## Useful Integrals

$$\int \sin^2\left(\frac{n\pi}{L}x\right) dx = \frac{x}{2} - \frac{L}{4n\pi} \sin\left(\frac{2n\pi}{L}x\right)$$

$$\int x \sin^2\left(\frac{n\pi}{L}x\right) dx = \frac{x^2}{4} - \frac{Lx}{4n\pi} \sin\left(\frac{2n\pi}{L}x\right) - \frac{L^2}{8n^2\pi^2} \cos\left(\frac{2n\pi}{L}x\right)$$

$$\int x^2 \sin^2\left(\frac{n\pi}{L}x\right) dx = \frac{x^3}{6} - \frac{Lx^2}{4n\pi} \sin\left(\frac{2n\pi}{L}x\right) - \frac{L^2x}{4n^2\pi^2} \cos\left(\frac{2n\pi}{L}x\right) + \frac{L^3}{8n^3\pi^3} \sin\left(\frac{2n\pi}{L}x\right)$$

$$\int_0^\infty x^m e^{-bx} dx = \frac{m!}{b^{m+1}}$$

## Gaussian Integrals

$$\int_{-\infty}^{+\infty} e^{-a(z-b)^2} dz = \sqrt{\frac{\pi}{a}} \qquad \int_{-\infty}^{+\infty} ze^{-a(z-b)^2} dz = b\sqrt{\frac{\pi}{a}}$$

$$\int_{-\infty}^{+\infty} e^{-az^2+bz} dz = e^{b^2/4a}\sqrt{\frac{\pi}{a}} \qquad \int_{-\infty}^{+\infty} z^2 e^{-az^2} dz = \frac{1}{2}\sqrt{\frac{\pi}{a^3}}$$

$$\int_{-\infty}^{+\infty} z^{n+2} e^{-az^2} dz = -\frac{d}{da} \int_{-\infty}^{+\infty} z^n e^{-az^2} dz$$

# Nonclassical Physics

## Beyond Newton's View

### Randy Harris
University of California, Davis

 **ADDISON-WESLEY**

An imprint of Addison Wesley Longman, Inc.

Menlo Park, California • Reading, Massachusetts • New York
Harlow, England • Don Mills, Ontario • Sydney • Mexico City
Madrid • Amsterdam

| | |
|---|---|
| Publisher | Robin Heyden |
| Acquisitions Editor | Sami Iwata |
| Production Editor | Lisa Weber |
| Market Development Manager | Andy Fisher |
| Director of Marketing | Stacy Treco |
| Channel Marketing Manager | Gay Meixel |
| Art Supervisor | Kelly Murphy |
| Artist | Karl Miyajima |
| Photo Editor | Kathleen Cameron |
| Composition and Film Buyer | Vivian McDougal |
| Manufacturing Supervisor | Holly Bisso |
| Copyeditor | Luana Richards |
| Proofreader | Kathy Lee |
| Cover and Text Designer | Mark Ong, Side by Side Studios |
| Cover Photograph | © Courtesy of Don Eigler, IBM Almaden Research Center |

For permission to use copyrighted material, grateful acknowledgment is made to the copyright holders on page 601, which is hereby made part of this copyright page.

Many of the designations used by manufacturers and sellers to distinguish their products are claimed as trademarks. Where those designations appear in this book, and the publisher was aware of a trademark claim, the designations have been printed in initial caps or all caps.

**Library of Congress Cataloging-in-Publication Data**

Harris, Randy.
    Nonclassical physics : beyond Newton's view / Randy Harris.
      p.   cm.
    Includes index.
    ISBN 0-201-83436-7
    1. Physics.  I. Title.
QC21.2.H38  1998
530—DC21
                                  98-7460
                                  CIP

2  3  4  5  6  7  8  9  10—RNT—02  01  00  99

Addison Wesley Longman, Inc.
2725 Sand Hill Road
Menlo Park, CA 94025

**Of Related Interest from Addison Wesley Longman**

*Essentials of Modern Physics,* T. R. Sandin, 1989

*Physics of the Atom,* fourth edition, M. R. Wehr, J. A. Richards, T. W. Adder, III, 1984

*The Lectures on Physics: Commemorative Issue,* Volume 3, R. P. Feynman, R. B. Leighton, M. L. Sands, 1965

*Introductory Quantum Mechanics,* R. Liboff, 1998

*Modern Quantum Mechanics,* revised edition, Sakurai, S. F. Tuan, 1994

*Quantum Mechanics,* S. McMurray, 1993

# Brief Contents

# Contents

*Sections marked with a ▼ are optional and at the same level as the main sections.*

*Sections marked with a ◆ are optional and more advanced.*

# Preface

This text is written for the student who has completed a calculus-based introductory course in classical physics. As such, it is appropriate for a course taught either in the sophomore year or the early part of the junior year. It is intended to acquaint the student with the wide range of fascinating physical principles that have developed in the past century. Through discussions, examples, and exercises involving real-world applications, it will enable students to apply their newfound understanding quantitatively, in all areas of natural science, as well as to further study in physics. Before discussing in detail the text's topics and the required student preparation, let me explain the philosophy behind this text.

Having taught this subject for some years, I know the excitement students feel about it. After classical physics, it is quite a shock. On the other hand, because it so often seems counterintuitive, explanations and discussions must be much more thorough than in classical physics, if the excitement is not to be overshadowed by confusion. Thus, one of my goals has been to clarify in the simplest possible way the points where student misconceptions arise, as soon as they are encountered. Students unencumbered by these misconceptions are best able to successfully build upon the foundation. The central role of nonclassical physics in our technological world led me to another goal—to connect the underlying theories to applications in other areas of science and technology.

An all-encompassing goal, motivated by experiences with other texts on modern physics, was to write a text with a consistent approach. The inconsistency of many of the current texts can be extremely distracting to students, so I have striven to avoid vague, oversimplified explanations on fundamental points that should be treated with rigor at this level, but also to avoid doggedly preserving rigor when students at this level have not had the preparation to follow it. Accordingly, on certain points this text will seem more sophisticated than others, expanding on concepts the student is fully prepared to grasp. If these fundamental ideas are not explained and understood in this course, they often never are, for they are frequently taken for granted in later courses. Thus, more than the usual amount of time is spent on questions such as: How can two observers each see the other aging slower? What is the true basis of the uncertainty principle? Why is the wave function a complex number? What really causes quantization? On other points, this text will seem more basic than others. For various reasons many modern physics texts tend to concentrate on complicated early theories that were long in the development, but that are no longer central to the subject. Little time is spent on such theories here, so as to keep the focus on the key topics of concern today.

This is, first and foremost, a textbook. Owing to the nature of nonclassical physics, some topics are unavoidably complex. Such topics are clearly delineated within the text and are treated qualitatively, so that the student may most readily see the important underlying concepts. Whenever possible, however, the text ties the discussion to the first principles and quantitative techniques introduced in the chapter. Finally, clarity is the supreme goal, so new nomenclature is introduced carefully, not capriciously. Explanations are in the most basic terms, centered on main ideas of the current material and fundamentals from earlier courses.

## Course Prerequisites

It is assumed that the student is acquainted with the standard classical physics topics of mechanics, waves, thermodynamics, and electricity and magnetism, the first two being particularly important. Regarding the math prerequisites, some familiarity with partial derivatives is assumed, and use is made of complex numbers, though not much beyond their definition and simple arithmetic. Quantum mechanics cannot be understood without some use of differential equations; however, the text assumes only a very basic understanding of them, and the student is guided through the few necessary steps. Above all, the topic of the text is physics, not math. The mathematics associated with differential equations, for example, is merely a tool, and if there is a known mathematical solution to an equation, it will generally be used without much concern about how it is obtained. If this is borne in mind, the reader should quickly become comfortable with words like "we now see that this is a differential equation of a standard form, and the mathematical solution is . . . ." Appendix K summarizes the various types of differential equations, providing solutions to those used most often.

## Content and Features of This Text

Chapter 1 is devoted to the subject of relativity, the study of objects and reference frames in relative motion at speeds comparable to that of light. While not as widely applicable as the material in later chapters, the topic is one of the most fascinating in physics, posing profound challenges to classical notions of space, time and energy. No study of modern physics could be complete without it.

Chapters 2 and 3 introduce an area of physics inescapable in modern technology: quantum mechanics. Chapter 2 presents the nonclassical view of light and other forms of electromagnetic radiation, as particles rather than waves. The other side of the coin is presented in Chapter 3, which examines evidence and explores the consequences of the nonclassical view of matter, not as particles, as in classical physics, but as waves. In Chapter 4 are found the basic quantitative applications of quantum mechanics. Here the student will learn why the word quantum comes up so often in the study of microscopic things—small particles in small confines behave as standing waves between which there are "quantum jumps." Chapter 5 discusses surprising consequences of quantum mechanics that arise when free particles encounter forces of one type or another. Among other things, we find that a particle that should rebound at a barrier according to classical laws may in fact pass through—via quantum mechanical "tunneling."

Rounding out basic quantum mechanics, Chapter 6 discusses a logical test: the simple, one-electron hydrogen atom. Here are found the explanations of the atom's stable electron orbits and its "quantized" energies, facts that cannot be understood classically. Chapter 7 continues the study of the atom, adding the concept of spin, exploring the consequences of having multiple electrons in the same atom, and establishing the foundation of chemistry.

An introduction to statistical mechanics, the physics that explains the behaviors of systems of countless interacting particles at different temperatures, is found in Chapter 8. Among the applications discussed is the laser, the workhorse of modern optics.

Chapter 9 follows with a discussion of bonding between atoms, first in individual molecules, and then in solids consisting of countless atoms. The characteristics of insulators, conductors, and semiconductors are explained, as well as the reasons semiconductors are indispensable in modern electronics. The chapter finishes with a discussion of superconductivity, a fascinating and important phenomenon being found at ever higher temperatures.

Chapter 10 introduces the physics of the atomic nucleus, which is held together by a force that was unknown to classical physics but is now accepted as one of nature's fundamental forces: the "strong force." Among the important topics here are radioactivity and nuclear energy. Chapter 11 presents the modern view of interparticle forces at the most fundamental level. The various categories of fundamental particles are discussed and the presently accepted fundamental forces are compared—their similarities, their differences, and the possibility that further research will show some if not all to be merely different aspects of the same force.

Appendices follow Chapter 11. Some cover optional material that supplements the material in the body of the chapters; others contain math-heavy derivations that would be distracting within the discussion of the physics. Appendix J is tabulated data on nuclear isotopes, while Appendix K covers frequently used math.

Each chapter is divided into sections. In addition to the main text sections there are two types of optional sections. Optional sections presenting material additional to the main text and at the same level are marked with a ▼ in the Table of Contents and within the chapters. Optional, more challenging sections are marked with a ◆. With few exceptions, material in optional sections is not crucial to the understanding of the main text. However, some optional sections rely on material from earlier optional sections. Recognizing that time and curriculum constraints may limit the material covered in a course, sections and chapters have been made as self-contained as possible. Toward the end of each chapter, recent advances and lingering mysteries related to the material in the chapter, as well as a few of the more interesting and important applications, are discussed in a section entitled Progress and Applications.

Each chapter closes with a summary of the chapter's important points, usually preceded by a list of frequently used equations. At the end of each chapter are exercises grouped into sections that parallel the sections of the chapter. In some chapters, additional general exercises encompass ideas from various areas. All exercises are either standard (ranging from easy to fairly challenging) or advanced (involving deeper physical principles and/or more complex math). Each advanced exercise is distinguished by a blue-highlighted exercise number. By making suitable choices, students at all levels of understanding can be appropriately challenged.

Accompanying the text is an Instructor's Solutions Manual. These detailed solutions to all of the end-of-chapter exercises comprise not only the bare mathematical steps, but usually include important words of explanation.

The text is up to date with current research at the publication date, but progress is constant in this active field. Many developments are newsworthy, and an eye on the media and the World Wide Web will ensure a familiarity with current discoveries in modern physics.

## Acknowledgments

I would like to acknowledge the hard work and dedication of the Chemistry, Physics, and Earth Sciences team at Addison Wesley Longman, particularly that of physics editor Sami Iwata and production editor Lisa Weber. Artist Karl Miyajima and art supervisor Kelly Murphy created a consistent and pleasing art program. The diligence and fine work of photo editor Kathleen Cameron and copyeditor Luana Richards also deserve recognition. And I greatly appreciate the painstaking efforts of Chris Ray of Saint Mary's College of California and Mark Coffey of the University of Colorado, Boulder, who acted as accuracy checkers for the text, exercises, and solutions.

My most sincere thanks go to all those who have reviewed the text:

**John Albright,** Florida State University

**Paul Avery,** University of Florida

**Arthur J. Braundmeier,** Southern Illinois University

**Anthony J. Buffa,** California Polytechnic State University

**Marshall Burns,** Tuskegee University

**Louis H. Cadwell,** Providence College

**Bernard Chasan,** Boston University

**Randall T. Dillingham,** Northern Arizona University

**Theodore Einstein,** University of Maryland

**Richard F. Haglund,** Vanderbilt University

**L. Michael Hayden,** University of Maryland

**Michael Lieber,** University of Arkansas

**Linda L. McDonald,** North Park College

**Herbert Muether,** SUNY Stony Brook

**Lawrence Pinsky,** University of Houston

**John A. Polo,** Edinboro University

**Patricia Rankin,** University of Colorado

**Andrew Sustich,** Arkansas State University

I am deeply indebted to those who have tested the text in the classroom, especially Louis Cadwell and Rajiv Singh, who provided many invaluable comments and suggestions. Thanks are also due Richard Scalettar and Gergely Zimanyi for numerous helpful chats. Above all I am profoundly grateful to my students! None could be better at catching mistakes, questioning explanations, discerning ambiguities, and so on, and no text dealing with ideas so challenging could be successful were it not shaped by such input.

Randy Harris
Davis, California
May, 1998

# Introduction

Classical physics, consisting principally of mechanics, thermodynamics, and electricity and magnetism, is the body of knowledge that explains those physical behaviors that can easily be observed. It might fairly be said that classical physics agrees with our intuition, borne of experience. Whether or not we are acquainted with it, classical physics "merely" provides the theoretical basis—a reason—for "normal" physical behaviors. As such, it is no threat to our beliefs.

Certainly what we now know as classical physics was unknown to the ancients. The discovery of pieces of the puzzle has been going on for as long as we humans have been around to ponder them. However, in the latter part of the nineteenth century, many reputable scientists believed that knowledge of all important areas of physics had attained virtual completion—dare it be said, stagnation. To be sure, there lingered important phenomena that defied explanation. All too often these disturbing problems were simply swept under the rug. Of course, it was expected that later decades would see certain measurements made more precise, and that technology would progress with the application of *existing* science. But few scientists suspected they were on the eve of a revolution.

Near the turn of the twentieth century, shocking new ideas emerged, theories regarding the fundamental nature of matter and the very nature of space and time. These theories explained many of the lingering problems, but they ran counter to "classical intuition." Indeed, they demanded a complete reassessment of our view of the physical universe. Nevertheless, though staunch opposition arose initially, the weight of decades of experimental investigation was eventually convincing.

One of the primary reasons these theories remained hidden for so long is that they involve behaviors not easily observed; and the reason they are distinguished from classical physics and often studied separately is that they do seem counterintuitive. These reasons are related. A behavior certainly would not be counterintuitive if it were subject to simple observation. Everyone knows that an object will fall if let go, whether or not the universal law of gravitation is judged to be the explanation. However, it is because beliefs are too strongly influenced by experience that nonclassical physics is treated separately. Perhaps it would not be necessary if we always bore in mind that *we cannot justifiably claim to possess intuition about a phenomenon we cannot observe!* Roughly speaking, nonclassical physics is the study of the small and the fast, but *we have never seen an individual atom with our own eyes nor have we ever traveled at a significant fraction of the speed of light.* We must not apply preexisting notions based on experience to situations in which we have no experience.

Two important areas of nonclassical physics are special relativity and quantum mechanics. Special relativity is the theory governing the behavior of things moving at any speed, while classical mechanics may be viewed as a special case correct only for slow-moving things. Quantum mechanics is the theory governing the behavior of things of all sizes, and classical mechanics is a special case correct only if we do not investigate small things too closely. But how can classical mechanics be a special case of two different things? In reality, special relativity and quantum mechanics too have limitations. Both are special cases of an all-encompassing theory: relativistic quantum mechanics. The diagram crudely represents the realms of applicability of the various theories. The special relativity studied in Chapter 1 is valid only for large things; it is not correct quantum-mechanically. Similarly, the quantum mechanics studied in subsequent chapters is, with a few noted exceptions, not "relativistically correct," but valid only for slow-moving things.

Often, however, using an advanced theory—applicable though it may be—would be like driving a thumbtack with a sledgehammer. We do not use relativistic quantum mechanics, or even quantum mechanics or special relativity, to analyze the motion of a baseball moving at 40 m/s, though all are applicable. We use classical mechanics, the simplest theory correct within the realm of interest. Relativistic quantum mechanics is too sophisticated for a text at this level, but its influence is pervasive. One manifestation is a theme that crops up in *seemingly* unrelated contexts: a curious four-way relationship between position and time and momentum and energy.

After special relativity and quantum mechanics have been introduced (roughly Chapters 1–6), several modern physics topics are discussed. These point the way to the frontiers of physics and the future of technology. In many ways, they are applications of the earlier material, but each presents its own difficulties. In most cases, the phenomenon of interest is so complex that the "problem cannot be solved" from first principles. Thus, its study rests upon certain fundamental assumptions, or a model, and applies first principles to the greatest extent possible within the resulting framework.

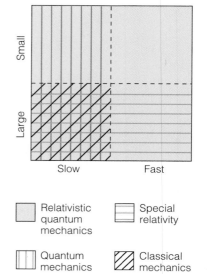

# Special Relativity

Special relativity is the theory governing physical phenomena when one object or reference frame moves relative to another at a speed comparable to that of light: $c = 3 \times 10^8$ m/s. At such high speeds, classical Newtonian mechanics has been found to be inadequate, and seemingly obvious truths about space, time, and energy are profoundly altered. It is very important to bear in mind that essentially all speeds with which we have experience are "low" speeds, much less than $c$. Spacecraft, for instance, travel no faster than a mere 25,000 mi/h, or about $10^4$ m/s. At such speeds, "relativistic effects" are imperceptibly small—classical mechanics is satisfactory. Because we have no experience with high speeds, we really have no reliable intuition against which to judge the predictions of special relativity. Startling though they are, it cannot be said that they are contrary to observation and therefore incorrect.

Given the high speed involved, it is little wonder that special relativity emerged so late; there was no obvious evidence of its effects. Its discovery by Albert Einstein in 1905 was precipitated by advances in the field of electromagnetism in the mid-19th century. Even so, Einstein's theory—a work of genius—was a tremendous leap forward and ran counter to many cherished beliefs. Not surprisingly, detractors abounded. However, with technological advances in the years since its discovery, the experimental evidence of its validity has mounted, until no serious opposition remains.

# 1.1  Basic Ideas

An important concept in special relativity is that of an **inertial frame of reference,** or "inertial frame." By definition, this is a frame in which an object experiencing zero net force—a "free object"—moves at constant velocity. In classical physics, this is any *frame* moving at constant velocity. Consider, for example, an object floating freely in space. Any observer moving at constant velocity, perhaps gliding by in an unpowered spacecraft, would determine the velocity of the object to be constant. An observer in an accelerating spacecraft would see an object whose velocity is changing. Accelerating frames are best understood using general (as opposed to special) relativity, the topic of Section 1.13.

Einstein's fundamental postulates of special relativity can be stated as follows:

> 1.  The form of each physical law is the same in all inertial frames.
> 2.  Light moves at the same speed relative to all observers.

According to the first postulate, an experiment should proceed identically in a spacecraft floating at rest relative to Earth as in a spacecraft moving at a constant 1,000,000 m/s relative to Earth. No amount of experimentation could prove that a given inertial frame is actually moving. Consider a simple example: If a man standing still drops a ball from shoulder height, it lands at his feet a certain span of time later. If he is inside a high-speed train moving horizontally at a constant 100 m/s and drops the ball, it lands at his feet after the same span of time. In each case, his observation agrees with a law of the same form, $\mathbf{F} = m\mathbf{a}$, without taking into account any motion relative to anything else. The physical behaviors are identical. In fact, were it not for giveaways like bumps, noise, windows, and so forth, the man simply could not prove that his frame is in motion relative to Earth. Since he has no evidence that it is *he* who moves, he is quite as justified in claiming that he is stationary and it is *Earth* that moves. According to Einstein's first postulate, then, it is impossible to prove that any one frame is truly in motion, or that any frame is truly at rest. We say that there is no "universal rest frame" relative to which all motion might be measured. No motion is absolute—all is relative.

Einstein's first postulate seems quite reasonable when applied to simple mechanical examples like the man on the train. But his second postulate can be very troubling. Imagine that Anna, a track star, runs to the right at $0.2c$ relative to Bob, who is standing still. Anna carries a flashlight and shines it in the direction in which she is running. According to the second postulate, Anna must determine that the beam moves at $c$ relative to herself, and Bob must also determine that the beam moves at $c$ *relative to himself.* The layman is tempted to claim either that the beam moves at $1.0c$ relative to Anna and $1.2c$ relative to Bob, or $1.0c$ relative to Bob and $0.8c$ relative to Anna. The layman's assumption is that one may add the velocity of the beam relative to Anna to the velocity of Anna relative to Bob to obtain the velocity of the beam relative to Bob. We shall see later that this assumption is invalid. But even if it were valid, the two claims are incompatible. Therefore, the layman faces the question: Relative to what or whom does light move at $c$? Besides the observer (i.e., besides Einstein's answer), there are only two alternatives: the source and a medium.

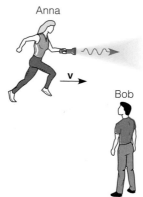

Anna

Bob

*Source*  Maxwell's equations, which govern all electromagnetic phenomena, predict that light propagates at a speed independent of the speed of the source (as do all mechanical waves). All experiments in which light from high-speed sources has been timed agree.

*Medium*  Mechanical waves require a medium whose deformation is communicated from one point to another via intermolecular forces, resulting in a propagating wave. The speed of the wave depends on properties of the medium, not on the motion of source or observer. An explosion causes pressure variations in the air around it. This "deformation" propagates outward as a sound wave at a speed dependent only on properties of the medium—air. It is reasonable to suppose that the propagation of light also requires a medium, and it was given a name: **aether.** What air is to sound, aether would be to light. Thus, light would always move at speed *c relative to the aether,* independent of the motion of source or observer. Furthermore, since light appears to travel throughout the universe, aether would have to permeate space. We might even go so far as to proclaim it the "universal rest frame"—the one thing "really" at rest—relative to which the motions of all other things could be judged.

Let us return to Anna running with the flashlight. (We now accept that since Anna is the source, her motion has no effect on the speed of the light.) To claim that the beam moves at *c* relative to Bob is to claim that the aether is at rest with respect to Bob and that Anna would see the light moving away from herself at 0.8*c*. By analogy, were Anna to shout while running through still air at $0.2v_s$, where $v_s$ is the speed of sound through air, the sound in front of Anna would be moving away at $0.8v_s$, while of course moving at $1.0v_s$ relative to Bob. On the other hand, to claim that the light beam moves at *c* relative to Anna is to claim that the aether moves with Anna, and would therefore carry the light away from Bob at 1.2*c*. Again by analogy, if the air happened to be stationary relative to Anna, blowing at $0.2v_s$ relative to Bob, the sound of Anna's shout would move at $1.0v_s$ relative to Anna and $1.2v_s$ relative to Bob.

If aether exists, we might expect it to be detectable. Certainly it is not as apparent as air. But by considering how we might "prove the existence of air," let us see how we might do the same for aether. Even if observers cannot feel the wind at their faces, they can determine the speed at which air moves—by timing echoes. Suppose the air is moving at $v_{air}$ relative to the stationary Bob (Figure 1.1). Bob shouts, and listens for the echo from an obstacle downwind, a distance *L* away. The sound moves at $v_s + v_{air}$ relative to Bob as it travels downwind, arriving at the obstacle after a time $L/(v_s + v_{air})$. After reflecting, the sound returns at $v_s - v_{air}$, and arrives back at Bob's location after an additional time $L/(v_s - v_{air})$. Adding these gives a total time of $2v_sL/(v_s^2 - v_{air}^2)$. Bob could calculate the speed $v_{air}$ relative to himself knowing only the speed $v_s$ of sound in air, the distance *L* and the round-trip time.

Since aether-wind cannot be felt, experiments to determine the speed of aether relative to Earth have involved timing round-trips of light beams. (The most famous, the **Michelson–Morley experiment,** is discussed in Appendix A.) *All have failed to find a nonzero speed.* Curiously, aether does not appear to be moving relative to Earth. Maybe this is true. But Earth and other planets orbit the Sun.

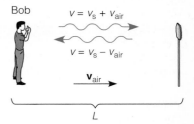

Bob

$v = v_s + v_{air}$

$v = v_s - v_{air}$

$\mathbf{v}_{air}$

$L$

**Figure 1.1** The round-trip time for sound depends on the motion of its medium—air.

Why should aether orbit with Earth? To salvage the belief in aether, some scientists suggested that perhaps moving objects (like Earth) "drag" aether along with them. The results of several experiments cast considerable doubt on this suggestion: One demonstrated no ability of water to drag aether; others failed to find variations in the apparent locations of stars as planets passed through the sky, even though a planet dragging light's medium should cause unusual refraction around it. On the contrary, the belief in aether became increasingly untenable with each new attempt to find it. There being no *evidence* of its existence, another possibility is that aether doesn't exist! This is the prevailing belief. *Light does not require a medium.*

Were aether to exist, then *only* in a frame stationary relative to the aether would light move at $c$ in all directions. In any other frame, light would move away from an observer at different speeds in different directions, like a shout on a windy day. Maxwell's equations predict that light will move at $c$, independent of direction. If aether existed, this could only be true in the one frame stationary relative to the aether, and Maxwell's equations would have to be of a different form in all other frames. Einstein's first postulate would not be met. Perhaps it is possible to patch up Maxwell's equations to include the effects of a frame's motion through aether, to generalize them so that those we now accept are a special case of a form correct for all frames. However, with no evidence of aether's existence, Einstein's second postulate is as plausible a solution to the "problem" as any. Essentially, it is a proclamation that Maxwell's equations *as we now have them* are correct in all frames. Consequently, light moves at $c$ in all directions relative to an observer regardless of his frame of reference.

*Casting Our Lot with Einstein*    Still the reader may find the postulate unpalatable, that though Anna and Bob move at $0.2c$ relative to one another, each determines a beam of light to be moving at $c$ relative to him or herself. Is something wrong with our concepts of distance and time, the laws of mechanics? The answer is full of irony. Originally, no one doubted the laws of mechanics. It was the laws of electromagnetism that were thought to be in error, because they predicted the same speed for light moving in all directions without reference to a medium. But with Einstein's postulate that Maxwell's equations are correct in all frames, we find that it is the laws of mechanics that must be reevaluated. Classical notions of space and time must be discarded.

In special relativity, distance and time become inextricably related. Accordingly, we find the following definition indispensable:

An **event** is anything with a location in space and a time.

The birth of a child is a single event, occurring at a certain time and at a certain location. A grain of sand embedded in an ancient mountain may be thought of as a *collection* of events, all with the same location but different times. In taking a snapshot of a friend, we record a collection of events, all with the same time but different locations.

Let us now discuss the startling consequences of Einstein's postulates. As we do so, we will often refer to observers Anna and Bob. Understanding special relativity is impossible if we can't keep straight *who* is doing the observing, and assigning names is the easiest way to do this.

# 1.2 Consequences of Einstein's Postulates

The consequences we now discuss are provocative. Distance and time intervals are no longer what our limited intuition tells us they should be. We will quantify these claims later, by applying Einstein's postulates to derive the all-important Lorentz transformation equations. But to gain the firmest grasp of their origin and meaning, let us begin with a qualitative discussion.

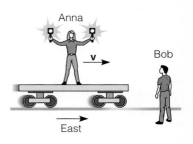

Anna
Bob
v
East

## Consequence I: Relative Simultaneity

Anna is riding on a railroad flatcar moving east at speed $v$. In each hand she holds a flashbulb, which she causes to flash at precisely the same instant. Bob is standing on the ground, watching Anna pass. We have two events (i.e., the flashes) with different locations. According to postulate 2, Anna will see both light beams move at $c$ *relative to herself* despite the fact that she moves *relative to Earth*. The flashes occur at the same distance from Anna, so the beams arrive simultaneously. This arrival is a single event—at a single location (her head) and a single time.

According to postulate 2, Bob must also see both beams travel at $c$ relative to himself. But according to Bob, Anna moves some distance to the east while the light beams travel, so the beam from the west flashbulb has farther to go. If Bob agrees that the beams originated simultaneously, he is forced to conclude that the beam from the east flashbulb will reach Anna first. *This is a contradiction!* A single event, the arrival of two light beams at the same time at a single location (i.e., Anna's head), either occurs or it doesn't. Though Anna and Bob might not agree on its location or its time, they must agree on the question of whether it occurs at all. So long as we accept Einstein's postulate, there is only one way out of the contradiction. If Bob is to agree that the beams arrive at Anna simultaneously, that the single final event even occurs, he must determine that the beams from the flashbulbs do not originate simultaneously, that these *two* events do not occur at the same time. In fact, he must determine that the west flashbulb goes off first, since light from it has farther to go and the beams move at the *same speed*. Later we will use the Lorentz transformation equations to quantify the time interval according to Bob, and to verify that the west flashbulb flashes first. Though the discrepancy in times is extremely small at ordinary speeds, the effect is real.

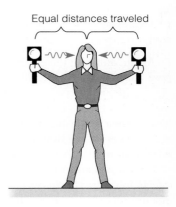

Equal distances traveled

**What Anna sees**

> Two events at different locations, simultaneous in one frame of reference, will not be simultaneous in a frame of reference moving relative to the first.

Before introducing our next consequence, we note a crucial point: The preceding arguments have *nothing at all* to do with Bob's specific location in his frame or the time it takes light to reach his eyes. The issue here is *not* which beam *reaches* Bob first, which does depend on where he stands, but rather which *begins* first, which has nothing to do with where he stands. Attempting to "blame" the effect on "optical illusions" is the comfortable yet incorrect way out of accepting a counterintuitive notion. The truth is more challenging. Bob could be standing right next to one flashbulb when it flashes, and have an assistant, Bob Jr., standing next to the other flashbulb when it flashes. Each records the time of his flash, and neither has to wait for the light to reach his eyes. *They record*

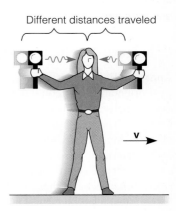

Different distances traveled

v

**What Bob sees**

*different times.* Anna, in her frame, causes the bulbs to flash simultaneously, but Bob and Bob Jr., from their frame, determine that she causes the bulbs to flash at different times.

## Consequence II: Time Dilation

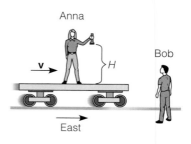

Anna

Bob

East

Anna is again on her flatcar and Bob is standing on the ground. Now Anna holds a flashlight and directs the beam to a mirror at her feet. The beam reflects back to the flashlight *at the same location* in Anna's frame, covering a total distance $2H$. The total time is $2H/c$. Bob, on the other hand, sees the beam reflect at a point east of where the beam originated (the flatcar is in motion) and return to the flashlight at a point still farther to the east. Moving diagonally, the beam travels a total distance *greater* than $2H$. By postulate 2, Bob must see the beam move at $c$ (not faster), so Bob will have to wait a total time *greater* than $2H/c$. We have two events, the creation of the light beam and its arrival back at the flashlight, occurring at the same location in Anna's frame, and the time that passes in Bob's frame is longer.

What Anna sees

> Two events occurring at the same location in one frame will be separated by a longer time interval in a frame moving relative to the first.

Again, although the effect is not noticeable at ordinary speeds, it is real. Suppose Anna brushes her teeth. It takes her 30 s by her watch. Bob, watching from the outside, is precisely by Anna's side when she starts brushing her teeth. Bob records this time—call it noon. Bob Jr., standing on the ground east of Bob, is beside Anna when she finishes. Bob Jr. will record a time of *more* than 30 s after noon. According to Bob (and Bob Jr. and all others in Bob's frame of reference), Anna is brushing her teeth slowly, doing all things slowly, aging slowly.

What Bob sees

It is important to note that all that is required for this and other relativistic effects to occur is *relative* motion. Anna is moving relative to Bob, but Bob is moving relative to Anna. Thus, Anna must see Bob brushing his teeth slowly, aging slowly. The reader may apprehend a paradox. How can each see the other aging more slowly than themselves? The resolution of this seeming paradox is intimately tied in with the principle that simultaneity is not absolute, as we shall see in Example 1.4.

## Consequence III: Length Contraction

Anna and Bob occupy their usual frames, but Bob now holds a plank in an east-west orientation. He could determine the length of the plank by finding the time interval between two events: Anna passing one end of the plank, and Anna passing the other. The length of the plank would be the time between these events multiplied by Anna's velocity relative to Bob. Though Anna sees a *moving* plank, she could determine its length by finding the time interval between the same two events: first one end of the plank passing herself, then the other. The plank's length is the time between these events multiplied by the plank's velocity relative to Anna. Now, the two events occur in the *same place* in Anna's frame of reference, namely Anna's very location. (According to Bob, the plank is stationary, so its

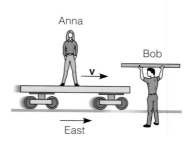

Anna

Bob

East

ends are at different locations.) Our discussion of time dilation then tells us that the time interval in Anna's frame will be shorter: Taking less time to pass, the plank is shorter according to Anna than it is according to Bob. The effect, an object occupying less space when moving than when stationary, is called **length contraction.**[1]

The same warning applies to this as to the two previous effects: It is a grave mistake to dismiss length contraction as an optical illusion caused by delays in light traveling to the observer from the moving object. The effect is real. Suppose the plank is of length $L$ according to Bob (who holds it). Now in Anna's frame, Anna and her young assistant Amy each happen to be aligned with an end of the passing plank *at precisely the same time.* In Anna's frame, she and Amy will be a distance apart *less than L!*

> The length of an object in a frame through which it moves, determined by viewing its ends at the same time (two events) in that frame, is smaller than the length of the object in the frame in which it is at rest.

Again, all that is required is relative motion. If Anna on her flatcar were holding a plank of length $L,$ it would occupy a distance less than $L$ in Bob's frame of reference, because Anna's plank moves relative to Bob.

We now consider a question that in one form or another is a common source of consternation for the beginning student of special relativity. In so doing, we will begin to see the interdependence of the preceding consequences. We construct paradoxes in applying one consequence of special relativity only by ignoring another. They are a "package deal," and we must buy the whole package.

Anna has developed a revolutionary new plane capable of attaining speeds near that of light. It is 40 m long when parked on the runway. Bob has a 20-m-long airplane hangar with open doorways at each end, Figure 1.2(a). Anna, a young and daring type, takes off, accelerates to high speed, then swoops through the hangar at constant speed. At a high enough speed, Anna's plane fits entirely within the hangar all at once according to Bob, occupying only 20 m, Figure 1.2(b). In other words, Bob sees the tail of Anna's plane at one doorway at precisely noon on his watch, while Bob Jr. sees the plane's nose at the other doorway, also at precisely noon. Now consider Anna's perspective: Anna is at rest relative to her plane and thus sees its length as 40 m. She sees the *hangar* as moving (relative to herself). It must occupy less space in Anna's frame than it does in Bob's frame, the one in which it is at rest. Thus, Anna will see the hangar as only 10 m long, Figure 1.2(c). Her 40-m plane cannot possibly fit in all at once!

How can we resolve this apparent paradox? Here is the connection: The two events—the two ends of the plane at the two doorways—occur *simultaneously* according to Bob, so they cannot be simultaneous according to Anna. In fact, Anna observes that her plane's nose reaches the doorway through which it exits *before* its tail enters the other doorway. According to Anna, the plane need not and does not fit in the hangar all at once.

A question often asked is, What would happen if the plane were to be stopped suddenly—will it fit in, or not? We note first that no object can be stopped without force being applied to it somewhere, and a sudden stopping of the plane would be a violent process certain to deform the object in either frame. A mechanism in

[1] The reader may be suspicious of basing a "proof" of length contraction on the concept of time dilation. Is there not a more direct way? It must be remembered that we investigate an effect of relative motion. It would not be fair to stop the object to measure it. We must consider some way of determining its length while moving. Distance = speed $\times$ time is as good as any. If a train moving at 10 m/s passes by in 100 s, its length is unquestionably 1000 m.

Anna's plane      Bob's hangar

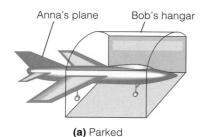

**(a)** Parked

**Figure 1.2** Whether one object fits inside another depends on the observer's frame of reference.

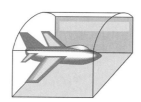

**(b)** What Bob sees

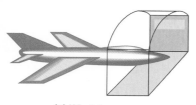

**(c)** What Anna sees

Bob's hangar reaches out and grabs Anna's plane, exerting sufficient force at many points on the plane to stop it essentially all at once (zero stopping distance would require infinite force). But simultaneous stopping of the different points of the plane in Bob's frame cannot be simultaneous in Anna's. Anna sees the nose stop first. Meanwhile, the other points are still moving, getting closer to the front, reducing the overall length. The last point to stop is the tail. When it is all over, the plane will have been squashed until it fits entirely within the hangar. Bob and Anna, then in the same reference frame, would agree that it fits, and that the experiment was a harrowing set of events.

## 1.3  Evidence of Relativistic Effects

The most famous experimental evidence supporting special relativity involves the behavior of muons produced in the upper atmosphere. Muons are subatomic particles (rare compared to protons, electrons, and neutrons) that are created when cosmic rays strike the upper atmosphere. It is well known that the muon is unstable; it will spontaneously "decay" (into something different) after a period of time. The time elapsing before decay varies greatly from muon to muon, but the average is known to be 2.2 $\mu$s, in a frame in which muons are at rest. Given the altitude at which the muons are produced and the speed at which they are found to travel thereafter, a classical calculation indicates that essentially no muons should reach Earth's surface—but many do! The explanation involves either time dilation or length contraction, *depending on the observer.*

Bob stands on Earth and watches a clock glued to a muon. Anna flies with the muon.

*Bob's View*  Bob observes events, ticks of a clock, that occur at the same place in the muon's (Anna's) frame. Due to *time dilation,* these events must take more time as measured by Bob's clock than by the muon's. On average, 2.2 $\mu$s will go by before the muon decays in the frame in which it is at rest (Anna's frame), so Bob's watch must register more than 2.2 $\mu$s. If according to Bob the muons last longer before decaying (age less rapidly), they will indeed reach Earth's surface in greater abundance.

*Anna's View*  Relative to Anna, it is Earth and the atmosphere that move. Due to *length contraction,* the distance between upper atmosphere and surface will be smaller in Anna's frame than in the frame in which they are at rest (Bob's frame). If according to Anna and her muon traveling companions the distance that passes by is shorter, more muons will survive to meet the surface.

Again we see the harmony of the various effects in reconciling the views of different observers. Notably, the prediction of muon flux at Earth's surface given by special relativity is in excellent agreement with the experimental findings.[2] We will treat the muon "problem" quantitatively in Example 1.2.

## 1.4  The Lorentz Transformation Equations

The Lorentz transformation equations are the fundamental equations relating position and time in one inertial frame to position and time in another. They are a modification of classical relationships, necessary to accommodate Einstein's postulates. We begin with the classical relationships.

First, however, the conventions we will use throughout our discussion of special relativity need to be introduced. In relating observations in one frame to those in another, we refer to two representative frames, frame $S$ and frame $S'$. Frame $S'$ moves at speed $v$ in the positive $x$-direction relative to frame $S$. Correspondingly, viewed from frame $S'$, frame $S$ moves in the *negative $x'$-direction.* (Their $x$-axes are coincident. They are shown otherwise in Figure 1.3 merely for the sake of clarity.) We reserve the symbol $v$ *exclusively* for the relative speed between the

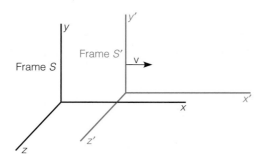

**Figure 1.3** Two reference frames: Frame $S'$ moves at velocity $+v$ relative to $S$, and frame $S$ at $-v$ relative to $S'$.

[2]It is practically impossible to follow an individual muon. Rather, we measure a *flux* at Earth's surface—number per unit time per unit area. Relativistically or not, muons do not all decay in the same amount of time; some invariably "live" much longer than others. But relativistic effects "stretch" the lives of all and so argue for a larger flux surviving to reach Earth's surface than expected classically. (See Exercise 14.)

two frames. The symbol $u$ is used for the velocity of an object moving relative to a frame. Quantities such as position, velocity, and time have different values in different frames. In frame $S'$, these would be denoted by $x'$, $u'$, and $t'$, while in frame $S$, they would be denoted by $x$, $u$, and $t$. Lastly, we define the instant at which the two origins pass as time zero in both frames.

## Classical (Galilean) Relativity

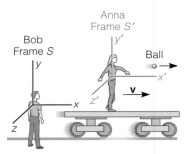

**Figure 1.4** The velocity of a ball depends on the frame of reference.

Once again, Anna is on her flatcar (frame $S'$ in Figure 1.4) moving east at speed $v$ relative to Bob, who stands on the ground (frame $S$). Anna now throws a ball eastward at speed $u'$ relative to herself. Relative to Anna in frame $S'$, the position of the ball (ignoring gravity and air resistance) is $x' = u't$. *Classically,* time is time; there is no reason to use separate $t$ and $t'$. Relative to Bob in frame $S$, the position of the ball is denoted by $x$ (without a prime) and is the sum of the distances from the ball to Anna and from Anna to Bob: $x = u't + vt$ or $x = x' + vt$. Taken together with our assumption that $t' = t$, we see that we have equations relating position and time in one frame to position and time in another:

$$x' = x - vt \qquad t' = t \qquad \text{(1-1)}$$

Classical transformation

We would say that equations (1-1) are a **transformation** between frames $S$ and $S'$. From these we may also relate the velocities. Recalling that, in any frame, the velocity of an object is the rate of change (time derivative) of its position, we obtain a result familiar from classical mechanics: $u' = dx'/dt' = d(x - vt)/dt = u - v$.

$$u' = u - v \qquad \text{(1-2)}$$

Classical velocity transformation

We have shown that both position and velocity have different values in different frames of reference. These results should not surprise the student of classical physics. It seems quite natural that an object's position relative to a moving reference frame could be expressed as a function of its position relative to the ground, the speed of the frame, and the time that has elapsed. But the assumption that time is absolute, *also quite natural,* dooms classical relativity. Equation (1-2), for instance, is at odds with Einstein's second postulate, for it says that if Anna were to "throw" a beam of light instead of a ball—that is, $u' = c$—its speed $u$ according to Bob would be $c + v$.

## Special Relativity

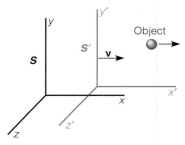

**Figure 1.5** Two frames of reference and a moving object.

The Lorentz transformation equations begin with the application of Einstein's postulates to the law of physics most students learn first: A free object moves at constant velocity. Imagine that at time zero in both frames, an object moving east happens to be at the exact point where the origins coincide (Figure 1.5). If the object experiences no net force, it must travel at constant velocity in both frames; its position must be a *linear* function of time in both frames.

$$x' = u't' \qquad x = ut \qquad \text{(1-3)}$$

Just as we did classically, we seek to relate positions, etc. in one frame to positions, etc. in the other, but we want to avoid making any unnecessary assumptions about time. Therefore, we allow that a given event in the "life" of the moving object may occur at different locations *and times* in the two frames. For instance, were the object a duck flying east while quacking regularly, we would like to be able to find the $x'$ *and* $t'$ of a given quack in frame $S'$, knowing its $x$ *and* $t$ in frame $S$. We seek the transformation between frames $S$ and $S'$—a functional relationship giving $(x', t')$ in terms of $(x, t)$. At this point, our only guidelines are that equations (1-3) must be obeyed. As it turns out, this is quite a restriction. The only functional relationship for which (1-3) are obeyed is the **linear transformation:** [3]

$$x' = Ax + Bt \qquad t' = Cx + Dt \qquad \text{(1-4)}$$

where *A, B, C,* and *D* are *constants*. It is known as a "linear" transformation because $x'$ and $t'$ depend on $x$ and $t$ to the *first* power only.

Though we have not derived equations (1-4), we may at least verify that they agree with equations (1-3). Suppose Bob in frame $S$ sees the object with position $x = ut$. We find the positions and times of events in the object's life according to Anna in frame $S'$ by inserting $x = ut$ into the transformation equations (1-4):

$$x' = A(ut) + Bt = (uA + B)t \qquad t' = C(ut) + Dt = (Cu + D)t$$

Eliminating $t$ between the equations, we obtain

$$x' = \frac{uA + B}{Cu + D} t'$$

Since *A, B, C, D,* and *u* are constant, so is the coefficient of $t'$, as equations (1-3) demand. In fact, it must be that $(uA + B)/(Cu + D)$ is the object's speed $u'$ in frame $S'$. The most important point here is that for a free object to move at constant velocity in all frames, all that is required is that the transformation between frames be linear. It is not required that time be absolute, that $t = t'$.

Now let us determine the constants *A, B, C,* and *D* in equations (1-4) with as few assumptions as possible. We need consider only three special cases of motion. Suppose the object were . . .

1. Fixed at the origin of frame $S'$: $x' = 0$.
   Since the origin of frame $S'$ moves in the positive $x$-direction at speed $v$ relative to frame $S$, an object fixed there would have $u = v$. Using equations (1-3), $x = vt$. Inserting this and $x' = 0$ into the first of equations (1-4) gives

   $$0 = A(vt) + Bt \quad \text{or} \quad \boxed{B = -Av}$$

2. Fixed at the origin of frame $S$: $x = 0$.
   The origin of frame $S$ moves in the negative $x'$-direction at speed $v$ relative to frame $S'$. An object fixed there would have $u' = -v$, so that $x' = -vt'$. Inserting this and $x = 0$ into equations (1-4) gives

   $$-vt' = Bt \quad \text{and} \quad t' = Dt \quad \text{or} \quad \boxed{D = -\frac{B}{v} = A}$$

[3]We omit the proof. Typically, it might be done in a class on linear algebra.

3.  A beam of light, heading to the right.
    Here is where Einstein's second postulate comes in: This object must have
    speed $c$ in both frames. Thus, $x = ct$ and $x' = ct'$. Inserting these into equations (1-4) and eliminating $D$ and $B$ using the preceding conditions,

$$ct' = A(ct) + (-Av)t \quad \text{and} \quad t' = C(ct) + At \quad \text{or} \quad C = -\frac{v}{c^2}A$$

We have found $B$, $C$, and $D$ in terms of $A$, and equations (1-4) become

$$x' = A(x - vt) \qquad t' = A\left(-\frac{v}{c^2}x + t\right) \tag{1-5}$$

To complete our derivation, we solve these for $x$ and $t$ in terms of $x'$ and $t'$.

$$x = \frac{1}{A\left[1 - \dfrac{v^2}{c^2}\right]}(x' + vt') \qquad t = \frac{1}{A\left[1 - \dfrac{v^2}{c^2}\right]}\left(+\frac{v}{c^2}x' + t'\right) \tag{1-6}$$

According to Einstein's first postulate, the transformation "law" to find $x$ and $t$ must be of the same form as that to find $x'$ and $t'$. Since $S'$ moves at $+v$ relative to frame $S$, while $S$ moves at $-v$ relative to $S'$, the different signs within the parentheses are acceptable as they are. To make equations (1-5) and (1-6) of the same form, then, we need only make the leading coefficients in the pairs of equations equal. Thus,

$$\frac{1}{A\left[1 - \dfrac{v^2}{c^2}\right]} = A \quad \text{or} \quad A = \frac{1}{\sqrt{1 - \dfrac{v^2}{c^2}}} \tag{1-7}$$

This factor is ubiquitous in special relativity, so it is given a special symbol, shown in the margin. As we shall see, this factor is often the measure of the departure of observed behavior from classical expectations. It increases continuously from essentially unity at small, "ordinary" speeds toward infinity as $v$ approaches $c$.

$$\gamma_v \equiv \frac{1}{\sqrt{1 - \dfrac{v^2}{c^2}}}$$

   With equations (1-5), (1-6), and (1-7), our derivation of the Lorentz transformation equations is complete.

$$x' = \gamma_v(x - vt) \tag{1-8a}$$
$$t' = \gamma_v\left(-\frac{v}{c^2}x + t\right) \tag{1-8b}$$
$$x = \gamma_v(x' + vt') \tag{1-9a}$$
$$t = \gamma_v\left(+\frac{v}{c^2}x' + t'\right) \tag{1-9b}$$

Lorentz transformation equations

Remember: These equations yield position and time of an event in one frame, knowing its position and time in another. [*Note:* Equations (1-8) and (1-9) are really equivalent, just solved for different quantities. We include both merely for convenience.]

   The case $v \ll c$ is known as the **classical limit**. In this limit, $\gamma_v$ is essentially unity and $v/c^2$ is negligible. As the reader may verify, the Lorentz transformation equations become the "classical transformation" of (1-1). Classical physics is valid in the realm of small speeds.

Though discussed in a later section, it is important to note here that the spatial coordinates perpendicular to the direction of relative motion are the same in both frames. That is, $y' = y$ and $z' = z$. A consequence is that *distances perpendicular to the direction of relative motion are not subject to length contraction.*

## Two Events: Consequences Revisited

Now we are ready to quantify the claims of Section 1.2. Each of the consequences we discussed there involved two different frames, but each also involved intervals between a *pair* of events—either spatial intervals or time intervals or both. Let event 1 have position and time $(x_1, t_1)$ in frame $S$ and $(x'_1, t'_1)$ in frame $S'$. Event 2 has positions and times $(x_2, t_2)$ and $(x'_2, t'_2)$. Inserting the positions and times for each event separately into equations (1-8) and (1-9), then subtracting, we obtain

$$x'_2 - x'_1 = \gamma_v [(x_2 - x_1) - v(t_2 - t_1)] \tag{1-10a}$$

$$t'_2 - t'_1 = \gamma_v \left[ -\frac{v}{c^2}(x_2 - x_1) + (t_2 - t_1) \right] \tag{1-10b}$$

$$x_2 - x_1 = \gamma_v [(x'_2 - x'_1) + v(t'_2 - t'_1)] \tag{1-11a}$$

$$t_2 - t_1 = \gamma_v \left[ +\frac{v}{c^2}(x'_2 - x'_1) + (t'_2 - t'_1) \right] \tag{1-11b}$$

We see that what is either strictly a time interval (two events with the same position) or strictly a distance interval (two events with the same time) in one frame is a combination of time and distance intervals in the other.

Let us examine the consequences of Einstein's postulates one at a time. The first—relative simultaneity—we quantify in an example.

### Example 1.1

Anna is on a flatcar moving east at $0.6c$ relative to Bob (Figure 1.6). She holds a flashbulb in each hand and causes them to flash simultaneously. Anna's hands are 2 m apart and her arms are oriented in an east-west direction. According to Bob, which flashbulb flashes first and by how much?

### Solution

Let Anna be $S'$ and Bob $S$. Event 1 is the east flash and event 2 the west flash. According to Anna, the two events occur simultaneously, so $t'_2 - t'_1 = 0$, and

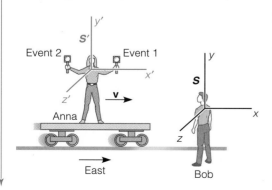

**Figure 1.6** Bob viewing two events that occur simultaneously according to Anna.

they are a distance apart $x_2' - x_1' = -2$ m. (This is negative because event 2 is west—i.e., on the negative side—of event 1.) To find the time interval according to Bob, we use (1-11b).

$$t_2 - t_1 = \frac{1}{\sqrt{1 - \frac{(0.6c)^2}{c^2}}}\left[\frac{0.6c}{c^2}(-2 \text{ m}) + 0\right] = 1.25\frac{-0.6 \times 2 \text{ m}}{3 \times 10^8 \text{ m/s}} = -5 \text{ ns}$$

Event 2 has the smaller (earlier) time; that is, the west bulb flashes 5 ns earlier.[4]

### Time Dilation

Consider two events, ticks of a clock, that occur *at the same location* in frame $S'$: $x_2' = x_1'$. From (1-11b), we see that $t_2 - t_1 = \gamma_v(t_2' - t_1')$, or, in more compact notation, $\Delta t = \gamma_v \Delta t'$. The symbol $\Delta t_0$ and special term **proper time** are used for the time that elapses in the frame in which events all occur at the same location. In the case we have considered here, that frame is $S'$, but it need not be in general, so it is better to write our result using the symbol $\Delta t_0$.

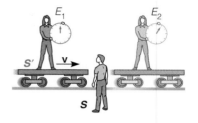

| $\Delta t = \gamma_v \Delta t_0$ | (1-12) |

The time interval $\Delta t$ that passes in another frame is longer by the factor $\gamma_v$ than the interval $\Delta t_0$ that passes in the frame in which events all occur at the same location.

For any series of events, there is always a frame where events all occur at the same place, a frame that "follows the events around." (For instance, a duck may dive and turn as it quacks in flight, but in the duck's frame of reference, the quacks all occur at the same location.) In Section 1.11, it is shown that observers in all other frames will agree on the time they see elapsing in this frame, although each sees a different time pass on their own clock. We might have guessed this, though, because the alternative makes no sense: Imagine the time passing on Bob's bathroom clock as he brushes his teeth depending on who happens to be observing him from a passing spaceship! The point is that relativity never causes us to feel as though we are moving in "slow-motion"; we each feel normal in our own frame, no matter who else might be watching.

### Length Contraction

Consider two events that occur *at the same time* in frame $S$, simultaneous observations from frame $S$ of the ends of an object fixed in frame $S'$: $t_2 = t_1$. Using (1-10a), we find that $x_2' - x_1' = \gamma_v(x_2 - x_1)$, or $L = L'/\gamma_v$.[5] The length of an object in the frame in which it is at rest is given the symbol $L_0$ and is known as **proper**

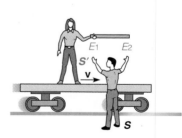

---

[4]We could, of course, have calculated the two times separately using (1-9b): Flash/event 1 occurs at $x_1' = +1$ m, giving $t_1 = +2.5$ ns, and flash/event 2 occurs at $x_2' = -1$ m, yielding $t_2 = -2.5$ ns. According to Bob, the west flash occurs (naturally, before actually reaching anyone's eyes) before the origins cross—that is, at negative time—while the east flash occurs afterward.

[5]An observer in frame $S'$ will not agree with the observer in frame $S$ that the observations of the ends of the object are made simultaneously. However, the observer in frame $S'$ does see that the observations are at least made *at the ends of the object,* a distance apart equal to the length of the object fixed in her frame.

**length.** Since the "rest frame" of an object need not in general be frame $S'$, we express our result using $L_0$ in place of $L'$.

$$L = \frac{L_0}{\gamma_v} \qquad\qquad (1\text{-}13)$$

The length $L$ of an object in a frame relative to which the object moves is shorter by the factor $\gamma_v$ than its length $L_0$ in the frame in which it is at rest.

As in the case of time dilation, no matter what frame he occupies or who else might be watching, an observer will never feel contracted along his direction of motion. He is always in his rest frame; he never moves relative to himself! It is only observers in other frames who will determine that he occupies less space than normal.

Now let us put what we have learned into practice.

---

### Example 1.2

A muon is created in the atmosphere 3 km above Earth's surface, heading downward at speed $0.98c$. It survives 2.2 $\mu$s *in its own frame of reference* before decaying. (a) *Classically,* how far would the muon travel before decaying and how much longer than 2.2 $\mu$s would it have to survive to reach the surface? (b) *Relativistically,* according to an observer on Earth, how long will the muon survive before decaying and will it reach the surface? (c) *Relativistically,* according to the muon, how far is it from the point in the atmosphere where the muon is created to Earth's surface, and how much time will it take this distance to pass the muon?

### Solution

(a) The muon would travel $d = ut = (0.98 \times 3 \times 10^8 \text{ m/s})(2.2 \times 10^{-6} \text{ s}) = 647$ m. To travel 3 km would require

$$t = \frac{d}{u} = \frac{3 \times 10^3 \text{ m}}{0.98 \times 3 \times 10^8 \text{ m/s}} = 10.2 \ \mu s$$

An extra 8 $\mu$s of "life" would be required. The muon should not reach the surface, for which there are two equivalent explanations: The distance is about five times too far; or the muon survives only about one-fifth the required time.

(b) The observer on Earth, watching a clock glued to the muon (fixed in that frame) ticking off 2.2 $\mu$s, will find that more time elapses on his own clock, by the factor $\gamma_v$.

$$\gamma_v = \frac{1}{\sqrt{1 - \dfrac{v^2}{c^2}}} = \frac{1}{\sqrt{1 - \dfrac{(0.98)^2 c^2}{c^2}}} = 5.03$$

*Warning:* Having found the ubiquitous $\gamma_v$, we may find correct time and space intervals. But "plugging in" blindly to a time-dilation or length-contraction formula will give the wrong answer 50% of the time. The student should reason beforehand whether the interval sought is larger or smaller than the known interval. Here, the observer on Earth watches events occurring at the same location in the muon's frame and must therefore see his own clock register the longer time—we must *multiply* (not divide) 2.2 $\mu$s by 5.03. Of course, the equation $\Delta t = \gamma_v \Delta t_0$ agrees; it comes with the condition that $\Delta t_0$ is the interval in the frame where the events occur at the same location. Thus, $\Delta t_0$ must be the 2.2 $\mu$s in the muon's frame and $\Delta t$ the interval in the Earth observer's frame (in which the atmosphere and surface of Earth are different locations).

$$\Delta t = \gamma_v \Delta t_0 = 5.03 \times 2.2 \ \mu s = 11.1 \ \mu s$$

We showed earlier that from the classical perspective, the muon would have to "live" at least 10.2 $\mu$s to reach the ground. This calculation was valid because the speed, distance, and time involved were all according to the *same* (Earth) observer. (The trap to avoid is using a distance according to an observer in one frame and a time according to an observer in a different frame.) Similarly, we can find a distance according to an earth observer if we use the "lifetime" of the muon as computed in the Earth frame: At $0.98c$, the muon could travel $(0.98 \times 3 \times 10^8 \text{ m/s})(11.1 \times 10^{-6} \text{ s}) = 3.25$ km. According to the Earth observer, the muon, surviving 11.1 $\mu$s and able to travel 3.25 km, will definitely reach the surface.

(c) The muon sees a moving "object," namely the region in the atmosphere from where it is "born" to Earth's surface. These points are 3 km apart according to an Earth observer. When viewed *simultaneously by the muon,* they will be separated by a smaller distance. We should expect to divide 3 km by 5.03. The equation $L = L_0/\gamma_v$ agrees; it is Earth's frame in which the "object" is at rest, so $L_0$ is 3 km.[6]

$$L = \frac{L_0}{\gamma_v} = \frac{3 \text{ km}}{5.03} = 597 \text{ m}$$

Length contraction is real. According to the muon, this passing "object" occupies only 597 m. Consequently, it will pass by in only

$$t = \frac{L}{u} = \frac{597 \text{ m}}{0.98 \times 3 \times 10^8 \text{ m/s}} = 2.03 \ \mu s$$

The muon, which by its own clock survives 2.2 $\mu$s, will hit the ground after 2.03 $\mu$s.

It is worth noting that according to the Earth observer, the muon could have been created at an altitude as high as 3.25 km and still strike the surface, although just barely. If it had been created at this height, the muon would have seen an "object" 3.25 km/5.03 = 647 m long, which would pass in 647 m/($0.98 \times 3 \times 10^8$ m/s) = 2.2 $\mu$s. The muon would agree that it barely survives to meet the surface. Both observers must agree on this! A collision between a muon and Earth is a single event. It cannot occur in one frame and not in another. The *explanations* may differ, however. The observer on Earth explains that time dilation is responsible, while the muon argues that it is length contraction.

[6]When we derived $\Delta t = \gamma_v \Delta t_0$ and $L = L_0/\gamma_v$, the frame in which the events occurred at the same locations (where $\Delta t_0$ passes) and the frame in which the object was at rest (where the length is $L_0$) were both $S'$. It might be said, then, that we defined Earth as frame $S$ in part (a) and the muon as frame $S$ in part (b). Since it cannot be said that any frame is truly at rest, we are always free to assign $S$ and $S'$ any way we find convenient. The only distinction between them is that $S'$ moves in the positive direction relative to $S$ and $S$ in the negative relative to $S'$. In Example 1.2, we are not concerned with direction.

**Example 1.3**

A proposed plane would carry intercontinental travelers at speeds many times those of today's fastest airliners. Imagine that such a plane is 50 m long (when parked) and cruises at 6000 m/s (13,400 mi/h, or about Mach 18). At precisely noon, the plane passes over Los Angeles and an observer on the ground with exceptional eyesight peers through its window. He sees that the clock inside also reads precisely noon. (a) How long is the moving plane according to this observer? (b) Later, a ground observer in Seattle, 1600 km north, checks the plane's clock as it passes overhead. By how much does the reading differ from that on her watch?

**Solution**

(a) The ground observer in Los Angeles, viewing the plane's ends *at the same time,* must see a length shorter than the plane's proper length of 50 m.

$$L = \frac{^{\text{!`}}L_0}{\gamma_v} = 50 \text{ m}\sqrt{1 - \left(\frac{6 \times 10^3}{3 \times 10^8}\right)^2} = 50 \text{ m}(1 - 4 \times 10^{-10})^{1/2}$$

Here we have a situation that crops up often when $v$ is much less than $c$. The quantity in parentheses differs from unity by only a very small amount. Raising it to any power requires either a calculator of great precision or some kind of approximation. The approximation shown in the margin works well.[7] To use it here, we replace $x$ with $-4 \times 10^{-10}$ and $a$ with $\frac{1}{2}$.

When $|x| \ll 1$:
$$(1 + x)^a \cong 1 + ax \quad \textbf{(1-14)}$$

$$L = 50 \text{ m }[1 + (-4 \times 10^{-10})]^{1/2}$$

$$\cong 50 \text{ m}\left[1 + \frac{1}{2}(-4 \times 10^{-10})\right] = 50 \text{ m} - 10^{-8} \text{ m}$$

The plane would occupy 10 nm less space than it does at rest.

(b) Both ground observers are watching a single clock ticking off events *at the same location* in the plane's frame, so this clock registers the proper time $\Delta t_0$ and the time passing on their own watches is the longer $\Delta t$. We calculate $\Delta t$ in the usual classical way, because both it and the 1600 km are according to the same (ground) observers. (The relative speed $v$ between frames is the same according to all observers.)

$$\Delta t = \frac{\Delta x}{v} = \frac{1600 \times 10^3 \text{ m}}{6000 \text{ m/s}} = 267 \text{ s} \quad \text{(L.A. to Seattle in 4.4 min!)}$$

$$\Delta t_0 = \frac{\Delta t}{\gamma_v} = 267 \text{ s}\sqrt{1 - \left(\frac{6 \times 10^3}{3 \times 10^8}\right)^2}$$

---

[7]It is the series expansion

$$(1 + x)^a = 1 + \frac{1}{1!}ax^1 + \frac{1}{2!}a(a - 1)x^2 + \cdots$$

truncated after the $x^1$ term.

Again using the approximation,

$$\Delta t_0 = 267 \text{ s}[1 + (-4 \times 10^{-10})]^{1/2} \cong 267 \text{ s}\left[1 + \frac{1}{2}(-4 \times 10^{-10})\right]$$

$$= 267 \text{ s} - 5.3 \times 10^{-8} \text{ s}$$

The plane's clock will read 53 ns earlier than the Seattle observer's watch. We see that even at speeds very high by ordinary standards, relativistic effects are small.

Now let us confront the perplexing question of two observers each seeing the other aging more slowly and contracted along the direction of relative motion.

## Example 1.4

Anna and Bob are on identical spaceships of length $\ell = 12 \times \sqrt{3}$ m $\cong$ 21 m.[8] Anna's (frame $S'$) moves to the right relative to Bob's (frame $S$). Clocks visible from the exterior are glued to the center, the nose, and the tail of each ship. Figure 1.7 shows what Bob sees at $t = 0$: The tails of the ships are aligned, Anna's tail-end clock reads zero, and the nose of Anna's spaceship is precisely aligned with the center of Bob's. According to Bob, (a) what is the speed of Anna's ship, and (b) what are the readings on Anna's clocks? (c) Show that the centers of the ships will be aligned 20 ns later and find the readings on Anna's clocks (again according to Bob) at this time. (d) Find the readings on Anna's clocks when Bob's read 40 ns and 60 ns. (e) Find pairs of events showing time dilation and length contraction from both travelers' perspectives.

## Solution

(a) Suppose Bob is at the tail end of his ship and Bob Jr. is at the center. At exactly the same time in their frame ($t = 0$), each sees one end of Anna's ship—its length $L$ is only $\frac{1}{2}\ell$! But in the frame in which it is at rest, $S'$, Anna's ship is of length $L_0 = \ell$. Thus, $L = L_0/\gamma_v \Rightarrow \gamma_v = 2$.

$$\frac{1}{\sqrt{1 - \dfrac{v^2}{c^2}}} = 2 \Rightarrow v = \frac{\sqrt{3}}{2}c$$

(b) At $t = 0$ in Bob's frame, Bob Jr. looks at the clock directly across from him on Anna's ship. This clock registers the time at that location in that frame (as do all clocks!). *We may find the time in Anna's frame when Bob Jr. makes his observation because we know when and where this observation/event occurs in Bob's frame.* It occurs at $t = 0$ and $x = \frac{1}{2}\ell$. Using (1-8b),[9]

$$t' = 2\left[-\frac{(\sqrt{3}/2)c}{c^2}\left(\frac{1}{2}\ell\right) + 0\right]$$

$$= -2\frac{\sqrt{3}/2}{c}\left(\frac{1}{2}\right)(12 \times \sqrt{3}) \text{ m}$$

$$= -\frac{(3/2) \times 12 \text{ m}}{3 \times 10^8 \text{ m/s}} = -60 \text{ ns}$$

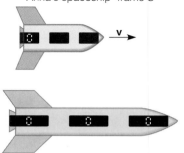

Anna's spaceship–frame $S'$

Bob's spaceship–frame $S$

**Figure 1.7** How Anna's passing ship appears to Bob.

[8]The reason for the weird value is so that later values work out to be "nice" numbers.

[9]Whenever any two of $x$, $t$, $x'$, and $t'$ are known, the other two can be found from equations (1-8) or (1-9), whichever is easiest. At times the reader may wonder whether it might be better to use the time dilation or length contraction equation. As a rule, if a single time or location (one event) is sought, the Lorentz transformation should be used. If a time or space *interval* is sought, time dilation or length contraction is usually fastest. In the question at hand, we consider a single event. The *passage* of time hasn't yet come into the picture.

Again, we are not studying optical illusions here; this is *the time* according to whoever stands at this spot in Anna's frame when Bob Jr. makes his observation! Now, concerning the center clock on Anna's ship, Bob might have an assistant on his ship across from that clock, too, whose observation occurs at $t = 0$ and $x = \frac{1}{4}\ell$. Equation (1-8b) then yields $t' = -30$ ns (Figure 1.8). Anna's clocks do not strike zero simultaneously according to Bob—but they shouldn't, for they *do* strike zero simultaneously according to Anna.

(c) Bob sees the center of Anna's ship move $\frac{1}{4}\ell$.

$$\Delta t = \frac{(1/4)(12\sqrt{3}\text{ m})}{(\sqrt{3}/2)(3 \times 10^8 \text{ m/s})} = 20 \text{ ns}$$

To find the new readings seen on Anna's clocks by observers in Bob's frame we merely insert 20 ns instead of zero. At the center of Bob's ship, $x = \frac{1}{2}\ell$ and $t = 20$ ns, and equation (1-8b) yields $t' = -20$ ns. Note that while those in Bob's frame have seen 10 ns go by on the center clock in Anna's frame, 20 ns have gone by in their own frame—time dilation according to Bob. We do not have to use the Lorentz transformation equations for the other readings; Bob must see all the clocks in Anna's frame advance at the same *rate*, 10 ns for every 20 ns of Bob's time. The nose-end clock must read $-50$ ns, and the tail-end one must read 10 ns.

(d) Another 20 ns later according to Bob, the nose of Anna's ship will be aligned with the nose of Bob's ship. Bob will have seen another 10 ns pass on the clocks in Anna's frame, thus reading 20, $-10$, and $-40$ ns. Still another 20 ns later, these will be 30, 0, and $-30$ ns, and the center of Anna's ship will be aligned with the nose of Bob's ship, as shown in Figure 1.9.

(e) We have already discussed length contraction and time dilation according to Bob, in parts (a) and (c). Anna also observes these effects. Consider two events: $E_1$, the center of Bob's ship passing the nose of Anna's, and $E_2$, the passing of the centers of the ships. In the first, an observer in Anna's frame, at $-60$ ns on her clock, sees the clock at the center of Bob's ship reading zero. In event $E_2$, an observer in Anna's frame 40 ns later looks at *the same clock* in Bob's frame and sees that it reads only 20 ns. Observers in Anna's frame have observed ticks of a clock fixed in Bob's frame, and see half the time elapse on that clock as on their own—time dilation according to Anna. (Since Bob and Anna *both* must see the other's clocks as unsynchronized, to verify time dilation they must always watch the *same* moving clock.) Consider a different pair of events: $E_3$, the passing of the tails of the spaceships, and $E_4$, the nose of Bob's ship passing the center of Anna's. Though these are separated by 60 ns in Bob's frame, they occur *simultaneously* according to Anna. At $t' = 0$, observers aboard Anna's ship, at the tail and center, simultaneously see the ends of Bob's ship. Accordingly, to Anna the length of Bob's ship is only $\frac{1}{2}\ell$.

Figure 1.10 shows Anna's complementary view of the whole affair. The important point is this: *It is the fact that simultaneity of events is not absolute that allows each observer to see the other's clocks running slowly and the other's distances contracted.*

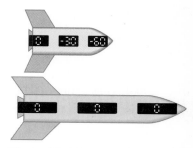

Figure 1.8 According to Bob, Anna's clocks aren't synchronized.

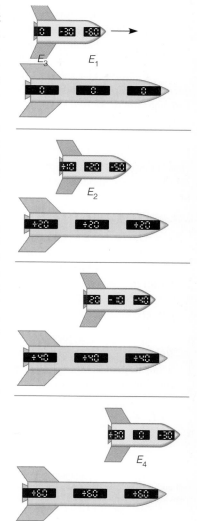

Figure 1.9 What Bob sees.

A final note: As we see in Figures 1.9 and 1.10, the *order* of events depends on the frame from which they are observed! In Bob's frame, the sequence of the events is $E_1$ and $E_3$ simultaneously, then $E_2$, then $E_4$. In Anna's frame it is $E_1$, then $E_2$, then $E_3$ and $E_4$ simultaneously. In particular, Anna and Bob see the order of $E_2$ and $E_3$ reversed![10] We shall return to this interesting point in Section 1.12.

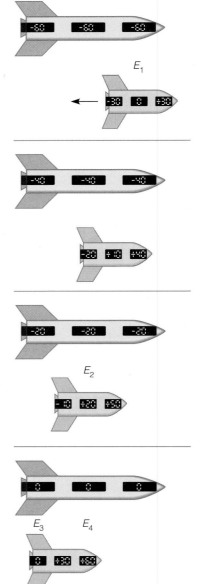

Figure 1.10  What Anna sees.

A simple rotation of coordinate axes in the *x-y* plane is a "transformation" between reference frames, and can be represented graphically. In much the same way, the Lorentz transformation can be represented graphically, with the two "space-time" axes plotted together. Appendix B shows how this can be done, and provides an alternative view of Example 1.4.

Using a situation similar to that in Example 1.4, Figure 1.11 clarifies Einstein's second postulate. It shows Bob's perspective of the passing of two identical ships. In this case, however, the ships' centers are the origins of the frames. (The speeds and lengths are also not the same as in the example. The goal here is simply to strengthen our qualitative understanding of Einstein's second postulate. It is the task of Exercise 11 to verify that the times shown are correct.) A burst of light occurs at the origins just as they cross. The pulse then spreads in all directions. Bob sees the pulse reach the ends of Anna's "moving" ship at different times, but it reaches the ends of his "stationary" ship simultaneously at 40 ns—he sees the pulse spreading symmetrically from his origin. Now consider Anna's view: Judging by the times we see in her frame, she sees the pulse reach the ends of her ship simultaneously, just as Bob does, and in the same amount of time, 40 ns. Each sees the spherical pulse spreading symmetrically at the same speed from the center of their own ship.

## Example 1.5

It takes light 40 years to reach Planet X from Earth. Anna has just been born. Can she get to Planet X by the time she is 30 years old? If so, what speed is required?

### Solution

It might seem that her speed would have to exceed *c*, since she wishes to be 40 light-years (ly) from Earth by the time she is 30.[11] But Planet X is 40 ly away according to an *Earth* observer—Bob, of course. To Anna in her spaceship moving at *v* relative to Earth, the Earth–Planet X "object" will be shorter:

$$L = \frac{40 \text{ ly}}{\gamma_v}$$

---

[10]Perhaps two more obvious time-reversed events are the passing of the ships' ends. As Figures 1.9 and 1.10 clearly show, and as must be the case if each observer sees the other ship as shorter, Bob sees the ships' tails pass first, Anna their noses.

[11]A light-year, or 1 ly, is $(3 \times 10^8 \text{ m/s})(3.16 \times 10^7 \text{ s}) = 9.5 \times 10^{15}$ m, but it is easier *not* to make the conversion here.

As $v$ approaches $c$, the distance approaches zero! The conclusion is provocative: *Anna may go anywhere she likes, in as little time as she pleases; she need only travel at a large enough fraction of c to contract the required distance to an arbitrarily small value.*[12] Here, the "object" must pass by in 30 years of Anna's time. Thus,

$$\text{speed} = \frac{\text{distance}}{\text{time}} \quad \rightarrow \quad v = \frac{40 \text{ ly}/\gamma_v}{30 \text{ yr}}$$

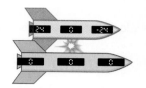

Dividing by $c$, we have[13]

$$\frac{v}{c} = \frac{40 \text{ ly}}{30 \text{ yr} \cdot c} \sqrt{1 - \frac{v^2}{c^2}}$$

Squaring and solving for $v$ yields

$$v = 0.8c$$

We also find that

$$\gamma_v = \frac{1}{\sqrt{1 - (0.8)^2}} = \frac{5}{3}$$

According to Anna, the Earth–Planet X "object" is only $40 \text{ ly} \div \frac{5}{3} = 24 \text{ ly}$ long. At $0.8c$, it passes in only 30 years.

How can Anna reach Planet X in 30 years when it takes light 40 years? One must be careful about *who* is doing the observing. Were a light beam to leave Earth at the same time as Anna, Anna and Bob would agree that the light beam will reach Planet X first. *According to Anna,* the beam, moving away from her at $c$, will meet Planet X, moving toward her at $0.8c$, in much less than 30 years ($13\frac{1}{3}$ years, as shown in the next section). *According to Bob,* the light does take 40 years, but Anna herself takes even longer, since she moves at "only" $0.8c$. In fact, according to Bob, Anna takes $\Delta t_B = 40 \text{ ly}/0.8c = 50 \text{ yr}$ to reach Planet X. Still he will agree that Anna is only 30 years old when she gets there, but by different reasoning than Anna (as in Example 1.2 on muons). While Anna's explanation of why she is able to reach Planet X is length contraction, Bob's explana-

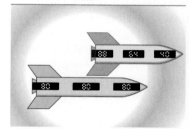

Figure 1.11 A light pulse appears to spread the same distance in all directions in the same time according to both observers.

---

[12]The effects on times and time intervals of any accelerations depend on the separation of the observers; see Exercise 27. To a good approximation they may be neglected at the journey's start, when Anna and Bob are close together. Still we overlook physiological limitations—too rapid an acceleration would be fatal. These are taken up in Exercise 84. We could avoid both problems by supposing that Anna is born aboard an already moving ship just as it passes Earth.

[13]Here is where the use of light years is so convenient: $1 \text{ ly} = c \cdot 1 \text{ yr}$, so $1 \text{ ly}/c = 1 \text{ yr}$. Note also that the same equation arises taking Bob's view: a distance of 40 ly and time of $\gamma_v(30 \text{ yr})$ gives

$$\text{speed} = \frac{\text{distance}}{\text{time}} = \frac{40 \text{ ly}}{\gamma_v(30 \text{ yr})}$$

tion is time dilation. Bob watches a clock fixed in Anna's frame register $\Delta t_A$. The interval $\Delta t_B$ that passes on his own will be longer.

$$\Delta t_B = \gamma_v \, \Delta t_A \quad \rightarrow \quad 50 \text{ yr} = \frac{5}{3} \Delta t_A \quad \Rightarrow \quad \Delta t_A = 30 \text{ yr}$$

As noted, a light beam leaving Earth at the same time as Anna would appear to her to be moving at $c$. Bob would see this beam moving at $c$ relative to himself, or $c - 0.8c = 0.2c$ *relative to Anna.*[14] But, seeing Anna aboard her ship making very lethargic observations of the light beam, Bob will certainly understand her "mistake."

A question to ponder: Because Anna moves relative to the Earth–Planet X system, though still at a speed less than $c$, she sees a distance shorter than 40 ly, requiring less than 40 years to travel. What must light "see"?

## ▼ 1.5 The Twin Paradox

Simply stated, the twin paradox is as follows: If one of two twins stays on Earth and the other travels away at high speed, and they each see the other aging more slowly, when the traveling twin returns, each twin would claim that the other is younger. Impossible! Two people cannot sit down together in a room and each see that the other is many years younger. Such a situation would be a violation of everyday experience.

There is really no paradox. The crucial factor overlooked is that in order for the two twins to get back together and discuss their beliefs, one of them must *accelerate.* In so doing, the accelerating twin does not remain in the same inertial frame, but rather ends up in a frame heading back toward Earth. An observer always in *an inertial frame* will observe relativistic effects in the usual way — moving objects contracted and moving clocks running slowly — *even if the object or clock he observes accelerates.* At any instant, the proper speed to use in $\gamma_v$ would be the instantaneous speed. In the twin paradox, the observations of the twin who remains on inertial Earth are the reliable ones. But if an *observer* accelerates, we cannot apply length contraction and time dilation simply. If Anna (Bob's fraternal twin sister) as observer accelerates *toward* Bob, she moves to a different inertial frame in which the truth is that Bob is *absolutely* older, regardless of any difference in aging *rates.* (An observer accelerating away from Bob moves to a frame in which Bob is absolutely *younger.* See Exercise 26.) It is the asymmetry caused by Anna's acceleration that allows her to agree with ever-reliable observer Bob when she gets back — that Bob is older.

Let us examine the twin paradox in the context of Example 1.5: Earthbound Bob waits 50 years by his own clock for Anna, traveling away from Earth at $0.8c$ $\left(\gamma_v = \frac{5}{3}\right)$, to reach Planet X, 40 ly away. As we know, Bob, watching a clock fixed

---

[14]This is no violation of Einstein's postulate; this is not a speed *relative to Bob as observer,* but a speed *according to Bob* of a light beam relative to something else — that is, Anna. See discussion following Example 1.7.

in Anna's frame of reference, determines that Anna ages 50 yr $\div \frac{5}{3} = 30$ yr. But according to Anna, *Bob* is moving. Anna watches a clock fixed in Bob's frame, and it registers less time than goes by on her own clock, by the same factor $\gamma_v$. In the 30 years passing on her watch as she travels to Planet X, Anna must claim that Bob ages only 30 yr $\div \frac{5}{3} = 18$ yr. Thus, when Anna reaches Planet X, Bob says "I'm 50 years old and Anna's 30 years old," and Anna says "I'm 30 years old and Bob's 18 years old." [15]

Thus far, we have no paradox. None of this is contrary to our experience, because no one has ever done this! What *would* be a paradox is if Anna were to return to Earth and sit down with Bob, with Bob saying "I'm 100, you're 60" and Anna saying "No, I'm 60 but you're 36." This would be contrary to our experience.

A paradox

To resolve the paradox, we introduce another observer, Carl, who happens to be in a ship heading toward Earth at 0.8$c$ and passing Planet X just as Anna does. Carl is in an inertial frame—always has been and always will be. When Carl passes Earth, *Bob will unarguably be 100 years old,* because Bob waits 50 years for Anna to reach Planet X and will wait another 50 for Carl to arrive. Now Carl will age 30 years in flight, and must say that Bob ages less, just as Anna does—the same relative speed is involved. Carl determines that Bob ages 18 years. If Bob is 100 when Carl gets to Earth, and Bob aged 18 years during the flight, then Bob must be 82 years old, according to Carl, when he passes Planet X. This is truth *in Carl's frame!*

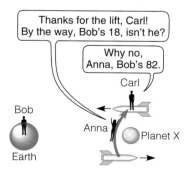

[15] In truth, neither will actually *see* this, because it does take light a certain time to reach each observer. But Bob's assistant on Planet X could bear witness to Anna's age when Bob is 50. Correspondingly, if Anna's ship were *very* long, such that the tail end is still at Earth while the nose is at Planet X, Anna's assistant at the back of her ship could testify to Bob's youth, 18 years, when Anna is 30.

To return, Anna must *accelerate*—that is, change from one inertial frame to another. (It makes no difference whether the *speed* could be kept constant; an inertial frame has constant *velocity*.) So far, we have not considered how this might affect Anna's perceptions, but Carl does not accelerate, so we may depend on his observations. In changing frames, Anna goes from one frame, where Bob is 18 years old, to another, where (as professed by Carl himself) Bob is 82 years old. Although on the return leg of the journey, Anna does indeed determine that Bob is aging more slowly than herself, it is too late. Bob's huge forward leap in age (according to Anna) has settled the matter: Bob will be much older than Anna when they reunite. The paradox is resolved: For the twins to be able to compare their ages, one of them must *accelerate,* and that makes all the difference.

Two questions may still haunt the reader: (1) How is it possible that when Anna and Carl are both at Planet X, Anna says Bob is 18 years old, while Carl says Bob is 82 years old? (2) Can Carl tell Anna about Bob's future? We begin by having Bob send a light signal from Earth toward Planet X so that it arrives at the precise instant Anna and Carl pass Planet X. If according to Bob it takes light 40 years to reach Planet X and it takes Anna 50 years, then Bob had better send this signal when he is 10 years old. Let the light signal be an image of Bob blowing out candles at his tenth birthday party. When this signal reaches Planet X, both Anna and Carl will see an image of Bob as a 10-year-old. Each will have intercepted all earlier such signals, so they will have knowledge of Bob's prior activities, but no knowledge beyond that. The answer to question 2 is, No, because neither has actually seen any of Bob's life after his tenth birthday party.

However, we do not answer the question "What is Bob's age?" by saying "The age he was when the light that I see originated." Neither Anna nor Carl will say that Bob is 10 years old when they reach Planet X, because each realizes that Bob will have aged while the light traveled. But Anna and Carl are scrupulous record keepers; from their records they can calculate what Bob's age must be when they pass Planet X, and the conclusions they reach differ markedly. Let us investigate this, and in so doing answer question 1.

Figure 1.12 shows how each of the three observers sees the situation as Bob's light signal leaves Earth. As we found in Example 1.5, Anna sees the distance between Earth and Planet X as 24 ly, and she sees a relative velocity between the light signal and Planet X of $1.8c$. Thus, only 24 ly/$1.8c$ = $13\frac{1}{3}$ yr passes for Anna. If $13\frac{1}{3}$ years go by on her clock, the time interval she determines to have passed on Bob's moving clock is smaller: $13\frac{1}{3}$ yr = $\gamma_v \, \Delta t_{\mathrm{B}}$  $\Rightarrow$  $\Delta t_{\mathrm{B}} = 13\frac{1}{3}$ yr $\div \frac{5}{3}$ =

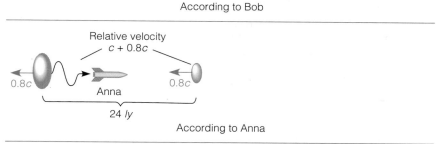

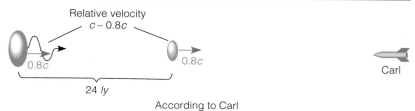

**Figure 1.12** A light beam traveling from Earth to Planet X according to three different observers.

8 yr. Anna concludes that when she reaches Planet X, Bob is 10 yr + 8 yr = 18 yr. Carl, on the other hand, must also see a separation between Earth and Planet X of 24 ly, but he sees a relative velocity between the light signal and Planet X of only $0.2c$. According to him, then, the time that passes as the light signal travels to Planet X is 24 ly/$0.2c$ = 120 yr. He too knows that less time will have passed on Bob's moving clock, and so judges that it marks off $\Delta t_B = $ 120 yr $\div \frac{5}{3}$ = 72 yr. Added to the 10-yr age indicated by the arriving light, Carl deduces that when he is at Planet X, Bob is 82 years old.

Another observation is of interest: If instead of Anna switching ships, Carl jumped to Anna's, accelerating *away* from Bob, Carl would indeed jump from a frame in which Bob is 82 years old to one in which Bob is 18 years old. But alas, this would be of little more than academic interest, again because the light both have seen bears information about only the first 10 years of Bob's life.

## The Light Signals Solution

We conclude our discussion of the twin paradox with an alternative way of arriving at the same conclusion—asymmetric aging—in Anna's round-trip to Planet X.

Figure 1.13 shows Anna's journey from Bob's point of view: Anna travels away at a speed/slope somewhat less than $c$, then reverses course and returns at the same speed. To keep each other fully informed, Anna and Bob by prearrangement send out light signals at 10-yr intervals. The signals Bob sends away from Earth have slope $+c$. If he sends one every 10 years, the first, Bob at his tenth birthday party,

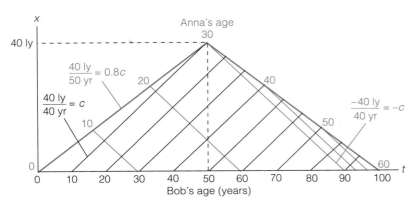

**Figure 1.13** The motion of Anna and of light signals exchanged between Anna and Bob on Anna's round-trip to Planet X.

will not reach the speedy Anna until she reaches Planet X, when she is 30 years old. The signals Anna sends back toward Earth have slope $-c$. So long as no one changes reference frames, symmetry demands that Bob also receive his first signal from Anna when he is 30 years old, and it must show Anna as a 10-year-old. Each *actually sees* images of the other aging only one-third as much as themselves. Immediately upon turning around, Anna begins to receive signals from Bob much more frequently—a new image of Bob another 10 years older every $3\frac{1}{3}$ years of her return trip. The images she now sees show Bob aging three times as rapidly as herself. As soon as Bob begins to receive signals Anna sends on her return trip, he too receives one every $3\frac{1}{3}$ years and sees Anna aging three times as rapidly as himself. But here is the asymmetry: While Anna begins to see Bob aging more rapidly at the *midpoint* of her journey, the instant she changes frames, Bob has to wait 90 years (50 years for Anna to reach Planet X and another 40 for the light from there to reach Earth). Now there is only another 10 years left. Anna sees Bob age one-third as fast for 30 years and three times as fast for 30 years [$(30 \text{ yr} \times \frac{1}{3}) + (30 \text{ yr} \times 3) = 100 \text{ yr}$], but Bob sees Anna age one-third as fast for 90 years and three times as fast for only 10 [$(90 \text{ yr} \times \frac{1}{3}) + (10 \text{ yr} \times 3) = 60 \text{ yr}$]. When it is all over, Bob has received a total of six signals, confirming that Anna has aged 60 years, and Anna has received ten, proving that poor Bob has aged 100 years.

In closing, we should note that the danger in this kind of solution to the twin paradox is that it emphasizes what one *sees,* rather than what is actually happening. During neither leg of the journey would the observers claim the other is aging only one-third as rapidly as themselves: $\gamma_v$ is not 3; it is $\frac{5}{3}$. Much less would either observer make the mistake of saying that the other is aging *more* rapidly (so long as the observer is not accelerating). Both realize that what they actually see is a combination of time dilation and the effect of their rapidly increasing or decreasing separation. The Doppler effect, discussed in Section 1.6, takes both changes in separation and time dilation into account. It shows that the rate at which signals would actually be received is decreased by the factor $\sqrt{(1 - v/c)/(1 + v/c)}$ when the source is moving away, and increased by the factor $\sqrt{(1 + v/c)/(1 - v/c)}$ when the source is approaching. For a speed $0.8c$, these are, respectively, $\frac{1}{3}$ and 3. The light signals approach to the twin paradox is straightforward: One twin need only count the light signals he or she sees to determine the age of the other twin—and seeing is believing. But it does skirt the point that the asymmetry is introduced instantly when the traveling twin changes frames.

# ▼ 1.6 The Doppler Effect

As is the case for sound, if a light source moves relative to an observer, the frequency received by the observer need not be that emitted by the source. The effects are not identical, though. Sound requires a medium, while light does not. Moreover, in the case of light we are usually interested in relative speeds comparable to *c,* in which case the time interval between the *production* (not simply *arrival*) of consecutive wave fronts in the observer's frame is not the same as in the source's.

An observer stands at the origin of frame *S* (Figure 1.14). A light source moves at an angle $\theta$ with respect to the line connecting the observer and the source, and emits a frequency $f_{source}$ in the frame in which it is at rest, frame *S'*. It emits consecutive wave fronts at point 1 and point 2. In the observer's frame, the time between the *production* of these fronts is $\Delta t_p$. They would also *arrive* at the observer separated by this time interval were it not for the motion of the source. If the source moves away, the second wave front has farther to travel. Thus, the time interval between the arrivals, $\Delta t_a$, differs by the amount of time required for the second front to travel the extra distance. As shown, the extra distance is $\sim v \Delta t_p \cos \theta$. (This is an approximate value since the angle labeled $\sim\theta$ is not exactly $\theta$. But even at near-light speeds, a source would not move far during the $10^{-14}$ s period of light, so the lines from the observer to points 1 and 2 are essentially parallel. Also, note that the result makes sense no matter what direction the source moves; if it moves toward the observer, $\theta = 180°$ and the "extra" distance is negative.) Dividing by *c,* we obtain the time to travel this extra distance: $(v/c)\Delta t_p \cos \theta$. Thus, the time between the arrivals of the two fronts is

$$\Delta t_a = \Delta t_p + \frac{v}{c}\Delta t_p \cos \theta = \Delta t_p \left( 1 + \frac{v}{c}\cos \theta \right)$$

The observed period is $\Delta t_a$. Were classical physics applicable, $\Delta t_p$ would be the source period, and we could merely invert this equation to find an equation relating observed frequency to source frequency. However, the source period $T_{source}$ is the time interval between production of the consecutive wave fronts *in frame S'*. Since the wave fronts are produced *at the same place* in frame *S'*, the time interval between their production in frame of reference *S* is longer: $\Delta t_p = \gamma_v T_{source}$. Now we have a relationship between the observed period ($\Delta t_a \equiv T_{obs}$) and source period.

$$T_{obs} = (\gamma_v T_{source})\left( 1 + \frac{v}{c}\cos \theta \right) = T_{source}\frac{\left( 1 + \frac{v}{c}\cos \theta \right)}{\sqrt{1 - \frac{v^2}{c^2}}}$$

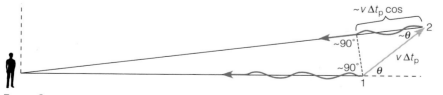

Frame *S*

**Figure 1.14** The production and propagation of two wave fronts emitted by a light source moving relative to the observer.

Inverting then gives a relationship between the source and observed frequencies.

Doppler effect

$$f_{obs} = f_{source} \frac{\sqrt{1 - \dfrac{v^2}{c^2}}}{1 + \left(\dfrac{v}{c}\right)\cos\theta} \qquad (1\text{-}15)$$

In conclusion, there are two factors governing the observed frequency: (1) Consecutive wave fronts produced by the source will in general have different distances to travel to the observer (taken into account in the classical Doppler effect in sound), and (2) time dilation.

If the source moves away from the observer, $\theta = 0$.

$$f_{obs} = f_{source} \frac{\sqrt{1 - \dfrac{v^2}{c^2}}}{1 + \dfrac{v}{c}} = f_{source} \frac{\sqrt{\left(1 - \dfrac{v}{c}\right)\left(1 + \dfrac{v}{c}\right)}}{\sqrt{\left(1 + \dfrac{v}{c}\right)\left(1 + \dfrac{v}{c}\right)}} = f_{source}\left(\frac{1 - \dfrac{v}{c}}{1 + \dfrac{v}{c}}\right)^{\frac{1}{2}}$$

$$(1\text{-}16)$$

The observed frequency is smaller, the period and wavelength larger. One of the important clues that the universe is expanding is that the light from distant stars is "red-shifted"—all of the wavelengths in the normal spectrum emitted by the hot gases are shifted toward the red (longer wavelength) end of the spectrum. Astronomers have documented shifts in which the observed wavelengths are many times the wavelengths usually emitted at rest, indicating that these stars are moving away very rapidly.

It is left as an exercise to show that if the source moves *toward* the observer ($\theta = 180°$), then $f_{obs}$ is found merely by switching the sign of the speed $v$ in equation (1-16). In this case, the observed frequency would be larger, and the period and wavelength shorter—the light is "blue-shifted." Time dilation alone would lead to a *longer* period, because the charges in the source indeed oscillate more slowly according to an observer in frame $S$ than they do in their "rest frame" frame $S'$. Since the observed period is actually smaller, we conclude that the time-dilation effect is more than compensated for by the decreasing separation between source and observer.

If the source gets neither closer to nor farther from the observer ($\theta = 90°$), equation (1-15) becomes $f_{obs} = f_{source}\sqrt{1 - v^2/c^2}$. The observed frequency is smaller. This is the so-called *transverse* red-shift. In the classical treatment of sound, there is *no* Doppler shift if the source neither gets closer to nor farther from the source. The sole reason for the shift here is time dilation. A simple example would be a source circling an observer. Waves all have the same distance to travel to the observer, but they are produced by charges that, according to the observer, are moving (oscillating) slowly.

**Example 1.6**

A star's spectrum is observed to have all wavelengths four times their ordinary values. How fast is the star receding from Earth?

## Solution

A longer wavelength corresponds to a smaller frequency. Thus, $f_{obs} = \frac{1}{4} f_{source}$.

$$f_{obs} = f_{source} \left( \frac{1 - \dfrac{v}{c}}{1 + \dfrac{v}{c}} \right)^{\frac{1}{2}} \rightarrow \frac{1}{4} f_{source} = f_{source} \left( \frac{1 - \dfrac{v}{c}}{1 + \dfrac{v}{c}} \right)^{\frac{1}{2}}$$

or

$$\frac{1}{4} \sqrt{1 + \frac{v}{c}} = \sqrt{1 - \frac{v}{c}}$$

Squaring and solving,

$$\frac{v}{c} = \frac{1 - (1/4)^2}{1 + (1/4)^2} = 0.882c$$

## 1.7 Velocity Transformation

With a firm grasp on position and time, we may now tackle velocity. We know that the classical transformation $u' = u - v$ is wrong. To find the correct transformation involves a straightforward application of the Lorentz transformation equations. We begin by considering an object moving only along the $x$-axis. By definition, the velocity of an object in frame $S'$ is the differential displacement *in that frame* divided by the differential time interval *in that frame*.

$$u' \equiv \frac{dx'}{dt'}$$

We may now use equations (1-10), but with the finite intervals replaced by infinitesimal ones (e.g., $x_2' - x_1' \rightarrow dx'$ and $t_2 - t_1 \rightarrow dt$).

$$u' = \frac{\gamma_v (dx - v\, dt)}{\gamma_v \left( -\dfrac{v}{c^2} dx + dt \right)}$$

Canceling, and dividing numerator and denominator by $dt$, we find that

$$u' = \frac{\dfrac{dx}{dt} - v}{-\dfrac{v}{c^2} \dfrac{dx}{dt} + 1}$$

But by definition $dx/dt$ is $u$, the velocity of the object in frame $S$. Thus,

$$u' = \frac{u - v}{1 - \dfrac{uv}{c^2}} \quad \textbf{(1-17a)} \quad \text{and} \quad u = \frac{u' + v}{1 + \dfrac{u'v}{c^2}} \quad \textbf{(1-17b)}$$

Equation (1-17b) is simply (1-17a) solved for $u$, but its form is no surprise: The same law must apply in both frames, the only difference being the sign of the relative velocity between them (i.e., $v$ is replaced by $-v$, as usual). If velocities are small compared to $c$, the denominators are essentially unity. As we would expect, the "classical velocity transformation," $u' = u - v$, is the result.

---

**Example 1.7**

A spaceship is traveling at $0.8c$ away from Earth. Further out, a meteorite is traveling toward Earth at $0.6c$. Both velocities are according to an observer on Earth. According to an observer on the spaceship, what is the velocity of the meteorite (relative to himself)?

Earth                                    Meteorite

**Solution**

Our first step is to determine what the symbols in equations (1-17) represent. They involve a velocity $u$ of an object according to frame $S$, a velocity $u'$ of that object according to frame $S'$ and, of course, a relative velocity $v$ between frames. The "object" here is the meteorite and the two reference frames are then Earth and the spaceship. The choice of which frame to designate as $S'$ is arbitrary, but once made, it must be followed carefully. Let us choose Earth as frame $S$ and the spaceship frame $S'$. Since by convention $S'$ moves in the positive direction relative to $S$, velocities away from Earth must be taken as positive. We know the object's velocity according to Earth/$S$ ($u = -0.6c$) and the relative velocity of the frames ($v = 0.8c$), and seek the object's velocity according to the spaceship/$S'$—that is, $u'$.

Inserting in equation (1-17a),

$$u' = \frac{-0.6c - 0.8c}{1 - \dfrac{(-0.6c)(0.8c)}{c^2}} \quad \Rightarrow \quad u' = \frac{-1.4c}{1.48} = -0.946c$$

Not surprisingly, the meteorite travels in the negative (toward Earth) direction according to an observer on the spaceship, *but at a speed less than c!* The layman is inclined to guess an answer of $-1.4c$ (i.e., the numerator).

---

A bit of valid classical physics should be noted. The meteorite's speed relative to the spaceship is indeed $1.4c$, *according to an observer on Earth*. If Bob on Earth sees a spaceship moving away at $0.8c = 2.4 \times 10^8$ m/s and the meteorite moving toward Earth at $0.6c = 1.8 \times 10^8$ m/s, the two are surely getting $4.2 \times 10^8$ m closer every second. This logic is sound because *all three of these velocities are according to the same (Earth) observer.* The *relativistic* velocity transformation, which relates velocities according to *different* observers, would *not* be appropriate. But the $1.4c$ would not be an example of something moving *relative to an observer* at a speed greater than $c$. It is the meteorite's speed *according* to the Earth observer but *relative* to the spaceship. The speed $0.946c$ is also the meteorite's speed relative to the spaceship, but *according to the spaceship*. Since in special relativity we are not usually interested in a relative velocity between two objects according to a third party, when we say simply "the velocity of B relative to A," it is understood to mean also *according to A.*

Provided that $v$ and $u$ (or $u'$) are less than $c$, the value of $u'$ (or $u$) given by equations (1-17) is *always* strictly less than $c$; no observer will ever see something moving relative to himself at a speed greater than $c$. On the other hand, if the "object" is a beam of light, of speed $c$ in one frame, equations (1-17) would show its speed to be $c$ in the other frame (independent of $v$), as Einstein's postulates demand. The proofs of these assertions are left as exercises.

We complete the relativistic velocity transformation by considering components of velocity *not* along the axis of relative motion. By definition, the y-component of velocity in frame $S'$ involves time and displacement (along the y'-axis) *in that frame.*

$$u'_y \equiv \frac{dy'}{dt'}$$

Using $y' = y$ and equation (1-10b) to rewrite the denominator, we have

$$u'_y = \frac{dy}{\gamma_v \left( -\dfrac{v}{c^2} dx + dt \right)}$$

Dividing by $dt$ and rearranging,

$$u'_y = \frac{\dfrac{dy}{dt}}{\gamma_v \left( -\dfrac{v}{c^2}\dfrac{dx}{dt} + 1 \right)} = \frac{u_y}{\gamma_v \left( 1 - \dfrac{u_x v}{c^2} \right)}$$

It may seem odd that the y-component of velocity in frame $S'$ should depend on the x-component in frame $S$. The connection is that, while the displacements along the y-axes are equal, the *time intervals* differ due to relative motion along $x$. The corresponding equation relating z-components follows identically. Thus,

$$u'_x = \frac{u_x - v}{1 - \dfrac{u_x v}{c^2}} \qquad u'_y = \frac{u_y}{\gamma_v \left( 1 - \dfrac{u_x v}{c^2} \right)} \qquad u'_z = \frac{u_z}{\gamma_v \left( 1 - \dfrac{u_x v}{c^2} \right)}$$

$$\text{(1-18a)} \qquad\qquad \text{(1-18b)} \qquad\qquad \text{(1-18c)}$$

Relativistic velocity transformation

As always, solving explicitly or merely switching the sign of $v$ would yield $(u_x, u_y, u_z)$ in terms of $(u'_x, u'_y, u'_z)$.

## 1.8 Dynamics

We now retrace a discovery that changed the world, contributing among other things to the unleashing of nuclear energy.

In a sense, we are at a familiar juncture. In classical physics, kinematics, in which we *describe* motion (via position, time, velocity, and so on), is usually followed by dynamics, a study of *why* things move. Though $\mathbf{F}_{net\,external} = m\mathbf{a}$ is often the first step, the student soon finds that for a system of particles the more fundamental and often more useful law is $\mathbf{F}_{net\,external} = d\mathbf{P}_{total}/dt$, where $\mathbf{P}_{total}$ is the total momentum of the system. From this we know that if the system is isolated, so that $\mathbf{F}_{net\,external} = 0$ and the only forces are those between particles in the system, the total momentum is conserved.

$$\mathbf{P}_{total} = \sum_i \mathbf{p}_i = \text{constant} \quad \text{(Isolated system)}$$

In common with all laws of physics, this law cannot be proved, but only experimentally verified. Had there ever been a case where it was violated, momentum conservation would have ceased being considered a true law of physics.

Classically, the momentum of an individual particle is $m\mathbf{u}$. (Again, since the symbol $v$ is reserved, we use a $\mathbf{u}$ for the velocity of a particle.) A question rarely considered is, Why should the conserved property we call "momentum" be given by the simple formula $m\mathbf{u}$? It has several things going for it: (1) Its conservation is experimentally verified (a must!), at least at low speeds. (2) It is reasonable that the vector property that is transferred but not lost as one object collides with another should depend on mass and velocity. But it has another interesting property: Classically, if the sum of $m\mathbf{u}$ for all particles in an isolated system is conserved in one frame of reference, then it is automatically conserved in another. We can show this easily via the classical velocity transformation. (For the sake of simplicity, we consider only one dimension.)

$$P'_{total} = \sum_i (m_i u_i') = \sum_i [m_i(u_i - v)] = \sum_i (m_i u_i) - \sum_i (m_i v)$$

or

$$P'_{total} = P_{total} - v \sum_i m_i \qquad \text{(1-19)}$$

Classical momentum transformed via classical velocity transformation

Given that the system's total mass doesn't change, and that the relative velocity $v$ between the frames is constant, it follows that if $P_{total}$ is constant, then so is $P'_{total}$.

But the classical velocity transformation is wrong! Let us see what happens when we use the relativistically correct transformation.

$$P'_{total} = \sum_i (m_i u_i') = \sum_i \left( m_i \frac{u_i - v}{1 - \dfrac{u_i v}{c^2}} \right)$$

$$= \sum_i (m_i u_i) - v \sum_i \left( m_i \frac{1 - u_i^2/c^2}{1 - u_i v/c^2} \right)$$

The last step involves algebraic rearrangement that the reader may easily verify. Thus,

$$P'_{total} = P_{total} - v \sum_i \left( m_i \frac{1 - u_i^2/c^2}{1 - u_i v/c^2} \right)$$

Classical momentum transformed via relativistic velocity transformation

Now it is no longer obvious that if $P_{total}$ is constant, then $P'_{total}$ is automatically constant. Their difference, the second term on the right, might not be constant, because the particle velocities $u$ change as the particles interact. In fact, it is not only not obvious—it is not true, as we see in the following example.

### Example 1.8

*If* momentum were given by $p = mu$, the collision depicted in frame $S$ would conserve momentum. Since the objects would have equal and opposite momentum, the total momentum would be zero, before and after. That is, $P_{total, before} = 0 = P_{total, after}$. (*Note:* Masses are in arbitrary units.) Using the relativistic velocity transformation, find the four velocities in frame $S'$, moving to the right at $0.6c$, and show that momentum (again, if given by $mu$) would not be conserved in this frame.

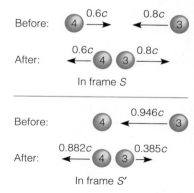

### Solution

With $v = 0.6c$, and using $u' = (u - v)/(1 - uv/c^2)$,

**Before:**
$$\frac{0.6c - 0.6c}{1 - \dfrac{(0.6c)(0.6c)}{c^2}} = 0 \qquad \frac{-0.8c - 0.6c}{1 - \dfrac{(-0.8c)(0.6c)}{c^2}} = -0.946c$$

**After:**
$$\frac{-0.6c - 0.6c}{1 - \dfrac{(-0.6c)(0.6c)}{c^2}} = -0.882c \qquad \frac{0.8c - 0.6c}{1 - \dfrac{(0.8c)(0.6c)}{c^2}} = 0.385c$$

The collision would be as shown at right. Summing over the particles,

$$P'_{total, before} = \sum_i m_i u_i = 4(0) + 3(-0.946c) = -2.84c$$

$$P'_{total, after} = \sum_i m_i u_i = 4(-0.882c) + 3(0.385) = -2.38c$$

According to Einstein's first postulate, a law of physics should hold in the same form in all frames. Thus, the example gives us theoretical grounds for claiming that the classical form for momentum must be incorrect! But correct theories invariably go hand in hand with experiment, so we shouldn't be too surprised that it also fails the experimental test: At high speeds, $\sum_i m_i u_i$ is not conserved in *any* frame! (Neither "after picture" is what would actually be observed.) *Theoretically and experimentally, m**u** is wrong.*

Equation (1-19) shows that—classically—momentum conservation in one frame implies momentum conservation in another; the *classical* velocity transformation reveals that the total momenta in the two frames *differ by a constant.* Relativistically, things are not so easy, but the "complication" leads to one of the most stunning claims of Einstein's theory. Given the *relativistic* velocity transformation, it can be proven that there is *no* candidate for a relativistic momentum (a function of mass and velocity) for which the total momenta in two frames would differ by a constant. The difference would invariably have some functional dependence on the particle's velocities $u$, which may vary as the particles interact. Therefore, the only way momentum conservation can be salvaged is if this other function of the particles' velocities is also conserved. *The requirement that momentum conservation hold in all frames implies that momentum conservation is tied to conservation of some other property.*

This theoretical line of reasoning would eventually lead to the correct function for momentum and for the mysterious "other property," but it would be very messy. Let us take a short-cut by temporarily switching to an experimental argument for momentum, then resuming the theoretical argument for the "other property."

By virtue of the experimental evidence of its conservation at all speeds, the relativistically correct expression for momentum is

$$\mathbf{p} = \gamma_u m\mathbf{u} \quad \text{where} \quad \gamma_u \equiv \frac{1}{\sqrt{1 - \dfrac{u^2}{c^2}}}. \tag{1-20}$$

Momentum

(In three dimensions, the speed $u$ is $\sqrt{u_x^2 + u_y^2 + u_z^2}$.) Note that (1-20) reduces to the classical momentum at ordinary speeds, where $\gamma_u \cong 1$.

To investigate the "other property," let us now transform the relativistic momentum from one frame to another via the relativistic velocity transformation. Using (1-18a) and the identity $\gamma_{u'} = (1 - u_x v/c^2)\gamma_v\gamma_u$, which follows directly from equations (1-18) (see Exercise 79), we obtain

$$P'_{\text{total}} = \sum_i (\gamma_{u_i} m_i u'_i) = \sum_i \left[ \left(1 - \frac{u_i v}{c^2}\right) \gamma_v \gamma_{u_i} m_i \left(\frac{u_i - v}{1 - \dfrac{u_i v}{c^2}}\right) \right] \tag{1-21}$$

$$= \gamma_v \sum_i (\gamma_{u_i} m_i u_i) - v\gamma_v \sum_i (\gamma_{u_i} m_i)$$

or

$$P'_{\text{total}} = \gamma_v P_{\text{total}} - v\gamma_v \sum_i (\gamma_{u_i} m_i)$$

Relativistic momentum transformed via relativistic velocity transformation

If it is true that momentum is conserved in both frames, then $\sum_i (\gamma_{u_i} m_i)$ must also be constant! What could it be? What besides momentum must be conserved in an isolated system? There is only one ready candidate: *total* energy. It cannot be just kinetic energy, because kinetic energy need not be conserved. The quantity $\gamma_{u_i} m_i$ must be related to the total energy of particle $i$, including both kinetic *and internal* energies. Even so, as it stands, it cannot be energy, because it has units of mass, but if we multiply it by $c^2$, it becomes $\gamma_{u_i} m_i c^2$, which has the proper units: kg·m$^2$/s$^2$.[16]

$$E = \gamma_u mc^2 \tag{1-22}$$

Total energy of an object moving at speed $u$

This claim follows directly from Einstein's postulates—through the Lorentz transformation, the requirement that the same law hold in all frames, and so on. It must not be underestimated. Nowhere in classical physics is the claim made that there is a simple way, if any at all, of determining the *total* energy of an object. The idea is revolutionary.

The most striking consequence is that mass measures energy; were an object *stationary,* its total energy would be $mc^2$. This includes all *internal* energy.

---

[16]Because $c$ is a universal constant, this doesn't really change the character of the quantity; it may be thought of as simply a conversion of units. The motivation might still seem a bit weak, but we soon see how well it fits.

$$E_{\text{internal}} = mc^2 \tag{1-23}$$

Internal energy of an object (measured by mass)

Classically or relativistically, kinetic energy is defined as the energy associated simply with the motion of an object. If we subtract from the energy of a moving object the (internal) energy it would have at rest, we have kinetic energy.

$$KE = (\text{energy moving}) - (\text{energy at rest})$$
$$= \gamma_u mc^2 - mc^2 = (\gamma_u - 1)mc^2 \tag{1-24}$$

Kinetic energy

We can begin to see the plausibility and consistency of Einstein's claim by considering the total energy of a slow-moving object.

$$\gamma_u mc^2 = \left(1 - \frac{u^2}{c^2}\right)^{-1/2} mc^2 \cong \left[1 + \left(-\frac{1}{2}\right)\left(-\frac{u^2}{c^2}\right)\right] mc^2$$

$$= mc^2 + \frac{1}{2}mu^2 \quad (u \ll c)$$

This fits perfectly: If $u$ is zero, the total energy is $mc^2$, simply the internal energy, and if $u$ is nonzero, the total energy exceeds the internal energy by the classical kinetic energy. Now consider the simple case of a completely inelastic head-on collision between two objects of the same mass and (low) speed (Figure 1.15). Momentum conservation requires that the objects be stationary after the collision. By symmetry, neither object would exchange net energy with the other. Therefore, conservation of total energy would require that for each object,

$$\left(mc^2 + \frac{1}{2}mu^2\right)_{\text{final}} = \left(mc^2 + \frac{1}{2}mu^2\right)_{\text{initial}}$$

But since $u_f = 0$,

$$(mc^2)_f = \left(mc^2 + \frac{1}{2}mu^2\right)_i$$

or

$$m_f c^2 = m_i c^2 + \frac{1}{2}m_i u_i^2$$

The meaning is clear: The amount by which the final internal/mass energy exceeds the initial internal/mass energy is the "lost" kinetic energy. This is in perfect harmony with the classical explanation that "lost" kinetic energy becomes internal thermal energy.

**Figure 1.15** A completely inelastic collision: Kinetic energy is converted to internal energy.

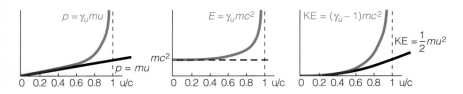

Figure 1.16 The variation of momentum, energy, and kinetic energy as an object's speed varies from zero to $c$.

Einstein's findings, expressed in (1-20) and (1-22) through (1-24), are profound. Momentum, classically a linear function, we now find to be a more complicated function of velocity. As shown in Figure 1.16, the classical and true momenta agree at low speeds, but the true momentum diverges as $u$ approaches $c$. In this limit, the kinetic energy and (therefore) total energy also diverge. For this reason alone, it should not be possible to accelerate a massive object to the speed of light; infinite energy would be required.

## Mass and Energy

According to (1-23), the energy of an object at rest depends solely on its mass. For example, heating (adding energy to) an object at rest increases its mass; cooling decreases it. In ordinary cases, these changes are immeasurably small, because $c^2$ is a very large number. A mass change of 1 kg corresponds to an energy difference of $9 \times 10^{16}$ J, more than enough to put the Empire State Building in a low orbit about Earth. However, one cannot simply make mass disappear arbitrarily. Generally speaking, exchange of energy requires a force.

*Chemical* reactions involve the *electrostatic force,* attractive between positive nucleus and orbiting electrons, repulsive between the electrons themselves. In chemical reactions, the electrons rearrange themselves, resulting in a change in the internal (electrostatic potential plus kinetic) energy. In an endothermic reaction, energy is absorbed from the surroundings, and the internal energy and therefore mass of the reacting sample increases. In an exothermic reaction, energy is lost to the surroundings and mass correspondingly decreases. For example, as two hydrogen atoms unite to form a molecule, the lowering of electrostatic potential energy results in approximately $7 \times 10^{-19}$ J being given off. Using $\Delta E = \Delta mc^2$, we see that the mass decreases by about $8 \times 10^{-36}$ kg. This is only about one-billionth of the molecule's mass. Understandably, such small changes went unnoticed for centuries.

*Nuclear* reactions, on the other hand, are much more energetic, for the simple reason that they involve a much stronger force, aptly named the *strong force.* We will discuss this force more extensively in Chapters 10 and 11. For the present, we note only that it is an attractive force shared by all constituents within the nucleus, neutrons and protons alike. (Clearly it must be "strong," for it is what prevents the nucleus from flying apart; it overwhelms the electrostatic repulsion between protons.) When a proton and neutron unite to form a deuterium (or "heavy hydrogen") nucleus, the lowering of the energy associated with the strong force results in approximately $4 \times 10^{-13}$ J being given off. The corresponding mass decrease is $4 \times 10^{-30}$ kg, about one-thousandth of the nuclear mass. Thus, the fractional change in mass is about six orders of magnitude larger than a typical chemical reaction. Pound for pound, nuclear reactions are much more energetic.

When discussing how the mass of an object may change, it is important to bear in mind that mass measures the internal energy of the whole system of particles. Consider a simple object: a helium balloon. If the balloon is heated, its internal thermal energy will increase and so will its mass. But what of the individual helium atoms? Their masses do not increase, because their internal energies have not changed. The mass of the balloon as a whole has increased because its individual constituents have gained *kinetic* energy and that kinetic energy is *internal* to the system. It may strike the reader odd that a system can gain mass without its constituents gaining mass, but *mass is not conserved; energy is*—and mass measures its internal forms. On the other hand, the mass of an object is not dependent on its *overall* motion relative to an observer (see the essay entitled "A Note of Caution"). Were a balloon thrown with a speed $u$, its *overall* kinetic energy would increase, but its *internal* energy—its mass—would not be affected. The mass of an object is the same when viewed from any reference frame. (See Section 1.11 for a more formal discussion.)

In physics, we often speak of fundamental particles. A fundamental particle is one that is not composed of other particles. Most objects of study are *not* fundamental. As we know, an atom is composed of electrons and a nucleus. Nuclei, in turn, comprise neutrons and protons; even neutrons and protons are combinations of more-basic particles. Nevertheless, there is widespread belief that certain particles are fundamental. The electron appears to be of this class. As far as is known, the electron has *no internal structure*. Even so, it does have mass. (Remarkably, it has internal energy without internal structure!) But with nothing inside to jiggle around more rapidly and thus absorb heat (or any other form of energy), this mass is not subject to change. Immutable properties that characterize a fundamental particle are known as "intrinsic" properties. Charge is one such property; **intrinsic mass** is another.

Incredible though they may seem, Einstein's claims have been borne out conclusively. Their validity is reaffirmed daily in such applications as nuclear reactors and the particle accelerators of high-energy physics (both discussed in detail in later chapters). Let us now solidify our grasp of the principles with several examples.

## A Note of Caution

In some introductory texts on special relativity, the notion of a so-called "relativistic mass" is introduced: $M_{rel} = \gamma_u m$. With this, the relativistic expression for momentum resembles the classical (i.e., $p = \gamma_u mu$ becomes $p = M_{rel}u$.) However, the association is not at all necessary. Worse yet, confusing interpretations are often the result. Quoting Albert Einstein himself: "It is not good to introduce the concept of the mass $M = m/\sqrt{1 - v^2/c^2}$ of a moving body for which no clear definition can be given. It is better to introduce no other mass concept than the 'rest mass' $m$. Instead of introducing $M$ it is better to mention the expression for the momentum and energy of a body in motion." Therefore, if reading phrases elsewhere such as "the (relativistic) mass of the electron is 5 times its *rest mass*," the reader should translate this to "gamma is 5," and proceed to calculate momentum and energy via $p = \gamma_u mu$ and $E = \gamma_u mc^2$.

## 1.9 Momentum and Energy: Applications

### Example 1.9

A proton has a kinetic energy half its internal energy, the energy it would have at rest. (a) What is the proton's speed? (b) What is its total energy? (c) Determine the potential difference $V$ through which the proton would have to be accelerated to attain this speed.

### Solution

(a) Kinetic energy is $(\gamma_u - 1)mc^2$, and the energy an object has at rest is $mc^2$.

$$(\gamma_u - 1)m_p c^2 = \frac{1}{2} \times m_p c^2 \quad \Rightarrow \quad \gamma_u = 1.5$$

$$\frac{1}{\sqrt{1 - (u/c)^2}} = 1.5 \quad \Rightarrow \quad u = 0.745c$$

Generally speaking, the internal/mass energy of an object is huge. For its kinetic energy to be any significant fraction of its internal energy, an object must be moving very fast (cf. Figure 1.16).

(b) The total energy of the proton is the sum of its kinetic and internal energies.

$$E = (\gamma_u - 1)mc^2 + mc^2 = \gamma_u m_p c^2$$

$$= 1.5(1.67 \times 10^{-27} \text{ kg})(3 \times 10^8 \text{ m/s})^2$$

$$= 2.25 \times 10^{-10} \text{ J} = 1409 \text{ MeV}$$

(c) Classically or relativistically, the kinetic energy gained by a charged particle is equal in magnitude to the potential energy change as the particle is accelerated through a potential difference.

$$\text{KE} = |q \, \Delta V|$$

$$(\gamma_u - 1)mc^2 = |q \, \Delta V|$$

$$(1.5 - 1)(1.67 \times 10^{-27} \text{ kg})(3 \times 10^8 \text{ m/s})^2 = (1.6 \times 10^{-19} \text{ C})|\Delta V|$$

$$\Rightarrow \quad \Delta V = 470 \text{ MV}$$

It should be noted that if we had used the classical expression for kinetic energy, while still desiring a final speed of $0.745c$, we would have obtained $\Delta V = 291$ MV. Kinetic energy increases very rapidly as speed approaches $c$, and a much larger potential difference is required than would be expected classically.

### Example 1.10

A 1.20-kg block heading east at $0.8c$ collides with a 1.60-kg block moving west at $0.6c$. They stick together. Find the mass and speed of the resulting combined object.

(*Note:* This example is unrealistic. Rather than forming a single object, the blocks would probably disintegrate. However, we are interested in merely studying how conservation laws govern the outcome, no matter how improbable it may be.)

### Solution

Let us choose our coordinate system so that east is the positive *x*-direction. All motion will be along this axis.

(*Note:* Were classical physics applicable, the combined mass would "of course" be 2.8 kg, and momentum conservation along the *x*-axis would then yield

$$(1.20 \text{ kg})(+0.8c) + (1.60 \text{ kg})(-0.6c) = (2.80 \text{ kg})u_{\text{final}} \quad \Rightarrow \quad u_{\text{final}} = 0$$

The total momentum would be zero, and the 2.8-kg object would be at rest.)

Relativistically, we must not assume that mass is constant, which would be to assume that the internal energy of the objects remains constant. Since the collision is completely inelastic, kinetic energy will be lost, so we should expect the internal energy (thermal energy in this case) to increase. With both mass and speed unknown, the single momentum conservation equation is not sufficient to solve the problem. But we have a powerful new tool: an expression for the total energy possessed by an object, and total energy must be conserved. (In what follows, a subscript f refers to the final combined object.)

Momentum ($\gamma_u mu$) is conserved:

$$\gamma_1 m_1 u_1 + \gamma_2 m_2 u_2 = \gamma_{u_f} m_f u_f$$

$$\frac{1}{\sqrt{1 - (0.8)^2}} (1.20 \text{ kg})(0.8c)$$

$$+ \frac{1}{\sqrt{1 - (0.6)^2}} (1.60 \text{ kg})(-0.6c) = \frac{1}{\sqrt{1 - (u_f/c)^2}} m_f u_f$$

$$(1.67)(1.20 \text{ kg})(0.8c) - (1.25)(1.60 \text{ kg})(0.6c) = \frac{1}{\sqrt{1 - (u_f/c)^2}} m_f u_f$$

$$+0.400 \text{ kg} \cdot c = \frac{1}{\sqrt{1 - (u_f/c)^2}} m_f u_f$$

(*Note:* The total momentum is not zero. The classical formula is in error.)

Energy ($\gamma_u mc^2$) is conserved:

$$\gamma_1 m_1 c^2 + \gamma_2 m_2 c^2 = \gamma_{u_f} m_f c^2$$

$$(1.67)(1.20 \text{ kg})c^2 + (1.25)(1.60 \text{ kg})c^2 = \frac{1}{\sqrt{1 - (u_f/c)^2}} m_f c^2$$

$$4.00 \text{ kg} \cdot c^2 = \frac{1}{\sqrt{1 - (u_f/c)^2}} m_f c^2$$

We have two equations in two unknowns. Dividing the momentum equation by the energy equation,

$$\frac{0.400 \text{ kg} \cdot c}{4.00 \text{ kg} \cdot c^2} = \frac{u_f}{c^2} \quad \Rightarrow \quad u_f = 0.100c$$

Reinserting in either equation yields $m_f = 3.98$ kg.

A calculation of total kinetic energy before and after the collision in the example verifies that kinetic energy has decreased. Using equation (1-24),

$$\begin{aligned} KE_f - KE_i &= (\gamma_{0.10c} - 1)(3.98 \text{ kg})c^2 \\ &\quad - [(1.67 - 1)(1.20 \text{ kg})c^2 + (1.25 - 1)(1.60 \text{ kg})c^2] \\ &= -1.18 \text{ kg} \cdot c^2 \end{aligned}$$

Not coincidentally, this is precisely $-\Delta mc^2$. Kinetic energy is not lost, but is converted into increased thermal energy, which registers as increased mass. The reverse of the example would be a single object suddenly breaking (exploding) into two pieces. Internal energy, measured by mass, would decrease as kinetic energy increases.

---

**Example 1.11**

The nucleus of a beryllium atom has a mass of 8.0031 u, where u is an **atomic mass unit:** $1.66 \times 10^{-27}$ kg. It is known to spontaneously fission (break up) into two identical pieces, each of mass 4.0015 u. Assuming the nucleus to be initially at rest, at what speed will its fragments move after the fission and how much kinetic energy is released?

**Solution**

Let us use a subscript Be to refer to the nucleus and subscripts 1 and 2 to refer to its fragments.
    Momentum ($\gamma_u mu$) is conserved:

$$1 m_{Be}(0) = \gamma_1 m_1 u_1 + \gamma_2 m_2 u_2$$

But $m_1 = m_2$ then implies that

$$\frac{1}{\sqrt{1 - (u_1/c)^2}} u_1 = -\frac{1}{\sqrt{1 - (u_2/c)^2}} u_2 \quad \text{or} \quad u_1 = -u_2$$

Although the relativistic expression for momentum is somewhat more complicated than the classical, the same conclusion is reached: The two fragments must move at the same speed but in opposite directions. Let us now refer to this speed as simply $u$ and the mass of a fragment as $m$.
    Energy ($\gamma_u mc^2$) is conserved:

$$m_{Be}c^2 = \gamma_1 m_1 c^2 + \gamma_2 m_2 c^2 = 2 \times (\gamma_u mc^2)$$

$$\gamma_u = \frac{m_{Be}}{2m} = \frac{8.0031\ u}{2 \times 4.0015\ u} = 1.0000125$$

$$\frac{1}{\sqrt{1 - (u/c)^2}} = 1.0000125 \quad \Rightarrow \quad u = 0.005c$$

The total kinetic energy is

$$2 \times (\gamma_u - 1)mc^2$$

$$= 2 \times (0.0000125)(4.0015 \times 1.66 \times 10^{-27}\ \text{kg})\,(3 \times 10^8\ \text{m/s})^2$$

$$= 1.5 \times 10^{-14}\ \text{J} = 93\ \text{keV}$$

This is in good agreement with the experimentally measured value.

There is a simple way to determine how much kinetic energy is released in a general reaction. Suppose we initially have particles 1 and 2. They interact, and particles 3 and 4 are the result.

$$E_3 + E_4 = E_1 + E_2$$

$$[m_3 c^2 + (\gamma_3 - 1)m_3 c^2] + [m_4 c^2 + (\gamma_4 - 1)m_4 c^2] =$$
$$[m_1 c^2 + (\gamma_1 - 1)m_1 c^2] + [m_2 c^2 + (\gamma_2 - 1)m_2 c^2]$$

Here we have broken each total energy explicitly into internal and kinetic parts. Rearranging,

$$[(\gamma_3 - 1)m_3 c^2 + (\gamma_4 - 1)m_4 c^2] - [(\gamma_1 - 1)m_1 c^2 + (\gamma_2 - 1)m_2 c^2]$$
$$= -[(m_3 + m_4) - (m_1 + m_2)]c^2$$

$$\text{KE}_{final} - \text{KE}_{initial} = -(m_{final} - m_{initial})c^2$$

$$\Delta \text{KE} = -\Delta mc^2 \qquad \text{(1-25)}$$

As we would expect, kinetic energy increases as internal energy decreases, and vice versa. This applies even for collisions/explosions at ordinary speeds, but because $c^2$ is so large, the differences in mass are negligible.

Another useful result is the following general relationship between energy, momentum, and mass in an arbitrary frame of reference:

$$E^2 = p^2 c^2 + m^2 c^4 \qquad \text{(1-26)}$$

Energy-momentum-mass relationship

Because this relationship is independent of the speed $u$, it is often very convenient, and we will have occasion to use it now and then. Even so, it is nothing new; it may be derived directly from $E = \gamma_u mc^2$ and $p = \gamma_u mu$. This is left as an exercise.

It may puzzle the reader that none of the examples in this section has involved a second reference frame; a pattern has been broken. But views from different frames are not so much the interest here as the fact that momentum and energy *in any frame* depend on mass and speed in ways different from the classical. The fascinating point is that it is the claim that conservation laws must be the same in all frames that led to this discovery.

## The Particle Accelerator

Einstein's famous mass-energy equivalence is central to the purpose of **particle accelerators.** These immense scientific apparatus are used to speed up particles such as protons to nearly the speed of light. They are the tools of "high-energy" physics, which seeks to identify the fundamental building blocks of the universe. Though the electron is the most familiar fundamental particle, many others have been found. Still others are at present mere theoretical predictions. Often, these undiscovered particles are expected to be fairly heavy. But how do we create mass? As illustrated in Example 1.10, Einstein's mass-energy relationship is the key. If particles are caused to collide inelastically, with kinetic energy decreasing, mass/internal energy must increase. In high-energy physics, however, we do not smash together macroscopic objects (e.g., 1.2-kg blocks), as it would be utterly impossible to ensure beforehand, or detect afterward, the microscopic reaction of interest. We smash together simple particles (e.g., protons, electrons) to restrict the outcome to but a few possibilities, and we smash fast particles together to make available a large initial kinetic energy from which to create massive new particles. The greater the mass expected of a particle, the greater must be the initial kinetic energy.

We now see why we might want to build a particle accelerator, but there are two types, and one is more desirable. Let us see why. In a "stationary-target accelerator" a moving particle is smashed into a stationary particle. In a "collider" (short for colliding beams apparatus), the two particles smashed together are both moving, toward a head-on collision. To produce the maximum mass with the minimum input of energy, the collider is preferable. Suppose two identical particles of total kinetic energy $KE_0$ collide. If both are moving at speed $u$ in opposite directions, the total momentum is zero. Therefore, if they stick together, the final particle must be at rest—*all* the initial kinetic energy will have been converted to mass. If, however, one particle is moving and one stationary, the initial total momentum is nonzero, so if they stick together, the final particle must be moving. There is necessarily some kinetic energy left over after the collision, not available to become mass. Although this simple example of a completely inelastic collision is not realistic, the underlying principle holds no matter what may result from the collision. In a collider, the final kinetic energy is a minimum and the final mass a maximum. The two types of collision are compared in Exercise 65.

## 1.10 Massless Particles

Experimental evidence indicates that there are particles for which $E = pc$: When these particles interact with other objects, the momentum and energy they carry are found to obey this relationship within experimental uncertainty. Comparing with (1-26), there is only one possible conclusion: they have no mass!

$$E = pc \quad (m = 0)$$

But shouldn't real objects, things that move about and smash into other things, have mass? There is no theoretical requirement. That applying $\mathbf{F} = m\mathbf{a}$ is problematic does not prove that $m = 0$ is impossible; it may simply indicate that this classical "law" does not apply to massless particles. Coincidentally, the topic of Chapter 2 is the most well-known massless particle: the particle of light, or **photon.** Many experiments have been done, but none have found a nonzero mass for the photon; if it is not zero, it is immeasurably small. In fact, the assumption that it is zero is central to our accepted theories of electromagnetic radiation. If a nonzero mass were ever to be found, these theories would have to be modified.

Our present interest is in seeing how far special relativity *alone* goes in characterizing massless particles. (We use the photon as an example, but it is representative of other massless particles.) First, having no mass, photons must have no internal energy; their energy is all kinetic. Secondly, if $m$ is zero, the only way $p = \gamma_u mu$ and $E = \gamma_u mc^2$ can be nonzero is for the speed $u$ to be $c$, in which case, both are undefined products of infinity and zero. This suggests that if a photon is to have at least *some* property that can be measured, it should move at the speed of light. This has been experimentally verified. Photons always move at $c$, from the instant of their creation to the instant of their disappearance. However, because it is also an experimental fact that photons do not all have the same energy and momentum, and the expressions for $p$ and $E$ are indeterminate, we have a mystery: Upon what property or properties do the momentum and energy of a photon depend? The answer awaits in Chapter 2.

## ◆ 1.11   The Fourth Dimension

In this section we shall see why one often hears of "space-time," as though time were no different than the three traditional spatial dimensions. First, though, we must justify a claim made earlier—that $y' = y$ and $z' = z$.

Recall that for a single spatial dimension, the transformation yielding constant velocity motion in all frames for a free object is given by the linear transformation (1-4). The logical generalization to three spatial dimensions is

$$x' = A_{11}x + A_{12}y + A_{13}z + A_{14}t$$
$$y' = A_{21}x + A_{22}y + A_{23}z + A_{24}t$$
$$z' = A_{31}x + A_{32}y + A_{33}z + A_{34}t$$
$$t' = A_{41}x + A_{42}y + A_{43}z + A_{44}t$$

Simple arguments show that most of the constants are zero. Were motion confined to the $x$-$y$ and $x'$-$y'$ planes, $z$ and $z'$ would have to be zero. The third equation would become $0 = A_{31}x + A_{32}y + 0 + A_{34}t$. But since $x$, $y$, and $t$ would still be arbitrary, all three constants must be zero. Consequently, for general three-dimensional motion, the third equation would be $z' = A_{33}z$. For the same law to hold in both frames, we would then have to have $z = A_{33}z'$, so the third equation must be $z' = z$. A similar argument involving motion confined to the $x$-$z$ and $x'$-$z'$ planes shows that the second equation becomes $y' = y$. Regarding the first and fourth equations: If we imagine rotating both coordinate frames by 180° about their coincident $x$- and $x'$-axes, the values of $x$, $t$, $x'$, and $t'$ for all events would be unaffected, but the values of $y$, $z$, $y'$, and $z'$ would change sign. Nevertheless, the transformation equations between the rotated frames would have to be the same as before, so the coefficients of $y$ and $z$—$A_{12}$, $A_{13}$, $A_{42}$, and $A_{43}$—must be zero. The remaining constants $A_{11}$, $A_{14}$, $A_{41}$, and $A_{44}$ are the $A$, $B$, $C$, and $D$ from Section 1.4.

Altogether, then, we see that to account for objects moving in three dimensions we need only add $y' = y$ and $z' = z$ to our earlier equations (1-8). Thus, with slight rearrangements to accentuate one of the most important features, the Lorentz transformation equations may be written as follows:

Lorentz transformation equations

$$x' = \gamma_v\left(x - \frac{v}{c}ct\right)$$
$$y' = y$$
$$z' = z \qquad \text{or}$$
$$ct' = \gamma_v\left(ct - \frac{v}{c}x\right)$$

$$\begin{bmatrix} x' \\ y' \\ z' \\ ct' \end{bmatrix} = \begin{bmatrix} \gamma_v & 0 & 0 & -\gamma_v\dfrac{v}{c} \\ 0 & 1 & 0 & 0 \\ 0 & 0 & 1 & 0 \\ -\gamma_v\dfrac{v}{c} & 0 & 0 & \gamma_v \end{bmatrix} \begin{bmatrix} x \\ y \\ z \\ ct \end{bmatrix} \qquad (1\text{-}27)$$

In the first and fourth equations, $ct'$ and $ct$ are interchangeable with $x'$ and $x$. *In special relativity, time behaves in exactly the same way as position.* It must be treated, in a real sense, as a fourth dimension. The three spatial dimensions together with time are referred to as a **four-vector.**

The matrix form of (1-27) is a way of writing the Lorentz transformation equations that emphasizes the four-dimensional nature of space-time.[17] Moreover, it is often a convenient way of using the equations all at once, as we see in the following example.

---

### Example 1.12

For the situation in Example 1.4, (a) determine the Lorentz transformation matrix from Bob's frame to Anna's frame—that is, yielding $(x', t')$ from $(x, t)$. (b) Using matrix multiplication, verify the initial time on the nose-end clock on Anna's ship.

### Solution

(a) Since $v/c = \sqrt{3}/2$ and $\gamma_v = 2$, the Lorentz transformation matrix in (1-27) is

$$\begin{bmatrix} 2 & 0 & 0 & -\sqrt{3} \\ 0 & 1 & 0 & 0 \\ 0 & 0 & 1 & 0 \\ -\sqrt{3} & 0 & 0 & 2 \end{bmatrix}$$

(b) The event is an observation by Bob Jr. at the center of Bob's ship ($x = \frac{1}{2}\ell = 6 \times \sqrt{3}$ m), at time zero in Bob's frame ($t = 0$).

$$\begin{bmatrix} x' \\ y' \\ z' \\ ct' \end{bmatrix} = \begin{bmatrix} 2 & 0 & 0 & -\sqrt{3} \\ 0 & 1 & 0 & 0 \\ 0 & 0 & 1 & 0 \\ -\sqrt{3} & 0 & 0 & 2 \end{bmatrix} \begin{bmatrix} 6 \times \sqrt{3}\text{ m} \\ 0 \\ 0 \\ 0 \end{bmatrix} = \begin{bmatrix} 2 \times 6 \times \sqrt{3}\text{ m} \\ 0 \\ 0 \\ -\sqrt{3} \times 6 \times \sqrt{3}\text{ m} \end{bmatrix}$$

[17]As it stands, it "mixes" the $x$- and $t$-dimensions between frames, leaving the $y$ and $z$ alone, because the relative motion is along the $x$-axes. The reader is encouraged to deduce what the matrix would be if the direction of relative motion were instead along the frames' $y$-axes.

Bob Jr.'s viewing of the clock at the nose of Anna's ship occurs at position $x' = 12 \times \sqrt{3}$ m—the nose of Anna's ship, as it should be—and time:

$$ct' = -18 \text{ m} \quad \text{or} \quad t' = -60 \text{ ns}$$

The matrix in (1-27) gives the space-time coordinates of an event in frame $S'$ from space-time coordinates of the event in frame $S$. As we know, to transform "the other way," we need only replace $v$ with $-v$. This merely switches the signs of the top-right and bottom-left elements. Let us see what we get by transforming the space-time coordinates in $S'$, the left-hand side of (1-27), with such a matrix. Of course, if the equation is still to hold, we must also multiply on the right-hand side.

$$\begin{bmatrix} \gamma_v & 0 & 0 & \gamma_v \dfrac{v}{c} \\ 0 & 1 & 0 & 0 \\ 0 & 0 & 1 & 0 \\ \gamma_v \dfrac{v}{c} & 0 & 0 & \gamma_v \end{bmatrix} \begin{bmatrix} x' \\ y' \\ z' \\ ct' \end{bmatrix} = \begin{bmatrix} \gamma_v & 0 & 0 & \gamma_v \dfrac{v}{c} \\ 0 & 1 & 0 & 0 \\ 0 & 0 & 1 & 0 \\ \gamma_v \dfrac{v}{c} & 0 & 0 & \gamma_v \end{bmatrix} \begin{bmatrix} \gamma_v & 0 & 0 & \gamma_v \dfrac{-v}{c} \\ 0 & 1 & 0 & 0 \\ 0 & 0 & 1 & 0 \\ \gamma_v \dfrac{-v}{c} & 0 & 0 & \gamma_v \end{bmatrix} \begin{bmatrix} x \\ y \\ z \\ ct \end{bmatrix}$$

As the reader may verify, the product of the two matrices on the right is the "identity matrix." Thus, as we would expect,

$$\begin{bmatrix} x \\ y \\ z \\ ct \end{bmatrix} = \begin{bmatrix} \gamma_v & 0 & 0 & \gamma_v \dfrac{v}{c} \\ 0 & 1 & 0 & 0 \\ 0 & 0 & 1 & 0 \\ \gamma_v \dfrac{v}{c} & 0 & 0 & \gamma_v \end{bmatrix} \begin{bmatrix} x' \\ y' \\ z' \\ ct' \end{bmatrix}$$

## Four-Vectors

In special relativity, many quantities that classically are vectors acquire a fourth component—they become four-vectors. By definition,

> Any four quantities $(A_x, A_y, A_z, A_t)$ that transform from one frame to another via a Lorentz transformation constitute a four-vector.

We already have one example: $(x, y, z, ct)$. One of the most important properties of a four-vector is that there is a quantity associated with it that is the same for observers in all frames. Following the form of (1-27), the Lorentz transformation equations for an arbitrary four-vector are

$$A'_x = \gamma_v \left( A_x - \frac{v}{c} A_t \right) \quad A'_y = A_y \quad A'_z = A_z \quad A'_t = \gamma_v \left( A_t - \frac{v}{c} A_x \right)$$

It is left as an exercise to show from these relationships that

$$A'^2_t - (A'^2_x + A'^2_y + A'^2_z) = A^2_t - (A^2_x + A^2_y + A^2_z) \tag{1-28}$$

Whatever this quantity may represent, it is the same in frame $S$ as in frame $S'$. Such a quantity is called an **invariant.** Identifying $A_x^2 + A_y^2 + A_z^2$ with the square of the three-component vector portion of $A$, we may write

$$A_t'^2 - \mathbf{A}' \cdot \mathbf{A}' = A_t^2 - \mathbf{A} \cdot \mathbf{A} = \text{"invariant" associated with } A$$

For every four-vector there is an invariant. Let us investigate the invariant associated with the position-time four-vector $A = (x, y, z, ct)$. For any collection of events,

$$A_t^2 - \mathbf{A} \cdot \mathbf{A} = (ct)^2 - (x^2 + y^2 + z^2)$$

Since $A_t^2 - \mathbf{A} \cdot \mathbf{A}$ is the same in all frames, we may find its value in the most convenient—the frame in which the events occur all at the origin. In this frame, $A_t^2 - \mathbf{A} \cdot \mathbf{A}$ is simply $(ct)^2$, where $t$ is the time in that frame. Perhaps the events are the moments in the life of an astronaut on board a spacecraft. We conclude that *all* observers must calculate the same amount of time passing on the clock fixed at the location of the traveling astronaut; all will agree on the question of how much the astronaut ages. As noted in Section 1.4, this is the whole idea behind "proper time."

## Momentum-Energy Four-Vector

We have encountered another four-vector, but to see this clearly, a little more work must be done. In equation (1-21), we used the relativistic velocity transformation to relate the *total* momentum in frame $S'$ to the total momentum in frame $S$. Let us do the same here, but this time considering just the momentum of an *individual object*. Again using (1-18a) and the identity $\gamma_{u'} = (1 - u_x v/c^2)\gamma_v \gamma_u$,

$$p_x' = \gamma_u' m u_x' = \left[ \left( 1 - \frac{u_x v}{c^2} \right) \gamma_v \gamma_u \right] m \left( \frac{u_x - v}{1 - \frac{u_x v}{c^2}} \right)$$

$$= \gamma_v \gamma_u m u_x - v \gamma_v \gamma_u m$$

Here we have been explicit in specifying this as the $x$-component of momentum. Noting that $p_x = \gamma_u m u_x$ and $E = \gamma_u mc^2$, the preceding may be written

$$p_x' = \gamma_v \left[ p_x - \frac{v}{c} \left( \frac{E}{c} \right) \right] \tag{1-29}$$

The object's momentum in frame $S'$ is a combination of its momentum *and energy* in frame $S$.

Now applying a similar procedure to the energy in frame $S'$, we obtain

$$E' = \gamma_u' mc^2 = \left[ \left( 1 - \frac{u_x v}{c^2} \right) \gamma_v \gamma_u \right] mc^2 = \gamma_v [(\gamma_u mc^2) - v(\gamma_u m u_x)]$$

Recognizing $E$ and $p_x$ in frame $S$, this becomes

$$E' = \gamma_v (E - v p_x) \tag{1-30}$$

Dividing this by $c$ and placing it and (1-29) beside the position-time Lorentz transformation equations, the similarity becomes irresistible.

$$x' = \gamma_v\left[x - \frac{v}{c}(ct)\right] \qquad p'_x = \gamma_v\left[p_x - \frac{v}{c}\left(\frac{E}{c}\right)\right]$$

$$ct' = \gamma_v\left[(ct) - \frac{v}{c}x\right] \qquad \frac{E'}{c} = \gamma_v\left[\left(\frac{E}{c}\right) - \frac{v}{c}p_x\right]$$

The relationship between momentum and energy is virtually identical to that between position and time. Just as do $x$, $y$, $z$, and $ct$, the quantities $p_x$, $p_y$, $p_z$, and $E/c$ obey a Lorentz transformation and thus constitute a four-vector. (It is left as an exercise to show that $p'_y = p_y$ and $p'_z = p_z$.) As time ($ct$ for proper units) must be considered an essential part of a position-time four-vector, or "four-space," energy ($E/c$) is an essential part of the momentum-energy four-vector, or "four-momentum." The momentum-energy transformation may be expressed exactly as the position-time transformation in equation (1-27):

$$
\begin{aligned}
p'_x &= \gamma_v\left(p_x - \frac{v}{c}\frac{E}{c}\right) \\
p'_y &= p_y \\
p'_z &= p_z \\
\frac{E'}{c} &= \gamma_v\left(\frac{E}{c} - \frac{v}{c}p_x\right)
\end{aligned}
\quad \text{or} \quad
\begin{bmatrix} p'_x \\ p'_y \\ p'_z \\ \dfrac{E'}{c} \end{bmatrix}
=
\begin{bmatrix} \gamma_v & 0 & 0 & -\gamma_v\dfrac{v}{c} \\ 0 & 1 & 0 & 0 \\ 0 & 0 & 1 & 0 \\ -\gamma_v\dfrac{v}{c} & 0 & 0 & \gamma_v \end{bmatrix}
\begin{bmatrix} p_x \\ p_y \\ p_z \\ \dfrac{E}{c} \end{bmatrix}
\qquad (1\text{-}31)
$$

Now let us determine the invariant associated with the momentum-energy four-vector. In this case, $A = (p_x, p_y, p_z, E/c)$, and the invariant is related not to events in position and time but to "dynamical properties" of an object.

$$A_t^2 - \mathbf{A} \cdot \mathbf{A} = (E/c)^2 - p^2$$

If we choose $S$ to be the frame in which the object is at rest, the momentum is zero and the energy is simply $mc^2$. The invariant is $m^2c^2$, where $m$ is the mass in the rest frame of the object. Observers in all frames will measure the same mass/internal energy of an object. Kinetic energy and (therefore) total energy depend on the speed of an object through a given frame of reference, but internal energy does not.

## ▾ 1.12 The Light Barrier

In the 1940s, it was widely believed that no airplane could be made to exceed the speed of sound, but would be shaken to pieces in a spectacular exhibition if the attempt were made. Many doubted this belief, there being no theoretical basis for it. We have, however, in the Lorentz transformation equations a theoretical basis for the claim that *no information may travel faster than the speed of light*.[18]

A technique often used to prove that something *cannot* occur is to assume that it can. We then investigate whether the assumption leads to a contradiction. Let us assume that there is a means by which information can be sent at a speed $u_0 > c$.

[18]It follows that *human beings*, bearers of information that they are, cannot travel faster than light.

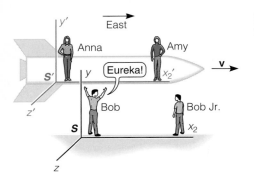

**Figure 1.17** Bob has an original thought.

Bob is at the origin of frame $S$ and Anna is at the origin of frame $S'$, a space-plane moving to the east (the $+x$-direction) at speed $v < c$ relative to Bob (Figure 1.17). At time zero, just as the origins cross, Bob has an original thought (event 1): In a spasm of inspiration, he has seized upon the perfect finale to the symphony he had been composing. This event has position and time $x_1 = 0$, $t_1 = 0$. Bob wishes to share his newly completed symphony with others as soon as possible. He sends it to Bob Jr., standing east of Bob at $x_2$. The completed symphony—the "information"—arrives at Bob Jr.'s location at time $t_2$ (event 2). Since Bob sent it via the assumed faster-than-light method, we must have

$$x_2 = u_0 t_2$$

Amy, Anna's assistant, is riding near the nose of Anna's spaceplane and happens to be precisely aligned with Bob Jr. just as he receives the information. Bob Jr. instantaneously communicates the information to Amy. Occurring at once at the same location, the arrival of the information at Bob Jr.'s location and its communication to Amy is but a single event (again, event 2). Since we have its position and time in Bob's frame, we may use the Lorentz transformation equations (1-8) to find its position (Amy's location) and time in Anna's frame. For reasons to become clear shortly, we express these in terms of $x_2$, the location of Bob Jr. in Bob's frame.

$$x_2' = \gamma_v(x_2 - vt_2) = \gamma_v\left(x_2 - v\frac{x_2}{u_0}\right)$$

$$= \gamma_v\left(1 - \frac{v}{u_0}\right)x_2$$

$$t_2' = \gamma_v\left(-\frac{v}{c^2}x_2 + t_2\right) = \gamma_v\left(-\frac{v}{c^2}x_2 + \frac{x_2}{u_0}\right)$$

$$= \gamma_v\left(-\frac{v}{c^2} + \frac{1}{u_0}\right)x_2 \qquad\qquad (1\text{-}32)$$

Amy urgently desires that Anna hear the symphony. Immediately upon receipt of the information, Amy sends it to Anna, also via the faster-than-light method. In Anna's frame, the time that passes as the information travels from $x_2'$ (Amy's location) to the origin of $S'$ (Anna's location) is $x_2'/u_0$. Since it leaves Amy at $t_2'$, it arrives at the origin (event 3) at time

$$t_3' = t_2' + \frac{x_2'}{u_0}$$

Now substituting from (1-32),

$$t_3' = \left[\gamma_v\left(-\frac{v}{c^2} + \frac{1}{u_0}\right)x_2\right] + \frac{1}{u_0}\left[\gamma_v\left(1 - \frac{v}{u_0}\right)x_2\right]$$

$$= \gamma_v x_2\left(-\frac{v}{c^2} + \frac{2}{u_0} - \frac{v}{u_0^2}\right)$$

$$= \gamma_v x_2\frac{2}{u_0}\left[1 - \left(\frac{u_0 v}{2c^2} + \frac{v}{2u_0}\right)\right]$$

For any $u_0 > c$, there is a $v < c$ for which the quantity in parentheses is greater than unity—in which case, $t_3'$ is negative! (For example, if $u_0$ were $2c$, any $v$ greater than $0.8c$ would cause $t_3'$ to be negative.) Recall that event 3 is the symphony arriving at the origin of frame $S'$ (Anna's location) and that the origins (Anna and Bob) pass at time zero on both clocks. Anna would be whistling the completed symphony *before* she passes Bob, who by assumption conceives of it at the instant of passing. This is a gross violation of the principle of cause and effect! If we accept that cause (conception of the symphony) must precede effect (its appreciation by others) and, naturally, that the Lorentz transformation equations are correct, there is only one candidate for an invalid assumption. Our proof is complete.

## Absolute Causality

The issue of cause and effect leads us back to a troubling point raised in Example 1.4: Events may occur in a different order in different frames. If this is true, might cause precede effect in one frame but effect precede cause in another? Such a thing would certainly be mind-boggling. But the Lorentz transformation equations do not allow it. Using equation (1-10b),

$$\Delta t' = \gamma_v\left(-\frac{v}{c^2}\Delta x + \Delta t\right) = \gamma_v \Delta t\left(1 - \frac{v}{c^2}\frac{\Delta x}{\Delta t}\right)$$

Suppose that $\Delta t$ and $\Delta t'$ are of opposite sign, that the time ordering of the events in the two frames is indeed reversed. The quantity in parentheses would have to be negative.

$$\left(1 - \frac{v}{c^2}\frac{\Delta x}{\Delta t}\right) < 0 \quad \text{or} \quad \Delta x > \frac{c^2}{v}\Delta t$$

Since $c^2/v$ is greater than $c$, the distance $\Delta x$ between the two events is farther than light could travel in the time interval $\Delta t$. Now, for one event to be a cause and the other an effect, information must travel between them, so if information cannot travel faster than light, there cannot be a cause-effect relationship. The order of the events can still be reversed; they just cannot be "causally related." (Of course, if information could travel faster than light, cause and effect would be reversed in the two frames, but we already know that to accept faster-than-light speed is to give up on cause and effect.)

## Light-Cones

We may visualize potentially causal events in three of the four space-time dimensions via the diagram of Figure 1.18. The focus of the diagram is an event at the "space-time origin." All events that may have had an effect on (been a cause of) this event—events of the **absolute past**—are confined within the **past light-cone,** of which the slope is the speed of light: $\Delta x/\Delta t = c$. For instance, an event on the time axis at $t < 0$ certainly could have had an effect on what occurs at the space-time origin at $t = 0$—and so could events spatially not too far from this axis. However, an event on the *surface* of the past light-cone, a distance $\Delta x = c\,\Delta t$ away, could have had an effect only if communicated to the space-time origin at the speed of light. Accordingly, events in the past but *outside* the past light-cone are too far away to have had an effect. This region is known as the **absolute elsewhere.** At time zero, it constricts to a single point, because no event anywhere else can have an *instantaneous* effect on the event at the space-time origin. The meaning of the **future light-cone** is complementary. It contains all events that are not too far away to be affected by the event at the space-time origin—events of the **absolute future.** Naturally, it spreads spatially as time progresses. The gray

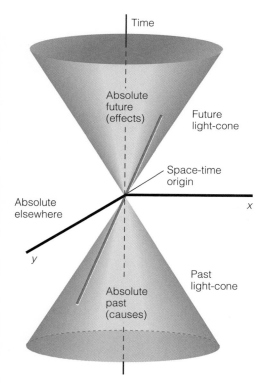

**Figure 1.18** Causes (events of the past) reside in the past light-cone, while effects (events of the future) reside in the future light-cone.

line is an example of a **world line,** a path showing the events in the "life" of some thing. This thing simply moves in the $+x$-direction at a constant speed $(dx/dt)$ not much less than $c$.

# ▾ 1.13   General Relativity and Cosmology

It might be said that *special* relativity begins with Einstein's deceptively simple postulate that the speed of light is the same in all frames. *General* relativity, a more general theory, begins with another of Einstein's "simple" postulates—that "inertial mass" and "gravitational mass" are the same. In introductory physics we learn Newton's universal law of gravitation: The force between objects of mass $m$ and $M$ separated by $r$ is $F = GMm/r^2$, where $G$ is the universal gravitational constant. Taking $m$ to be an object near Earth's surface, with $M$ and $r$ being Earth's mass and radius, the force on $m$ is

$$F_{\text{grav}} = \frac{GM_{\text{Earth}}}{r_{\text{Earth}}^2} m \cong 9.8 \text{ m/s}^2 \times m \qquad\qquad \text{(1-33a)}$$

In introductory mechanics we also learn the second law of motion: The acceleration of an object of mass $m$ is proportional to the net force on the object and inversely proportional to $m$.

$$\mathbf{a} = \mathbf{F}_{\text{net}} \frac{1}{m} \qquad\qquad \text{(1-34a)}$$

On the face of it, these two properties of the mass $m$ are entirely different. There is no fundamental reason that the property governing how hard gravity pulls on a given object should have anything to do with the property governing its re-luctance to accelerate when a net force is applied! (After all, the net force might be purely electrostatic, unrelated to gravity.) Accordingly, it might be safer to use an $m_g$ in the first equation, signifying a gravitational property, and an $m_i$ in the second, signifying an inertial property.

$$F_{\text{grav}} = 9.8 \text{ m/s}^2 \times m_g \quad \text{(1-33b)} \qquad a = F_{\text{net}} \frac{1}{m_i} \quad \text{(1-34b)}$$

Now consider what happens when an object is dropped from shoulder height. In this case, the net force is simply the gravitational force. Thus,

$$a = (9.8 \text{ m/s}^2 \times m_g) \frac{1}{m_i} = 9.8 \text{ m/s}^2 \frac{m_g}{m_i}$$

Were $m_g$ and $m_i$ truly different properties, there is no reason why we could not have an object 1 with $m_i = 1.1 m_g$ and an object 2 with $m_i = 0.9 m_g$, in which case they would accelerate at different rates. This is certainly not what we expect. In fact, the equivalence of $m_g$ and $m_i$ has been experimentally verified to better than 1 part in $10^{12}$. It is natural to assume that the properties are the same. But with the unthinking assumption comes a certain blindness.

Albert Einstein was the first to discover the startling possibilities that arise by postulating that $m_g$ and $m_i$ are the same. In particular, if $m_g = m_i$, it would be impossible to determine whether we are in an inertial frame permeated by a uniform gravitational field, or in a frame in which there is no field, but which accelerates at a constant rate.

Suppose Bob stands in a closet on Earth, frame $B$ (Figure 1.19). He is in a frame of reference that is inertial and in which there is a uniform gravitational field of $g = 9.8$ m/s² downward.[19] Anna is in an identical closet, frame $A$, but one that is out in space, far from any gravitational fields. By means of a rocket engine, Anna's closet is accelerating in a straight line at 9.8 m/s². For Bob to remain stationary, the floor must push upward on his feet with a force whose magnitude is equal to the downward force, $F_N = m_g \times 9.8$ m/s². For Anna to accelerate along with her rocket-powered closet, the floor must push "upward" on her feet with a force sufficient to give her an acceleration of 9.8 m/s². By the second law of motion, this force is $F_N = m_i \times 9.8$ m/s². Now, if $m_g$ and $m_i$ are equal, the forces in the two frames would be equal, and would provide no clue to distinguish whether an observer is in frame $A$ or frame $B$.

The normal force is only the simplest example of something we might use; the fact is that no mechanical experiment would be able to distinguish the frames. In the linearly accelerating frame $A$, all things appear to be affected by a downward force just as they are in frame $B$: the floor must push "up" on objects; "dropped" objects to appear to accelerate downward (because, once let go, they do not accelerate along with the frame). All these effects could be attributed to a fictitious "inertial force" $-m_i\mathbf{a}$ opposite the direction of acceleration. *Provided that $m_i$ equals $m_g$, this force would mimic a gravitational force $m_g\mathbf{g}$ in all respects, as shown in Figure 1.20.*

By the same token, no mechanical experiment would be able to distinguish a frame that is accelerating in "free-fall" in a uniform gravitational field from one

[19] Actually, Earth revolves, so the frame does accelerate, and the field is not exactly uniform, because all field lines point toward Earth's center and so are not exactly parallel. But it is inertial and of a uniform field to a reasonably good approximation.

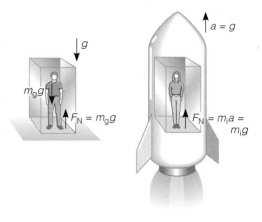

Frame *B*     Frame *A*

**Figure 1.19** A stationary frame with gravity and an accelerating frame without gravity.

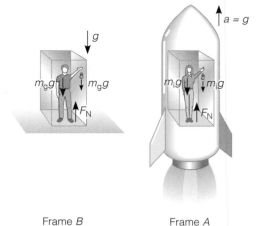

Frame *B*     Frame *A*

**Figure 1.20** All forces appear to be the same in both frames.

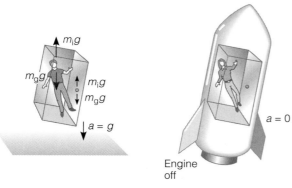

Free-falling frame                    Inertial frame

that is inertial and without a gravitational field. An observer in the inertial frame would see all objects floating or moving at constant velocity, because no forces act. As shown in Figure 1.21, an observer in the free-falling frame would also see objects seemingly moving at constant velocity (though all would actually be accelerating with the frame), because the gravitational force $m_g\mathbf{g}$ is exactly canceled by the "inertial force" $-m_i\mathbf{a}.$

Einstein's customary leap forward was to postulate that *all* physical phenomena, not just mechanical ones, occur identically in a frame accelerating in gravitational free-fall as in an inertial frame without gravity. No experiment could distinguish the frames. Accordingly, he generalized the concept of an inertial frame by defining a "locally inertial frame": a frame that is falling freely in a gravitational field (which includes ordinary inertial frames—no gravity, no acceleration—as a special case.) [20] With this definition, Einstein's fundamental postulate of general relativity, known as the **principle of equivalence,** is

> The form of each physical law is the same in all locally inertial frames.

Principle of equivalence

This postulate is only the basis of general relativity. Just as the Lorentz transformation equations follow from the postulates of special relativity, a mathematical framework follows from this postulate. Unfortunately, general relativity theory is too sophisticated to discuss quantitatively here (it involves the mathematics of tensors and differential geometry). Nevertheless, some of its astonishing predictions can be understood qualitatively just from the principle of equivalence. Three have attracted particular attention: (1) gravitational red-shift; (2) the deflection of light by the sun, and (3) the precession of the perihelion of Mercury.

## Gravitational Red-Shift

According to the equivalence principle, light *emitted* at one point in a gravitational field will have a different frequency if *observed* at a different point. We see this by analyzing not a fixed light source and observer in a gravitational field $g$, but the equivalent case of a source and observer in a frame accelerating at $g$ (without gravity). In Figure 1.22(a), a source in an accelerating frame emits a wave front when the frame has zero speed. An observer a distance $H$ "above" the source

[20]The acceleration of freely falling objects in a frame would not be the same unless the gravitational field is uniform. Therefore, the frame must also be small enough that any nonuniformities in the field within it are negligible.

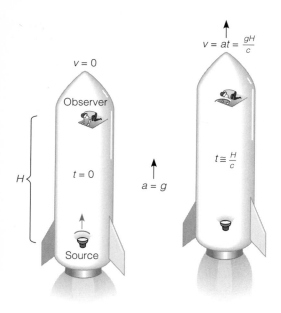

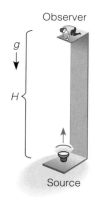

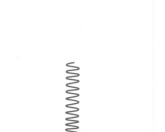

**(a)** Accelerating frame

**(b)** Equivalent frame

**Figure 1.22** A light beam traveling from source to observer (a) in an accelerating frame without gravity, and (b) in the equivalent stationary frame with gravity.

observes the wave front after a time $H/c$. But by this time the observer is moving "upward" at speed $v = gH/c$. Thus, the light was emitted in a frame that moves at velocity $gH/c$ *away* from the frame in which the light is observed. According to the observer, the light will be red-shifted. By the arguments of Section 1.6, the observed period will be longer than the source period by the factor $1 + v/c = 1 + gH/c^2$.[21]

$$T_{obs} = \left(1 + \frac{gH}{c^2}\right) T_{source}$$

Inverting to obtain frequencies,

$$f_{obs} = \left(1 + \frac{gH}{c^2}\right)^{-1} f_{source} \cong \left(1 - \frac{gH}{c^2}\right) f_{source}$$

The observed frequency is lower by the factor $gH/c^2$. The fractional change in frequency is given by

$$\frac{\Delta f}{f_{source}} \equiv \frac{|f_{obs} - f_{source}|}{f_{source}} = \frac{gH}{c^2}$$

Now since a gravity-free frame accelerating at $g$ is equivalent to a fixed frame in a gravitational field $g$, the same conclusion must apply to the latter case, Figure 1.22(b). Thus, as light moves upward, its frequency becomes smaller and its wavelength longer (Figure 1.23). Two interesting conclusions follow.

First conclusion: This is a time-dilation effect, but it has nothing to do with being in different reference frames. In Figure 1.22(b), there is no relative motion between source and observer. Assuming light of frequency $5 \times 10^{14}$ Hz (600 nm), as 1 s passes at the source, $5 \times 10^{14}$ wave fronts are emitted. It might be said that

**Figure 1.23** A light beam's wavelength grows longer as it rises in a gravitational field.

[21] This is a "lowest-order" result, correct only when the acceleration and distance traveled are small.

### Does Gravity Really Affect Time?

Is time really passing slower on Earth's surface than in the surrounding space, or is it just that gravity interferes with the behavior of light sources, clocks, and so on? In introductory physics we learn that time is based upon an accepted standard unit. Today we define time in terms of the period of apparently regular oscillations of cesium-133 atoms. There does seem to be a predictable relationship between this unit of time and the rate of all other microscopic physical processes, such as those that govern the affairs of the human body. If all processes occur at a slower rate on Earth's surface than at altitude, as it appears that they do, it certainly seems reasonable to say that time passes more slowly at the surface. Of course, if we cannot equate "real" time with the rate at which physical processes occur, the question remains unanswered.

$5 \times 10^{14}$ distinct events in the life of the source pass. But in a given second, the observer receives fewer wave fronts, because he observes a smaller frequency. To witness all $5 \times 10^{14}$ events in the life of the source, the observer has to wait more than 1 s. Relative to the observer, the source, "deeper" in the gravitational field, is aging slowly.

Does time really pass more slowly on the surface of Earth than at some altitude above? The weakness of most gravitational fields ($gH/c^2 \ll 1$) had long made verification of gravitational time dilation difficult. But modern atomic clocks of extremely high precision have changed this. Atomic clocks use as their basic unit of time the very short period of certain atomic oscillations. By comparing the frequency of such a clock on the surface to that of one aboard a high-altitude rocket, strong confirmation of Einstein's equivalence principle has been obtained. (See the essay entitled "Does Gravity Really Affect Time?")

Before we discuss the second conclusion, let us venture a little further with the present train of thought: If a gravitational field somehow "warps" time intervals, even when there is no relative motion, why not space intervals? Indeed, one of the tenets of general relativity is that a massive heavenly body warps space-time nearby. Representing warped space-time in three dimensions is difficult. It is easier in two dimensions, in which space is area. Figure 1.24 shows a massive

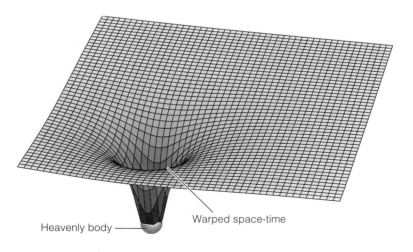

Heavenly body

Warped space-time

**Figure 1.24** A two-dimensional representation of a massive object warping space.

heavenly body disturbing the regularity of a two-dimensional space. Inhabitants of this two-dimensional universe *expect* all "cells" to be of equal area. We "outside observers," however, can see that the cells near the heavenly body are really larger—but only from our extra-dimensional viewpoint. Similarly, the warpage of real three-dimensional space is not apparent to human beings; we are creatures of our space of three dimensions, and are unable to stand back and view our universe on *four*-dimensional axes. Nevertheless, even reduced-dimensional views such as Figure 1.24 provide a qualitative understanding of some features of general relativity, as we shall soon see.

Second conclusion: Gravity has an effect on light! Now light has no mass—its energy is all kinetic—so it must be that gravity pulls on forms of energy besides mass (or internal) energy. As it moves away, the light's kinetic energy must decrease as potential energy increases. But this leads to the question, How can the universal law of gravitation, $F = GMm/r^2$, be used to account for the effect of gravity on light without an $m$ for light? The answer is that Newton's "universal law" is really a special case, correct only when the gravitational field is very weak—the so-called "classical limit." [22] In this limit, the effect on light is negligible and there is no need to consider an $m$. For a strong field, however, general relativity comes into play and Newton's law is replaced by a different view—warped space-time. Near a massive heavenly body, the regularity in space and time intervals is disturbed, and light naturally changes frequency as it passes through.

Most heavenly bodies produce only a very small gravitational red-shift. They are not dense enough to warp nearby space-time significantly. The gravitational red-shift of our sun is only about 2 parts in $10^6$. (Earth's is much smaller still.) It has been measured, though, and agrees quite well with what Einstein's principle of equivalence predicts. On the other hand, it is theoretically possible for an object to be so dense that light simply cannot escape its gravitational potential energy at all. We discuss such objects, known as "black holes," later in this section.

## Deflection of Light by the Sun

We have seen that gravity "pulls" on light moving directly away from a heavenly body. But what if the light is traveling laterally—does it curve? It must! Since a laterally moving light beam would appear to curve toward the floor in Anna's rocket-powered closet, as shown in Figure 1.25, it must curve toward the floor in Bob's Earth-bound closet.

Again, however, we need not attempt to reconcile light's curvature with Newton's law of gravitation; rather, the light simply moves in the most natural way given the warped space through which it travels. A guiding classical principle applies even in warped space-time: Light always takes the minimum time to travel from one point to another. If we combine this with the idea of warped space-time, it becomes clear why light should "bend" near massive heavenly bodies. Figure 1.26 shows two possible paths of a light beam originating at one point in space, passing through the warped space-time near a large heavenly body, then observed at another point. The darker path follows what inhabitants of the flat, two-dimensional space might believe is a straight line, one of the lines in a "grid" that would be regular if space were not warped. Smugly observing from our

[22]Newton's gravitation law $F = Gm_1m_2/r^2$ bears somewhat the same relationship to general relativity theory that Coulomb's Law $F = (1/4\pi\epsilon_0)q_1q_2/r^2$ does to Maxwell's equations. Coulomb's law is a special case, correct only for *static* electric fields. It cannot explain electromagnetic *waves*, as can Maxwell's equations. Similarly, general relativity, not Newton's static law, is needed to understand gravitational waves (see Progress and Applications on pp. 60–61).

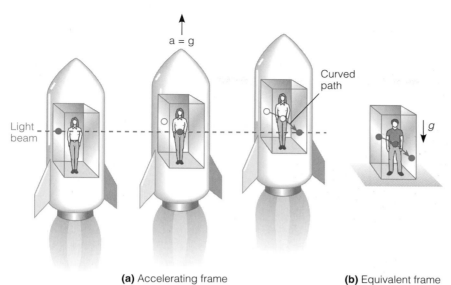

a = g

Light beam

Curved path

g

**(a)** Accelerating frame

**(b)** Equivalent frame

**Figure 1.25** A light beam curving (a) in an accelerating frame without gravity, and (b) in the equivalent stationary frame with gravity.

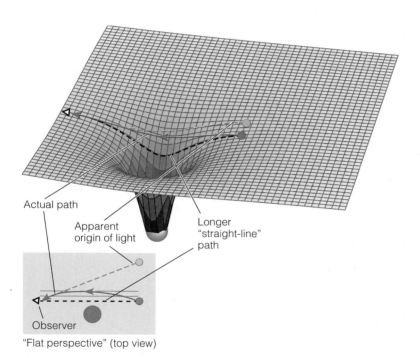

Actual path

Apparent origin of light

Longer "straight-line" path

Observer

"Flat perspective" (top view)

**Figure 1.26** A light beam takes the shortest path, which to inhabitants of warped space does not appear straight.

greater-dimensional perspective, however, we see that this path is rather long. The other path, though appearing not to follow a straight line from the "flat perspective," is shorter. Since this is the path the light actually takes, an observer with a "flat perspective" sees a beam that curved as it passed near the heavenly body and thus *seems* to have originated at some other point in space.

As noted, our Sun does not warp space much. To have any hope of detecting deflection of light by our Sun, observations have to be made of light rays passing very close to it, where its field is strongest. This light, of course, comes from other stars. Their positions should appear to shift slightly as the Sun passes between them and Earth. The problem is that the stars would ordinarily be completely obscured by the brightness of the Sun. Therefore, such observations have to be made when the Sun is "darkened"—during a solar eclipse. One of the first, during an eclipse in 1919, showed a deflection of about 2 s of arc for light barely grazing the Sun. As in the case of gravitational red-shift, agreement with the prediction of general relativity is very good.[23]

## Precession of the Perihelion of Mercury

The orbits of the planets about the Sun are not exactly circular, but slightly ellip-tical. At one point in its orbit, called aphelion, a planet will be slightly farther than average from the Sun, and at another, called perihelion, slightly closer. So long as a system is simply one object orbiting another, it is a direct prediction of classical Newtonian gravitation that the same path in space is retraced indefinitely. But if anything interferes with the simple interaction, the orbit will *precess;* the points of aphelion and perihelion progressively creep around in a circular fashion. One source of interference is the presence of other heavenly bodies. The slightly el-liptical orbit of Mercury, for instance, precesses due to the perturbing effects of the other planets. Now Newtonian theory can account for these effects, predicting a rate at which precession should occur. But, much to the consternation of early astronomers, the predicted rate did not agree precisely with observation. The problem was that Newton's law of gravitation is correct only for a weak gravita-tional field. A strong field might well cause observation to deviate from the clas-sical expectation. If any planet shows deviation, it should be Mercury, for it orbits where the Sun's field is strongest. Using general relativity, a correction to the clas-sically expected precession rate of Mercury may be calculated. The result of 43 s of arc per century is in good agreement with observation.

We have seen how successful general relativity has been in explaining some slight discrepancies between classical expectation and actual observation. Al-though this evidence took decades to accumulate, its weight was irresistible, lead-ing to near-universal acceptance of Einstein's "new" approach to gravitation. Let us now explore an important topic in which general relativity plays a signifi-cant role.

## Cosmology

Cosmology is the study of the behaviors of heavenly bodies, both individually and collectively. The topic is invariably coupled to gravitation, because gravitation is the only fundamental force in nature that is both long-range and only attractive. Thus, it is by far the most important force in the evolution of, and interaction between, huge heavenly bodies.[24] Much of cosmology can be understood via clas-sical Newtonian gravitation, but certain behaviors require general relativity.

**Stellar collapse** is the term describing the fate of individual stars. The energy source of stars is nuclear fusion (cf. Chapter 11). Early in the "life" of a star,

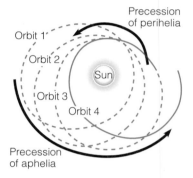

[23]The warpage of space caused by an immense mass allows for *multiple* short-est distances between two points. Light from a star may curve around opposite sides of an intervening massive object, producing a double image at Earth. As-tronomers have indeed observed large galaxy clusters that produce this effect. They are referred to as "gravitational lenses."

[24]Large objects tend to be electrically neutral, and so do not exert significant electrostatic forces on one another. The other fundamental forces (the "weak" and the "strong," see Chapter 11) are sig-nificant only for microscopic particles at very close range, typically $10^{-15}$ m and less.

the enormous gravitational attraction of its mass is balanced by the constant generation of energy, which would tend to scatter the material—a constant size is maintained. When a star's fuel begins to burn out, the gravitational attraction predominates and gravitational collapse begins. If the star is not much more massive than our Sun, the final result is a **white dwarf,** and ultimately a cold, dead chuck of matter typically no larger than Earth! For stars several times the Sun's mass, the extra gravitational pressure is able to force protons and electrons to combine, forming neutrons and tiny particles known as **neutrinos** (see Chapter 11). The neutrinos carry away a huge amount of energy in a cataclysmic explosion known as a supernova. What remains is a cold **neutron star** of fantastic density—a typical radius is only tens of kilometers. The density of a star collapsed so small would be roughly $10^{17}$ kg/m$^3$, approximately equal to that of the atomic nucleus, $10^{13}$ times that of lead![25] For even larger stars there is a third possible fate—becoming a **black hole,** and here is where general relativity comes into play.

One of the most startling features of the theory of general relativity is the possibility of singularities in space.[26] A black hole would be such a singularity. A singularity would occur if a body of mass $M$ becomes so compact that its radius drops below the so-called **Schwartzchild radius:** $r_S = 2GM/c^2$. (Assuming that the critical radius depends only on the mass of the body and the fundamental constants $G$ and $c$, simple dimensional analysis yields this equation to within a multiplicative constant; see Exercise 74.) The singularity would be separate from the universe as we know it. It would simply be a silent "hole" in space. Moreover, its gravitational field would be so strong that not even light could leave it, thus, the name "*black* hole."[27] Its presence would be betrayed only by its warpage of space—that is, its external gravitational field—and its ability to gobble up things from the outside. A black hole would be uncharted territory, literally not on the map, and an unlucky passerby would be in for quite a ride. It should be noted that not even the atomic nucleus—trillions of times denser than the densest of ordinary materials—comes close to qualifying as a black hole. The diameter of a typical nucleus is $10^{-15}$ m, but its Schwartzchild radius would be roughly $10^{-52}$ m. Only under conditions such as the tremendous gravitational pressure in immense heavenly bodies might a black hole occur. Since black holes would be light-years away and dark, with gravitational effects serving as our only clues, we expect them to be difficult to detect. Although there are several candidates, there is understandable hesistancy in declaring the evidence conclusive.

### The Evolution of the Universe

The most apparent change in the universe as a whole is that it is expanding. Furthermore, the way in which it is expanding suggests that it began with a "bang," from the explosion of an initial, supremely compact body of energy. Let us view the universe as consisting of galaxies, each galaxy being a distinct frame of reference.[28] If all galaxies began moving away from the origin at the same time, after an arbitrary time they would be spread out according to their speeds; speed should increase linearly with distance from the origin. After one unit of time, those moving a unit distance per time would be one unit away, those moving two units of distance per time would be two units away, and so forth. Perhaps surprisingly, this would also appear to be the case no matter which frame of reference (galaxy) an observer happened to occupy (see Exercise 76 for a simple

[25] Chapters 7 and 8 provide the background necessary to understand why white dwarfs and neutron stars should be so small. Calculations of their radii are taken up in the Chapter 8 exercises. The term pulsar refers to a neutron star that has a strong magnetic field and is spinning so as to send regular pulses of electromagnetic radiation toward the earth.

[26] Mathematically, a "singularity" is a point at which a discontinuity or divergence occurs.

[27] See Exercise 75 for a crude "derivation" of the Schwartzchild radius from this assumption.

[28] The reason for this arbitrary choice is that galaxies are the smallest units that move *more or less* independently. Stars within a given galaxy definitely do not. Our Sun, for instance, orbits about the center of its galaxy, the Milky Way.

argument). Thus, if the universe did begin with a "big bang," any galaxy (including our own, of course) should serve as a valid base from which to observe, and the speeds of the galaxies should increase linearly with their distances from that frame of reference.

This is just what we observe! This evidence comes from the Doppler shifts in the light spectra from distant galaxies; the farther a galaxy is from Earth, the greater the red-shift in its spectrum, which implies a greater recessional speed. We find that the speed increases linearly with distance:

$$v = H_0 r$$

This relationship is known as Hubble's law, and the constant $H_0$ as the **Hubble constant.** Since an object moving a distance $r$ at constant speed $v$ would travel for a time $r/v$, and this is the same constant $1/H_0$ for all galaxies, $1/H_0$ is often referred to as "the age of the universe." Modern Doppler evidence gives a value of approximately $2 \times 10^{-18}$ s$^{-1}$ for $H_0$, or an age of the universe on the order of 10 billion years.

Unfortunately, the Hubble constant is not known with great precision, mostly because it is difficult to be certain of the distances to faraway galaxies. Even if distances could be known precisely, a guess at the age of the universe based on the Hubble constant is approximate at best. Among the problems is that galaxies really haven't moved out independently at constant velocity since an initial "big bang." All forms of energy in the universe share a gravitational attraction, which necessarily decreases the expansion rate. Indeed, one of the most compelling questions in cosmology is whether this gravitational attraction is sufficient to cause the universe at some time in the future to stop expanding and then reconverge. The answer hinges on the energy density in the universe.

General relativity predicts a **critical density** $\rho_c$. If the actual density $\rho$ is less than $\rho_c$, the universe will expand forever; if greater, it will eventually reconverge. The big question is, How does $\rho$ compare to $\rho_c$? Unfortunately, the energy density in the universe is not known well enough to answer the question. One of the most important factors is the amount of "dark matter." It is fairly easy to estimate the energy density of things we can see (i.e., stars), and this suggests that $\rho < \rho_c$. But there is much out there that we cannot see. This dark matter may take several forms: intergalactic dust, collapsed galaxies, black holes, background neutrinos, or unknown particles or radiation. The density of neutrinos in space is difficult to determine, but of particular interest. Though once thought to be massless (like light), neutrinos now appear to have mass, and this may greatly affect the value of $\rho$. Much research has to be done before we can say whether the universe will or will not expand forever.

## Progress and Applications

Special relativity is now well matured. In modern applications of physics, the need to observe the relativistic rules at high speeds goes without saying, relegating classical mechanics to the role of special case. Most current research in the field is in the area of general relativity, and most of its applications are in cosmology.

■ Gravitation theory predicts how the rotation rate of stars and other orbiting matter in spiral galaxies should depend on the galaxy's mass/energy density. Recent Doppler-shift data from hydrogen clouds surrounding such galaxies clearly show that these rotations cannot be explained by luminous matter alone—strong evidence for the existence of dark matter, though not for its specific form. The evidence suggests that the density of dark matter may be ten times that of the luminous matter. The search for dark matter is one of the most active in physics.

■ A promising candidate for dark matter is the **brown dwarf:** a heavenly body of mass intermediate between a large planet like Jupiter and a star. Too small to produce its own light via nuclear fusion (see Chapter 10), it is still large enough to have a significant effect on the universe's energy density. Brown dwarfs are being sought through infrared "signatures" and gravitational "light-bending." Though not yet conclusive, the evidence for their existence is strong and mounting.

■ Other candidates are the so-called weakly interacting massive particles, or WIMPs. If they do exist, the very trait of being "weakly interacting" would make them difficult to detect. (While the smell of chlorine is hard to ignore, nitrogen passes undetected through the nose, with which it "weakly interacts.") Still, no evidence has been found.

■ The most conspicuous possible contributors to dark matter, neutrinos (see Chapters 10 and 11), permeate space, and so are constantly bombarding our world. The possibility that they have significant mass is the big question. Theory predicts that if neutrinos are able to "oscillate"—that is, change from one type to another—they must have mass. Oscillations have been detected recently, but the actual masses are not yet well known.

■ Probably the most fascinating prediction of general relativity is the black hole. Theories uniting the prediction with the ideas of quantum field theory continue to make progress (see Chapter 11), and experimental evidence mounts daily. Among the tell-tale signs being studied are unusual concentrations of starlight in, and movements of gases about, certain galaxies; unexplained creations of subatomic particles known as positrons in our Milky Way galaxy; and conspicuous polarization of x rays (see Chapter 2) from various objects in the heavens, all thought to be caused by the effects of nearby black holes.

■ Just as Maxwell's equations, describing electromagnetic phenomena, predict the existence of electromagnetic waves, general relativity, underlying gravity, predicts the existence of **gravitational waves.** But comparatively speaking, electromagnetism is strong and gravity is weak. (Everyone has seen static electric forces between ordinary objects, but the only gravitational forces readily apparent to anyone involve Earth, of immense mass.) To have any hope of success, efforts to detect gravitational waves have involved apparatus of extreme sensitivity. Indeed, so sensitive must they be that they impinge upon the theoretical limits to our knowledge—the most important being the "uncertainty principle" (see Chapters 3 and 4). Unfortunately, sorting out the weak effects among the various kinds of noise still proves elusive.

# Basic Equations

## Lorentz transformation equations

$$x' = \gamma_v(x - vt) \qquad \text{(1-8a)}$$

$$t' = \gamma_v\left(-\frac{v}{c^2}x + t\right) \qquad \text{(1-8b)}$$

$$x = \gamma_v(x' + vt') \qquad \text{(1-9a)}$$

$$t = \gamma_v\left(+\frac{v}{c^2}x' + t'\right) \qquad \text{(1-9b)}$$

$$\gamma_v \equiv \frac{1}{\sqrt{1 - \dfrac{v^2}{c^2}}}$$

## Time dilation

$$\Delta t = \gamma_v \, \Delta t_0 \qquad \text{(1-12)}$$

## Length contraction

$$L = \frac{L_0}{\gamma_v} \qquad \text{(1-13)}$$

## Doppler effect

$$f_{\text{obs}} = f_{\text{source}} \frac{\sqrt{1 - \dfrac{v^2}{c^2}}}{1 + \left(\dfrac{v}{c}\right)\cos\theta} \qquad \text{(1-15)}$$

## Relativistic velocity transformation

$$u'_x = \frac{u_x - v}{1 - \dfrac{u_x v}{c^2}} \qquad u'_y = \frac{u_y}{\gamma_v\left(1 - \dfrac{u_x v}{c^2}\right)} \qquad u'_z = \frac{u_z}{\gamma_v\left(1 - \dfrac{u_x v}{c^2}\right)}$$

$$\text{(1-18a)} \qquad\qquad \text{(1-18b)} \qquad\qquad\qquad \text{(1-18c)}$$

## Momentum

$$\mathbf{p} = \gamma_u m\mathbf{u} \quad \text{where} \quad \gamma_u \equiv \frac{1}{\sqrt{1 - \dfrac{u^2}{c^2}}} \qquad \text{(1-20)}$$

## Energy

$$E = \gamma_u mc^2 \qquad \text{(1-22)}$$

## Internal energy

$$E_{\text{internal}} = mc^2 \qquad \text{(1-23)}$$

## Kinetic energy

$$\text{KE} = (\gamma_u - 1)mc^2 \qquad \text{(1-24)}$$

$$E^2 = p^2c^2 + m^2c^4 \qquad \text{(1-26)}$$

# Summary

No experiment has been successful in demonstrating the *absolute* motion of any inertial frame. There is apparently no fixed universal frame of reference; all motion is relative. Arguing that an experiment should thus proceed identically in any frame, irrespective of its motion relative to any other, Einstein postulated that the laws of physics must be the same in all inertial frames. In particular, Maxwell's equations, which yield the speed of light $c$, are the same in all inertial frames. Accordingly, light will move at $c$ relative to an observer no matter how his frame may move relative to another.

One of the most important results of Einstein's postulates is that an event's position *and time* in one frame will in general be different than its position *and time* in another. The equations relating these positions and times are the Lorentz transformation equations.

$$x' = \gamma_v(x - vt) \qquad y' = y \qquad z' = z \qquad t' = \gamma_v\left(-\frac{v}{c^2}x + t\right)$$

They may be used to quantify several startling relativistic effects: Time dilation, in which an observer in an inertial frame will always determine time to pass more slowly in a moving frame than in his own; length contraction, in which an object in a moving frame is shorter along the direction of relative motion than it would be if at rest with respect to the observer; and relative simultaneity, in which two events simultaneous in one frame are separated by a nonzero time interval in another. These effects are imperceptibly small for relative velocities very much less than $c$.

The demand that momentum conservation hold in all frames leads to the relativistically correct expression for momentum *and,* remarkably, to an expression for the total energy of a moving object.

$$\mathbf{p} = \gamma_u m\mathbf{u} \qquad E = \gamma_u mc^2$$

Momentum and energy are inextricably related; the laws governing their conservation are inseparable.

Special relativity shows that associated with some classical vectors is a fourth quantity—a fourth component. For the position vector it is time, and for the momentum vector it is energy. Four-component quantities are known as four-vectors.

The speed of light is an absolute upper limit on the speed at which anything carrying information may move relative to an observer.

# E X E R C I S E S

These exercises are either standard (ranging from easy to fairly challenging) or advanced (involving deeper physical principles and/or more complex math). Each advanced exercise is distinguished by a blue-highlighted exercise number.

## Section 1.4

1. Appearing in the time-dilation and length-contraction formulas, $\gamma_v$ is a reasonable measure of the size of relativistic effects. Roughly speaking, at what speed would observations deviate from classical expectations by 1%?

2. Through a window in Carl's spaceplane, passing at $0.5c$, you watch Carl doing an important physics calculation. By your watch it takes him 1 min. How much time did Carl spend on his calculation?

3. According to an observer on Earth, a spacecraft whizzing by at $0.6c$ is 35 m long. What is the length of the spacecraft according to passengers on board?

4. According to Bob on Earth, Planet Y (uninhabited) is 5 ly away. Anna is in a spaceship moving away from Earth at $0.8c$. She is bound for Planet Y, to study its geology. Unfortunately, Planet Y explodes. According to Bob, this occurred 2 yr after Anna passed Earth. (Bob, of course, has to wait a while for the light from the explosion to arrive, but reaches his conclusion by "working backward.") Call the passing of Anna and Bob time zero for both. (a) According to Anna, how far away is Planet Y when it explodes? (b) At what time does it explode?

5. Anna is on a railroad flatcar moving at $0.6c$ relative to Bob. (Their clocks read zero as Anna's center of mass passes Bob's.) Anna's arm is outstretched in the direction the flatcar moves, and in her hand is a flashbulb. When the flashbulb goes off, the reading on Anna's clock is 100 ns. The reading on Bob's differs by 27 ns. (a) Is it earlier or later than 100 ns? (b) How long is Anna's arm (i.e., from her hand to her center of mass)?

6. A pole-vaulter holds a 16-ft pole. A barn has doors at both ends, 10 ft apart. The pole-vaulter on the outside of the barn begins running toward one of the open barn doors, holding the pole level in the direction he's running. When passing through the barn, the pole fits (barely) entirely within the barn all at once. (a) How fast is the pole-vaulter running? (b) According to whom, the pole-vaulter or an observer stationary in the barn, does the pole fit in all at once? (c) According to the other person, which occurs first, the front end of the pole leaving the barn, or the back end entering, and (d) what is the time interval between these two events?

7. Anna and Bob are in identical spaceships 100 m long each, as shown, with distances from the back labeled along the sides. Prior to taking up space travel in retirement, Bob and Anna owned a clock shop, and they glued the leftovers all over the walls of their ships. The diagram shows Bob's view as Anna's ship passes at $0.8c$. Just as the backs of the ships pass one another, both clocks there read zero. At the instant shown, Bob Jr., on board Bob's ship, is aligned with the very front edge of Anna's ship. He peers through a window in Anna's ship and looks at the clock. (a) In relation to his own ship, where is Bob Jr., and (b) what does the clock he sees read?

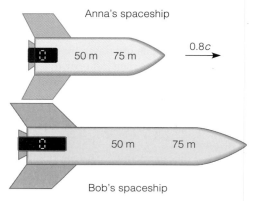

Anna's spaceship

50 m    75 m

0.8c →

50 m    75 m

Bob's spaceship

8. Bob is watching Anna fly by in her new high-speed plane, which Anna knows to be 60 m in length. As a greeting, Anna turns on two lights simultaneously, one at the nose, one at the tail. According to Bob, the lights come on at different times, 40 ns apart. (a) Which comes on first? (b) How fast is the plane moving?

9. Bob and Bob Jr. stand at open doorways at opposite ends of an airplane hangar 25 m long. Anna owns a spaceplane, 40 m long as it sits on the runway. Anna takes off in her spaceplane, then swoops through the hangar at constant velocity. At precisely zero time on both Bob's clock and Anna's, Bob sees the nose of Anna's spaceplane reach his doorway. At time zero on his clock, Bob Jr. sees the tail of Anna's spaceplane at his doorway. (a) How fast is Anna's spaceplane moving? (b) What will Anna's clock read when she sees the tail of her spaceplane at the doorway where Bob Jr. is standing? (c) How far will Anna say the nose of her spaceplane is from Bob at this time?

10. The following diagram shows Bob's view of the passing of two identical spaceships, Anna's and his own, where $\gamma_v = 2$. The length of either spaceship in its rest frame is $L_0$. What are the readings on Anna's two unlabeled clocks?

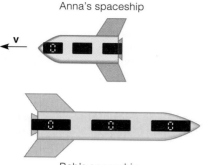

Anna's spaceship

Bob's spaceship

11. Refer to Figure 1.11. (a) How long is a spaceship? (b) At what speed do the ships move relative to one another? (c) Show that Anna's times are in accord with the Lorentz transformation equations. (d) Sketch a set of diagrams showing Anna's complementary view of the passing of the ships. Include times in both frames.

12. You are in a bus traveling on a straight road at 20 m/s. As you pass a gas station, your clock and a clock in the station read precisely zero. You pass another gas station 900 m farther down the road. (In the frame of reference of the gas stations, all gas station clocks are synchronized.) (a) As you pass the second station, do you find its clock to be ahead of, or behind your own clock, and (b) by how much?

13. A spaceplane travels at $0.8c$. As this plane covers the 4000 km from coast to coast, by how much will the time interval registered on a clock on board the spaceplane differ from the time interval measured on the ground?

14. In the frame in which they are at rest, the number of muons at time $t$ is given by

$$N = N_0 e^{-t/\tau}$$

where $N_0$ is the number at $t = 0$ and $\tau$ is the mean life time 2.2 $\mu$s. (a) If muons are produced at a height of 4.0 km, heading toward the ground at $0.93c$, what fraction will survive to reach the ground? (b) What fraction would reach the ground if classical mechanics were valid?

15. A supersonic plane travels at 420 m/s. As this plane passes two markers a distance of 4.2 km apart on the ground, how will the time interval registered on a very precise clock on board the plane differ from 10 s?

16. How fast must a plane 50 m long travel to be found by observers on the ground to be 0.10 nm shorter than 50 m?

17. According to Bob, on Earth, it is 20 ly to Planet Y. Anna has just passed Earth, moving at a constant speed $v$ in a spaceship. When Anna passes Planet Y, she is 20 years older than when she passed Earth. Calculate $v$.

18. A plank is stationary in frame $S$. It is of length $L_0$ and makes an angle of $\theta_0$ with the $x$-axis. It is then caused to move relative to frame $S$ at a constant speed $v$ parallel to the $x$-axis. Show that according to an observer who remains at rest in frame $S$, the length of the plank is now

$$L = L_0 \sqrt{1 - \frac{v^2}{c^2} \cos^2 \theta_0}$$

and the angle it makes with the $x$-axis is

$$\theta = \tan^{-1}(\gamma \tan \theta_0)$$

19. Bob, in frame $S$, is observing the moving plank of Exercise 18. He quickly fabricates a wall, fixed in his frame, that has a hole of length $L$ and is slanted at angle $\theta$, such that the plank will completely fill the hole as it passes through. This occurs at the instant $t = 0$. According to Anna, moving with the plank, the plank is of course not of length $L$, but of length $L_0$. Moreover, because Bob's wall moves relative to her, Anna sees a hole that is less than $L$ in length; a plank longer than $L$ is headed toward a hole shorter than $L$. Can the plank pass through the hole according to Anna? If so, at what time(s)? Explain.

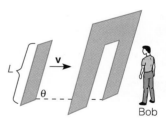

Bob

20. An experimenter determines that a particle created at one end of the laboratory apparatus moved at $0.94c$ and "survived" for 0.032 $\mu$s, decaying just as it reached the other end. (a) According to the experimenter, how far did the particle move? (b) In its own frame of reference, how long did the particle survive? (c) According to the particle, what was the length of the laboratory apparatus?

21. A muon has a mean lifetime of 2.2 $\mu$s in its rest frame. Suppose muons are traveling at $0.92c$ relative to Earth. What is the mean distance a muon will travel as measured by an observer on Earth?

22. A pion is an elementary particle that on average disintegrates $2.6 \times 10^{-8}$ s after creation, in a frame at rest relative to the pion. An experimenter finds that pions created in the laboratory travel 13 m on average before disintegrating. How fast are the pions traveling through the lab?

23. Anna and Bob have identical spaceships 60 m long. The diagram shows Bob's observations of Anna's ship, which passes at a speed of $c/\sqrt{2}$. Clocks at the back of both ships read zero just as they pass. Bob is at the center of his ship and, at $t = 0$ on his wristwatch, peers at a second clock on Anna's ship.
    (a) What does this clock read?
    (b) Later, the back of Anna's ship passes Bob, who is at the center of his ship. At what time does this occur according to Bob?
    (c) What will observers in Bob's frame see on Anna's two clocks at this time?
    (d) Identify two events that show time dilation and two that show length contraction *according to Anna*.

Anna's spaceship

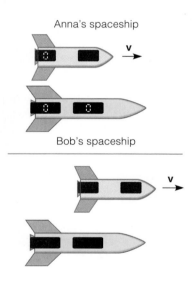

Bob's spaceship

**Section 1.5**

24. Anna and Bob are both born just as Anna's spaceship passes Earth at $0.9c$. According to Bob on Earth, Planet Z is a fixed 30 ly away. As Anna passes Planet Z on her continuing outward journey, what will be (a) Bob's age according to Bob, (b) Bob's age according to Anna, (c) Anna's age according to Anna, and (d) Anna's age according to Bob?

25. Planet W is 12 ly from Earth. Anna and Bob are both 20 years old. Anna travels to Planet W at $0.6c$, quickly

turns around, and returns to Earth at $0.6c$. How old will Anna and Bob be when Anna gets back?

26. You stand at the center of your 100-m spaceship and watch Anna's identical ship pass at $0.6c$. At $t = 0$ on your wristwatch, Anna, at the center of her ship, is directly across from you and her wristwatch also reads zero.
    (a) A friend on your ship, 24 m from you in a direction toward the tail of Anna's passing ship, looks at a clock directly across from him on Anna's ship. What does it read?
    (b) Your friend now steps onto Anna's ship. By this very act he moves from a frame where Anna is one age to a frame where she is another. What is the difference in these ages? Explain. (*Hint:* Your friend moves to Anna's frame, where the time is whatever the clock at the location reads.)
    (c) Answer parts (a) and (b) for a friend 24 m from you but in a direction toward the nose of Anna's passing ship.
    (d) What happens to the reading on a clock when you accelerate toward it? Away from it?

Anna's spaceship

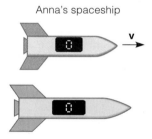

Your spaceship

27. From a standstill you begin jogging at 5 m/s directly toward the galaxy Centaurus A, on the horizon $2 \times 10^{23}$ m away.
    (a) There is a clock in Centaurus A. According to you, how will readings on this clock differ before and after you begin jogging? (Remember: You change frames.)
    (b) The planet Neptune is between Earth and Centaurus A ($4.5 \times 10^9$ m from Earth). How much would readings on a clock there differ?
    (c) What would be the time differences if you had instead begun jogging in the opposite direction?
    (d) What do these results tell us about the observations of a traveling twin who accelerates toward his Earth-bound twin, and how do these observations depend on the distance between the twins?

## Section 1.6

28. By what factor would a star's characteristic wavelengths of light be shifted if it were moving away from Earth at $0.9c$?

29. Show that for a source moving toward an observer, equation (1-15) becomes

$$f_{\text{obs}} = f_{\text{source}} \sqrt{\frac{1 + v/c}{1 - v/c}}$$

30. The light from galaxy NGC 221 consists of a recognizable spectrum of wavelengths. However, all are shifted toward the shorter-wavelength end of the spectrum. In particular, the calcium "line" ordinarily observed at 396.85 nm is observed at 396.58 nm. Is this galaxy moving toward, or away from Earth, and at what speed?

31. At rest, a light source emits 532-nm light. (a) As it moves along the line connecting it and Earth, observers on Earth see 412 nm. What is the source's velocity (magnitude and direction)? (b) Were it to move in the opposite direction at the same speed, what wavelength would be seen? (c) Were it to circle Earth at the same speed, what wavelength would be seen?

32. With reckless disregard for safety and the law, you set your high-performance rocket cycle on course to streak through an intersection at top speed. Approaching the intersection, you *observe* green (540-nm) light from the traffic signal. After passing through, you look back to *observe* red (650-nm) light. Actually, the traffic signal never changed color—it didn't have time! What is the top speed of your rocket cycle, and what was the color of the traffic signal (according to an appalled bystander)?

33. A space probe has a powerful light beacon that emits 500-nm light in its own rest frame. Relative to Earth, the space probe is moving at $0.8c$. An observer on Earth is viewing the light arriving from the distant beacon and detects a wavelength of 500 nm. Is this possible? Explain.

34. To catch speeders, a police "radar gun" detects the beat frequency between the signal it emits and that which reflects off a moving vehicle. What would be the beat frequency for an emitted signal of 900 MHz reflected from a car moving at 30 m/s?

35. For reasons having to do with quantum mechanics, a given kind of atom can emit only certain wavelengths of light. These "spectral lines" serve as a "fingerprint." For instance, hydrogen's only visible spectral lines are 656, 486, 434, and 410 nm. Were spectral lines of absolutely precise wavelength, they would be very difficult to discern. Fortunately, two factors broaden them: the uncertainty principle (discussed in Chapter 3) and Doppler broadening. Atoms in a gas are in motion, so some light will arrive having been emitted by atoms moving toward the observer and some from atoms moving away. Thus, the light reaching the observer will cover a range of wavelengths. (a) Making the assumption that atoms move no faster than their rms speed, $v_{\text{rms}} = \sqrt{3k_B T/m}$, where $k_B$ is the Boltzmann constant, obtain a formula in terms of the wavelength $\lambda$ of the spectral line, atomic mass $m$, and temperature $T$ for the range of wavelengths. (*Note:* $v_{\text{rms}} \ll c$.) (b) Evaluate this range for the 656-nm hydrogen spectral line, assuming a temperature of $10^5$ K.

## Section 1.7

36. According to Anna, on Earth, Bob is on a spaceship moving at $0.8c$ toward Earth, and Carl, a little farther out, is on a spaceship moving at $0.9c$ toward Earth. (a) According to Bob, how fast and in what direction is Carl moving relative to himself (Bob)? (b) According to Bob, how fast is Carl moving relative to Earth?

37. Bob is on Earth. Anna is on a spacecraft moving away from Earth at $0.6c$. At some point in Anna's outward travel, Bob fires a projectile loaded with supplies out to Anna's ship. Relative to Bob, the projectile moves at $0.8c$. (a) How fast does the projectile move relative to Anna? (b) Bob also sends a light signal, "Greetings from Earth," out to Anna's ship. How fast does the light signal move relative to Anna?

38. In a particle collider experiment, particle 1 is moving to the right at $0.99c$ and particle 2 to the left at $0.99c$, both relative to the laboratory. What is the relative velocity of the two particles according to (an observer moving with) particle 2?

39. Prove that if $v$ and $u'$ are less than $c$, it is impossible for a speed $u$ greater than $c$ to result from equation (1-17b). [*Hint:* the product $(c - u')(c - v)$ is positive.]

40. A light beam moves at an angle $\theta$ with the $x$-axis as seen from frame $S$. Using the relativistic velocity transformation, find the components of its velocity when viewed from frame $S'$, and from these verify explicitly that its speed is $c$.

## Sections 1.8 and 1.9

41. What is the ratio of the relativistically correct expression for momentum to the classical expression, and under what condition does the deviation become significant?

42. What are the momentum, energy, and kinetic energy of a proton moving at $0.8c$?

43. What would be the internal energy, kinetic energy, and total energy of a 1-kg block moving at $0.8c$?

44. By how much (in picograms) does the mass of 1 mol of ice at 0°C differ from that of 1 mol of water at 0°C?

45. A spring has a force constant of 18 N/m. If it is compressed 50 cm from its equilibrium length, how much mass will it have gained?

46. A typical household uses 500 kWh of energy in 1 month. How much mass is converted to produce this energy?

47. Determine the momentum of an electron moving (a) at speed $2.4 \times 10^4$ m/s (about three times escape velocity), and (b) at speed $2.4 \times 10^8$ m/s. (c) In each case, by how much is the classical formula in error?

48. In the collision shown, energy is conserved, because both objects have the same speed and mass after as before the collision. Since the collision merely reverses the velocities, the final (total) momentum is opposite the initial. Thus, momentum can be conserved only if it is zero.

    (a) Using the relativistically correct expression for momentum, show that the total momentum is zero—that momentum is conserved. (Masses are in arbitrary units.)

    (b) Using the relativistic velocity transformation, find the four velocities in a frame moving to the right at 0.6c.

    (c) Verify that momentum is conserved in the new frame.

    Before:

    After:

49. Is it possible for the momentum of an object to be $mc$? If not, why not? If so, under what condition?

50. Show that the relativistic expression for kinetic energy $(\gamma_u - 1)mc^2$ is equivalent to the classical $\frac{1}{2}mu^2$ when $u \ll c$.

51. At Earth's location, the intensity of sunlight is $1.5 \text{ W/m}^2$. If no energy escaped Earth, by how much would Earth's mass increase in 1 day?

52. The weight of the Empire State Building is 365 kilotons. Show that the complete conversion of 1 kg of mass would provide sufficient energy to put this rather large object in a low Earth orbit (orbit radius $\cong$ Earth's radius).

53. Radiant energy from the Sun, approximately $1.5 \times 10^{11}$ m away, arrives at Earth with an intensity of $1.5 \text{ W/m}^2$. At what rate is mass being converted in the Sun to produce this radiant energy?

54. (a) A high-explosive material employing chemical reactions has an "explosive yield" of $10^6$ J/kg, measured in joules of energy released per kilogram of material. By what fraction does its mass change when it explodes? (b) What is the explosive yield of a material that pro-

duces energy via *nuclear* reactions in which its mass decreases by 1 part in 10,000?

55. How fast must an object be moving for its kinetic energy to equal its internal energy?

56. How much work must be done to accelerate an electron (a) from 0.3c to 0.6c, and (b) from 0.6c to 0.9c?

57. An electron accelerated from rest through a potential difference $V$ acquires a speed of 0.9998c. Find the value of $V$.

58. What is the momentum of a proton accelerated through 1 gigavolt (GV)?

59. A proton is accelerated from rest through a potential difference of 500 MV. (a) What is its speed? (b) Classical mechanics indicates that quadrupling the potential difference would double the speed. Were a classical analysis valid, what speed would result from such an increased potential difference? (c) What speed actually results?

60. A particle of mass $m_0$ moves through the lab at 0.6c. Suddenly it explodes into two fragments. Fragment 1, mass $0.66m_0$, moves at 0.8c in the same direction the original particle had been moving. Determine the velocity (magnitude and direction) and mass of fragment 2.

61. The boron-14 nucleus (mass: 14.02266 u) "beta-decays," spontaneously becoming an electron (mass: 0.00055 u) and a carbon-14 nucleus, of mass 13.99995 u. What will be the speeds and kinetic energies of the carbon-14 nucleus and the electron? (*Note:* A neutrino is also produced. We consider the case in which its momentum and energy are negligible. Also, since the carbon-14 nucleus is much more massive than the electron, it recoils "slowly"; $\gamma_u \cong 1$.)

62. A 3.000-u object moving to the right through a laboratory at 0.8c collides with a 4.000-u object moving to the left through the laboratory at 0.6c. Afterward, there are two objects, one of which is a 6.000-u mass at rest. (a) What are the mass and speed of the other object? (b) Determine the change in kinetic energy in this collision.

63. A 10-kg object is moving to the right at 0.6c. It explodes into two pieces, one of mass $m_1$ moving to the left at 0.6c, and one of mass $m_2$ moving right at 0.8c. (a) Find $m_1$ and $m_2$. (b) Find the change in kinetic energy in this explosion.

64. Particle 1, of mass $m_1$, moving at 0.8c relative to the lab, collides head-on with particle 2, of mass $m_2$, moving at 0.6c relative to the lab. Afterwards, there is a single stationary object. Find, in terms of $m_1$, (a) $m_2$, (b) the mass of the final stationary object, and (c) the change in kinetic energy in this collision.

65. Consider the collisions of two identical particles, each of mass $m_0$. In experiment A, a particle moving at 0.9c strikes a stationary particle.

(a) What is the total kinetic energy before the collision?

(b) In experiment B, both particles are moving at a speed $u$ (relative to the lab), directly toward one another. If the total kinetic energy before the collision in experiment B is the same as that in experiment A, what is $u$?

(c) In both experiments, the particles stick together. Find the mass of the resulting single particle in each experiment. In which is more of the initial kinetic energy converted to mass?

Experiment A

Experiment B

66. A kaon (denoted $K^0$) is an unstable particle of mass $8.87 \times 10^{-28}$ kg. One of the means by which it decays is by spontaneous creation of two pions, a $\pi^+$ and a $\pi^-$. The decay process may be represented as

$$K^0 \rightarrow \pi^+ + \pi^-$$

Both the $\pi^+$ and $\pi^-$ have mass $2.49 \times 10^{-28}$ kg. A kaon moving in the $+x$-direction decays by this process, the $\pi^+$ moving off at speed $0.9c$ and the $\pi^-$ at $0.8c$. (a) What was the speed of the kaon before decay? (b) In what directions do the pions move after the decay?

67. Show that $E^2 = p^2c^2 + m^2c^4$ follows from expressions (1-20) and (1-22) for momentum and energy in terms of $m$ and $u$.

## Section 1.11

68. For the situation given in Exercise 4, find the Lorentz transformation matrix from Bob's frame to Anna's frame, then solve the problem via matrix multiplication.

69. (a) Determine the Lorentz transformation matrix giving position and time in frame $S'$ from those in frame $S$ in the classical limit $v \ll c$. (b) Show that it yields equations (1-1).

70. Show that equation (1-28) follows from the arbitrary four-vector Lorentz transformation equations immediately preceding it.

71. From $\mathbf{p} = \gamma_u m\mathbf{u}$ (whose components are $p_x = \gamma_u m u_x$, $p_y = \gamma_u m u_y$, and $p_z = \gamma_u m u_z$), the relativistic velocity transformation (1-18), and $\gamma_{u'} = (1 - u_x v/c^2)\gamma_v\gamma_u$, show that $p'_y = p_y$ and $p'_z = p_z$.

72. A 1-kg object moves at $0.8c$ relative to Earth. (a) Calculate the momentum and energy of the object. (b) Determine the Lorentz transformation matrix from Earth's frame to the object's frame. (c) Find the momentum and energy of the object in this "new" frame via matrix multiplication.

## Section 1.12

73. In a television picture tube, a beam of electrons is sent from the back to the front (screen) by an electron "gun." When an electron strikes the screen, it causes a phosphor to glow briefly. To produce an image across the entire screen, the beam is electrically deflected up and down and left and right. The beam may sweep from left to right at a speed greater than $c$. Why is this not a violation of the claim that no information may travel faster than the speed of light?

## Section 1.13

74. If it is fundamental to nature that a given mass has a "critical" radius at which something extraordinary happens (i.e., a black hole forms), we might guess that this radius should depend only on the mass and fundamental constants of nature. Assuming that $r_{\text{critical}}$ depends only on $M$, $G$, and $c$, show that dimensional analysis gives the equation for the Schwartzchild radius to within a multiplicative constant.

75. A projectile is a distance $r$ from the center of a heavenly body and is heading directly away. Classically, if the sum of its kinetic and potential energies is positive, it will escape the gravitational pull of the body, but if negative, it cannot escape. Now imagine that the projectile is a pulse of light of energy $E$. Since light has no internal energy, $E$ is also the kinetic energy of the light pulse. Suppose that the gravitational potential energy of the light pulse is given by Newton's classical formula, $U = -(GMm/r)$, where $M$ is the mass of the heavenly body and $m$ is an "effective mass" of the light pulse. Assume that this effective mass is given by $m = E/c^2$.

Show that the critical radius for which light could not escape the gravitational pull of a heavenly body is within a factor of 2 of the Schwartzchild radius given in the chapter. (This kind of "semiclassical" approach to general relativity is sometimes useful but always vague. To be reliable, predictions must be based from beginning to end on the logical, but unfortunately complex, fundamental equations of general relativity.)

76. Suppose particles begin moving in one dimension away from the origin at $t = 0$ with the following velocities: $0, \pm1, \pm2, \pm3$ m/s, and so on. (a) After 1 s, how will the velocities of the particles depend on distance from the origin? (b) Now consider an observer on one of the moving particles *not* at the origin. How will the *relative* velocities of the other particles depend on distance from the observer?

## General Exercises

**77.** Derive the following expressions for the components of acceleration of an object, $a'_x$, $a'_y$, $a'_z$, in frame $S'$ in terms of its components of acceleration and velocity in frame $S$.

$$a'_x = \frac{a_x}{\gamma_v^3 \left(1 - \frac{u_x v}{c^2}\right)^3}$$

$$a'_y = \frac{a_y}{\gamma_v^2 \left(1 - \frac{u_x v}{c^2}\right)^2} + \frac{a_x \frac{u_y v}{c^2}}{\gamma_v^2 \left(1 - \frac{u_x v}{c^2}\right)^3}$$

$$a'_z = \frac{a_z}{\gamma_v^2 \left(1 - \frac{u_x v}{c^2}\right)^2} + \frac{a_x \frac{u_z v}{c^2}}{\gamma_v^2 \left(1 - \frac{u_x v}{c^2}\right)^3}$$

**78.** (a) Determine the Lorentz transformation matrix giving position and time in frame $S'$ from those in frame $S$ for the case $v = 0.5c$. (b) If frame $S''$ moves at $0.5c$ relative to frame $S'$, the Lorentz transformation matrix is the same as the previous one. Find the product of the two matrices, which gives $x''$ and $t''$ from $x$ and $t$. (c) To what single speed does the transformation correspond? Explain this result.

**79.** Using equations (1-18), show that

$$\gamma_{u'} = \left(1 - \frac{u_x v}{c^2}\right)\gamma_v \gamma_u$$

where

$$\gamma_{u'} \equiv \frac{1}{\sqrt{1 - \frac{u'^2}{c^2}}} = \frac{1}{\sqrt{1 - \frac{u'^2_x + u'^2_y + u'^2_z}{c^2}}}$$

**80.** Show that, relativistically as classically, the net work done on an initially stationary object equals the final kinetic energy of the object. Consider only one-dimensional motion. It will be helpful to use the expression for $p$ as a function of $u$ in the following:

$$W = \int F\,dx = \int \frac{dp}{dt}\,dx = \int \frac{dx}{dt}\,dp = \int u\,dp$$

**81.** Both classically and relativistically, the force on an object is what causes a time rate of change of its momentum: $F = dp/dt$.

(a) Using the relativistically correct expression for momentum, show that

$$F = \gamma_u^3 m \frac{du}{dt}$$

(b) Under what condition does the classical equation $F = ma$ hold?

(c) Assuming a constant force and that the speed is zero at $t = 0$, separate $t$ and $u$, then integrate to show that

$$u = \frac{1}{\sqrt{1 + (Ft/mc)^2}}\frac{F}{m}t$$

(d) Plot $u$ versus $t$. What happens to the velocity of an object when a constant force is applied for an indefinite length of time?

**82.** A rocket maintains a constant thrust $F$, giving it an acceleration of $g$ (i.e., 9.8 m/s$^2$). (a) If classical physics were valid, how long would it take for its speed to reach $0.99c$? (b) Using the result of Exercise 81(c), how long will it really take to reach $0.99c$?

**83.** Exercise 81 gives the speed $u$ of an object accelerated under a constant force. Show that the distance it travels is given by

$$x = \frac{mc^2}{F}\left[\sqrt{1 + \left(\frac{Ft}{mc}\right)^2} - 1\right]$$

**84.** In Example 1.5, we noted that Anna could go wherever she wished in as little time as desired, by going fast enough to length-contract the distance to an arbitrarily small value. This overlooks a physiological limitation. Accelerations greater than about $30g$ are fatal, and there are serious concerns about what the effects might be of prolonged accelerations greater than $1g$. Here we see how far a person could go under a constant acceleration of $1g$, producing a comfortable artificial gravity.

(a) Though traveler Anna accelerates, Bob, being on near-inertial Earth, is a reliable observer and will see less time go by on Anna's clock ($dt'$) than on his own ($dt$). Thus, $dt' = (1/\gamma_u)dt$, where $u$ is Anna's instantaneous speed relative to Bob. Using the result of Exercise 81(c), with $g$ replacing $F/m$, substitute for $u$, then integrate to show that

$$t = \frac{c}{g}\sinh\frac{gt'}{c}$$

(b) How much time goes by for observers on Earth as they "see" Anna age 20 years?

(c) Using the result of Exercise 83, show that when Anna has aged a time $t'$, she is a distance from Earth (according to Earth observers) of

$$x = \frac{c^2}{g}\left(\cosh\frac{gt'}{c} - 1\right)$$

(d) If Anna accelerates away from Earth while aging 20 years, then slows to a stop while aging another 20, how far away from Earth will she end up and how much time will have passed on Earth?

# Waves and Particles I: Electromagnetic Radiation Behaving as Particles

W e now begin our study of quantum mechanics. In some sense, quantum mechanics is the study of "small" things— so small that it is essentially impossible to observe their behavior without altering the very behavior we wish to observe! For example, the simplest way to observe an object is to look at it, but to do so, light must be bounced off it, and light carries energy. Some of this energy will unavoidably be transferred to the object itself. Ordinarily this transference is inconsequential, but if the object is very small, such as a single electron, the effect might be significant. We should not be too surprised, then, that the behavior we observe might vary, depending on how the observation is made.

A central idea in quantum mechanics is **wave-particle duality:** Things may behave as waves *or* as discrete particles depending on the situation. The "situation" might be an experiment designed to reveal a property of some thing, or it might be determined simply by the dimensions of a region to which the thing is confined. Two of the most important things we study are massive objects and electromagnetic radiation. In classical situations, our observations (and any observation is, after all, an experiment) reveal electromagnetic radiation behaving as waves and massive objects as discrete particles.[1] We now look at the nonclassical side of the

---

[1]Even the behaviors of mechanical waves and classical fluids can be understood as collective motions of discrete particles.

coin. In this chapter we investigate electromagnetic radiation behaving as a collection of discrete particles. In Chapter 3, we study the complementary topic—massive objects behaving as waves.

## 2.1  The Photoelectric Effect

Classically, electromagnetic radiation is a wave. The energy it carries is diffuse, distributed continuously along a broad wave front, and its intensity (energy per unit time per unit area) is proportional to $E_0^2$, where $E_0$ is the amplitude of the electric field oscillations. Long before the advent of quantum physics, it was known that a beam of light directed at the surface of a metal could eject electrons. This is called the **photoelectric effect**—light producing a flow of electricity. It was also known that a certain minimum energy is required simply to free an electron from the metal. The electron is bound to the metal, and pulling it loose requires a certain amount of energy, with any surplus energy becoming the freed electron's kinetic energy. The amount of energy required to free an electron, the **work function** $\phi$, is a characteristic of the metal (see Table 2.1).

If light is a wave, several definite things should be observed. First, if light of one wavelength is able to eject electrons, then light of any wavelength should be able to do so; independent of the wavelength, the rate at which energy arrives (i.e., the intensity), and thus the rate at which electrons are ejected, could be increased to any arbitrarily large value simply by increasing $E_0$. Secondly, if the intensity is low, although electrons might still be ejected, a measurable time lag should be evident; a wave being diffuse, considerable time may be required for enough energy to accumulate in the electron's vicinity. Finally, at any given frequency, if the intensity is increased, the departing electrons should be more energetic; a stronger electric field should produce a larger acceleration.

Imagine the surprise of the experimenter whose weak light of 520-nm wavelength ejects electrons from sodium, with no time lag, while light of 550 nm cannot, even at *many times* the intensity, and the energy of the electrons liberated by the 520-nm light is completely independent of the intensity. Classically, this cannot be explained!

Albert Einstein in 1905 proposed the following explanation: The light is behaving as a collection of particles, called **photons,** each with energy given by

$$E = hf \tag{2-1}$$

where $h$ is **Planck's constant,** $6.63 \times 10^{-34}$ J·s, and $f$ is the frequency. (See the essay entitled "A New Fundamental Constant.") The ejection of a given electron is accomplished by a *single* photon, the photon transferring all its energy to the electron then disappearing. If the frequency of the incoming light is too low, so that the photon energy is less than the work function, there is simply insufficient energy in any given photon to liberate an electron. *Thus, no electron can be ejected, no matter what the intensity*—no matter how abundant the photons. (In this case, each photon's energy becomes some combination of heat and reflected light before another can strike; it is improbable that multiple photons could strike the electron simultaneously.) However, if the frequency of the light is high enough,

**Table 2.1**

| Metal | Work Function $\phi$ (in eV) |
|---|---|
| Potassium | 2.2 |
| Sodium | 2.3 |
| Magnesium | 3.7 |
| Zinc | 4.3 |
| Chromium | 4.4 |
| Tungsten | 4.5 |

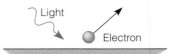

Energy of a photon

$h = 6.63 \times 10^{-34}$ J·s

## A New Fundamental Constant

**E**instein did not originate the idea that electromagnetic energy might be discrete in nature. This we attribute to Max Planck. Planck was attempting to explain the electromagnetic spectrum emitted by a so-called "blackbody," an object that *reflects* no electromagnetic radiation, but does emit radiation due to the thermal motion of its charges. Charcoal, for instance, reflects little light but certainly emits light when heated "red-hot." Classical electromagnetic wave theory predicted that the intensity emitted by a blackbody would increase without bound as the frequency increases! Obviously the theory was flawed, for the experimental evidence is that intensity increases with increasing frequency, reaches a maximum, then decreases to zero.

In the year 1900, Planck found that he could match the experimental curve perfectly by augmenting the existing theory with a curious assumption—that the energy in the electromagnetic radiation at a given frequency may take on values restricted to $E = nhf$, where $n$ is an integer, $f$ is the frequency, and $h$ is a constant. He was thus the first to obtain a value for $h$, which now bears his name. For his discovery he was awarded the 1918 Nobel prize.

Interestingly, Planck quoted $h$ as simply the value that caused his theory to match the experimental curve. The question arises, Why couldn't he theoretically *derive* it? On the other hand, that it was even possible to

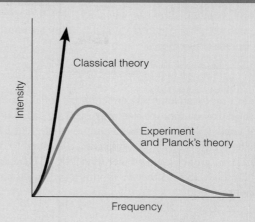

match the curve is significant. Even with the freedom to adjust one constant, it would be a fantastic coincidence if the functional form of the curve predicted by an incorrect theory were to match the experimental curve. With the benefit of hindsight, we now realize that Planck's constant cannot be derived; it is one of the fundamental constants of nature (e.g., the universal gravitational constant $G$), all of which must be determined experimentally. In the end, although Planck's $E = nhf$ might suggest that electromagnetic radiation is a collection of particles of energy $hf$, the ramifications were pursued not by Planck, but by Einstein. For a detailed discussion of Planck's **black-body radiation** work, see Appendix C.

so that the photon energy $hf$ is greater than the work function, an electron can be ejected. The kinetic energy given to the electron would then be the difference between the photon's energy and the energy required to free the electron from the metal.

$$KE_{max} = hf - \phi \qquad (2\text{-}2)$$

Photoelectric effect

(The subscript "max" is necessary because the electrons are not all bound equally strongly to the metal. The work function is the energy required to free those least strongly bound, and it is only these that leave with $KE_{max}$.)

Einstein's interpretation of the photoelectric effect explains not only the observation that a certain minimum frequency is required, but the other classically unexpected results as well. If a single photon—a *particle of concentrated energy* rather than a diffuse wave—does have enough energy, electron ejection should be immediate, with no time lag. Then too, the electron's kinetic energy should depend only on the energy of the single photon (i.e., the frequency), not on how many strike the metal per unit time (the intensity). In all respects, Einstein's explanation agrees with the experimental evidence. This accomplishment won Einstein the 1921 Nobel prize in physics.

### Example 2.1

Light of 380-nm wavelength is directed at a metal plate, plate 1 (Figure 2.1). To determine the energy of electrons ejected, a second metal plate, plate 2, is placed parallel to the first, and a potential difference is established between them. Photoelectrons ejected from plate 1 are found to reach plate 2 as long as the potential difference is no greater than 2.00 V. Determine (a) the work function of the metal, and (b) the maximum-wavelength light that can eject electrons from this metal.

### Solution

(a) Electrons are ejected from plate 1 with a certain maximum kinetic energy. If none have enough kinetic energy to surmount the electrostatic potential energy difference, no electrons will reach plate 2. We are told that a potential energy difference of $qV = (1.6 \times 10^{-19} \text{ C})(2.00 \text{ V}) = 3.20 \times 10^{-19} \text{ J} = 2.00 \text{ eV}$ is the maximum that can be surmounted, so the maximum kinetic energy must be 2.00 eV. (*Note:* Requiring them to cross a potential difference is a common way of determining the energy of photoelectrons. The potential difference at which electrons cease to reach the other plate is called the **stopping potential**.) Using equation (2-2),

$$3.20 \times 10^{-19} \text{ J} = (6.63 \times 10^{-34} \text{ J·s}) \left( \frac{3 \times 10^8 \text{ m/s}}{380 \times 10^{-9} \text{ m}} \right) - \phi$$

$$\Rightarrow \quad \phi = 2.03 \times 10^{-19} \text{ J} = 1.27 \text{ eV}$$

(b) As the wavelength of incoming light is increased to $\lambda'$ (frequency decreased to $f'$) the energy in the photons is decreased. The limit for ejecting electrons occurs when an incoming photon has only enough energy to free an electron from the metal, with none left for kinetic energy. Again using equation (2-2),

$$0 = hf' - \phi = (6.63 \times 10^{-34} \text{ J·s}) \left( \frac{3 \times 10^8 \text{ m/s}}{\lambda'} \right) - 2.03 \times 10^{-19} \text{ J}$$

$$\Rightarrow \quad \lambda' = 978 \text{ nm}$$

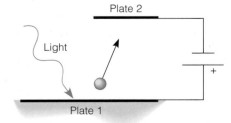

**Figure 2.1** An opposing potential difference is used to determine the maximum kinetic energy of photoelectrons liberated by photons.

Wavelengths longer than 978 nm have insufficient energy per photon, and no photoelectrons will be produced. The maximum wavelength for which electrons can be freed is commonly called the **threshold wavelength,** and the corresponding minimum frequency the **threshold frequency.**

The central point in Einstein's explanation of the photoelectric effect is that electromagnetic radiation appears to be behaving as a collection of *particles,* each with a discrete energy. Something that is discrete (as opposed to continuous) is said to be **quantized.** In the photoelectric effect, the energy in light is quantized.

**Example 2.2**

How many photons per second emanate from a 10-mW 633-nm laser?

Solution

For each photon,

$$E = hf = h\frac{c}{\lambda} = (6.63 \times 10^{-34} \text{ J·s})\left(\frac{3 \times 10^{8} \text{ m/s}}{633 \times 10^{-9} \text{ m}}\right) = 3.14 \times 10^{-19} \text{ J}$$

To find number of particles per unit time, we divide energy per unit time by energy per particle:

$$\frac{\text{number of particles}}{\text{time}} = \frac{10 \times 10^{-3} \text{ J/s}}{3.14 \times 10^{-19} \text{ J/particle}}$$

$$= 3.18 \times 10^{16} \text{ particles/s}$$

Clearly, photons are "small," and it is easy to see how a light beam might appear continuous. By analogy, a stream of water strikes a wall via countless collisions of individual molecules, but to a casual observer its behavior appears continuous.

## 2.2   The Production of x Rays

We use the name **x ray** for electromagnetic radiation whose wavelength is in the $10^{-2}$ to 10 nm region of the spectrum. The name was coined by Wilhelm Röentgen, who first studied this type of radiation. Among other things, he found that it could expose photographic film after having passed through a solid object, such as a human being. For his work he received the first Nobel prize in physics (1901). Our interest is in studying how the production of x rays demonstrates the particle nature of electromagnetic radiation.

As shown in Figure 2.2, x rays may be produced by smashing high-speed electrons into a metal target. Any charged particle radiates electromagnetic energy when it accelerates, and smashing electrons into a target is an unusually violent acceleration, so much radiation is produced. (Do not confuse this with the photoelectric effect, in which essentially the reverse occurs—electromagnetic radiation produces freely moving electrons.) The radiation associated with the abrupt

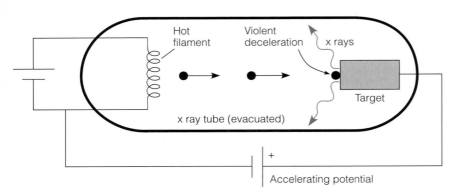

**Figure 2.2** To produce x rays, electrons "boiled" off a hot filament are accelerated into a metal target.

stopping of a charged particle is called ***bremsstrahlung,*** a German word meaning "braking radiation." Physicists were not surprised that rapid deceleration of electrons produced electromagnetic radiation, but they expected to see radiation covering the entire spectrum of wavelengths. Although the total energy is, of course, limited by the number of electrons arriving per unit time, physicists saw no reason that there should not be some amplitude of waves produced at all frequencies, from the very low to the very high.

Experiment differed. Figure 2.3 depicts the spectrum of wavelengths produced when electrons of kinetic energy 25 keV strike a molybdenum target. A broad range is apparent,[2] but there is no radiation of wavelength less than $\lambda_c = 0.050$ nm. This is called the **cutoff wavelength,** and there is no classical explanation for so sharp a termination of the spectrum.

A nonclassical explanation is that at a given frequency, electromagnetic radiation cannot be of arbitrarily small amplitude. If the radiation is quantized, the minimum energy that can be produced at frequency $f$ is $hf$, that of a *single photon.* We cannot produce a half of a photon. Suppose that each electron smashing into the target may produce any number of photons, but that multiple electrons cannot *combine* to produce a photon. If so, no photon could ever be produced with an energy greater than the kinetic energy of a single electron. Is this the case? Setting the kinetic energy of an incoming electron equal to the energy of one photon,

$$25 \text{ keV} = h\frac{c}{\lambda}$$

$$\Rightarrow \quad \lambda = \frac{(6.63 \times 10^{-34} \text{ J·s})(3 \times 10^8 \text{ m/s})}{(25 \times 10^3 \text{ eV})(1.6 \times 10^{-19} \text{ J/eV})} = 5.0 \times 10^{-11} \text{ m} = 0.050 \text{ nm}$$

[2]Not shown are characteristic x-ray "spikes" due to an interaction specific to the particular element in the target material rather than nonspecific deceleration of the electron. Characteristic x rays are discussed in Section 7.10.

**Figure 2.3** The x-ray spectrum produced when 25-keV electrons strike a molybdenum target.

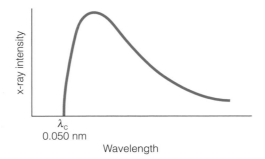

This is fairly convincing evidence! The cutoff wavelength is that for which the entire kinetic energy of a single incoming electron goes into a single photon. No single electron has enough energy to produce a photon of shorter wavelength, and multiple electrons apparently cannot combine their energies to do so.

## 2.3  The Compton Effect

Once again, we consider a phenomenon that could not be explained by the classical view of electromagnetic radiation as waves. In so doing, we will uncover another important property of photons. The situation is the scattering of electromagnetic radiation from free stationary electrons. According to classical electromagnetic wave theory, free electrons would be caused to oscillate, scattering (or "re-radiating") electromagnetic energy in all directions *initially at the same frequency as the incoming radiation.* The electrons would also slowly begin moving in the direction of the incident waves, so there would be a slowly developing Doppler shift in the wavelength they radiate. The energy scattered backward would be of progressively longer wavelength.

Arthur Holly Compton investigated this phenomenon with x rays, using carbon atoms as the source of electrons.[3] He found that some radiation scattered backward immediately with a wavelength significantly longer than that of the incoming x rays! His explanation rested upon treating the x rays as a collection of photons, each with a discrete energy as well as another property we ordinarily associate with a particle—momentum.

Even accepting that a massless particle can carry energy, it may seem a great leap to accept that such a particle carries momentum. But it appears to be true that anything with kinetic energy—and, having no mass, photons have only kinetic energy—must have momentum. Signs point this way from two different directions:

1. According to special relativity, an object with zero mass should have momentum related to its energy by

$$E = pc$$

2. Classical electromagnetic wave theory shows that electromagnetic waves do carry momentum, although for a diffuse wave, we speak of momentum "density." It is related to the energy density by

$$\frac{\text{energy}}{\text{volume}} = \frac{\text{momentum}}{\text{volume}} \times c$$

With two such compatible clues, it seems reasonable that the momentum of a photon should be given by

$$p = \frac{E}{c} = \frac{hf}{c} = \frac{h}{\lambda}$$

But the claim must be verified experimentally. Compton provided the first experimental evidence that a photon does carry this momentum, and the phenomenon now bears his name.

[3] Strictly speaking, electrons bound to carbon atoms are neither stationary nor free, but they are bound to fixed nuclei so weakly (compared to the x-ray photon energy) as to be regarded both stationary and free.

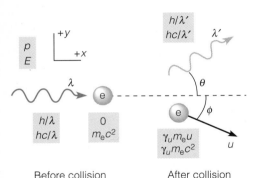

**Figure 2.4** Momentum and energy when a photon strikes a free electron.

In the "particle view" of electromagnetic radiation, the interaction of x rays and electrons is simply a collection of two-particle collisions between photon and electron. Let us set up equations for momentum and energy conservation. As shown in Figure 2.4, a stationary electron is struck by an x-ray photon of wavelength $\lambda$. Afterward, the electron moving at speed $u$ and a scattered photon of wavelength $\lambda'$ depart at angles $\phi$ and $\theta$, respectively. (For convenience we use $hc/\lambda$ for photon energy, rather than the equivalent $hf$.)

Momentum conserved:

$$x\text{-component:} \quad \frac{h}{\lambda} = \frac{h}{\lambda'} \cos\theta + \gamma_u m_e u \cos\phi \tag{2-3a}$$

$$y\text{-component:} \quad 0 = \frac{h}{\lambda'} \sin\theta - \gamma_u m_e u \sin\phi \tag{2-3b}$$

Energy conserved:

$$h\frac{c}{\lambda} + m_e c^2 = h\frac{c}{\lambda'} + \gamma_u m_e c^2 \tag{2-3c}$$

(Because it happens that the electron often moves very fast after the collision, we must use the relativistically correct expressions for its momentum and energy. The corresponding expressions for the photon are necessarily relativistically correct; *non*relativistic ones don't exist for things that always move at the speed of light.)

Equations (2-3) have been found to agree completely with the experimental observations when a photon collides with a free electron. All tests since Compton's original one have reaffirmed the conclusion: The momentum of a photon is given by

Momentum of a photon

$$p = \frac{h}{\lambda} \tag{2-4}$$

(See the essay entitled "Why Not $hf/c$?")

The Compton effect's most striking departure from classical expectation is the large and immediate wavelength shift in the scattered radiation. This is made most clear by eliminating the electron speed $u$ and scattering angle $\phi$ from the three

## Why Not *hf/c*?

It may bother the reader that the momentum of a photon is always expressed as $h/\lambda$, rather than $hf/c$. The expressions are equivalent for a photon. However, we shall see in later chapters that they are not equivalent for a massive object with a wave nature. It is found that the first expression is correct while the second is not. There is for all phenomena a fundamental relationship between momentum and wavelength, not momentum and frequency. There is a complementary fundamental relationship between energy and frequency (not energy and wavelength), so energy is expressed as $E = hf$, rather than $E = hc/\lambda$, although they are again equivalent for a photon.

equations (2-3). (The somewhat lengthy algebra is left to Exercise 22.) What remains is

$$\lambda' - \lambda = \frac{h}{m_e c}(1 - \cos \theta) \tag{2-5}$$

Compton effect

The difference in wavelength between the incident and scattered photons depends only on the angle of scatter, not on the time of exposure to the x-ray source. For this discovery, Compton was awarded the 1927 Nobel prize in physics.[4] Note also that the scattered photon's wavelength is always greater, the frequency and energy smaller, than the incident's. This must be the case because the electron gains kinetic energy. In particular, the maximum increase in wavelength is for backward scatter of the photon ($\theta = 180°$), because a head-on collision imparts the maximum possible energy to the electron.

### Example 2.3

An x-ray photon of 0.0500-nm wavelength strikes a free, stationary electron. A photon scatters at 90°. Determine the momenta of the incident photon, the scattered photon, and the electron.

### Solution

For the incident photon:

$$p = \frac{h}{\lambda} = \frac{6.63 \times 10^{-34} \text{ J·s}}{0.05 \times 10^{-9} \text{ m}} = 1.33 \times 10^{-23} \text{ kg·m/s}$$

Inserting what we know, we solve (2-5) for the scattered photon's wavelength:

$$\lambda' - 0.0500 \times 10^{-9} \text{ m} = \frac{(6.63 \times 10^{-34} \text{ J·s})}{(9.11 \times 10^{-31} \text{ kg})(3 \times 10^8 \text{ m/s})}(1 - \cos 90°)$$

$$\Rightarrow \quad \lambda' = 0.0524 \text{ nm}$$

Thus,

$$p' = \frac{h}{\lambda'} = \frac{6.63 \times 10^{-34} \text{ J·s}}{0.0524 \times 10^{-9} \text{ m}} = 1.26 \times 10^{-23} \text{ kg·m/s}$$

[4]It should be noted that in Compton's experiment some radiation *of the incident wavelength* was indeed scattered at all angles. Some of carbon's electrons are so tightly bound to their atoms as to prevent their being stripped away by the incident photon. The photon instead interacts essentially with the atom as a whole. In effect, $m_e$ in (2-5) is replaced by the atomic mass, many thousands of times larger, giving a negligible shift in wavelength.

To find the momentum of the electron, we may use equations (2-3a) and (2-3b).

$$\frac{h}{\lambda} = \frac{h}{\lambda'} \cos 90° + \gamma_u m_e u \cos \phi \quad \rightarrow \quad \frac{h}{0.0500 \text{ nm}} = \gamma_u m_e u \cos \phi$$

$$0 = \frac{h}{\lambda'} \sin 90° - \gamma_u m_e u \sin \phi \quad \rightarrow \quad \frac{h}{0.0524 \text{ nm}} = \gamma_u m_e u \sin \phi$$

Dividing the latter ($y$) equation by the former ($x$),

$$\frac{0.0500}{0.0524} = \tan \phi \quad \Rightarrow \quad \phi = 43.6°$$

Reinserting in either equation then yields the magnitude of the electron's momentum.

$$0 = \frac{6.63 \times 10^{-34} \text{ J·s}}{0.0524 \times 10^{-9} \text{ m}} \sin 90° - \gamma_u m_e u \sin 43.6°$$

$$\Rightarrow \quad \gamma_u m_e u = 1.83 \times 10^{-23} \text{ kg·m/s}$$

Thus,

$$p_e = 1.83 \times 10^{-23} \text{ kg·m/s at 43.6° from direction of incident photon}$$

The addition of the momenta is shown in Figure 2.5. Even though massless, x rays are very energetic particles—the electron is "bumped" to $2 \times 10^7$ m/s, or about $0.07c!$

If treating x rays as a collection of photons predicts so well what is experimentally observed, why is the *wave* theory of electromagnetic radiation still around? The answer is quite simple: The behavior of electromagnetic radiation is just as wave theory predicts—*if* the wavelength is long. Particle and wave behaviors converge at long wavelengths. Of course, this "answer" raises an even thornier question: Can a line be drawn between wave and particle? This we take up in Sec-

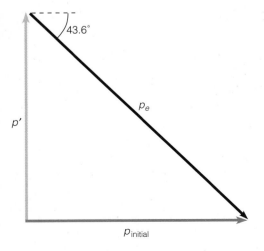

**Figure 2.5** Momentum conservation in photon-electron collision.

tion 2.5. For the present, let us concentrate on how the two behaviors agree for long wavelengths.

Suppose instead of a 0.050-nm wavelength x ray, visible light of wavelength 500 nm "collides" with the electron in the previous example. The wavelength *change* predicted by equation (2-5) would be the same as for the x-ray photon. The scattered photon's wavelength would thus be 500.0024 nm. Rather than a change of approximately 5% in wavelength, the change would be only 0.0005%. If the apparatus were not of great precision, it might well seem that the scattered radiation is of the same wavelength as the incident—the prediction of classical wave theory. Furthermore, when an energetic x-ray photon strikes, the electron departs speedily; no further interaction with x rays is noticed. But when a feeble 500-nm photon strikes, the electron acquires so little speed ("only" about 2 km/s, as opposed to the 0.07$c$ of the example; see Exercise 19) that it is subject to repeated collisions with incident photons. It would be slowly sped up and the photons scattered from it would begin to exhibit the Doppler shift expected classically.

These arguments provide just one example of the **correspondence principle.** This principle, which should be viewed as a guideline rather than a quantitative law of physics, states that there is always some limit in which a nonclassical theory should agree with the preexisting classical one. For example, special relativity is a nonclassical theory that agrees with classical mechanics in the limit of small velocities. To this we add that the nonclassical particle theory of electromagnetic radiation agrees with classical wave theory in the limit of long wavelengths: As the wavelength of a beam of electromagnetic radiation is increased, the energy per photon decreases. A given intensity would then comprise a larger number of less-energetic photons and would begin to exhibit the behavior expected of a continuous wave.[5]

In connection with the correspondence principle, it is often said that classical behavior follows "in the limit that $h$ goes to zero." It is true that photon energies and momenta tend to be small because $h$ is so small ($\sim 10^{-33}$ J·s). Were $h$ indeed zero, $E$ and $p$ (i.e., $hf$ and $h/\lambda$) would be zero, and light could never behave as a granular collection of particles. It would always behave as a continuous classical wave. However, it is important to remember that Planck's constant is a fundamental *constant* of nature—the constant associated with quantum phenomena—and as such never "goes" anywhere. Thus, it is better to say that classical behavior follows when the frequency is small and the wavelength long.

## An Inelastic Collision

The collision between the photon and the electron in the Compton effect is necessarily elastic. Since the mass/internal energy of a truly fundamental particle such as the electron cannot change, neither can the overall kinetic energy. Indeed, we may rearrange the energy-conservation equation (2-3c) as $E = E' + (\gamma_u - 1)m_e c^2$. Photons have only kinetic energy (no mass/internal energy), so

---

[5]The photoelectric effect might appear to violate the correspondence principle. The classical expectation was that all wavelengths should be capable of ejecting electrons, if any are. The photoelectric effect occurs—agrees with this expectation—not at long wavelengths, but only at *shorter* wavelengths! The "if any are" is the catch; the binding of electrons in a metal simply cannot be understood classically, so the classical expectation is really of little value.

this is a statement of *kinetic* energy conservation. Let us consider a collision of a photon with a massive object in which mass *is* subject to change.

---

**Example 2.4**

A neutron and proton bound together by the "strong force" is called a **deuteron,** the nucleus of the hydrogen isotope **deuterium** (also known as "heavy hydrogen"). A helium nucleus is two protons and two neutrons bound together by the same force. Suppose an extremely energetic photon strikes a helium nucleus and breaks it into two deuterons, each of which departs at 0.6c. The photon vanishes in the process. (a) What was the photon's wavelength? (b) In what directions do the deuterons depart? (The mass of a deuteron is 2.01355 u and of a helium nucleus is 4.00151 u.)

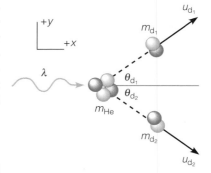

**Solution**

(a) Let us use the subscripts He, $d_1$, and $d_2$ to refer to the helium nucleus and the two deuterons. Before the collision, we have a photon and a stationary helium nucleus. Afterward, we have two moving deuterons. Energy is conserved:

$$h\frac{c}{\lambda} + m_{\text{He}}c^2 = \gamma_{d_1}m_{d_1}c^2 + \gamma_{d_2}m_{d_2}c^2$$

Because their speeds and masses are the same, the deuterons have the same energy, so we may simply double either term. Noting that $\gamma_{0.6c} = 1.25$, we have

$$h\frac{c}{\lambda} = c^2(2 \times 1.25m_d - m_{\text{He}})$$

$$= c^2[2 \times 1.25(2.01355 \text{ u}) - 4.00151 \text{ u}]$$

$$= c^2(1.0324 \text{ u})$$

A very energetic photon indeed! Its energy equals approximately the entire mass-energy of a proton. Such an energetic photon might be created when a proton and antiproton engage in "pair annihilation" (see Section 2.4). Now, cancelling a factor of $c$ on both sides,

$$6.63 \times 10^{-34} \text{ J·s}\frac{1}{\lambda} = (3 \times 10^8 \text{ m/s})(1.0324 \text{ u})(1.66 \times 10^{-27} \text{ kg/u})$$

$$\lambda = 1.29 \times 10^{-15} \text{ m}$$

(b) Directions are involved only in (vector) momentum-conservation equations, not the scalar energy-conservation equation.

x-component conserved:  $\dfrac{h}{\lambda} = (\gamma_{d_1}m_{d_1}u_{d_1})\cos\theta_{d_1} + (\gamma_{d_2}m_{d_2}u_{d_2})\cos\theta_{d_2}$

y-component conserved:  $0 = (\gamma_{d_1}m_{d_1}u_{d_1})\sin\theta_{d_1} - (\gamma_{d_2}m_{d_2}u_{d_2})\sin\theta_{d_2}$

Since $\gamma_u mu$ is the same for both deuterons, it cancels in the y-component equation, so that $\theta_{d_1} = \theta_{d_2}$ (as symmetry demands). Simplifying the x-component equation,

$$\frac{h}{\lambda} = 2(\gamma_d m_d u_d)\cos \theta_d$$

$$\frac{6.63 \times 10^{-34} \text{ J·s}}{1.29 \times 10^{-15} \text{ m}} = 2[1.25(2.01355 \text{ u} \times 1.66 \times 10^{-27} \text{ kg/u}) \times (0.6 \times 3 \times 10^8 \text{ m/s})]\cos \theta_d$$

$$\theta_d = 70°$$

The deuterons depart at 70° from the photon's initial direction of motion.

In the example, the initial mass is less than the final. It is left as an exercise to verify that the photon's initial kinetic energy exceeds the deuterons' final kinetic energy by an amount that agrees with equation (1-25), $\Delta KE = -\Delta mc^2$.

## 2.4 Pair Production

The photoelectric effect and the Compton effect are two important ways in which electromagnetic radiation interacts as a particle with matter. We now discuss a third—**pair production.**

In 1932, a new particle was discovered. Carl D. Anderson (Nobel prize 1936) was studying the effects of "cosmic rays," electromagnetic radiation from space that constantly bombards Earth's surface, when he noticed a particle behaving like an electron, but of positive charge. It curved the right amount but the "wrong" way in a magnetic field. This positively charged electron was termed the **positron.** We now know that its creation is quite common in the process known as pair production. Figure 2.6 depicts the process, which occurs whenever very high energy photons, such as cosmic-ray photons, are present. Typically, it would be revealed using a **bubble chamber** detector—in which charged particles leave visible trails of bubbles—immersed in a magnetic field. From an initial absence of charged particles, there suddenly appear two charged particles that deflect in similar paths but opposite directions. The energy to produce the massive electron-positron pair

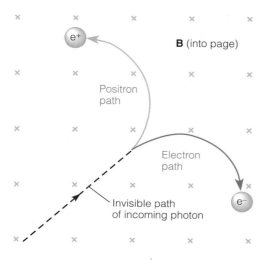

**Figure 2.6** In pair production, a gamma-ray photon becomes an electron, which curves one way if a magnetic field is present, and a positron, which curves the other way.

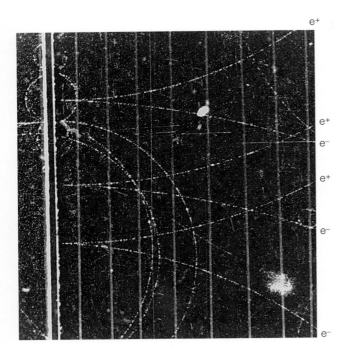

e+

e+

e−

e+

e−

e−

**Figure 2.7** Trails in a bubble chamber showing the sudden appearance of electron-positron pairs, the three pairs of arcs diverging toward right, from photons incident at left.

comes from a high-energy photon, which, being uncharged, leaves no trail. Charge is conserved because the total charge of the pair is zero.[6] Figure 2.7 shows actual bubble-chamber trails of several electron-positron pairs.

## Example 2.5

Calculate the energy and wavelength of the least-energetic photon capable of producing an electron-positron pair.

### Solution

The energy in the photon becomes the energy of the massive particles, internal/mass energy plus any kinetic energy. The minimum energy required is that which is barely able to produce the pair, with *no* kinetic energy. In this case, the photon energy equals just the internal energy of the pair:

$$2 \cdot m_e c^2 = 2(9.11 \times 10^{-31} \text{ kg})(3 \times 10^8 \text{ m/s})^2$$

$$= 1.64 \times 10^{-13} \text{ J} \quad (\cong 1 \text{ MeV})$$

Thus,

$$h\frac{c}{\lambda} = 2m_e c^2 \quad \Rightarrow \quad \lambda = \frac{hc}{2m_e c^2}$$

$$= \frac{(6.63 \times 10^{-34} \text{ J·s})(3 \times 10^8 \text{ m/s})}{1.64 \times 10^{-13} \text{ J}}$$

$$= 1.21 \times 10^{-12} \text{ m}$$

[6]The positron and other antiparticles are discussed in Chapter 11, but for the positron, life is short and its end dramatic. After a very short time, it will find an electron (not necessarily its original pair electron), and the two will "annihilate" together—erased—their entire energy suddenly converted to photons. Accordingly, positrons are comparatively rare. **Pair annihilation** is addressed in Exercise 25.

The term often used for electromagnetic radiation of wavelength shorter than those in the x-ray range is **gamma rays.** In the preceding process, a gamma-ray photon becomes an electron-positron pair.

Judging from Example 2.5, it might seem that pair production is not really an example of the *interaction* of electromagnetic radiation with (existing) matter, but is instead simply a way of *producing* matter. On the contrary, a gamma ray cannot become an electron-positron pair in a vacuum! Conspicuously missing in the example was any consideration of momentum. Momentum is not conserved if a (moving) photon becomes two stationary massive particles. Even if the photon were more energetic, allowing the electron and positron some kinetic energy after their creation, momentum could not be conserved. We can always choose to consider the process from a frame of reference in which the newly created particles have opposite velocities. The final momentum would again be zero, but we would have begun with a nonzero initial momentum (i.e., a single photon). Impossible!

What actually happens is that the gamma ray passes by a preexisting massive particle, such as the nucleus of an atom, the two interact via the electromagnetic force and a pair is created.[7] Afterward, the nucleus is in motion. Although momentum cannot be conserved without it, the nucleus is not affected much by the whole affair. It has momentum, but "steals" relatively little energy. Let us add to the example a stationary lead nucleus of mass $3.5 \times 10^{-25}$ kg. Suppose the momentum of the photon were transferred completely to this nucleus; its speed would be

$$ v = \frac{p_{photon}}{m_{nucleus}} = \frac{h/\lambda}{m_{nucleus}} = \frac{6.63 \times 10^{-34} \text{ J·s}/1.21 \times 10^{-12} \text{ m}}{3.5 \times 10^{-25} \text{ kg}} \cong 1600 \text{ m/s} $$

Its kinetic energy would be $\frac{1}{2}(3.5 \times 10^{-25} \text{ kg})(1600 \text{ m/s})^2 \cong 4 \times 10^{-19}$ J. (Clearly, nonrelativistic formulas are adequate for this slow nucleus.) Since the total energy involved in the process is about six orders of magnitude larger, we see that the nucleus can indeed ensure momentum conservation and yet have a negligible effect on the energy.

## Momentum Conservation: Small and Large Particles

The preceding discussion raises a point that crops up often in physics. Whenever "small" particles interact with "large" ones, the small ones tend to possess essentially all the kinetic energy. There is a simple argument: For a nonrelativistic massive particle, kinetic energy may be written as $p^2/2m$. If the interacting particles have comparable momenta (numerator), then a particle with a much larger mass (denominator) will have a much smaller kinetic energy. This conclusion usually holds for massless particles as well, with the massless particle filling the "small" role and the massive particle the "large" role. For instance, when an atom emits a photon, the recoiling atom and photon have equal momenta, but the photon carries essentially all the kinetic energy.

---

[7]The pair created need not be an electron-positron pair. Another possibility is a proton-antiproton pair. In any case, the photon must have an energy of at least $2mc^2$, where $m$ is the mass of either "half" of the pair.

## 2.5  Is It a Wave, or a Particle?

This may be the most perplexing question for the student of quantum mechanics, but it is important that we confront it early, because it lies at the heart of the subject. The simplest answer is that the object of study has no predetermined nature; the observation itself—whether the experimenter bounces light off the object of study, places something in its path, or interacts with it in any way—determines whether a wave nature or a particle nature will be observed.

Much of what we discuss here applies not only to electromagnetic radiation but also to massive objects. Accordingly, the principles we discuss serve as the crucial link between this chapter—the *particle* nature of electromagnetic radiation—and the next—the *wave* nature of massive objects.

Will it behave as a wave, or a particle? We begin with an admittedly vague criterion: It depends on how the object's wavelength $\lambda$ compares to the "relevant dimensions" of the experimental apparatus, which we represent simply as $D$. Every object may be described by a **wave function.** The reader is probably familiar with the sinusoidal functions describing the electric and magnetic fields in a plane wave of light. As we find in Chapter 3, a massive object may also be described by a sinusoidal wave function. If its wavelength is much smaller than $D$, an object will exhibit a particle behavior, and if comparable to or larger than $D$, a wave behavior.

Figure 2.8 gives a crude picture of how the behavior an object exhibits might depend upon a "relevant dimension" of the apparatus chosen for its study. The fellows in the boat are conducting an experiment—blindfolded. If the wavelength

$\lambda \ll D$:   particle
$\lambda \gtrsim D$:   wave

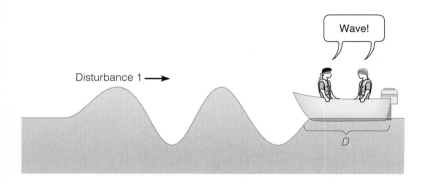

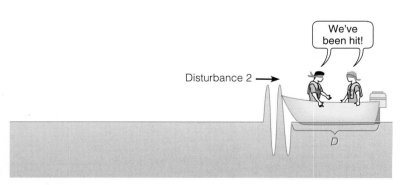

**Figure 2.8** An "experiment" in which a disturbance behaves as a wave or a particle depending on the relative size of the wavelength and the relevant dimensions of the apparatus.

of the disturbance is larger than $D$ (the width of the boat), the experimenters will determine the disturbance to be a wave; the dimension $D$ is small enough to respond to different regions of the wave. If the wavelength is much smaller than $D$, the boat responds not to different regions of the wave, but rather to all regions at once. The experimenters are likely to conclude that some *thing*, a particle rather than a diffuse wave, has struck. By the same logic, passengers on an immense ocean liner would see even the first disturbance as a particle, because $\lambda \ll D$, while ants floating on a popcorn kernel would see the second one as a wave, because $\lambda > D$.

Let us see how the criterion applies to electromagnetic radiation. Passing light through a single slit is a good way to reveal its wave nature. But to see the wave phenomenon of diffraction, the slit has to be narrow compared to the wavelength. (If the slit is wide, the wave behavior is indistinguishable from the particle; both would pass straight through.) On the other hand, as we found in Section 2.3, having electromagnetic radiation interact with weakly bound atomic electrons is a good way to reveal its particle nature. Bumping into electrons and transferring discrete bits of momentum and energy is definitely a particlelike behavior. But it is apparent only for short-wavelength x-ray photons, whose wavelengths are fairly small compared to the relevant atomic dimensions. (In this case, the particle behavior merges with the wave behavior if the wavelength becomes too large.)

It is worthwhile to reiterate that for a given intensity, or "flow rate," of some thing—electromagnetic radiation or massive objects—the shorter the wavelength, the higher the energy per particle and the coarser, more particlelike the behavior. Conversely, a longer wavelength means a smaller energy per particle and therefore a smoother, wavelike behavior. Nevertheless, it is dangerous to view wave-particle duality too rigidly, as wave *versus* particle. The natures are not incompatible, but complementary, two faces of the same phenomenon. We cannot fully explain the behavior of the phenomenon with either alone, and there are close relationships between the two. The equations $E = hf$ and $p = h/\lambda$, which *quantitatively* link the particle properties $E$ and $p$ to the wave properties $f$ and $\lambda$, are one part of the relationship. We now discuss another.

## A Two-Slit Experiment

There is no more direct way to see the link between wave and particle natures than the double-slit experiment. We discuss it now in connection with electromagnetic radiation and will again in Chapter 3 when we discuss matter waves.

Suppose light of wavelength $\lambda$ is directed at a double slit (in which both slits are "narrow"[8]) and the light passing through is detected on photographic film placed beyond the slits. At high intensities, we would observe a typical interference pattern: the intensity varying from a maximum (constructive interference) at the center of the film, to zero (destructive), then back to a maximum (constructive again), and so on. The light is exhibiting the wave nature we attribute to it in classical physics. Particles don't interfere, or "cancel"; waves do.

Now suppose the intensity is greatly reduced, to the point that the unaided eye can no longer see light on the screen. The film still registers the arrival of light, but sporadically at scattered locations. Evidently, the light is being detected one photon (one particle) at a time. Figure 2.9 shows what would be found if the film

[8]As in physical optics, "narrow" means that the individual slits are much narrower than the wavelength $\lambda$ being studied. This assumption simplifies analysis of the *double* slit because the effects of *single*-slit diffraction may be largely ignored. If an individual slit is much narrower than $\lambda$, the wave emerges as though from a point source. It spreads essentially uniformly in all directions and there are no diffraction minima.

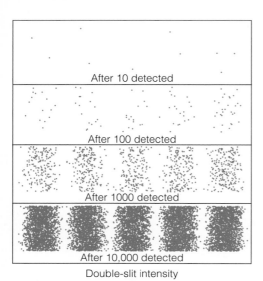

**Figure 2.9** Photons producing a double-slit interference pattern— one particle at a time.

After 10 detected

After 100 detected

After 1000 detected

After 10,000 detected

Double-slit intensity

were to register the arrival of light by producing a visible spot. Although the locations at which spots are recorded seem at first almost completely unpredictable, there soon appears a regular pattern to the density of spots. Understanding the link between wave and particle rests on two key observations: First, the exact location where the next photon will be found clearly cannot be known, but the probability of detecting the next photon in a given region is surely proportional to the density of spots—high where density is high, low where density is low. Therefore, if the density of spots assumes a pattern, the *probability* assumes a pattern. Secondly, careful study reveals that the density of spots in a region is directly proportional to what wave theory (physical optics) predicts should be the intensity in that region, which is in turn proportional to the square of the amplitude of the wave. In particular, no photons are ever detected at the locations where there should be points of destructive interference. *Although the light is being detected one particle at a time, its wave nature is still apparent.*

Combining these two observations, we conclude that since both are proportional to the density of spots, there is a proportionality between the probability of detecting the particle and the square of the amplitude of the wave. This connection between particle and wave natures is a cornerstone of quantum mechanics, crucial to its understanding:

When a phenomenon is detected as *particles,* it cannot be predicted with certainty where a given particle will be found. The most that can be determined is a probability of finding it in a given region, and this "probability density" is proportional to the square of the amplitude of the associated *wave* in that region.

$$\begin{array}{c}\text{probability density}\\ \text{of finding } particle\end{array} \propto (\text{amplitude of } wave)^2$$

In the case of electromagnetic radiation, the "associated wave" is the oscillating electromagnetic field. Electromagnetic fields, then, have multiple traits to their personality. We learn in electricity and magnetism that they exert forces on charges, and we now see that for an electromagnetic wave they also measure the probability of finding the "associated particle" (i.e., the photon). We point this out partly to prepare the reader for a possible shock in the next chapter. The wave associated with massive objects is a measure of the probability of finding a massive particle, but it does not appear to have any readily observable "other side" to its personality!

Before leaving the double slit, we address a point central to wave-particle duality. In the double slit, it is not proper to ask, "Through which slit did the 17th photon pass?" This cannot be determined in the *specific* experiment assumed, and by altering the experiment so as to "observe" a particle passing through a slit—an experiment in itself, requiring some interaction with light in one slit alone—we would alter the very behavior we wish to observe. Since interference requires two coherent waves, the pattern would be disturbed. We can't have it both ways! Wavelike interference can only be observed *by allowing each "particle" to behave as a wave—passing through both slits simultaneously.*

---

### Example 2.6

Light of wavelength 633 nm is directed at a double slit, and the interference pattern is viewed on a screen. The intensity at the center of the screen is 4.0 W/m$^2$. (a) At what rate are photons detected at the center of the screen? (b) At what rate are photons detected at the first interference minimum? (c) At what rate are photons detected at a point on the screen where the waves from the two sources are out of phase by one-third of a cycle? (*Note:* From physical optics, the double-slit intensity varies according to $I = I_0 \cos^2(\frac{1}{2}\phi)$, where $\phi$ is the phase difference between the waves from the two slits and $I_0$ is the intensity when $\phi = 0$—i.e., at the center of the screen.)

#### Solution

(a) Each photon has energy

$$h\frac{c}{\lambda} = (6.63 \times 10^{-34} \text{ J·s})\frac{3 \times 10^8 \text{ m/s}}{633 \times 10^{-9} \text{ m}} = 3.14 \times 10^{-19} \text{ J}$$

We must therefore have

$$\frac{4.0 \text{ J/s·m}^2}{3.14 \times 10^{-19} \text{ J/photon}} = 1.27 \times 10^{19} \text{ photons/s·m}^2$$

(b) The first interference minimum is a point of destructive interference because the difference in distances traveled to the screen creates a phase difference of $\pi$ (one-half cycle) between the waves from the two sources. Since the net (electric field) wave is zero, its square is zero and so is the intensity. No photons are detected here.

(c) Somewhere between the center of the screen and the first minimum, the waves will be out of phase by one-third cycle: $\phi = \frac{1}{3}2\pi$. Given that $I_0$ is 4.0 W/m$^2$, then, we have

$$I = (4.0 \text{ W/m}^2)\cos^2\left[\frac{1}{2}\left(\frac{1}{3}2\pi\right)\right] = 1.0 \text{ W/m}^2$$

Since this is one-fourth the value at the center, the probability of detecting photons here is one-fourth as large, so only one-fourth as many will be detected per unit time in a given area around this point.

$3.18 \times 10^{18}$ photons/s·m$^2$

The important point: Although particles may be detected, the probability of their detection is governed by the behavior of the associated electromagnetic wave.

## Diffraction: Uncertainties in Particle Properties

A cornerstone of quantum mechanics is the **uncertainty principle.** We will have more to say about it in Chapter 3, but the underlying idea is this: The mere fact that something has a *wave* nature implies that there are inherent uncertainties in some of its *particle* properties. We see this most easily via the single slit (Figure 2.10).[9]

Suppose a plane wave of light moving in the z-direction strikes a barrier in which there is a single slit of width $\Delta x$. As we know, the wave emerging from the slit will diffract, moving not only in the z-direction, but also in the x-direction. Even classical electromagnetic waves carry momentum, so in passing through the slit the waves have acquired a significant x-component of momentum. But what of the particles? To produce, one photon at a time, this spread-out single-slit pattern, their x-components of momentum must fluctuate significantly. The "experiment" of *sending the* wave *through a slit that restricts its location to within a width $\Delta x$ causes an uncertainty in the* particles' *x-components of momentum detected in a subsequent experiment* (e.g., at a screen). If the wavelength is very small compared to the slit width, little diffraction occurs; no wave nature is apparent and the particles do not seem to acquire any unexpected sideways momentum. But if the slit is narrow, the sideways momentum we observe might be any value within a large range. This dependence of an uncertainty in particle momentum on a width to which the particles are restricted is intimately tied to the wave behavior of diffraction.

Is it a wave, or a particle? It has the potential to behave as either, depending on the situation, but the two natures are inextricably related.

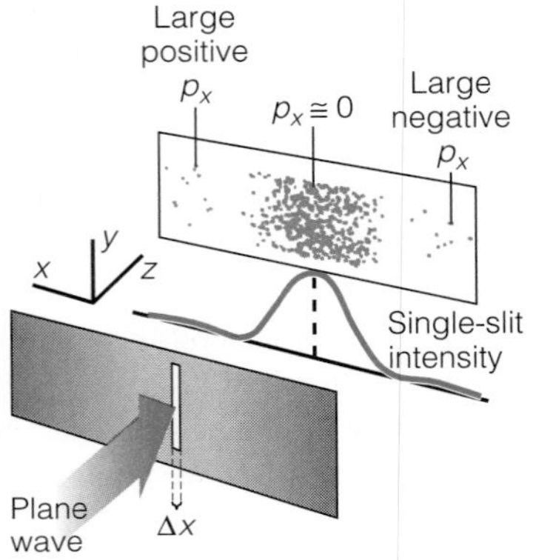

**Figure 2.10** Single-slit diffraction: Better knowledge of particle position along x (a narrower slit) implies decreased knowledge of x-component of momentum (a wider pattern), and vice versa.

## Progress and Applications

Consequences of light's particle nature crop up in all areas of physics, indeed in all of science.

■ The quest to transmit information as rapidly as possible has accelerated tremendously in recent decades. Were light strictly a wave, more information could be transmitted per unit time simply by increasing its frequency; roughly speaking, more bumps (crests) per unit time would translate to more bits of

[9]Because we are now interested in the single-slit diffraction pattern, we do *not* make the assumption, as we did in the double slit, that the slit width is much less than $\lambda$.

information per unit time. Due to $E = hf$, however, for a given energy per time, an increased frequency implies a smaller *number* of photons per unit time, and there are fundamental limits on how much information can be carried in a single particle of light. At one level of analysis, the many-photons-but-too-low-frequency and the high-frequency-but-too-few-photons effects should partly cancel, leaving a maximum transmission rate independent of frequency, but inversely proportional to $h$. Planck's constant being nonzero is the root of the "problem." Another quantum mechanical complication arises as we try to push our communication abilities to the limit: The uncertainty principle is far-reaching, and ties together inherent unpredictabilities in different aspects of light. We can exploit one aspect for "noiseless" communication only by scrupulously maintaining a hands-off policy with regard to the other. Novel ways to circumvent some seeming limits on communication are being investigated. One is communication via **solitons,** light pulses that, due to special properties in certain media, propagate without distortion.

■ Whereas high-energy photons may become electron-positron pairs via pair production, in the reverse process, pair annihilation, electron-positron pairs become photons. The latter effect is being exploited to probe the microscopic structure of solids such as metals and semiconductors. A positron is introduced, eventually pairing up and annihilating with an electron in the material, and the resulting high-energy photons are detected and used to produce a "picture" of the sample's structure. A related technique, positron emission tomography, or **PET,** produces images of the human body's interior.

■ A photon-based experimental technique much in use nowadays involves the phenomenon of **Raman scattering.** In Compton scattering, the photon's energy changes simply due to the energy transfer to the recoiling electron, a rather lightweight target. If a whole atom recoils, the photon's energy shift is negligible—like a Ping-Pong ball bouncing off a bowling ball. However, a photon may "bounce off" an entire molecule and yet suffer a significant energy shift—if in the process the molecule either loses or gains *internal* energy. As discussed in Chapter 9, the energies that molecules may possess are quantized and depend on the kind of molecule, like a fingerprint. Thus, the amount by which the photon's energy increases or decreases tells something about the molecule it has struck. Raman spectroscopy, watching and studying these shifts, is now indispensable in physics, chemistry, biology, and many other areas where the identification of molecular species is essential.

■ In recent years, x rays have enjoyed increased interest. Their ability to "see through" objects (being attenuated in proportion to the density of the region through which they pass) and their wavelike utility in revealing crystalline structures of solids via diffraction have long been exploited. However, we learn much about solids' thermal, electrical, and magnetic properties by analyzing how x rays inelastically collide with them as particles, and x-ray beams have recently become available from a source superior in many ways to the time-honored smashing of electron beams into targets: Electrons circularly accelerated by magnetic fields in particle accelerators known as synchrotrons continuously emit electromagnetic radiation. Such **synchrotron radiation** gives more ordered x rays, in a narrow beam, polarized and with tunable frequencies, and above all at much higher power.

## Basic Equations

**Photon properties:**

$$E = hf \tag{2-1}$$

$$p = \frac{h}{\lambda} \tag{2-4}$$

**Photoelectric effect:**

$$\text{KE}_{\text{max}} = hf - \phi \tag{2-2}$$

**Compton effect:**

$$\lambda' - \lambda = \frac{h}{m_e c}(1 - \cos\theta) \tag{2-5}$$

## Summary

Electromagnetic radiation behaves in some situations as a collection of particles—photons. These have the particle properties of discrete energy and momentum related to the wave properties of frequency and wavelength via

$$E = hf \qquad p = \frac{h}{\lambda}$$

where $h$ is Planck's constant, $6.63 \times 10^{-34}$ J·s. Electromagnetic radiation is more likely to exhibit a particle nature when the wavelength is small compared to the relevant dimensions of the experimental apparatus. For a given intensity, a short wavelength corresponds to a large energy per particle and a correspondingly small particle flux (number per unit time per unit area). The radiation is thus relatively discontinuous and particlelike. For long wavelengths, the radiation is more likely to behave as a continuous wave.

Electromagnetic radiation may exhibit its particle nature in several ways. In the photoelectric effect, a photon gives up a discrete amount of energy to an electron in a metal. In the Compton effect, a short-wavelength photon "scatters" from an essentially free electron, and both electron and photon obey the usual conservation laws for particles with discrete momentum and energy. In pair production, these conservation laws are also obeyed as a high-energy photon disappears, creating and giving its discrete energy to a pair of massive particles.

When electromagnetic radiation is detected as particles, the exact location where the next particle will be found cannot be predicted, but the probability of finding a particle within a region is proportional to the square of the associated wave, the oscillating electromagnetic field, in that region.

## EXERCISES

### Section 2.1

1. Light of 300-nm wavelength strikes a metal plate, and photoelectrons are produced moving as fast as $0.002c$. (a) What is the work function of the metal? (b) What is the threshold wavelength for this metal?

2. What is the stopping potential when 250-nm light strikes a zinc plate?

3. What wavelength of light is necessary to produce photoelectrons of speed $2 \times 10^6$ m/s with a magnesium target?

4. What is the wavelength of a 2.0-mW laser from which $6 \times 10^{15}$ photons emanate every second?

5. A 940-kHz radio station broadcasts 40 kW of power. How many photons emanate from the transmitting antenna every second?

6. To expose photographic film, photons of light dissociate silver bromide (AgBr) molecules, which requires an energy of 1.2 eV. What limit does this impose on the wavelengths that may be recorded by photographic film?

7. Light of wavelength 590 nm is barely able to eject electrons from a metal plate. What would be the speed of the fastest electrons ejected by light of one-third the wavelength?

8. With light of wavelength 520 nm, photoelectrons are ejected from a metal surface with a maximum speed of $1.78 \times 10^5$ m/s. (a) What wavelength would be needed to give a maximum speed of $4.81 \times 10^5$ m/s? (b) Can you guess what metal it is?

9. You are an early 20th century experimental physicist, and do not know the value of Planck's constant. By a suitable plot of the following data, and using Einstein's explanation of the photoelectric effect (KE = $hf - \phi$, where $h$ is *not* known), determine Planck's constant.

| Wavelength of light (nm) | Stopping potential (V) |
|---|---|
| 550 | 0.060 |
| 500 | 0.286 |
| 450 | 0.563 |
| 400 | 0.908 |

10. A sodium vapor light emits 10 W of light energy. Its wavelength is 589 nm and it spreads in all directions. How many photons pass through your pupil, diameter 4 mm, in 1 s if you stand 10 m from the light?

### Section 2.2

11. When a beam of mono-energetic electrons is directed at a tungsten target, x rays are produced with wavelengths no shorter than 0.062 nm. How fast are the electrons in the beam moving?

12. A television picture tube accelerates electrons through a potential difference of 30,000 V. Find the minimum wavelength to be expected in x rays produced by this tube.

### Section 2.3

13. Verify that $\Delta KE = -\Delta mc^2$ in Example 2.4.

14. A typical ionization energy—the energy needed to remove an electron—for the elements is 10 eV. Explain why the energy binding the electron to its atom can be ignored in Compton scattering involving an x-ray photon whose wavelength is about a tenth of a nanometer.

15. In the Compton effect, we may always choose the electron to be at the origin and the initial photon's direction of motion to be in the $+x$-direction.
    (a) We may also choose the $xy$-plane so that it contains the velocities of the outgoing electron and photon. Why?
    (b) The wavelength $\lambda$ of the incoming photon is assumed known. The unknowns after the collision are the outgoing photon's wavelength and direction, $\lambda'$ and $\theta$, and the speed and direction of the electron, $u_e$ and $\phi$. With only three equations—two components of momentum conservation and one of energy—we may not find all four; equation (2-5) gives $\lambda'$ in terms of $\theta$. Our lack of knowledge of $\theta$ *after* the collision (without an experiment) is directly related to a lack of knowledge of something *before* the collision. What is it? (*Hint:* Imagine the two objects are spheres.)
    (c) Is it reasonable to suppose that we *could* know this quantity? Explain.

16. A 0.065-nm x-ray source is directed at a sample of carbon. Determine the maximum speed of scattered electrons.

17. A 0.057-nm x-ray photon "bounces off" an initially stationary electron and scatters with a wavelength of 0.061 nm. Find the directions of scatter of (a) the photon, and (b) the electron.

18. An x-ray source of unknown wavelength is directed at a carbon sample. An electron is scattered with a speed of

4.5 × $10^7$ m/s at an angle of 60°. Determine the wavelength of the x-ray source.

19. Show that if a 500.00000-nm photon were to scatter at 90° off a free electron, the electron would acquire a speed of about 2 km/s.

20. A photon scatters off a free electron. (a) What is the maximum possible change in wavelength? (b) Suppose a photon scatters off a free proton. What is now the maximum possible change in wavelength? (c) Which, collision with an electron or with a proton, more clearly demonstrates the particle nature of electromagnetic radiation?

21. Determine the wavelength of an x-ray photon that can impart at most 80 keV of kinetic energy to a free electron.

22. From equations (2-3) obtain equation (2-5). It is easiest to start by eliminating $\phi$ between equations (2-3a) and (2-3b) using $\cos^2 \phi + \sin^2 \phi = 1$. The electron speed $u$ may then be eliminated between the remaining equations.

23. Show that the angles of scatter of the photon and electron in the Compton effect are related by the following formula:

$$\cot \frac{\theta}{2} = \left(1 + \frac{h}{mc\lambda}\right) \tan \phi$$

## Section 2.4

24. A gamma-ray photon changes into a proton-antiproton pair. Ignoring momentum conservation, what must have been the wavelength of the photon (a) if the pair is stationary after creation, and (b) if each moves off at 0.6c perpendicular to the motion of the photon? (c) Assume these interactions occur as the photon encounters a lead plate, and that a lead nucleus participates in momentum conservation. In each case, what fraction of the photon's energy must be absorbed by a lead nucleus?

25. *Pair Annihilation.* A stationary muon $\mu^-$ annihilates with a stationary antimuon $\mu^+$ (same mass, 1.88 × $10^{-28}$ kg, but opposite charge). The two disappear, replaced by electromagnetic radiation. (a) Why is it not possible for a single photon to result? (b) Suppose two photons result. Describe their possible directions of motion and wavelengths.

## Section 2.5

26. A beam of 500-nm light strikes a barrier in which there is a narrow single slit. At the very center of a screen beyond the single slit, $10^{12}$ photons are detected per square millimeter per second. (a) What is the intensity of the light at the center of the screen? (b) A second slit is now added very close to the first. How many photons will be detected per square millimeter per second at the center of the screen now?

27. Electromagnetic "waves" strikes a single slit of 1-$\mu$m width. Determine the "angular full width" (angle from first minimum on one side of center to first minimum on the other) in degrees of the central diffraction maximum if the waves are (a) visible light of wavelength 500 nm, and (b) x rays of wavelength 0.05 nm. (c) Which more clearly demonstrates a wave nature?

28. A beam of light moving in the $z$-direction is directed at a single slit of width $\Delta x$, as shown in Figure 2.10. As we see, the $x$-component of momentum of a photon after passing through the slit is uncertain. It may take on various values on either side of zero. Accordingly, its *uncertainty* $\Delta p_x$ should be comparable to a typical *value* $p_x$. If the diffraction angle $\theta$ is small, $p_x$ would be roughly $p \sin \theta$. With this in mind, use the single-slit diffraction equation $m\lambda = \Delta x \sin \theta$ to show that $\Delta p_x$ is inversely proportional to $\Delta x$.

## General Exercises

29. A photon has the same momentum as an electron moving at $10^6$ m/s. (a) Determine the photon's wavelength. (b) What is the ratio of the kinetic energies of the two? (*Note:* A photon is *all* kinetic energy.)

30. A photon and an object of mass $m$ have the same momentum $p$.
    (a) Assuming that the massive object is moving slowly, so that nonrelativistic formulas are valid, find in terms of $m$, $p$, and $c$, the ratio of the massive object's kinetic energy to the photon's kinetic energy and argue that it is small.
    (b) Find the same ratio found in part (a) but using relativistically correct formulas for the massive object. (*Note:* $E^2 = p^2c^2 + m^2c^4$ may be helpful.)
    (c) Show that the low-speed limit of the ratio of part (b) agrees with part (a) and that the high-speed limit is unity.
    (d) Show that at *very* high speed, the kinetic energy of a massive object is $pc$, just as it is for a photon.

31. Radiant energy from the Sun arrives at Earth with an intensity of 1.5 W/$m^2$. Making the rough approximation that all photons are absorbed, find (a) the radiation pressure, and (b) the total force experienced by Earth due to this "solar wind."

32. A flashlight beam produces 2.5 W of electromagnetic radiation in a narrow beam. Although the light it produces is white (all visible wavelengths), make the simplifying assumption that the wavelength is 550 nm, the middle of the visible spectrum. (a) How many photons

per second emanate from the flashlight? (b) What force would the beam exert on a "perfect" mirror (i.e., one that reflects all light completely)?

33. The average intensity of an electromagnetic wave is $\frac{1}{2}\epsilon_0 c E_0^2$, where $E_0$ is the amplitude of the electric-field portion of the wave. Find a general expression for the photon flux $j$ (measured in photons/s·m$^2$) in terms of $E_0$ and wavelength $\lambda$.

34. Show that the laws of momentum and energy conservation forbid the complete *absorption* of a photon by a free electron. (*Note:* This is not the photoelectric effect. In the photoelectric effect, the electron is not free; the metal participates in momentum and energy conservation.)

35. An electron moving to the left at $0.8c$ collides with an incoming photon moving to the right. After the collision, the electron is moving to the right at $0.6c$ and an outgoing photon moves to the left. What was the wavelength of the incoming photon?

36. An object moving to the right at $0.8c$ is struck head-on by a photon of wavelength $\lambda$ moving to the left. The object absorbs the photon (i.e., the photon disappears), and is afterward moving to the right at $0.6c$. (a) Determine the ratio of the object's mass after the collision to its mass before the collision. (*Note:* The object is not a "fundamental particle," and its mass is therefore subject to change.) (b) Does kinetic energy increase or decrease?

37. Cosmic-ray photons from space are bombarding your laboratory and smashing massive objects to pieces! Your detectors indicate that two fragments, each of mass $m_0$, depart such a collision moving at $0.6c$ at $60°$ to the photon's original direction of motion. In terms of $m_0$, what are the energy of the cosmic-ray photon and the mass $M$ of the particle being struck (assumed initially stationary)?

# Waves and Particles II: Matter Behaving as Waves

W e have seen that electromagnetic radiation, classically a wave, also has a particle nature. We now begin our study of a complementary and fascinating truth: Matter, classically particlelike, has a wave nature. Although a challenging notion, it is the foundation of our understanding of the physical world on the submicroscopic level.

To make electromagnetic radiation reveal its wave nature, we must study it with an apparatus of small dimensions, comparable to the wavelength ($D \lesssim \lambda$). We must also do this to reveal the wave nature of matter. But as we shall see, the wavelengths of matter waves are typically very small, far too small for human vision, relying on "unwieldy" visible light, to resolve any wavelike detail. It is understandable, then, that we tend to regard matter as particles; only a situation involving *very* small dimensions could show matter's other side. Perhaps the situation that best reveals this other side is the atom, in which electrons are confined within dimensions so small that their wave nature must come to the fore. In a study of waves and sound, we learn that there are only certain, discrete standing waves possible on a stretched string, in an organ pipe, and so on. We shall see in the coming chapters that electron energies in an atom are quantized for the very same reasons—the electrons are behaving as standing waves.

Being diffuse, waves are analyzed differently from discrete particles. In introductory mechanics, objects are particles and the equations governing their behavior are relatively simple (kinematic equations, $\mathbf{F} = m(d\mathbf{v}/dt)$, etc.). We find that waves, on the other hand, obey equations involving more sophisticated calculus—

for instance, the differential "wave equation" for waves on a string and Maxwell's equations for light waves. This difference also shows up in what we seek to determine. For particles, the question is often Where is it going? For waves, we ask: What is its amplitude? What is its wavelength? How spread out is it? Where is it zero?

We shall soon consider such questions with regard to matter waves. Afterward, we introduce the Schrödinger equation, the equation obeyed by matter waves, and then discuss one of the most important consequences of matter's wave nature—the uncertainty principle. We begin, however, with the question probably foremost in the reader's mind: Just what is a matter wave and how do we know they even exist? Nothing illustrates matter's wave nature more convincingly than the double-slit experiment.

## 3.1 A Two-Slit Experiment

Imagine a beam of mono-energetic electrons[1] striking a barrier in which there is a slit. Beyond the slit is a phosphorescent screen that registers each electron's arrival by producing a flash. With a "wide" slit, as shown in Figure 3.1, the beam passes straight through and produces—electron by electron—a stripe on the screen the same width as the slit. With a narrow slit, however, we find electrons registering sporadically *over the entire screen*. While this spreading of the beam is difficult to reconcile with the notion of electrons as strictly particles, if a second slit is added, the conclusion is inescapable.

Suppose, then, that we add a second narrow slit. Again, with either slit open alone, electrons are detected sooner or later at *all* points on the screen. But when both are open at the same time, at certain locations, locations *where electrons had been detected with either slit open separately,* electrons are *never* detected. Opening a second "door" *decreases* to zero the number arriving per unit time at specific, regularly spaced locations on the screen! This is impossible to explain if electrons are simply particles passing through one slit or the other; a particle passing through one slit would not suddenly have reason to avoid specific locations on the screen just because a slit had been opened elsewhere. Rather, this is destructive interference, and since interference requires multiple coherent waves, *each* electron must be behaving as a wave passing through *both* slits at once.

Figure 3.2 shows how the electron flashes would accumulate with both slits open. It should look familiar—it is the double-slit photon-detection pattern of Figure 2.9. The point is that both electromagnetic radiation and massive objects exhibit the same kind of wave-particle duality (see also Figure 3.3): Associated with the particle of light (photon) is a wave of oscillating electromagnetic field; we must now accept that there is also a wave associated with a massive particle. Cer-

[1] Electrons are our preferred objects of study for two reasons: First, among the familiar massive particles—protons, neutrons, and electrons—the electron is the simplest. As far as is known, it is truly fundamental, having no internal structure and occupying no space. Secondly and more importantly, at a given speed its wavelength is much longer. (As we shall see, this is due to its small mass.) With a relatively large wavelength, its wave nature should be more readily apparent.

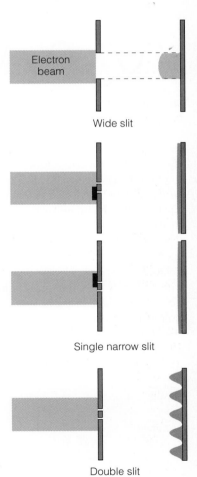

**Figure 3.1** Intensity patterns when an electron beam strikes various slits.

Electron beam

Wide slit

Single narrow slit

Double slit

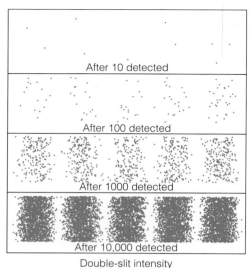

After 10 detected

After 100 detected

After 1000 detected

After 10,000 detected

Double-slit intensity

**Figure 3.2** Electrons producing a double-slit interference pattern—one particle at a time.

tainly it is different, and it is natural to wonder what is analogous to the electromagnetic fields—that is, what is oscillating.

To identify at least one property of this wave, we note that the same two observations we made about electromagnetic waves also apply here: (1) Although it is obviously impossible to know where the next electron will be detected, it is surely true that the density of spots in a given region *is proportional to the probability* that it will be found in that region. (2) If we concede that coherent waves of equal amplitude are emanating from the two slits, we should expect an interference pattern. In particular, we would expect the amplitude at the center of the screen (constructive interference) to be twice what it would be if only one slit were open—the square of the amplitude, proportional to an intensity, should be 4 times as large. Conspicuously, the density of spots is indeed 4 times larger at the center with both slits open than with either slit alone. In fact, at all points on the screen we find that the density of spots *is proportional to the square of the total wave* obtained in a standard physical-optics analysis of the double slit. Combining these observations—that the square of the wave's amplitude and the probability of finding the particle are proportional to the same thing (i.e., the density of spots)—we arrive at the same conclusion as for electromagnetic radiation:

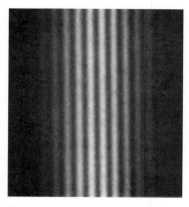

**Figure 3.3** Actual double-slit pattern produced by electrons.

> When a phenomenon is detected as *particles,* it cannot be predicted with certainty where a given particle will be found. The most that can be determined is a probability of finding it in a given region, and this "probability density" is proportional to the square of the amplitude of the associated *wave* in that region.
>
> probability density
> of finding *particle*    $\propto$ (amplitude of *wave*)$^2$

Though the wave-particle relationship for electromagnetic radiation and massive objects is the same, matter waves differ in an important way: They cannot be directly detected! Electric and magnetic fields can be isolated and caused to exert forces on objects; they may be directly detected. But we have not found any analogous way of directly detecting matter waves; no one has ever "seen" one. What then is the answer to the question, What is oscillating? We have no other readily identifiable property—*probability* is oscillating.[2]

## Matter Wave Interference: Evidence and Application

The double slit is conceptually the simplest experiment in which to show matter wave interference, but it was not the first. The first evidence that matter has a wave nature was obtained by Clinton J. Davisson and Lester H. Germer in 1927. Investigating properties of metallic surfaces by observing how a nickel crystal scatters a beam of electrons, Davisson and Germer were perplexed to discover that the electrons seemed to scatter preferentially at only certain discrete angles. Particles wouldn't do this—but waves would.

Microscopically, a crystal is simply an arrangement of regularly spaced atoms. If a wave is incident upon it, each individual atom on the surface will reflect the wave in all directions. In essence, each becomes a point source of waves. These sources will produce an interference pattern just as do the multiple slits of a diffraction grating. The similarities are shown in Figure 3.4. In each case, constructive interference occurs at angles where the wave from one point source has an integral number of wavelengths farther to travel than that from the adjacent source. In the Davisson–Germer experiment, electrons in a normally incident beam encounter the surface *one at a time,* and the wave associated with each reflects from the many surface atoms and *interferes with itself.* At those angles where we would expect this interference to be constructive, electrons are detected in great abundance; at angles where the interference would be destructive, no electrons are to be found. The experimentally observed electron detection rate versus angle is in perfect agreement with the theoretical prediction based upon the assumption of wave interference.

In the Davisson–Germer experiment, whose primary apparatus is shown in Figure 3.5, a fairly low accelerating potential was used (54 V). At higher energies,

[2]Beginning students of quantum mechanics sometimes assume that *mass* is oscillating, that an electron wave is somehow the mass of the electron, or some portion thereof, jiggling back and forth. The wave is *not* the particle, nor is it a collection of pieces of particles. Massive particles do not oscillate in a matter wave any more than photons oscillate in an electromagnetic wave. It is probably best, as a default, to view all things (e.g., electrons, electromagnetic radiation) as a wave of oscillating *probability*; where that wave is large there is a large probability of finding the particle.

**Figure 3.4** Diffraction of light by a grating and of matter waves by a crystal surface.

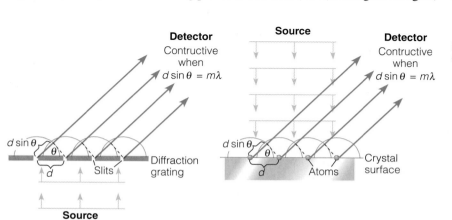

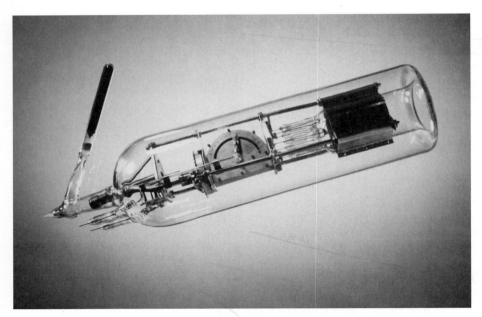

**Figure 3.5** Davisson and Germer's original electron diffraction apparatus, showing the mechanism for varying the sample's angle with the incident beam.

the matter wave is able to penetrate to atomic planes *beneath* the surface, and a different experimental setup is preferred: The beam is directed at the crystal's surface *at an angle $\theta$* with the normal, and the detector is placed at the same angle on the other side of the normal. An atom in the top plane reflects a portion of the wave, ray 1 in Figure 3.6, which scatters in all directions. But much of the wave penetrates deeper, so an atom in the second atomic plane also scatters a portion in all directions. As shown in the figure, ray 2 has $2D \cos \theta$ farther to travel than ray 1 to reach the detector. Ray 3 travels the same distance farther than ray 2, and ray 4 farther than ray 3, and so on. Having the incident beam and detector at equal angles ensures that waves scattering from adjacent *columns* of atoms always have the same distance to travel from source to detector. Therefore, the interference is only between different atomic *planes*. Constructive interference occurs at angles

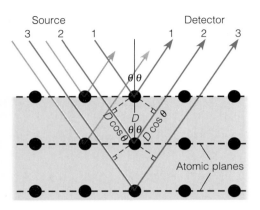

**Figure 3.6** Diffraction of a deeply penetrating beam.

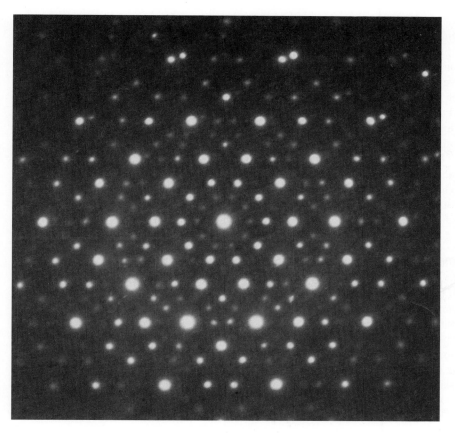

**Figure 3.7** Electron wave diffracted by AlMn quasicrystal.

where $2D \cos \theta = m\lambda$. Significantly, the same equation gives the interference maxima when x rays scatter from a crystal. (It is known as the **Bragg law,** for father-son team W. H. and W. L. Bragg, who were awarded the 1915 Nobel prize for their x-ray diffraction work.) We saw in Section 2.3 that x rays may behave as particles where visible light (of longer wavelength) would behave as waves. The spacing $D$ in a crystal is so small that even x rays behave as waves. However, not only electrons and x rays behave as waves; diffraction patterns have been produced by beams of neutrons (see Progress and Applications) and even whole atoms. Crystal diffraction is an indispensable tool in the study of solids; the details of the diffraction patterns provide much information about the crystal's microscopic geometry. Figure 3.7 is a good example. Produced by electrons diffracting from a single grain of an aluminum–manganese alloy, it shows an unexpected symmetry: the pattern is identical if rotated by one-*fifth* of a whole rotation. Study of such "quasicrystals" has been quite active since this discovery of the mid-1980s.

## 3.2 Properties of Matter Waves

There are several properties we expect of any wave: It should have a wavelength, a frequency, and a speed, and it should have an amplitude that varies with position and time. The generic term for the function giving the amplitude is **wave**

**function.** Since what actually oscillates depends on the kind of wave, we use different symbols for the wave functions of different kinds of waves: For transverse waves on a string, we might use the symbol $y(x, t)$; the string's transverse displacement $y$ varies as a function of the position $x$ along the string and time $t$. For an electromagnetic plane wave moving along the $x$-axis, we have two wave functions, $\mathbf{E}(x, t)$ and $\mathbf{B}(x, t)$, which describe how the oscillating electric and magnetic fields vary with position and time. For the wave function for matter waves, we choose the symbol $\Psi(x, t)$. Describing oscillating probability, it is properly referred to as the **probability amplitude.** In practice, however, it is usually simply referred to as "the wave function." In the next section we discuss how wave functions in general are determined; each obeys its own "wave equation." But for the time being we concentrate on important matter wave behaviors that can be understood without having an explicit formula for the wave function.

*Matter wave function: $\Psi(x, t)$
"probability amplitude"*

## Wavelength

The basic properties of wavelength, frequency, and velocity we might expect to be related in the usual way: $v = f\lambda$. They are, but this is not immediately useful, for the $v$ here is *not* the velocity of the particle. (Although it *is* the velocity of the matter wave, the two are not the same. We return to this point later in the section.) But if we cannot relate $v$ to the particle velocity, can we relate *any* of the wave properties $v$, $f$, or $\lambda$ to familiar properties of the particle? Is there any quantitative relationship we can use as a starting point?

Louis de Broglie submitted the following hypothesis: The wavelength of the matter wave associated with a massive particle of momentum $p$ is given by

*Wavelength of matter wave*

$$\lambda = \frac{h}{p}$$

This relationship has been indisputably confirmed, even for relativistic speeds, by experiments such as crystal diffraction. The momentum of the electrons in a beam is known, analysis of the diffraction pattern establishes the wavelength, and together the relationship is verified. The discovery won de Broglie the 1929 Nobel prize. (In recognition, we often refer to the wavelength of a matter wave as the "de Broglie wavelength.") The reader may have noticed that this is the same equation as for electromagnetic radiation—relating the wave property of wavelength to the particle property of discrete momentum. *There is for all phenomena a fundamental relationship between wavelength and momentum.*

---

**E x a m p l e   3 . 1**

If moving at 300 m/s, what would be the wavelength (a) of an 18,000-kg airplane, and (b) of an electron? (c) Which is more likely to exhibit a wave nature?

Solution

(a) $\lambda_{\text{airplane}} = \dfrac{6.63 \times 10^{-34} \text{ J·s}}{(18,000 \text{ kg})(300 \text{ m/s})} = 1.23 \times 10^{-40} \text{ m}$

(b) $\lambda_{\text{electron}} = \dfrac{6.63 \times 10^{-34} \text{ J·s}}{(9.11 \times 10^{-31} \text{ kg})(300 \text{ m/s})} = 2.43 \times 10^{-6} \text{ m}$

(c) The wavelength of the airplane is 25 orders of magnitude smaller than the atomic nucleus ($\sim 10^{-15}$ m). This being so much smaller than any conceivable investigative apparatus, it is absurd to speak of the airplane as anything but a particle. An electron might well display a wave nature under very close scrutiny.

In ordinary situations, the wavelength of matter waves is short enough to ensure that matter behaves as particles. This is true because Planck's constant, the fundamental constant of quantum mechanics, is so small. Even so, as an object's momentum approaches zero, its wavelength would become arbitrarily large. We might draw the unsettling conclusion that a stationary particle, with an infinite wavelength, should certainly behave as a wave. As we shall see in Example 3.5, the way out of this predicament rests upon the fact that it is difficult to guarantee that a particle is indeed *stationary*.

With a way of quantifying wavelength at our disposal, we can now quantify one of the most important things matter waves tell us: how probability varies with position.

---

**Example 3.2**

A beam of electrons each of 1.0-eV kinetic energy strikes a barrier in which there are two narrow slits separated by 0.020 $\mu$m. Beyond the barrier is a bank of electron detectors. At the center detector, directly in the path the beam would follow if unobstructed, 100 electrons per second are detected. As the detector angle varies, the number per unit time varies in a typical two-slit pattern between the maximum of 100 s$^{-1}$ and the minimum of zero. The first minimum occurs at detector X, an angle $\theta_X$ off center. (a) Determine $\theta_X$. (b) How many electrons would be detected per second at the center detector if one of the slits were blocked? (c) How many electrons would be detected per second at the center detector and at detector X if one of the slits were narrowed to 36% of its original width?

Solution

(a) At detector X, the first point of destructive interference, the wave arriving from one of the slits has $\frac{1}{2}\lambda$ farther to travel than that from the other slit. From physical optics we know that the difference in distances traveled by the two waves in a double-slit experiment is $d \sin \theta$, where $d$ is the slit separation. Thus,

$$d \sin \theta_X = \frac{1}{2}\lambda$$

According to de Broglie's hypothesis, $\lambda$ is related to $p$. We first determine $v$:

$$\mathrm{KE} = \frac{1}{2}mv^2$$

$$\rightarrow \quad (1\text{ eV}) \times (1.6 \times 10^{-19}\text{ J/eV}) = \frac{1}{2}(9.11 \times 10^{-31}\text{ kg})v^2$$

$$\Rightarrow \quad v = 5.93 \times 10^5 \text{ m/s}$$

Therefore,

$$p = mv = (9.11 \times 10^{-31} \text{ kg})(5.93 \times 10^5 \text{ m/s})$$

$$= 5.40 \times 10^{-25} \text{ kg·m/s}$$

and

$$\lambda = \frac{h}{p} = \frac{6.63 \times 10^{-34} \text{ J·s}}{5.40 \times 10^{-25} \text{ kg·m/s}} = 1.23 \times 10^{-9} \text{ m}$$

Reinserting,

$$(0.020 \times 10^{-6} \text{ m})\sin\theta_X = \frac{1}{2}(1.23 \times 10^{-9} \text{ m}) \quad \Rightarrow \quad \theta_X = 1.76°$$

(b) With both slits open, the electron flux (electrons per second) is $100 \text{ s}^{-1}$. But the electron flux *is proportional to the probability of detection and therefore to the square of the amplitude* of the total matter wave (from both slits).

$$|\Psi_T|^2 \propto 100 \text{ s}^{-1} \quad \Rightarrow \quad |\Psi_T| \propto 10$$

Since the waves from the two slits *add* equally at this point of constructive interference, the amplitude of either individual wave must be half the total.

$$|\Psi_1| \propto 5 \quad \Rightarrow \quad |\Psi_1|^2 \propto 25 \text{ s}^{-1}$$

The electron flux would be $25 \text{ s}^{-1}$ at the center detector. It is important to note that without a second slit/wave to interfere, 25 electrons per second would arrive at *all* detectors. With two slits, twice as many electrons should be detected each second, and this is indeed the case, but only *on average:* At points of constructive interference, 100 are detected per second, and at points of destructive interference, none. The average is $50 \text{ s}^{-1}$, but they are distributed in an interference pattern that cannot be understood by a strict particle interpretation.

(c) If only one slit were open and its width were reduced to 0.36 of its original value, all detectors would register an electron flux 0.36 times $25 \text{ s}^{-1}$, or $9 \text{ s}^{-1}$.

$$|\Psi_1'|^2 = 0.36 \times |\Psi_1|^2 \propto 0.36 \times 25 \text{ s}^{-1} = 9 \text{ s}^{-1}$$

The amplitude of the wave from the reduced-width slit would then be

$$|\Psi_1'|^2 \propto 9 \text{ s}^{-1} \quad \Rightarrow \quad |\Psi_1'| \propto 3$$

This is 60% of the original amplitude. Sensibly, an amplitude only 60% of the original corresponds to a *square* of the amplitude only 36% of the original.

With both slits open, we have two waves of different amplitudes, one proportional to 5 (original width) and one to 3 (reduced width). At points of constructive interference, such as the center detector, where the waves *add,* the total amplitude will be proportional to $5 + 3$.

$$|\Psi_T'|_{\text{constr}} \propto 8 \quad \Rightarrow \quad |\Psi_T'|^2_{\text{constr}} \propto 64 \text{ s}^{-1}$$

At points previously of destructive interference, where the two waves are 180° out of phase, as at detector X, the interference would no longer be totally destructive because the waves are not of equal amplitude. Even so, the waves *subtract,* and the total amplitude would therefore be proportional to $5 - 3$.

$$|\Psi'_T|_{destr} \propto 2 \quad \Rightarrow \quad |\Psi'_T|^2_{destr} \propto 4 \text{ s}^{-1}$$

The average electron flux is $\frac{1}{2}(64 \text{ s}^{-1} + 4 \text{ s}^{-1}) = 34 \text{ s}^{-1}$, the sum of the $9 \text{ s}^{-1}$ and $25 \text{ s}^{-1}$ expected from each slit alone. Again, however, to find the probability (or flux) at a location, we do not add the *probabilities* (or fluxes) from each slit at that location; these are always positive, so they cannot cancel. Rather, we add the *waves,* which may add constructively or destructively, to find the total wave, and then square the total wave to find the probability.

Figure 3.8 shows how electron flux would vary with angle for the equal-width, one-slit, and unequal-width cases in the example. Note that, except for the method of finding the electron momentum, Example 3.2 could have involved light. The words "matter wave" would be replaced by "electric field," and "electron flux" or "probability" by "light intensity." We have analyzed electron behavior via standard wave theory.

## Frequency

Matter waves, of course, have a frequency. The evidence is not quite so direct as the interference pattern evidence that clearly demonstrates a wavelength. But once again, the relationship is the same as for electromagnetic radiation:

$$f = \frac{E}{h}$$

Frequency of matter wave

The wave property of frequency is related to the particle property of discrete energy.

We now have ways of quantifying two of the properties we would expect matter waves to have:

$$\lambda = \frac{h}{p} \qquad f = \frac{E}{h}$$

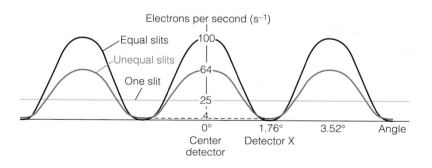

**Figure 3.8** Electron detection rate versus angle for equal-width slits, unequal-width slits, and one slit.

In many cases it is more convenient to express these in terms of the **wave number** and **angular frequency,** defined as follows:

$$k \equiv \frac{2\pi}{\lambda} \qquad \omega \equiv \frac{2\pi}{T}(=2\pi f)$$

wave number · · · · · · angular frequency

Because they are so simply related, it might seem a mystery why wave number would be more convenient than wavelength. One reason is that it is often easier to work with a frequency rather than a period—and wave number is a "spatial frequency." Whereas angular frequency is inversely proportional to the temporal period $T$, wave number is inversely proportional to the spatial period $\lambda$. Secondly, while wavelength is strictly a scalar, wave number can be made a vector, which conveniently points in the direction the wave travels. Another very convenient definition is

$$\hbar \equiv \frac{h}{2\pi} = 1.055 \times 10^{-34} \text{ J·s}$$

With this we may write the preceding wave-particle relationships as

$$p = h/\lambda = \hbar k \quad \text{(3-1)} \qquad E = hf = \hbar\omega \quad \text{(3-2)}$$

Fundamental wave-particle relationships

*These are the fundamental relationships between wave and particle properties for all phenomena. Their importance cannot be overstated.*

It is interesting to note the curious connection between these relationships and the idea of "space-time" in special relativity. Here, momentum is proportional to the spatial frequency and energy to the temporal frequency. In special relativity, the relationship between momentum and energy is essentially identical to that between position and time, with momentum filling the role of position and energy of time. (See Section 1.11.) The connection is not an accident, and to formulate a relativistically correct quantum mechanics, space and time must be treated symmetrically. The topic of **relativistic quantum mechanics** is beyond the scope of the text, but a small piece of the puzzle is discussed in Section 11.2.

Still we have side-stepped the issue of velocity. From our wave-particle relationships,

$$v_{\text{wave}} = f\lambda = \frac{E}{h}\frac{h}{p} = \frac{E}{p}$$

Knowing the particle momentum and energy, we could find the wave velocity, referred to as the **phase velocity.** However, this is not in general equal to the particle velocity, known as **group velocity.** For electromagnetic radiation in a vacuum, the two *are* equal—both electromagnetic particles (photons) and waves move at $c$. But for massive objects, the two are definitely *not* equal. In fact, relating particle velocity to wave properties requires much more sophistication than we have at this point. (It is discussed in Chapter 5.) For the present, we merely stress that $f\lambda \neq v_{\text{particle}}$. Consequently, while for electromagnetic radiation, $E$ can easily be related to $\lambda$ ($E = hf = hc/\lambda$) and $p$ to $f$ ($p = h/\lambda = hf/c$), for massive objects,

$E = hf \neq hv/\lambda$, and $p = h/\lambda \neq hf/v$. The wave-particle relationships fundamental to all phenomena are between momentum and wavelength, and energy and frequency.

## 3.3  The Free-Particle Schrödinger Equation

Let us now take up the all-important question: How do we determine the wave function $\Psi(x, t)$ of a matter wave? In one sense, we do this the same way for all waves; for each kind of wave, there is an underlying **wave equation,** of which the **wave function** must be a solution. To see exactly what this means, it will be helpful to see how it applies to kinds of waves with which we are already familiar.

### Waves on a String

For transverse waves on a stretched string, the wave equation is

$$v^2 \frac{\partial^2 y(x, t)}{\partial x^2} = \frac{\partial^2 y(x, t)}{\partial t^2}$$

where $v$ is the speed of the wave. The wave function (Figure 3.9) is $y(x, t)$, giving the string's transverse displacement as a function of position and time, and it must be a solution of this wave equation. All wave equations are ultimately based upon a fundamental law, and this one may be derived from the second law of motion: $\mathbf{F}_{\text{net}} = m\mathbf{a}.$

The simplest wave to analyze is a sinusoidal, or harmonic, wave. A sinusoidal solution of the above wave equation is

$$y(x, t) = A \sin(kx - \omega t) \quad \text{where} \quad \frac{\omega}{k} = v$$

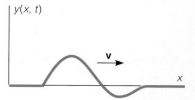

Figure 3.9  A wave disturbance on a string.

### Electromagnetic Waves

As waves on a string must obey a wave equation derivable from the fundamental mechanical law $\mathbf{F} = m\mathbf{a},$ electromagnetic waves must obey wave equations (one for $\mathbf{E}$ and one for $\mathbf{B}$) derivable from the fundamental laws of electricity and magnetism: Maxwell's equations. In a vacuum, with neither charges nor currents present, these are

$$\oint \mathbf{E} \cdot d\mathbf{A} = 0 \tag{3-3a}$$

$$\oint \mathbf{B} \cdot d\mathbf{A} = 0 \tag{3-3b}$$

$$\oint \mathbf{E} \cdot d\boldsymbol{\ell} = -\frac{\partial}{\partial t} \int \mathbf{B} \cdot d\mathbf{A} \tag{3-3c}$$

$$\oint \mathbf{B} \cdot d\boldsymbol{\ell} = \frac{1}{c^2} \frac{\partial}{\partial t} \int \mathbf{E} \cdot d\mathbf{A} \tag{3-3d}$$

Maxwell's equations (in vacuum)

The general wave equations for $\mathbf{E}$ and $\mathbf{B}$ are rather involved. Our main purpose here is to point out *similarities* between wave equations, so we will be content

---

### Amplitude

**U**nfortunately, some ambiguity has crept into the jargon. We say that a plane wave has a constant amplitude, and yet we say that the amplitude of a wave varies with position and time. The first usage of "amplitude" is proper—a wave's amplitude is its *maximum* magnitude (absolute value) as it oscillates. The second usage is loose; rather than saying that "the electric and magnetic fields" or "the displacement of the string" or "the pressure in the air" (for sound) varies with position and time, we use "amplitude" as a generic term for the actual instantaneous value of the thing that is oscillating. Quantum mechanics has not helped the situation: $\Psi(x, t)$ is known as the "probability amplitude," and varies with position and time even for a plane wave whose maximum magnitude is constant. As a rule, the loose usage applies in quantum mechanics: "Amplitude" means the actual instantaneous value of the oscillating thing, and this varies with position and time.

---

with merely noting that the wave equations we derive from Maxwell's equations are partial differential equations.[3] This was the case for waves on a string and will be the case for matter waves. An obvious difference between electromagnetic waves and waves on a string is that electromagnetic waves *by nature* have two parts, **E** and **B,** although Maxwell's equations (3-3c) and (3-3d) couple them together inextricably.

Our simplest solution (i.e., wave) for the string was a sine wave. The three-dimensional analog is a **plane wave** (Figure 3.10). By definition, a plane wave moves in one direction and has a constant amplitude (see the essay entitled "Amplitude"); it doesn't spread out. It is shown in almost all introductory electricity and magnetism texts that the following sinusoidal electromagnetic plane wave is solution of Maxwell's equations:

$$\mathbf{E}(x,\ t) = A\ \sin(kx - \omega t)\hat{\mathbf{y}}$$
$$\mathbf{B}(x,\ t) = \frac{1}{c}A\ \sin(kx - \omega t)\hat{\mathbf{z}} \qquad \text{where} \quad \frac{\omega}{k} = c \qquad (3\text{-}4)$$

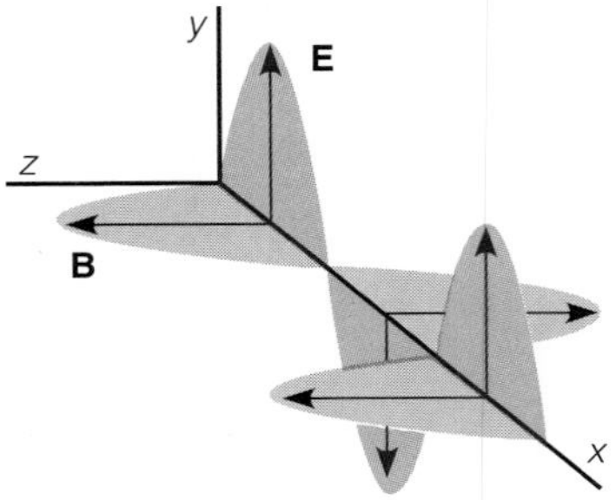

**Figure 3.10**  An electromagnetic plane wave.

## Matter Waves

The wave equation obeyed by matter waves is the **Schrödinger equation.** In this chapter we consider only the special case of "free particles," taking up the effects of external forces in later chapters. In the absence of external forces, the Schrödinger equation is

$$-\frac{\hbar^2}{2m}\frac{\partial^2 \Psi(x,\ t)}{\partial x^2} = i\hbar\frac{\partial \Psi(x,\ t)}{\partial t} \qquad (3\text{-}5)$$

Schrödinger equation
(free particle)

The equation often troubles beginning students of quantum mechanics on two accounts: First, its form is certainly not intuitively obvious, so it is natural to hope for a derivation from first principles. However, the Schrödinger equation *is* the first principle. Classical wave equations rest upon a fundamental assumption or physical law, but there simply is no underlying, more basic principle upon which the Schrödinger equation is built. Its acceptance as, in essence, a "law" arises from the fact that it renders correct predictions about the behavior of matter (probabili-

[3]For a wave moving in the *x*-direction, with **E** along *y* and **B** along *z*, they are:

$$c^2\frac{\partial^2 E_y}{\partial x^2} = \frac{\partial^2 E_y}{\partial t^2} \quad \text{and}$$

$$c^2\frac{\partial^2 B_z}{\partial x^2} = \frac{\partial^2 B_z}{\partial t^2}$$

ties of finding particles, for instance). By analogy, Maxwell's equations, which cannot be derived or proved, will remain accepted laws only so long as we discover no situation in which they are violated. Although the Schrödinger equation cannot be derived, we may argue that it is at least plausible. We will do this shortly, when we discuss the simplest possible matter wave—the plane wave.

The second concern is that the Schrödinger equation is complex (involving $i$, that is, $\sqrt{-1}$). This distresses and alienates some students of quantum mechanics. It might be incorrectly concluded that a matter wave is not "real." Unfortunately, this would also seem to fit perfectly with the point we made earlier—that the wave function is not directly observable. (Why should it be, since it isn't real anyway?) The truth is that matter waves cannot be represented by a single real function; like electromagnetic waves, they *by nature* have two inextricably related parts. And though *quantum mechanics could be formulated without them, complex numbers are the easiest way to handle the two parts—it is simply a matter of convenience.*

Table 3.1 provides some insight. Column 1 shows the equations of electromagnetic waves. Column 2 shows the simplification that results merely by making the definition: $\mathbf{F} \equiv \mathbf{E} + ic\mathbf{B}$. Maxwell's four equations become two, and the plane wave functions for $\mathbf{E}$ and $\mathbf{B}$ become one wave function.[4] This change does not make $\mathbf{E}$ or $\mathbf{B}$ complex; it is merely a mathematical "trick" by which we could handle these two real parts simultaneously. Similarly, column 3 shows the equations with which matter waves could be analyzed completely without complex numbers. The two real waves are $\Psi_1$ and $\Psi_2$, and they obey a pair of real wave equations that link them together. Completing the picture, column 4 shows what happens when we make the definition $\Psi \equiv \Psi_1 + i\Psi_2$: We cut in half both the

[4]The tasks of showing that columns 2 and 1 are equivalent and that columns 3 and 4 are equivalent are left as exercises.

**Table 3.1**

| | Electromagnetic Waves | | Matter Waves | |
|---|---|---|---|---|
| | **1. Real Approach** | **2. Complex Approach** | **3. Real Approach** | **4. Complex Approach** |
| **Wave** | $\left.\begin{array}{c}\mathbf{E}\\\mathbf{B}\end{array}\right\}$ | $\mathbf{F} \equiv \mathbf{E} + ic\mathbf{B}$ | $\left.\begin{array}{c}\Psi_1\\\Psi_2\end{array}\right\}$ | $\Psi \equiv \Psi_1 + i\Psi_2$ |
| | *Maxwell's Equations* | | *Matter Wave Equations* | |
| **Underlying Laws** | $\left.\begin{array}{l}\oint \mathbf{E}\cdot d\mathbf{A} = 0\\\oint \mathbf{B}\cdot d\mathbf{A} = 0\end{array}\right\}$ $\oint \mathbf{E}\cdot d\boldsymbol{\ell} = -\dfrac{\partial}{\partial t}\int \mathbf{B}\cdot d\mathbf{A} \left.\vphantom{\begin{array}{c}a\\b\end{array}}\right\}$ $\oint \mathbf{B}\cdot d\boldsymbol{\ell} = \dfrac{1}{c^2}\dfrac{\partial}{\partial t}\int \mathbf{E}\cdot d\mathbf{A}$ | $\oint \mathbf{F}\cdot d\mathbf{A} = 0$ $\oint \mathbf{F}\cdot d\boldsymbol{\ell} = \dfrac{i}{c}\dfrac{\partial}{\partial t}\int \mathbf{F}\cdot d\mathbf{A}$ | $\left.\begin{array}{l}-\dfrac{\hbar^2}{2m}\dfrac{\partial^2\Psi_1}{\partial x^2} = -\hbar\dfrac{\partial\Psi_2}{\partial t}\\[2mm]-\dfrac{\hbar^2}{2m}\dfrac{\partial^2\Psi_2}{\partial x^2} = \hbar\dfrac{\partial\Psi_1}{\partial t}\end{array}\right\}$ | $-\dfrac{\hbar^2}{2m}\dfrac{\partial^2\Psi}{\partial x^2} = i\hbar\dfrac{\partial\Psi}{\partial t}$ |
| **Plane Wave Function** | $\left.\begin{array}{l}\mathbf{E}(x,t) = A\sin(kx - \omega t)\hat{\mathbf{y}}\\[2mm]\mathbf{B}(x,t) = \dfrac{1}{c}A\sin(kx - \omega t)\hat{\mathbf{z}}\end{array}\right\}$ | $\mathbf{F}(x,t) = A\sin(kx - \omega t)(\hat{\mathbf{y}} + i\hat{\mathbf{z}})$ | $\left.\begin{array}{l}\Psi_1(x,t) = A\cos(kx - \omega t)\\\Psi_2(x,t) = A\sin(kx - \omega t)\end{array}\right\}$ | $\Psi(x,t) = A e^{i(kx - \omega t)}$ |
| **Square of the Wave** $\left(\begin{array}{l}\propto \text{intensity,}\\\text{probability}\end{array}\right)$ | $E^2 + (cB)^2 = |F|^2 = F^*F$ | | $\Psi_1^2 + \Psi_2^2 = |\Psi|^2 = \Psi^*\Psi$ | |

number of wave equations and wave functions. Since electric and magnetic fields do have different "personalities," we keep them separate by taking the real approach to electromagnetic waves. But there is no similar distinction between $\Psi_1$ and $\Psi_2$, so we streamline the math somewhat by taking the complex approach.

It may be convenient, but can we *physically interpret* a complex wave function? There is no need, because the wave function itself cannot be physically detected. What *is* open to experimental scrutiny is its square, the probability density, and this is the same real quantity whether the real or complex approach is taken.

## Probability Density

As we have noted for both electromagnetic radiation and matter, the probability of detecting the particle is proportional to the square of the wave's amplitude. But what does this mean if the wave has two parts (**E** and **B** or $\Psi_1$ and $\Psi_2$)? For electromagnetic waves, experiment verifies that it is $E^2 + (cB)^2$. Notably, the same factor appears in the classical expression for electromagnetic wave intensity—this is the natural way of combining contributions from both **E** and **B**. For massive objects, experiment verifies that the probability of finding the particle depends on the wave function's two parts in a very similar way—a sum of squares:

$$\Psi_1(x,\ t)^2\ +\ \Psi_2(x,\ t)^2\ =\ \Psi^*(x,\ t)\Psi(x,\ t)\ =\ \big|\Psi(x,\ t)\big|^2$$

where $\Psi = \Psi_1 + i\Psi_2$ is a solution of the Schrödinger equation. Note that this is a real, positive quantity, as any probability must be.

Thus far, we have been careful to say that the probability and square of the wave's amplitude are "proportional." Suppose we "look" for the particle within a region of width $\delta$ surrounding point $P$. For a given wave amplitude, there is a given probability of finding the particle in this region. But if $\delta$ is so small that the wave's amplitude is essentially constant within the region, then the probability of finding the particle in a region of width $2\delta$ must be twice as large. Thus, the square of the wave function's amplitude must give a probability *per unit length*. In three dimensions (as we see in Chapter 6), it is probability *per unit volume*. The generic term for probability per unit length/volume is **probability density.**

$$\text{probability density} = \big|\Psi(x,\ t)\big|^2 \tag{3-6}$$

## The Plane Wave

A plane wave solution of the Schrödinger equation is the **complex exponential:**

$$\Psi(x,\ t)\ =\ Ae^{i(kx - \omega t)} \tag{3-7}$$

We can gain important insight into the Schrödinger equation by explicitly verifying that the complex exponential is a solution of equation (3-5). The question is

$$-\frac{\hbar^2}{2m}\frac{\partial^2 Ae^{i(kx-\omega t)}}{\partial x^2} \overset{?}{=} i\hbar\,\frac{\partial Ae^{i(kx-\omega t)}}{\partial t}$$

Taking the partial derivatives on both sides, we have

$$-\frac{\hbar^2}{2m}(ik)^2 A e^{i(kx-\omega t)} = i\hbar(-i\omega)A e^{i(kx-\omega t)}$$

and canceling,

$$\frac{\hbar^2 k^2}{2m} = \hbar\omega \qquad\qquad\qquad\qquad (3\text{-}8\text{a})$$

The functional dependence on position and time cancels, so (3-7) obeys the Schrödinger equation for all values of $x$ and $t$, provided $k$ and $\omega$ are related as in (3-8a). The reader is strongly encouraged to verify that this is not the case for the familiar plane waves: $A \sin(kx - \omega t)$ and $A \cos(kx - \omega t)$ (appropriate for waves on a string, sound waves, etc.). They are *not* solutions of the Schrödinger equation. *The fundamental matter wave is a complex exponential.* Because $A e^{i(kx-\omega t)} = A \cos(kx - \omega t) + iA \sin(kx - \omega t)$, there is a certain similarity between the complex exponential and the more familiar plane waves. But the complex exponential has two parts, and they are "out of phase" by one-quarter cycle.

The requirement that (3-8a) hold gives us an important link between the Schrödinger *wave* equation and the *particle* it is supposed to represent. Using $p = \hbar k$ and $E = \hbar\omega$, condition (3-8a) becomes

$$\frac{p^2}{2m} = E \qquad\qquad\qquad\qquad (3\text{-}8\text{b})$$

Since $p^2/2m$ is the kinetic energy,[5] this "condition" merely requires that kinetic energy and total energy be equal, precisely what is true for the associated particle. (A "free particle," subject to no external force, has no potential energy.[6]) We see then that *the Schrödinger equation is based upon energy.*

Figure 3.11 may help in visualizing the plane matter wave:

$$\Psi(x, t) = \begin{cases} A e^{i(kx-\omega t)} \\ A \cos(kx - \omega t) + iA \sin(kx - \omega t) \\ \Psi_1(x, t) + i\Psi_2(x, t) \end{cases}$$

At $t = 0$, the real part $\Psi_1$ is a cosine function and the imaginary part $\Psi_2$ a sine. (*Always bear in mind: $e^{ir}$*, where $r$ is real, is an *oscillatory* function.) It should not be concluded from Figure 3.11 that there is something about the wave function that swirls in space. The axes other than the $x$-axis are not the $y$ and $z$, and do not point in any direction in space; the wave function is not a vector. The diagram is simply meant to show the two-part nature of the wave function, and that its two parts are out of phase in such a way that its *magnitude* (absolute value) is *constant*—it varies neither in position nor time.

$$|\Psi(x, t)|^2 = \Psi^*(x, t)\Psi(x, t) = [A e^{-i(kx-\omega t)}][A e^{+i(kx-\omega t)}] = A^2$$

$$|\Psi(x, t)| = A$$

That the wave's magnitude is a constant means that the probability per unit length is constant. Were we to "look" for it, *a particle represented by a plane wave would be equally likely to be found anywhere!*

[5] In quantum mechanics, $p^2/2m$ is used *much* more often than $\frac{1}{2}mv^2$.

$$\frac{1}{2}mv^2 = \frac{(mv)^2}{2m} = \frac{p^2}{2m}$$

[6] A massive particle does possess *internal* energy. But unless it were struck or otherwise affected by something external, its internal energy wouldn't change. The inclusion of a constant term to account for internal energy would not really change the arguments here; it would serve only to complicate the math. In nonrelativistic quantum mechanics, as in nonrelativistic classical mechanics, the essentially constant internal energy of an object may be largely ignored.

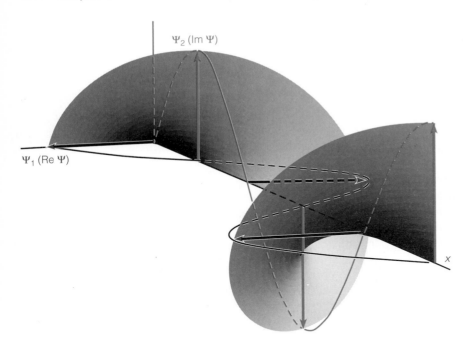

$\Psi_2$ (Im $\Psi$)

$\Psi_1$ (Re $\Psi$)

$x$

**Figure 3.11** A plane matter wave: $Ae^{i(kx-\omega t)}$.

Summarizing what we have learned from the plane-wave solution of the free-particle Schrödinger equation:

The Schrödinger equation is based upon energy. The simplest solution is a plane wave, a complex exponential with two sinusoidal parts. The wave's magnitude and the probability density vary neither in time nor in position.

**Example 3.3**

A certain free particle is represented by the wave function

$$\Psi(x,\, t) = Ae^{i[(1.58\times10^{12}\ \mathrm{m}^{-1})x - (7.91\times10^{16}\ \mathrm{s}^{-1})t]}$$

Determine the momentum, energy, and mass of the particle.

**Solution**

From the fundamental wave-particle relationships,

$$p = \hbar k = (1.055\times10^{-34}\ \mathrm{J\cdot s})(1.58\times10^{12}\ \mathrm{m}^{-1})$$

$$= 1.67\times10^{-22}\ \mathrm{kg\cdot m/s}$$

$$E = \hbar\omega = (1.055\times10^{-34}\ \mathrm{J\cdot s})(7.91\times10^{16}\ \mathrm{s}^{-1})$$

$$= 8.35\times10^{-18}\ \mathrm{J}$$

Knowing the momentum and the energy (which is solely kinetic), we may find the mass directly from the energy relationship (3-8b).

$$E = \frac{p^2}{2m} \rightarrow 8.35 \times 10^{-18} \text{ J} = \frac{(1.67 \times 10^{-22} \text{ kg·m/s})^2}{2m}$$

$$\Rightarrow \quad m = 1.67 \times 10^{-27} \text{ kg}$$

(*Note:* We find a wave, or "phase," velocity via $\lambda f$. The values in the example would give us $f\lambda = \omega/k = E/p = 5 \times 10^4$ m/s. This is not the particle's velocity. Dividing $p$ by $m$ gives $10^5$ m/s, which is correct, but the proper way of finding a particle, or "group," velocity is really more involved. These ideas are discussed fully in Chapter 5.)

Representing a particle equally likely to be found anywhere, the plane wave might seem unrealistic. Indeed, a true plane wave is always a theoretical ideal, but it is still very useful. In physical optics, we often speak of plane waves of light, because they are easily analyzed and are a sufficiently good approximation of the true wave. The same is true of matter waves. But even when the wave is more complicated, the plane wave is important, for a complicated wave can always be treated as an algebraic sum of plane waves; plane waves are easily analyzed "building blocks." We discuss specific wave functions of more complicated matter waves in later chapters. Now, however, we investigate an important phenomenon where we need not know the specific wave function. Its understanding is based in part on the familiar wave behavior of single-slit diffraction, which occurs the same way for all kinds of waves—light, sound, matter waves, and so forth.

## 3.4 The Uncertainty Principle

In Section 2.5, we noted that the mere fact that a phenomenon has a wave nature (whether obvious or not) implies inherent uncertainties in its particle properties. Passing through a single slit causes an electromagnetic wave to spread out, and so must also cause uncertainty in the momenta of the particles detected afterward. As shown in Figure 3.12, the same thing occurs when an electron matter wave passes through a slit.[7]

But what do we mean by "uncertainty" in momentum? We mean that if the experiment were repeated many times *identically,* the momentum detected after passing through the slit would still vary over a range of values. The question then is, How do we quantify uncertainty? Suppose that in a single-slit experiment the $p_x$-values found as electrons reach the screen fall within the range $-1$ kg·m/s to $+1$ kg·m/s, except for one value of 500 kg·m/s. What value do we assign to the uncertainty? 1 kg·m/s? 250 kg·m/s? 501 kg·m/s?

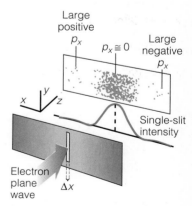

Figure 3.12 Single-slit diffraction of an electron matter wave.

---

[7]We now choose the wave to be moving in the $z$-direction, because we prefer to use $x$ for the focus of attention, which is now *perpendicular* to the direction of motion.

The definition of uncertainty is always an arbitrary choice, but it obviously should measure how far deviations are from some mean (average) value. In physics, uncertainty is defined as **standard deviation.** For example, suppose repeated experiments are carried out to determine a quantity $Q$, where $Q$ might represent position, charge, $x$-component of momentum, or any other measurable quantity. The value $Q_1$ is obtained $n_1$ times, the value $Q_2$ is obtained $n_2$ times, and so on. The **mean** $\bar{Q}$ is found by multiplying a value $Q_i$ by the number of times $n_i$ that value is obtained, summing over all values, then dividing by the total number of values obtained. The standard deviation $\Delta Q$ is the square root of the mean of the *squares of the deviations* of the values from the mean value.[8] It is a good definition because it is the simplest possible one that is zero *if and only if* all $Q_i$ values are equal (necessarily to the mean $\bar{Q}$), and increases as deviations increase.

We shall discuss the use of this definition later. For now, we simply note that as the electrons are detected one at a time in the single-slit experiment, there is an uncertainty—a nonzero standard deviation—in $p_x$. On the other hand, were we to conduct a different experiment, to find exactly where within the width $\Delta x$ a given electron passed through the slit, there would also be uncertainty. The electron wave front is spread over the entire width $\Delta x$, so there would be a probability of finding the electron anywhere within. There would be an uncertainty in $x$, and the narrower the slit, the smaller would be this uncertainty. Above all, there is a link between the uncertainty in $p_x$ and the uncertainty in $x$: The width of a diffraction pattern is inversely proportional to the slit width, so the uncertainty $\Delta p_x$ in the electron's $x$-component of momentum is inversely proportional to the uncertainty $\Delta x$ in the electron's position.

$$\Delta p_x \propto \frac{1}{\Delta x}$$

The specific experiment (i.e., the single slit) is not to "blame" for this state of affairs. On the contrary, it is a direct consequence of matter's wave nature, that increased precision in the knowledge of position implies decreased precision in the knowledge of momentum, and vice versa. Figure 3.13 provides a simplified,

Mean: $\bar{Q} = \dfrac{\Sigma_i Q_i n_i}{\Sigma_i n_i}$

Standard deviation:

$$\Delta Q = \sqrt{\frac{\Sigma_i (Q_i - \bar{Q})^2 n_i}{\Sigma_i n_i}}$$

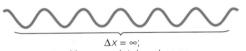

$\Delta x = \infty$;
position completely unknown;
$\lambda$ and $p$ well defined

$\Delta x$ finite;
position better known;
$\lambda$ and $p$ less well defined

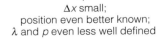

$\Delta x$ small;
position even better known;
$\lambda$ and $p$ even less well defined

**Figure 3.13** As a wave becomes more compact, its overall wavelength and momentum are less well defined.

---

[8] Standard deviation is sometimes referred to by the more descriptive term "rms deviation." The "rms" has the same meaning as in alternating current ($I_{\mathrm{rms}}$) and molecular speeds ($v_{\mathrm{rms}}$).

qualitative argument. The top wave is infinite and regular. Though there is no doubt of its wavelength, there would be a probability of finding the particle anywhere. Wavelength and thus momentum $h/\lambda$ are certain ($\Delta p = 0$), but the position where the particle would be found is completely unknown ($\Delta x = \infty$). The center wave is regular over only a finite region. The cost of obtaining a wave for which the probable whereabouts of the particle are narrowed down ($\Delta x \neq \infty$) is that the wave is not regular everywhere. In any fair way of taking into account all of space, the wavelength cannot be said to be simply $\lambda$, nor the momentum $h/\lambda$ ($\Delta p \neq 0$). The bottom wave gives an even better known position, but only by further restricting the region over which the wave is regular. Accordingly, it is even less fair to say that the wavelength of this wave as a whole is $\lambda$, so $\Delta p$ is larger still. The rigorous proof that $\Delta p$ is inversely proportional to $\Delta x$ is a topic in itself (see Section 3.6). Here we concern ourselves only with the conclusion, known as the uncertainty principle, and its ramifications.

> If a phenomenon has a wave nature, it is theoretically impossible to know precisely the position along an axis and the component of momentum along that axis *simultaneously;* $\Delta x$ and $\Delta p_x$ cannot be simultaneously zero. Rather, there is a strict theoretical lower limit on their product:
>
> $$\Delta p_x \Delta x \geq \frac{\hbar}{2} \tag{3-9}$$

Momentum-position uncertainty principle

The uncertainty principle was discovered by Werner Heisenberg (Nobel prize 1932)[9]. It is a shocking revelation. There exists a *theoretical limit* on the precision with which some dynamical quantities can be known simultaneously. If we know position *exactly* ($\Delta x = 0$), *nothing* is known about momentum ($\Delta p_x = \infty$). If momentum is known exactly, position is completely unknown. The plane wave is a good example. This fundamental matter wave has a precise wave number $k$ and momentum $p = \hbar k$, but represents a particle that is equally as likely to be found anywhere. A property in which there is no uncertainty is said to be **well defined.** Thus, for the plane wave, momentum is well defined ($\Delta p_x = 0$), but position clearly is not ($\Delta x = \infty$).

Well defined: No uncertainty

Do not be troubled by the inequality in (3-9)—there is no uncertainty about the uncertainty principle. The "$\geq$" reflects the fact that there is a particular wave shape, called a **gaussian,** for which the product of uncertainties is a minimum, $\frac{1}{2}\hbar$. If a particle's matter wave were of this shape, its position and momentum would be known simultaneously as precisely as theoretically possible. For any other shape, simultaneous knowledge of the two is less precise: $\Delta p_x \Delta x > \frac{1}{2}\hbar$. The gaussian wave form (at $t = 0$) is shown in Figure 3.14. It contains a periodic $\cos(kx)$,

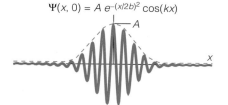

$\Psi(x, 0) = A\, e^{-(x/2b)^2} \cos(kx)$

**Figure 3.14** Gaussian wave form ($\Delta x\, \Delta p = \frac{1}{2}\hbar$).

[9]In recognition, it is often referred to as the "Heisenberg uncertainty principle."

but it is modulated by the "gaussian factor" $e^{-(x/2b)^2}$, where $b$ is a constant. The gaussian factor is maximum at $x = 0$, falls off toward zero symmetrically as $x$ becomes large, and the rate of fall-off depends on $b$. If $b$ is large the wave form is broad, falling off very slowly, while if $b$ is small, the wave form is narrow.

The whole idea behind the uncertainty principle is rather upsetting to a student of classical physics. Classically, we claim that we can calculate a particle's position and velocity for all time, via $\mathbf{F} = m\mathbf{a}$ and kinematics. We need only know the forces acting and the initial position and velocity. But now we see that even starting such a calculation is problematic, for it is theoretically impossible to know initial position and velocity simultaneously! Fortunately, as we shall soon see, the uncertainty principle is of little consequence for "large" things.

On the other hand, it is of great consequence for the small things we study in quantum mechanics. As we saw in Example 3.1, while wavelengths of macroscopic objects are often ridiculously small, an electron might well have a wavelength measured in nanometers or larger. Undoubtedly it would behave as a particle in a situation where distances are measured in meters, as in a computer monitor's cathode-ray tube. But it should definitely show its wave side when confined within a system measured in fractions of a nanometer. Such a system is the atom, probably the most logical test of quantum mechanics, and the simplest atom is hydrogen—essentially an electron orbiting a stationary proton. In such small confines, the electron must be treated not as an orbiting particle but as a bound three-dimensional wave surrounding the proton. Because the wave is spread diffusely, the probability of finding the electron is spread diffusely. Our knowledge of the atom's approximate size sets a rather small *maximum* possible value for the position uncertainty, and there is correspondingly a significant *minimum* theoretical uncertainty in momentum.

---

### Example 3.4

Much evidence indicates that the hydrogen atom is about 0.1 nm in radius. That is, the electron's orbit (regardless of whether it is circular) extends to about this far from the proton. Accordingly, the uncertainty in the electron's position is no larger than about 0.1 nm. What is the minimum theoretical uncertainty in its velocity?

(*Note:* An electron in an atom moves in three dimensions; the values in the example apply to the component of its motion along a given axis.)

#### Solution

$$\Delta p_x \, \Delta x \geq \frac{\hbar}{2} \quad \rightarrow \quad \Delta p_x (0.1 \times 10^{-9} \text{ m}) \geq \frac{1.055 \times 10^{-34} \text{ J·s}}{2}$$

$$\Rightarrow \quad \Delta p_x \geq 5.3 \times 10^{-25} \text{ kg·m/s}$$

Now using $p = mv$,

$$\Delta v_x = \frac{\Delta p_x}{m} \geq \frac{5.3 \times 10^{-25} \text{ kg·m/s}}{9.11 \times 10^{-31} \text{ kg}} = 5.8 \times 10^5 \text{ m/s}$$

From Example 3.4, we learn that, whatever might be the nature of the electron's orbit (circular, elliptical, etc.), repeated experiments dedicated to determining its speed must find a range of values covering more than $\frac{1}{10}$% of $c$! Indeed, careful study of the hydrogen atom reveals slight relativistic effects. Actually, the theoretical minimum $\Delta p_x$ of $5.3 \times 10^{-25}$ kg·m/s would apply only if the wave function were a gaussian, which it is not, so the true uncertainty is somewhat larger. However, to be too concerned with this point is to overlook much of the power of the uncertainty principle: It governs all wave phenomena, regardless of their specific form, and so may be used for order-of-magnitude calculations *in complete ignorance of the wave function*. In Example 3.6, we shall see just how useful this can be. First, however, let us return to the problem of reconciling the uncertainty principle with classical mechanics.

## The Classical Limit

The uncertainty principle places no significant limitation on the use of classical mechanics in classical situations, for the simple reason that $\hbar$ is small. In the following example, we verify this and at the same time see that the uncertainty principle is the way out of what would otherwise be a very disturbing conclusion: Any *stationary* particle, with a corresponding infinite wavelength, should behave as a wave!

### Example 3.5

A stationary 1-mg grain of sand is found to be at a given location within an uncertainty of 550 nm. (a) What is the minimum uncertainty in its velocity? (b) Were it moving at this speed, how long would it take to travel 1 $\mu$m? (c) Can classical mechanics be applied reliably? (d) What is a reasonable wavelength of the grain of sand, and will it behave as a wave, or as a particle?

(*Note:* From physical optics, we know that we cannot resolve details smaller than the wavelength of the light used to view them. Thus, 550 nm, the middle of the visible spectrum, is a reasonable limit on the precision with which location could be established by an experiment using light, such as a simple visual inspection.)

### Solution

(a) $\Delta p_x \Delta x \geq \dfrac{\hbar}{2} \rightarrow \Delta p_x(5.5 \times 10^{-7} \text{ m}) \geq \dfrac{1.055 \times 10^{-34} \text{ J·s}}{2}$

$\Rightarrow \Delta p_x \geq 9.59 \times 10^{-29}$ kg·m/s

In essence, $\Delta p_x$ is small because $\hbar$ is small.

$\Delta v = \dfrac{\Delta p}{m} \geq \dfrac{9.59 \times 10^{-29} \text{ kg·m/s}}{10^{-6} \text{ kg}} = 9.59 \times 10^{-23}$ m/s

It is a relief to find that this $\Delta v$ is much smaller than the $\sim10^6$ m/s of the electron in Example 3.4. The disparity is due mostly to the disparity in the particles' masses. Quantum-mechanically speaking, a grain of sand is huge.

(b) The grain of sand is within about $10^{-22}$ m/s of stationary. Were it moving at the greatest speed that might be expected, it would travel 1 $\mu$m in

$$t = \frac{\text{distance}}{\text{speed}} = \frac{10^{-6} \text{ m}}{9.59 \times 10^{-23} \text{ m/s}} = 1.04 \times 10^{16} \text{ s}$$

$$= 3.3 \text{ million centuries}$$

(c) Never is precision as fine as $10^{-22}$ m/s required in classical physics. Indeed, it is unrealistic to expect that we would, or even could, determine a speed with such precision. Thus, the uncertainty principle places no significant limitation on the use of classical mechanics for macroscopic (as opposed to microscopic) objects. Both position and velocity may be known with much finer precision than needed.

(d) What is a reasonable *momentum?* We have found that the object's momentum may take on any of a range of values within at least $10^{-28}$ kg·m/s of zero. *It is therefore extremely unlikely that the momentum would be zero— we cannot know that the object is stationary!* Suppose its momentum is the largest consistent with the uncertainty.

$$\lambda = \frac{h}{p} = \frac{6.63 \times 10^{-34} \text{ J·s}}{9.59 \times 10^{-29} \text{ kg·m/s}} = 6.9 \times 10^{-6} \text{ m}$$

A wavelength of 7 $\mu$m is smaller than most apparatus we might use to study macroscopic objects. Even if the grain of sand's momentum happened to be only one-hundredth of its largest possible value, its wavelength would be only about 0.7 mm, also rather small. Still, these wavelengths are not so small as to make us completely confident that the grain of sand should behave as a particle. However, we have considered only the *theoretical minimum* momentum uncertainty. It is ridiculously small (classically speaking) and would in any case apply only for a gaussian wave function. No technique of classical mechanics applied to macroscopic objects could verify this tiny uncertainty.

Suppose, then, that the momentum uncertainty we are able to verify is a less unrealistic value—that the velocity is found to be zero within an uncertainty of 1 mm per century.

$$\Delta p = m \, \Delta v = (10^{-6} \text{ kg}) \left( \frac{10^{-3} \text{ m}}{100 \text{ yr} \times 3.16 \times 10^7 \text{ s/yr}} \right)$$

$$= 3.2 \times 10^{-19} \text{ kg·m/s}$$

Might the grain of sand have a momentum small enough, a wavelength long enough, to behave as a wave in some cases? Even if its momentum were as small as a millionth of its maximum possible value, its wavelength would be

$$\lambda = \frac{h}{p} = \frac{6.63 \times 10^{-34} \text{ J·s}}{\frac{1}{1,000,000} \times 10^{-19} \text{ kg·m/s}} \cong 7 \times 10^{-9} \text{ m}$$

In no case in which we would ever conceive of using classical mechanics would the grain of sand behave as a wave.

At $\sim 10^{-34}$ J·s, Planck's constant is certainly small by classical standards, and this alone is responsible for several classical observations: (1) Electromagnetic radiation studied long ago tended to be of fairly long wavelength. With $h$ being so small, $p = h/\lambda$ would be very small. Classically, with long wavelengths and small particle momenta, electromagnetic radiation is a wave. (2) Macroscopic massive objects, with large mass, tend to have large particle momenta. The small value of $h$ then ensures that $\lambda = h/p$ is minute. Classically, massive objects are particles. (As we are learning, however, *microscopic* massive objects are a different matter.) (3) Because $h$ is so small, the link between uncertainties in particle properties is invisible. For all classical purposes, $\Delta x$ and $\Delta p$ may be zero simultaneously.

## A Practical Application

In Chapter 6, we discuss electron wave functions in simple atoms, but the uncertainty principle alone explains a characteristic of the atom for which classical arguments fail: Classically, there is no lower limit on the energy a small particle may have as it orbits a large body. For instance, a satellite may be positioned at any distance from a planet, and, if given the proper velocity, will maintain a circular orbit. The smaller the orbit radius, the lower would be the energy, though there is *practical* lower limit in which the satellite simply rests on the planet's surface. By analogy, if an electron orbits a proton in a hydrogen atom, there should be no lower limit on the energy it may have. But there *is* a minimum energy, called the "ground-state" energy, and it is inconsistent with the electron simply "resting on the proton." The electron's wave nature, specifically the inverse relation between the uncertainties in momentum and position, is the answer to the mystery.

---

### Example 3.6

An electron is held in orbit about a proton by electrostatic attraction.

(a) Assume that an "orbiting electron wave" has the same energy an orbiting particle would have if at radius $r$ and of momentum $mv$. Write an expression for this energy.

(b) If the electron behaves as a classical particle, it must obey $F = ma$. Assuming circular orbit, apply $F = ma$ to eliminate $v$ in favor of $r$ in the energy expression.

(c) Suppose instead that the electron is an orbiting wave, and that the product of the uncertainties in radius $r$ and momentum $p$ is governed by an uncertainty relation of the form: $\Delta p \, \Delta r \approx \hbar$. Also assume that a typical radius of this orbiting wave is roughly equal to the uncertainty $\Delta r$, and that a typical magnitude of the momentum is roughly equal to the uncertainty $\Delta p$,[10] so that the uncertainty relation becomes $pr \approx \hbar$. Use this to eliminate $v$ in favor of $r$ in the energy expression.

(d) Sketch on the same graph the expressions from parts (b) and (c).

(e) Find the minimum possible energy for the orbiting electron wave, and the value of $r$ to which it corresponds.

[10]These are reasonable assumptions. Both position and momentum *vectors* must be zero on average, since the electron is bound to the proton/origin and it goes in no one direction consistently. The magnitudes of $r$ and $p$, therefore, should be comparable to their respective fluctuations about zero—that is, their uncertainties.

Solution

(a) Kinetic energy is $\frac{1}{2}mv^2$ and electrostatic potential energy is $(1/4\pi\epsilon_0)[(+e)(-e)/r]$

$$E = \frac{1}{2}mv^2 - \frac{1}{4\pi\epsilon_0}\frac{e^2}{r}$$

(b) The electrostatic force is $(1/4\pi\epsilon_0)(e^2/r^2)$, and the acceleration of a particle in circular motion is $v^2/r$.

$$F = ma \quad \rightarrow \quad \frac{1}{4\pi\epsilon_0}\frac{e^2}{r^2} = m\frac{v^2}{r} \quad \Rightarrow \quad v^2 = \frac{e^2}{4\pi\epsilon_0 mr}$$

Thus, the energy may be written

$$E_{\substack{\text{classical} \\ \text{particle}}} = \frac{1}{2}m\left(\frac{e^2}{4\pi\epsilon_0 mr}\right) - \frac{1}{4\pi\epsilon_0}\frac{e^2}{r} = \underbrace{\frac{e^2}{8\pi\epsilon_0 r}}_{\text{kinetic}} - \underbrace{\frac{e^2}{4\pi\epsilon_0 r}}_{\text{potential}} = -\frac{e^2}{8\pi\epsilon_0 r}$$

The negative electrostatic potential energy is always of greater magnitude than the positive kinetic, so the total energy strictly decreases as $r$ decreases—there is no minimum energy.

(c) Assuming $pr = \hbar$, we have $p = \hbar/r$ or $v = \hbar/mr$. Thus,

$$E_{\substack{\text{matter} \\ \text{wave}}} = \frac{1}{2}m\left(\frac{\hbar}{mr}\right)^2 - \frac{1}{4\pi\epsilon_0}\frac{e^2}{r} = \frac{\hbar^2}{2mr^2} - \frac{e^2}{4\pi\epsilon_0 r}$$

In this case, as $r$ decreases and the wave becomes more compact, the likely speed increases inversely. The kinetic energy increases faster than the potential decreases, and the total energy at some point must increase.

(d) The two plots are shown in Figure 3.15. While the energy of a classical particle would monotonically decrease as $r$ decreases, the energy of the matter wave reaches a minimum then increases.

(e) To find the minimum, we set the derivative to zero.

$$\frac{dE_{\substack{\text{matter} \\ \text{wave}}}}{dr} = -\frac{\hbar^2}{mr^3} + \frac{e^2}{4\pi\epsilon_0 r^2} = 0$$

$$\Rightarrow \quad r = \frac{4\pi\epsilon_0\hbar^2}{me^2} = \frac{4\pi(8.85 \times 10^{-12} \text{ N} \cdot \text{m}^2/\text{C}^2)(1.055 \times 10^{-34} \text{ J·s})^2}{(9.11 \times 10^{-31} \text{ kg})(1.6 \times 10^{-19} \text{ C})^2}$$

$$= 5.3 \times 10^{-11} \text{ m}$$

Reinserting,

$$E_{\substack{\text{matter} \\ \text{wave}}} = \frac{\hbar^2}{2m}\left(\frac{me^2}{4\pi\epsilon_0\hbar^2}\right)^2 - \frac{e^2}{4\pi\epsilon_0}\left(\frac{me^2}{4\pi\epsilon_0\hbar^2}\right) = -\frac{me^4}{32\pi^2\epsilon_0^2\hbar^2}$$

$$= -\frac{(9.11 \times 10^{-31} \text{ kg})(1.6 \times 10^{-19} \text{ C})^4}{32\pi^2(8.85 \times 10^{-12} \text{ N·m}^2/\text{C}^2)^2(1.055 \times 10^{-34} \text{ J·s})^2}$$

$$= -2.2 \times 10^{-18} \text{ J} = -13.6 \text{ eV}$$

The energy happens to equal the correct, experimentally determined value, and the radius is indeed the most probable radius at which the electron would be found. That these agree *so closely* is an accident; many approximations

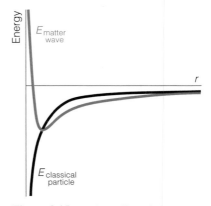

**Figure 3.15** As the radius of an orbiting matter wave approaches zero, its momentum uncertainty and thus kinetic energy approach infinity.

have been made. Nevertheless, the uncertainty principle does impose a lower limit on the energy, and it is no accident that the value we obtained is of the correct order of magnitude.

## The Uncertainty Principle in Three Dimensions

So far, we have really only discussed the uncertainty principle in one dimension. The only hitch in three dimensions is relating momentum, a vector, to wavelength, a scalar. However, since $\mathbf{p} = h/\lambda\,\hat{\mathbf{v}} = \hbar(2\pi/\lambda)\hat{\mathbf{v}}$, where $\hat{\mathbf{v}}$ is a unit vector in the wave's direction of motion, we may define a vector wave number $\mathbf{k} \equiv (2\pi/\lambda)\hat{\mathbf{v}}$, and the fundamental wave-particle relationship is then $\mathbf{p} = \hbar\mathbf{k}$. As we know, wave number is a spatial frequency, so the three-dimensional generalization of the uncertainty principle is quite sensible: The more compact a wave is *along a given axis,* the less definite we can be about its spatial frequency and thus momentum component along that axis (cf. Figure 3.13). Accordingly, in three dimensions we have three uncertainty relations:

$$\Delta p_x\,\Delta x \geq \frac{\hbar}{2} \qquad \Delta p_y\,\Delta y \geq \frac{\hbar}{2} \qquad \Delta p_z\,\Delta z \geq \frac{\hbar}{2}$$

An uncertainty relation holds only between the width of a wave along an axis and the component of momentum *along that axis.* Therefore, it is theoretically possible for the $x$-component of position and $y$-component of momentum of a particle to be precisely known simultaneously (i.e., $\Delta x$ and $\Delta p_y$ may both be zero). The independence of the three axes explains why in Figure 3.12 the passage of the wave through the slit, narrow along the x-axis, affects only $p_x$, not $p_y$ or $p_z$.

## The Energy-Time Uncertainty Principle

The root of the momentum-position uncertainty principle is that as an oscillatory function of position becomes more compact, its wavelength (spatial period) becomes less well defined. It follows *mathematically* that as an oscillatory function of *time* becomes more "compact," its (temporal) period becomes less well defined. But period is related to energy precisely as wavelength is to momentum ($E = h/T$ and $p = h/\lambda$). Apparently, a *physical* consequence is

$$\Delta E\,\Delta t \geq \frac{\hbar}{2} \tag{3-10}$$

Energy-time uncertainty principle

How do we interpret this? If a wave function oscillating in time were to exist, or merely to be subject to study, for a *limited* span of time, its energy would be uncertain. A good example is an atomic electron undergoing a transition from a higher to a lower energy state. During the transition, the electron's wave function temporarily oscillates in an unusual way, generating a photon that carries away the energy difference. Since an electron somehow gotten into the higher-energy state drops to the lower-energy one in a finite time interval $\Delta t$, the energy given to the photon is uncertain by an amount $\Delta E$, inversely related to $\Delta t$. The mathematical basis of both uncertainty principles is discussed in Section 3.6.

## 3.5 The Not-Unseen Observer

Let us spend a little time summarizing the limitations quantum mechanics places on our knowledge. Loosely speaking, an object small enough to require a quantum mechanical analysis of its behavior is too small to be unaffected by an observation of it. The observer alters the very behavior he wishes to observe. Even if the observer employs only a single photon of light in his observation, it is likely to have a significant effect on the object of study, if that object is very small.

Suppose the object of our study is a single electron, about which we wish to learn as much as possible. There is no way to do so without interacting with it in some way via a mutual force. We study the effect of interactions/forces in Chapter 4. Here we need only note that such an interaction would affect the electron's wave function *in some way*.[11] In an attempt to determine the location of the electron, we choose to bounce light off it. Wishing to affect the electron as little as possible, we reduce the intensity to the point that the light strikes the electron one photon at a time. In physical optics, we learn that resolution increases with decreasing wavelength. Therefore, photons of smaller wavelength would provide better resolution in locating the electron. However, photons of smaller wavelength have greater momentum ($p = h/\lambda$), and accordingly may impart a larger momentum to the electron. (There is no phenomenon for which this fundamental relationship does not hold, so the same difficulty arises no matter what one "throws" at the electron.) Moreover, since our knowledge of the photon's motion after the collision is necessarily blurred by diffraction when we detect it, we cannot be certain how much of its momentum it transferred to the electron. Thus, the more precisely we attempt to localize the electron, the more uncertain will be its momentum. *This is the uncertainty principle!* Not only is a wave function of simultaneously precise momentum and position a *theoretical* impossibility, but we now see how *any experiment* intended to minimize the uncertainty in position would necessarily alter the electron's wave function so that there is large uncertainty in its momentum.

We learn what we may about a particle by interacting with it. Given knowledge of the forces involved, the Schrödinger equation may in principle be solved for the wave function, which contains all information that may be known. But we see that there are limitations on what may be known *simultaneously*.

### What *Can* Be Known?

Suppose we have carried out an experiment, experiment A, and have determined both the position and momentum of an electron as precisely as possible: $\Delta x \, \Delta p = \hbar/2$. We have caused the wave function to be a gaussian. Let us assume that $\Delta x$ is 1 $\mu$m, implying that $\Delta p = \hbar/(2 \ \mu\text{m})$, and refer to this wave function as $\Psi_A$. We have found the wave function, yet we are unsatisfied, for we have not really "found" the electron (its "location"); we have found only the mysterious "probability amplitude."

We conduct another experiment, experiment B, in which the electron registers its presence at a detector at a definite location. We rejoice—we have found the electron. If causing a detector to respond is our criterion, then we have found the electron *by definition*. However, there are no "point detectors." If the width of our detector is smaller than the 1-$\mu$m uncertainty in position of $\Psi_A$, then we have

[11] This is not contrary to the claim that the wave function cannot be directly observed. There is a difference between an oscillating wave simply interacting with something, and actually being able to isolate and control the thing that oscillates. We may produce constant electric and magnetic fields and directly observe their effects, but we cannot produce and observe the effects of a constant $\Psi$.

indeed narrowed down the possible locations, but we have not established the electron's location with complete certainty. We have reduced, not eliminated, the uncertainty in position, and if this is so, experiment B has changed the wave function—increased $\Delta p$, if nothing else.

Were we to repeat this pair of experiments many times, experiment A to establish an initial wave function and experiment B to "find" the electron, experiment B would find the electron at various locations within the $1$-$\mu$m uncertainty of wave function $\Psi_A$, and the number detected at a given location would be proportional to $|\Psi_A|^2$. In essence, we would simply verify that $|\Psi_A|^2$ is proportional to the probability of finding the electron. But since experiment B changes the wave function, it is not possible to "watch," to repeatedly find, the *same* electron while preserving a single wave function $\Psi_A$.

The double-slit experiment (Figure 3.16) is a good example of these ideas. In effect, the slits are an experiment A, establishing a wave function $\Psi_A$ beyond the slits, and experiment B is the detection of an electron at a point beyond the slits. By sending in a beam of electrons one at a time, we are carrying out experiment A then experiment B repeatedly. Where $\Psi_A$ is large, experiment B will register electrons abundantly; where $\Psi_A$ is zero, experiment B will register no electron. We cannot conduct an intermediate experiment, determining which slit a given electron passes through, and yet hope to observe the interference pattern exhibited by $\Psi_A$, for this intermediate experiment would itself alter the wave function. To observe interference at the screen, we must allow each electron's wave function to pass through both slits simultaneously, or there would not be two coherent waves to interfere.

"What can be known" are probabilities and corresponding uncertainties based upon the most recent observation of the "particle"—that is, the most recent determination of the wave function.

### An Interesting Observation

The discussion raises an interesting point: If we cannot know the location of a particle *until* we actually look for it, it is rather an act of faith to claim that it even *has* a location before we look for it. Early in the "quantum age," many eminent physicists, most notably Einstein himself, asserted that theories of wave-particle duality must be incomplete, that some modification is necessary to make allowance for particles having definite locations at all times. However, although the

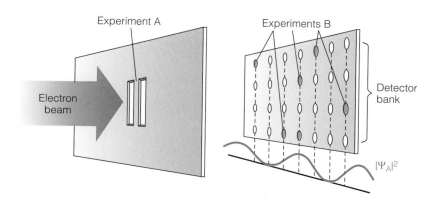

Figure 3.16 Experiment A establishes $\Psi_A$, which experiment B verifies.

theoretical arguments and experimental tests are unfortunately beyond the scope of the text, the strong consensus now is that a particle indeed does not have a location until it is explicitly looked for.

# ◆ 3.6 Mathematical Basis of the Uncertainty Principle—The Fourier Transform

The basis of the uncertainty principle is a mathematical truth completely independent of any physical application: *The more spatially compact a wave is, the less well its wavelength may be specified.*

To begin to understand this, we consider whether it is even fair to speak of "the" wavelength of an arbitrary periodic function. The left-hand plots in Figure 3.17 show periodic functions of position. All repeat within the same interval along the x-axis, indicated by $\lambda_1$. The top wave form is a pure sinusoidal wave. The other two are not pure, but are algebraic sums of pure wave forms of different wavelengths, as shown by the center plots in the figure. These sums have the same fundamental wavelength as that in the top wave form, plus differing amounts of wavelengths half as long and one-third as long. Wave forms rich in harmonics are common; the wave on a guitar string consists of a large amplitude of the fundamental wavelength coexisting with shorter-wavelength harmonics of various amplitudes. (It is largely by the harmonics' relative amplitudes that we distinguish one musical instrument from another playing the same fundamental frequency.) The important point here is that, even if it is not obvious, a complicated periodic

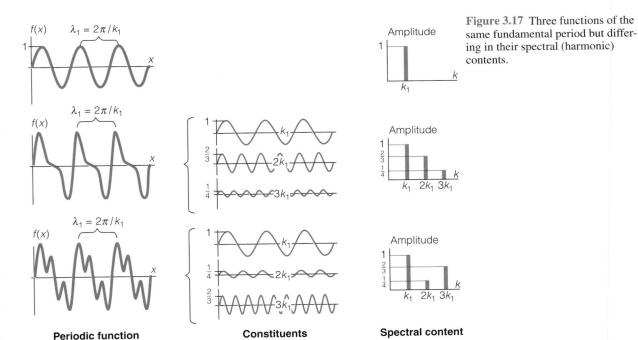

**Figure 3.17** Three functions of the same fundamental period but differing in their spectral (harmonic) contents.

Periodic function          Constituents          Spectral content wave form is really an algebraic sum of pure sine waves of different wavelengths. To know the amplitudes of these wavelengths is to know the wave form's **spectral content.**

Adding pure sine waves whose wave numbers are multiples of a fundamental ($\lambda_n = \lambda_1/n \Rightarrow k_n = nk_1$) always yields a *periodic* function.[12,13] But what of a non-periodic wave form? A wave pulse, such as that shown in the margin, a common object of interest in quantum mechanics, is not periodic, but is no less a wave. Can it too be considered as a sum of pure sine waves? The answer is yes, but not if the sum is restricted to multiples of a fundamental.

Figure 3.18 shows what is needed. Wave form (a) is a pure sine wave whose wave number is the pulse's *apparent* wave number $k_0$. Obviously, it is a poor approximation of the pulse. Wave form (b), a sum of just three sine waves of different amplitudes and wave numbers, does considerably better. Wave forms (c) and (d) add wave numbers more densely spaced and covering a greater range above and below $k_0$. Here we begin to see a trend: The periodic "impostors" move farther from our desired wave form. They can be eliminated completely only by including an *infinite* number of waves! We shall soon see how we knew what amplitudes and wavelengths to add together, but, as illustrated by wave form (e), the main point is this: A nonperiodic wave can be considered as a sum of sine

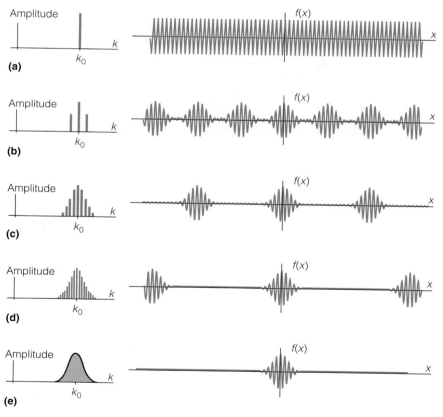

**Figure 3.18** Building a single isolated pulse from pure sine waves requires a continuum of wave numbers.

[12]Because it is easier to work with a spatial frequency than with a spatial period, we choose to express functions in terms of wave number, $k = 2\pi/\lambda$ rather than wavelength.

[13]Such a sum, $\Sigma A_n \sin[n(2\pi/\lambda_1)x]$, will repeat whenever $x$ increases by $\lambda_1$. That is, it has the same overall period as the fundamental, as shown in Figure 3.17.

waves of different amplitudes covering a *continuous* range of wave numbers. It is not a sum, but an integral.[14]

In quantum mechanics, our basic sinusoidal plane waves have two parts (real and imaginary). As we know, they are of the form $e^{ikx}$ (assuming $t = 0$). Therefore, a general complex wave form may be considered as a sum of such pure waves over a continuum of wave numbers:

$$f(x) = \int_{-\infty}^{+\infty} \tilde{f}(k)\, e^{ikx}\, dk \qquad (3\text{-}11)$$

Constituent plane wave — (above $\tilde{f}(k)\ e^{ikx}$)

Amplitude of plane wave — (below)

Giving the amplitude as a function of $k$ of each constituent plane wave, $\tilde{f}(k)$ is the spectral content. Clearly, it is intimately related to the $f(x)$ itself, and it might be nice to be able to determine it. In Appendix D, it is shown that we do so as follows:

$$\tilde{f}(k) = \frac{1}{2\pi} \int_{-\infty}^{+\infty} f(x) e^{-ikx}\, dx \qquad (3\text{-}12)$$

Spectral content of $f(x)$

At this point the reader may ask, Why is it important that a general wave can be thought of as a continuous sum of pure plane waves, and what good is knowing "spectral content"? The answer is that *if* the wave is equivalent to a sum, *it is a sum!* It is not a wave of pure wave number, but a combination of them, and the spectral content is what makes clear the wave numbers of which it is composed. A quick glance reveals whether the wave is pure, nearly pure, or contains a broad range of wave numbers of significant amplitudes. The quantum-mechanical application is that a compact wave such as a pulse does not have a well-defined wave number. It is really a combination of waves of various wave numbers *and hence momenta*. Let us investigate this by analyzing a typical wave pulse.

## Gaussian Wave Packet

In quantum mechanics it is often necessary to treat a moving object (a photon, an electron, etc.) as a traveling wave pulse. Figure 3.19 depicts such a pulse. The particle associated with it would most likely be found in the region where the pulse is large, and the region moves at the speed of the particle. The most common such wave form is the so-called **gaussian wave packet:** [15]

$$\psi(x) = A e^{-(x/2\epsilon)^2} e^{ik_0 x} \qquad (3\text{-}13)$$

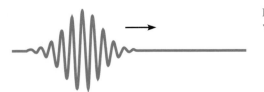

**Figure 3.19** A traveling gaussian wave pulse.

[14]Because the spacing is thus infinitesimal, it might be said that the "wave number of the fundamental" is infinitesimal, corresponding to infinite wavelength and allowing the overall period of the wave to be infinite—that is, the pulse *doesn't* repeat.

[15]Presently, we are interested in the relationship between the wave number and spatial extent of the wave *at some instant*, not how the wave propagates in time. It is a universal convention to use the lower case $\psi(x)$ for a wave function at a particular instant, and the upper case $\Psi(x, t)$ when explicit time dependence is considered.

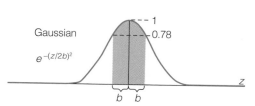

**Figure 3.20** A gaussian falls to 78% of its maximum when $z = \pm b$.

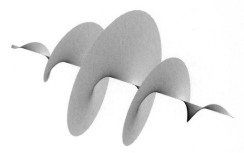

**Figure 3.21** A gaussian matter wave.

What qualifies this as "gaussian" is the $e^{-(x/2\epsilon)^2}$. Any factor of the form $e^{-(z/2b)^2}$, shown in Figure 3.20, is called a gaussian. As noted in Section 3.4, it has a maximum at $z = 0$ and falls off toward zero as $|z|$ increases. The rate at which it falls off is governed by the constant $b$. At $z = \pm b$, it falls off to $e^{-0.25}$—about 78%—of its maximum value. If $b$ is small, the gaussian falls off rapidly; if large, it falls off slowly. (The limits are $b \to 0$, for which it would be nonzero only for $z$ identically equal to zero, and $b \to \infty$, for which the gaussian is everywhere unity.) Since the width of its "bulge" is proportional to $b$, we find it convenient to simply define $b$ as the "width." Thus, the $\epsilon$ in the gaussian factor in (3-13) is the spatial width $\Delta x$ of the packet.

Multiplying the gaussian and giving $\psi(x)$ its oscillatory character is a complex exponential of wave number $k_0$: $e^{ik_0x}$. Accordingly, the gaussian wave packet shares much with the plane wave of Section 3.3. Figure 3.21 represents both the real and imaginary parts of the gaussian packet in the same way Figure 3.11 does for the plane wave. But there is more to the gaussian wave, as we now see.

---

**Example 3.7**
Given $\psi(x) = Ae^{-(x/2\epsilon)^2}e^{ik_0x}$, (a) find the spectral content, and (b) interpret the result.

**Solution**

(a) Using (3-12),

$$\tilde{\psi}(k) = \frac{1}{2\pi}\int_{-\infty}^{+\infty} \psi(x)e^{-ikx}\,dx$$

$$= \frac{1}{2\pi}\int_{-\infty}^{+\infty} (Ae^{-(x/2\epsilon)^2}e^{ik_0x})e^{-ikx}\,dx$$

$$= \frac{A}{2\pi}\int_{-\infty}^{+\infty} e^{-(1/4\epsilon^2)x^2 + i(k_0-k)x}\,dx$$

The integral is of a standard form:

$$\int_{-\infty}^{+\infty} e^{-az^2 + bz} \, dz = e^{b^2/4a} \sqrt{\frac{\pi}{a}}$$

where $a = 1/4\epsilon^2$ and $b = i(k_0 - k)$. It is known not coincidentally as a **gaussian integral.** Thus,

$$\tilde{\psi}(k) = \frac{A}{2\pi} e^{-\epsilon^2(k-k_0)^2} \sqrt{4\epsilon^2 \pi}$$

$$= \frac{A\epsilon}{\sqrt{\pi}} e^{-\epsilon^2(k-k_0)^2}$$

$$= \frac{A\epsilon}{\sqrt{\pi}} e^{-\{(k-k_0)/[2(2\epsilon)^{-1}]\}^2}$$

The final rearrangement is done to emphasize the fact that $\tilde{\psi}(k)$ happens also to be a gaussian, centered at $k = k_0$ and with a "wave number width" (whatever multiplies the 2 in the exponent's denominator) of $\Delta k = (2\epsilon)^{-1}$. It is plotted in Figure 3.22.

(b) According to equation (3-11), any $f(x)$ can be considered as a sum of many plane waves, the amplitudes given by $\tilde{f}(k)$. In this example, $\tilde{\psi}(k)$ has a maximum at $k = k_0$, *as we should expect.* Since $k_0$ is the wave number of the oscillatory part $e^{ik_0 x}$, the "spectral content" should be large there. But this is not the only value of $k$ where $\tilde{\psi}(k)$ is nonzero; the wave form $\psi(x)$ "contains" other wave numbers. Most are within a width $\Delta k = (2\epsilon)^{-1}$ of $k_0$, and because this width is inversely proportional to the width $\Delta x = \epsilon$ of $\psi(x)$, $\tilde{\psi}(k)$ becomes narrow as $\psi(x)$ becomes wide (and vice versa). In fact, $\psi(x)$ has a unique, well-defined wave number if and only if its width $\epsilon$ is infinite. In this limit, the gaussian factor $e^{-(x/2\epsilon)^2}$ would become unity, $\psi(x)$ would accordingly be just the pure $e^{ik_0 x}$, and $\tilde{\psi}(k)$ would be a spike at $k_0$.

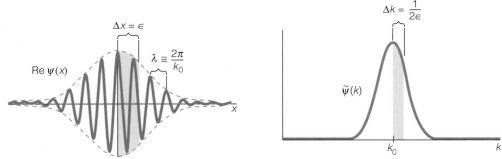

**Figure 3.22**  A gaussian wave packet $\psi(x)$ and its spectral content $\tilde{\psi}(k)$.

Example 3.7 shows why we chose the waves that we did to add together in Figure 3.18. All the spectral contents in that figure have gaussian shapes. Again, though, the single isolated pulse is reproduced perfectly only by including the entire continuous spectrum of its constituent waves.

Now let us turn from strictly math to the central physical consequence revealed by the example. The physical link is $p = \hbar k$. If a general wave is really a combination of a range $\Delta k$ of wave numbers, it is a combination of a range $\Delta p = \hbar \, \Delta k$ of *momenta! Were we to measure its momentum, the result is not certain, but would most likely be one of the values within this range.* Since the uncertainty in momentum is *inversely* proportional to $\epsilon$ and the uncertainty in the particle's position is *directly* proportional to $\epsilon$, we have $\Delta p \propto 1/\Delta x$. This is the uncertainty principle![16]

We introduced the uncertainty principle in connection with the single slit. We may now understand the connection even better. By assumption, the wave incident on a single slit (Figure 3.12) is a plane wave; it does not vary along the width of the slit, which we designate the *x*-direction. (It does oscillate with a given wavelength in the *z*-direction, the direction in which it moves.) So far as the *x*-dimension goes, a wave passing through a slit is simply a constant value $C$ that ends abruptly at the edges of the slit, as shown in Figure 3.23. Such a function may not seem very wavelike, but its spectral content $\tilde{\psi}(k)$ is nontrivial and tells us much. (It is left as an exercise to actually calculate it.) First, it is centered on the point $k = 0$. Since the incident wave's *x*-component of momentum is zero, it is reasonable that it should have no *average* $p_x$ after passing through a slit. Secondly, it says that $\psi(x)$ is not a simple wave of zero $p_x$, but rather a combination of waves

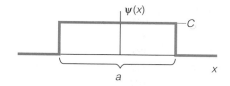

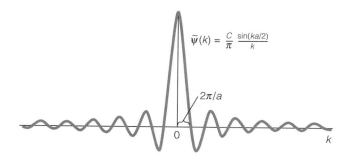

Figure 3.23 A not very wavelike function $\psi(x)$ and its interesting spectral content $\tilde{\psi}(k)$.

[16]It should be noted that the uncertainty principle involves the *uncertainty* in $k$, not the particular *value* of $k_0$. A stationary particle would have $k_0 = 0$, and $\psi(x)$ would not be oscillatory, because only the gaussian $e^{-x^2/(2\epsilon)^2}$ would remain. But the only effect on $\tilde{\psi}(k)$ would be that it would be centered at $k = 0$. It would still show a range of wave numbers inversely proportional to $\epsilon$, so a measurement of the momentum of such a "stationary" particle would not likely find a value of precisely zero.

with positive and negative wave numbers/momenta. This is precisely what we see in the single slit: There is a probability of detecting particles with positive and negative $p_x$. Even the fact that in Figure 3.12 some values of $p_x$ are apparently never observed (i.e., the diffraction minima) we see to be a consequence of $\tilde{\psi}(k)$ being zero at certain values of $k$.[17] Finally and perhaps most importantly, the width of $\tilde{\psi}(k)$ is (again!) inversely proportional to the width of $\psi(x)$. The central maximum of $\tilde{\psi}(k)$ ends at its first zeros on either side of $k = 0$, where $ka/2 = \pm\pi$ or $k = 2\pi/a$. Thus, most $k$-values are within $2\pi/a$ of zero: The range of $k$ and hence $p_x$-values is inversely proportional to the slit width $a$. Once again, the better the position is known, the less well-known the momentum, and vice versa.

By now, the reader should see that the inverse relationship between $\Delta x$ and $\Delta k$, based on the mathematical link between a function $\psi(x)$ and its spectral content $\tilde{\psi}(k)$, is the true basis of the uncertainty principle. We have considered two different functions for which $\Delta x \propto 1/\Delta k$ holds, and we have certainly made the most of it. Still the question may be asked, is this inverse relationship merely a mathematical peculiarity of the functions we have chosen to consider? Absolutely not! It always holds: *A more compact wave is necessarily a wave of less well-defined wave number* (further examples may be found in the exercises). Consequently, the uncertainty principle is quite general. As noted in Section 3.4, a gaussian is the function for which the product $\Delta k\,\Delta x$ is a minimum. From Example 3.7, we see that it is $\Delta x\,\Delta k = \epsilon(2\epsilon)^{-1} = \frac{1}{2}$. For any other function, though the inverse relation still holds, the product is *strictly greater* than $\frac{1}{2}$. (Unfortunately, the rigorous mathematical proof is somewhat involved and must be omitted.) Thus, for an arbitrary wave function,

$$\Delta k\,\Delta x \geq \frac{1}{2}$$

which, combined with $p = \hbar k$, yields the physical consequence, the uncertainty principle.

$$\Delta p\,\Delta x \geq \frac{\hbar}{2}$$

Summarizing our findings somewhat, Figure 3.24 plots $\psi(x)$ and $\tilde{\psi}(k)$ for a gaussian wave, the wave for which the product $\Delta x\,\Delta k$ is as small as theoretically possible. All plots are of the same approximate wavelength, but $\Delta x$ varies from very large at the top to very small at the bottom. The inverse relationship is clear. An infinitely broad wave ($\Delta x \to \infty$)—a simple plane wave—is the only one for which $\Delta k = 0$, for which wave number and thus momentum are truly well defined. Conversely, a supremely compact wave ($\Delta x \to 0$)—a spike at a definite position—is a combination of an infinitely broad range of wave numbers/momenta.

## The Fourier Transform

The strictly mathematical procedure of finding the spectral content of $f(x)$, given in equation (3-12), has a more common name: "finding the Fourier transform of $f(x)$." The function $\tilde{f}(k)$ is the **Fourier transform** of $f(x)$. Correspondingly, since

[17]There is an obvious difference: $\tilde{\psi}(k)$ takes on negative values, while the plot of Figure 3.12 is strictly positive. The reason is that for the sum of $\tilde{\psi}(k)e^{ikx}$ to properly add up to $\psi(x)$, some of the coefficients $\tilde{\psi}(k)$ must be negative. The function $\tilde{\psi}(k)$ is an amplitude, while Figure 3.12 plots intensity, proportional to the *square* of an amplitude.

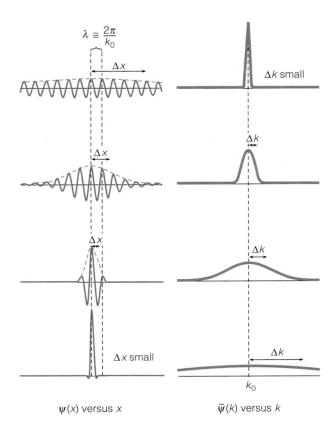

$\psi(x)$ versus $x$      $\tilde{\psi}(k)$ versus $k$

**Figure 3.24** The spatial width $\Delta x$ of a function $\psi(x)$ is inversely proportional to the wave number width $\Delta k$ of its spectral content $\tilde{\psi}(k)$.

equation (3-11) gives the function $f(x)$ from its Fourier transform $\tilde{f}(k)$, it is known as the inverse Fourier transform.

$$f(x) = \int_{-\infty}^{+\infty} \tilde{f}(k)e^{ikx}\,dk \quad \textbf{(3-11)} \qquad \tilde{f}(k) = \frac{1}{2\pi}\int_{-\infty}^{+\infty} f(x)e^{-ikx}\,dx \quad \textbf{(3-12)}$$

**Inverse Fourier transform** $(k \rightarrow x)$     **Fourier transform** $(x \rightarrow k)$

In short, the inverse Fourier transform represents an arbitrary function of position $f(x)$ as a combination of sinusoidal functions, and the Fourier transform yields the amplitude of each such function that goes into $f(x)$.[18]

As a mathematical procedure, the Fourier transform takes a function of one variable and returns a related function of another, while its inverse of course performs the reverse. Thus far, we have given it a function of $x$ and received back a function of $k$, $2\pi$ over the "spatial period" $\lambda$. Logically, were we to give it a function of $t$, it would have to return a function of $\omega$, $2\pi$ over the temporal period

---

[18]A crude analogy: The composition of a building of varying numbers of different-size bricks might be difficult to see at a glance. The Fourier transform of the building would be a tidy row of all the bricks, stacked in order of increasing size. The building is the wave $f(x)$, the bricks are the $e^{ikx}$, varying $k$ corresponding to varying brick size, and the tidy row of bricks is $\tilde{f}(k)$. The inverse Fourier transform acting on the row of bricks would rebuild the building.

*T*. It should be useful to reveal a relationship between time and frequency,[19] just as it does between position and wave number. This direct correspondence is represented in the margin. Thus, for oscillatory functions of time rather than position, we may write

$$f(t) = \int_{-\infty}^{+\infty} \tilde{f}(\omega) e^{i\omega t}\, d\omega \qquad\qquad \tilde{f}(\omega) = \frac{1}{2\pi} \int_{-\infty}^{+\infty} f(t) e^{-i\omega t}\, dt$$

$$x \leftrightarrow k \quad \left( \equiv 2\pi \frac{1}{\lambda} \right)$$

$$t \leftrightarrow \omega \quad \left( \equiv 2\pi \frac{1}{T} \right)$$

**Inverse Fourier transform** $(\omega \rightarrow t)$     **Fourier transform** $(t \rightarrow \omega)$

Given the direct correspondence, if it is true mathematically that $\Delta k\, \Delta x \geq \frac{1}{2}$, then it follows that $\Delta \omega\, \Delta t \geq \frac{1}{2}$. Combining this with $E = \hbar \omega$, the energy-time uncertainty principle $\Delta E\, \Delta t \geq \frac{1}{2}\hbar$ is a direct physical consequence.

The Fourier transform is an indispensable mathematical tool in many areas of modern technology involving waves—for example, quantum optics (lasers), signal transmission, acoustics, quantum mechanics. One example is the problem of transmitting a laser pulse $f(t)$ whose duration $\Delta t$ is short. Regardless of the intended "central" frequency $\omega_0$, the Fourier transform $\tilde{f}(\omega)$ would reveal that the pulse really contains a range of frequencies $\Delta \omega$ around $\omega_0$. Although the central frequency has the largest amplitude, others, intended or not, are present. Moreover, since $\Delta \omega \propto 1/\Delta t$, the shorter the pulse, the greater is the range of frequencies around $\omega_0$ that must be transmitted to accurately reproduce it. If any are left out, the pulse will be distorted. A similar situation arises in transmitting any signal that is not a pure sine wave. Any modulation or higher-harmonic detail superimposed on a pure sine wave—as radio superimposes voice upon a "carrier frequency"—results in a signal comprising a range of frequencies within a nonzero "bandwidth" of the central frequency. Both transmitter and receiver must be able to process the entire range if the complex signal is to be reproduced faithfully.

## 3.7 The Bohr Model of the Atom

It is instructive at this point to take a look at an early attempt to solve the mysteries of the atom through application of some principles of quantum mechanics. This work, for which Niels Bohr won the 1922 Nobel prize, became known as the **Bohr model of the atom,** or simply the "Bohr atom."

When the fundamental workings of something are so obscure as to defy formulation of a comprehensive theory, we construct a "model." We observe that a system behaves in a certain way, and our model is a simplified theory that (hopefully) explains the behavior. If it agrees with further experimental observation, we have evidence that the assumptions upon which it is based are valid; quite likely, we will have learned something about the subject's underlying nature. But if our model's predictions are at odds with further experiment, the model must be modified. Even so, something is learned—that at least one of the assumptions in the existing model is invalid.

The Bohr model of the atom predicted that the electron orbiting the proton in a hydrogen atom may take on only certain, discrete energies, and the predicted values agreed with experimental findings. Viewed in retrospect, this agreement was largely accidental, and the model failed in numerous other ways. But its

[19] Strictly speaking, $\omega$ should be called "angular frequency"; we refer to it simply as "frequency" in cases where the distinction is not significant or context is sufficient to eliminate confusion.

shortcomings are understandable, because at the time the model was proposed, quantum mechanics simply had not been developed sufficiently to allow a more complete description. In any case, we gain valuable training in the application of knowledge at hand through study of such early attempts. The Bohr model combines primitive quantum principles with classical physics. It is based upon (1) the classical second law of motion, applied to an electron assumed held in circular orbit by its electrostatic attraction to a proton; (2) a classical expression for the energy of the orbiting electron; and (3) a postulate involving the quantization of the electron's angular momentum.

The magnitude of the force on an electron a distance $r$ from a proton is given by $F = (1/4\pi\epsilon_0)(e^2/r^2)$, and in circular orbit the acceleration of a classical particle is $v^2/r$. Thus, $F = ma$ gives us

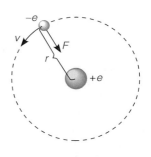

$$\frac{1}{4\pi\epsilon_0}\frac{e^2}{r^2} = m_e\frac{v^2}{r} \quad \text{or} \quad v^2 = \frac{e^2}{4\pi\epsilon_0 m_e}\frac{1}{r} \qquad (3\text{-}14)$$

The energy of an electron a distance $r$ from a proton is given by

$$E = \mathrm{KE} + U_{\mathrm{elec}} = \frac{1}{2}m_e v^2 + \frac{1}{4\pi\epsilon_0}\frac{-e^2}{r} \qquad (3\text{-}15)$$

Substituting equation (3-14) into (3-15) gives the energy solely in terms of $r$:

$$E(r) = -\frac{1}{8\pi\epsilon_0}\frac{e^2}{r} \qquad (3\text{-}16)$$

(These steps were also carried out in Example 3.6. As noted there, the negative sign indicates that the negative potential energy is of greater magnitude than the kinetic.) Equation (3-16) is purely classical. Total energy varies continuously, as $r$ may take on any of a continuum of values. Energy is not quantized.

Bohr added one more condition—he postulated that the electron's angular momentum $L$ may take on only the values

$$L = n\hbar \quad \text{(where } n = 1, 2, 3, \ldots)$$

Since $L = mvr$ in circular orbit, this condition may also be written

$$m_e vr = n\hbar \quad (n = 1, 2, 3, \ldots) \qquad (3\text{-}17)$$

A plausible basis for Bohr's postulate is shown in the margin. If we assume that the orbiting electron behaves as a wave wrapped around a circle, and that the circumference of the circle must be an integral number of wavelengths, the condition $\lambda = h/p$ implies that the product $mvr$ may take on only the values $n\hbar$.

Equation (3-17) may now be combined with (3-14) to eliminate $v$ and obtain a condition restricting $r$ to only certain values.

$$r = \frac{(4\pi\epsilon_0)\hbar^2}{m_e e^2}n^2 \quad (n = 1, 2, 3, \ldots) \qquad (3\text{-}18)$$

or

$$r = a_0 n^2 \quad \text{where } a_0 \equiv \frac{(4\pi\epsilon_0)\hbar^2}{m_e e^2} = 0.0529 \text{ nm}$$

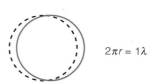

$2\pi r = 1\lambda$

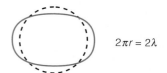

$2\pi r = 2\lambda$

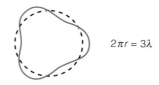

$2\pi r = 3\lambda$

$2\pi r = n\lambda$ but $\lambda = \dfrac{h}{p} = \dfrac{h}{mv}$

Thus $2\pi r = n\dfrac{h}{mv}$

$mvr = n\dfrac{h}{2\pi}$

$L = n\hbar$

According to Bohr's theory, the electron may orbit at only certain multiples of the so-called **Bohr radius** $a_0$.

Energy in turn is also quantized. Inserting equation (3-18) into (3-16),

$$E = -\frac{m_e e^4}{2(4\pi\epsilon_0)^2 \hbar^2} \frac{1}{n^2} = -13.6 \text{ eV} \left(\frac{1}{n^2}\right) \qquad (n = 1, 2, 3, \ldots) \qquad (3\text{-}19)$$

Energy is quantized according to the value of the integer $n$, known as the "principal quantum number." As noted, the predicted energies agree with the experimental evidence.

The Bohr model of the atom is an excellent example of working with what theory is at hand—Bohr didn't even have the Schrödinger equation at his disposal! However, the model *is* flawed. While orbiting electrons are *most likely* to be found at certain distances from the proton, they must really be treated as diffuse waves, which do not orbit at precise radii. This casts doubt on (3-18), which in turn calls into question the model's prediction of energies. In reality, "orbiting" electrons have not only *rotational* kinetic energy due to orbital motion about the origin, but *radial* kinetic energy due to motion toward and away from the origin. A fully quantum-mechanical treatment of the electron's behavior in the hydrogen atom, based upon the Schrödinger equation, is found in Chapter 6.

## Progress and Applications

The fundamental truth of matter's wave nature, and its consequences, continue to bear directly on our efforts to learn more about our universe and to apply the benefits of science to our civilization.

■ The uncertainty principle is everywhere, and as technology advances, we will run into it more and more. Not only does it tie together inherent uncertainties in position and momentum, and in time and energy, the physical world is full of pairs of quantities, known as "conjugate quantities," inherently linked in the same way. To know one quantity exactly is to be completely ignorant of its conjugate. Even so, if one quantity *is* known, an inadvertent "observation" of its conjugate will disturb that knowledge. This is the root of the interest in quantum-nondemolition, or **QND,** measurements, now in their infancy: To control a quantity as precisely as possible, we must be careful not to "touch" its conjugate. Conjugate quantities having to do with light are currently of great interest, being at the root of high-speed communication. If QND techniques are not used, "quantum fluctuations" in the quantities' values—the inevitable result of the uncertainty principle—are introduced, adding undesirable noise. Light in which fluctuations in one aspect have been reduced at the cost of uncontrolled fluctuations in its conjugate aspect is known as **squeezed light.** By its unusually "smooth" nature, squeezed light has application in many fields— studying delicate goings-on in the atom, for instance. The desire to observe restrictions imposed by the uncertainty principle carries over into gravitation. Gravitational waves should exist, but are so weak that quantum fluctuations figure prominently in obscuring their observation. Controlling the fluctuations offers hope of success.

■ Technological advances have allowed use of **"ultracold" neutrons,** not only to study neutrons themselves, but to probe the structures of materials and to substantiate fundamental ideas of quantum mechanics. "Thermal" neutrons, owing their speeds to the ambient temperature, travel at thousands of meters per second and pass through a typical apparatus in the wink of an eye. An ultracold neutron moves so slowly that it can be "held" for seconds. How a neutron scatters in a solid tells something about the nature of vibrations therein, a very important topic in this age of solid-state electronics. Probabilities of neutron absorption into nuclei are also of interest. The use of very slow neutrons provides information on these quantities in a realm previously unexplored. Causing neutrons to move so slowly may also help in the search for a possible electric dipole moment for the neutron. This is an intriguing possibility, for if true, it would imply a violation on a microscopic scale of "time-reversal symmetry"; in other words, things would not obey the same physical laws if viewed in a motion picture run backward (see Chapter 11). Ultracold neutrons are also noteworthy for their straightforward adherence to one of quantum mechanics' most fascinating predictions: quantum-mechanical "tunneling" (see Chapter 5).

■ A technique being used to "see"—again within quantum-mechanical limits— wave functions of electrons in atoms and molecules is **(e,2e) spectroscopy.** The idea is fairly simple: An electron whose wave function is to be probed is struck by a second of known wave number/momentum. The result is two electrons whose momenta (wave numbers) are detected simultaneously as they depart the site of the collision. Repeated trials determine the first electron's Fourier transform $\tilde{\psi}(k)$, and from this its wave function is computed. The experimental technique has provided decisive insight into the complicated multiple-electron models where theoretical calculations are difficult (see Chapter 7).

■ Matter waves diffract, and this fact continues to be exploited in many applications. Figure 3.25 shows the result of a neutron diffraction experiment revealing clearly a square pattern of magnetic field vortices (see Chapter 9) in the superconductor $YNi_2B_2C$. Another use is **helium atom scattering (HAS),** in which beams of helium atoms are directed at crystal surfaces. Helium atoms are particularly good for viewing surfaces because they neither stick, being "noble" elements that do not chemically bond, nor do they penetrate beyond the surface, as do smaller "particles."

■ Even at temperatures we would consider very cold, a freely moving particle the size of an atom still may have so large a momentum that its wavelength is smaller than the physical dimensions of the atom itself. Correspondingly, the momentum *uncertainty* is large, so that the position uncertainty may be no larger than atomic dimensions. It is rather silly to speak of it as a wave. Advances in **laser cooling** (see Chapter 8), however, have allowed atoms to be "cooled" to previously unknown low speeds. Quantum behaviors then come to the fore. A particularly nice recent example slowed a beryllium ion to a very low speed, then coaxed it into a combination of two different quantum states, with "bumps" separated by about 80 nm. No longer considered microscopic, this "mesoscopic" separation is hundreds of times the "size of the atom." Clearly, this "particle" was capable of being found over a much broader region

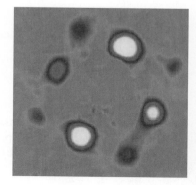

**Figure 3.25** Neutron diffraction pattern showing the arrangement of magnetic flux vortices in a superconductor.

than a naive particle view would expect (C. Monroe, et al., *Science*, May 24, 1996, pp. 1131–1136). Such experiments offer to help in understanding the coherence of different quantum states (important to "quantum computing"; see Chapter 4), and may serve to uncover secrets that quantum mechanics might yet harbor.

## Basic Equations

| Wave-particle relationships | Probability density |
|---|---|
| $p = \dfrac{h}{\lambda} = \hbar k \qquad$ (3-1) | $\lvert \Psi(x,\ t) \rvert^2 \qquad$ (3-6) |
| $E = hf = \hbar\omega \qquad$ (3-2) | |

**Free-Particle Schrödinger equation**

$$-\frac{\hbar^2}{2m}\frac{\partial^2 \Psi(x,\ t)}{\partial x^2} = i\hbar\frac{\partial \Psi(x,\ t)}{\partial t} \qquad (3\text{-}5)$$

| Momentum-position uncertainty relation | Energy-time uncertainty relation |
|---|---|
| $\Delta p_x\,\Delta x \geq \dfrac{\hbar}{2} \qquad$ (3-9) | $\Delta E\,\Delta t \geq \dfrac{\hbar}{2} \qquad$ (3-10) |

| Inverse Fourier transform | Fourier transform |
|---|---|
| $f(x) = \displaystyle\int_{-\infty}^{+\infty} \tilde{f}(k)e^{ikx}\,dk \qquad$ (3-11) | $\tilde{f}(k) = \dfrac{1}{2\pi}\displaystyle\int_{-\infty}^{+\infty} f(x)e^{-ikx}\,dx \quad$ (3-12) |

## Summary

As electromagnetic radiation, classically a wave, may behave as particles, a massive object, classically a discrete particle, may behave as a wave. Its wave number and angular frequency are given by

$$p = \hbar k \qquad E = \hbar\omega$$

The symbol for a matter wave is $\Psi(x,\ t)$. It is a solution of the Schrödinger equation, which is based upon energy. The equation and its solutions are complex by choice; a matter wave has two parts, and with complex numbers we treat both parts at once. The complex square of the wave gives the probability per unit length/volume of finding the particle:

$$\text{probability density} = \lvert \Psi(x,\ t) \rvert^2$$

Because a massive object has a wave nature, its momentum and position cannot be precisely known simultaneously; there is a theoretical lower limit on the product of the uncertainties:

$$\Delta p_x \, \Delta x \geq \frac{\hbar}{2}$$

This is known as the uncertainty principle, and is a direct consequence of $p = \hbar k$ and the mathematical properties of waves. A wave function for which momentum is known precisely ($\Delta p = 0$) is one for which the position is completely unknown ($\Delta x = \infty$)—the probability density would be spread uniformly throughout space. Conversely, a wave function for which the position is known precisely is one for which the momentum is completely unknown—any momentum might be measured.

Given the wave nature of matter, the behavior of massive objects cannot be described as completely as is assumed possible in classical physics.

## E X E R C I S E S

### Section 3.1

1. A beam of electrons strikes a barrier with two narrow but equal-width slits. A screen is located beyond the barrier, and electrons are detected as they strike the screen. The "center of the screen" is the point on the screen equidistant from the slits. When either slit alone is open, ten electrons arrive per second in a 0.1-mm-wide band at the center of the screen. When both slits are open, how many electrons will arrive per second in a 0.1-mm-wide band at the center of the screen?

2. A beam of low-energy electrons is directed normally at a metal surface, and strong reflection is detected only at an angle of 35°. What is the ratio $\lambda/D$, where $D$ is the atomic spacing at the surface?

### Section 3.2

3. Determine the Compton wavelength of the electron, defined to be the wavelength it would have if its momentum were $m_e c$.

4. Show that the wavelength of a particle of mass $m$ and charge $q$ accelerated through a potential difference $V$ is given by

$$\lambda = \frac{h}{\sqrt{2mqV}}$$

Assume that the potential difference is insufficient to accelerate the particle to relativistic speeds.

5. How slow would an electron have to be traveling for its wavelength to be at least 1 $\mu$m?

6. A particle is "thermal" if it is in equilibrium with its surroundings—its average kinetic energy would be $\frac{3}{2}k_B T$. Show that the wavelength of a "thermal" particle is given by

$$\lambda = \frac{h}{\sqrt{3mk_B T}}$$

7. The average kinetic energy of a particle at temperature $T$ is $\frac{3}{2}k_B T$. (a) What is the wavelength of a room temperature (22°C) electron? (b) Of a room temperature proton? (c) In what circumstances should each behave as a wave?

8. What is the wavelength of a neutron of kinetic energy 1 MeV? Of kinetic energy 20 eV? (The difference is important for fission of uranium, as a neutron will be more easily absorbed when its wavelength is very large compared to the dimensions of the uranium nucleus, ~15 fm. In effect, a neutron of long wavelength behaves more as a diffuse wave than as a localized particle.)

9. (a) What is the range of possible wavelengths for a neutron corresponding to a range of speeds from "thermal" at 300 K (see Exercise 6) to 0.01$c$? (b) Repeat part (a), but with reference to an electron. (c) For this range of speeds, what range of dimensions $D$ would reveal the wave nature of a neutron? Of an electron?

10. The size of the smallest detail resolvable by a microscope is approximately equal to the wavelength of the radiation used to view it. Since the minimum wavelength of visible light is approximately 400 nm, a microscope employing light can resolve details no smaller than approximately 400 nm. An electron microscope directs electron waves rather than light toward a specimen. If an electron microscope is to resolve details as small as 1 nm, (a) what must be the speed of the electrons, and (b) what must be the accelerating potential $V$?

11. A beam of electrons, kinetic energy = 54 eV, is directed normally at a nickel surface, and strong reflection is detected only at an angle of 50°. Determine the spacing of nickel atoms on the surface.

12. The Moon orbits Earth at a radius of $3.84 \times 10^8$ m. To do so as a classical particle, its wavelength should be small. But small relative to what? Being a rough measure of the region where it is confined, the orbit radius is certainly a relevant dimension against which to compare the wavelength. Compare the two. Does the Moon indeed orbit as a classical particle? ($m_{\text{Earth}} = 5.98 \times 10^{24}$ kg and $m_{\text{Moon}} = 7.35 \times 10^{22}$ kg)

13. In the hydrogen atom, the electron's orbit, not necessarily circular, extends to a distance of about an angstrom (1 Å = 0.1 nm) from the proton. If it is to move about *as a compact classical particle* in the region where it is confined, the electron's wavelength had better *always* be much smaller than an angstrom. Is it? How large might be the electron's wavelength—how small might be its speed? If orbiting as a particle, its speed at 1 Å could be no faster than that for *circular* orbit at that radius. (Why?) Find the corresponding wavelength and compare it to 1 Å. Can the atom be treated classically?

14. A beam of electrons accelerated through an 80-V potential difference is directed at a single slit of 5.0-$\mu$m width, then detected at a screen 10 m beyond the slit. How far from a point directly in the line of the beam is the first location where no electrons are ever detected?

15. Electrons are accelerated through a 50-V potential difference, producing a mono-energetic beam. This is directed at a two slit apparatus of 0.010-mm slit separation. A bank of electron detectors is 12 m beyond the double slit. With slit 1 alone open, 100 electrons per second are detected at all detectors. With slit 2 alone open, 900 electrons per second are detected at all detectors. Now both slits are open.
    (a) The first minimum in the electron count occurs at detector X. How far is it from the center of the interference pattern?
    (b) How many electrons per second will be detected at the center detector?

    (c) How many electrons per second will be detected at detector X?

16. A beam of particles, each of mass $m$ and (nonrelativistic) speed $v$, strikes a barrier in which there are two narrow slits, and beyond which is a bank of detectors. With slit 1 alone open, 100 particles are detected per second at all detectors. Now slit 2 is also opened. An interference pattern is noted in which the first minimum, 36 particles per second, occurs at an angle of 30° from the initial direction of motion of the beam.
    (a) How far apart are the slits?
    (b) How many particles would be detected (at all detectors) per second with slit 2 alone open?
    (c) There are multiple answers to part (b). For each, how many particles would be detected at the center detector with both slits open?

## Section 3.3

17. Show that the "underlying laws" given in column 2 of Table 3.1 are equivalent to those given in column 1 (i.e., Maxwell's equations). Also verify that the single complex plane wave function given in column 2 is equivalent to the two real plane wave functions given in column 1.

18. Verify that the pair of real wave functions $\Psi_1$ and $\Psi_2$ given in column 3 of Table 3.1 are solutions of the "underlying laws," the two "coupled" wave equations, in that column.

19. Show that the Schrödinger equation given in column 4 of Table 3.1 is equivalent to the pair of real wave equations given in column 3. Also verify that the single complex plane wave function given in column 4 is equivalent to the two real plane wave functions given in column 3.

20. An electron moves along the $x$-axis with a well-defined momentum of $5 \times 10^{-25}$ kg·m/s. Write an expression describing the matter wave associated with this electron. Include numerical values where appropriate.

21. Determine the mass of a free particle whose wave function is the plane wave

$$\Psi(x, t) = \hat{A}e^{i(2.5 \times 10^{11}x - 2.1 \times 10^{13}t)}$$

where distance is in meters and time in seconds.

## Section 3.4

22. A visual inspection of an ant (mass: 0.5 mg) verifies that it is within an uncertainty of 0.7 $\mu$m of a given point, apparently stationary. How fast might the ant actually be moving?

23. The uncertainty in the position of a baseball of mass 0.145 kg is 1 $\mu$m. What is the minimum uncertainty in its speed?

24. A mosquito of mass 0.15 mg is found to be flying at a speed of 50 cm/s within an uncertainty of 0.5 mm/s. (a) How precisely may its position be known? (b) Does this inherent uncertainty present any hindrance to the application of classical mechanics?

25. The position of a neutron in a nucleus is known within an uncertainty of $\sim 5 \times 10^{-15}$ m. At what speeds might we expect to find it moving?

26. To how small a region must an electron be confined for borderline relativistic speeds—say, $0.05c$—to become reasonably likely? On the basis of this, would you expect relativistic effects to be prominent for hydrogen's electron, of orbit radius near $10^{-10}$ m? For a lead atom "inner-shell" electron, of orbit radius $10^{-12}$ m?

27. A 65-kg man walks at 1 m/s, known to within an uncertainty (unrealistically small) of 1 $\mu$m/h. (a) Compare the minimum uncertainty in his position to his actual physical dimension in his direction of motion, 25 cm from front to back. (b) Is it sensible to apply the uncertainty principle to the man?

28. One of the cornerstones of quantum mechanics is that bound particles cannot be stationary—even at zero absolute temperature! A "bound" particle is one that is confined in some finite region of space, as is an atom in a solid. There is a nonzero lower limit on the kinetic energy of such a particle. Suppose a particle is confined in one dimension to a region of width $L$. Obtain an approximate formula for its minimum kinetic energy.

29. A crack between two walls is 10 cm wide. What is the angular width of the central diffraction maximum when (a) an electron moving at 50 m/s passes through? (b) When a baseball of mass 0.145 kg and speed 50 m/s passes through? (c) In each case an uncertainty in momentum is introduced by the "experiment" (i.e., passing through the slit). Specifically, what aspect of the momentum becomes uncertain, and how does this uncertainty compare to the initial momentum of each?

30. If things really do have a dual wave-particle nature, then if the wave spreads, the probability of finding the particle should spread *proportionally*, independent of the degree of spreading, mass and speed, and even Planck's constant.

     Imagine that a beam of particles of mass $m$ and speed $v$, moving in the $x$-direction, passes through a single slit of width $a$. Show that the angle $\theta_1$ at which the first diffraction minimum would be found ($m\lambda = a \sin \theta_m$, from "wave theory") is proportional to the angle at which the particle would likely be deflected $\theta \cong \Delta p_y/p$, and that the proportionality factor is a pure number, independent of $m$, $v$, $a$, and $h$. (Assume small angles: $\sin \theta \cong \tan \theta \cong \theta$.)

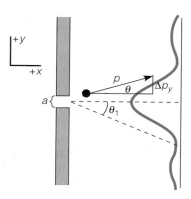

31. In Exercise 27, the case is made that the position uncertainty for a typical macroscopic object is generally so much smaller than its actual physical dimensions that applying the uncertainty principle would be absurd. Here we gain some idea of how small an object would have to be before quantum mechanics might rear its head. The density of aluminum, $2.7 \times 10^3$ kg/m$^3$, is typical of solids and liquids around us. Suppose we could narrow down the velocity of an aluminum sphere to within an uncertainty of 1 $\mu$m per decade. How small would it have to be for its position uncertainty to be at least as large as $\frac{1}{10}\%$ of its radius?

32. A particle is connected to a spring and undergoes one-dimensional motion.
    (a) Write an expression for the total (kinetic plus potential) energy of the particle in terms of its position $x$, its mass $m$, its momentum $p$, and the force constant $\kappa$ of the spring.
    (b) Now treat the particle as a wave. Assume that the product of the uncertainties in position and momentum is governed by an uncertainty relation $\Delta p\, \Delta x \approx \frac{1}{2}\hbar$. Also assume that since $x$ is on average zero, the uncertainty $\Delta x$ is roughly equal to a typical value of $|x|$. Similarly assume that $\Delta p \cong |p|$. Eliminate $p$ in favor of $x$ in the energy expression.
    (c) Find the minimum possible energy for the wave.

33. The energy of a particle of mass $m$ bound by an unusual spring is $p^2/2m + bx^4$.
    (a) Classically, it can of course have zero energy. Quantum-mechanically, however, while both $x$ and $p$ are "on average" zero, its energy cannot be zero. Why?
    (b) Roughly speaking, $\Delta x$ is a typical value of the particle's position. Making a reasonable assumption

about a typical value of its momentum, find the particle's minimum possible energy.

## Section 3.6

34. A 1-ns pulse of electromagnetic waves would be 30 cm long. (a) Consider such a pulse of 633-nm wavelength laser light. Find its central wave number and the range of wave numbers it comprises. (b) Repeat part (a), but for a 1-ns pulse of 100-MHz radio waves.

35. If a laser pulse is of short enough duration, it becomes rather superfluous to refer to its specific color. How short a duration must a light pulse be for its range of frequencies to cover the entire visible spectrum? (The visible spectrum covers frequencies of ~4.5 to 7.5 × $10^{14}$ Hz; 1 fs = $10^{-15}$ s.)

36. What is the range of frequencies in a 1-ns pulse of (a) 1060-nm infrared laser light, and (b) 100-MHz radio waves? (c) For which is the "uncertainty" in frequency, relative to its approximate value, larger?

37. A 1-fs pulse of laser light would be 0.3 $\mu$m long. What is the range of wavelengths in a 0.3-$\mu$m-long pulse of (approximately) 600-nm laser light? (1 fs = $10^{-15}$ s.)

38. Determine the spectral content $\tilde{f}(k)$ of the function $f(x)$ and interpret the result.

$$f(x) = \begin{cases} C & -\frac{1}{2}a \le x \le +\frac{1}{2}a \\ 0 & |x| > \frac{1}{2}a \end{cases}$$

39. Determine the spectral content $\tilde{f}(k)$ of the oscillatory function $f(x)$ and interpret the result.

$$f(x) = \begin{cases} e^{ik_0 x} & -\frac{1}{2}L \le x \le +\frac{1}{2}L \\ 0 & |x| > \frac{1}{2}L \end{cases}$$

40. Determine the spectral content $\tilde{f}(k)$ of the oscillatory function $f(x)$ and interpret the result. [The identity $\cos(k_0 x) = \frac{1}{2}(e^{+ik_0 x} + e^{-ik_0 x})$ is useful.]

$$f(x) = \begin{cases} \cos(k_0 x) & -\frac{1}{2}L \le x \le +\frac{1}{2}L \\ 0 & |x| > \frac{1}{2}L \end{cases}$$

41. A particle is described by

$$\psi(x) = \begin{cases} C & |x| < \frac{1}{2}w \\ 0 & |x| > \frac{1}{2}w \end{cases}$$

What momenta can never be measured?

42. A function $f(\alpha)$ is nonzero only in the region of width $2\delta$ centered at $\alpha = 0$.

$$f(\alpha) = \begin{cases} C & |\alpha| \le \delta \\ 0 & |\alpha| > \delta \end{cases}$$

where $C$ is a constant.

(a) Find and plot versus $\beta$ the Fourier transform $\tilde{f}(\beta)$ of this function.

(b) The function $f(\alpha)$ might represent a pulse occupying either finite distance ($\alpha$ = position) or finite time ($\alpha$ = time). Comment on the wave number spectrum if $\alpha$ is position, and the frequency spectrum if $\alpha$ is time. Specifically address the dependence of the "width" of the spectrum on $\delta$.

43. A signal is described by the function

$$D(t) = Ae^{-|t|/\tau}$$

(a) Calculate the Fourier transform $\tilde{D}(\omega)$. Sketch and interpret your result.

(b) How are $D(t)$ and $\tilde{D}(\omega)$ affected by a change in $\tau$?

44. Determine the Fourier transform of the function $D(t)$ and interpret the result. [The identity $\sin(\omega_0 t) = -\frac{1}{2}i(e^{+i\omega_0 t} - e^{-i\omega_0 t})$ is useful.]

$$D(t) = \begin{cases} i \sin(\omega_0 t) & -\frac{1}{2}\tau \le t \le +\frac{1}{2}\tau \\ 0 & |t| > \frac{1}{2}\tau \end{cases}$$

45. Consider the following function:

$$f(x) = \begin{cases} Ae^{+\alpha x} & -\infty < x < 0 \\ Be^{-\alpha x} & 0 \le x < +\infty \end{cases}$$

(a) Sketch this function. (Without loss of generality, it may be assumed that $A$ is greater than $B$.) Calculate the Fourier transform $\tilde{f}(k)$.

(b) Show that for large $k$, $\tilde{f}(k)$ is proportional to $1/k$.

(c) In general, $f(x)$ is not continuous. Under what condition will it be, and how does $\tilde{f}(k)$ behave at large values of $k$ if this condition holds?

(d) How does a discontinuity in a function affect the spectral content for large values of $k$?

## Section 3.7

46. The allowed electron energies predicted by the Bohr model of the hydrogen atom are correct. (a) Determine the three lowest. (b) The electron can "jump" from a higher to lower energy, with a photon carrying away the energy difference. From the three energies found in part (a), determine three possible wavelengths of light emitted by a hydrogen atom.

47. In the Bohr model of the hydrogen atom, the electron can have only certain velocities. Obtain a formula for the allowed velocities, then obtain a numerical value for the highest speed possible.

48. What is the density of a solid? Although the *mass* densities of solids vary greatly, the *number* densities (in mol/m³) vary surprisingly little. The value, of course, hinges on the separation between the atoms—but where does a theoretical prediction start? The electron in hy-

drogen must be treated not as a particle orbiting at a strict radius, as in the Bohr atom, but as a diffuse orbiting wave. Given a diffuse probability, identically repeated experiments dedicated to "finding" the electron would obtain a range of values—with a mean and standard deviation (i.e., uncertainty). Nevertheless, the allowed radii predicted by the Bohr model are very close to the true mean values.

(a) Assuming that atoms are packed into a solid typically $j$ Bohr radii apart, what would be the number of moles per cubic meter?

(b) Compare this with the typical mole density in a solid of $10^5$ mol/m$^3$. What would be the value of $j$?

## General Exercises

49. (a) Find the wavelength of a proton whose kinetic energy is equal to its internal energy. (b) The proton is generally regarded as being roughly of radius $10^{-15}$ m. Would this proton behave as a wave, or a particle?

50. According to the energy-time uncertainty principle, the lifetime $\Delta t$ of a transition between states of different energy is inversely proportional to the uncertainty $\Delta E$ in the energy difference. Therefore, in real cases in which a photon carries off the energy difference between two states, both the duration of the transition and the uncertainty in the energy of the photon are nonzero. Hydrogen's 656-nm red spectral line is the result of a transition between quantum states of the electron in the hydrogen atom. Such transitions occur within approximately $10^{-8}$ s.

(a) What inherent uncertainty in the energy of the emitted photon does this imply? (*Note:* Unfortunately, we might use the symbol $\Delta E$ for the energy difference—i.e., the energy of the photon—but

here it means the *uncertainty* in that energy difference.)

(b) To what range in wavelengths does this correspond? (As noted in Exercise 1.35, the uncertainty principle is one contributor to the broadening of spectral lines.)

(c) Obtain a general formula relating $\Delta\lambda$ to $\Delta t$.

51. The proton and electron were known to exist years before the discovery of the neutron, and the structure of the nucleus accordingly was baffling. Helium's nucleus, for example, has a mass approximately four times the proton mass but a charge of only twice that of the proton. It was suggested that it consisted of four protons plus two electrons, accounting for the mass (electrons are "light") and the total charge. (Today, of course, we know that the helium nucleus is two protons and two *uncharged* neutrons whose mass is very nearly that of the protons.) Quantum mechanics makes the electrons-in-the-nucleus theory untenable. A *confined* electron is a *standing* wave, whose wavelength in the nucleus could be no longer than approximately $4R_{nuc}$. (By analogy, the "fundamental" standing wave on a string has a wavelength satisfying $L = \frac{1}{2}\lambda$. In the case of the nucleus, the "length of the string" $L$ is the diameter, $2R_{nuc}$.)

Assuming a nuclear charge of $2e$ and a typical nuclear radius of $5 \times 10^{-15}$ m, show that the kinetic energy of an electron standing wave confined in the nucleus would be much greater than the magnitude of the attractive potential energy when the electron is at the surface of the nucleus. (*Warning:* Is the electron moving "slow," or "fast"?) Since it would be the responsibility of the electrostatic potential energy to prevent the electron from escaping, this calculation effectively proves that an electron cannot be confined in the nucleus.

# Bound States: Simple Cases

To make quantum mechanics useful in real applications, we must be able to account for the effects of external forces.[1] We begin this chapter by adapting the Schrödinger equation to include these effects. Afterward, we will study applications to **bound states.** A bound state is one in which a particle's motion is restricted by an external force/interaction to a finite region of space. A mass on a spring, subject only to a "Hooke's law" restoring force, is an example of a bound state; no matter how much energy is given to the mass, the spring will invariably cause it to oscillate between finite extremes of travel ("turning points") equidistant from the spring's equilibrium position. Were the spring to break, the mass would no longer be bound; it would move at constant speed in one direction indefinitely. We take up the study of such "unbound states" in Chapter 5. It may seem backward to study the behavior of particles bound by external forces before studying presumably simpler unbound states, but the study of bound states turns out to be the simplest, and therefore best, framework for grasping most of the basic ideas of quantum mechanics.

The standard problems we consider here may seem oversimplified and therefore unrealistic and of little applicability, but an understanding of them is crucial

---

[1] The word *interaction,* encompassing both force and potential energy, would be a better choice. We will be concerned exclusively with conservative forces, for each of which there is a potential energy dependent solely on position. Thus, a particle experiencing an external force has potential energy. Because force may be found from potential energy (e.g., $F = -dU/dx$), and vice versa, they are really two aspects of the same thing.

for two reasons. First, as in all of science, we must understand the simple if we are to have any hope of understanding the complicated; much is to be learned about the nature of quantum mechanics by studying the so-called simple cases. Secondly, most of the real applications confronting scientists cannot be solved exactly. They are either too complicated mathematically, or are plagued by incomplete knowledge of the interactions, or (most often) they are both. A simple, tractable model is required. Our simple cases are indeed used as models of more complicated systems.

## 4.1 The Schrödinger Equation

In Section 3.3, we discussed the Schrödinger equation in the absence of external forces:

$$-\frac{\hbar^2}{2m}\frac{\partial^2\Psi(x,\,t)}{\partial x^2} = i\hbar\frac{\partial\Psi(x,\,t)}{\partial t} \tag{4-1}$$

By substitution, we verified that the plane wave $\Psi(x,\,t) = Ae^{i(kx-\omega t)}$ is a solution:

$$-\frac{\hbar^2}{2m}(ik)^2 Ae^{i(kx-\omega t)} = i\hbar(-i\omega)Ae^{i(kx-\omega t)}$$

or

$$\frac{\hbar^2 k^2}{2m}\Psi(x,\,t) = \hbar\omega\Psi(x,\,t)$$

Using $p = \hbar k$ and $E = \hbar\omega$, this becomes

$$\frac{p^2}{2m}\Psi(x,\,t) = E\Psi(x,\,t) \quad\rightarrow\quad KE\,\Psi(x,\,t) = E\Psi(x,\,t)$$

An important conclusion is that the Schrödinger equation is based upon energy. There would seem to be an obvious way to generalize it to include the effects of an external force—merely "tack on" the potential energy associated with that force:

$$(KE + U(x))\Psi(x,\,t) = E\Psi(x,\,t)$$

Guided by this, it is reasonable that the appropriate partial differential equation to replace (4-1) should be

$$-\frac{\hbar^2}{2m}\frac{\partial^2\Psi(x,\,t)}{\partial x^2} + U(x)\Psi(x,\,t) = i\hbar\frac{\partial\Psi(x,\,t)}{\partial t} \tag{4-2}$$

Time-dependent Schrödinger equation[2]

The predictions yielded by equation (4-2) have been borne out experimentally, and on this basis we accept it as the correct "law" by which the wave function may be found. To determine the behavior of a particle in classical mechanics, we solve $\mathbf{F} = m(d^2\mathbf{r}/dt^2)$ for $\mathbf{r}(t)$, given knowledge of the net force on the particle. The analogous task in quantum mechanics is to solve the Schrödinger equation for $\Psi(x,\,t)$, given knowledge of the potential energy $U(x)$.

[2]The "time-dependent Schrödinger equation" is the general form of the Schrödinger equation, so it is not really necessary to qualify it as "time-dependent." We do this only to distinguish it from the soon-to-be-discussed "time-independent Schrödinger equation," a special case of the general form.

## 4.2 Stationary States

The first step in solving (4-2) is the use of a standard mathematical technique: **separation of variables.** We assume that the wave function may be expressed as a product of a spatial part and a temporal part.

$$\Psi(x,\ t)\ =\ \psi(x)\phi(t) \tag{4-3}$$

Inserting this in (4-2), then factoring out terms constant with respect to the partial derivatives, we have

$$-\frac{\hbar^2}{2m}\phi(t)\frac{\partial^2\psi(x)}{\partial x^2}\ +\ U(x)\psi(x)\phi(t)\ =\ i\hbar\psi(x)\frac{\partial\phi(t)}{\partial t}$$

Finally, dividing by $\psi(x)\phi(t)$ yields

$$-\frac{\hbar^2}{2m}\frac{1}{\psi(x)}\frac{\partial^2\psi(x)}{\partial x^2}\ +\ U(x)\ =\ i\hbar\frac{1}{\phi(t)}\frac{\partial\phi(t)}{\partial t}$$

The left-hand side depends solely on position, the right solely on time—the variables are separate. Position and time are both independent variables; a wave varies with position at a given time, and with time at a given position. Therefore, the equation can hold for all values of $x$ and $t$ *only* if both sides happen to be the same constant. Let us use the symbol $C$ for this constant.

$$-\frac{\hbar^2}{2m}\frac{1}{\psi(x)}\frac{d^2\psi(x)}{dx^2}\ +\ U(x)\ =\ i\hbar\frac{1}{\phi(t)}\frac{d\phi(t)}{dt}\ =\ C \tag{4-4}$$

We now have two separate ordinary differential equations, one involving the spatial part of the wave function and one the temporal. Since the temporal part is simpler, being independent of the potential energy, we start there.

## The Temporal Part of $\Psi(x, t)$

$$i\hbar\frac{1}{\phi(t)}\frac{d\phi(t)}{dt}\ =\ C\ \rightarrow\ \frac{d\phi(t)}{dt}\ =\ -\frac{iC}{\hbar}\phi(t)$$

Substitution will verify that the solution of this first-order differential equation is

$$\phi(t)\ =\ Ae^{-i(C/\hbar)t}$$

This oscillatory function has a well-defined angular frequency, that is, the coefficient multiplying $t$, of $\omega = C/\hbar$. (As the *spatial* frequency $k$ of $e^{ikx}$ is well defined, so is the *temporal* frequency $\omega$ of $e^{-i\omega t}$.) Thus, it has the well-defined energy

$$E\ =\ \hbar\omega\ =\ C$$

*In separating variables, we are considering only those solutions for which the energy is well defined, and the separation constant is that energy.*

The symbol $A$ is an arbitrary multiplicative constant. The spatial part of the wave function $\psi(x)$ will also have such a constant. Because $\psi(x)$ and $\phi(t)$ are ultimately multiplied, we adopt the simplifying convention of treating the product of the two constants as a single one, which is kept with $\psi(x)$. Therefore,

$$\phi(t) = e^{-i(E/\hbar)t} \tag{4-5}$$

Temporal part of $\Psi(x, t)$

and the total wave function is

$$\Psi(x, t) = \psi(x)e^{-i(E/\hbar)t} \tag{4-6}$$

Wave functions whose energy is well defined have a very important property: The probability density doesn't vary with time.

$$\Psi^*(x, t)\Psi(x, t) = [\psi^*(x)e^{+i(E/\hbar)t}][\psi(x)e^{-i(E/\hbar)t}] = \psi^*(x)\psi(x) \tag{4-7}$$

Probability density is subject to experimental scrutiny, but the wave function itself is not. So we see that for wave functions of form (4-6), the particle's whereabouts do not change in any *observable* way. Accordingly, they are called **stationary states.** This is a pivotal idea in quantum mechanics: *Classically,* an electron orbiting in an atom should constantly lose energy in the form of electromagnetic radiation—the atom should be unstable. *Quantum-mechanically,* it need not be; the electron is not an accelerating charged particle, but rather a stationary "cloud," whose charge density is constant. Quantum mechanics explains the stability of the atom!

### The Spatial Part of $\Psi(x, t)$

Turning now to the "spatial half" of equation (4-4) and replacing $C$ by $E$, we obtain what is known as the **time-independent Schrödinger equation.**

$$-\frac{\hbar^2}{2m}\frac{d^2\psi(x)}{dx^2} + U(x)\psi(x) = E\psi(x) \tag{4-8}$$

Time-independent
Schrödinger equation

This is as far as we can go without knowledge of $U(x)$. All stationary-state wave functions are of the form (4-6), but for each $U(x)$, there is a different $\psi(x)$ given by (4-8). Thus, it is the potential energy that distinguishes one application from another ( just as the net force does classically). Unfortunately, the functions $U(x)$ for which the time-independent Schrödinger equation has a simple solution are few indeed, and it is this fact that dictates what we regard as a "simple" case.

Before concentrating on the spatial part of the wave function, we note that even though the time-*independent* Schrödinger equation contains no complex numbers, and $\psi(x)$ is thus real, the (total) wave function is still complex, always containing $e^{-i(E/\hbar)t}$. The temporal part of the wave function must not be forgotten.[3]

## 4.3 Well-Behaved Functions and Normalization

Each simple case we consider has its own unique wave functions describing a given "particle," but certain conditions must be met in all cases. For one thing, the total probability of finding the particle must be unity. Perhaps the particle is an electron; whether or not it has a definite location, it has a definite charge with

[3]Even so, it is common habit to refer to $\psi(x)$, the *spatial* part of the wave function, as simply "the wave function." Not without drawbacks, this habit in turn leads to the coinage of new descriptive terms, such as "total" and "overall" wave function. The reader should find that little confusion actually results from the loose usage.

a strict value—we cannot have half an electron. The procedure by which we en-sure that the wave function gives a unit probability is called **normalization.** Another condition is that a wave function be **smooth.**

## Normalization

*To be acceptable, a wave function must be normalizable.* If the probability per unit distance $|\Psi(x, t)|^2$ is multiplied by a small distance $dx$, then summed over all space, the result must be unity.

$$\int_{\text{all space}} |\Psi(x, t)|^2 \, dx = 1 \tag{4-9}$$

We might liken the probability to a feather pillow. It may be squashed flat, fluffed into a rounded form, or compressed to a small volume, but although the density and extent may vary, the total amount of feathers is fixed.

The requirement that the probability integral be a finite number places two specific restrictions on the wave function: (1) It must not diverge (i.e., approach $\infty$). (2) If it does extend to $\pm\infty$ (as does a gaussian), it must fall to zero faster than $|x|^{-1/2}$. The first is sensible. Let us consider the second: Suppose the wave function were proportional to $x^{-a}$ as $x$ becomes large. Its square would be proportional to $x^{-2a}$, so the probability of finding the particle between some arbitrary $x_0$ and $\infty$ would be proportional to $\int_{x_0}^{\infty} x^{-2a} \, dx$. As we know from calculus, $\int_{z_0}^{\infty} z^{-b} \, dz$ diverges unless $b > 1$. Therefore, the probability would not be normalizable unless $2a > 1$, or $a > \frac{1}{2}$. (In three dimensions, where the square of the wave function is probability *per unit volume,* the corresponding requirement is that the wave function fall off faster than $|r|^{-3/2}$.) (See the essay entitled "Two Exceptional Wave Functions.")

## Smoothness

*To be acceptable, a wave function must be smooth.* There are two aspects to smoothness: continuity of the wave function, and continuity of its derivative. Under no circumstance may the wave function itself be discontinuous. Not only

---

### Two Exceptional Wave Functions

There are two notable exceptions to the normalization restrictions on the wave function: (1) The function representing a particle whose position is known precisely ($\Delta x = 0$) *does* diverge, but only at a single point, at which the probability density should indeed be infinite. Its name is the Dirac delta function. The spectral analysis of Section 3.6 tells us that such a function comprises an infinitely broad range of wave numbers; $\Delta k = \Delta p = \infty$. (2) The complex exponential $\psi(x) = Ae^{ikx}$, representing a particle whose wave number and momentum are known precisely

($\Delta p = 0$), has a *constant* amplitude throughout space. As we know, such a particle is equally likely to be found anywhere; $\Delta x = \infty$. Of course, for the integral of $|\psi(x)|^2$ over all space to be finite, the amplitude $A$ would have to be infinitesimal. Whether or not these two special cases are realistic, they are convenient for representing a particle whose position or momentum is reasonably certain. Significantly, there is a link between them: The Dirac delta function and the complex exponential are Fourier transforms of one another (see Section. 3.6).

would this be contrary to what we expect of wavelike phenomena, it would represent a particle of infinite kinetic energy. Figure 4.1 shows a wave with an abrupt deviation. Near the deviation it resembles a wave of very short wavelength. We may say that it possesses a significant "amount" of a very large momentum. Were we to measure its momentum, there would be a significant probability of finding a very large value. In the limit that the abrupt deviation becomes a discontinuity, the "short wavelength" becomes zero wavelength, with infinite momentum and kinetic energy. Thus, a discontinuity is unacceptable. The mathematical form of the Schrödinger equation argues for the same restriction: Its kinetic energy term involves a second derivative of $\psi(x)$ with respect to position. Were the wave function discontinuous, its first derivative would be infinite at the point of discontinuity; its second derivative would be decidedly pathological. No sense could be made of such a wave function.

With one exception, the first derivative of the wave function must also be continuous. If the first derivative is continuous, the second derivative and thus the kinetic energy are finite. If the potential energy $U(x)$ and total energy $E$ are both finite, the Schrödinger equation couldn't hold otherwise. The exception arises when this argument breaks down. If the potential energy does jump to infinity at some point, we cannot demand a continuous derivative. Obviously, this is not completely realistic, but the notion of a solid wall or barrier is a useful one. In confining a particle to some region of space, we often imagine that it is confined between "walls" stout enough to prevent its escape, no matter how much kinetic energy it might have. If we assume that a wall exerts a strong repulsive force—electrostatic, for instance—the particle encountering the wall will lose kinetic energy rapidly, gaining electrostatic potential energy. The strongest wall would be one where the potential energy jumps to an arbitrarily large value immediately; a particle of any finite speed would instantly rebound. Thus, in the idealized but useful case of walls of infinite potential energy, the derivative of the wave function is discontinuous.

## 4.4 A Review of Classical Bound States

There is one last thing to do before we consider our first simple case of a quantum-mechanical bound particle. We must understand what goes on when a classical particle is bound. In classical mechanics, we learn that when nonconservative forces are absent, total mechanical energy, $E = \text{KE} + U$, is conserved. Plots of energy versus position show this clearly. Figure 4.2 gives the simple example of a

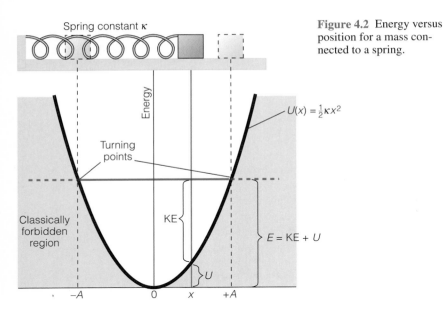

**Figure 4.2** Energy versus position for a mass connected to a spring.

mass on a spring of spring constant $\kappa$. The potential energy of a spring is $\frac{1}{2}\kappa x^2$—a parabolic plot. The total mechanical energy never varies from whatever value is initially given to the system—it is a horizontal line. The kinetic energy is the "distance" above the potential energy parabola but below the total energy line. These "heights" are labeled for an arbitrary value of $x$ and arbitrary value of $E$. Increasing $E$ would increase the amplitude $A$ of the motion, but at any finite $E$, the mass is bound within the region $-A \leq x \leq +A$.

Now suppose that at the arbitrary $x$-value shown in Figure 4.2, the mass is moving to the right. Its kinetic energy is large compared to the potential, but as it moves, potential increases as kinetic decreases—it slows down. When it reaches the point $x = +A$, the energy is all potential; the kinetic energy is zero. This is a "turning point." The mass is stationary, there is a force on it to the left, and it must return toward the origin. It cannot proceed to values of $x$ larger than $+A$. Were it to do so, the potential energy would of course increase, but since the total must be constant, the kinetic energy would have to be negative! Because for any given $E$, the mass is confined to values of $x$ "inside" the potential energy plot, the shaded region outside is known as the **classically forbidden region.**

Figure 4.3 shows a different system in which bound states are possible. Atom 1 is fixed at the origin, while atom 2 moves in response to their shared electrostatic interaction. (We won't study the origin of this potential energy; we simply note that atoms, due their dispersed orbiting electrons and compact positive nuclei, do not attract or repel as simple point charges.) The potential energy is positive and large for small interatomic separation $x$, has a minimum value that is negative, then asymptotically approaches zero from the negative side as $x \to \infty$. At small values of $x$, atom 2 is strongly repelled by atom 1 [$F = -(dU/dx) > 0$], and at large $x$, it is attracted [$F = -(dU/dx) < 0$]. Accordingly, a bound, oscillatory state (a diatomic molecule) is possible—but it depends on the total energy. If the total energy is negative (labeled $E_{bound}$), atom 2 will be bound, oscillating between turn-

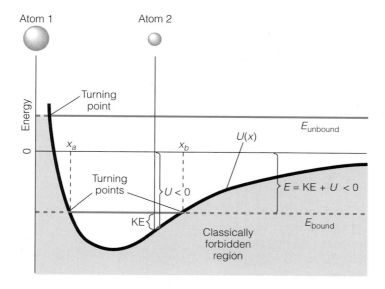

**Figure 4.3** Energy versus position for the interatomic force between a large atom at the origin and a small one free to move.

ing points $x_a$ and $x_b$.[4] If the total energy is positive (labeled $E_{unbound}$), atom 2 has only a single turning point. Were it moving toward atom 1, it would momentarily stop at this point, then move away, never to return. Although the potential energy does increase as $x$ becomes very large, so that atom 2 slows down as it moves off, there is no outer turning point where the potential energy rises to meet the total energy. By contrast, there is no total energy large enough that a mass on a spring would be unbound, since the harmonic-oscillator potential energy $\frac{1}{2}\kappa x^2$ increases without bound in both directions. In summary, a bound state results when a particle is "caught" between two turning points, where the potential energy rises to meet the total energy.

## Bound States in Quantum Mechanics

In common with the classical bound states, quantum-mechanical bound states have well-defined, constant energy, which is somehow divided between kinetic and potential. But in quantum mechanics we don't speak of a particle's position at a given time—a classical notion. Rather, we seek the wave function $\Psi(x, t)$. We will find that in quantum mechanics, bound states are *standing waves*. Accordingly, just as there are only certain discrete standing waves possible for waves on a string stretched between fixed ends, there are only certain discrete states possible for a matter wave bound by a potential energy. Figure 4.4 shows the first four possible standing waves obtained in a quantum-mechanical treatment of the harmonic oscillator (the topic of Section 4.8). They are plotted on separate horizontal axes whose heights correspond to their respective energies. Different energies are possible, but not all energies. The "fundamental" (one antinode) standing wave is the lowest in energy; the "second harmonic" (two antinodes) is next, and so forth.

Besides being restricted to only certain energies, the standing waves of quantum-mechanical bound states differ from classical bound states in another way: While a classical particle is restricted to locations "inside" the potential

[4]Energy may be negative due to the usual freedom to choose zero potential energy. We choose it at $x = \infty$. Although this makes the *potential* energy negative for certain values of $x$, *kinetic* energy is still positive between $x_a$ and $x_b$, as it must be. The reason for the choice is convenience; it divides bound from unbound states simply according to the sign of the total energy.

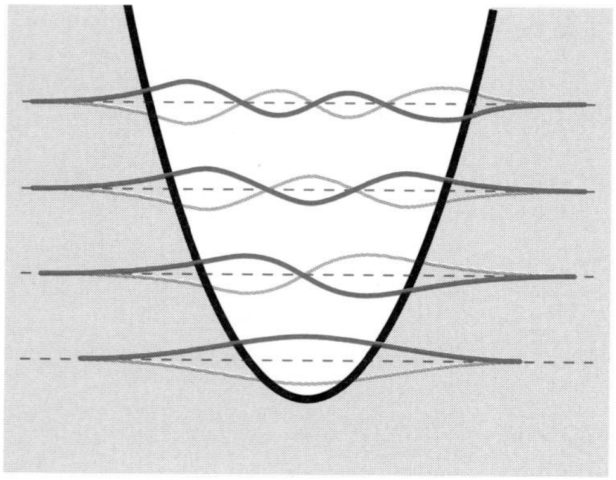

**Figure 4.4** In quantum mechanics, bound states are standing waves.

energy plot, wave functions for quantum-mechanical bound states often extend beyond the classical turning points—into the "classically forbidden region." Consequently, there is a probability of finding the particle where, classically, it cannot be!

# 4.5  Case 1: Particle in a Box—Infinite Well

The situation in which the particle-confining potential energy $U(x)$ allows the simplest solution of the time-independent Schrödinger equation is the so-called **particle in a box,** or **infinite well,** in which a particle is confined between (infinitely) rigid walls. Classically, it would simply bounce back and forth, but quantum-mechanically it is a standing wave.

The forces exerted by ordinary walls of wood, brick, and so on, are complicated; we simplify the problem by choosing the particle to be an electron and the "walls" to be electrostatic potential energy barriers. As we proceed, we shall see what further assumptions are needed to give the true simplest case of an infinite well.

Batteries are connected to two parallel-plate capacitors separated by a distance $L$, and an electron moves in the region between them, as shown in Figure 4.5. Small holes allow the electron entry. The batteries' positive terminals are connected to the "inside" plates, so once inside, the electron experiences an electrostatic force back the way it came.[5] Since (ideally) the electric field is confined to the space within each capacitor, the electron would move with constant speed and kinetic energy everywhere else. Correspondingly, the electrostatic potential and the electron's potential energy are *constants* everywhere but in the space within each capacitor. Defining the batteries' positive terminals as $V = 0$, and their negative terminals as $V = -V_0$, the electron's potential energy would be zero when between the capacitors and $U_0 = eV_0$ when outside the capacitors (on either side). The potential energy is shown in Figure 4.6.

Suppose the electron is between the capacitors, with kinetic energy greater than $eV_0$. Once inside a capacitor, it would of course slow down, but it would nonetheless pass completely through, never to return. It would not be bound. If, on the other hand, its kinetic energy were less than $eV_0$, it could not escape. Its kinetic

[5]Remember, the force on an electron is $\mathbf{F} = (-e)\mathbf{E}$; it is *opposite* the direction of the electric field. Similarly, the sign of an electron's potential energy is opposite that of the potential; $U = (-e)V$.

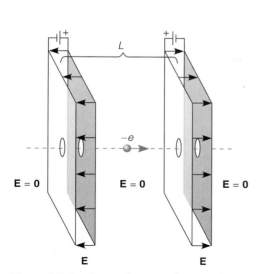

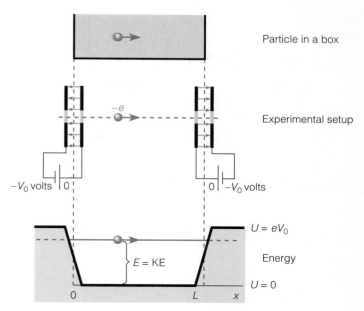

Particle in a box

Experimental setup

$-V_0$ volts    $-V_0$ volts

$U = eV_0$

Energy

$U = 0$

**Figure 4.5**  An electron between electrostatic "walls."

**Figure 4.6**  Confining a particle.

energy would fall to zero before it reached the negative plate, and it would therefore return the way it came. It is such a bound state that we consider.

We now further simplify our problem. To avoid addressing the electron's behavior in the space within each capacitor, we assume the capacitors to be of negligible width. The potential energy that results is known as a **potential well,** Figure 4.7. With this assumption, the particle would rebound at the wall instantly—very simple. It is true that the assumption is unrealistic, but it is a good approximation of more realistic cases, and it simplifies the solution of the Schrödinger equation greatly. It divides the one-dimensional space into (*only*) three distinct regions in which $U(x)$ is constant: $0 < x < L$ (region I), $x < 0$ (region II) and $x > L$ (region III). The all-important potential energy is thus

$$U(x) = \begin{cases} 0 & 0 < x < L \\ U_0 & x < 0, \; x > L \end{cases} \tag{4-10}$$

In cases where the potential energy changes abruptly from one region to another, we solve the Schrödinger equation for $\psi(x)$ independently in each region ("piecewise"). Then, by adjusting the arbitrary constants accompanying each $\psi(x)$, we

II          I          III

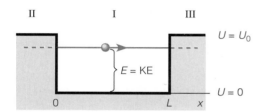

**Figure 4.7**  A "potential well."

$U = U_0$

$E = \text{KE}$

$U = 0$

ensure that the wave function as a whole obeys the required smoothness conditions and is normalizable.

---

## Solving the Schrödinger Equation: Potential Well

### Region I ($0 < x < L$)

Since $U(x) = 0$ here, the time-independent Schrödinger equation (4-8) is

$$-\frac{\hbar^2}{2m}\frac{d^2\psi(x)}{dx^2} = E\psi(x) \quad \text{or} \quad \frac{d^2\psi(x)}{dx^2} = -\frac{2mE}{\hbar^2}\psi(x)$$

For convenience, let us make the following definition (which we very soon see is a wave number, thus the symbol):

$$k \equiv \sqrt{\frac{2mE}{\hbar^2}} \tag{4-11}$$

Thus,

$$\frac{d^2\psi(x)}{dx^2} = -k^2\psi(x)$$

As with all *second*-order differential equations, the solution will involve *two* independent functions and *two* arbitrary constants. Substitution will verify that a pair of independent solutions[6] is

$$\psi(x) = A \sin kx \quad \text{and} \quad \psi(x) = B \cos kx$$

(As discussed in the essay entitled "Other Solutions," there is an alternative mathematical solution, but it ends up being equivalent physically.) The general solution is the sum:

$$\psi(x) = A \sin kx + B \cos kx$$

This function is "smooth" and it diverges nowhere, so it is "well behaved." But to be assured that the wave function as a whole (i.e., for all space) is well behaved, we must consider the other regions.

### Region II ($x < 0$)

Here, $U(x) = U_0$, so the time-independent Schrödinger equation is

$$-\frac{\hbar^2}{2m}\frac{d^2\psi(x)}{dx^2} + U_0\psi(x) = E\psi(x) \quad \text{or} \quad \frac{d^2\psi(x)}{dx^2} = \frac{2m(U_0 - E)}{\hbar^2}\psi(x)$$

Again we make a simplifying definition:

$$\alpha \equiv \sqrt{\frac{2m(U_0 - E)}{\hbar^2}} \tag{4-12}$$

Thus,

$$\frac{d^2\psi(x)}{dx^2} = \alpha^2\psi(x)$$

[6]In physics, we are concerned with physical consequences more than mathematical procedure. Thus, the most common approach to differential equations is to guess the solution and verify that it works. Of course, this still requires some familiarity with the particular differential equation. As it turns out, most of the differential equations we "solve" are of the same two basic forms: $y' = Cy$ and $y'' = Cy$. Solutions are summarized in Appendix K.

A crucial difference in this region is that the constant $\alpha^2$ is positive, where it was negative before (i.e., $-k^2$). A pair of independent solutions is

$$\psi(x) = Ce^{+\alpha x} \quad \text{and} \quad \psi(x) = De^{-\alpha x}$$

and the general solution is the sum:

$$\psi(x) = Ce^{+\alpha x} + De^{-\alpha x}$$

To be normalizable, the wave function must not diverge. But no matter what the value of $U_0$, $e^{-\alpha x}$ diverges as $x \to -\infty$. (*Note:* $x = +\infty$ is not part of this region, so nowhere in this region does the other term diverge.) Therefore, although it is an acceptable solution mathematically, the physical requirement that the wave function be normalizable makes this term physically unacceptable: *D must be zero.*

### Region III $(x > L)$

Since $U(x)$ is the same, the mathematical solution of the Schrödinger equation in this region is the same as in region II. Thus,

$$\psi(x) = Fe^{+\alpha x} + Ge^{-\alpha x}$$

However, the physically unacceptable term in this case is $e^{+\alpha x}$, which diverges as $x \to +\infty$ for any value of $U_0$: *F must be zero.*

Altogether, we have

$$\psi(x) = \begin{cases} Ce^{+\alpha x} & x < 0 \\ A \sin kx + B \cos kx & 0 < x < L \\ Ge^{-\alpha x} & x > L \end{cases} \qquad \text{(4-13)}$$

where $k \equiv \sqrt{\dfrac{2mE}{\hbar^2}}$ and $\alpha \equiv \sqrt{\dfrac{2m(U_0 - E)}{\hbar^2}}$

Having thrown out divergent terms, and given that exponential functions do fall to zero faster than $|x|^{-1/2}$, our wave function is normalizable. Also, its pieces are separately smooth. All that remains is to make sure that the pieces obey the continuity requirements at the points where the separate regions meet.

## The Infinite Well

We shall see in Section 4.7 just how difficult it would be to proceed with the potential well without still further simplification. Indeed, it can't be solved "in closed form" (i.e., without iterative numerical techniques). In the interest of finding a "problem" that *is* solvable, then, we make one final approximation: We imagine the capacitors in Figure 4.5 to be charged to infinite(!) potential—that is, $U_0 \to \infty$. Perhaps we wish to be able to confine the electron no matter how high its kinetic energy. In any case, the problem is simplified because $\alpha \to \infty$ as $U_0 \to \infty$, so both exponentials in the solutions *outside* the well ($x < 0$ and $x > L$) go to zero. The

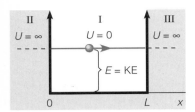

**Figure 4.8** Infinite well.

conclusion is general: *The wave function cannot extend into a region where the potential energy is infinite.* (See Exercise 4.2.) Thus, the probability of finding a particle anywhere beyond infinitely rigid walls is zero. The name given to this $U(x)$ is the **infinite well** potential energy, Figure 4.8. (Though the assumption is unrealistic, the results, as we shall see, are really very close to the case in which $U_0$ is finite but much larger than $E$. The infinite well may be viewed as a convenient approximation of that case.) For an infinite well, equation (4-13) is replaced by

$$\psi(x) = \begin{cases} A \sin kx + B \cos kx & 0 < x < L \\ 0 & x < 0,\; x > L \end{cases} \qquad (4\text{-}14)$$

Now let us finish our solution by ensuring that $\psi(x)$ has the required smoothness.

---

## Solving the Schrödinger Equation (continued): Infinite Well

### Continuity for All Space[7]

If the wave function as a whole is to be continuous, its pieces must have the same value at $x = 0$. Thus,

$$\psi_{II}(0) = \psi_I(0)$$

or

$$0 = A \sin k0 + B \cos k0 \quad \Rightarrow \quad \boxed{B = 0}$$

The cosine term must be zero if the solution in region I is to be continuous at $x = 0$ with that in region II. (See the essay entitled "Other Solutions.") Similarly, the pieces must have the same value at $x = L$:

$$\psi_I(L) = \psi_{III}(L)$$

or

$$A \sin kL = 0$$

Now we have a problem: The constant $A$ must not be zero. If it were, the entire wave function would be zero—there would be zero probability, no particle. The only way the wave function as a whole can be continuous and yet nonzero is if the following condition holds:

$$\boxed{kL = n\pi} \qquad (4\text{-}15)$$

where $n$ is an integer.

[7]Recall from Section 4.3 that since the potential energy is infinite at the walls, we do not require that the *derivative* of the wave function be continuous. In fact, it must be discontinuous.

## Other Solutions

**A**n alternative solution of the Schrödinger equation in region I is $\psi(x) = A'e^{+ikx} + B'e^{-ikx}$. Why are there two seemingly different solutions, and which is "correct"? It is left as an exercise to show that merely by redefining the constants $A'$ and $B'$ this solution is identical to the one we chose earlier. We chose sines and cosines only because we prefer to work with real functions rather than complex ones. More importantly, though, *when the conditions we require of well-behaved wave functions are imposed,* *the differences vanish.* The condition that $\psi(0) = 0$ requires that $A' + B' = 0$, so the "complex" solution would become $\psi(x) = A'(e^{+ikx} - e^{-ikx}) = 2A' \sin kx$, the same as the "real" one we chose.

Similarly, an alternative solution in the region $x > L$ is $\psi(x) = A' \sinh \alpha x + B' \cosh \alpha x$. But this diverges as $x \to +\infty$ unless $A' + B' = 0$. Again, an essentially equivalent solution results: $\psi(x) = A'(\sinh \alpha x - \cosh \alpha x) = -A'e^{-\alpha x}$.

We have "solved" the Schrödinger equation; we have found the physically acceptable wave function $\psi(x)$ for the given potential energy $U(x)$. But more importantly, the final, seemingly innocuous condition (4-15) heralds one of the most fascinating predictions of quantum mechanics; it implies that the energy of a particle is quantized. Using equation (4-11),

$$\sqrt{\frac{2mE}{\hbar^2}}\, L = n\pi \quad \Rightarrow \quad E = \frac{n^2\pi^2\hbar^2}{2mL^2}$$

The particle's energy is not arbitrary; it must be one of these discrete values.

Before seeing how we might reconcile this with our classical (and therefore limited) experience, there is a central point to be made regarding the particle's possible wave functions. Inserting $kL = n\pi$ into (4-14), the expression for $\psi(x)$, we obtain for each integer $n$—that is, for each energy—a corresponding wave function.

$$\psi_n(x) = A \sin \frac{n\pi}{L} x \quad (0 < x < L) \qquad E_n = \frac{n^2\pi^2\hbar^2}{2mL^2}$$

The wave functions are of typical standing-wave form. *Quantum-mechanically, bound states are standing waves.*

All that remains is to normalize the wave functions. In any of the allowed states, the total probability of finding the particle must be unity, and this requirement fixes the value of the remaining arbitrary constant $A$. Inserting the preceding wave function into the normalization condition (4-9), we have

$$\int_{\text{all space}} |\Psi(x, t)|^2 \, dx = \int_{-\infty}^{+\infty} |\psi(x)|^2 \, dx = \int_0^L \left( A \sin \frac{n\pi}{L} x \right)^2 dx = 1$$

[*Note:* We have used the fact, shown in equation (4-7), that the probability density is independent of time for a stationary state.] Regardless of the value of $n$, the integral $\int_0^L \sin^2[(n\pi/L)x]\, dx$ is $L/2$. Consequently,

$$A^2 \frac{L}{2} = 1 \quad \Rightarrow \quad A = \sqrt{\frac{2}{L}}$$

Thus, the normalized, continuous wave functions representing the allowed states of a particle in an infinite well, and their corresponding energies, are

$$\psi_n(x) = \begin{cases} \sqrt{\dfrac{2}{L}} \sin \dfrac{n\pi x}{L} & 0 < x < L \\ 0 & x < 0,\ x > L \end{cases} \qquad E_n = \frac{n^2\pi^2\hbar^2}{2mL^2} \qquad (4\text{-}16)$$

Figure 4.9 shows the four lowest-energy wave functions $\psi(x)$ plotted on axes whose heights are proportional to the respective energies $E_n$ (ratios $1:4:9:16$). Because the temporal parts of the wave functions oscillate in time (the meaning of the faint inverted wave at each level), they are in many ways typical standing waves.[8] In fact, the quantum energy levels may be obtained solely from classical standing-wave conditions and $p = \hbar k$: Wavelengths of standing waves are given by $\lambda_n = 2L/n$, and wave numbers by $k_n = 2\pi/\lambda = \pi n/L$, so that

$$E = \mathrm{KE} = \frac{p^2}{2m} = \frac{(\hbar k)^2}{2m} = \frac{\hbar^2 \pi^2 n^2}{2mL^2}$$

Where there are nodes in the standing waves, the probability density $|\psi(x)|^2$ is zero (Figure 4.10). But how, we might ask, can the particle get from one place in the well to another if there is no possibility of it ever being found at a point in between? The answer is simple: It is a wave, not a particle. As we know from a study of classical waves, standing waves may be thought of as two waves passing one another in opposite directions. Although they bounce back and forth, there are nodes where the displacement is always zero. We cannot demand that the "par-

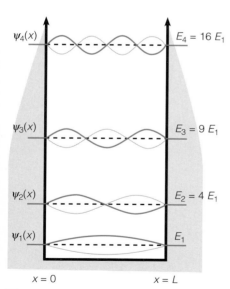

Figure 4.9 Infinite-well wave functions.

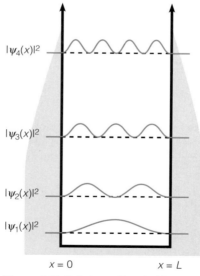

Figure 4.10 Infinite-well probability densities.

[8] While standing waves on a string are flat periodically, this is *never* the case for quantum-mechanical standing waves; the total probability must never vanish. The time-varying function multiplying $\psi(x)$ is not a sine or cosine, as for waves on a string, but a complex exponential $\phi(t) = e^{-i\omega t}$, which is never zero. When Re $\Psi(x, t)$ is zero, Im $\Psi(x, t)$ is maximum, and vice versa. One might visualize a quantum-mechanical standing wave as a portion of a sine wave spinning about the $x$-axis in the Re-Im plane.

ticle" behave as a classical particle. For one thing, it cannot be "watched"; the photons that would have to be bounced off it introduce potential energies $U(x)$ that have not been taken into account in our solution of the Schrödinger equation. The assumed simplicity of the situation forbids our having detailed knowledge of the particle's activities. On the other hand, lack of knowledge of precise whereabouts is a fact in all the real situations to which quantum mechanics is applied. We cannot, for example, say exactly what an electron orbiting a nucleus is doing. But despite these limitations, in microscopic applications, the predictions of quantum mechanics have succeeded, while those of classical physics have failed. Sight unseen, the "particle" is a wave.

The standing wave of minimum energy is known as the **ground state** and its energy as the **ground-state energy** (or **zero-point energy**). It is in the ground state that quantum-mechanical behavior deviates most from the classical expectation. Most importantly, the kinetic energy is not zero—*a bound particle cannot be stationary.* As we learned in Chapter 3, this would violate the uncertainty principle; to have a position uncertainty comparable to $L$ and a certain momentum of zero is impossible. Another deviation from the classical is that in its ground state the particle is most likely to be found near the center. Classically, if we were to "turn on the lights" suddenly and catch the particle somewhere in the course of its constant-speed back-and-forth motion, it is equally likely to be found anywhere within the well. On the other hand, the correspondence principle (Section 2.3) says that there should be a limit in which the particle indeed behaves as a classical particle. This limit is at the "other end"—*large n.* Larger $n$ correspond to shorter wavelengths ($\lambda_n = 2L/n$), and the shorter the wavelength, the more particlelike and classical the behavior should be. As we see in Figure 4.10, the larger the $n$, the more evenly the probability of finding the particle is spread over the well— the classical expectation. But we still have those perplexing nodes. Let us see what sense we can make of this.

---

**Example 4.1**

An electron is confined in an infinite well of 30 cm width. (a) What is the ground-state energy? (b) In this state, what is the probability that the electron would be found within 10 cm of the left-hand wall? (c) If the electron instead has an energy of 1.0 eV, what is the probability that it would be found within 10 cm of the left-hand wall? (d) For the 1-eV electron, what is the distance between nodes and the minimum possible fractional decrease in energy?

Solution

(a) The electron's ground-state energy is

$$E_1 = \frac{(1)^2 \pi^2 (1.055 \times 10^{-34} \text{ J·s})^2}{2(9.11 \times 10^{-31} \text{ kg})(0.3 \text{ m})^2} = 6.70 \times 10^{-37} \text{ J}$$

(b) The probability density is $|\psi(x)|^2$.

$$|\psi_n(x)|^2 = \left( \sqrt{\frac{2}{L}} \sin \frac{n\pi x}{L} \right)^2 = \frac{2}{L} \sin^2 \frac{n\pi x}{L}$$

Thus, since 10 cm is one-third the width of the well, we have

$$\text{probability} = \int \frac{\text{probability}}{\text{length}} \, dx = \int |\psi_n(x)|^2 \, dx = \frac{2}{L} \int_0^{L/3} \sin^2 \frac{n\pi x}{L} \, dx$$

$$= \frac{2}{L} \left[ \frac{x}{2} - L \frac{\sin(2n\pi x/L)}{4n\pi} \right] \Bigg|_0^{L/3} = \frac{2}{L} \left\{ \frac{L}{6} - L \frac{\sin[2n\pi(L/3)/L]}{4n\pi} \right\}$$

$$= \frac{1}{3} - \frac{1}{2n\pi} \sin \frac{2n\pi}{3}$$

The width of the well cancels. This is reasonable since we seek the probability of the electron being found within a certain *fraction* of the width and the shape of the probability density is independent of $L$. For an electron in the $n = 1$ state,

$$\text{probability} = \frac{1}{3} - \frac{1}{2\pi} \sin \frac{2\pi}{3} = \frac{1}{3} - 0.137 = 0.196$$

The probability is less than the classical expectation of one-third. This agrees with the probability density plots of Figure 4.10. In the ground state, the probability density is largest in the center. In fact, the probability of finding the electron in the *center* third must be $1 - 2 \times 0.196 = 0.609$.

(c)  To carry out the integral, we need to know $\psi(x)$, specifically the value of $n$.

$$E_n = \frac{n^2 \pi^2 \hbar^2}{2mL^2}$$

$$1.0 \text{ eV} \cdot 1.6 \times 10^{-19} \text{ J/eV} = \frac{n^2 \pi^2 (1.055 \times 10^{-34} \text{ J·s})^2}{2(9.11 \times 10^{-31} \text{ kg})(0.3 \text{ m})^2}$$

$$\Rightarrow \quad n = 4.89 \times 10^8$$

The previous result may now be used, up to the point where $n = 1$ was inserted.

$$\text{probability} = \frac{1}{3} - \frac{1}{2n\pi} \sin \frac{2n\pi}{3}$$

$$= \frac{1}{3} - \frac{1}{2(4.89 \times 10^8)\pi} \sin \frac{2(4.89 \times 10^8)\pi}{3}$$

Whatever may be the precise value of $n$, the probability is practically one-third.

(d)  From the study of classical standing waves, and as shown in Figure 4.10, the distance between nodes is $L/n$. For $n = 4.89 \times 10^8$, this is

$$\frac{0.3 \text{ m}}{4.89 \times 10^8} \cong 0.6 \text{ nm}$$

The probability density has $4.89 \times 10^8$ "bumps," each separated by 0.6 nm.

The electron's energy is quantized, so the smallest possible decrease from the $n$th state would be to the $(n - 1)$th state. Thus,

$$\text{fractional change} = \frac{E_{\text{final}} - E_{\text{initial}}}{E_{\text{initial}}} = \frac{\dfrac{(n-1)^2 \pi^2 \hbar^2}{2mL^2} - \dfrac{n^2 \pi^2 \hbar^2}{2mL^2}}{\dfrac{n^2 \pi^2 \hbar^2}{2mL^2}}$$

$$= \frac{(n-1)^2 - n^2}{n^2} = -\frac{2}{n} + \frac{1}{n^2}$$

Since $n$ is so large, the $1/n^2$ may be ignored.

$$\text{fractional change} \cong -\frac{2}{n} = -\frac{2}{4.89 \times 10^8} \cong -4 \times 10^{-9}$$

Parts (c) and (d) of Example 4.1 clarify important points. First, even though a 1 eV electron is not very energetic by ordinary standards (requiring only a meager 1-V accelerating potential), in this large—not microscopic—container, it is in a high energy state quantum-mechanically speaking: $n > 10^8$. The relative change in energy from one state to the next is less than one part in a hundred million! Quantization would be extremely difficult to substantiate; the "energy spectrum" would appear continuous. Secondly, the electron has a wavelength $(2L/n)$ very short compared to the dimension $L$ of its container. We should expect classical, particlelike behavior, and this is what we find: Not only do the electron's allowed energies appear continuous, but it is equally likely to be found in a given width $\Delta x$ anywhere within the box. To see this most clearly, we rewrite the probability density as follows:

$$|\psi_n(x)|^2 = \frac{2}{L}\sin^2\frac{n\pi x}{L} = \frac{2}{L}\left[\frac{1 - \cos(2n\pi x/L)}{2}\right] = \frac{1}{L} - \frac{\cos(2n\pi x/L)}{L}$$

The first term is what we would expect classically: A uniform probability corresponds to a probability per unit length of $1/L$. The second term oscillates between $+1/L$ and $-1/L$, $n$ times over the well's width. As a whole, then, the probability density oscillates $n$ times between 0 and $2/L$ (cf. Figure 4.10). If $n$ is large, the oscillation is so rapid that there are countless bumps in any reasonable $\Delta x$ in the well. If $n$ were $10^8$, for instance, within an interval $\Delta x$ of $\frac{1}{1000}L$, there would be $10^5$ bumps. Whether the interval included all or only part of the two bumps at its edges is immaterial; the probability of finding the particle within $\Delta x$ would be a constant. The nodes would be practically unresolvable.

The example also allows us to judge the plausibility of the infinite well's assumptions. We assumed that the potential energy is infinite outside the well, and we found that even a poky 1-eV electron would be far "above" the well's ground state. Since the capacitors (i.e., the walls) might readily be charged to thousands of volts, we see that the electron could occupy any of a huge number of states "near the bottom," yet the well would still appear very deep. We also assumed the walls to be of zero width. It is quite possible to have a capacitor charged to thousands of volts, yet thinner than 1 mm. Compared to 30 cm, this is rather small.

The remarkable evidence shown in Figure 4.11 may dispel lingering doubts about quantum-mechanical bound states. Electrons on the surface of a piece of

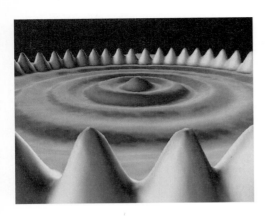

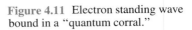

**Figure 4.11** Electron standing wave bound in a "quantum corral."

copper are bound within a circular "corral" of approximately 14-nm diameter, formed by 48 iron atoms (the spikes at the perimeter). Though more complicated than our simple one-dimensional cases, the result is qualitatively the same: We see a clear pattern of concentric antinodes—a two-dimensional standing wave. The image was produced by a scanning tunneling electron microscope (cf. Section 5.2). Importantly, it represents a probability density, not a wave function, and is the combined result of many electron detections.

## 4.6 Expectation Values, Uncertainties, and Operators

Quantum-mechanically, we cannot know all that we might hope to know. Besides simple probabilities, two important things we can determine are **expectation values** and **uncertainties.** Mathematically, an expectation value is an average. The probability *per unit length* is $|\psi(x)|^2$, so we may write

probability of being found in the interval $dx = |\psi(x)|^2\ dx$

Multiplying a position $x$ by the probability of finding the particle there, then summing (integrating) over all values of $x$, would yield an average $x$. The symbol for this average, called an expectation value, is a bar over the quantity being averaged.

$$\bar{x} \equiv \int_{\text{all space}} x\ |\psi(x)|^2\ dx \qquad (4\text{-}17)$$

The words "expectation value" are carefully chosen. To refer to (4-17) as the "average value" of the position would be to suggest that the particle always *has* a position. Moreover, as we know, a small particle cannot be "watched" without altering its state, so it would be impossible to verify an "average position" in any case. The expectation value is the value we would obtain if we were to begin with a particle in state $A$, do an experiment to "find" the particle, *begin again* with a particle in state A, "find" it, and so on, repeating the identical experiment many times and averaging the locations. Importantly, the expectation value of $x$ is a numerical value, not an explicit function of $x$, since the integral has been taken over that variable.

By the same logic, the expectation value of the square of the position is

$$\overline{x^2} \equiv \int_{\text{all space}} x^2 |\psi(x)|^2 \, dx$$

The reason for our concern about $\overline{x^2}$ is that it is part of how we choose to define uncertainty. In quantum mechanics, we define uncertainty as standard deviation. In the case of position, this is given by

$$\Delta x \equiv \sqrt{\int_{\text{all space}} (x - \overline{x})^2 |\psi(x)|^2 \, dx} \tag{4-18}$$

It is the square root of the mean of the squares of the deviations of $x$ from its expectation value—the rms deviation. It is the simplest possible quantity that is zero if and only if $x$ *never* deviates from its expectation value, and increases as deviations increase. We see this more clearly by rewriting it as

$$\Delta x = \sqrt{\int_{\text{all space}} |(x - \overline{x})\psi(x)|^2 \, dx} \tag{4-19}$$

The integral will be zero if and only if the *integrand* is everywhere zero, since the integrand is manifestly positive definite. Imagine a wave function $\psi(x)$ that is centered on the value $x = x_0$ and falls off symmetrically in both directions. Clearly, its expectation value $\overline{x}$ would be $x_0$. But for the integrand in (4-19) to be zero, $\psi(x)$ itself would have to be zero everywhere except at the single point $x = \overline{x} = x_0$, where multiplication by $(x - \overline{x})$ would cause the integrand to be zero. The wave function would have to be infinitesimally narrow.[9] Thus, (4-19) is zero if and only if there is but a single point where the particle might be found. When (4-19) is *non*zero, it increases as $\psi(x)$ becomes broader—a good definition of uncertainty.

### Example 4.2

An electron is found to be in a state given by the wave function

$$\psi(x) = A e^{-[(x-a)/2\epsilon]^2}$$

Determine (a) the value of $A$, (b) the expectation value of the position, and (c) the uncertainty in position.

### Solution

(a) The arbitrary constant $A$ is determined via normalization.

$$\int_{\text{all space}} |\psi(x)|^2 \, dx = \int_{-\infty}^{+\infty} [A e^{-[(x-a)/2\epsilon]^2}]^2 \, dx$$

$$= A^2 \int_{-\infty}^{+\infty} e^{-(x-a)^2/2\epsilon^2} \, dx = 1$$

[9]This is again the so-called Dirac delta function, discussed briefly in the "Two Exceptional Wave Functions" essay.

This is a standard **gaussian integral** (see Appendix K). Its value is $\sqrt{2\pi\epsilon^2}$.

$$A^2\sqrt{2\pi\epsilon^2} = 1 \quad \Rightarrow \quad A = \frac{1}{\sqrt{\epsilon}(2\pi)^{1/4}} \tag{4-20}$$

(b) $\bar{x} = \displaystyle\int_{\text{all space}} x|\psi(x)|^2\, dx = \int_{-\infty}^{+\infty} x\,[Ae^{-[(x-a)/2\epsilon]^2}]^2\, dx$

$\qquad = A^2 \displaystyle\int_{-\infty}^{+\infty} xe^{-(x-a)^2/2\epsilon^2}\, dx$

This is also a gaussian integral. Its value is $a\sqrt{2\pi\epsilon^2}$.

$$\bar{x} = A^2 a\sqrt{2\pi\epsilon^2} = \left(\frac{1}{\sqrt{\epsilon}(2\pi)^{1/4}}\right)^2 a\sqrt{2\pi\epsilon^2} = a$$

The result makes sense, since the function is symmetric about the point $x = a$.

(c) $\Delta x = \sqrt{\displaystyle\int_{\text{all space}} (x-\bar{x})^2\,|\psi(x)|^2\, dx} = \sqrt{A^2\displaystyle\int_{-\infty}^{+\infty} (x-\bar{x})^2 e^{-(x-a)^2/2\epsilon^2}\, dx}$

Noting that $\bar{x} = a$, and using the change of variables $x - a \equiv z$, all dependence on $a$ (or $\bar{x}$) drops out.

$$\Delta x = \sqrt{A^2\int_{-\infty}^{+\infty} z^2 e^{-(z^2/2\epsilon^2)}\, dz}$$

Yet again we are left with a standard gaussian integral: $\frac{1}{2}\sqrt{\pi(2\epsilon^2)^3}$.

$$\Delta x = \sqrt{\left[\frac{1}{\sqrt{\epsilon}(2\pi)^{1/4}}\right]^2 \frac{1}{2}\sqrt{\pi(2\epsilon^2)^3}} = \epsilon$$

We have seen gaussian wave functions in Chapter 3, where we noted that the factor multiplying the 2 in the denominator of the exponent is a measure of the wave function's width. Example 4.2 proves that this factor is in fact the uncertainty as we define it—the standard deviation. Another feature of the gaussian wave function also becomes more clear. Rewriting $\psi(x)$ in terms of $\bar{x}$ and $\Delta x$ rather than $a$ and $\epsilon$,

$$\psi(x) = \frac{1}{\sqrt{\Delta x}(2\pi)^{1/4}} e^{-[(x-\bar{x})/2\Delta x]^2}$$

As the width $\Delta x$ of $\psi(x)$ becomes small, its maximum value $1/[\sqrt{\Delta x}(2\pi)^{1/4}]$ becomes large. This must be so if the total probability is to remain fixed.

Uncertainty is not usually expressed as in equation (4-18), but in a simpler form. Noting that $\bar{x}$ is a value, not a function of $x$, we may write (4-18) as

$$\Delta x = \sqrt{\int_{\text{all space}} (x^2 - 2x\overline{x} - \overline{x}^2) |\psi(x)|^2 \, dx}$$

$$= \sqrt{\int_{\text{a.s.}} x^2 |\psi(x)|^2 \, dx - 2\overline{x} \int_{\text{a.s.}} x |\psi(x)|^2 \, dx + \overline{x}^2 \int_{\text{a.s.}} |\psi(x)|^2 \, dx}$$

The first integral is the expectation value of $x^2$, the second the expectation value of $x$, and the third, due to normalization, unity. Thus,

$$\Delta x = \sqrt{\overline{x^2} - 2\overline{x}^2 + \overline{x}^2} = \sqrt{\overline{x^2} - \overline{x}^2}$$

Although it is no longer obvious that the quantity in the square root is positive, this form for uncertainty is usually more convenient. It also makes clear that to calculate uncertainty/standard deviation we need only two pieces of information: the average of the square and the square of the average.

We may calculate the expectation value of any function of position in the same way as for $x$ and $x^2$.

$$\overline{f(x)} = \int_{\text{all space}} f(x) |\psi(x)|^2 \, dx$$

A note of caution: The expectation value of the function is *not* in general the function evaluated at the point $\overline{x}$, that is, $\overline{f(x)} \neq f(\overline{x})$. Consider a simple counterexample: the potential energy function $U(x) = \frac{1}{2}\kappa x^2$ for a mass on a spring. By symmetry, the expectation value of the position is zero; $\overline{x} = 0$. At this point in space, the potential energy is also zero; $U(\overline{x}) = 0$. But $\overline{U(x)}$ is an average of potential energy values times probabilities at all locations. Since the position $x$ of the particle varies and the potential energy is positive definite, this average potential energy is certainly not zero; $\overline{U(x)} \neq U(\overline{x})$.

Now let us generalize expectation values, uncertainties, and so on, to things besides position. The generic name **observable** is used for any dynamical property that can be measured, such as position, momentum, energy, or angular momentum. The expectation value of an observable $Q$ is found as follows:

$$\overline{Q} = \int_{\text{all space}} \Psi^*(x, t) \hat{Q} \Psi(x, t) \, dx \tag{4-21}$$

Expectation value

The symbol $\hat{Q}$ is the **operator** associated with the observable $Q$; for each observable there is a unique operator. For position, it is simply $x$. But in many cases $\hat{Q}$ is a differential operator; it takes a derivative of the function to its right. Consequently, the location of $\hat{Q}$ in the probability integral is not arbitrary—it *must* be immediately to the left of the wave function $\psi(x)$, the function upon which it operates.

While it is reasonable that $x$ is the proper operator to put in the probability integral to calculate the expectation value of position, it is by no means obvious what we would use for any other observable. Each case is unique, and "derivations" often involve subtle arguments and nontrivial math. The momentum operator is a good example. The details of its derivation are found in Appendix E. Since we are concerned mostly with its use, we simply present the result. The operator

**Table 4.1**

| Observable | Momentum | Position | Energy | Time * |
|---|---|---|---|---|
| Operator | $\hat{p} = -i\hbar\dfrac{\partial}{\partial x}$ | $\hat{x} = x$ | $\hat{E} = i\hbar\dfrac{\partial}{\partial t}$ | $\hat{t} = t$ |

\* Never in nonrelativistic quantum mechanics does the fact that there is an "operator" for time come into play; time is always treated in the usual way, as a variable in an oscillatory function, for instance. The operator is included merely to show the recurring similarities between time-energy relationships and position-momentum relationships, notable also in special relativity and the uncertainty principle. The true origin of these similarities is *relativistic* quantum mechanics, touched upon briefly in Section 11.2, where time and space are treated on equal footing.

associated with momentum—which when put in the proper location in the probability integral gives the expectation value—is given by

$$\hat{p} = -i\hbar\frac{\partial}{\partial x}$$

Momentum operator

*Functions* of an operator are also allowed. For example, the operator for kinetic energy we obtain as follows:

$$KE = \frac{p^2}{2m} \;\rightarrow\; \hat{KE} = \frac{1}{2m}\hat{p}^2 = \frac{1}{2m}\left(-i\hbar\frac{\partial}{\partial x}\right)\left(-i\hbar\frac{\partial}{\partial x}\right) = \frac{-\hbar^2}{2m}\frac{\partial^2}{\partial^2 x}$$

If not basic themselves, most quantum-mechanical operators are combinations of those for the basic dynamical properties: momentum, position, energy, and time. In Table 4.1, we complete the list of basic operators by adding those for energy (total energy, as opposed to kinetic) and time.[10] Note that with the operators now in hand we may write the Schrödinger equation (4-2) in a form clearly illustrating its foundation—energy.

$$\hat{KE}\,\Psi(x,\, t) + \hat{U}(x)\,\Psi(x,\, t) = \hat{E}\,\Psi(x,\, t)$$

Having a general method for calculating expectation values, we may now generalize the definition of uncertainty to arbitrary observables:

$$\Delta Q = \sqrt{\overline{Q^2} - \overline{Q}^2} \tag{4-22}$$

Uncertainty

Let us investigate uncertainties in the now somewhat familiar setting of the infinite well.

---

**Example 4.3**

For a particle in the ground state of an infinite well, find (a) $\Delta x$, (b) $\Delta p$, and (c) the product $\Delta x\,\Delta p$.

**Solution**

(a) Uncertainties involve the expectation value of the square and the square of the expectation value. Using general form (4-21)—where the temporal part of $\Psi$ drops out, as in (4-7)—the expectation value of position is

$$\overline{x} = \int_{\text{all space}} \psi^*(x)\, x\, \psi(x)\, dx$$

[10]The nature of operators is discussed further in Section 4.10.

In the ground state, $\psi_1(x) = \sqrt{2/L}\,[\sin(\pi x/L)]$ for $0 < x < L$, so

$$\bar{x} = \int_0^L \left(\sqrt{\frac{2}{L}}\sin\frac{\pi x}{L}\right) x \left(\sqrt{\frac{2}{L}}\sin\frac{\pi x}{L}\right) dx = \frac{2}{L}\int_0^L x\sin^2\frac{\pi x}{L}\,dx$$

$$= \frac{2}{L}\int_0^L x\frac{1-\cos(2\pi x/L)}{2}\,dx = \frac{2}{L}\left(\frac{L^2}{4} - \frac{1}{2}\int_0^L x\cos\frac{2\pi x}{L}\,dx\right)$$

Integration by parts reveals the remaining integral to be zero. Therefore,

$$\bar{x} = \frac{1}{2}L$$

This expectation value is certainly expected. Even classically, the average of the positions we would find if we were to repeatedly turn on the lights and catch the particle in its back-and-forth motion is the center of the box.

$$\overline{x^2} = \int_0^L \left(\sqrt{\frac{2}{L}}\sin\frac{\pi x}{L}\right) x^2 \left(\sqrt{\frac{2}{L}}\sin\frac{\pi x}{L}\right) dx = \frac{2}{L}\int_0^L x^2\sin^2\frac{\pi x}{L}\,dx$$

$$= \frac{2}{L}\int_0^L x^2\frac{1-\cos(2\pi x/L)}{2}\,dx = \frac{2}{L}\left(\frac{L^3}{6} - \frac{1}{2}\int_0^L x^2\cos\frac{2\pi x}{L}\,dx\right)$$

Integration by parts may be used to carry out the remaining integral, yielding

$$\overline{x^2} = \frac{2}{L}\left(\frac{L^3}{6} - \frac{L^3}{4\pi^2}\right) = L^2\left(\frac{1}{3} - \frac{1}{2\pi^2}\right)$$

Now using the general form (4-22),

$$\Delta x \sqrt{\overline{x^2} - \bar{x}^2} = \sqrt{L^2\left(\frac{1}{3} - \frac{1}{2\pi^2}\right) - \left(\frac{1}{2}L\right)^2} = (0.181)L$$

The result doesn't mean that the particle may never be found farther than $0.181L$ from the center. Standard deviation is a statistical definition, only a measure of the range where the *large majority* of values should be found. Logically, this range increases in proportion to the width of the box.

(b) Again using (4-21), we have

$$\bar{p} = \int_{\text{all space}} \psi^*(x)\,\hat{p}\psi(x)\,dx$$

Inserting the momentum operator and $\psi(x)$ gives

$$\bar{p} = \int_0^L \left(\sqrt{\frac{2}{L}}\sin\frac{\pi x}{L}\right)\left(-i\hbar\frac{\partial}{\partial x}\right)\left(\sqrt{\frac{2}{L}}\sin\frac{\pi x}{L}\right) dx$$

$$= \int_0^L \left(\sqrt{\frac{2}{L}}\sin\frac{\pi x}{L}\right)(-i\hbar)\frac{\pi}{L}\left(\sqrt{\frac{2}{L}}\cos\frac{\pi x}{L}\right) dx$$

$$= -i\hbar\frac{2\pi}{L^2}\int_0^L \sin\frac{\pi x}{L}\cos\frac{\pi x}{L}\,dx$$

The integral is zero, and with it $\bar{p}$. We might have guessed this also. The symmetry of the situation, quantum-mechanical or not, demands that it should be equally likely to find the particle traveling to the right as to the left. Since momentum is a vector, the "average" must be zero.

$$\overline{p^2} = \int_{\text{all space}} \psi^*(x)\, \hat{p}^2 \psi(x)\, dx$$

$$= \int_0^L \left( \sqrt{\frac{2}{L}} \sin \frac{\pi x}{L} \right) \left( -i\hbar \frac{\partial}{\partial x} \right)^2 \left( \sqrt{\frac{2}{L}} \sin \frac{\pi x}{L} \right) dx$$

$$= \int_0^L \left( \sqrt{\frac{2}{L}} \sin \frac{\pi x}{L} \right) (-\hbar^2) \frac{\partial^2}{\partial x^2} \left( \sqrt{\frac{2}{L}} \sin \frac{\pi x}{L} \right) dx$$

$$= \int_0^L \left( \sqrt{\frac{2}{L}} \sin \frac{\pi x}{L} \right) (-\hbar^2) \left( \frac{-\pi^2}{L^2} \right) \left( \sqrt{\frac{2}{L}} \sin \frac{\pi x}{L} \right) dx$$

$$= \frac{\pi^2 \hbar^2}{L^2} \int_0^L \left( \sqrt{\frac{2}{L}} \sin \frac{\pi x}{L} \right)^2 dx$$

The integrand is just $|\psi(x)|^2$, so the remaining integral is merely the total probability, which is unity.

$$\overline{p^2} = \frac{\pi^2 \hbar^2}{L^2}$$

Now using (4-22):

$$\Delta p = \sqrt{\overline{p^2} - \overline{p}^2} = \sqrt{\frac{\pi^2 \hbar^2}{L^2} - 0} = \frac{\pi \hbar}{L}$$

The uncertainty is nonzero—but it must be! Although the average of this vector is zero, the particle may at least be found moving in different directions. This alone is enough to see that there is uncertainty in the momentum.

(c) The product of the uncertainties is

$$\Delta x\, \Delta p = (0.181) L \frac{\pi \hbar}{L} = 0.568 \hbar$$

According to the uncertainty principle, $\Delta x\, \Delta p \geq \frac{1}{2}\hbar$. Here we have verification. While the uncertainty in momentum decreases with increasing width of the box (a longer wavelength implies less kinetic energy), the uncertainty in position increases (as the wave is more spread out), and the product is independent of $L$. The reason the product is not *equal to* $\frac{1}{2}\hbar$ is simple: The wave function is not a gaussian, the function for which the product is minimum.

It is left as an exercise to show that in the infinite well both $\Delta x$ and $\Delta p$ increase with increasing $n$; the ground state is the least uncertain combination of position and momentum. It is also an exercise to show that $\Delta x$ approaches the classical value $L/\sqrt{12}$ as $n \rightarrow \infty$. Again, this is in accord with the correspondence principle: Quantum-mechanical predictions approach the classical as $n$ becomes large.

## 4.7 Case 2: The Finite Well

In the infinite well, we analyzed the simplest bound-state problem. We now turn to a more realistic case: the potential well *before* the assumption of infinite potential energy walls. This is known as the **finite well** (Figure 4.12). Although more realistic, it is also more difficult to analyze. It is mathematically impossible to solve for the quantum energy levels in closed form—it is not as clean a problem. However, it actually differs in few ways from the infinite well. Our major concern in this section is to understand the important physical consequences of these differences.

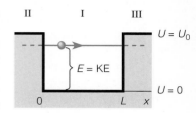

Figure 4.12 Finite well.

The wave function prior to the assumption $U_0 \to \infty$ is given by (4-13):

$$\psi(x) = \begin{cases} Ce^{+\alpha x} & x < 0 \\ A \sin kx + B \cos kx & 0 < x < L \\ Ge^{-\alpha x} & x > L \end{cases}$$ (4-13)

where $k = \sqrt{\dfrac{2mE}{\hbar^2}}$ and $\alpha \equiv \sqrt{\dfrac{2m(U_0 - E)}{\hbar^2}}$

Since $\alpha$ is not infinite, as it was in the $U_0 \to \infty$ case, we may not argue as we did then that the exponentials outside the box are zero. They are perfectly valid solutions of the Schrödinger equation in their respective regions. Therefore, we must take them into account in the overall smoothness of $\psi(x)$. But the potential energy is now finite, so both $\psi(x)$ *and its derivative* must be continuous.

---

### Solving the Schrödinger Equation (continued): Finite Well

**Smoothness for all space**

$\psi(x)$ **continuous at $x = 0$:** $\quad \psi_{II}(0) = \psi_I(0)$

$Ce^{+\alpha 0} = A \sin k0 + B \cos k0 \quad \Rightarrow \quad \boxed{C = B}$

$\dfrac{\partial \psi(x)}{\partial x}$ **continuous at $x = 0$:** $\quad \left. \dfrac{\partial \psi_{II}}{\partial x} \right|_{x=0} = \left. \dfrac{\partial \psi_I}{\partial x} \right|_{x=0}$

$\alpha Ce^{+\alpha 0} = kA \cos k0 - kB \sin k0 \quad \Rightarrow \quad \boxed{\alpha C = kA}$

$\psi(x)$ **continuous at $x = L$:** $\quad \psi_I(L) = \psi_{III}(L)$

$\boxed{A \sin kL + B \cos kL = Ge^{-\alpha L}}$

$\dfrac{\partial \psi(x)}{\partial x}$ **continuous at $x = L$:** $\quad \left. \dfrac{\partial \psi_I}{\partial x} \right|_{x=L} = \left. \dfrac{\partial \psi_{III}}{\partial x} \right|_{x=L}$

$\boxed{kA \cos kL - kB \sin kL = -\alpha Ge^{-\alpha L}}$

Although it is certainly not obvious, these four conditions do lead to quantization of the particle's energy, just as the continuity conditions did in the infinite well. From the first two conditions, we see that $A = \alpha C/k$ and $B = C$. Inserting these in the last two conditions,

$$\frac{\alpha}{k} C \sin kL + C \cos kL = Ge^{-\alpha L}$$

$$k \frac{\alpha}{k} C \cos kL - kC \sin kL = -\alpha Ge^{-\alpha L}$$

Dividing the last equation by $-\alpha$, the right-hand sides become equal. Thus,

$$\frac{\alpha}{k} \cancel{C} \sin kL + \cancel{C} \cos kL = -\cancel{C} \cos kL + \frac{k}{\alpha} \cancel{C} \sin kL$$

All dependence on arbitrary constants has canceled. Finally, dividing by $\sin kL$ and rearranging a bit, we have

$$2 \cot kL = \frac{k}{\alpha} - \frac{\alpha}{k} \tag{4-23a}$$

(*Note*: In the limit $U_0 \rightarrow \infty$, $\alpha \rightarrow \infty$, and equation (4-23a) would be $\cot kL = -\infty$. This would require that $kL = n\pi$, precisely condition (4-15) from the infinite well.)

It is mathematically impossible to solve equation (4-23a) for $k$. (It is a so-called "transcendental equation.") Even gaining the barest idea of what it implies is not a simple matter. Nevertheless, it involves $E$ and $U_0$ through $k$ and $\alpha$, and for a given $U_0$ can be solved for only certain values of $E$. Graphing is the least complicated way to see this.

It is left as an exercise to show that (4-23a) may be solved for $\alpha$, yielding

$$\alpha = \begin{cases} k \tan \frac{1}{2} kL & 2n\pi < kL < 2n\pi + \pi \\[2ex] -k \cot \frac{1}{2} kL & 2n\pi - \pi < kL < 2n\pi \end{cases} \tag{4-23b}$$

and that, using the definitions of $k$ and $\alpha$, this in turn leads to

$$U_0 = \begin{cases} \dfrac{\hbar^2 k^2}{2m} \sec^2 \frac{1}{2} kL & 2n\pi < kL < 2n\pi + \pi \\[3ex] \dfrac{\hbar^2 k^2}{2m} \csc^2 \frac{1}{2} kL & 2n\pi - \pi < kL < 2n\pi \end{cases} \tag{4-23c}$$

In Figure 4.13, we plot the right-hand sides of (4-23c) as functions of $k$. A plot of the left-hand sides would simply be a horizontal line at $U_0$, the depth of the well. We see then that allowed values of $k$ are intersections of the right-hand-side plots with a horizontal "well-depth" line. *Of primary importance is that they cross at only certain points; equation (4-23) holds for only certain discrete values of $k$,*

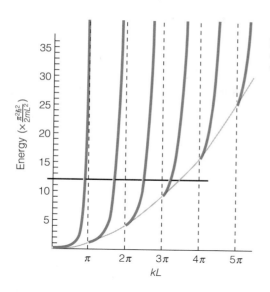

**Figure 4.13** Allowed wave numbers in the finite well: intersections of a horizontal "well-depth" line with curves.

with corresponding discrete values of $E (= \hbar^2 k^2/2m)$. A representative $U_0$ line is shown. Intersecting four curves, it represents a potential well with four allowed values of $k$, four energy levels. Clearly, even the shallowest $U_0$ line would intersect the leftmost curve; a finite well always has at least one bound state, one standing wave. As the potential well becomes deeper, the $U_0$ line would intersect more curves and the number of allowed values of $k$ and $E$ would increase; the well would "hold" more bound states. However, the number of discrete energy levels is finite for any finite $U_0$.

## Comparison: Finite and Infinite Wells

Figure 4.14 shows wave functions and probability densities for the finite and infinite wells. (See Appendix F for a detailed discussion of the finite-well wave functions.) In the infinite well, the confining potential energy is, of course, infinite. For the finite well, we have chosen one that corresponds to the $U_0$ line in Figure 4.13. Because all involve standing waves, certain characteristics are common

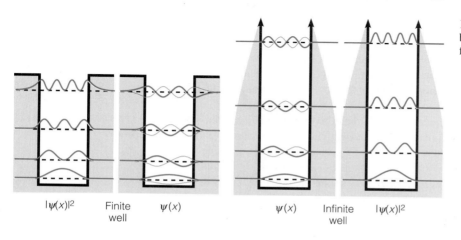

**Figure 4.14** Wave functions, probability densities, and energies in the finite and infinite wells.

**Table 4.2**

| n | Finite Well $U_0 = 12\dfrac{\pi^2\hbar^2}{2mL^2}$ | | Infinite Well $U_0 = \infty$ | |
|---|---|---|---|---|
|   | $k*$ | $E^\dagger$ | $k*$ | $E^\dagger$ |
| 1 | 0.84 | 0.71 | 1 | 1 |
| 2 | 1.68 | 2.82 | 2 | 4 |
| 3 | 2.49 | 6.20 | 3 | 9 |
| 4 | 3.23 | 10.5 | 4 | 16 |

\* $\left(\times\dfrac{\pi}{L}\right)$, from numerical solution of (4-23) in finite-well case.

$\dagger$ $\left(\times\dfrac{\hbar^2}{2m}\dfrac{\pi^2}{L^2}\right)$

to all bound-state situations: Energy is quantized, and there is a ground state of minimum total energy in which the kinetic energy is not zero—a "fundamental" standing wave whose wave number is nonzero. But what about the differences between the two wells?

One distinction is in the values of the allowed energies. The energy levels of a finite well are lower than those of an infinite well of the same width. We see this in Figure 4.14, and the corresponding data in Table 4.2 agree. Since for both wells $k \equiv \sqrt{2mE/\hbar^2}$, energy is directly related to the value of $k$. As we know, the allowed $k$-values in the infinite well are $n\pi/L$. For the finite well, a well depth $U_0 = 12(\pi^2\hbar^2/2mL^2)$ was chosen[11] and equation (4-23) was then solved numerically for allowed values of $k$ (where the $U_0$ line intersects curves in Figure 4.13). As shown in the table, these $k$-values are smaller than those in the infinite well, and so are the energies. But even without a calculation, we should expect longer wavelengths ($\lambda = 2\pi/k$) in the finite well: The waves do not stop abruptly, but penetrate somewhat into the classically forbidden region. As Figure 4.14 clearly shows, the penetration increases as the energy approaches the top of the well. Because the ground state penetrates the walls very little, its wave number is nearly the same as in the infinite well, so the energies of their ground states are nearly equal. By the fourth energy levels, however, there is a marked difference. Due to its significant penetration, the finite-well wave function is of considerably longer wavelength and smaller wave number and energy. In fact, the wavelength and energy of the fourth finite-well function are closer to those of the *third* infinite-well function.

Penetration of the classically forbidden region is the other important distinction between the two wells. Although exponentially decaying, the wave functions outside the finite well are not zero; there is a nonzero probability of the particle being found outside the well (i.e., outside the capacitors in Figure 4.5). This is remarkable! In these regions, the total energy $E$ is less than the potential. Apparently, the kinetic energy is negative. To a "classical physicist," this would be nonsense.

[11]Twelve times the ground-state energy of the infinite well.

The measure of how far the wave function extends into the classically forbidden region is the constant $\alpha$. Using the symbol $|x|$ for the magnitude of the distance from a "wall," the wave function in both classically forbidden regions obeys

$$\psi(x) \propto e^{-\alpha|x|}$$

The exponential will have decayed by a fraction $1/e$ at the point where $\alpha|x| = 1$, corresponding to a distance from the wall of $|x| = 1/\alpha$. This distance is known as the **penetration depth,** and is given the symbol $\delta$.[12]

$$\delta = \frac{1}{\alpha} = \frac{\hbar}{\sqrt{2m(U_0 - E)}} \qquad (4\text{-}24)$$

Penetration depth

The penetration becomes deeper as the energy $E$ nears the value of the confining potential $U_0$, in agreement with Figure 4.14.

Upon hearing that a particle may be found where it "cannot be" (with negative kinetic energy!), many students hunger for proof. Although there is clear evidence of matter waves passing through classically forbidden regions (alpha decay, for instance; cf. Section 5.2), actually detecting a particle with negative kinetic energy is problematic. Position and momentum cannot be determined simultaneously. To detect a particle in the classically forbidden region, the experiment would have to have an uncertainty in position no *larger* than $\sim\delta$, the interval within which the probability is significant. (We wouldn't want our experiment to have a much larger uncertainty, certainty not comparable to $L$, for we might then be detecting a particle actually *inside* the well.) This very experiment would introduce a momentum uncertainty no *smaller* than $\sim\hbar/\delta$. Since KE $=p^2/2m$, this would imply kinetic energy fluctuations no smaller than $\sim\hbar^2/2m\delta^2$. But according to (4-24), this in turn is $U_0 - E$, the magnitude of the (negative) kinetic energy. Thus, no experiment could be certain that the particle it finds had negative kinetic energy.

Furthermore, we must be careful in interpreting the penetration of the classically forbidden region. We cannot conclude that a particle is constantly passing outside the well, then going back inside. Once outside, what reason would it have to turn around? More to the point, we could not watch the particle undisturbed anyhow! We must be content with the meager knowledge that there is a probability of finding the particle outside the well *if* an attempt is made to find it there. If no attempt is made, we simply have an undisturbed standing wave that happens to penetrate the walls.

Matter-wave penetration into the classically forbidden region has an electromagnetic analog. When light waves totally internally reflect at an interface between two media, no sinusoidal wave propagates into the "second" medium. All energy is reflected. But there is an oscillating electromagnetic field that decays exponentially with distance from the interface. Furthermore, if the second medium is thin enough that the field does not decay too much within, a significant amount of light intensity will get through. If light waves may do this, might matter waves be able to escape through a thin wall? Indeed they can; the phenomenon is called tunneling and is discussed in Chapter 5.

[12]It might be argued that the "penetration depth" is really infinite, because any exponential takes an infinite distance to fall to zero. The definition we choose is a practical one—measuring the degree of *significant* penetration into the wall.

## 4.8 Case 3: The Simple Harmonic Oscillator

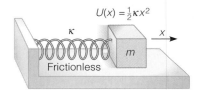

Of our simple cases, the harmonic oscillator is the most realistic. Once again, the potential energy defines the problem. The harmonic oscillator potential energy is $U(x) = \frac{1}{2}\kappa x^2$, where $\kappa$ is the spring constant. It is more realistic because it is a continuous function. Moreover, it is a good approximation of the actual potential energy whenever particles undergo small oscillations about an equilibrium position. In the immediate vicinity of a local minimum, all continuous potential energy functions "look" parabolic (i.e., there is a parabola that closely approximates the function about its local minimum). The bond between two atoms in a diatomic molecule is a good example. Figure 4.15 shows the potential energy shared by two atoms in a diatomic molecule, plotted versus their separation $x$. The force is strongly repulsive (large slope) for small separations and weakly attractive (small slope) at large separations. But in the immediate vicinity of the equilibrium separation $x_0$, $U(x)$ is nearly parabolic. Therefore, the lowest energy vibrations about the equilibrium separation will be essentially those of a simple harmonic oscillator.

With the harmonic oscillator potential energy, the time-independent Schrödinger equation is

$$-\frac{\hbar^2}{2m}\frac{d^2\psi(x)}{dx^2} + \frac{1}{2}\kappa x^2\psi(x) = E\psi(x) \tag{4-25}$$

There are *mathematical* solutions for all values of $E$, but except for certain discrete values the solutions all diverge as $|x| \to \infty$. These discrete values are the only ones for which the corresponding solutions $\psi(x)$ are *physically* acceptable. Energy is quantized.

Equation (4-25) is one of the few cases where the Schrödinger equation can be solved "in closed form" for the wave functions and corresponding energies. Unfortunately, it is still rather complicated. Let us take a different approach.

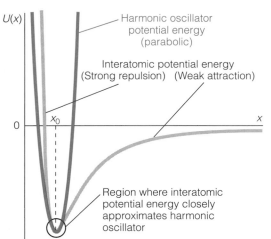

**Figure 4.15** For small oscillations about an equilibrium point, all potential energies resemble a parabola—a simple harmonic oscillator.

**Example 4.4**

For a simple harmonic oscillator of spring constant $\kappa$ and mass $m$, one solution of the Schrödinger equation is of the form $\psi(x) = Ae^{-ax^2}$, a gaussian centered at the origin. Determine fully the wave function and energy in this state.

Solution

The given $\psi(x)$ must be a solution of (4-25). Let us see what this tells us:

$$-\frac{\hbar^2}{2m}\frac{d^2Ae^{-ax^2}}{dx^2} + \frac{1}{2}\kappa x^2 Ae^{-ax^2} = EAe^{-ax^2}$$

The multiplicative constant cancels. Carrying out the differentiation,

$$-\frac{\hbar^2}{2m}(-2a + 4a^2x^2)e^{-ax^2} + \frac{1}{2}\kappa x^2 e^{-ax^2} = Ee^{-ax^2}$$

The exponential now also cancels. After rearranging a bit,

$$\left(\frac{\hbar^2 a}{m} - E\right) - \left(\frac{2\hbar^2 a^2}{m} - \frac{1}{2}\kappa\right)x^2 = 0$$

If this is to hold for all values of $x$, both the constant term and the constant coefficient of $x^2$ must be zero. (*Note:* It is no accident that the conditions to be met are fairly simple. Were the function not a solution, we would have quite a mess at this point.)

$$\frac{2\hbar^2 a^2}{m} = \frac{1}{2}\kappa \quad \Rightarrow \quad a = \frac{\sqrt{m\kappa}}{2\hbar} \quad \text{so that} \quad E = \frac{\hbar^2 a}{m} = \frac{\hbar}{2}\sqrt{\frac{\kappa}{m}}$$

To fully determine the wave function, we must find its normalization constant,

$$\int_{-\infty}^{+\infty} |\psi(x)|^2\, dx = \int_{-\infty}^{+\infty} (Ae^{-ax^2})^2\, dx = A^2 \int_{-\infty}^{+\infty} e^{-2ax^2}\, dx = A^2\sqrt{\frac{\pi}{2a}}$$

where Appendix K has been used to evaluate the gaussian integral. Setting this equal to unity, we see that

$$A = \left(\frac{2a}{\pi}\right)^{1/4} = \left(\frac{2\sqrt{m\kappa}/2\hbar}{\pi}\right)^{1/4} = \left(\frac{m\kappa}{\pi^2\hbar^2}\right)^{1/8}$$

Thus,

$$E = \frac{\hbar}{2}\sqrt{\frac{\kappa}{m}} \qquad \psi(x) = \left(\frac{m\kappa}{\pi^2\hbar^2}\right)^{1/8} e^{-(\sqrt{m\kappa}/2\hbar)x^2}$$

The rigorous mathematical derivation of energies and wave functions shows that the allowed energies of the harmonic oscillator are given by

**Table 4.3   Harmonic Oscillator Solutions**

$[b \equiv (m\kappa/\hbar^2)^{1/4}]$

| $n$ | $\psi_n(x)$ |
|-----|-------------|
| 0 | $\left(\dfrac{b}{\sqrt{\pi}}\right)^{1/2} e^{-(1/2)b^2x^2}$ |
| 1 | $\left(\dfrac{b}{2\sqrt{\pi}}\right)^{1/2} (2bx)e^{-(1/2)b^2x^2}$ |
| 2 | $\left(\dfrac{b}{8\sqrt{\pi}}\right)^{1/2} (4b^2x^2 - 2)e^{-(1/2)b^2x^2}$ |
| 3 | $\left(\dfrac{b}{48\sqrt{\pi}}\right)^{1/2} (8b^3x^3 - 12bx)e^{-(1/2)b^2x^2}$ |
| $n$ | $\left(\dfrac{b}{2^n n!\sqrt{\pi}}\right)^{1/2} H_n(bx)e^{-(1/2)b^2x^2}$ |

$$E = (n + \tfrac{1}{2})\hbar\omega_0 \quad (n = 0, 1, 2, 3, \ldots) \tag{4-26}$$

$$\text{where } \omega_0 \equiv \sqrt{\frac{\kappa}{m}}$$

Note that the energies are equally spaced.[13] Moreover, for each one there is a unique wave function. Several are given in Table 4.3. Each is the product of a normalization constant, a gaussian factor, and one of a set of related functions known as the **Hermite polynomials,** given the symbol $H_n(bx)$. From the table and equation (4-26), we see that the simple gaussian function considered in the example is, in fact, the ground state.

Several wave functions and probability densities are plotted on the energy-level diagram of Figure 4.16. As in the particle in a box, an oscillator in its ground state is most likely to be found at the center, contrary to the classical expectation. Classically, a harmonic oscillator is most likely to be found where its speed is smallest—at the extremes of its travel. However, we see once again that as the particle energy becomes large, the predictions of quantum mechanics and those of classical mechanics converge. For large $n$, the probability density is largest near the extremes of travel. Even so, there is penetration into the classically forbidden region at all energies.

As noted, the harmonic oscillator is a good model for lowest-energy oscillations of a diatomic molecule. Let us see what use we can make of this.

---

**Example 4.5**

The mass $m$ of a hydrogen atom is $1.67 \times 10^{-27}$ kg. Classically or quantum-mechanically, when two are bound together, they should oscillate about their

[13]That $n$ is allowed to be zero is not cause for alarm. This state does not correspond to zero *kinetic* energy, so it does not violate the uncertainty principle. The energies could certainly be written equivalently as $E_n = (n - \frac{1}{2})\hbar\omega_0$, with $n$ allowed to be any *positive* integer, but equation (4-26) is the conventional way of expressing them. Also, $\omega_0$ is the angular frequency of the *classical* harmonic oscillator, not a wave function frequency, as in $e^{-i\omega t}$; $\omega$ is always $E/\hbar$, and so would be $(n + \frac{1}{2})\omega_0$.

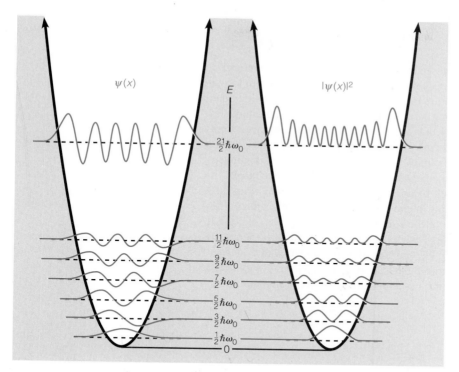

**Figure 4.16** Wave functions, probability densities, and energies of a harmonic oscillator.

equilibrium separation as a simple harmonic oscillator of mass $\frac{1}{2}m$. (This is the so-called "reduced mass," the effective mass when a spring connects two equal-mass *free* objects, rather than a single mass to a fixed point. See Exercise 6.50.) The effective spring constant is 573 N/m. Should vibrational energy contribute to hydrogen's heat capacity at room temperature? That is, will there be energy exchange between the molecules' translational (kinetic) and vibrational energies?

## Solution

The classical angular frequency of vibration is

$$\omega_0 = \sqrt{\frac{\kappa}{m}} = \sqrt{\frac{573 \text{ N/m}}{\frac{1}{2} \times 1.67 \times 10^{-27} \text{ kg}}} = 8.28 \times 10^{14} \text{ rad/s}$$

Quantum-mechanically, oscillator energy levels are separated by $\hbar\omega_0$, or

$$\Delta E = (1.055 \times 10^{-34} \text{ J·s})(8.28 \times 10^{14} \text{ rad/s}) = 8.74 \times 10^{-20} \text{ J}$$

Therefore, to "jump" from one vibrational state to another requires an energy input of $8.74 \times 10^{-20}$ J. Let us estimate the temperature required for the translational kinetic energy to be comparable. From an introductory study of gases, we know that the average translational kinetic energy of a gas particle is $\frac{3}{2}k_B T$, where $k_B$ is the Boltzmann constant, $1.38 \times 10^{-23}$ J/K.

$$\tfrac{3}{2}k_B T = \Delta E \quad \rightarrow \quad \tfrac{3}{2}(1.38 \times 10^{-23}\ \text{J/K})T$$
$$= 8.74 \times 10^{-20}\ \text{J} \quad \Rightarrow \quad T \cong 4200\ \text{K}$$

Rough though our estimate is, we see that the translational energy available in a collision between molecules surely should not be able to cause them to *gain* vibrational energy if the temperature is only about 300 K. (It is no coincidence that the specific heat of hydrogen gas rather abruptly increases at around a few thousand kelvins; see Figure 8.24.) On the other hand, molecules can certainly *lose* vibrational energy to other forms, so essentially all of them will be in their ground vibrational states. As we know, the ground-state energy is nonzero; however, because it cannot be taken away, it is not available to exchange with other forms. What is really meant by storing energy in vibrational motion is that vibrational energy participates in energy *exchange* with other forms, that it on average manifests *its share* of the equipartition of energy in the equilibrium state. This requires that energy be continually stored and returned at energy levels *above* ground state. We see that this is quite unlikely at room temperature. The fact that a diatomic gas can effectively store no vibrational energy below a certain temperature was a mystery prior to the advent of quantum mechanics.

## Absolute Zero

Not only do *individual molecules* have "springs"; at low temperatures, essentially all materials form solids—and solids are full of oscillators. An important consequence of the ground state's nonzero kinetic energy is that even at absolute zero temperature, motion does not stop. To cool a collection of oscillating particles to absolute zero, we would remove all energy *possible*. But the result would not be a state of zero motion; it would only be one in which all oscillators are in their ground states. Still there is *something* that would be zero, and that is entropy. In such a state, there could be no random, disordered distribution of energy. All is order.

## ◆ 4.9 Math, Physics, and Quantization

Having finished our simple cases, let us take a closer look at how the *mathematics* of the Schrödinger equation combines with the *physical* requirements we impose on the wave function to give quantized standing waves.

The time-independent Schrödinger equation may be written

$$\psi''(x) = \frac{2m}{\hbar^2}(U(x) - E)\psi(x)$$

Primes indicate derivatives with respect to $x$. We see that the sign of the second derivative of $\psi(x)$ must be the same as that of the product $(U(x) - E)\psi(x)$. Therefore, at points where $E$ is greater than $U(x)$, the wave function and its second derivative must be of opposite sign. *Necessarily this describes an oscillatory, wave-*

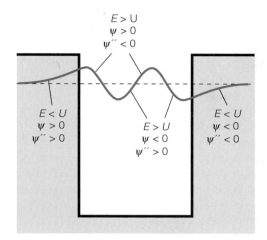

$E > U$
$\psi > 0$
$\psi'' < 0$

$E < U$
$\psi > 0$
$\psi'' > 0$

$E > U$
$\psi < 0$
$\psi'' > 0$

$E < U$
$\psi < 0$
$\psi'' < 0$

**Figure 4.17**  The behavior of a wave function depends on which is larger —$E$ or $U$.

*like function;* if $\psi(x)$ is always concave-down when it is positive and concave-up when negative, it must continually return to the axis. Thus, when $E > U$, $\psi(x)$ is oscillatory.[14] In classically forbidden regions, on the other hand, where $U(x)$ is greater than $E$, $\psi(x)$ and its second derivative must be of the same sign. Oscillatory functions do not meet this criterion,[15] but functions like decaying exponentials do. The finite-well wave function in Figure 4.17 clearly shows the change in the behavior of $\psi(x)$ from inside ($U < E$) to outside ($U > E$).

The time-independent Schrödinger equation is a *second*-order differential equation. Mathematically, its general solution should involve *two* arbitrary constants. Our rigorous solution to the infinite well had none! With the imposition of the physical conditions, the constants are no longer arbitrary. But the physical conditions lead to an even further restriction. Not only are there no arbitrary constants in the *solution* of the equation, there is no acceptable solution at all unless the constant $E$ in *the equation itself* is one of a discrete set of values. Let us investigate why and how.

In the finite well, we expended much effort ensuring smoothness.[16] Choosing to investigate an "unrealistic" discontinuous potential energy required us to solve the Schrödinger equation piece-wise, then match the pieces. However, had we not been concerned with normalizability, we would have had no reason to discard increasing exponentials outside the well, and by including them we could have fashioned smooth solutions for any value of $E$ (see Exercise 33). On the other hand, with a continuous potential energy such as the harmonic oscillator, we don't have

---

[14]The conclusion is valid only for *stationary states*, where the time-*independent* Schrödinger equation applies and probability density is constant in time. It is only in such states that there indeed is a well-defined $E$. In Chapter 3, we discussed the gaussian wave function $\psi(x) = Ae^{-(x/b)^2}$. It might well describe a free particle—for which $U$ (being zero) is less than $E$—at some instant; and it is *not* oscillatory. But, it is not a solution of the time-independent Schrödinger for equation for $U = 0$, it does not have a well-defined energy, and its probability density spreads with time. This is shown in Section 5.3.

[15]Roughly speaking, to be oscillatory is to have a wavelength and thus a momentum. In a classically forbidden region, kinetic energy is negative and momentum, proportional to the square root of kinetic energy, is not "real." Accordingly, we should not expect the wave to have a "real" wavelength.

[16]We do not address the infinite well specifically here. It may be thought of as a limiting case of a finite well.

to be concerned with smoothness at all; the Schrödinger equation automatically gives smooth solutions for all values of $E$. Since there are smooth solutions for any $E$ whether $U(x)$ is continuous or not, how is it that the physical conditions restrict $E$ to only certain values? The only requirement left is normalizibility, which may be viewed as three conditions: The wave function must fall to zero at *both* ends—these two are called **boundary conditions**—and the total probability must be unity. The unit probability condition fixes the wave function's overall multiplicative constant, while the two boundary conditions fix the other arbitrary constant *and* restrict $E$ to certain values. (This is exactly what we found in the infinite well.) Unless $E$ is just right, the solution diverges at one end or both.

Verifying this, and in so doing finding the allowed values of $E$, is a different matter altogether. In most real applications, $U(x)$ is such that we cannot solve the differential equation (with pencil and paper), so it is worthwhile to spend some time studying how we can show that normalization leads to quantization *without* solving it. The procedure is trial and error: We assume a value of $E$, then determine how $\psi(x)$ varies with $x$. If it diverges anywhere, then $E$ is "wrong" and we try another $E$. (A computer is very handy.) We find how $\psi(x)$ varies with $x$ by the following prescription:

$$\psi(x + \Delta x) = \psi(x) + \psi'(x)\,\Delta x$$

where $\psi'(x) = \psi'(x - \Delta x) + \psi''(x - \Delta x)\,\Delta x$

The next value of $\psi$ is found from a known value by a simple linear projection; it is the known value plus the rate of change times an incremental distance $\Delta x$. The rate of change $\psi'$ we find by linear projection from its previous value.

The procedure raises two questions: Where does it all start, and where do we get the previous *second* derivative? As we soon see, it starts with an initial assumption of $\psi$ and $\psi'$ at some value of $x$. The second question is where the Schrödinger equation and potential energy come into play: The previous second derivative comes directly from the previous $\psi$, through the time-independent Schrödinger equation.

$$\psi''(x) = -\frac{2m[E - U(x)]}{\hbar^2}\psi(x)$$

Starting only with $U(x)$ and initial assumptions for $\psi(0)$, $\psi'(0)$, and $E$, we could find a sequence of values for $\psi$ and $\psi'$ at points $\Delta x$ apart as follows:

(1) $\psi(\Delta x + 0) = \psi(0) + \psi'(0)\,\Delta x$

$$\psi'(\Delta x + 0) = \psi'(0) + \left[-\frac{2m(E - U(0))}{\hbar^2}\psi(0)\right]\Delta x$$

(2) $\psi(\Delta x + \Delta x) = \psi(\Delta x) + \psi'(\Delta x)\,\Delta x$

$$\psi'(\Delta x + \Delta x) = \psi'(\Delta x) + \left[-\frac{2m(E - U(\Delta x))}{\hbar^2}\psi(\Delta x)\right]\Delta x$$

(3) etc.

Notice how at each step, values needed at the next are determined.[17]

[17]The size of the increment $\Delta x$ is a matter of choice, but since the true solution of the Schrödinger equation is being generated, point by point, the behavior, or shape, of $\psi(x)$ should not depend on it. It won't, so long as $\Delta x$ is much smaller than the typical distance over which $\psi$ changes significantly.

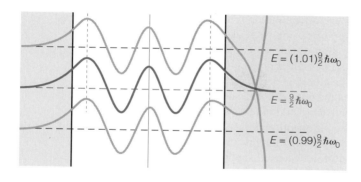

$E = (1.01)\frac{9}{2}\hbar\omega_0$

$E = \frac{9}{2}\hbar\omega_0$

$E = (0.99)\frac{9}{2}\hbar\omega_0$

**Figure 4.18** Unless the energy is just right, a solution of the Schrö-dinger equation will diverge at infinity. (*Note:* There is no signifi-cance to the functions seeming to cross at the same point.)

Applying this procedure to a given $U(x)$, we would find that for almost any choice of $E$, $\psi(x)$ diverges somewhere, no matter what values are chosen for $\psi(0)$ and $\psi'(0)$. We would have a valid mathematical solution, but an energy not physi-cally allowed. However, at certain values of $E$, *for any value of* $\psi(0)$, there would be a value of $\psi'(0)$ for which $\psi(x)$ would be finite everywhere, falling to zero at $x = \pm\infty$. These are the allowed quantum energy levels.

Figure 4.18 shows a "close-up" of the harmonic oscillator potential energy near its fifth allowed energy, $E_4 = \frac{9}{2}\hbar\omega_0$. (The range of energies shown is so small that the "walls" of the potential energy parabola appear straight.) The three func-tions are mathematical solutions of the time-independent Schrödinger equation for three different values of $E$. In each case, an arbitrary value of $\psi(0)$ is chosen, then $\psi'(0)$ is chosen so that the solution approaches zero as $x \to -\infty$. Although the energies differ by only 1%, the problem is obvious: The top and bottom func-tions diverge at $x = +\infty$; "nailing down" one end of the function causes the other to misbehave.

Closer inspection shows why. The leftmost crests of all three waves nearly co-incide, but because wavelength decreases with increasing energy, they are decid-edly out of step by the time they reach the right wall. Unless a wave hits the wall at just the right "height" and slope, it will diverge, one way or the other. As noted earlier, $\psi''$ and $\psi$ will be of the same sign in the classically forbidden region. The bottom wave is indeed positive and concave-up, but, hitting the wall too flat, it di-verges to $+\infty$. The top wave is too steep. While $\psi''$ is initially positive, the wave crosses the axis, is then negative, with negative $\psi''$, and diverges to $-\infty$. Only the middle wave hits the wall just right, so that $\psi$ asymptotically approaches zero. For an arbitrarily chosen $\psi(0)$, the value of $\psi'(0)$ is determined—its arbitrariness used up—by the requirement that the wave fall to zero at $x = -\infty$. Being one of the correct $E$-values, however, the wave falls to zero at $x = +\infty$ *naturally*. The con-stant $\psi(0)$ is thus still arbitrary, and available to ensure a total probability of unity.

## Quantization of Classical Waves

Although there are obvious differences, we should not overlook the many similar-ities between matter waves and classical waves—for example, sound waves in air and waves on strings. All interfere, diffract, form standing waves, and so on. One reason is that they all obey second-order differential equations. Of course, we do not speak of "unit probability" for classical waves; the overall amplitude is usually left arbitrary. But analogous boundary conditions are often imposed—solid walls at the ends, for instance—and quantization of standing waves is invariably the result. In a study of sound, we analyze standing waves in the air in organ pipes, and only certain wavelengths fit the conditions. For a pipe closed at one end only, the conditions are that there be a displacement antinode at the open end and a node at the closed end. As a result, only resonant frequencies that are odd multiples of a fundamental (ground state) are allowed. Because a pipe open at both ends has different boundary conditions, its allowed standing waves are related differently (both odd and even multiples). A uniform string stretched between two fixed supports has resonant frequencies that are all integral multiples of a fundamental. Now if its density were to vary along the string in a complicated way, the resonant frequencies would not be so simply related—but they would still exist! The allowed standing waves would depend on the boundary conditions *and* $\rho(x)$. For matter waves, the density has its analog in the potential energy. *If a particle is bound by a potential energy function U(x), no matter how complicated, discrete quantum energy levels will result.*

# ◆ 4.10 Well-Defined Observables: Eigenvalues

As we know, a "well-defined" observable is one for which the uncertainty is zero, for which an experiment designed to measure that observable can obtain only a single value. There is a mathematical tool that allows us to determine easily whether a wave function describes a state for which a given observable is well defined. Before introducing it, however, let us consider what wave functions we would *expect* to have well-defined values for the basic observables.

**Momentum:** A wave function for which the momentum/wave number is well defined is the basic quantum-mechanical plane wave—the complex exponential $e^{ikx}$.

**Energy:** By analogy, a wave function for which energy/frequency is well defined is the complex exponential $e^{-i\omega t}$ (see also Section 4.2).

**Position:** A wave function for which position is well defined would be one for which there is no probability of finding the particle anywhere but at a single point. Calling this point $x_0$, $\psi(x)$ would have to be zero everywhere but at $x_0$. We give this particular wave function the special symbol $\psi_{x_0}(x)$. It is crudely pictured in Figure 4.19.

Now the mathematical tool:

**Definition:** Given an operator $\hat{Q}$, the function $f(x)$ is an **eigenfunction** of the operator if and only if the operator acting on the function yields a constant times the same function.

$$\underset{\text{Operator}}{\hat{Q}} \quad \underset{\text{Eigenfunction}}{f(x)} \quad = \quad \underset{\text{Eigenvalue}}{\lambda} \quad \underset{\text{Eigenfunction}}{f(x)}$$

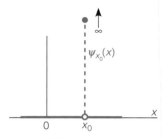

**Figure 4.19** A function for which position is well defined.

The constant $\lambda$ is said to be an **eigenvalue** of the operator. Each operator has a unique, characteristic set of eigenvalues and corresponding eigenfunctions for which the preceding is true.

Its utility in quantum mechanics is as follows:

A wave function $\Psi$ is one for which observable $Q$ is well defined, for which there is no uncertainty in the outcome of a measurement of $Q$, if and only if it is an eigenfunction of the operator $\hat{Q}$; and the eigenvalue is the well-defined value of the observable. Thus,[18]

$$\Delta Q = 0 \iff \hat{Q}\Psi = \overline{Q}\Psi$$

Before discussing the proof of this claim, let us see if it agrees with our previous thoughts about which functions are well defined for the basic observables.

**Momentum:** If the function $e^{ikx}$ is one of well-defined momentum ($\Delta p = 0$), it should be an eigenfunction of the momentum operator.

$$\hat{p}e^{ikx} = -i\hbar\frac{\partial}{\partial x}e^{ikx} = \hbar k e^{ikx}$$

It is—and the eigenvalue is indeed the well-defined momentum, $p = \hbar k$. (Not all functions behave this way! In particular, a gaussian wave does not, and we already know that its momentum is *not* well defined. It has a finite $\Delta x$ and a finite $\Delta p$.)

**Energy:** Similarly, $e^{-i\omega t}$ should be an eigenfunction of the energy operator.

$$\hat{E}e^{-i\omega t} = i\hbar\frac{\partial}{\partial t}e^{-i\omega t} = \hbar\omega e^{-i\omega t}$$

This works the same way; $E$ is indeed $\hbar\omega$.

**Position:** We seek to show that the somewhat unrealistic $\psi_{x_0}(x)$ is an eigenfunction of the position operator $x$ and that the eigenvalue is the well-defined position $x_0$.[19] Thus,

$$x\psi_{x_0}(x) \stackrel{?}{=} x_0\psi_{x_0}(x)$$

Since the function $\psi_{x_0}(x)$ is zero everywhere but at $x_0$, both sides of the equation are also zero. At the only other point, $x = x_0$, the variable $x$ multiplying $\psi_{x_0}(x)$ on the left-hand side is (trivially) $x_0$, the same as the value on the right. The equation holds, and this unusual function is the only one for which it does.

To prove that $\hat{Q}\Psi = \overline{Q}\Psi \implies \Delta Q = 0$ is straightforward.

$$\overline{Q^2} = \int_{\text{all space}} \Psi^*\hat{Q}^2\Psi\,dx = \int_{\text{all space}} \Psi^*\hat{Q}(\hat{Q}\Psi)\,dx$$
$$= \int_{\text{all space}} \Psi^*\hat{Q}(\overline{Q}\Psi)\,dx = \overline{Q}\int_{\text{all space}} \Psi^*\hat{Q}\Psi\,dx = \overline{Q}^2$$

[18]Using the symbol $\overline{Q}$ (i.e., a mean) for the eigenvalue might suggest to the reader that there is some variation in the observable, when in fact there is not. We use it because (1) it makes clear that the eigenvalue is indeed a value, not an operator, and (2) it is not incorrect—if there is no variation, then the mean value and the *only* value are one and the same.

[19]The reader may quail at such a glib discussion of an obviously pathological function. Things must be much more precisely defined before mathematicians are happy. However, precise justifications do exist. For better or for worse, physicists often charge forward without much attention to them.

Thus, $\Delta Q = \sqrt{\overline{Q^2} - \overline{Q}^2} = 0$. A general proof that $\Delta Q = 0 \Rightarrow \hat{Q}\Psi = \overline{Q}\Psi$ involves math too sophisticated to present here. It may, however, be verified fairly easily in the case of the momentum operator. This is left as an exercise.

---

**Example 4.6**

Which of the following wave functions have a well-defined momentum, and for those that do, what is its value? (a) $Ae^{-(x/2\epsilon)^2}$, (b) $A\cos kx$, (c) $A(\cos kx - i\sin kx)$, (d) $A(\cos kx - \sin kx)$.

**Solution**

(a) $\hat{p}\psi(x) = -i\hbar\dfrac{\partial}{\partial x}Ae^{-(x/2\epsilon)^2} = (-i\hbar)Ae^{-(x/2\epsilon)^2}\left(-\dfrac{x}{2\epsilon^2}\right)$

This is not a constant times $Ae^{-(x/2\epsilon)^2}$, so the momentum of this wave function is not well defined. Of course, we would not expect a gaussian to have zero $\Delta p$, because $\Delta x$ is not infinite.

(b) $-i\hbar\dfrac{\partial}{\partial x}A\cos kx = i\hbar kA\sin kx$

Again, since this is not a constant times $A\cos kx$, the momentum is not well defined. The reader may wonder why an infinitely long cosine function does not have a well-defined momentum; it would seem to qualify as having a well-defined wavelength. The answer is basic to quantum mechanics and is intimately tied in with the momentum operator's definition: The fundamental plane waves are not sines or cosines, but complex exponentials (cf. Section 3.3). Only these have well-defined momenta.

(c) $-i\hbar\dfrac{\partial}{\partial x}A(\cos kx - i\sin kx) = -i\hbar kA(-\sin kx - i\cos kx)$

$$= -\hbar kA(-i\sin kx + \cos kx)$$

This is $-\hbar k$ times the original wave function. The constant (eigenvalue) is the well-defined value of the observable, so the momentum is a well-defined $-\hbar k$. This function is, in fact, $Ae^{-ikx}$. Since the momentum is negative, we see that this complex exponential represents a wave moving in the negative direction, whereas $Ae^{+ikx}$ represents one moving in the positive.[20] We also see why $A\cos kx$ does *not* have a well-defined momentum. Writing it as $\cos kx = \frac{1}{2}(e^{+ikx} + e^{-ikx})$, we see that it is really equal parts of two waves of opposite momentum. Its momentum expectation value would be zero, but the particle could be found moving in different directions. The same arguments apply to the sine functions of the infinite well.

(d) $-i\hbar\dfrac{\partial}{\partial x}(A\cos kx - A\sin kx) = i\hbar k(A\sin kx + A\cos kx)$

Momentum is not well defined; the "missing" $i$ is crucial.

---

[20]We may draw an analogy with traveling electromagnetic waves. The direction of motion of an electromagnetic wave may be seen from a "snapshot"; it is the direction of $\mathbf{E} \times \mathbf{B}$. If $\mathbf{E}$ were to be shifted relative to $\mathbf{B}$ by 180°, the result would be a wave moving in the opposite direction. Similarly, if the real part of a positive-moving matter wave is shifted 180° relative to the imaginary, an $e^{+ikx}$ is changed to an $e^{-ikx}$. The result is a negative-moving matter wave.

Categorizing wave functions as eigenfunctions of operators has many applications. It is widely used in more advanced analysis, such as "matrix quantum mechanics." Probably its most useful application, though, is the one to which we have put it: *It is the simplest way of determining whether a function has a well-defined value of a given observable.*

# ◆ 4.11 Nonstationary States

In separating variables to solve the Schrödinger equation, we restricted our attention to states for which the energy is well defined and the probability density $\Psi^*\Psi$ independent of time. We now consider a more general case.

Consider a bound particle. As we know, its energy will be quantized. Let us refer to its allowed energies as $E_1, E_2, E_3, \ldots$, and the corresponding solutions of the time-independent Schrödinger equation as $\psi_1, \psi_2, \psi_3, \ldots$. The total wave function in stationary state $n$ is given by equation (4-6).

$$\Psi_n(x,\ t) = \psi_n(x)e^{-i(E_n/\hbar)t}$$

Now consider a mathematical sum of two such solutions, but of different energies.

$$\Psi(x,\ t) = \psi_n(x)e^{-i(E_n/\hbar)t} + \psi_m(x)e^{-i(E_m/\hbar)t} \tag{4-27}$$

Since the Schrödinger equation is a *linear* differential equation, any sum of solutions is also a solution. Both of the functions added above are solutions of the time-*dependent* Schrödinger equation, so their sum is also a solution. However, the sum does not have a well-defined energy, because $\psi_n$ and $\psi_m$ involve different values of $E$. (Neither is a solution of the time-*independent* Schrödinger equation *of the other.*) We can prove this by showing that this $\Psi(x, t)$ is not an eigenfunction of the energy operator.

$$\hat{E}\Psi(x,\ t) = i\hbar\frac{\partial}{\partial t}\left[\psi_n(x)e^{-i(E_n/\hbar)t} + \psi_m(x)e^{-i(E_m/\hbar)t}\right]$$

$$= \psi_n(x)(i\hbar)\left(-i\frac{E_n}{\hbar}\right)e^{-i(E_n/\hbar)t} + \psi_m(x)(i\hbar)\left(-i\frac{E_m}{\hbar}\right)e^{-i(E_m/\hbar)t}$$

$$= \psi_n(x)E_n e^{-i(E_n/\hbar)t} + \psi_m(x)E_m e^{-i(E_m/\hbar)t}$$

Since the *same* constant cannot be factored out of both terms, the expression on the right-hand side is *not* a constant times $\Psi(x, t)$ itself. Accordingly, $\Psi(x, t)$ is not an eigenfunction, and the uncertainty in the energy must be nonzero. (We might guess that $\Delta E$ is proportional to $E_n - E_m$. It is left as an exercise to verify this for the infinite well.)

Not only is its energy not certain, this $\Psi(x, t)$ lacks another property of stationary states. Its probability density varies with time.

$$\Psi^*(x,\ t)\Psi(x,\ t) = [\psi_n(x)e^{+i(E_n/\hbar)t} + \psi_m(x)e^{+i(E_m/\hbar)t}]$$

$$\times [\psi_n(x)e^{-i(E_n/\hbar)t} + \psi_m(x)e^{-i(E_m/\hbar)t}] \tag{4-28}$$

$$= \psi_n^2(x) + \psi_m^2(x) + 2\psi_n(x)\psi_m(x)\cos\left(\frac{E_n - E_m}{\hbar}t\right)$$

We see that the variation is a periodic one, of angular frequency $(E_n - E_m)/\hbar$.

The wave function $\Psi(x, t)$ given in (4-27) may be thought of as crudely representing a transitional state between two stationary states. Perhaps it represents an electron in an infinite well as it "jumps down" from a high-energy state, the $n$th, to a lower-energy one, the $m$th. During the transition, the wave function and energy are combinations of those of the initial and final states. On the other hand, if energy is to be conserved, there must be something missing; a system cannot by itself start with one energy and end with another.

The most common result from such a transition is the emission of a photon. Initially, the system is in a stationary state of well-defined energy $E_n$. It then enters a transitional phase, during which a photon is being generated and the energy of the system is not well defined. Finally, the system is in a stationary state of well-defined energy $E_m$, and a photon of energy $E_n - E_m$ emerges. Energy is conserved. A detailed study of the photon's generation would not be appropriate here, but our crude picture contains the key feature: In a *non*stationary state, probability density does vary with time, and if the particle making the transition is charged, its *charge* density should vary with time. Electromagnetic radiation *should* result!

Is the association of charge density with probability density valid? Classically, the hydrogen atom should be unstable, the orbiting electron accelerating and therefore radiating away its energy into electromagnetic waves. The quantum-mechanical explanation of the atom's stability is that the electron is in a stationary state, so its probability density *and charge density* do not vary with time in any way. If no charge accelerates, no radiation is generated. This alone would make the association of the two densities plausible. What we have found in this section further strengthens the case: To generate a photon of energy $(E_n - E_m)$, it is necessary that *charge* oscillate at angular frequency $\omega = (E_n - E_m)/\hbar$, and the transitional state *probability* density (4-28) oscillates at exactly this frequency.

## Progress and Applications

The essential correctness of quantum mechanics is not doubted. It is the proper description of small things, and its differences from classical mechanics continue to be revealed in modern technology in new and fascinating ways.

■  What often had to be taken on faith in the past is now being verified. A number of experiments have been successful in recent years in "feeling the shape" of matter-wave functions. A lone observation can gain only a speck of information, while destroying the delicate wave function. Accordingly, these experiments have involved many thousands of repetitions on identically prepared objects, the data then being knitted together to yield the complete picture (as in Figure 4.11). One picture showing the matter wave's expected complex interference over a fairly broad region of space has been obtained by sending "excited" helium atoms through a double slit (Figure 4.20). Their ejection of electrons upon striking a metal target is monitored as the target is moved progressively farther from the slits, in a process known as "phase-space tomography" (C. Kurtsiefer, et al., *Nature*, March 13, 1997, pp. 150–153). The results confirm our "blind" theoretical expectations.

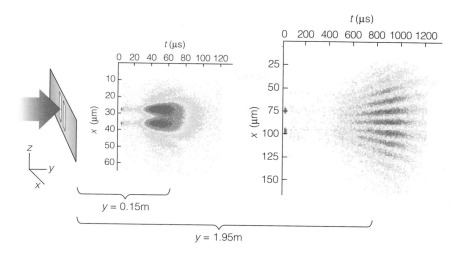

**Figure 4.20** A helium atom double-slit interference pattern. The spread in arrival times is due to a range of helium atom speeds (and thus wavelengths).

- Due to improved technology in manipulating materials on a microscopic scale, much interest has lately been focused on so-called **quantum dots,** from which a host of novel devices springs. A "quantum well" confines a particle in one dimension; but in a real three-dimensional system, the particle would still be free to move in two dimensions, so its energy would not strictly speaking be confined to certain discrete values. A "quantum wire" confines a particle in two dimensions, leaving it free in one, again without fully discrete energies. Bound in all dimensions, free in none, a particle in a "quantum dot" is indeed restricted to only certain discrete energies. (The three-dimensional particle in a box is discussed in Chapter 6.) Today's experimental quantum dots confine electrons in various ways inside semiconductors or metals and are typically 10–100 nm on a side. One possible application is in data storage, where the well-defined energy level serves as the information, and photon emission and absorption as the particles jump among levels serves as the "read and write" mechanism. In another possible application, leads are attached in a particular way to produce a "quantum-dot transistor," a minute circuit element in which extremely small electrical signals control the flow of other signals, conceivably allowing another great leap in the miniaturization of electronic circuitry. A collection of quantum dots has also been explored as a possible laser medium, but the difficulty (among others) of producing the required identical, regularly spaced dots has been a major challenge.

- The idea of **quantum computing** has occupied many scientists lately. In the simplest analysis, it is based on an alternative to the standard "data bit," the two-state 1 or 0 (on or off) of conventional digital computing. A quantum system may also have "only" two states, wave functions 1 and 0. Two different vibrational levels of a harmonic oscillator have been used, as well as two different "spin" states (see Chapter 7) of an atom's orbiting electron. But in a quantum system the states may also coexist in an infinite number of combinations, by varying the phase relationship between them. Quantum computing deals intimately with the wave function $\psi$, not just a probability $|\psi|^2$. As a rough analogy, adding two incoherent (irregular) light beams merely increases the intensity, but carefully adding two coherent light beams can give a continuum of interference results. Such an increase in information holds the potential for "massive parallel processing." Quantum computing is being applied to finding the prime factors of large whole numbers quickly, a problem heretofore renowned for the tedious algorithms it required, and accordingly used in information encryption schemes. Still, debate surrounds the topic. Preserving the crucial coherent phase relationship in practical applications is a formidable hurdle.

- A particularly nice example of penetration of the classically forbidden region is being studied in the atomic nucleus (the subject of Chapter 10). In certain special nuclei, neutrons and protons that are only barely bound to the nucleus have wave functions extending significantly beyond its "normal" confines. The result is a so-called nuclear halo. The effective cross-sectional areas of these nuclei are much larger as a result of their halos. In one method of study, a beam containing the halo nuclei is directed at stationary target atoms, the details of the halos being gleaned from the scattering results.

## Basic Equations

**Time-dependent Schrödinger equation**

$$-\frac{\hbar^2}{2m}\frac{\partial^2\Psi(x,\,t)}{\partial x^2} + U(x)\Psi(x,\,t) = i\hbar\,\frac{\partial\Psi(x,\,t)}{\partial t} \tag{4-2}$$

**Temporal part of $\Psi(x,t)$**

$$\phi(t) = e^{-i(E/\hbar)t} \tag{4-5}$$

**Total wave function**

$$\Psi(x,\,t) = \psi(x)e^{-i(E/\hbar)t} \tag{4-6}$$

**Time-independent Schrödinger equation**

$$-\frac{\hbar^2}{2m}\frac{d^2\psi(x)}{dx^2} + U(x)\psi(x) = E\psi(x) \tag{4-8}$$

**Normalization**

$$\int_{\text{all space}} |\Psi(x,\,t)|^2\,dx = 1 \tag{4-9}$$

**Potential well**

$$U(x) = \begin{cases} 0 & 0 < x < L \\ U_0 & x < 0, \, x > L \end{cases} \tag{4-10}$$

**Infinite well ($U_0 \to \infty$)**

$$\psi_n(x) = \begin{cases} \sqrt{\dfrac{2}{L}}\,\sin\dfrac{n\pi x}{L} & 0 < x < L \\ 0 & x < 0, \, x > L \end{cases} \qquad E_n = \frac{n^2\pi^2\hbar^2}{2mL^2} \tag{4-16}$$

| Expectation value | Uncertainty |
|---|---|
| $$\overline{Q} = \int_{\substack{all \\ space}} \Psi^*(x, t)\hat{Q}\Psi(x, t)\, dx \quad \text{(4-21)}$$ | $$\Delta Q = \sqrt{\overline{Q^2} - \overline{Q}^2} \qquad \text{(4-22)}$$ |

**Momentum operator**

$$\hat{p} = -i\hbar\frac{\partial}{\partial x}$$

**Penetration depth**

$$\delta = \frac{\hbar}{\sqrt{2m(U_0 - E)}} \qquad \text{(4-24)}$$

**Harmonic oscillator energy levels**

$$E = (n + \tfrac{1}{2})\hbar\omega_0 \quad (n = 0, 1, 2, 3, \ldots) \quad \text{where } \omega_0 \equiv \sqrt{\frac{\kappa}{m}} \qquad \text{(4-26)}$$

## Summary

To understand the behavior of a massive object, it is necessary to know the wave function $\Psi(x, t)$. The wave function is a solution of the Schrödinger equation, which in the presence of external interactions, represented by $U(x)$, is

$$-\frac{\hbar^2}{2m}\frac{\partial^2}{\partial x^2}\Psi(x, t) + U(x)\Psi(x, t) = i\hbar\frac{\partial}{\partial t}\Psi(x, t)$$

or

$$\hat{KE}\Psi(x, t) + \hat{U}(x)\Psi(x, t) = \hat{E}\Psi(x, t)$$

Solving the Schrödinger equation for $\Psi(x, t)$ is roughly analogous to the classical procedure of solving $\mathbf{F} = m(d^2\mathbf{r}/dt^2)$ for $\mathbf{r}(t)$.

Employing separation of variables, the wave function $\Psi(x, t)$ becomes $\psi(x)\phi(t)$, and the resulting solutions are called stationary states. The energy is well defined, and the probability density becomes $|\psi(x)|^2$; it does not vary with time.

A quantum-mechanical treatment is required in "small systems," whenever the dimensions of the system are comparable to or less than the wavelength of $\Psi(x, t)$. Such a small system cannot be constantly "watched," so as to keep track of position, momentum, and so on, since an observer must somehow interact with the system, which would disturb the very behavior being observed. If no attempt is made to find it, a massive object in a small system is not a particle but a wave.

When a massive object is bound, as when a classical wave is confined to a finite region of a medium, only certain waves satisfy the boundary conditions and may form standing waves. This is the root of energy quantization. The energy of a bound particle is restricted to a certain discrete set of allowed values, each with corresponding wave function. In all cases of particles bound by a potential energy, the minimum-energy state has nonzero kinetic energy; if a particle is bound, it cannot be stationary. Three simple cases of bound particles, often used as models for more complicated systems, are the infinite well, the finite well, and the harmonic oscillator. In the limit of large energies, the predictions of quantum mechanics and classical mechanics converge.

Observations are governed by probabilities. An experiment yielding a value for an observable such as position or momentum would, in general, yield a different value if repeated identically. The experiment cannot simply be a later observation of the initial system, as the first observation will have disturbed the system. The result of many identical repetitions would be an average value and a standard deviation. These may be predicted from knowledge of the wave function, and are known, respectively, as the expectation value and the uncertainty.

$$\overline{Q} = \int_{\text{all space}} \Psi^*(x,\ t)\hat{Q}\Psi(x,\ t)\ dx \qquad \Delta Q = \sqrt{\overline{Q^2} - \overline{Q}^2}$$

$$\text{Expectation value} \qquad\qquad \text{Uncertainty}$$

where $\hat{Q}$ is the operator associated with the observable. For a particle whose wave function is $\Psi(x,\ t)$, the following are equivalent:

$$\text{the observable } Q \text{ is well defined} \quad \Leftrightarrow \quad \Delta Q = 0 \quad \Leftrightarrow \quad \hat{Q}\Psi = \overline{Q}\Psi$$

# E X E R C I S E S

## Section 4.4

1. A classical particle experiences a force whose potential energy is

   $$U(x) = \frac{1}{x^2} - \frac{2}{x} + 1 \quad \text{(in SI units)}$$

   (a) By finding its minimum value and determining its behaviors at $x = 0$ and $x = +\infty$, sketch this potential energy.
   (b) Suppose the particle has an energy of 0.5 J. Find any turning points. Would the particle be bound?
   (c) Suppose the particle has an energy of 2.0 J. Find any turning points. Would the particle be bound?

## Section 4.5

2. In the time-independent Schrödinger equation, if $\psi(x)$ were nonzero and $U(x)$ infinite, what else would have to be infinite? Would such a situation be realistic?
3. Write out the total wave function $\Psi(x, t)$ for an electron in the $n = 3$ state of a 10-nm-wide infinite well. Besides the symbols $x$ and $t$, the function should include only numerical values.
4. An electron in the $n = 4$ state of a 5-nm-wide infinite well makes a transition to the ground state, giving off

energy in the form of a photon. What is the photon's wavelength?

5. An electron is trapped in a quantum well (practically infinite). If the lowest-energy transition is to produce a photon of 450-nm wavelength, what should be the well's width?
6. What is the probability that a particle in the first excited $(n = 2)$ state of an infinite well would be found in the middle third of the well? How does this compare with the classical expectation? Why?
7. Show that $\psi(x) = A'e^{ikx} + B'e^{-ikx}$ is equivalent to the expression $\psi(x) = A \sin kx + B \cos kx$, provided that $A' = \frac{1}{2}(A - iB)$ and $B' = \frac{1}{2}(A + iB)$.
8. A particle is bound by a potential energy of the form

$$U(x) = \begin{cases} 0 & |x| < \frac{1}{2}a \\ \infty & |x| > \frac{1}{2}a \end{cases}$$

Determine the allowed energies and corresponding normalized wave functions.

## Section 4.6

9. Classically, if a particle is not observed, the probability of finding it in a box is a constant $1/L$ per unit length along the entire length of the box. With this, show that the "classical expectation value" of the position is $\frac{1}{2}L$, that the expectation value of the square of the position is $\frac{1}{3}L^2$, and that the uncertainty in position is $L/\sqrt{12}$.

10. Show that the uncertainty in the position of a particle in an infinite well in the general case of arbitrary $n$ is given by

$$L\sqrt{\frac{1}{12} - \frac{1}{2n^2\pi^2}}$$

Discuss the dependence; in what circumstance does it agree with the "classical uncertainty" of $L/\sqrt{12}$ (see Exercise 9)?

11. Show that the uncertainty in the momentum of a particle in an infinite well in the general case of arbitrary $n$ is given by $n\pi\hbar/L$.

12. What is the product of the uncertainties determined in the previous two exercises? Discuss the result.

13. If a particle in a stationary state is *bound,* the expectation value of its momentum must be zero. (a) In words, why? (b) Prove it! Starting from the general expression (4-21) with $\hat{p}$ in place of $\hat{Q}$, integrate by parts, then argue that the result is identically zero. (Be careful that your argument is somehow based on the particle being *bound;* a free particle certainly may have a nonzero momentum. *Note:* $\psi(x)$ may always be chosen to be real in a stationary state.)

## Section 4.7

14. A 50-eV electron is trapped between negligible-width capacitors charged to 200 V (each with an exit hole). How far does its wave function extend beyond the capacitors?

15. An electron is trapped in a finite well. How "far" (in eV) is it from being free if the penetration length of its wave function into the classically forbidden region is 1 nm?

16. While an infinite well has an infinite number of bound states, a finite well does not. By relating the well height $U_0$ to the kinetic energy, and the kinetic energy (through $\lambda$) to $n$ and $L$, show that the number of bound states is given roughly by $\sqrt{8mL^2U_0/h^2}$. (Assume that the number is large.)

17. Obtain expression (4-23c) from equation (4-23a). (*Hint:* Using $\cos\theta = \cos^2\frac{1}{2}\theta - \sin^2\frac{1}{2}\theta$ and $\sin\theta = 2\sin\frac{1}{2}\theta \cos\frac{1}{2}\theta$, first convert the argument of the cotangent from $kL$ to $\frac{1}{2}kL$. Put the resulting equation in quadratic form, then factor. Note that $\alpha$ is positive by definition.)

**Exercises 18–21 refer to a particle of mass $m$ trapped in a "half-infinite" well, with potential energy given by**

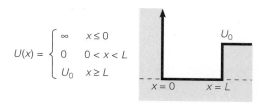

$$U(x) = \begin{cases} \infty & x \leq 0 \\ 0 & 0 < x < L \\ U_0 & x \geq L \end{cases}$$

18. Advance an argument based upon $p = h/\lambda$ that there is no bound state in a half-infinite well unless $U_0$ is at least $h^2/32mL^2$. (*Hint:* What is the maximum wavelength possible within the well?)

19. As noted in Section 4.7, a finite well always has at least one bound state. Why does the argument of the previous exercise fail in the case of a finite well?

20. Write solutions to the Schrödinger equation appropriate in the various regions, impose required continuity conditions, and obtain the energy quantization condition:

$$\sqrt{E} \cot\left(\frac{\sqrt{2mE}}{\hbar}L\right) = -\sqrt{U_0 - E}$$

21. Using the result of Exercise 20 and a numerical technique of your choice, find the two bound-state energies for a well in which $U_0 = 4(\pi^2\hbar^2/2mL^2)$.

## Section 4.8

22. A 2-kg block oscillates with an amplitude of 10 cm on a spring of force constant 120 N/m. (a) In which quantum state is the block? (b) The block has a slight electric charge and may therefore drop to a lower energy level by generating a photon. What is the minimum energy decrease possible, and what would be the corresponding fractional change in energy?

23. Air is mostly $N_2$, diatomic nitrogen, with an effective spring constant of $2.3 \times 10^3$ N/m. At room temperature, air's heat capacity is what we would expect of a collection of "rigid dumbbells." That is, each molecule seems to behave as two point masses connected by a rigid rod along which they cannot vibrate. Argue that according to quantum mechanics this is to be expected. (The effective oscillating mass is half the atomic mass. *Note:* Diatomic oxygen, $O_2$, making up the remaining $\frac{1}{5}$ of the air, behaves the same way for the same reasons.)

24. What is the most probable location to find a particle that is in the "first excited" ($n = 1$) state of a harmonic oscillator potential energy?

25. Determine the expectation value of the position of a harmonic oscillator in its ground state.

26. Show that the uncertainty in the position of a ground-state harmonic oscillator is $(1/\sqrt{2})(\hbar^2/m\kappa)^{1/4}$.

27. Show that the uncertainty in the momentum of a ground state harmonic oscillator is $(\sqrt{\hbar/2})(m\kappa)^{1/4}$.

28. What is the product of the uncertainties determined in the two previous exercises? Explain.

29. Repeat Exercises 26-28 for the "first excited" ($n = 1$) state of a harmonic oscillator.

30. The potential energy shared by two atoms in a diatomic molecule, depicted in Figure 4.15, is often approximated by the fairly simple function $U(x) = (a/x^{12}) - (b/x^6)$, where constants $a$ and $b$ depend on the atoms involved. In Section 4.8, it is said that near its minimum value it can be approximated by an even simpler function—it should "look like" a parabola.
    (a) In terms of $a$ and $b$, find the minimum potential energy $U(x_0)$ and the separation $x_0$ at which it occurs.
    (b) The parabolic approximation $U_p(x) = U(x_0) + \frac{1}{2}\kappa(x - x_0)^2$ has the same minimum value at $x_0$ and the same first derivative there (i.e., zero). Its second derivative is $\kappa$, the spring constant of this "Hooke's law" potential energy. In terms of $a$ and $b$, what is the spring constant of $U(x)$?

## Section 4.9

31. For a Hooke's law potential energy of $U = \frac{1}{2}\kappa x^2$, the ground-state wave function is $\psi(x) = Ae^{-(\sqrt{m\kappa/2\hbar})x^2}$ and its energy is $\frac{1}{2}\hbar\sqrt{\kappa/m}$.
    (a) Find the classical turning points for a particle with this energy.
    (b) By the arguments in Section 4.9, the second derivative of $\psi(x)$ should be negative between the classical turning points, where $E > U$, and positive outside the turning points, where $E < U$. Verify this. (*Hint:* Look for the inflection points.)

32. Consider a particle bound in an infinite well where the potential inside is not constant, but a linearly varying function. Suppose that the particle is in a fairly high energy state, so that its wave function stretches across the entire well; that is, it isn't caught in the "low spot."

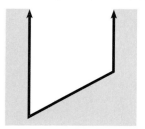

(a) Would the wave function have a constant wavelength? Explain.
(b) Sketch a plausible wave function.

33. In Section 4.9, it is said that for a given $U(x)$ there may be mathematical solutions to the Schrödinger equation for *any* E. In fact, in the finite well there are *smooth* solutions for any energy. Show this as follows:
    (a) Do not throw out any mathematical solutions. That is, in region II ($x < 0$) assume that $\psi(x) = Ce^{+\alpha x} + De^{-\alpha x}$, and in region III ($x > L$), assume that $\psi(x) = Fe^{+\alpha x} + Ge^{-\alpha x}$. Write the smoothness conditions.
    (b) In Section 4.7, we combined these four conditions to eliminate A, B, and D in favor of C. In the remaining equation, C canceled, leaving an equation involving only $k$ and $\alpha$, solvable for only certain values of E. Why can't this be done here?
    (c) Our solution is smooth. What is still wrong with it physically?
    (d) Show that
    $$D = \frac{1}{2}\left(B - \frac{k}{\alpha}A\right) \quad \text{and} \quad F = \frac{1}{2}e^{-\alpha L} \times$$
    $$\left[\left(A - B\frac{k}{\alpha}\right)\sin kL + \left(A\frac{k}{\alpha} + B\right)\cos kL\right]$$
    and that setting these offending coefficients to zero reproduces quantization condition (4-23a).

## Section 4.10

34. To describe a matter wave, does the function $A\sin(kx)\cos(\omega t)$ have a well-defined energy? Explain.

35. Does the wave function $\psi(x) = A(e^{+ikx} + e^{-ikx})$ have a well-defined momentum? Explain.

36. The operator for angular momentum about the $z$-axis in spherical polar coordinates is $-i\hbar(\partial/\partial\phi)$. Find a function $f(\phi)$ that would have a well-defined $z$-component of angular momentum.

37. Show that $\Delta p = 0 \Rightarrow \hat{p}\psi(x) = \overline{p}\psi(x)$. That is, show that unless the wave function is an eigenfunction of the momentum operator, there will be a nonzero uncertainty in momentum. The following procedure may be followed. Show first that the quantity
    $$\int_{\text{all space}} \psi^*(x)(\hat{p} - \overline{p})^2\psi(x)\, dx$$
    is $(\Delta p)^2$. Then, using the differential operator form of $\hat{p}$ and integration by parts, show that it is also
    $$\int_{\text{all space}} [(\hat{p} - \overline{p})\psi(x)]^*[(\hat{p} - \overline{p})\psi(x)]\, dx$$

(*Note:* Since momentum is real, $\bar{p}$ is real.) Together these show that if $\Delta p$ is zero, then the preceding quantity must be zero. However, the integral of the complex square of a function (the quantity in brackets) can only be zero if the function is identically zero. The assertion is proved.

### Section 4.11

**38.** Consider a wave function that is a combination of two different infinite-well stationary states, the $n$th and the $m$th.

$$\Psi(x, t) = \frac{1}{\sqrt{2}} \psi_n(x) e^{-i(E_n/\hbar)t} + \frac{1}{\sqrt{2}} \psi_m(x) e^{-i(E_m/\hbar)t}$$

(a) Show that $\Psi(x, t)$ is properly normalized.
(b) Show that the expectation value of the energy is the average of the two energies: $\bar{E} = \frac{1}{2}(E_n + E_m)$.
(c) Show that the expectation value of the square of the energy is given by

$$\overline{E^2} = \frac{1}{2}(E_n^2 + E_m^2)$$

(d) Determine the uncertainty in the energy.

### General Exercises

**39.** The harmonic oscillator potential energy is proportional to $x^2$, and the energy levels are equally spaced: $E_n \propto (n + \frac{1}{2})$. The energy levels in the infinite well become farther apart as energy increases: $E_n \propto n^2$. Since the function $\lim_{b \to \infty} |x/L|^b$ is zero for $|x| < L$ and infinitely large for $|x| > L$, the infinite well potential energy may be thought of as proportional to $|x|^\infty$.

$U \propto x^2$    $U \propto |x|^\infty$    $U \propto |x|^1$    $U \propto -|x|^{-1}$

How would you expect energy levels to be spaced in a potential well that is (a) proportional to $|x|^1$, and (b) proportional to $-|x|^{-1}$? For the harmonic oscillator and infinite well, the number of bound-state energies is infinite, and arbitrarily large bound-state energies are possible. Are these characteristics shared (c) by the $|x|^1$ well, and (d) by the $-|x|^{-1}$ well?

**40.** The energy levels for the harmonic oscillator potential $U(x) = \frac{1}{2}\kappa x^2$ are equally spaced: $E_n \propto (n + \frac{1}{2})$. Consider the following crude argument as to why: Energy is roughly proportional to $k^2$ (*kinetic* energy $= p^2/2m$

$= \hbar^2 k^2/2m$), and $k = n\pi/L$ for a standing wave, so that $E \propto n^2/L^2$. The argument is completely valid for the infinite well, in which the potential energy is zero inside the well and $L$ is strictly constant. For the harmonic oscillator, the length $L$ allowed the wave should be roughly proportional to the classical amplitude $A$ of the particle's motion. At the extremes of travel, $E = U = \frac{1}{2}\kappa A^2$, so that $A \propto \sqrt{E}$. Inserting this for $L$, $E \propto n^2/\sqrt{E}^2$ or $E \propto n$. Use a similar procedure to argue that the energy levels in a potential energy where $U \propto 1/x$ should be proportional to $1/n^2$. (Although three-dimensional, the potential energy in the hydrogen atom is of this form, and the quantum energy levels are indeed proportional to $1/n^2$.)

**41.** Consider the differential equation $d^2f(x)/dx^2 = bf(x)$.
(a) Suppose that $f_1(x)$ and $f_2(x)$ are solutions, that is,

$$\frac{d^2f_1(x)}{dx^2} = bf_1(x) \quad \text{and} \quad \frac{d^2f_2(x)}{dx^2} = bf_2(x)$$

Show that the equation also holds when the linear combination $A_1f_1(x) + A_2f_2(x)$ is inserted. (b) Suppose that $f_3(x)$ and $f_4(x)$ are solutions of $d^2f(x)/dx^2 = bf^2(x)$. Is $A_3f_3(x) + A_4f_4(x)$ a solution? Justify your answer.

### Exercises 42–52 refer to a particle of mass $m$ described by the wave function:

$$\psi(x) = \begin{cases} 2\sqrt{a^3}\,xe^{-ax} & x > 0 \\ 0 & x < 0 \end{cases}$$

**42.** Verify that the normalization constant $2\sqrt{a^3}$ is correct.
**43.** Sketch the wave function. Is it smooth?
**44.** Determine the particle's most probable position.
**45.** What is the probability that the particle would be found between $x = 0$ and $x = 1/a$?
**46.** Calculate the expectation value of the position of the particle.
**47.** Calculate the uncertainty in the particle's position.
**48.** Determine the expectation value of the momentum of the particle. Explain.
**49.** Calculate the uncertainty in the particle's momentum.
**50.** What is the product of $\Delta x$ and $\Delta p$ (obtained in Exercises 47 and 49)? How does it compare with the minimum theoretically possible? Explain.
**51.** The particle has $E = 0$. (a) Show that the potential energy for $x > 0$ is given by

$$U(x) = -\frac{\hbar^2 a}{m}\frac{1}{x} + \frac{\hbar^2 a^2}{2m}$$

(b) What is the potential energy for $x < 0$?

**52.** The potential energy in the case where the particle has $E = 0$ is given in Exercise 51. (a) On the same axes, sketch the wave function and the potential energy. (b) What is the probability that the particle would be found in the classically forbidden region?

**53.** A particle is described by the wave function

$$\psi(x) = \frac{\sqrt{2/\pi}}{x^2 - x + 1.25}$$

    **(a)** Show that the normalization constant $\sqrt{2/\pi}$ is correct.

    **(b)** A measurement of the position of the particle is to be made. At what location is it most probable that the particle would be found?

    **(c)** What is the probability per unit length of finding the particle at this location?

**Exercises 54–56 refer to a particle described by the wave function:**

$$\psi(x) = \sqrt{\frac{2}{\pi}}\, a^{3/2} \frac{1}{x^2 + a^2}$$

**54.** Show that the normalization constant is correct.

**55.** Calculate the uncertainty in the particle's position.

**56.** (a) Taking the particle's total energy to be zero, find the potential energy. (b) On the same axes, sketch the wave function and the potential energy. (c) To what region would the particle be restricted classically?

**57.** Consider the "delta well" potential energy:

$$U(x) = \begin{cases} 0 & x \neq 0 \\ -\infty & x = 0 \end{cases}$$

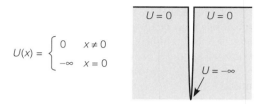

Although not completely realistic, this potential energy is often a convenient approximation to a *very* strong, *very* narrow attractive potential energy well. It has only one allowed bound-state wave function and, since the top of the well is defined as $U = 0$, the corresponding bound-state energy is negative. Call its value $-E_0$.

    **(a)** Applying the usual arguments and required continuity conditions, show that the wave function is given by

$$\psi(x) = \left(\frac{2mE_0}{\hbar^2}\right)^{1/4} e^{-(\sqrt{2mE_0}/\hbar)\,|x|}$$

    **(b)** Sketch $\psi(x)$ and $U(x)$ on the same diagram. Does this wave function exhibit the expected behavior in the classically forbidden region?

**58.** A harmonic oscillator has its minimum possible energy. What is the probability of finding it in the classically forbidden region? (*Note:* At some point in the calculation, a computer will be needed.)

# Unbound States: Obstacles, Tunneling, and Particle-Wave Propagation

Common to all topics in this chapter is the absence of a force capable of confining a particle to a region of space. Since the particle is not bound, standing waves are not possible and energy is not quantized. Of course, classical mechanics is full of applications in which particles are not bound, but we are interested in what quantum mechanics has to say in such cases. Among other things, we find that a "particle" may be turned back by a force when classical mechanics says it should proceed, and that a particle may prevail against a force where classical physics says it should be turned back. We begin in Section 5.1 by seeing what effect simple forces have on simple matter waves. In Section 5.2, we apply these ideas to a classically baffling phenomenon—radioactive alpha decay. Section 5.3 is devoted to understanding the richer behavior of not-so-simple, but more realistic free-particle wave functions. Here we clarify the relationship between the wave function and the motion of the particle. In the final section, we make an important connection between force in quantum mechanics and force in classical mechanics.

## 5.1 Obstacles and Tunneling

As a light wave striking a glass plate divides—some transmitted, some reflected—so does a matter wave divide when it encounters an abrupt change in conditions. This doesn't mean that a massive particle divides, any more than photons divide in the case of light. But since in either case the wave is related to the probability of finding the particle, the division of the wave governs the probability of a given incident particle (photon or massive object) being transmitted or reflected.

To say anything about this, we must be able to distinguish between particles moving one way and particles moving the other, and to do this we must be careful about the wave functions we use. In Section 3.3, we obtained a plane-wave solution of the free-particle Schrödinger equation:

$$\Psi(x,\,t) = A e^{+ikx-i\omega t} \quad \textbf{(Right-moving)}$$

This represents a particle whose momentum is in the positive $x$-direction. For a particle whose momentum is in the negative $x$-direction, we would have

$$\Psi(x,\,t) = A e^{-ikx-i\omega t} \quad \textbf{(Left-moving)}$$

Since we will be concentrating on stationary states, in which the same $e^{-i\omega t}$ would appear in all $\Psi(x,\,t)$, the spatial parts are our main concern: $\psi(x) = e^{+ikx}$ moves to the right and $\psi(x) = e^{-ikx}$ moves left. It is important to note that we *could* choose otherwise. Sums of solutions are still solutions, and by adding or subtracting these two, we would obtain $\cos(kx)$ and $\sin(kx)$. The latter pair are thus combinations of right-moving and left-moving. Inside the infinite well, where the particle does move freely, we found $\sin(kx)$ to be the proper solution. It is appropriate because a bound particle can be found to be moving in either direction. But if we now wish to distinguish right-moving from left-moving, we must use $e^{+ikx}$ and $e^{-ikx}$.

$e^{+ikx}$       $e^{-ikx}$

Right-moving    Left-moving

## Potential Step: $E > U_0$

We now consider a free particle encountering a simple force, an abrupt change in potential energy known as a **potential step,** shown in Figure 5.1. A free electron moves in a region ($x < 0$) of zero potential energy. At $x = 0$, it encounters a very

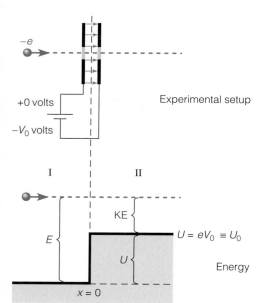

**Figure 5.1** A free electron encounters a "potential step" that is classically surmountable.

narrow region where it experiences a large force opposite its direction of motion—a backward "kick." Passing through, its potential energy jumps abruptly to $U_0$ and stays there. Because a force corresponds to a *change* in potential energy, the narrow region near $x = 0$ is the only place where the electron experiences a force.

$$U(x) = \begin{cases} 0 & x < 0 \\ U_0 & x > 0 \end{cases}$$

For the present, we consider only the case where the particle initially has kinetic energy greater than the potential energy jump. (In Figure 5.1, $E$ is greater than $U_0$.) Classically, it should merely slow down abruptly at $x = 0$, its kinetic energy dropping as potential jumps, then proceed at constant reduced speed. There is no classical turning point where its kinetic energy would fall to zero and the force cause a reversal of the motion; the particle would not rebound.

But classical physics won't do. To understand the behavior of something so small as an electron, we must apply quantum mechanics. We assume that the electron's energy is well defined, so that we solve the time-independent Schrödinger equation for $\psi(x)$.

$$\frac{d^2\psi(x)}{dx^2} = \left[\frac{2m(U(x) - E)}{\hbar^2}\right]\psi(x) \qquad (5\text{-}1)$$

Just as for the potential wells, we obtain solutions in the various regions, then impose proper matching conditions at points where the regions meet.

## $E > U_0$ Potential Step

### Region I ($x < 0$)

In this region, the quantity in brackets is the negative constant $-2mE/\hbar^2$. As noted, to distinguish right-moving from left-moving particles, we do not choose $\sin(kx)$ and $\cos(kx)$ as we did in the potential well, but rather

$$\psi_1(x) = \underset{\text{Incident}}{Ae^{+ikx}} + \underset{\text{Reflected}}{Be^{-ikx}} \qquad \text{where } k \equiv \sqrt{\frac{2mE}{\hbar^2}}$$

The complex square of the first term is a probability density—the probability density of finding a particle moving to the right. The complex square of any complex exponential is unity (i.e., $e^{+iz} \times e^{-iz}$), so this probability density is particularly simple: $A*A$. Similarly, the probability density of finding a particle moving to the left is $B*B$.

$$|\psi|^2_{\text{inc}} = A*A \qquad |\psi|^2_{\text{refl}} = B*B \qquad (5\text{-}2)$$

Classically, the allowance for reflection is absurd; a particle would not be "reflected" in the case $E > U_0$. But knowing that light waves divide upon encountering another medium, whether the speed (governed by refractive index) decreases or increases, we are careful to allow for the possibility.

**Region II ($x > 0$)**

In this region, the quantity in brackets in equation (5-1) is the negative constant $-[2m(E - U_0)/\hbar^2]$. (It has been rearranged merely as a convenience.) An appropriate solution would seem to be $Ce^{+ik'x} + De^{-ik'x}$, where $k' \equiv \sqrt{2m(E - U_0)}/\hbar^2$. But there is a physical argument against one of these functions. Beyond $x = 0$, there is no change in potential energy—no force to cause a wave moving to the right to reflect. Because there can thus be no left-moving wave in this region, $e^{-ik'x}$ is physically unacceptable. Therefore,

$$\psi_{II}(x) = Ce^{+ik'x} \qquad \text{where } k' \equiv \sqrt{\frac{2m(E - U_0)}{\hbar^2}}$$

Transmitted

The probability density of finding a right-moving particle in this region is $C^*C$.

$$|\psi|^2_{trans} = C^*C \tag{5-3}$$

**Smoothness**

The combined wave function must be smooth. We have found its pieces, and we must ensure that they match as they should.

**$\psi(x)$ continuous at $x = 0$:**    $\psi_I(0) = \psi_{II}(0)$

$$Ae^{+ik0} + Be^{-ik0} = Ce^{+ik'0} \quad \Rightarrow \quad \boxed{A + B = C} \tag{5-4}$$

$$\frac{\partial\psi(x)}{\partial x} \text{ continuous at } x = 0: \qquad \left.\frac{\partial\psi_I}{\partial x}\right|_{x=0} = \left.\frac{\partial\psi_{II}}{\partial x}\right|_{x=0}$$

$$ikAe^{+ik0} - ikBe^{-ik0} = ik'Ce^{+ik'0} \quad \Rightarrow \quad \boxed{k(A - B) = k'C} \tag{5-5}$$

In the case of a bound particle, imposing boundary conditions on the wave function leads to quantization. However, no restriction on the value of $k$ (hence $E$) arises from conditions (5-4) and (5-5). Since the particle is not bound, we do not have the boundary conditions at *two* ends that would inevitably lead to a standing wave. An incident particle may have any energy.

What the boundary conditions give us here are reflection and transmission probabilities. If a single electron is incident upon the step, it will later be found either moving to the left in region I ($x < 0$) or to the right in region II ($x > 0$). Quantum mechanics predicts the probabilities of these two outcomes. The easiest approach is based upon the fact that a continuous incident *beam* of monoenergetic electrons would divide according to the same probabilities. A coherent beam would still be described by $Ae^{+ikx}$, but $A$ would be arbitrary; instead of a *probability* per unit distance, $|\psi|^2$ would be proportional to an arbitrary *number* per unit distance in the incident beam. A similar statement would apply to the reflected and transmitted beams. To find reflection and transmission probabilities, we need only find ratios of numbers per unit time in the corresponding beams, and the number per unit time follows directly from $|\psi|^2$:

$$\frac{\text{number}}{\text{time}} = \frac{\text{number}}{\text{distance}}\frac{\text{distance}}{\text{time}} \propto |\psi|^2 v$$

It will be convenient to use $v = p/m = \hbar k/m \propto k$ to express this in terms of $k$:

$$\frac{\text{number}}{\text{time}} \propto |\psi|^2 k$$

The transmission probability, given the symbol $T$, is thus

$$T = \frac{\dfrac{\text{number transmitted}}{\text{time}}}{\dfrac{\text{number incident}}{\text{time}}} = \frac{|\psi|^2_{\text{trans}}\, k_{\text{II}}}{|\psi|^2_{\text{inc}}\, k_{\text{I}}} = \frac{C^*C}{A^*A}\frac{k'}{k}$$

Similarly, the reflection probability $R$ is

$$R = \frac{\dfrac{\text{number reflected}}{\text{time}}}{\dfrac{\text{number incident}}{\text{time}}} = \frac{|\psi|^2_{\text{refl}}\, k_{\text{I}}}{|\psi|^2_{\text{inc}}\, k_{\text{I}}} = \frac{B^*B}{A^*A}$$

Conditions (5-4) and (5-5) may now be used to express $C$ and $B$ separately in terms of $A$. Eliminating $B$ yields $C = [2k/(k + k')]A$, and eliminating $C$ yields $B = [(k - k')/(k + k')]A$. Thus,

$$T = \frac{\left(\dfrac{2k}{k + k'}A\right)^*\left(\dfrac{2k}{k + k'}A\right)}{A^*A}\frac{k'}{k} = \frac{4kk'}{(k + k')^2}$$

and

$$R = \frac{\left(\dfrac{k - k'}{k + k'}A\right)^*\left(\dfrac{k - k'}{k + k'}A\right)}{A^*A} = \frac{(k - k')^2}{(k + k')^2}$$

We could have deduced one probability from the other, for the transmission and reflection probabilities *must* add to unity. (The reader is encouraged to verify that they do.) Finally, we express the probabilities in terms of $U_0$ and $E$. Using the definitions of $k$ and $k'$, we have

$$R + T = 1$$

$$T = 4\frac{\sqrt{E(E - U_0)}}{(\sqrt{E} + \sqrt{E - U_0})^2} \qquad R = \frac{(\sqrt{E} - \sqrt{E - U_0})^2}{(\sqrt{E} + \sqrt{E - U_0})^2} \qquad (5\text{-}6)$$

Notably, expressions (5-6) are essentially identical to those in physical optics that give reflected and transmitted intensities of a light wave incident normally on the interface between two media. Those for light differ only in that $\sqrt{E}$ and

$\sqrt{E - U_0}$, proportional to the speeds in the two regions, are replaced by the speeds $c/n$, where $n$ is the refractive index. The behaviors are completely analogous. The most important point here is that, contrary to the classical expectation, *the reflection probability is not in general zero.*

---

**Example 5.1**

An electron whose kinetic energy is 5 eV encounters a 2-eV potential step. What is the probability that it will be reflected?

Solution

The electron's total energy is its initial kinetic energy of 5 eV (since it starts by convention where $U = 0$), and $U_0$ is given to be 2 eV.

$$R = \frac{(\sqrt{5} - \sqrt{5 - 2})^2}{(\sqrt{5} + \sqrt{5 - 2})^2} = 0.016$$

---

Before moving on, we should confront an issue that might otherwise cause the reader some consternation. Planck's constant is nowhere to be found in expressions (5-6), so the result obtained in the example apparently would also apply for a bowling ball with a kinetic energy of 50 J encountering a potential jump of 20 J. But never would we expect to see the bowling ball reflect; it is a classical case, and the bowling ball should simply slow down to a kinetic energy of 30 J. There must be something wrong with our analysis!

Our simple assumption was that the region in which the potential energy jumps is of infinitesimal width. This is unrealistic. But it is valid as long as this region is much narrower than the wavelength of the incoming particle. For a given $E = p^2/2m$, a subatomic particle has a comparatively small momentum, because of its small mass. Accordingly, its wavelengths $\lambda = h/p$ is comparatively large. Circumstances in which the assumption is valid are common. As we have come to expect, a wave nature is exhibited because the wavelength is large compared to the relevant dimension of the apparatus (i.e. the step's abrupt jump). For "large" moving objects, on the other hand, as shown in Example 3.1, wavelengths are typically smaller than atomic dimensions! Obviously, it would be absurd to speak of so abrupt a potential energy jump and thus any wavelike nature to the bowling ball, even if we could treat it as a simple large "particle." The quantum mechanics we study, however, is in any case not able to handle such complex real objects.

Another issue is noteworthy: What would happen if a particle were to encounter a potential *drop?* Some beginning students of quantum mechanics assume that a particle may reflect upon encountering a potential jump simply because the potential energy increases. However, a "particle" may reflect even if the potential drops; *a force in a particle's direction of motion may cause it to reflect.* This shouldn't be too surprising if it is borne in mind that, since $E$ is greater than the potential energy at all points, reflection is at odds with classical mechanics *in either case.* Reflection and transmission probabilities for a potential drop may be found by essentially identical analysis, merely replacing $U_0$ by $-U_0$. In particular, the reflection probability for a 5-eV electron encountering a 2-eV drop is not zero

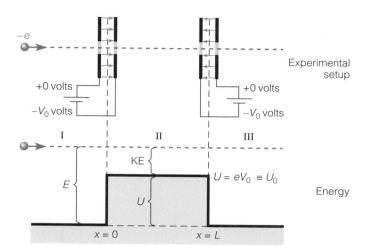

**Figure 5.2** A classically surmountable "potential barrier."

(see Exercise 1). Just as any increase *or* decrease in refractive index causes partial reflection of a light wave, *any* change in potential energy results in partial reflection of a matter wave.

## Potential Barrier: $E > U_0$

We now consider a **potential barrier,** in which the potential energy jump is only temporary, as shown in Figure 5.2. At $x = L$, it drops abruptly back to zero, during which the particle would experience a kick in the forward direction.

$$U(x) = \begin{cases} 0 & x < 0,\ x > L \\ U_0 & 0 < x < L \end{cases}$$

Again, classically, the particle would not reflect. It would merely slow down temporarily, between $x = 0$ and $x = L$, then return to its initial speed. Let us see what quantum mechanics has to say.

---

### $E > U_0$ Potential Barrier

**Region I ($x < 0$)**

The solution is the same here as for the potential step.

$$\psi_I(x) = \underset{\textbf{Incident}}{Ae^{+ikx}} + \underset{\textbf{Reflected}}{Be^{-ikx}} \qquad \text{where} \quad k \equiv \sqrt{\frac{2mE}{\hbar^2}}$$

**Region II ($0 < x < L$)**

Here the mathematical solution is also the same as for region II of the potential step. But since a wave may reflect at the $x = L$ interface (capacitor), we must allow for both left and right movement "over" the barrier. Both solutions are physically acceptable.

$$\psi_{\mathrm{II}}(x) = \underset{\text{Right-moving}}{Ce^{+ik'x}} + \underset{\text{Left-moving}}{De^{-ik'x}} \quad \text{where} \quad k' \equiv \sqrt{\frac{2m(E - U_0)}{\hbar^2}}$$

### Region III ($x > L$)

The mathematical solution here is the same as for region I, since the potential energy is zero in both. But beyond $x = L$, there is no change in potential energy that might reflect a right-moving wave, so the left-moving $e^{-ikx}$ is physically unacceptable.

$$\psi_{\mathrm{III}}(x) = \underset{\text{Transmitted}}{Fe^{+ikx}} \quad \text{where} \quad k \equiv \sqrt{\frac{2mE}{\hbar^2}}$$

### Smoothness

$\psi(x)$ **continuous at $x = 0$:**  $\psi_{\mathrm{I}}(0) = \psi_{\mathrm{II}}(0)$

$$Ae^{+ik0} + Be^{-ik0} = Ce^{+ik'0} + De^{-ik'0}$$

$$\Rightarrow \qquad \boxed{A + B = C + D}$$

$\dfrac{\partial\psi(x)}{\partial x}$ **continuous at $x = 0$:**  $\left.\dfrac{\partial\psi_{\mathrm{I}}}{\partial x}\right|_{x=0} = \left.\dfrac{\partial\psi_{\mathrm{II}}}{\partial x}\right|_{x=0}$

$$ikAe^{+ik0} - ikBe^{-ik0} = ik'Ce^{+ik'0} - ik'De^{-ik'0}$$

$$\Rightarrow \qquad \boxed{k(A - B) = k'(C - D)}$$

$\psi(x)$ **continuous at $x = L$:**  $\psi_{\mathrm{II}}(L) = \psi_{\mathrm{III}}(L)$

$$\boxed{Ce^{+ik'L} + De^{-ik'L} = Fe^{+ikL}}$$

$\dfrac{\partial\psi(x)}{\partial x}$ **continuous at $x = L$:**  $\left.\dfrac{\partial\psi_{\mathrm{II}}}{\partial x}\right|_{x=L} = \left.\dfrac{\partial\psi_{\mathrm{III}}}{\partial x}\right|_{x=L}$

$$ik'Ce^{+ik'L} - ik'De^{-ik'L} = ikFe^{+ikL}$$

$$\Rightarrow \qquad \boxed{k'(Ce^{+ik'L} - De^{-ik'L}) = kFe^{+ikL}}$$

Again, from the smoothness conditions come reflection and transmission probabilities. However, while the reflection probability is found as before, a "transmission" in this case is finding a right-moving particle in *region III*.

$$R = \frac{B^*B}{A^*A} \qquad T = \frac{|\psi|^2_{\text{trans}}\, k_{\mathrm{III}}}{|\psi|^2_{\text{inc}}\, k_{\mathrm{I}}} = \frac{F^*F}{A^*A}$$

Using the smoothness conditions to eliminate C, D, and F leaves an equation for B in terms of A. If this is inserted in the numerator of the preceding expression for R, A*A cancels, as it did in the case of the potential step. Similarly, A*A cancels when the smoothness conditions are solved for F in terms of A, which is then inserted in the expression for T. What remains in each case depends only on the constants $k'$ and $k$, and consequently on E and $U_0$. The lengthy algebraic details are left as an exercise. The results are

$$R = \frac{\sin^2(k'L)}{\sin^2(k'L) + 4\dfrac{k'^2 k^2}{(k^2 - k'^2)^2}} \qquad T = \frac{4\dfrac{k'^2 k^2}{(k^2 - k'^2)^2}}{\sin^2(k'L) + 4\dfrac{k'^2 k^2}{(k^2 - k'^2)^2}} \qquad (5\text{-}7)$$

or

$$R = \frac{\sin^2\left[\dfrac{\sqrt{2m(E - U_0)}}{\hbar} L\right]}{\sin^2\left[\dfrac{\sqrt{2m(E - U_0)}}{\hbar} L\right] + 4\dfrac{E}{U_0}\left(\dfrac{E}{U_0} - 1\right)}$$

$$T = \frac{4\dfrac{E}{U_0}\left(\dfrac{E}{U_0} - 1\right)}{\sin^2\left[\dfrac{\sqrt{2m(E - U_0)}}{\hbar} L\right] + 4\dfrac{E}{U_0}\left(\dfrac{E}{U_0} - 1\right)} \qquad (5\text{-}8)$$

As they should, R and T add to unity.

With the width L now a factor, the probabilities for a barrier are more complicated than those for a step. But this richer detail admits an interesting phenomenon: **resonant transmission.** Although the reflection and transmission probabilities are in general nonzero, the numerator of the reflection probability involves a sine. When the sine is zero, there is no reflection whatsoever (and T = 1). The condition is

$$\frac{\sqrt{2m(E - U_0)}}{\hbar} L = n\pi \quad \text{or} \quad E = U_0 + \frac{n^2 \pi^2 \hbar^2}{2mL^2} \qquad (5\text{-}9)$$

This is not energy quantization. The energy of incident particles may be any value, but resonant (complete) transmission occurs at only certain energies. Resonant transmission is also found in physical optics. Nonreflective coatings exploit the same wave properties. If the width and refractive index are chosen properly, a thin film will pass light without reflection. In both applications, reflections within the film ("over" the barrier) are caused to precisely cancel the wave reflected from the first interface. As always, it is important to remember that waves, not particles, interfere. A particle (photon or massive) encountering an obstacle may later be detected as a *particle,* having been either reflected or transmitted, but the process involves a probability that comes only from knowledge of the *wave* function.

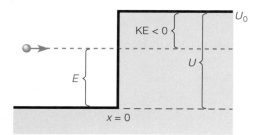

**Figure 5.3** A classically insurmountable potential step.

## Potential Step: $E < U_0$

Thus far, we have studied forces that, classically, should not be capable of reflecting a particle. We now turn to the other possibility: forces that "should" cause reflection. We begin with the simple case of a particle encountering a potential step whose height $U_0$ is greater than the particle's energy (Figure 5.3). The experimental setup now would be the same as before, except that the potential difference across the capacitor is higher. An electron *behaving classically* would slow to zero speed in the narrow region between the plates, then reverse its direction and return to $x < 0$; it would "reflect."

---

### $E < U_0$ Potential Step

**Region I ($x < 0$)**

The wave function here is the same as before:

$$\psi_I(x) = \underbrace{Ae^{+ikx}}_{\text{Incident}} + \underbrace{Be^{-ikx}}_{\text{Reflected}} \qquad \text{where } k \equiv \sqrt{\frac{2mE}{\hbar^2}}$$

**Region II ($x > 0$)**

The solution in this region is clearly different than in previous cases, even mathematically. The quantity in brackets in equation (5-1) is the *positive* constant $2m(U_0 - E)/\hbar^2$. The sign of this constant is crucial! If it is negative, as it has been so far, (5-1) has oscillatory, complex exponential solutions; if positive, solutions are real exponentials.

$$\psi(x) = Ce^{+\alpha x} + De^{-\alpha x} \qquad \text{where } \alpha \equiv \sqrt{\frac{2m(U_0 - E)}{\hbar^2}}$$

The solution and the constant $\alpha$ are the same as those *outside* the potential well [cf. equation (4-13)], because in both cases $E$ is less than $U$. Here it is important to note that neither $e^{-\alpha x}$ nor $e^{+\alpha x}$ can be interpreted as representing particles moving right and/or left. Real exponentials do not have a wavelength, and with no wavelength there is no real momentum.[1] Even so, we have the same physical grounds as in the potential well to discard one of the

[1] As in the region outside the potential well, the kinetic energy $p^2/2m = E - U_0$ is negative, and it would seem that the momentum is a purely imaginary number. Since the use of complex numbers is a matter of choice, we need not spend time wondering what a momentum such as $1.5i$ kg·m/s might mean. We simply accept that there is no "real" momentum in this region.

functions: As $x \rightarrow +\infty$, $e^{+\alpha x}$ diverges, which makes it physically unacceptable.

$$\psi_{\mathrm{II}}(x) = De^{-\alpha x} \quad \text{where } \alpha \equiv \sqrt{\frac{2m(U_0 - E)}{\hbar^2}}$$

**Smoothness**

**$\psi(x)$ continuous at $x = 0$:** $\quad \psi_{\mathrm{I}}(0) = \psi_{\mathrm{II}}(0)$

$$Ae^{+ik0} + Be^{-ik0} = De^{-\alpha 0} \quad \Rightarrow \quad \boxed{A + B = D}$$

**$\dfrac{\partial \psi(x)}{\partial x}$ continuous at $x = 0$:** $\quad \left.\dfrac{\partial \psi_{\mathrm{I}}}{\partial x}\right|_{x=0} = \left.\dfrac{\partial \psi_{\mathrm{II}}}{\partial x}\right|_{x=0}$

$$ikAe^{+ik0} - ikBe^{-ik0} = -\alpha De^{-\alpha 0} \quad \Rightarrow \quad \boxed{ik(A - B) = -\alpha D}$$

To determine the reflection probability, we eliminate $D$ between the smoothness conditions: $B = -[(\alpha + ik)/(\alpha - ik)]A$. As it turns out, simply considering the magnitude of this complex number tells us much.

$$|B| = \sqrt{B^*B} = \sqrt{\left(-\frac{\alpha + ik}{\alpha - ik}A\right)^* \left(-\frac{\alpha + ik}{\alpha - ik}A\right)}$$

$$= \sqrt{\left(\frac{\alpha - ik}{\alpha + ik}A^*\right)\left(\frac{\alpha + ik}{\alpha - ik}A\right)} = \sqrt{A^*A} = |A|$$

We see that $B$ is of the same magnitude as $A$; it merely points in a different "direction" in the complex plane. It follows that

$$R = \frac{B^*B}{A^*A} = 1$$

Because the transmission and reflection probabilities must add to unity, it would seem that the transmission probability is zero. Indeed it is. Were a particle to be transmitted, it should continue moving to the right at constant speed indefinitely, since it experiences no force beyond $x = 0$. However, the wave function falls exponentially to zero beyond the step; there is no remnant that survives to be found infinitely far beyond the step. Although it does not do so abruptly at $x = 0$, the incoming wave does reflect completely. Inferring its value from $R$ is the best way to determine $T$, because it is not proper to *calculate* a transmission probability from $D^*D$. Our ratios have involved the number *per unit time*: $|\psi|^2 v$. In the region $x > 0$, the speed $v$ is not "real" (again, $\mathrm{KE} = \frac{1}{2}mv^2$ is negative). Since no number per unit time can be specified, neither can a transmission probability.

On the other hand, a calculation of number *per unit distance* is still justified. The quantity $|\psi|^2$ is still real, and that the wave penetrates the step does mean that there is a probability of finding an electron on the "wrong" side of the capacitor.

But it cannot be concluded that particles are constantly getting through the capacitor to $x > 0$, then going back through to $x < 0$, for there is no force beyond the step that could reflect anything. Rather, we accept that, as long as no attempt is made to find a particle, there is simply an undisturbed wave, which, though reflecting completely, happens to penetrate the classically forbidden step. The extent of this penetration is governed by constant $\alpha$. With penetration depth (see Section 4.7) defined as

$$\delta = \frac{1}{\alpha} = \frac{\hbar}{\sqrt{2m(U_0 - E)}}$$

we see that the wave function is proportional to $e^{-\alpha x} = e^{-x/\delta}$. The closer $E$ is to $U_0$, the greater is $\delta$ and the slower is the decay of the wave function. But in any case the wave function becomes very small at values of $x$ much larger than $\delta$.

Once again there is an electromagnetic analog. Just as a $U_0 > E$ step completely reflects massive particles, a smooth metal surface completely reflects light—metals make good mirrors for this reason. The light wave does not propagate through the metal, and no photons are transmitted. However, there is an electromagnetic field within the metal, oscillating with time and decaying exponentially with depth. For an electromagnetic wave, the penetration depth is often called "skin depth."

Figure 5.4 shows reflection and transmission probabilities plotted versus $E/U_0$ for a potential step. If $E$ is less than $U_0$, the wave is totally reflected. When $E$ exceeds $U_0$, the reflection probability falls rapidly with increasing $E$. At $E = 3U_0$, the wave is almost completely transmitted.

## Potential Barrier: $E < U_0$—Tunneling

Our final case is perhaps the most startling: A particle may escape confinement that classical mechanics says it cannot. Our simple test case, shown in Figure 5.5, is a potential barrier with $U_0 > E$, so high that a classical particle would rebound

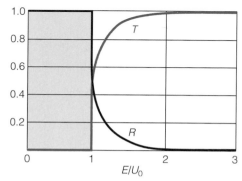

**Figure 5.4** Reflection and transmission probabilities for a potential step.

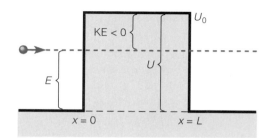

**Figure 5.5** Tunneling.

at first encounter—that is, at $x = 0$. The quantum mechanical solution is similar to the others.

### $E < U_0$ Potential Barrier

#### Region I ($x < 0$)

$$\psi_{\text{I}}(x) = \underbrace{Ae^{+ikx}}_{\text{Incident}} + \underbrace{Be^{-ikx}}_{\text{Reflected}}$$

#### Region II ($0 < x < L$)

Mathematically, the solution here is the same as in region II of the potential step with $E < U_0$. But we may not discard either function, since neither diverges in this region.

$$\psi_{\text{II}}(x) = Ce^{+\alpha x} + De^{-\alpha x}$$

#### Region III ($x > L$)

$$\psi_{\text{III}}(x) = \underbrace{Fe^{+ikx}}_{\text{Transmitted}}$$

#### Smoothness

$\psi(x)$ **continuous at** $x = 0$:    $\psi_{\text{I}}(0) = \psi_{\text{II}}(0)$

$$\boxed{A + B = C + D}$$

$\dfrac{\partial\psi(x)}{\partial x}$ **continuous at** $x = 0$:    $\left.\dfrac{\partial\psi_{\text{I}}}{\partial x}\right|_{x=0} = \left.\dfrac{\partial\psi_{\text{II}}}{\partial x}\right|_{x=0}$

$$ikAe^{+ik0} - ikBe^{-ik0} = \alpha Ce^{+\alpha 0} - \alpha De^{-\alpha 0}$$

$$\Rightarrow \quad \boxed{ik(A - B) = \alpha(C - D)}$$

$\psi(x)$ **continuous at** $x = L$:    $\psi_{\text{II}}(L) = \psi_{\text{III}}(L)$

$$\boxed{Ce^{+\alpha L} + De^{-\alpha L} = Fe^{+ikL}}$$

$\dfrac{\partial\psi(x)}{\partial x}$ **continuous at** $x = L$:    $\left.\dfrac{\partial\psi_{\text{II}}}{\partial x}\right|_{x=L} = \left.\dfrac{\partial\psi_{\text{III}}}{\partial x}\right|_{x=L}$

$$\alpha Ce^{+\alpha L} - \alpha De^{-\alpha L} = ikFe^{+ikL}$$

$$\Rightarrow \quad \boxed{\alpha(Ce^{+\alpha L} - De^{-\alpha L}) = ikFe^{+ikL}}$$

The algebra yielding $R$ and $T$ is nearly identical to that in the $E > U_0$ barrier. (In fact, replacing $k'$ by $-i\alpha$, it is *completely* identical.) The result is

$$R = \frac{\sinh^2(\alpha L)}{\sinh^2(\alpha L) + 4\dfrac{\alpha^2 k^2}{(k^2 + \alpha^2)^2}} \qquad T = \frac{4\dfrac{\alpha^2 k^2}{(k^2 + \alpha^2)^2}}{\sinh^2(\alpha L) + 4\dfrac{\alpha^2 k^2}{(k^2 + \alpha^2)^2}}$$

or

$$R = \frac{\sinh^2\left[\dfrac{\sqrt{2m(U_0 - E)}}{\hbar}L\right]}{\sinh^2\left[\dfrac{\sqrt{2m(U_0 - E)}}{\hbar}L\right] + 4\dfrac{E}{U_0}\left(1 - \dfrac{E}{U_0}\right)}$$

$$T = \frac{4\dfrac{E}{U_0}\left(1 - \dfrac{E}{U_0}\right)}{\sinh^2\left[\dfrac{\sqrt{2m(U_0 - E)}}{\hbar}L\right] + 4\dfrac{E}{U_0}\left(1 - \dfrac{E}{U_0}\right)}$$

(5-10)

Both are in general nonzero; it is entirely possible for a particle to escape through a barrier that it cannot surmount classically. It is important to grasp that a particle is able to do this *not* due to some mysterious fluctuation in its energy, allowing it to go "over the top." On the contrary, we assumed at the outset that the energy was well defined and less than the barrier height. *The particle never has sufficient energy to surmount the barrier!* Instead, it "tunnels" through, and the principles of quantum mechanics demand that there be such a possibility. The solution to the left of the barrier is some combination of incident and reflected waves, of positive and negative momentum. This is matched at $x = 0$ to a monatonically decreasing wave (without a real momentum) inside the barrier, and to this is matched at $x = L$ a transmitted wave of positive momentum. There is no smooth, physically acceptable solution that is identically zero in region III.

Figure 5.6 shows wave functions (real part) for particles of different energies incident from the left upon a barrier. Note that the wavelengths decrease as kinetic

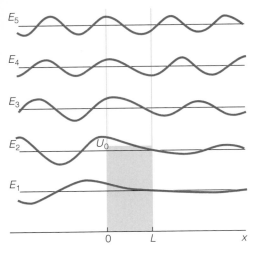

**Figure 5.6** Wave functions for particles of different energies incident from left upon a potential barrier.

energy increases. At energy $E_1$, about half of $U_0$, the exponential decay within the barrier is rapid and little transmission occurs. At energy $E_2$, barely below the barrier height, the wave decays less rapidly, leaving a larger transmitted "tail." Energy $E_3$ is above the barrier, but we see evidence of significant reflection; the wave is of smaller amplitude to the right of the barrier than to the left. (Also, the wavelength is longer when it is "over" the barrier, because its speed is smaller there.) Energy $E_4$ happens to be the first transmission resonance—the wave on the right is of precisely the same amplitude as that on the left. (Notably, the width of the barrier is $\frac{1}{2}\lambda$. This is just the condition we would expect for maximum transmission of a light wave through a thin film surrounded by air, an analogous situation.) Energy $E_5$ is not a transmission resonance, but even so, little reflection occurs— the amplitude on the right is only slightly less than on the left.

Reflection and transmission probabilities for this barrier are shown in Figure 5.7 for all energies, with the energies of Figure 5.6 indicated by dashed lines. Although reflection predominates for $E < U_0$, and transmission for $E > U_0$, both occur at all energies (except transmission resonances). In the limit $L \to \infty$, the barrier becomes a step and Figure 5.7 becomes Figure 5.4.

The electromagnetic analog to tunneling is light striking a thin metal film. If the film is thin enough, the exponentially decaying electromagnetic field within the metal will still be significant at the far side of the film, and some fraction of the incident intensity will emerge. A half-silvered mirror operates on this principle. Another electromagnetic analog to quantum-mechanical tunneling is the phenomenon of "frustrated total internal reflection," represented schematically in Figure 5.8. In the top diagram, an electromagnetic wave (blue) traveling through a prism strikes the interface with the outside air, of lower refractive index. The incident angle exceeds the critical angle, so the wave is totally internally reflected (black wave). But though there is no light *per se* transmitted into the air, there is a time-varying, exponentially decaying electromagnetic field in the space beyond. To this point, the behavior is analogous to the $E < U_0$ potential step (except that the incident and reflected waves move in perpendicular directions). If a second prism is brought near, so that the decaying field is intercepted before it falls to a negligible value, the step is effectively replaced by a narrow barrier (i.e., the

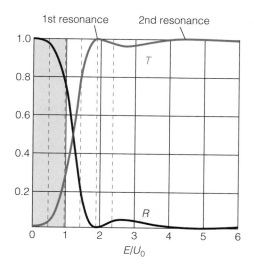

Figure 5.7 Reflection and transmission probabilities for a potential barrier.

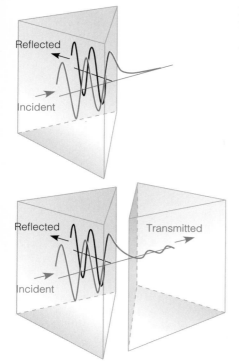

**Figure 5.8** Frustrated total internal reflection—an electromagnetic analog of quantum-mechanical tunneling.

air-space). We find that the internal reflection is no longer total—it is "frustrated." The incident wave is partially transmitted, and the narrower the barrier, the greater is the transmission.

## Tunneling Through Wide Barriers

In many applications, the transmission probability for tunneling is very small. Inside a barrier, the wave function is essentially proportional to $e^{-\alpha x}$, or $e^{-x/\delta}$, where $\delta \equiv 1/\alpha$ is the penetration depth.[2] If $L \gg \delta$, very little of the wave will "survive" to $x = L$. The condition for a "wide" barrier is thus

$$\frac{1}{\delta}L = \alpha L = \frac{\sqrt{2m(U_0 - E)}}{\hbar}L \gg 1 \tag{5-11}$$

It is wide if $L$ is large, or $E$ is much less than $U_0$, or both.

It is left as an exercise to show that if (5-11) holds, the transmission probability of (5-10) is given by the approximation

$$T \cong 16\frac{E}{U_0}\left(1 - \frac{E}{U_0}\right)e^{-2[\sqrt{2m(U_0 - E)}/\hbar]L} \tag{5-12}$$

**Example 5.2**

An electron encounters a barrier of height 0.100 eV and width 15 nm. What is the transmission probability if its energy is (a) 0.040 eV? (b) 0.060 eV?

[2]The solution within the barrier is not strictly a decaying exponential, but contains an exponentially *increasing* part $e^{+\alpha x}$. In most cases, this part is inconsequentially small.

Solution

First we see if condition (5-11) holds.

For 0.040 eV:

$$\frac{L}{\delta} = \frac{\sqrt{2(9.11 \times 10^{-31} \text{ kg})(0.100 - 0.040) \times 1.6 \times 10^{-19} \text{ J}}}{1.055 \times 10^{-34} \text{ J·s}} 15 \times 10^{-9} \text{ m}$$

$$= 18.8$$

For 0.060 eV:

$$\frac{L}{\delta} = \frac{\sqrt{2(9.11 \times 10^{-31} \text{ kg})(0.100 - 0.060) \times 1.6 \times 10^{-19} \text{ J}}}{1.055 \times 10^{-34} \text{ J·s}} 15 \times 10^{-9} \text{ m}$$

$$= 15.4$$

In both cases, the barrier is wide, many times the penetration depth. Now, noting that we have just calculated the arguments of the exponential in (5-12),

$$T_{0.040 \text{ eV}} = 16 \frac{0.04}{0.1}\left(1 - \frac{0.04}{0.1}\right)e^{-2 \times 18.8} = 1.8 \times 10^{-16}$$

$$T_{0.060 \text{ eV}} = 16 \frac{0.06}{0.1}\left(1 - \frac{0.06}{0.1}\right)e^{-2 \times 15.4} = 1.8 \times 10^{-13}$$

As expected, for both barriers the transmission probability *for a single event* is very small. But in many real situations, barriers are constantly bombarded by particles. If electrons at either of these energies were to strike the barrier $10^{20}$ times each second, there would be a significant flux of escaping particles. Such high frequency is not at all unrealistic; alpha-particles *almost* trapped in an atomic nucleus typically get $10^{20}$ chances to tunnel out every second. Alpha decay is discussed in the next section.

The example illustrates another important point: When transmission probabilities are very small, they vary sharply with energy. Both probabilities are small, but a modest 50% increase in particle energy results in a transmission probability about a thousand times larger. This sensitivity is due to the exponential dependence in (5-12), and the smaller the probability, the more pronounced is the variation.[3]

## 5.2 Alpha Decay and Other Applications

Imagine: Bob is standing on ground that is flat except for an isolated hill. Anna, on the far side of the hill, wishes to roll a large ball over the hill to Bob. The ball would have a gravitational potential energy of 900 J if perched on the top of the hill. Anna rolls the ball, which loses no mechanical energy as it travels; friction is negligible. When the ball reaches Bob, its kinetic energy is 800 J. Classically impossible! The ball's total mechanical energy at the top of the hill must have

[3] E.g., $e^{-2 \times 18.8} \cong 10^3 \times e^{-2 \times 15.4}$, but $e^{-2 \times 1.88}$ is only $2 \times e^{-2 \times 1.54}$, while $e^{-2 \times 188}$ is about $10^{30} \times e^{-2 \times 154}$.

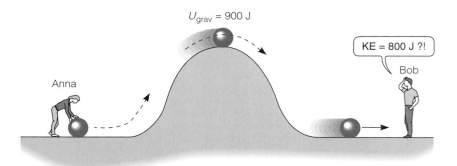

been *at least* 900 J—its potential energy at that point—and this should have been added to its kinetic energy as it rolled down. Classically, the ball can have no less than 900 J of kinetic energy when it reaches Bob. Since mechanical energy is conserved, for the ball to arrive with a kinetic energy of 800 J, its kinetic energy as it passed over (through?) the hill would have had to be *negative* 100 J!

Something quite like this occurs in the radioactive process known as **alpha decay.** Certain atomic nuclei are known to be unstable; they will eliminate excess energy by spontaneously emitting a high-energy particle. The energy carried away by this particle leaves the remaining nucleus in a lower-energy, more stable state. The original nucleus is usually called the "parent nucleus," and the nucleus remaining after decay the "daughter nucleus." One of the particles that may be emitted is two protons and two neutrons stuck together tightly, known as an **alpha-particle.** An example is the alpha decay of uranium-238, depicted in Figure 5.9. A uranium-238 nucleus contains 92 protons and 146 neutrons, for a total of 238 "nucleons." After emitting an alpha-particle, what remains is a nucleus of 90 protons (92−2) and 144 neutrons (146−2), a total of 234 nucleons: thorium-234. Both being positively charged, the thorium nucleus and alpha particle repel one another. As the alpha-particle moves away from the thorium, electrostatic potential energy is converted to kinetic; the alpha-particle speeds up. However, experiment reveals a deficiency—classically speaking—in the final kinetic energy. Let us calculate the minimum energy we might *classically* expect the alpha-particle to have.

To understand alpha decay, we must introduce certain ideas belonging to nuclear physics (the topic of Chapter 10). The constituents of the nucleus are held together not by electrostatic force, for this alone would violently scatter the positively charged protons, but by the altogether different **strong force.** All nucleons, including neutrons, are found to attract one another via this very strong force, but only when very close together. It may loosely be said that they attract one another only when essentially in contact. Thus, the nucleus may be viewed crudely as a collection of nucleons stuck as close together as possible in an approximately spherical arrangement. The radius may be determined via scattering experiments (discussed in Section 10.1). Here, we need only note that beyond this radius the only significant force is the electrostatic repulsion, while inside this

Uranium-238

Thorium-234    $\alpha$-particle

**Figure 5.9** Alpha decay.

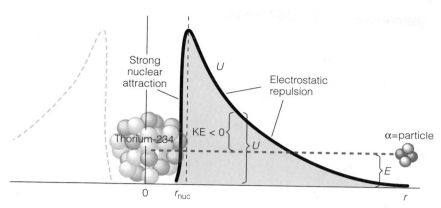

**Figure 5.10** To escape the nucleus, an alpha-particle must tunnel.

radius, the strong, attractive nuclear force is overwhelming (see Figure 5.10). An alpha-particle brought progressively closer to a nucleus against the electrostatic repulsion would experience an *increase* in electrostatic potential energy. But upon reaching the surface of the nucleus, it would experience a strong attractive force and therefore a rapidly *decreasing* potential energy. Accordingly, the alpha-particle is at a point of maximum potential energy when at the nuclear surface.[4]

Experiment has found the radius of the thorium-234 nucleus to be approximately $7.4 \times 10^{-15}$ m. At the surface of a thorium nucleus (90 protons), an alpha-particle (2 protons) would thus possess an electrostatic potential energy

$$U_{\mathrm{elec}} = \frac{q_1 q_2}{4\pi\epsilon_0 r} = \frac{(2 \times 1.6 \times 10^{-19}\ \mathrm{C})(90 \times 1.6 \times 10^{-19}\ \mathrm{C})}{4\pi(8.85 \times 10^{-12}\ \mathrm{C^2/N \cdot m^2})(7.4 \times 10^{-15}\ \mathrm{m})}$$

$$= 5.6 \times 10^{-12}\ \mathrm{J} = 35\ \mathrm{MeV}$$

When a uranium nucleus alpha-decays, the alpha-particle at the surface of the daughter thorium nucleus would have 35 MeV of electrostatic potential energy, and would speed up as it is repelled away. Classically, the kinetic energy can be no less than zero at the nuclear surface, so it can be no less than 35 MeV when very far away, where the potential is negligible—*an alpha-particle with less than 35 MeV of kinetic energy should never be found.* The experimental evidence is that an alpha-particle emitted from uranium-238 and detected far from the thorium-234 daughter nucleus has a kinetic energy of 4.3 MeV! Quantum mechanics provides the answer: The alpha-particle never possesses more than 4.3 MeV of energy; it tunnels through the 35-MeV barrier.

Figure 5.10 is a schematic representation of the phenomenon. Initially, an alpha particle moves relatively freely within a uranium nucleus, pulled equally in all directions by the 234 other nucleons. Approaching the barrier at the surface, however, it experiences a net force back toward the center (due to its attraction to the other nucleons). Although it does not have sufficient energy to surmount this barrier, it may quantum-mechanically tunnel.

A calculation of the transmission probability for our multifarious potential energy is too involved to present here. But calculated **decay rates** (decays per

[4]The behavior of a water molecule in a water drop is very similar. Due to the intermolecular forces attracting it to all others, much energy is needed to bring a given water molecule to the surface. The water drop is thus in a state of lowest overall energy when its surface area is a minimum. This is the basis of surface tension. Not coincidentally, one of the most successful models of the nucleus, aptly named the "liquid drop model" and discussed in Chapter 10, assumes that the nucleons in the nucleus behave like molecules in a liquid drop.

second) do agree quite well with experimentally determined values. In simplified form, the calculation is as follows:

$$\frac{\text{number of decays}}{\text{time}} = \frac{\text{number of times } \alpha \text{ strikes barrier}}{\text{time}}$$

$$\times \text{ transmission probability}$$

$$= \frac{\text{one strike}}{\text{time to cross diameter}} \times T$$

$$= \frac{1}{\text{diameter of nucleus/speed}} \times T$$

$$= \frac{v}{2r_{\text{nuc}}} T$$

Obviously, once the alpha-particle escapes, alpha decay *of the original nucleus* is no longer possible, since the element has changed (e.g., uranium becomes thorium); to refer to a number per unit time for a single nucleus is nonsense. But with the huge number of nuclei in even a (macroscopically) small sample of a radioactive material, simply multiplying by the total number yields a proper average decay rate. As noted, agreement with experiment is very good. Even so, we must not overlook the simple fact that the probabilistic nature of alpha decay is good evidence of its quantum-mechanical basis; watching *individual* nuclei, an observer would find great variation in the time required for decay, and this is exactly what we would expect if the decay is governed by quantum-mechanical probabilities.

Example 5.2 provides the key to understanding another (classical) mystery of alpha decay: Most alpha-particles have energies $E$ in the range $4-9$ MeV, yet their decay rates differ by more than 20 orders of magnitude (see Table 5.1). The probability of the alpha-particle tunneling in a single encounter with the nuclear surface is very small.[5] As we found in Example 5.2, when they are very small, transmission probabilities are extremely sensitive to the value of $E$.

In conclusion: Were classical physics applicable, alpha decay would be—in the unexpected low energy of alpha-particles, in its probabilistic nature, and in the vast range of decay rates—completely baffling. The explanation of alpha decay is a triumph of quantum mechanics.

**Table 5.1**

| $\alpha$-emitting nucleus | $\alpha$-particle energy (MeV) | Mean time to decay |
|---|---|---|
| Po-212 | 8.8 | $4.4 \times 10^{-7}$ seconds |
| Rn-220 | 6.3 | 79 seconds |
| Ra-224 | 5.7 | 5.3 days |
| Ra-226 | 4.8 | 2300 years |
| U-238 | 4.3 | $6.5 \times 10^9$ years |

[5]Even so, the alpha-particle in the nucleus prior to decay may, depending on its speed, encounter the barriers at the surface so often that decay occurs in a very short time.

## The Tunnel Diode

A tunnel diode is an electronic circuit element whose response to applied voltages is unusual and very fast. In a narrow region between the device's two ends (leads), there is a change in the material's physical properties that prevents simple conduction of electrons from one end to the other. In essence, the electrons at the ends are separated by an electrostatic potential barrier they cannot classically surmount, crudely depicted in Figure 5.11. By design, however, the barrier is not wide and significant tunneling occurs. With no applied voltage, tunneling occurs equally in both directions; there is no *net* flow. When a potential difference is applied, the situation becomes asymmetric. Right- and left-tunneling transmission rates differ, and a net current flows.

The tunnel diode may seem to have little utility, as an applied voltage will induce current flow in many materials. But one of its distinctive features is *how* the current varies with voltage. It does not always increase as applied voltage is increased; at some points it decreases (see Exercise 9.32). Moreover, changing the applied voltage changes the transmission rates almost instantly, and quick response is very desirable at high frequencies. The more common devices that control current via voltage changes rely upon relatively slow thermal diffusion of the charge carriers (see Section 9.9). Although its early promise as a high-frequency switch in integrated circuits has dimmed, the tunnel diode has found use in a variety of modern electronic circuits.

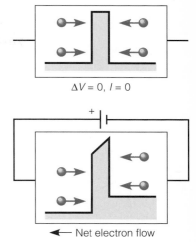

$\Delta V = 0, I = 0$

◄— Net electron flow

**Figure 5.11** Tunnel diode.

## Field Emission

As we know, to remove an electron from a given kind of metal requires a certain minimum amount of energy: the work function $\phi$. In effect, the metal's electrons reside in a potential well, the result of their attraction to the positive ions, and they are repelled at its walls by a potential step they cannot surmount, as shown in Figure 5.12(a). Due to the random thermal distribution of speeds, at any given temperature a small fraction of the electrons have sufficient kinetic energy to escape the metal. Those that do participate in an equilibrium exchange of electrons with the surrounding space. Heating a metal filament to enhance this effect, known as "thermionic emission," has long been used as a source for electron beams, such as those in the conventional (elongated) cathode-ray tubes used in televisions and computer monitors.

But there is another way to coax electrons from their potential well. If a positive electrode is brought nearby, the potential step may effectively be changed to a potential barrier. In Figure 5.12(b), the positive electrode modifies the potential energy function "seen" by the electrons—lowering it outside the metal. (Remember: Negatively charged electrons are at low potential *energy* where the *potential* is high.) Now electrons moving too slowly to surmount the work function barrier may tunnel. The technique may be used to generate an electron beam, and is known as **field emission.** (Between the metal and the positive electrode, where the potential changes, there is an electric field.) In many applications, it is preferable to thermionic emission, since heating a metal filament not only wastes power but often produces a great deal of electrical noise.

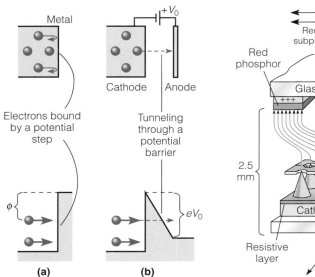

**Figure 5.12** Electrons in a metal (a) behave as though in a finite well. An electric field (b) alters the "wall," so that tunneling may occur.

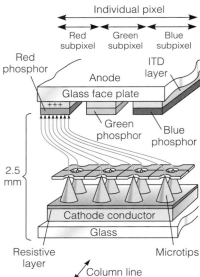

**Figure 5.13** One pixel of a field emission display. Applying a positive bias turns on any of the three different colors of subpixel.

Field emission is now being tried as the source of illumination in a new kind of flat-screen display, the aptly named Field Emission Display (FED) shown in Figure 5.13. Its potential advantages over the liquid crystal displays commonly used in laptop computers include wider viewing angle and quicker response.

## The Scanning Tunneling Microscope

The extreme sensitivity of tunneling probability to width in a wide barrier is the root of the utility of the **scanning tunneling microscope** (STM). In this device, a slender metal tip is positioned near the sample under study. Since tip and sample are not actually in contact, the free electrons they harbor cannot pass between them in the usual (classical) way; a potential barrier intervenes. However, in the STM the separation is made small enough that significant quantum-mechanical tunneling occurs, though still large enough that the barrier qualifies as "wide." Accordingly, extremely small variations in the tip-sample separation (barrier width) translate to easily measurable changes in tunneling current. (Tunneling preferentially from sample to tip is ensured by biasing the tip at a higher potential than the sample.) In fact, under typical conditions, a change in tunneling current of a few percent may be produced by a separation change of only about $\frac{1}{1000}$ nm (see Exercise 19). This is much smaller than typical atomic dimensions. It is little wonder, then, that as the tip of a tunneling microscope is scanned laterally over a sample's surface, as shown in Figure 5.14, it is able to "see" individual atoms! Furthermore, calling the tunneling direction the $z$ and the scan direction the $x$, by

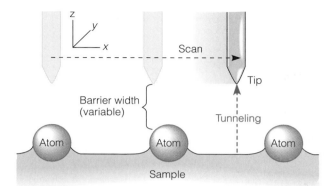

**Figure 5.14** Schematic diagram of an STM in operation.

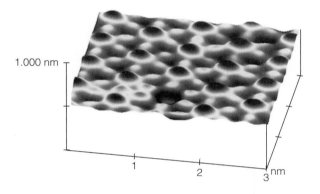

**Figure 5.15** A missing atom, as seen by an STM.

repeating the scan but with the tip displaced ever so slightly in the *y*-direction, an entire "topological map" of the sample is generated. Phenomenal results have been obtained. Figure 5.15 shows a lone atom missing from a pattern of iodine atoms adsorbed on a platinum surface. The hexagonal geometry that characterizes the iodine bonding is quite clear.

The STM must be ranked as one of the most indispensable tools in modern technology. Its uses are already legion—studying geometry and composition of a seeming endless list of surfaces; locating important biological molecular groups, such as the fundamental building blocks of DNA; mapping microscopic "vortices" in certain kinds of superconductors; nudging atoms from one point on a surface to another—and they continue to expand.

# ◆ **5.3** Particle-Wave Propagation

Thus far, our study of unbound states has concentrated on the effect of simple forces on plane waves. A plane wave is not the most realistic representation of the matter wave of a single particle. To represent a reasonably compact moving particle, a traveling wave pulse is much better: It is broad in one region and essentially zero elsewhere. But the behavior of a wave pulse is more complex, even with *no* external force. Accordingly, we have enough to keep us busy without considering the effects of steps, barriers, and so forth; we thus restrict our attention to wave pulses moving in vacuum or homogeneous media.

To understand the behavior of a wave pulse, we must consider the behavior of its parts. As discussed in Section 3.6, a wave pulse may be thought of as a sum (a linear combination) of plane waves. Each constituent plane wave moves at its own speed, or **phase velocity.** There are two important consequences: (1) The speed of the point where the probability density is largest—the "speed of the particle"—may differ distinctly from the speeds of the individual plane waves, and (2) an initially well-localized wave pulse will spread out with time, as its constituents get progressively out of step.

## Phase and Group Velocities

Figure 5.16 shows a traveling wave pulse—which we might associate with a traveling particle—at several successive times. It may be thought of as a sum of plane waves, so we refer to it as a **wave group.** The gray arrows indicate the motion of a particular crest. The thin blue outline traces the "envelope" of the wave group, a shape that moves with the group and defines the maximum displacement at each point. The blue arrows indicate the motion of the envelope, which clearly doesn't move at the same speed as the crests. The speed of the envelope is known as the **group velocity.** If we are interested in knowing when the particle is likely to arrive at some destination, it is the speed of the envelope, where the probability is large—the group velocity—that is of interest. The speed at which an individual crest appears, passes through the envelope, then disappears—phase velocity—is rather unimportant. Nevertheless, an understanding of the group's motion comes only through study of the motion of its parts. Let us begin by analyzing the simplest possible wave group: one comprising only two constituent plane waves.

### A Simple Wave Group

Consider a wave group consisting of just two plane waves of equal amplitude A:[6]

$$\Psi(x, t) = Ae^{i(k_1 x - \omega_1 t)} + Ae^{i(k_2 x - \omega_2 t)}$$

We wish our group to have reasonably well-defined wave number and frequency, so we choose $k$- and $\omega$-values that deviate by only very small amounts $dk$ and $d\omega$ above and below central values $k_0$ and $\omega_0$.

$$k_1 = k_0 + dk \qquad k_2 = k_0 - dk$$

$$\omega_1 = \omega_0 + d\omega \qquad \omega_2 = \omega_0 - d\omega$$

Thus,

$$\Psi(x, t) = A\left\{e^{i[(k_0 + dk)x - (\omega_0 + d\omega)t]} + e^{i[(k_0 - dk)x - (\omega_0 - d\omega)t]}\right\}$$

---

[6]Although most of the conclusions in this section apply to all wave phenomena, including electromagnetic radiation, we retain the symbol $\Psi$ for the wave.

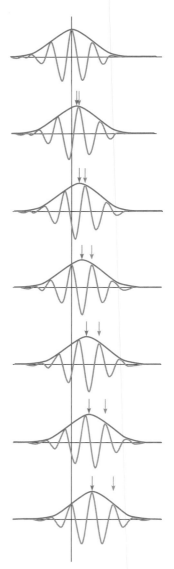

**Figure 5.16** Wave group. The crests and envelope move at different speeds.

Factoring out common terms, we obtain

$$\Psi(x, t) = Ae^{i(k_0 x - \omega_0 t)} \left\{ e^{+i[(dk)x - (d\omega)t]} + e^{-i[(dk)x - (d\omega)t]} \right\}$$

and thus,[7]

$$\Psi(x, t) = Ae^{i(k_0 x - \omega_0 t)} 2 \cos[(dk)x - (d\omega)t]$$

We have a complex exponential that moves at speed $\omega_0/k_0$, the phase velocity appropriate to the central values $k_0$ and $\omega_0$, modulated by a cosine function that moves at speed $d\omega/dk$. As shown in Figure 5.17, it is the latter term that defines the envelope; the group velocity is $d\omega/dk$. We see this more clearly by calculating the probability density:

$$\Psi^*(x, t)\Psi(x, t) = 4A^2 \cos^2[(dk)x - (d\omega)t]$$

The phase velocity has disappeared. Since a particle must move as the probability of finding it moves, the speed of any particle described by this wave must be the speed of $|\Psi|^2$: $d\omega/dk$. Perhaps surprisingly, the group/particle velocity does not depend on the actual central values $\omega_0$ and $k_0$, but rather on *how $\omega$ varies with $k$* from one constituent wave to another.

Although the simple wave group we have considered exhibits an important feature—that the probability density moves at speed $d\omega/dk$—it is not a "pulse." Rather, it is periodic. (A *finite* sum of periodic functions will always be periodic; only a *continuous* sum—an integral—can represent a pulse.) Let us turn to the more particlelike case of a wave pulse.

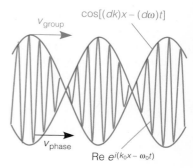

Figure 5.17 Simple wave group.

### A Wave Pulse

The most general way of expressing a wave group—a sum of plane waves—is

$$\Psi(x, t) = \int_{-\infty}^{+\infty} \tilde{\psi}(k)e^{i(kx - \omega t)} \, dk \tag{5-13}$$

This is of the form of the inverse Fourier transform of equation (3-11); each plane wave of wave number $k$ has an amplitude given by the value of $\tilde{\psi}(k)$. There is an important difference here, however. We are interested in the explicit time behavior of the wave group, so we must include the explicit time dependence of each plane wave. Each has a temporal part involving an angular frequency $\omega$. Since the sum includes plane waves of various $k$ ($p/\hbar$) and $\omega$ ($E/\hbar$) values, neither the momentum nor the energy of the group is well defined.

The conclusions we draw about wave group (5-13) apply to both matter and electromagnetic waves, because for each the underlying wave equation is *linear;* a sum of planes waves is a solution so long as each plane wave is a solution. But if each plane wave indeed is a solution of an underlying wave equation, $\omega$ and $k$ are not independent variables. *Built in to each wave equation is a relationship between $\omega$ and $k$.* From equations (3-4), we see that for plane-wave solutions to Maxwell's equations in vacuum, $\omega = ck$. For plane-wave solutions to the free-particle Schrödinger equation, equation (3-8a) shows that $\omega = \hbar k^2/2m$. Actually, this relationship is flawed. Using $E = \hbar\omega$ and $p = \hbar k$, we find that it is equivalent to $E = p^2/2m$, equation (3-8b). However, this is only the kinetic energy. Strictly

---

[7]Not coincidentally, the result is similar to a typical beat frequency function—a *sum* of sinusoidal functions whose frequencies are close becomes a *product* of sinusoidal functions whose frequencies are vastly different.

speaking, $\omega$ is related to *total* energy: kinetic plus internal energy. Thus, even at low speeds (where KE $= p^2/2m$ is valid), the total energy is given by $E = mc^2 + (p^2/2m)$ and the true relationship between $\omega$ and $k$ is therefore $\omega = (mc^2/\hbar) + (\hbar k^2/2m)$.[8] (As we shall see, to understand most behaviors the extra term is irrelevant, but in one aspect it is vitally important.) A relationship between frequency and wave number is known as a **dispersion relation**. Table 5.2 lists the dispersions relations for matter and electromagnetic waves in vacuum. If a wave passes through a medium other than vacuum, the wave equation will be affected (matter waves by new potential energies $U(x)$, electromagnetic waves by new charge densities), and $\omega$ and $k$ would be related in still other ways. Let us consider $\omega$ to be some unspecified function of $k$.

**Table 5.2    Dispersion Relations**

| EM waves in vacuum | Matter waves in vacuum |
|---|---|
| $\omega = ck$ | $\omega = \dfrac{mc^2}{\hbar} + \dfrac{\hbar k^2}{2m}$ |

$$\omega = \omega(k) \tag{5-14}$$

Dispersion relation

In this way we will be able to apply our results to different cases simply by substituting the appropriate relationship.

Our study of the group velocity and wave-group behavior will make more sense if we first consider the phase velocity of plane waves in the familiar cases. In general, the phase velocity of a plane wave is given by

Phase velocity

$$v_{\text{phase}} = \lambda f = \frac{\omega}{k} \tag{5-15}$$

Using the dispersion relations of Table 5.2, we see that

Electromagnetic waves:    $$v_{\text{phase}} = \frac{\omega}{k} = c \tag{5-16}$$

Matter waves:    $$v_{\text{phase}} = \frac{\omega}{k} = \frac{mc^2}{\hbar k} + \frac{\hbar k}{2m} \tag{5-17}$$

While electromagnetic plane waves all move through vacuum at the same speed $c$, matter plane waves of different wave number move at different speeds.

With a general relationship between $\omega$ and $k$, our wave group (5-13) is

$$\Psi(x, t) = \int_{-\infty}^{+\infty} \tilde{\psi}(k) e^{i(kx - \omega(k)t)} \, dk \tag{5-18}$$

While it is possible to obtain the group velocity of (5-18) for a general wave pulse, it will better serve our purposes to consider a commonly occurring specific type: a gaussian wave pulse. In Section 3.6, it is noted that (at a time arbitrarily chosen to be zero) a gaussian packet is described by

$$\psi(x) = A e^{-(x/2\epsilon)^2} e^{ik_0 x} \tag{5-19}$$

and its Fourier transform/spectral content[9] is

$$\tilde{\psi}(k) = \frac{A\epsilon}{\sqrt{\pi}} e^{-\epsilon^2(k - k_0)^2}$$

[8] Although we have added rest energy, this expression still applies only to "slow-moving" particles; it is still based upon KE $= p^2/2m$. It is left as an exercise to show that the relativistically correct expression is $\omega = \sqrt{k^2 c^2 + m^2 c^4/\hbar^2}$, and that the two agree at low speeds.

[9] These were obtained with matter waves in mind. But since the relationship between a function and its Fourier transform is strictly mathematical, a similar relationship would hold for electromagnetic waves.

Inserting this in (5-18), we obtain

$$\Psi(x,\ t) = \int_{-\infty}^{+\infty} \left[ \frac{A\epsilon}{\sqrt{\pi}} e^{-\epsilon^2(k-k_0)^2} \right] e^{i(kx-\omega(k)t)}\ dk \qquad \textbf{(5-20)}$$

Strictly speaking, without knowing the specific form of $\omega(k)$, we can proceed no further. However, our wave pulse has a reasonably well-defined wave number/momentum. Being itself a gaussian, $\tilde{\psi}(k)$ is a fairly sharply peaked function, with its maximum occurring at $k_0$. Accordingly, most of the contribution to the integral occurs near $k_0$, and if we replace $\omega(k)$ with a power series expanded about the point $k_0$, little error should be introduced. Thus, we write

$$\omega(k) \cong \omega(k_0) + s(k - k_0) + \tfrac{1}{2} d(k - k_0)^2$$

$$\text{where} \quad s \equiv \left. \frac{d\omega(k)}{dk} \right|_{k_0} \quad \text{and} \quad d \equiv \left. \frac{d^2\omega(k)}{dk^2} \right|_{k_0} \qquad \textbf{(5-21)}$$

Keeping terms higher than the quadratic would of course be more accurate, but as we shall see, the behaviors of interest reside in the terms we have kept. Equation (5-20) now becomes

$$\Psi(x,\ t) = \int_{-\infty}^{+\infty} \left[ \frac{A\epsilon}{\sqrt{\pi}} e^{-\epsilon^2(k-k_0)^2} \right] e^{i\{kx-[\omega_0 + s(k-k_0) + \frac{1}{2}d(k-k_0)^2]t\}}\ dk \quad \textbf{(5-22)}$$

This integral may be evaluated by completing the square in the exponential's argument, leaving a standard gaussian integral (see Exercise 28). The result is

$$\Psi(x,\ t) = \frac{A}{\sqrt{1 + \dfrac{idt}{2\epsilon^2}}} \exp\left[ -\frac{(x-st)^2}{4\epsilon^2\left(1 + \dfrac{idt}{2\epsilon^2}\right)} \right] e^{i(k_0 x - \omega_0 t)} \qquad \textbf{(5-23)}$$

[*Note:* When legibility dictates, we will use "exp(argument)" style for the exponential function.] At $t = 0$, this is the initial gaussian wave pulse of (5-19), but we now see how it propagates in time. The complex exponential on the right is the "oscillatory part" of the initial wave function, including its explicit time dependence. It is a plane wave of the approximate wave number and frequency of the pulse (i.e., $k_0$ and $\omega_0$), and moves at the corresponding phase velocity $\omega_0/k_0$. The other exponential is the "gaussian part," which modulates the plane wave. Its maximum occurs at the point $x = st$, so it moves at the speed $s$. The probability density is

$$|\Psi(x,\ t)|^2 = \Psi(x,\ t)^*\Psi(x,\ t)$$

$$= \left\{ \frac{A}{\sqrt{1 - \dfrac{idt}{2\epsilon^2}}} \exp\left[ -\frac{(x-st)^2}{4\epsilon^2\left(1 - \dfrac{idt}{2\epsilon^2}\right)} \right] e^{-i(k_0 x - \omega_0 t)} \right\}$$

$$\times \left\{ \frac{A}{\sqrt{1 + \dfrac{idt}{2\epsilon^2}}} \exp\left[ -\frac{(x-st)^2}{4\epsilon^2\left(1 + \dfrac{idt}{2\epsilon^2}\right)} \right] e^{+i(k_0 x - \omega_0 t)} \right\}$$

or

$$|\Psi(x,\ t)|^2 = \frac{A^2}{\sqrt{1 + \dfrac{d^2t^2}{4\epsilon^4}}} \exp\left[-\frac{(x-st)^2}{2\epsilon^2\left(1 + \dfrac{d^2t^2}{4\epsilon^4}\right)}\right] \tag{5-24}$$

The "oscillatory part" and its phase velocity have vanished, unimportant to the propagation of the particle or the energy it conveys. The probability of finding the particle is a function of $(x - st)$. It is thus a shape that moves at velocity $s$, and with (5-21) we have our result.

Group velocity

$$v_{\text{group}} = \frac{d\omega(k)}{dk}\bigg|_{k_0} \tag{5-25}$$

Let us calculate the group velocity for the familiar cases. Using Table 5.2,

Electromagnetic waves: $v_{\text{group}} = \dfrac{d}{dk}(ck)\bigg|_{k_0} = c$ (5-26)

Matter waves: $v_{\text{group}} = \dfrac{d}{dk}\left(\dfrac{mc^2}{\hbar} + \dfrac{\hbar k^2}{2m}\right)\bigg|_{k_0} = \dfrac{\hbar k_0}{m}$ (5-27)

The group velocity of an electromagnetic pulse in vacuum is the speed of light, because all its constituent waves share that phase velocity; they must move as one. Equation (5-17), on the other hand, shows that a matter-wave pulse's constituent plane waves would move at different speeds. Nevertheless, the group velocity is just what we should expect: $\hbar k_0/m = p_0/m$ is indeed the velocity of a *particle* whose approximate momentum in $p_0$. No matter what may be the phase velocity of constituent waves, the region of high probability moves at the speed expected of the associated particle.

In summary, the phase velocity of an individual plane wave of a wave group depends on its wave number and frequency in the usual way: $\omega/k$. If frequency is directly proportional to wave number (e.g., $\omega = ck$), then all phase velocities are equal and the group necessarily moves at that phase velocity. But if its constituents move at different speeds, then the group (and with it the particle) moves at a speed dependent on how the speeds differ—specifically, how frequency varies with wave number, $d\omega/dk$.

### Application: An Electromagnetic Pulse

Let us consider the behavior of an electromagnetic wave pulse traveling through a medium where plane waves of different frequencies move at different speeds.

**Example 5.3**

In the earth's ionosphere, for waves in the region of the electromagnetic spectrum used for "short-wave" radio communication, the refractive index varies according to

$$n(\omega) = \sqrt{1 - \frac{b}{\omega^2}}$$

where $\omega$ is angular frequency and $b$ is a constant. (a) Find the dispersion relation. (b) For a pulse of approximate frequency $\omega_0$, determine the phase and group velocities.

(*Note:* This variation in refractive index arises from the detailed interaction between the electromagnetic wave and the medium. Although the reader may quail at the notion of a refractive index less than unity, and therefore a phase velocity greater than $c$, situations in which it occurs are not uncommon. We confront the seeming violation of special relativity afterward.)

## Solution

(a) By definition, the refractive index of a material is the ratio of the speed of light in vacuum to the speed of a pure electromagnetic plane wave in the material—that is, to the phase velocity.

$$n = \frac{c}{v_{\text{phase}}} \quad \Rightarrow \quad v_{\text{phase}} = \frac{c}{\sqrt{1 - b/\omega^2}}$$

But it is also true that the speed of a plane wave is $\omega/k$. Thus,

$$\frac{\omega}{k} = \frac{c}{\sqrt{1 - b/\omega^2}}$$

Solving for $\omega$, we obtain the dispersion relation

$$\omega(k) = \sqrt{b + (kc)^2} \tag{5-28}$$

(b) Using (5-25),

$$v_{\text{group}} = \left.\frac{d\omega(k)}{dk}\right|_{k_0} = \frac{k_0 c^2}{\sqrt{b + (k_0 c)^2}}$$

We now reexpress this in terms of the given $\omega_0$. Using (5-28), we obtain

$$\omega_0 = \sqrt{b + (k_0 c)^2} \quad \Rightarrow \quad k_0 = \frac{1}{c}\sqrt{\omega_0^2 - b}$$

Thus,

$$v_{\text{group}} = \frac{c\sqrt{\omega_0^2 - b}}{\omega_0} = c\sqrt{1 - b/\omega_0^2}$$

Evaluating the phase velocity also at the approximate frequency, we have

$$v_{\text{phase}} = \frac{c}{\sqrt{1 - b/\omega_0^2}} \qquad v_{\text{group}} = c\sqrt{1 - b/\omega_0^2}$$

In the example, the pulse's group velocity is less than $c$, but its phase velocity is greater than $c$! We do not have to look far for this "problem" to recur: The same holds true for matter waves—even in vacuum! Since equation (5-17) is correct at low speeds, where $\hbar k = p$ must be much less than $mc$, its first term is greater than $c$.[10] *Is special relativity violated?* It is true that any individual plane wave may travel faster than $c$. But a true plane wave—of infinite extent in space and infinite duration in time—cannot transmit *information*. It does not vary in any significant way—that is, in a way capable of conveying information from one place to another. On the contrary, to transmit any intelligence, the wave must be modulated

[10]This is where the internal energy term is (as noted earlier) "vitally important." All other uses of $\omega(k)$ involve taking derivatives with respect to $k$, in which case the constant term is irrelevant. It should also be noted that the relativistically correct dispersion relation agrees on the point considered here: $v_{\text{phase}} > c$. See Exercise 20.

in some way, perhaps varying amplitude or frequency. If it is modulated (even so simply as turning a beam on and off), the wave is no longer a single plane wave; it becomes a combination of plane waves, and for any such wave group the important speed is that of the probability density—the group velocity. This is never greater than $c$.

## Dispersion

The simplest wave pulse is an electromagnetic wave in vacuum. Because all of its constituent plane waves move at the same phase velocity $c$, the group propagates as an unchanging shape indefinitely. In other cases—electromagnetic waves moving through a medium and matter waves in general—there is invariably **dispersion,** the spreading of a wave pulse due to nonuniform speeds of its constituents.

Dispersion arises whenever the dispersion relation is nonlinear; that is, when $d$ in equation (5-21)—the *second* derivative of $\omega$ with respect to $k$—is nonzero. Let us take another look at (5-24), the probability density for a gaussian pulse.

$$|\Psi(x,\ t)|^2 = \frac{A^2}{\sqrt{1 + \dfrac{d^2 t^2}{4\epsilon^4}}} \exp\left[-\frac{(x-st)^2}{2\epsilon^2\left(1 + \dfrac{d^2 t^2}{4\epsilon^4}\right)}\right]$$

If $\omega$ is a linear function of $k$, as it is for electromagnetic waves in vacuum ($\omega = ck$), $d$ is zero. In this case, $|\Psi(x,\ t)|^2$ becomes simply $A^2 e^{-(x-st)^2/2\epsilon^2}$, an unchanging shape moving at speed $s$. When $d$ is nonzero, the pulse spreads. For this reason, $d$ is called the **dispersion coefficient.** In the denominator of the exponential's argument, it causes the moving gaussian to become broader, ultimately a constant of unity. In the factor multiplying the exponential, it causes the probability density as a whole to decrease. Thus, the pulse flattens out; the probability of finding the particle spreads over an ever larger region, while the probability per unit length diminishes.

Dispersion would occur for the electromagnetic pulse in the previous example, since the dispersion relation (5-28) for the medium is nonlinear (see Exercise 26). It occurs for a matter wave *even in vacuum.* Let us determine the dispersion coefficient. Using the dispersion relation in Table 5.2, we obtain

$$d \equiv \left.\frac{d^2\omega(k)}{dk^2}\right|_{k_0} = \left.\frac{d^2}{dk^2}\left(\frac{mc^2}{\hbar} + \frac{\hbar k^2}{2m}\right)\right|_{k_0} = \frac{\hbar}{m} \tag{5-29}$$

Therefore,

$$|\Psi(x,\ t)|^2 = \frac{A^2}{\sqrt{1 + \dfrac{\hbar^2 t^2}{4m^2\epsilon^4}}} \exp\left[-\frac{(x-st)^2}{2\epsilon^2\left(1 + \dfrac{\hbar^2 t^2}{4m^2\epsilon^4}\right)}\right] \tag{5-30}$$

Gaussian matter wave in vacuum

Note that since the factor $\hbar^2 t^2/4m^2\epsilon^4$ that causes the spreading is proportional to $\epsilon^{-4}$, the narrower the initial width of the pulse, the more rapidly it will spread in time. Exercise 27 shows just how quickly a pulse might spread in a typical case.

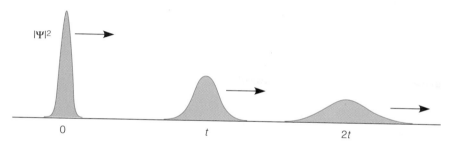

**Figure 5.18** Dispersion causes a matter wave to spread.

It is worth reiterating that the gaussian pulse we have considered is a solution of the free-particle time-dependent Schrödinger equation (verification is left as an exercise). It is not an unrealistic plane wave, but is better representative of a reasonably well localized particle. Figure 5.18 illustrates the effect of dispersion on the probability density of a moving free particle. Dispersion invariably leads to increasing uncertainty in a particle's position.

## ▼ **5.4** The Classical Limit: A Familiar Law

We now investigate what happens when a massive particle described by an arbitrary wave function encounters an arbitrary force! Of course, this is too ambitious; we can't expect to determine quantum-mechanical behavior in a completely general case. But we are able to show—after a fair amount of rather messy math—that in the classical limit, behaviors of all massive objects are governed by the same rule.

We begin with the time derivative of the expectation value of momentum.

$$\frac{d}{dt}\overline{p} = \frac{d}{dt}\int_{-\infty}^{+\infty}\Psi^*(x,\ t)\left(-i\hbar\frac{\partial}{\partial x}\right)\Psi(x,\ t)\ dx$$

The integral is a function of time only, but the integrand is a function of time and position. Therefore, to bring the derivative over time inside the integral, we change the ordinary derivative with respect to time to a partial derivative.

$$\frac{d}{dt}\overline{p} = \int_{-\infty}^{+\infty}\frac{\partial}{\partial t}\left[\Psi^*(x,\ t)\left(-i\hbar\frac{\partial}{\partial x}\right)\Psi(x,\ t)\right]dx$$

$$= \int_{-\infty}^{+\infty}\frac{\partial\Psi^*(x,\ t)}{\partial t}\left(-i\hbar\frac{\partial}{\partial x}\right)\Psi(x,\ t)\ dx$$

$$+ \int_{-\infty}^{+\infty}\Psi^*(x,\ t)\left(-i\hbar\frac{\partial}{\partial x}\right)\frac{\partial\Psi(x,\ t)}{\partial t}\ dx$$

We now eliminate the time derivative by using the Schrödinger equation, expressed as

$$\frac{\partial\Psi(x,\ t)}{\partial t} = \frac{-\dfrac{\hbar^2}{2m}\dfrac{\partial^2\Psi(x,\ t)}{\partial x^2} + U(x)\Psi(x,\ t)}{i\hbar}$$

Inserting this in the second term and its complex conjugate in the first,

$$\frac{d}{dt}\overline{p} = \int_{-\infty}^{+\infty} \frac{-\dfrac{\hbar^2}{2m}\dfrac{\partial^2 \Psi^*(x, t)}{\partial x^2} + U(x)\Psi^*(x, t)}{-i\hbar}\left(-i\hbar\frac{\partial}{\partial x}\right)\Psi(x, t)\, dx$$

$$+ \int_{-\infty}^{+\infty} \Psi^*(x, t)\left(-i\hbar\frac{\partial}{\partial x}\right)\frac{-\dfrac{\hbar^2}{2m}\dfrac{\partial^2 \Psi(x, t)}{\partial x^2} + U(x)\Psi(x, t)}{i\hbar}\, dx$$

Now canceling factors of $i\hbar$ and separating terms,

$$\frac{d}{dt}\overline{p} = \left(-\frac{\hbar^2}{2m}\right)\int_{-\infty}^{+\infty}\frac{\partial^2 \Psi^*(x, t)}{\partial x^2}\left(\frac{\partial}{\partial x}\right)\Psi(x, t)\, dx$$

$$+ \int_{-\infty}^{+\infty} U(x)\Psi^*(x, t)\left(\frac{\partial}{\partial x}\right)\Psi(x, t)\, dx$$

$$- \left(-\frac{\hbar^2}{2m}\right)\int_{-\infty}^{+\infty}\Psi^*(x, t)\left(\frac{\partial}{\partial x}\right)\frac{\partial^2 \Psi(x, t)}{\partial x^2}\, dx$$

$$- \int_{-\infty}^{+\infty}\Psi^*(x, t)\left(\frac{\partial}{\partial x}\right)(U(x)\Psi(x, t))\, dx$$

Combining the integrands in the first and third terms yields

$$\frac{\partial^2 \Psi^*(x, t)}{\partial x^2}\left(\frac{\partial}{\partial x}\right)\Psi(x, t) - \Psi^*(x, t)\left(\frac{\partial}{\partial x}\right)\frac{\partial^2 \Psi(x, t)}{\partial x^2}$$

$$= \frac{\partial}{\partial x}\left\{\frac{\partial \Psi^*(x, t)}{\partial x}\frac{\partial \Psi(x, t)}{\partial x} - \Psi^*(x, t)\frac{\partial^2 \Psi(x, t)}{\partial x^2}\right\}$$

Thus, the integral of these two terms is trivial, merely the quantity in braces evaluated between $x = -\infty$ and $x = +\infty$. But a particle's wave function must drop to zero at $x = \pm\infty$, so this integral is zero. Combining the second and fourth terms,

$$\frac{d}{dt}\overline{p} = \int_{-\infty}^{+\infty}\Psi^*(x, t)\left[U(x)\left(\frac{\partial}{\partial x}\right)\Psi(x, t) - \left(\frac{\partial}{\partial x}\right)(U(x)\Psi(x, t))\right]dx$$

Finally, using the product rule on the partial derivative of $U(x)\Psi(x, t)$, and then canceling, our result is

$$\frac{d}{dt}\overline{p} = \int_{-\infty}^{+\infty}\Psi^*(x, t)\left(-\frac{\partial U(x)}{\partial x}\right)\Psi(x, t)\, dx$$

or

Second law of motion

$$\frac{d}{dt}\overline{p} = -\overline{\frac{\partial U}{\partial x}} \qquad\qquad (5\text{-}31)$$

Classically, we identify the negative derivative of a particle's potential energy with the force it experiences. As we found in Section 3.4, theoretical minimum uncertainties are negligible, classically speaking. Therefore, the expectation values in (5-31) become simply the precisely known momentum and force. We see then that

Newton's second law of motion $dp/dt = F$ is a special-case prediction of the Schrödinger equation. According to the correspondence principle, we should expect the quantum-mechanical law to agree with the classical law in an appropriate limit. That limit occurs when quantum-mechanical uncertainties are negligible.

## Progress and Applications

■ It might seem that disagreements about presumably basic points of quantum mechanics should by now be fully resolved. Most are, but not all. The time required for a particle to tunnel is one such point. Several experiments seem to have measured it; the difficulty is the theoretical prediction. It can be shown that it is not always correct to associate the maximum of a wave pulse with the location of the particle; given certain conditions, the transmitted pulse's maximum can emerge *before* the incident pulse's maximum strikes a barrier! While some claim that even the fundamental question "What is the tunneling time?" has no meaning, most debate centers on the question "What theoretical definition of the time agrees with observation?" The answer lies ahead.

■ The so-called **tunnel junction** is a key element in many existing and proposed electronic devices. It consists of two electrical conductors separated by an insulating barrier. Many electrons move freely in each conductor, and if a potential difference is applied, they would tend to move from one side to the other. The insulator, were it true to its name, would of course prevent this. However, electrons may tunnel through it. Tunnel junctions are used in an experimental device known as the single-electron transistor. When electrons tunnel, they do so one particle at a time—an (entire) electron either tunnels or it doesn't. This imposes a natural granularity on the device's response and allows for extremely fine control of current flow—qualitatively varying due to the passage of a *single* electron! In a different application, a tunnel junction in which the conductors are ferromagnetic is being studied as a possible element of magnetic computer data storage. In perhaps the most important application, when the insulator separates two superconductors (see Chapter 9), it is known as a Josephson junction, and electrons tunnel in pairs. The Josephson junction is at the heart of devices known as **SQUIDS:** superconducting quantum interference devices. The insulator serves as a "weak link" between the charges' motions on the two sides, coupling them together as would a weak spring connecting two pendula. With this kind of coupling, energy/current flow is intimately dependent on the phase relationship between the things that oscillate (pendula or charge-carrier wave functions) on the two sides; there may be constructive or destructive interference. In the case of charges separated by the weak link in a SQUID, the phase relationship is also an extremely sensitive function of any nearby magnetic field. Depicted in Figure 5.19, a SQUID is in essence a loop with two Josephson junctions. Interference caused by small changes in the magnetic flux passing through it produces easily detected changes in the current $I$. SQUIDS may be used to detect extremely small magnetic fields, such as those produced by the human heart and brain.

■ Understanding the optical properties of materials is one of the more active pursuits in modern physics. Of particular interest are their "nonlinear responses."

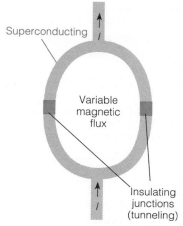

**Figure 5.19** Elements of a SQUID—superconducting quantum interference device.

In general, the speed of a light wave varies not only with frequency, which we know as dispersion, but also with the wave amplitude, usually referred to as a nonlinear response of the medium to the disturbance. The first effect tends to broaden a wave pulse, such as might carry a "data bit" of digital information; the second tends to compress a pulse. The effects may be pitted against one another, so that for a pulse of the proper size and shape they effectively cancel, leaving the pulse to propagate undistorted! Such a pulse is known as a **soliton.** Use of soltions in high-speed long-distance communication is now under vigorous investigation.

## Basic Equations

$E > U_0$ step:

$$T = 4\frac{\sqrt{E(E - U_0)}}{(\sqrt{E} + \sqrt{E - U_0})^2} \qquad R = \frac{(\sqrt{E} - \sqrt{E - U_0})^2}{(\sqrt{E} + \sqrt{E - U_0})^2} \qquad (5\text{-}6)$$

$E > U_0$ barrier:

$$R = \frac{\sin^2\left[\dfrac{\sqrt{2m(E - U_0)}}{\hbar}L\right]}{\sin^2\left[\dfrac{\sqrt{2m(E - U_0)}}{\hbar}L\right] + 4\dfrac{E}{U_0}\left(\dfrac{E}{U_0} - 1\right)}$$

$$T = \frac{4\dfrac{E}{U_0}\left(\dfrac{E}{U_0} - 1\right)}{\sin^2\left[\dfrac{\sqrt{2m(E - U_0)}}{\hbar}L\right] + 4\dfrac{E}{U_0}\left(\dfrac{E}{U_0} - 1\right)} \qquad (5\text{-}8)$$

**Resonant transmission:**

$$E = U_0 + \frac{n^2\pi^2\hbar^2}{2mL^2} \qquad (5\text{-}9)$$

$E < U_0$ Barrier (tunneling):

$$R = \frac{\sinh^2\left[\dfrac{\sqrt{2m(U_0 - E)}}{\hbar}L\right]}{\sinh^2\left[\dfrac{\sqrt{2m(U_0 - E)}}{\hbar}L\right] + 4\dfrac{E}{U_0}\left(1 - \dfrac{E}{U_0}\right)}$$

$$T = \frac{4\dfrac{E}{U_0}\left(1 - \dfrac{E}{U_0}\right)}{\sinh^2\left[\dfrac{\sqrt{2m(U_0 - E)}}{\hbar}L\right] + 4\dfrac{E}{U_0}\left(1 - \dfrac{E}{U_0}\right)} \qquad (5\text{-}10)$$

**Dispersion relation:**

$$\omega = \omega(k) \tag{5-14}$$

**Wave group:**

$$\Psi(x, t) = \int_{-\infty}^{+\infty} \tilde{\psi}(k) e^{i(kx - \omega(k)t)} \, dk \tag{5-18}$$

| **Phase velocity:** | **Group (particle) velocity:** |
|---|---|
| $$v_{phase} = \lambda f = \frac{\omega}{k} \tag{5-15}$$ | $$v_{group} = \frac{d\omega(k)}{dk}\bigg|_{k_0} \tag{5-25}$$ |

**Second law of motion:**

$$\frac{d}{dt}\overline{p} = -\overline{\frac{\partial U}{\partial x}} \tag{5-31}$$

## Summary

Just as it has an effect on a massive particle, any change in potential energy (i.e., force) has an effect on a matter wave function. Under certain conditions, a particle subject to a force will manifest its wave nature, doing what a classical particle would not. It may be reflected by a force that would not classically reflect it, even by a force in the particle's direction of motion. Conversely, it may pass through a region despite a force that classically should cause it to rebound. It may escape where classically it should be confined, a process known as tunneling. The probabilities of a particle being transmitted or reflected are governed by the solution $\Psi(x, t)$ of the Schrödinger equation for the given potential energy.

To understand the relationship between the motion of a particle and the motion of its matter wave, it is useful to represent the particle as a wave group—a sum of plane waves. The speed of an individual plane wave is known as phase velocity. It varies from one plane wave to another in the group, depending on the dispersion relation governing the wave. It may also exceed the speed of light. However, the probability density moves at a different speed, the group velocity, which is never greater than $c$ and is the same as the speed of the associated particle. Unless all plane waves in a group travel at the same speed, a wave group will spread with time—the phenomenon of dispersion. Dispersion occurs in matter waves even in vacuum, and leads to increasing uncertainty in a particle's position.

The effects of changes in potential energy on particles and waves are linked, so that the classical second law of motion may be viewed as a special-case prediction of the Schrödinger equation.

# E X E R C I S E S

## Section 5.1

1. Calculate the reflection probability for a 5-eV electron encountering a step in which the potential drops by 2 eV.

2. A particle moving in a region of zero force encounters a precipice—a sudden drop in the potential energy to an arbitrarily large negative value. What is the probability that it will "go over the edge"?

3. What fraction of a beam of 50-eV electrons would get through a 200-V, 1-nm-wide electrostatic barrier?

4. A beam of particles of energy $E$ and incident upon a potential step of $U_0 = \frac{3}{4}E$ is described by the wave function

$$\psi_{inc}(x) = 1 \ e^{ikx}$$

The amplitude of the wave (related to the number incident per unit distance) is arbitrarily chosen as unity. (a) Determine completely the reflected and transmitted waves by enforcing the required continuity conditions to obtain their (possibly complex) amplitudes. (b) Verify that the ratio of reflected probability density to the incident probability density agrees with equation (5-6).

5. The equations for $R$ and $T$ in the $E > U_0$ barrier are essentially the same as for light passing through a transparent film. It is possible to fabricate a thin film that reflects no light. Is it possible to fabricate one that transmits no light? Why or why not?

6. In the $E > U_0$ potential barrier, there should be no reflected wave when the incident wave is at one of the transmission resonances. Prove this: Assuming that a beam of particles is incident at the first transmission resonance, $E = U_0 + (\pi^2\hbar^2/2mL^2)$, combine the continuity conditions to show that $B = 0$. (*Note:* $k'$ is particularly simple in this special case, and this should be used to streamline your work as much as possible.)

7. As we learn in physical optics, thin film interference can cause some wavelengths of light to be strongly reflected while others are not reflected at all. Neglecting absorption, all light has to go one way or the other, so wavelengths not reflected are strongly transmitted. (a) For a film of thickness $t$ surrounded by air, what wavelengths $\lambda$ (while they are within the film) will be strongly transmitted? (b) What wavelengths (while they are "over" the barrier) of matter waves satisfy condition (5-9)? (c) Comment on the relationship between the answers in (a) and (b).

8. Consider a potential barrier of height 30 eV. (a) Find a width around 1.000 nm for which there will be no reflection of 35-eV electrons incident upon the barrier. (b) What would be the reflection probability for 36-eV electrons incident upon the same barrier? (*Note:* This corresponds to a difference in speed of less than $1\frac{1}{2}\%$.)

9. A particle of mass $m$ and energy $E$ moving in a region where there is initially no potential energy encounters a potential dip of width $L$ and depth $U = -U_0$.

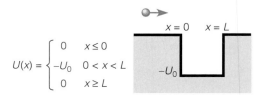

$$U(x) = \begin{cases} 0 & x \leq 0 \\ -U_0 & 0 < x < L \\ 0 & x \geq L \end{cases}$$

Show that the reflection probability is given by

$$R = \frac{\sin^2\left[\frac{\sqrt{2m(E + U_0)}}{\hbar}L\right]}{\sin^2\left[\frac{\sqrt{2m(E + U_0)}}{\hbar}L\right] + 4\frac{E}{U_0}\left(\frac{E}{U_0} + 1\right)}$$

(*Hint:* This may be inferred via an appropriate substitution in a known reflection probability.)

10. Given the situation of Exercise 9, show (a) that as $U_0 \to \infty$, the reflection probability approaches unity, and (b) that as $L \to 0$, the reflection probability approaches zero. (c) Consider the limit in which the well becomes infinitely deep and infinitesimally narrow—that is, $U_0 \to \infty$ and $L \to 0$—but the product $U_0 L$ is constant. (This "delta well" model approximates reasonably well the effect of a narrow but strong attractive potential, such as that experienced by a free electron encountering a positive ion.) Show that the reflection probability becomes

$$R = \frac{1}{1 + \dfrac{2\hbar^2 E}{m(U_0 L)^2}}$$

11. For the $E > U_0$ potential barrier, the reflection and transmission probabilities are the ratios

$$R = \frac{B*B}{A*A} \quad \text{and} \quad T = \frac{F*F}{A*A}$$

where $A$, $B$, and $F$ are the multiplicative coefficients of the incident, reflected and transmitted waves respectively. From the four smoothness conditions, solve for $B$ and $F$ in terms of $A$, insert them in the $R$ and $T$ ratios, and thus derive equations (5-7).

12. *Jump to Jupiter:* The gravitational potential energy of a 1-kg object is plotted versus position from Earth's surface to the surface of Jupiter. Mostly it is due to the Sun, but there are downturns at each end, due to the attractions to the two planets.

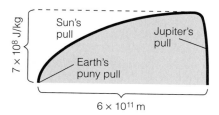

Make the crude approximation that this is a rectangular barrier of width $6 \times 10^{11}$ m and approximate (average) height $4 \times 10^8$ J/kg. Your mass is 65 kg, and you launch yourself from Earth at an impressive 4 m/s. What is the probability that you can jump to Jupiter?

13. A beam of particles of energy $E$ and incident upon a potential step of $U_0 = \frac{5}{4}E$ is described by the wave function

$$\psi_{inc}(x) = 1\ e^{ikx}$$

(a) Determine completely the reflected wave and the wave inside the step by enforcing the required continuity conditions to obtain their (possibly complex) amplitudes.

(b) Verify by explicit calculation that the ratio of reflected probability density to the incident probability density is unity.

14. Obtain equation (5-12) from (5-10) and (5-11).

15. It is shown in Section 5.1 that for the $E < U_0$ potential step, $B = -[(\alpha + ik)/(\alpha - ik)]A$. Use this to calculate the probability density (including both incident and reflected waves) to the left of the step:

$$|\psi_I|^2 = |Ae^{+ikx} + Be^{-ikx}|^2$$

(a) Show that the result is $4|A|^2\sin^2(kx - \theta)$, where $\theta = \tan^{-1}(k/\alpha)$. Because the reflected wave is of the same amplitude as the incident, this is a typical standing wave pattern, varying periodically between zero and $4A^*A$.

(b) Determine $\theta$ and $D$ (the amplitude of the $x > 0$ wave inside the step) in the limits $k \to 0$ and $\alpha \to 0$, and interpret the results.

## Section 5.2

16. Could the situation depicted in the following diagram represent a particle in a bound state? Explain.

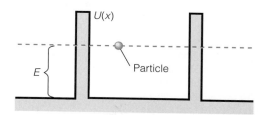

17. A particle experiences a potential energy given by $U(x) = (x^2 - 3)e^{-x^2}$ (in SI units). (a) Make a sketch of $U(x)$ including numerical values at the minima and maxima. (b) What is the maximum energy the particle could have and yet be bound? (c) What is the maximum energy the particle could have and yet be bound for a considerable length of time? (d) Is it possible for the particle to have an energy greater than that in part (c) and still be "bound" for some period of time? Explain.

18. To obtain a rough estimate of the mean time required for uranium-238 alpha decay, let us approximate the combined electrostatic and strong nuclear potential energies by a rectangular potential barrier half as high as the actual 35-MeV maximum potential energy. Alpha-particles (mass 4 u) of 4.3-MeV kinetic energy are incident. Let us also assume that the barrier extends from the radius of the nucleus, 7.4 fm, to the point where the electrostatic potential drops to 4.3 MeV (i.e., the classically forbidden region). Since $U \propto 1/r$, this point is 35/4.3 times the radius of the nucleus, the point at which $U(r)$ is 35 MeV.

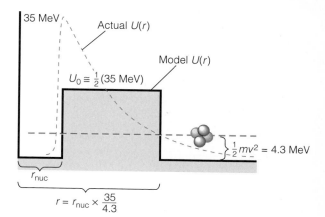

(a) Use these crude approximations, the method suggested in Section 5.2, and the wide barrier approximation to obtain a value for the time to decay.

(b) To gain some appreciation of the difficulties in a theoretical prediction, work the exercise "backward." Rather than assuming a value for $U_0$, use the known value of the mean time to decay for uranium-238 and infer the corresponding value of $U_0$. Retain all other assumptions.

(c) Comment on the sensitivity of the decay time to the height of the potential barrier.

19. As we learned in Chapter 2, a certain minimum amount of energy, the work function $\phi$, is needed to remove an electron from a given metal, such as the tip of a scanning tunneling microscope. The same is true of samples we might study with an STM. The two work functions are not in general equal, and in the STM a potential difference is applied between tip and sample, giving a slope to the potential energy function felt by electrons moving between them. Nevertheless, merely to gain some idea of the variation of tunneling current with barrier width/separation $L$, it is a fair approximation to treat the space between tip and sample as a simple square barrier that is 3 eV (a typical work function) higher than the electron energy. Given a nominal separation of $L = 0.2$ nm, by how much would the tunneling probability, and thus tunneling current, change due to a 0.001-nm change in barrier width? (Remember: This is a "wide" barrier.)

## Section 5.3

20. For a general wave pulse, neither $E$ nor $p$ (i.e., neither $\omega$ nor $k$) are well defined, but they have approximate values $E_0$ and $p_0$. Although it comprises many plane waves, the general pulse has an overall phase velocity corresponding to these values.

$$v_{\text{phase}} = \frac{\omega_0}{k_0} = \frac{E_0/\hbar}{p_0/\hbar} = \frac{E_0}{p_0}$$

If the pulse describes a "large" massive particle, the uncertainties are reasonably small, and the particle may be said to have energy $E_0$ and momentum $p_0$. Using the relativistically correct expressions for energy and momentum, show that the overall phase velocity is greater than $c$ and given by

$$v_{\text{phase}} = \frac{c^2}{u_{\text{particle}}}$$

Note that the phase velocity is greatest for particles whose speed is least!

21. The matter-wave dispersion relation in Table 5.2 is correct only at low speed.

(a) Using the relativistically correct relationship between energy, momentum, and mass, show that the correct dispersion relation is

$$\omega = \sqrt{k^2 c^2 + \frac{m^2 c^4}{\hbar^2}}$$

(b) Show that in the limit of low speed (small $p$ and $k$) this expression agrees with that of Table 5.2.

22. Equation (5-27) shows that the group velocity of a matter-wave group equals the velocity of the massive particle with which it is associated. However, both the dispersion relation used to show that $v_{\text{group}} = \hbar k_0/m$ and the formula $v_{\text{particle}} = p/m$ used to relate this to the particle velocity are relativistically incorrect. It might be argued that we proved what we wished to prove by making an even number of mistakes.

(a) Using the relativistically correct dispersion relation given in Exercise 21, show that the group velocity of a wave pulse is actually given by

$$v_{\text{group}} = \frac{\hbar k_0 c^2}{\sqrt{(\hbar k_0)^2 c^2 + m^2 c^4}}$$

(b) The fundamental relationship $p = \hbar k$ is universally correct, so $\hbar k_0$ is indeed the particle momentum $p$. (It is not well defined, but this is its approximate, or "central," value.) Making this substitution in the expression for $v_{\text{group}}$ from part (a), then using the relativistically correct relationship between momentum $p$ and particle velocity $v$, show that the group velocity again is equal to the particle velocity.

23. For wavelengths less than about a centimeter, the dispersion relation for waves on the surface of water is $\omega = \sqrt{(\gamma/\rho)k^3}$, where $\gamma$ and $\rho$ are the surface tension and density of water. Given $\gamma = 0.072$ N/m and $\rho = 10^3$ kg/m$^3$, calculate the phase and group velocities for a wave of 5-mm wavelength.

24. For waves on the surface of water, the behavior of long wavelengths is dominated by gravitational effects—a liquid "seeking its own level." Short wavelengths are dominated by surface tension effects. Taking both into account, the dispersion relation is $\omega = \sqrt{gk + (\gamma/\rho)k^3}$, where $\gamma$ is the surface tension, $\rho$ is the density of water, and $g$ is of course the gravitational acceleration.

(a) Make a qualitative sketch of group velocity versus wave number. How does it behave for very large $k$? For very small $k$?

(b) Find the minimum possible group velocity and the wavelength at which it occurs. (Use $\gamma = 0.072$ N/m, $\rho = 10^3$ kg/m$^3$, and $g = 9.8$ m/s$^2$.)

25. For wavelengths greater than about 20 cm, the disper-

sion relation for waves on the surface of water is $\omega = \sqrt{gk}$. (a) Calculate the phase and group velocities for a wave of 5-m wavelength. (b) Will the wave spread as it travels? Justify your answer.

26. Calculate the dispersion coefficient for the electromagnetic wave of Example 5.3. Which spreads more rapidly—a pulse of short or long wavelength?

27. As we learned in Example 4.2, in a gaussian function of the form $\psi(x) \propto e^{-(x/2\epsilon)^2}$, $\epsilon$ is the standard deviation, or uncertainty, in position. The probability density for a gaussian wave function would be proportional to $\psi(x)$ squared: $e^{-(x^2/2\epsilon^2)}$. Comparing with the time-dependent gaussian probability of equation (5-30), we see that the uncertainty in position of the time-evolving gaussian wave function of a free particle is given by

$$\Delta x = \epsilon \sqrt{1 + \frac{\hbar^2 t^2}{4m^2 \epsilon^4}}$$

That is, it starts at $\epsilon$ and increases with time. Suppose the wave function of an electron is initially determined to be a gaussian of 500-nm uncertainty. How long will it take for the uncertainty in the electron's position to reach 5 m, the length of a typical automobile?

28. In this exercise, equation (5-23) is obtained from (5-22). Making the substitution $k - k_0 \equiv z$, express the integral in equation (5-22) in terms of the new variable of integration $z$. The integral may then be put in the standard gaussian form:

$$\int_{-\infty}^{+\infty} e^{-az^2 + bz} \, dz = \sqrt{\frac{\pi}{a}} \, e^{b^2/4a}$$

from which equation (5-23) follows.

29. Verify that the gaussian matter wave given in equation (5-23) is indeed a solution of the time-dependent free-particle Schrödinger equation. (*Note:* $s = \hbar k_0/m$ and $d = \hbar/m$.)

# Quantum Mechanics in Three Dimensions and the Hydrogen Atom

Nature provides few more obvious failures of classical physics than the simple hydrogen atom. According to classical physics, the atom should be unstable. Anytime a charged particle accelerates, it emits electromagnetic radiation. Thus, the hydrogen atom's orbiting electron, due to its centripetal acceleration, should lose energy continuously, and so spiral into the nucleus. Our observation, however, is that atoms are usually supremely stable, emitting no radiation. Furthermore, when they are *induced* to radiate electromagnetic energy, they do so at only certain frequencies. Although easily demonstrated with a simple diffraction grating, this defied classical explanation. Quantum mechanics' first great triumph was its explanation of these observations. But before we can understand the hydrogen atom quantum-mechanically, we must first extend the Schrödinger equation to govern waves in real, three-dimensional space.

## 6.1 The Schrödinger Equation in Three Dimensions

In one dimension, the Schrödinger equation is

$$-\frac{\hbar^2}{2m}\frac{\partial^2}{\partial x^2}\Psi(x,\ t) + U(x)\Psi(x,\ t) = i\hbar\frac{\partial}{\partial t}\Psi(x,\ t)$$

In three dimensions, the potential energy will in general be a function of three spatial coordinates, $U(x, y, z)$, and so too the wave function, $\Psi(x, y, z, t)$. The first term on the left in the Schrödinger equation involves the operator for kinetic energy. In the one-dimensional case, we see this as follows:

$$\frac{p_x^2}{2m} \rightarrow \frac{\left(-i\hbar \dfrac{\partial}{\partial x}\right)^2}{2m} = -\frac{\hbar^2}{2m}\frac{\partial^2}{\partial x^2}$$

To generalize this, we should expect to replace it by an operator for kinetic energy in three dimensions:

$$\frac{p_x^2 + p_y^2 + p_z^2}{2m} \rightarrow \frac{\left(-i\hbar \dfrac{\partial}{\partial x}\right)^2 + \left(-i\hbar \dfrac{\partial}{\partial y}\right)^2 + \left(-i\hbar \dfrac{\partial}{\partial z}\right)^2}{2m}$$

$$= -\frac{\hbar^2}{2m}\left(\frac{\partial^2}{\partial x^2} + \frac{\partial^2}{\partial y^2} + \frac{\partial^2}{\partial z^2}\right)$$

With these assumptions, the Schrödinger equation becomes

$$-\frac{\hbar^2}{2m}\left(\frac{\partial^2}{\partial x^2} + \frac{\partial^2}{\partial y^2} + \frac{\partial^2}{\partial z^2}\right)\Psi(x, y, z, t) + U(x, y, z)\Psi(x, y, z, t)$$

$$= i\hbar\frac{\partial}{\partial t}\Psi(x, y, z, t)$$

Once again, this equation cannot be proved or derived. Its validity rests upon the agreement of its predictions with experiment.

To specify position in three-dimensional space, we are not required to use rectangular Cartesian coordinates. Indeed, in many cases (notably, the hydrogen atom) it would be very awkward to do so. To avoid bias, the Schrödinger equation is usually expressed in coordinate-independent form. Using the generic symbol $\mathbf{r}$ (boldface) to represent position—Cartesian $(x, y, z)$, cylindrical $(r, \theta, z)$, spherical polar $(r, \theta, \phi)$, and so on—$U(x, y, z)$ becomes $U(\mathbf{r})$ and $\Psi(x, y, z, t)$ becomes $\Psi(\mathbf{r}, t)$. All that remains, then, is to adopt a generic symbol for the differential operator in the kinetic energy term. Vector calculus provides this symbol: $\nabla^2$. Expressed in Cartesian coordinates,

$$\nabla^2 = \frac{\partial^2}{\partial x^2} + \frac{\partial^2}{\partial y^2} + \frac{\partial^2}{\partial z^2}$$

Although the operator may have a quite different appearance in other coordinate systems, *its mathematical function is the same no matter what coordinates are chosen*. Altogether, we have

$$-\frac{\hbar^2}{2m}\nabla^2\Psi(\mathbf{r}, t) + U(\mathbf{r})\Psi(\mathbf{r}, t) = i\hbar\frac{\partial}{\partial t}\Psi(\mathbf{r}, t) \qquad (6\text{-}1)$$

Time-dependent
Schrödinger equation

## Probability Density and Normalization

In one dimension, the complex square of the wave function is the probability per unit *length*. It is naturally the probability per unit *volume* in three dimensions. Both are called "probability density."

$$\text{probability density} = \frac{\text{probability}}{\text{volume}} = |\Psi(\mathbf{r}, t)|^2$$

If a wave function is to describe a single particle, the total probability of finding the particle somewhere in three-dimensional space must be unity.

*Normalization*

$$\int_{\text{all space}} |\Psi(\mathbf{r}, t)|^2 \, dV = 1 \qquad (6\text{-}2)$$

From this equation it is clear that the dimensions of a wave function in three dimensions are $(\text{volume})^{-1/2}$ or $(\text{length})^{-3/2}$.

## Stationary States

We will usually be concerned with states in which the energy is well defined and unchanging. As in one dimension, to single out such states, we separate spatial dependence from temporal by expressing the wave function as a product of spatial and temporal parts.

$$\Psi(\mathbf{r}, t) = \psi(\mathbf{r})\phi(t)$$

Following the standard procedure of separation of variables, we insert this in the Schrödinger equation (6-1), divide by it, then cancel terms. (*Note:* $\nabla^2$ is a differential operator involving *spatial* coordinates, not time.)

$$-\frac{\hbar^2}{2m} \frac{1}{\psi(\mathbf{r})} \nabla^2 \psi(\mathbf{r}) + U(\mathbf{r}) = i\hbar \frac{1}{\phi(t)} \frac{\partial}{\partial t} \phi(t)$$

The right-hand side is the same as in the one-dimensional case. Setting it equal to a constant $E$ (the "separation constant"), we have

*Temporal part of $\Psi(\mathbf{r}, t)$*

$$\phi(t) = e^{-i(E/\hbar)t}$$

Again $E$ is the well-defined energy and, since $\phi(t)*\phi(t)$ is unity, the probability density is a constant in time. Thus far, all is as before. The remaining spatial equation is the three-dimensional, time-independent Schrödinger equation, the generalization of (4-8).

*Time-independent Schrödinger equation*

$$-\frac{\hbar^2}{2m} \nabla^2 \psi(\mathbf{r}) + U(\mathbf{r})\psi(\mathbf{r}) = E\psi(\mathbf{r}) \qquad (6\text{-}3)$$

This is as far as we can go until the potential energy $U(\mathbf{r})$ is known.

## Bound States

As in Chapter 4, we concentrate here on bound states—potential energies capable of confining a particle to some region of space. As in one dimension, the physical conditions we impose on the wave function lead to quantized standing waves. However, with conditions in *three* dimensions, *three* **quantum numbers** arise, related to quantization of *three* things. Invariably energy is quantized, but the additional quantized properties depend on the particular $U(\mathbf{r})$.

## ▼ **6.2** The 3D Infinite Well

The simplest example of bound states in three dimensions is the 3D infinite well. It is important as a model of the so-called "quantum dot" (see Progress and Applications in Chapter 4). Our main purpose in studying it here is that other cases are more complicated mathematically and tend to obscure the crucial insight we gain by considering this simple case.

As always, the starting point in solving the Schrödinger equation is the potential energy. In the 3D infinite well (Figure 6.1), a particle is bound in a three-dimensional rectangular region by flat, infinitely rigid walls at $x = 0$ and $L_x$, $y = 0$ and $L_y$, and $z = 0$ and $L_z$. The potential energy is therefore

$$U(\mathbf{r}) = \begin{cases} 0 & 0 < x < L_x,\ 0 < y < L_y,\ 0 < z < L_z \\ \infty & \text{otherwise} \end{cases}$$

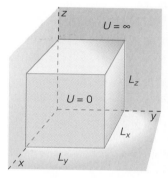

**Figure 6.1** A 3D infinite well.

### 3D Infinite Well

The symmetry of the potential energy makes Cartesian coordinates the logical choice, so that equation (6-3) is

$$-\frac{\hbar^2}{2m}\left(\frac{\partial^2}{\partial x^2} + \frac{\partial^2}{\partial y^2} + \frac{\partial^2}{\partial z^2}\right)\psi(x,y,z) + U(x,y,z)\psi(x,y,z) = E\psi(x,y,z)$$

or

$$-\frac{\hbar^2}{2m}\left(\frac{\partial^2\psi(x,y,z)}{\partial x^2} + \frac{\partial^2\psi(x,y,z)}{\partial y^2} + \frac{\partial^2\psi(x,y,z)}{\partial z^2}\right) + U(x,y,z)\psi(x,y,z)$$
$$= E\psi(x,y,z)$$

Now we are faced with handling not one but three spatial variables. To cope with them independently, we use the same technique used to separate time and space—separation of variables. We express the spatial function as a product of three functions, each of a different independent variable.

$$\psi(x,y,z) = F(x)G(y)H(z)$$

As usual, we insert this, distribute the terms among the partial derivatives, then divide by it. Suppressing the arguments of the functions, we arrive at

$$\frac{-\frac{\hbar^2}{2m}\left(GH\frac{\partial^2 F}{\partial x^2} + FH\frac{\partial^2 G}{\partial y^2} + FG\frac{\partial^2 H}{\partial z^2}\right) + U(x, y, z)(FGH)}{FGH} = \frac{E(FGH)}{FGH}$$

Now canceling $F$, $G$, and $H$ where possible,

$$-\frac{\hbar^2}{2m}\left(\frac{1}{F}\frac{\partial^2 F}{\partial x^2} + \frac{1}{G}\frac{\partial^2 G}{\partial y^2} + \frac{1}{H}\frac{\partial^2 H}{\partial z^2}\right) + U(x, y, z) = E$$

The potential energy is infinite at the walls, so the wave function must be zero outside the "box." We seek solutions inside, where $U$ is zero. Rearranging,

$$\frac{1}{F(x)}\frac{\partial^2 F(x)}{\partial x^2} + \frac{1}{G(y)}\frac{\partial^2 G(y)}{\partial y^2} + \frac{1}{H(z)}\frac{\partial^2 H(z)}{\partial z^2} = -\frac{2mE}{\hbar^2}$$

With this step, the spatial variables are now in separate terms. By the usual argument,[1] *each* term must be a constant. Thus,

$$\frac{d^2 F(x)}{dx^2} = C_x F(x), \qquad \frac{d^2 G(y)}{dy^2} = C_y G(y), \qquad \frac{d^2 H(z)}{dz^2} = C_z H(z), \quad \text{(6-4)}$$

and

$$C_x + C_y + C_z = -\frac{2mE}{\hbar^2} \tag{6-5}$$

We now show that, just as in one dimension, the conditions we impose on the wave function lead to energy quantization—quantum numbers creep into the picture. Consider the "x-equation" of (6-4). By process of elimination, the constant $C_x$ must be negative: Were $C_x$ zero, $F(x)$ would be a linear function of $x$. We rule this out because the wave function must be zero at both walls. Were $C_x$ positive, the general mathematical solution would be $F(x) = Ae^{+\sqrt{C_x}x} + Be^{-\sqrt{C_x}x}$. It is left as an exercise to show that this function also cannot be zero at two points (i.e., the walls). Thus, $C_x$ must be negative. It will streamline things to redefine it:

$$C_x \rightarrow -k_x^2$$

The mathematical solution to the "x-equation" is thus

$$\frac{d^2 F(x)}{dx^2} = -k_x^2 F(x) \quad \Rightarrow \quad F(x) = A \sin k_x x + B \cos k_x x$$

The continuity conditions are the same as in the one-dimensional infinite well, and identical conclusions follow:

$$F(0) = 0 \quad \rightarrow \quad A \sin k_x 0 + B \cos k_x 0 = 0 \quad \Rightarrow \qquad \boxed{B = 0}$$

[1]The "usual argument" in separation of variables is that if the entire equation holds for $x_0$, $y_0$, $z_0$, it cannot hold for $x'$, $y_0$, $z_0$ if the term involving $x$ indeed *varies* with $x$. See Section 4.2.

$$F(L_x) = 0 \quad \rightarrow \quad A \sin k_x L_x = 0 \quad \Rightarrow \quad \boxed{k_x L_x = n_x \pi}$$

We see that $k_x L_x$ must be an integral multiple of $\pi$. Thus,

$$F(x) = A \sin \frac{n_x \pi x}{L_x} \quad \text{and} \quad C_x = -k_x^2 = -\frac{n_x^2 \pi^2}{L_x^2}$$

The integer $n_x$ carries a subscript for a very important reason: There are independent integers, quantum numbers, for the other two dimensions. By identical arguments,

$$G(y) = A \sin \frac{n_y \pi y}{L_y} \quad \text{and} \quad C_y = -k_y^2 = -\frac{n_y^2 \pi^2}{L_y^2}$$

$$H(z) = A \sin \frac{n_z \pi z}{L_z} \quad \text{and} \quad C_z = -k_z^2 = -\frac{n_z^2 \pi^2}{L_z^2}$$

Altogether, our three-dimensional standing-wave solutions $F(x)G(y)H(z)$ are given by

$$\psi_{n_x, n_y, n_z}(x, y, z) = A \sin \frac{n_x \pi x}{L_x} \sin \frac{n_y \pi y}{L_y} \sin \frac{n_z \pi z}{L_z} \tag{6-6}$$

Each set of the three quantum numbers $(n_x, n_y, n_z)$ yields a unique wave function. (*Note:* The product of the multiplicative constants from each term has been replaced simply by $A$.) Finally, substituting into (6-5) gives the allowed energies.

$$-\frac{n_x^2 \pi^2}{L_x^2} - \frac{n_y^2 \pi^2}{L_y^2} - \frac{n_z^2 \pi^2}{L_z^2} = -\frac{2mE}{\hbar^2}$$

or

$$E_{n_x, n_y, n_z} = \left( \frac{n_x^2}{L_x^2} + \frac{n_y^2}{L_y^2} + \frac{n_z^2}{L_z^2} \right) \frac{\pi^2 \hbar^2}{2m} \tag{6-7}$$

One of the most important lessons here is that, whereas in one dimension the boundary conditions lead to energy quantization according to the value of a single quantum number, conditions in each of *three* dimensions lead to *three* quantum numbers. In the infinite well, they are $n_x$, $n_y$, $n_z$, but it is true in general that *for each dimension there is an independent quantum number*.

As we might expect, the smaller the quantum numbers, the smaller is the corresponding energy. Consider the case $L_x = 1$, $L_y = 2$, and $L_z = 3$ (in arbitrary units). None of the three quantum numbers may be zero, since this would make the wave function zero. The lowest energy is the case $(n_x, n_y, n_z) = (1, 1, 1)$.

$$E_{1,1,1} = \left( \frac{1^2}{1^2} + \frac{1^2}{2^2} + \frac{1^2}{3^2} \right) \frac{\pi^2 \hbar^2}{2m} = \frac{49 \pi^2 \hbar^2}{72m}$$

The corresponding wave function is

$$\psi_{1,1,1}(x, y, z) = A \sin \frac{1\pi x}{1} \sin \frac{1\pi y}{2} \sin \frac{1\pi z}{3}$$

With three quantum numbers rather than one, it is not so easy to see which state is the next-higher energy. It is left as a mental exercise to verify that it is the case $(n_x, n_y, n_z) = (1, 1, 2)$.

## Degeneracy

When there is much symmetry, multiple states can have the same energy. Suppose the box in the 3D infinite well is a cube.

$$L_x = L_y = L_z \equiv L$$

The energy levels (6-7) become

$$E_{n_x, n_y, n_z} = (n_x^2 + n_y^2 + n_z^2) \frac{\pi^2 \hbar^2}{2mL^2} \qquad (6\text{-}8)$$

Table 6.1 shows sets of the three quantum numbers for the 14 lowest allowed energies. Of those shown, only the energy levels $3(\pi^2 \hbar^2/2mL^2)$ and $12(\pi^2 \hbar^2/2mL^2)$ correspond to unique sets of quantum numbers, respectively (1, 1, 1) and (2, 2, 2). There are, for instance, four different sets of quantum numbers for the energy $27(\pi^2 \hbar^2/2mL^2)$. But although the energies are the same, *each set of quantum numbers corresponds to a different wave function.* The wave functions for the four sets are

$$\psi_{3,3,3} = A \sin \frac{3\pi x}{L} \sin \frac{3\pi y}{L} \sin \frac{3\pi z}{L} \qquad \psi_{5,1,1} = A \sin \frac{5\pi x}{L} \sin \frac{1\pi y}{L} \sin \frac{1\pi z}{L}$$

$$\psi_{1,5,1} = A \sin \frac{1\pi x}{L} \sin \frac{5\pi y}{L} \sin \frac{1\pi z}{L} \qquad \psi_{1,1,5} = A \sin \frac{1\pi x}{L} \sin \frac{1\pi y}{L} \sin \frac{5\pi z}{L}$$

If nothing else, these have their maximum values at different coordinates $(x, y, z)$. The coincidence—different wave functions having the same energy—is called **degeneracy,** and energy levels for which it is true are said to be **degenerate.** The energy $27(\pi^2 \hbar^2/2mL^2)$ is said to be 4-fold degenerate. Levels $3(\pi^2 \hbar^2/2mL^2)$ and $12(\pi^2 \hbar^2/2mL^2)$ are said to be **nondegenerate;** each corresponds to but a single wave function.

---

**Example 6.1**

An electron is in the $(n_x, n_y, n_z = 1, 2, 1)$ state of a cubic 3D infinite well of side length 1 nm. (a) Find the electron's energy and wave function. (b) Where is the probability of finding the electron the largest? (c) If there are other states of the same energy, where would an electron most likely be found in each?

Solution

(a) We may use equation (6-8) for the energy and (6-6) for the wave function.

**Table 6.1**

| $n_x, n_y, n_z$ | $E_{n_x, n_y, n_z}$* |
|---|---|
| 1, 1, 1 | 3 |
| 2, 1, 1 | 6 |
| 1, 2, 1 | 6 |
| 1, 1, 2 | 6 |
| 1, 2, 2 | 9 |
| 2, 1, 2 | 9 |
| 2, 2, 1 | 9 |
| 3, 1, 1 | 11 |
| 1, 3, 1 | 11 |
| 1, 1, 3 | 11 |
| 2, 2, 2 | 12 |
| 1, 2, 3 | 14 |
| 2, 1, 3 | 14 |
| 1, 3, 2 | 14 |
| 2, 3, 1 | 14 |
| 3, 1, 2 | 14 |
| 3, 2, 1 | 14 |
| 3, 2, 2 | 17 |
| 2, 3, 2 | 17 |
| 2, 2, 3 | 17 |
| 4, 1, 1 | 18 |
| 1, 4, 1 | 18 |
| 1, 1, 4 | 18 |
| 1, 3, 3 | 19 |
| 3, 1, 3 | 19 |
| 3, 3, 1 | 19 |
| 1, 2, 4 | 21 |
| 2, 1, 4 | 21 |
| 1, 4, 2 | 21 |
| 2, 4, 1 | 21 |
| 4, 1, 2 | 21 |
| 4, 2, 1 | 21 |
| 2, 3, 3 | 22 |
| 3, 2, 3 | 22 |
| 3, 3, 2 | 22 |
| 4, 2, 2 | 24 |
| 2, 4, 2 | 24 |
| 2, 2, 4 | 24 |
| 1, 3, 4 | 25 |
| 3, 1, 4 | 25 |
| 1, 4, 3 | 25 |
| 3, 4, 1 | 25 |
| 4, 1, 3 | 25 |
| 4, 3, 1 | 25 |
| 3, 3, 3 | 27 |
| 5, 1, 1 | 27 |
| 1, 5, 1 | 27 |
| 1, 1, 5 | 27 |

$$^* \left( \times \frac{\pi^2 \hbar^2}{2mL^2} \right)$$

$$E_{1,2,1} = (1^2 + 2^2 + 1^2)\frac{\pi^2(1.055 \times 10^{-34} \text{ J·s})^2}{2(9.11 \times 10^{-31} \text{ kg})(10^{-9} \text{ m})^2}$$

$$= 3.62 \times 10^{-19} \text{ J} = 2.26 \text{ eV}$$

$$\psi_{1,2,1}(x, y, z) = A \sin\frac{1\pi x}{10^{-9} \text{ m}} \sin\frac{2\pi y}{10^{-9} \text{ m}} \sin\frac{1\pi z}{10^{-9} \text{ m}}$$

[To be complete, the normalization constant $A$ would have to be determined. It is $(2/L)^{3/2}$, and is the same for all sets $n_x, n_y, n_z$. The proof is left as an exercise.]

(b) To answer this, we find where the probability density has its maximum value(s).

$$|\psi_{1,2,1}(x, y, z)|^2 = A^2 \sin^2\frac{1\pi x}{L} \sin^2\frac{2\pi y}{L} \sin^2\frac{1\pi z}{L}$$

In general, to find the maximum of a function of several variables, we set equal to zero the partial derivative with respect to each—one equation for each variable. But we are familiar enough with trigonometric functions to say that their maxima occur where they are unity: $x = \frac{1}{2}L$, $y = \frac{1}{4}L$ or $y = \frac{3}{4}L$, and $z = \frac{1}{2}L$. Thus, there are two points where the electron is most likely to be found.

Most probable $(x, y, z)$ in $\psi_{1,2,1}$ state:    $(\frac{1}{2}L, \frac{1}{4}L, \frac{1}{2}L)$ and $(\frac{1}{2}L, \frac{3}{4}L, \frac{1}{2}L)$

(c) By symmetry, $E_{2,1,1}$ and $E_{1,1,2}$ are of the same energy as $E_{1,2,1}$, but they correspond to different wave functions, with maxima at different locations.

Most probable $(x, y, z)$ in $\psi_{2,1,1}$ state:    $(\frac{1}{4}L, \frac{1}{2}L, \frac{1}{2}L)$ and $(\frac{3}{4}L, \frac{1}{2}L, \frac{1}{2}L)$

Most probable $(x, y, z)$ in $\psi_{1,1,2}$ state:    $(\frac{1}{2}L, \frac{1}{2}L, \frac{1}{4}L)$ and $(\frac{1}{2}L, \frac{1}{2}L, \frac{3}{4}L)$

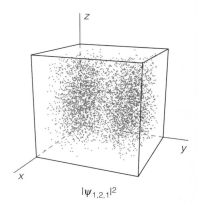

$|\psi_{1,2,1}|^2$

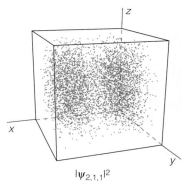

$|\psi_{2,1,1}|^2$

Figure 6.2 shows the probability densities (proportional to density of shading) for the three states in the example. Note the two regions of maximum probability in each case. The states have the same energy, and the shapes of their probability densities are similar, but they are different states.

## Splitting of Energy Levels

*Degeneracy results from symmetry.* In the 3D infinite well, there will be less coincidental equality of different wave functions' energies if the sides of the box are of unequal lengths. For example, if $L_x$ and $L_y$ were $L$, but $L_z$ were $0.9L$, much of the *cubic* well's symmetry would be destroyed. The result would be **splitting** of the formerly degenerate energy levels. From equation (6-7), we see that all energy levels would be raised, but energy $E_{1,1,2}$ would no longer equal $E_{2,1,1}$ and $E_{1,2,1}$. It would be disproportionately higher.

With equal-length sides, equation (6-8) shows that

$$E_{2,1,1} \qquad = \qquad E_{1,2,1} \qquad = \qquad E_{1,1,2}$$

$$(2^2 + 1^2 + 1^2)\frac{1}{L^2} = (1^2 + 2^2 + 1^2)\frac{1}{L^2} = (1^2 + 1^2 + 2^2)\frac{1}{L^2}$$

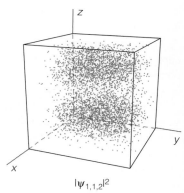

$|\psi_{1,1,2}|^2$

**Figure 6.2** Degenerate states— different, but of the same energy— in a 3D infinite well.

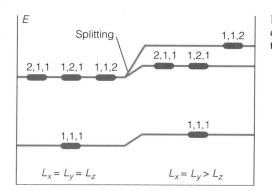

**Figure 6.3** Splitting of formerly degenerate states, caused by the introduction of an asymmetry.

But with unequal lengths, equation (6-7) gives us

$$\frac{2^2}{L^2} + \frac{1}{L^2} + \frac{1}{(0.9L)^2} = \frac{1}{L^2} + \frac{2^2}{L^2} + \frac{1}{(0.9L)^2} < \frac{1}{L^2} + \frac{1}{L^2} + \frac{2^2}{(0.9L)^2}$$

$$E_{2,1,1} \qquad = \qquad E_{1,2,1} \qquad < \qquad E_{1,1,2}$$

Figure 6.3 illustrates this energy-level splitting.

## 6.3  Toward the Hydrogen Atom

In the simplest analysis, the hydrogen atom is nothing more than an electron and a proton moving under the influence of their mutual electrostatic attraction, with insufficient kinetic energy to escape one another. Being about 2000 times as massive, however, the proton moves very little. As an approximation, we assume that it is stationary.[2] Thus, we study merely the behavior of an electron whose electrostatic potential energy is given by

$$U(r) = -\frac{1}{4\pi\epsilon_0}\frac{e^2}{r} \qquad (6\text{-}9)$$

where $e$ is the fundamental charge and $r$ the distance from the proton, which is at the origin.

Figure 6.4 plots this potential energy in two of the three spatial dimensions. It approaches negative infinity at zero distance and zero at infinite distance. To be bound, then, the electron must have a total energy $E$ less then zero—the maximum height of the potential energy "walls."

As already noted, imposing the physical requirements on $\psi(\mathbf{r})$ in three dimensions will give us three quantized properties. We shall soon find out what they are in the case of hydrogen. But the simple fact that the potential energy is spherically symmetric allows standing waves that differ in one of these properties, related only to angular orientation in space, to have the same energy. Different wave functions having the same energy is what we know as **degeneracy.**

Since the $r$ in equation (6-9) is $\sqrt{x^2 + y^2 + z^2}$, solving the Schrödinger equation in Cartesian coordinates would be extremely difficult. Spherical polar coor-

[2]This is, of course, an approximation. As it turns out, the noninfinite mass of the nucleus can be taken into account fairly easily. The net effect is that the electron behaves as a particle of "reduced mass" $m_e\, m_{nucleus}/(m_{nucleus} + m_e)$ orbiting the center of mass, which is very close to the location of the nucleus. For a hydrogen atom, where $m_{nucleus} = m_{proton}$, the electron's reduced mass is about $0.9995 m_e$. For the derivation of the reduced mass, see Exercise 50.

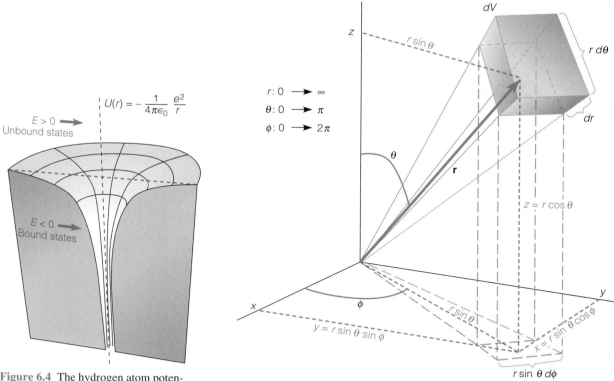

$$U(r) = -\frac{1}{4\pi\epsilon_0}\frac{e^2}{r}$$

$E > 0$
Unbound states

$E < 0$
Bound states

$r: 0 \longrightarrow \infty$

$\theta: 0 \longrightarrow \pi$

$\phi: 0 \longrightarrow 2\pi$

$dV$

$r\sin\theta$

$r\,d\theta$

$dr$

$z = r\cos\theta$

$z$

$\theta$

$\mathbf{r}$

$y$

$x$

$\phi$

$r\sin\theta$

$y = r\sin\theta\sin\phi$

$x = r\sin\theta\cos\phi$

$r\sin\theta\,d\phi$

**Figure 6.4** The hydrogen atom potential energy.

**Figure 6.5** Spherical polar coordinates.

dinates are simplest, because $r$ is one of the independent variables. Let us briefly review this coordinate system.

In spherical polar coordinates, all of three-dimensional space is represented by values of the three coordinates $r$, $\theta$, and $\phi$, shown in Figure 6.5. Associated with any position in space is a position vector $\mathbf{r}$, which begins at the origin and ends at that position. The coordinate $r$ is the length of the position vector (the *scalar* distance from the origin, not to be confused with the vector $\mathbf{r}$) and is called the radius. The coordinate $\theta$ is the angle between the position vector and the $z$-axis. It is called the polar angle (the "pole" being the $z$-axis). The coordinate $\phi$ is the angle, about the $z$-axis, between the $x$-axis and the projection of the position vector in the $x$-$y$ plane. It is called the azimuthal angle. Let us verify that these three cover all space: For a given radius $r$ and polar angle $\theta$, varying the azimuthal angle $\phi$ through $2\pi$ would inscribe a latitude at angle $\theta$ on the surface of a sphere of radius $r$. Allowing the polar angle $\theta$ to vary from zero to $\pi$ (not $2\pi$) would cover all latitudes from the "north pole" to the "south pole" of that sphere. Finally, allowing $r$ to vary from zero to infinity would include all such concentric spheres—covering all space. The equations relating $(x, y, z)$ and $(r, \theta, \phi)$ may be deduced from the diagram, important dimensions marked in blue, and are summarized in Table 6.2.

To calculate probabilities, as in equation (6-2), we need the expression for the

**Table 6.2**

| | |
|---|---|
| $x = r \sin \theta \cos \phi$ | $r = \sqrt{x^2 + y^2 + z^2}$ |
| $y = r \sin \theta \sin \phi$ | $\theta = \cos^{-1} \dfrac{z}{\sqrt{x^2 + y^2 + z^2}}$ |
| $z = r \cos \theta$ | $\phi = \tan^{-1} \dfrac{y}{x}$ |

infinitesimal volume element $dV$ in spherical polar coordinates. This element is the shaded region in Figure 6.5, with its dimensions indicated by gray lines. It has one side of length $r\, d\theta$ (an arc length due to a change in $\theta$), another of length $r \sin \theta\, d\phi$ (an arc length at distance $r \sin \theta$ in the x-y plane due to a change in $\phi$) and a thickness $dr$. The volume is the product:

$$dV = r^2 \sin \theta\, dr\, d\theta\, d\phi$$

## Schrödinger Equation in Spherical Polar Coordinates

The time-independent Schrödinger equation (6-3) is

$$-\frac{\hbar^2}{2m} \nabla^2 \psi(\mathbf{r}) + U(\mathbf{r})\psi(\mathbf{r}) = E\psi(\mathbf{r})$$

To express this in spherical polar coordinates, we replace the generic symbol $\mathbf{r}$ with $(r, \theta, \phi)$, and also use the appropriate expression for the operator $\nabla^2$. We know its form in Cartesian coordinates, so we could find $\nabla^2$ in spherical polar coordinates using the transformations in Table 6.2. Most standard texts on analytic geometry provide the details of this messy task. We merely use the result, expressed in spherical polar coordinates:

$$\nabla^2 = \frac{1}{r^2}\left[\frac{\partial}{\partial r}\left(r^2 \frac{\partial}{\partial r}\right) + \csc \theta \frac{\partial}{\partial \theta}\left(\sin \theta \frac{\partial}{\partial \theta}\right) + \csc^2 \theta \frac{\partial^2}{\partial \phi^2}\right]$$

While it is more complicated, it should be borne in mind that this expression is equivalent to the simpler kinetic energy operator in Cartesian coordinates. We have

$$-\frac{\hbar^2}{2m}\frac{1}{r^2}\left[\frac{\partial}{\partial r}\left(r^2 \frac{\partial}{\partial r}\right) + \csc \theta \frac{\partial}{\partial \theta}\left(\sin \theta \frac{\partial}{\partial \theta}\right) + \csc^2 \theta \frac{\partial^2}{\partial \phi^2}\right]\psi(r, \theta, \phi)$$

$$+ U(r, \theta, \phi)\psi(r, \theta, \phi) = E\psi(r, \theta, \phi) \qquad \text{(6-10)}$$

We attack this equation via separation of variables.

## 6.4 Central Forces

In the hydrogen atom, the electron's potential energy depends only on $r$, the scalar distance from the origin, not on $\theta$ or $\phi$. Thus, $U(\mathbf{r}) = U(r)$. Although this might seem to result in only a minor simplification of equation (6-10), it is very important, for without it the variables could not be separated. In this section, we will use simply $U(r)$ for the potential energy, rather than its specific form (6-9) for the hydrogen atom. Accordingly, *the conclusions we draw are valid in all cases in which U depends only on r.* Because the force associated with such a potential energy is always directed radially, it is called a **central force**.

Substituting $U(r)$ for $U(r, \theta, \phi)$ in (6-10) and rearranging a bit,

$$\left[\frac{\partial}{\partial r}\left(r^2 \frac{\partial}{\partial r}\right) + \csc\theta \frac{\partial}{\partial\theta}\left(\sin\theta \frac{\partial}{\partial\theta}\right) + \csc^2\theta \frac{\partial^2}{\partial\phi^2}\right]\psi(r, \theta, \phi)$$

$$= -r^2 \frac{2m(E - U(r))}{\hbar^2}\psi(r, \theta, \phi)$$

We separate variables in the usual way—by writing the wave function as a product of three functions, each of a different independent variable:

$$\psi(r, \theta, \phi) = R(r)\Theta(\theta)\Phi(\phi)$$

Here we adopt a helpful means of bookkeeping: The variable is in lowercase, while the corresponding *function* of that variable is in uppercase. Now inserting, distributing the functions among the partial derivatives that act upon them, then dividing as usual by $\psi$ (and suppressing functions' arguments), we obtain

$$\frac{\Theta\Phi\dfrac{\partial}{\partial r}\left(r^2 \dfrac{\partial R}{\partial r}\right) + R\Phi\csc\theta \dfrac{\partial}{\partial\theta}\left(\sin\theta \dfrac{\partial\Theta}{\partial\theta}\right) + R\Theta\csc^2\theta \dfrac{\partial^2\Phi}{\partial\phi^2}}{R\Theta\Phi}$$

$$= \frac{-r^2\dfrac{2m(E - U(r))}{\hbar^2}R\Theta\Phi}{R\Theta\Phi}$$

After canceling, we find that one of the variables is separate.

$$\frac{1}{R}\frac{\partial}{\partial r}\left(r^2 \frac{\partial R}{\partial r}\right) + \frac{1}{\Theta}\csc\theta \frac{\partial}{\partial\theta}\left(\sin\theta \frac{\partial\Theta}{\partial\theta}\right) + \csc^2\theta \underbrace{\left[\frac{1}{\Phi}\frac{\partial^2\Phi}{\partial\phi^2}\right]}_{\phi\text{—separate}}$$

$$= -r^2 \frac{2m(E - U(r))}{\hbar^2}$$

(6-11)

All dependence on $\phi$ resides in the term in brackets; it must be constant! Let us concentrate on this piece of the puzzle, by solving this "$\phi$-equation."

## The $\phi$-equation

Let us use the symbol $C_\phi$ for the constant associated with the $\phi$-equation.

$$\frac{d^2\Phi(\phi)}{d\phi^2} = C_\phi\Phi(\phi)$$

(6-12a)

Note that $C_\phi$ cannot be positive. If it were, $\Phi(\phi)$ would be $Ae^{+\sqrt{C_\phi}\phi} + Be^{-\sqrt{C_\phi}\phi}$. This is unacceptable simply because it is not periodic (approaching infinity as $\phi \to \pm\infty$). Whatever may be the wave function's value at azimuthal angle $\phi_0$, it must have the same value at $\phi_0 + 2\pi$, *because it is the same location.* Were the "$\phi$-part" of the wave function not periodic, the wave function would be multivalued, which makes no sense. Since it cannot be positive, we may redefine $C_\phi$ as follows:

$$C_\phi = -m_\ell^2$$

(As we shall see, the "weird" choice of symbol follows a convention.) Equation (6-12a) becomes

$$\frac{d^2\Phi(\phi)}{d\phi^2} = -m_\ell^2\Phi(\phi) \qquad\qquad \text{(6-12b)}$$

with general solution

$$\Phi(\phi) = Ae^{+im_\ell\phi} + Be^{-im_\ell\phi} \qquad\qquad \text{(6-13)}$$

As we know, complex exponentials are periodic, but this $\Phi(\phi)$ will have the *correct* period, repeating when $\phi$ increases by $2\pi$, only if $m_\ell$ is an integer— quantization! Imposing the boundary condition of periodicity on the $\phi$-part of the Schrödinger equation has given us a quantum number, $m_\ell$. One step remains: In an isolated atom, the electron is in a spherically symmetric environment, so its probability density should be spherically symmetric. But $\Phi(\phi)*\Phi(\phi)$ varies with $\phi$ unless *either A or B is zero* (see Exercise 11).[3] We may account for the two cases by simply letting $m_\ell$ take on positive and negative values. Thus,

$$\Phi_{m_\ell}(\phi) = e^{im_\ell\phi} \quad (m_\ell = 0, \pm1, \pm2, \pm3, \dots) \qquad \text{(6-14)}$$

This function has a subscript as a reminder that there are solutions for only certain values of a quantum number, namely $m_\ell$.[4]

The physical property that is quantized according to the value of $m_\ell$ is the $z$-component of angular momentum, $L_z$. We see this by rewriting (6-12b). The operator for $z$-component of angular momentum is given by $\hat{L}_z = -i\hbar(\partial/\partial\phi)$. (See the essay entitled "Combining Basic Operators.") Multiplying both sides of (6-12b) by $(-i\hbar)^2$, we obtain

$$\hat{L}_z^2\Phi_{m_\ell}(\phi) = (m_\ell\hbar)^2\Phi_{m_\ell}(\phi)$$

---

[3] As we know, $|e^{ia}|^2 = e^{+ia}e^{-ia} = 1$, so each of $e^{-im_\ell}$ and $e^{+im_\ell}$ *alone* gives a spherically symmetric probability. The function $A'\sin(m_\ell\phi) + B'\cos(m_\ell\phi)$ is equivalent to (6-13), but we discard it because it too is not spherically symmetric. It should be noted, however, that if the atom were not isolated— were it affected by other atoms in a molecule, for instance—we should not expect spherical symmetry. Sines and/or cosines are often valid choices in such cases. See Section 9.2.

[4] Note that $m_\ell = 0$ is allowed, and that the corresponding $\Phi(\phi)$ is constant. This may seem odd, since the infinite well obeys the same kind of second-order differential equation, yet $n = 0$ is not allowed. The difference is in the boundary conditions. In the infinite well, the wave function must be zero outside the well. But a circle *has* no outside, so it violates no boundary conditions for $\Phi(\phi)$ to be constant. Moreover, that $m_\ell$ may be zero does not imply that kinetic energy may be zero, because the quantized property here is not energy.

## Combining Basic Operators

Claiming that an operator is a combination of the basic operators is one thing—proving it is quite another. The $L_z$ operator, however, is a nontrivial yet not too difficult case, and so is worthwhile considering. Using the $(x, y, z) \leftrightarrow (r, \theta, \phi)$ transformations of Table 6.2 and rules for partial derivatives,

$$-i\hbar \frac{\partial}{\partial \phi} = -i\hbar \left( \frac{\partial x}{\partial \phi} \frac{\partial}{\partial x} + \frac{\partial y}{\partial \phi} \frac{\partial}{\partial y} + \frac{\partial z}{\partial \phi} \frac{\partial}{\partial z} \right)$$

$$= -i\hbar \left( -r \sin \theta \sin \phi \frac{\partial}{\partial x} + r \sin \theta \cos \phi \frac{\partial}{\partial y} + 0 \frac{\partial}{\partial z} \right)$$

$$= -i\hbar \left( -y \frac{\partial}{\partial x} + x \frac{\partial}{\partial y} \right)$$

$$= -y \left( -i\hbar \frac{\partial}{\partial x} \right) + x \left( -i\hbar \frac{\partial}{\partial y} \right)$$

$$= \hat{x} \, \hat{p}_y - \hat{y} \, \hat{p}_x$$

Classically, $\mathbf{L} = \mathbf{r} \times \mathbf{p}$ and $L_z = xp_y - yp_x$. Our $L_z$ operator is thus the proper combination of basic operators. The operator we began with is thus the proper combination of fundamental operators. See Appendix G for similar arguments involving the operators for $L_x$, $L_y$, and, particularly, $L^2$, which is used later in this section.

Now we have an equation involving the $L_z$ operator and a separation constant that is quantized and has proper units of angular momentum.[5] Thus, $\Phi_{m_\ell}(\phi)$ represents states for which $L_z$ may take on only the values

$$L_z = m_\ell \hbar \quad (m_\ell = 0, \pm 1, \pm 2, \pm 3, \dots) \tag{6-15}$$

The diagrams in Figure 6.6 provide a crude means of visualizing the $\phi$-part of the wave function $e^{im_\ell \phi}$, and an additional argument that $L_z$ is the quantized property. The dashed circle represents a "$\phi$-axis." For $m_\ell = 0$, $e^{i0\phi}$ is unity. It is thus the same distance "above" the $\phi$-axis around the entire circle. For $m_\ell = 1$, $e^{i1\phi}$ is $\cos \phi + i \sin \phi$. The real part passes through one cycle, varying from positive to negative and back, as $\phi$ goes from 0 to $2\pi$. For $m_\ell = 2$, $e^{i2\phi} = \cos 2\phi + i \sin 2\phi$, and passes through two cycles in an interval of $2\pi$. Similarly, the $m_\ell = 3$ wave passes through three cycles. In general, the circumference of the circle would be an integral number of "wavelengths": $2\pi r = m_\ell \lambda$. Assuming that a "wavelength around the $\phi$-axis" is inversely related to a "tangential momentum" $mv_t$ in the usual way ($\lambda = h/p$), we might also claim that $\lambda = h/mv_t$. Putting these two observations together, we have $2\pi r = m_\ell h/mv_t$, which if rearranged becomes the quantization condition $m_\ell \hbar = mv_t r = L_z$. Although somewhat vague, these arguments certainly reinforce the claim that the quantization associated with the $\phi$-part of the Schrödinger equation is that of $L_z$.

On the other hand, Figure 6.6 has its limitations. There really is no "$\phi$-axis." That the *real part* of $\Phi(\phi)$ in the $m_\ell = 1$ diagram is positive on the right side and

[5]The ideas of Section 4.10 provide the more precise argument: The function $\Phi_{m_\ell}(\phi) = e^{im_\ell \phi}$ is an eigenfunction of the operator $\hat{L}_z = -i\hbar(\partial/\partial\phi)$, and the eigenvalue is $m_\ell \hbar$, which is thus the well-defined value of $L_z$.

Re $\Phi_{m_\ell}(\phi)$

**Figure 6.6**  Standing waves on "the $\phi$-axis."

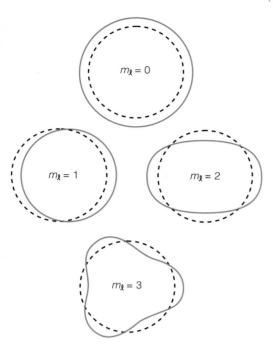

$m_\ell = 0$

$m_\ell = 1$

$m_\ell = 2$

$m_\ell = 3$

**Figure 6.6**  Standing waves on "the $\phi$-axis."

negative on the left does not mean that the wave function sticks out on the right. In fact, $\Phi(\phi)$ is complex; as $\phi$ increases, $e^{im_\ell\phi}$ undergoes a continuous "rotation" in the real-complex plane ($+1 \rightarrow +i \rightarrow -1 \rightarrow -i$, etc.). It is a unit-magnitude complex number that multiplies $R(r)\Theta(\theta)$ at all points in space. This complication disappears, however, when the probability density is considered, because $\Phi(\phi)^*\Phi(\phi)$ is unity.

## Separating $\theta$

Let us proceed with our solution. Reinserting (6-12b) into (6-11), we separate dependence on $\theta$ from dependence on $r$:

$$\frac{1}{R}\frac{\partial}{\partial r}\left(r^2\frac{\partial R}{\partial r}\right) + \underbrace{\left[\frac{1}{\Theta}\csc\theta\frac{\partial}{\partial\theta}\left(\sin\theta\frac{\partial\Theta}{\partial\theta}\right) + \csc^2\theta\,(-m_\ell^2)\right]}_{\theta\text{—separate}} \qquad (6\text{-}16)$$

$$= -r^2\frac{2m(E - U(r))}{\hbar^2}$$

Since all of the $\theta$ dependence is within brackets, the expression must be a constant. Let us use the symbol $C_\theta$ for this separation constant.

## The $\theta$-equation

We have

$$\frac{1}{\Theta} \csc \theta \frac{d}{d\theta}\left(\sin \theta \frac{d\Theta}{d\theta}\right) - m_\ell^2 \csc^2 \theta = C_\theta$$

or

$$\csc \theta \frac{d}{d\theta}\left(\sin \theta \frac{d\Theta(\theta)}{d\theta}\right) - m_\ell^2 \csc^2 \theta \, \Theta(\theta) = C_\theta \Theta(\theta) \qquad \text{(6-17a)}$$

This is a much more complicated differential equation than those to which we are accustomed. We omit the details,[6] but the solution is similar to that of the one-dimensional harmonic oscillator, and they share an important characteristic. Although mathematical solutions exist for all values of the constant ($C_\theta$ or $E$), all diverge except those for certain discrete values. For (6-17a), it is found that nondivergent solutions result only when $C_\theta$ and $m_\ell$ obey the following conditions:

$$\begin{array}{llllll} C_\theta & 0 & -2 & -6 & \cdots & -\ell(\ell+1) \quad (\ell = 0,1,2,\ldots) \\ m_\ell & 0 & 0,\pm 1 & 0,\pm 1,\pm 2 & \cdots & \mathbf{0,\pm 1,\pm 2,\ldots,\pm \ell} \end{array} \qquad \text{(6-18)}$$

We have uncovered another quantization, and the quantum number is $\ell$. Besides this, however, we find that *for each allowed value of $\ell$, there are only certain allowed values of $m_\ell$.* (This is why we attached the subscript to $m_\ell$.) We shall soon see the physical reason for this mathematical curiosity. Let us simplify things by dispensing with $C_\theta$, replacing it with its allowed values. Equation (6-17a) becomes

$$\csc \theta \frac{d}{d\theta}\left(\sin \theta \frac{d\Theta(\theta)}{d\theta}\right) - m_\ell^2 \csc^2 \theta \, \Theta(\theta) = -\ell(\ell+1)\Theta(\theta) \qquad \text{(6-17b)}$$

For each allowed set ($\ell, m_\ell$), there is a solution $\Theta_{\ell,m_\ell}(\theta)$. (The subscripts remind us that are *two* quantum numbers required to specify a particular function.) We write these in a compact form as follows:

$$\Theta_{\ell,m_\ell}(\theta) = P_{\ell,|m_\ell|}(\cos \theta) \qquad \begin{pmatrix} \ell = 0,\, 1,\, 2,\, \ldots \\ m_\ell = 0,\, \pm 1,\, \pm 2,\, \ldots,\, \pm \ell \end{pmatrix} \qquad \text{(6-19)}$$

The symbol $P_{\ell,m_\ell}(\cos \theta)$ represents a set of related functions known as the associated Legendre functions. Table 6.3 lists the first ten. The reader is strongly encouraged to verify by substitution (for at least the simpler cases) that the functions given in the table are indeed solutions of (6-17b) with the corresponding values of $\ell$ and $m_\ell$.

---

[6] They may be found in any dedicated quantum-mechanics text.

**Table 6.3  Associated Legendre Functions**

| $\ell, m_\ell$ | $P_{\ell,\,m_\ell}(\cos \theta)$ |
|---|---|
| 0, 0 | 1 |
| 1, 0 | $\cos \theta$ |
| 1, 1 | $\sin \theta$ |
| 2, 0 | $\frac{1}{2}(3\cos^2 \theta - 1)$ |
| 2, 1 | $3 \cos \theta \sin \theta$ |
| 2, 2 | $3 \sin^2 \theta$ |
| 3, 0 | $\frac{1}{2}(5\cos^3 \theta - 3\cos \theta)$ |
| 3, 1 | $\frac{3}{2}(5\cos^2 \theta - 1)\sin \theta$ |
| 3, 2 | $15 \cos \theta \sin^2 \theta$ |
| 3, 3 | $15 \sin^3 \theta$ |

Let us now attach meaning to the quantum number $\ell$, by revealing the physical property that is quantized according to its value. To do this, we must consider both angular parts together. It is shown in Appendix G that the operator for the square of the angular momentum is given by

$$\hat{L}^2 = -\hbar^2 \left[ \csc\theta \frac{\partial}{\partial\theta} \left( \sin\theta \frac{\partial}{\partial\theta} \right) + \csc^2\theta \frac{\partial^2}{\partial\phi^2} \right] \qquad (6\text{-}20)$$

Let us operate with this on the product $\Theta_{\ell,m_\ell}(\theta)\Phi_{m_\ell}(\phi)$.

$$\hat{L}^2(\Theta_{\ell,m_\ell}(\theta)\Phi_{m_\ell}(\phi))$$

$$= -\hbar^2 \left[ \csc\theta \frac{\partial}{\partial\theta} \left( \sin\theta \frac{\partial}{\partial\theta} \right) + \csc^2\theta \frac{\partial^2}{\partial\phi^2} \right] (\Theta_{\ell,m_\ell}(\theta)\Phi_{m_\ell}(\phi))$$

$$= -\hbar^2\Phi_{m_\ell}(\phi) \csc\theta \frac{\partial}{\partial\theta} \left( \sin\theta \frac{\partial\Theta_{\ell,m_\ell}(\theta)}{\partial\theta} \right) - \hbar^2\Theta_{\ell,m_\ell}(\theta) \csc^2\theta \frac{\partial^2\Phi_{m_\ell}(\phi)}{\partial\phi^2}$$

Using the $\phi$-equation (6-12b), the right-hand side becomes

$$-\hbar^2\Phi_{m_\ell}(\phi) \csc\theta \frac{\partial}{\partial\theta} \left( \sin\theta \frac{\partial\Theta_{\ell,m_\ell}(\theta)}{\partial\theta} \right) - \hbar^2\Theta_{\ell,m_\ell}(\theta) \csc^2\theta \left[ (-m_\ell^2)\Phi_{m_\ell}(\phi) \right]$$

Thus,

$$\hat{L}^2(\Theta_{\ell,m_\ell}(\theta)\Phi_{m_\ell}(\phi))$$

$$= -\hbar^2\Phi_{m_\ell}(\phi) \left[ \csc\theta \frac{\partial}{\partial\theta} \left( \sin\theta \frac{\partial\Theta_{\ell,m_\ell}(\theta)}{\partial\theta} \right) - m_\ell^2 \csc^2\theta \, \Theta_{\ell,m_\ell}(\theta) \right]$$

The expression in brackets, however, is the same as the left-hand side of (6-17b). Thus, the angular parts of the Schrödinger equation are equivalent to the following:

$$\hat{L}^2(\Theta_{\ell,m_\ell}(\theta)\Phi_{m_\ell}(\phi)) = \hbar^2\ell(\ell+1)(\Theta_{\ell,m_\ell}(\theta)\Phi_{m_\ell}(\phi)) \qquad (6\text{-}21)$$

We have an equation involving the operator for $L^2$, and a separation constant that is quantized and has units of angular momentum squared. We conclude that the solutions represent states for which the square of the angular momentum is quantized.[7] Since the square of a vector is the square of its magnitude, this means that the magnitude of the angular momentum may take on only the values

$$|L| = \sqrt{L^2} = \sqrt{\ell(\ell+1)}\,\hbar \qquad (\ell = 0, 1, 2, \ldots) \qquad (6\text{-}22)$$

The case $\ell = 0$ is troubling. How can a particle orbit without angular momentum? In general, the motion of the orbiting "particle" is a combination of rotational and radial motion. The $\ell = 0$ case is one in which there is no average rotational motion, but there may well be radial motion. The electron might be crudely pictured as oscillating back and forth through the origin, as would a ball dropped down a tunnel through Earth's center. Nevertheless, we must not throw out fundamental principles of quantum mechanics. If not "watched," the electron is not a particle following a definite path; it is a diffuse standing wave occupying three dimensions.

As shown in (6-18), no physically acceptable $\Theta(\theta)$ exists unless $|m_\ell| \leq \ell$. With this restriction, equation (6-15) becomes

[7] Again, in the more precise language of Section 4.10, the product of the two angular parts of the wave function is an eigenfunction of the $L^2$ operator, with eigenvalue $\hbar^2\ell(\ell+1)$.

$$L_z = m_\ell \hbar \qquad (m_\ell = 0, \pm 1, \pm 2, \ldots, \pm \ell) \qquad (6\text{-}23)$$

Let us now look at the physical reason that $m_\ell$ should depend on $\ell$. Since a component of a vector may never be greater than the magnitude of the vector, the maximum $L_z$ must be no greater than $|L|$. Equations (6-22) and (6-23) agree! That is, $\ell\hbar$, the maximum $L_z$, is strictly less than $\sqrt{\ell(\ell+1)}\,\hbar$. Due to its relationship to orbital angular momentum, the quantum number $\ell$ has been given the name **orbital quantum number,** while (for reasons discussed in Chapter 7) $m_\ell$ is known as the **magnetic quantum number.**

The first property we found to be quantized was the $z$-component of angular momentum; the second was the magnitude of the angular momentum. The reader may have expected the second instead to be quantization of a second component, then perhaps a third. Even accepting that only one is quantized, why is the $z$-direction "special"? The surprising answer to all these questions is that it is impossible to know all components of angular momentum at once! What results naturally from the Schrödinger equation for a central force is quantization of $|L|$ and one component of **L,** which we call the $z$-component by arbitrary convention. But this is all that *should* result, because to know any more than this would constitute simultaneous knowledge of position and momentum that would violate the uncertainty principle. For **L** to point with certainty along a given axis, the wave function (the "motion of the particle") would have to be confined to the plane perpendicular to that axis. Calling this the $x$-$y$ plane, both $z$ and $p_z$ would have to be zero, with zero uncertainty. Impossible! (In effect, this says that the wave function must spread out to some degree in all dimensions open to it.) If, however, only the magnitude and one component are fixed, **L** may be in any of an infinite number of directions, though not all directions (Figure 6.7). Furthermore, it is not only *theoretically* impossible to know all components of angular momentum simultaneously; it is impossible to devise an *experiment* that would do so. Experimental apparatus (discussed in Chapter 7) inherently introduce directionality along some axis. That axis becomes the "special" axis, and quantization is demonstrated only along it. It naturally exhibits the peculiarities of the axis we have chosen to call the $z$-axis. Indeed, without an externally imposed "$z$-axis," it makes little sense to speak of orientations in space.

The form of (6-22) is closely linked to these restrictions. Comparing $L_z = m_\ell \hbar$ and $|L| = \hbar\sqrt{\ell(\ell+1)}$, it may seem as though a mistake has been made. Should it not be $|L| = \ell\hbar$? On the contrary, if $L_z$ were its maximum value of $\ell\hbar$, and the magnitude of **L** were also $\ell\hbar$, **L** would certainly be pointing along $z$—the uncertainty principle would be violated. That the magnitude of **L** is strictly greater than its maximum $z$-component $[\sqrt{\ell(\ell+1)}\,\hbar > \ell\hbar]$ guarantees that **L** may not ever point along that axis. Its direction cannot be known with certainty. Figure 6.8 shows the possibilities for $\ell = 1$ and $\ell = 2$. Arrows represent angular momentum vectors, and the sphere's radii are thus $\hbar\sqrt{\ell(\ell+1)}$.[8] In the $\ell = 2$ case, for instance, all arrows are $\sqrt{6}\,\hbar$. According to (6-23), there would be five allowed

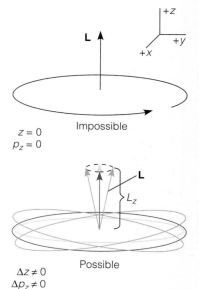

$z = 0$
$p_z = 0$

Impossible

$L_z$

$\Delta z \neq 0$
$\Delta p_z \neq 0$

Possible

**Figure 6.7** Planar motion violates the uncertainty principle.

---

[8]There is no diagram for $\ell = 0$ because the angular momentum vector is of zero length. Though **L** is known precisely, this case does not violate the uncertainty principle. In fact, if **p** were radial, **r** could be *anything* and still give zero angular momentum.

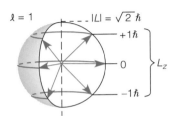

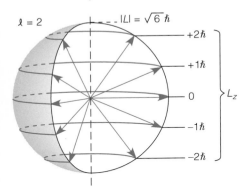

**Figure 6.8** Angular momentum quantization for $\ell = 1$ and $\ell = 2$.

values of $m_\ell$ ($-2$, $-1$, $0$, $+1$, $+2$) and five corresponding values of $L_z$. But in each case, with only the one component fixed, there are an infinite number of directions in which **L** may point.

---

**Example 6.2**

What is the minimum angle the angular momentum vector may make with the $z$-axis (a) in the case $\ell = 3$, and (b) in the case $\ell = 1$?

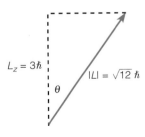

**Solution**

(a) When $\ell = 3$, the angular momentum is $\sqrt{12}\,\hbar$, with seven possible $z$-components. The vector will be most nearly parallel to the $z$-axis when its $z$-component is as large as possible: $3\hbar$. The angle between a vector and its $z$-component is the polar angle in spherical polar coordinates.

$$\cos\theta = \frac{L_z}{|L|} = \frac{3\hbar}{\sqrt{12}\,\hbar} \quad\Rightarrow\quad \theta = \cos^{-1}\frac{3}{\sqrt{12}} = 30°$$

(b) For $\ell = 1$, $|L| = \sqrt{2}\,\hbar$ and the largest $z$-component is $1\hbar$. Thus,

$$\cos\theta = \frac{L_z}{|L|} = \frac{1\hbar}{\sqrt{2}\,\hbar} \quad\Rightarrow\quad \theta = \cos^{-1}\frac{1}{\sqrt{2}} = 45°$$

---

From Example 6.2, we see that the greater the angular momentum, the more nearly parallel it can be to the axis. In the following example, we consider an extreme case.

**Example 6.3**

A 200-kg satellite orbits Earth at a radius of 42,300 km and speed of 3.07 km/s. What might be its angular momentum quantum numbers?

**Solution**

(*Note:* This example may seem absurd; the situation is one in which classical physics is completely adequate. Nevertheless, as has been noted, nowhere in our discussion have we taken into account the functional form of $U(r)$. Thus, what may be said in the electrostatic case $U(r) \propto e^2/r$ may also be said in the gravitational case $U(r) \propto m_1 m_2/r$.)

The angular momentum of a point mass in circular orbit is *mvr*.

$$L = (200 \text{ kg})(3.07 \times 10^3 \text{ m/s})(4.23 \times 10^7 \text{ m}) = 2.60 \times 10^{13} \text{ kg·m}^2/\text{s}$$

Thus,

$$L = \sqrt{\ell(\ell + 1)}\,\hbar \quad \Rightarrow \quad \sqrt{\ell(\ell + 1)} = \frac{2.60 \times 10^{13} \text{ kg·m}^2/\text{s}}{1.055 \times 10^{-34} \text{ J·s}}$$

$$= 2.46 \times 10^{47}$$

Clearly, $\ell$ is a very large number, so we are justified in replacing $\ell + 1$ by $\ell$:

$$\ell = 2.46 \times 10^{47}$$

Of course, without more precise data we cannot find $\ell$ exactly, but the satellite would be in one of the allowed quantum states near this value. Obviously, it is in a very high angular momentum state quantum-mechanically speaking. From our experience with one-dimensional quantum mechanics, we should not be surprised that classical situations correspond to large quantum numbers. Since it is so large, a change in $\ell$ of unity would represent negligible fractional change in the angular momentum. The range of angular momentum values would seem continuous.

The range of allowed components along an axis would also seem continuous.

$$m_\ell = 0, \pm 1, \pm 2, \ldots, \pm 2.46 \times 10^{47}$$

As we know, $L_z$ would be allowed any value from $-\ell\hbar$ to $+\ell\hbar$. Since $\ell \cong \sqrt{\ell(\ell + 1)}$, this range is $-|L|$ to $+|L|$, the classical expectation. For all classical purposes, the angular momentum may make any angle with the $z$-axis. (This could be established by an "experiment," such as bouncing radio waves off the satellite or a visual observation, without affecting the quantum state enough to be classically significant, and $L_x$ and $L_y$ can be known adequately for all classical purposes.)

Let us see what quantum mechanics says about the angular locations where a particle orbiting in a central force might likely be found, by looking at the complex square of the wave function's angular parts. Since the complex square of

$\Phi(\phi) = e^{im_\ell\phi}$ is unity and $\Theta(\theta)$ is real, this is simply $(\Theta_{\ell,m_\ell}(\theta))^2$. Still, it is fair to ask why this should depend on *either* angle. After all, we have said that probability density should be spherically symmetric for an isolated atom. The answer is that unless an experiment is actually conducted to determine $L_z$, *any* value of $L_z$ (hence $m_\ell$) is equally likely. It may be shown that for a given $\ell$, if we add the probabilities $(\Theta_{\ell,m_\ell}(\theta))^2$ for all allowed $m_\ell$-values from $-\ell$ to $+\ell$, the result is indeed independent of $\theta$ (see Exercise 20). In other words, if we are ignorant of the $m_\ell$-value, the probability of finding the particle *is* spherically symmetric. However, in nearly all real applications there is some kind of external influence that would distinguish the different states, so it is important to have some idea of how they differ.

Figure 6.9 shows $(\Theta_{\ell,m_\ell}(\theta))^2$ (from the functions in Table 6.3) plotted versus polar angle for all cases through $\ell = 3$. (Again, these curves represent only the contribution to the overall probability density from its angular parts.[9]) Several features are noteworthy:

1. In the zero angular momentum, $\ell = 0$, case the particle's energy is due solely to radial motion. But while a particle with no angular momentum would oscillate along a line through the origin, we see a probability density that is the same in any direction. What we have is not an orbiting *particle*, but a bound *standing wave*.

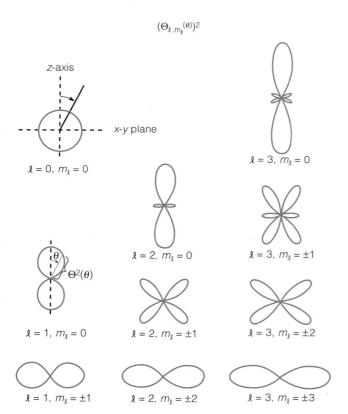

$(\Theta_{\ell,m_\ell}(\theta))^2$

**Figure 6.9** Angular probability densities for a central force.

[9]The curve's distance from the origin is $\Theta^2(\theta)$, not $r$. We can specify a probable region in space only after we find the wave function's *radial* part. Since the probability does not vary with $\phi$, its contribution may be included simply by imagining the plots rotated about the z-axis.

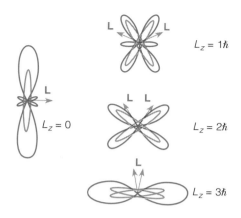

$L_z = 1\hbar$

$L_z = 0$

$L_z = 2\hbar$

$L_z = 3\hbar$

**Figure 6.10** A crude correspondence to orbital motion.

2. In all $\ell \neq 0$ cases, as the magnitude of $m_\ell$ increases, the probability density changes progressively from predominantly along the $z$-axis to predominantly in the $x$-$y$ plane. Correspondingly, as its plane of rotation becomes more nearly parallel to the $x$-$y$ plane, the angular momentum of an orbiting *particle* would also be more nearly parallel to the $z$-axis. (This crude view is depicted in Figure 6.10 for the $\ell = 3$ cases.) Again, however, it must not be thought of as a particle orbiting in a well-defined path. In particular, the separated lobes are typical standing-wave antinodes. They are irreconcilable with a particle view, which would have motion strictly confined to a plane.

3. In all $m_\ell \neq 0$ cases, the probability density is zero at the $z$-axis. If a particle is to have nonzero $L_z$, it must never be found on the $z$-axis.

4. The $m_\ell = \pm\ell$ states look similar, with equatorial lobes suggesting orbit near the $x$-$y$ plane. That they become flatter as $\ell$ increases is expected; in Example 6.2, we found that for large $\ell$ the angular momentum vector may point more nearly parallel to the $z$-axis, in which case the orbital motion should be more closely restricted to the $x$-$y$ plane. In the classical limit $\ell \to \infty$, **L** can be precisely along $z$, and the probability density would be planar.

## 6.5 The Hydrogen Atom

We have dealt with the angular parts of the wave function for central forces in general. It remains to complete our treatment of the hydrogen atom by studying the radial part for the specific hydrogen atom central force. Using (6-17b), equation (6-16) may be rewritten as

$$\frac{1}{R}\frac{d}{dr}\left(r^2\frac{dR}{dr}\right) - \ell(\ell + 1) = -r^2\frac{2m(E - U(r))}{\hbar^2} \tag{6-24}$$

Multiplying both sides by the radial wave function $R(r)$, then rearranging, we obtain an enlightening form.

$$\underbrace{\frac{-\hbar^2}{2m} \frac{1}{r^2} \frac{d}{dr}\left(r^2 \frac{d}{dr}\right) R(r)}_{\text{KE}_{\text{radial}}} + \underbrace{\frac{\hbar^2 \ell(\ell+1)}{2mr^2} R(r)}_{\text{KE}_{\text{rotational}}} + \underbrace{U(r)R(r)}_{\substack{\text{Potential} \\ \text{energy}}} = \underbrace{ER(r)}_{\substack{\text{Total} \\ \text{energy}}}$$

(6-25)

Rotational kinetic energy may be written $L^2/2I$, and the moment of inertia of a point particle is $I = mr^2$. Thus, since $\hbar^2\ell(\ell+1) = L^2$, the second term on the left accounts for rotational kinetic energy. The first term must account for radial kinetic energy, toward and away from the nucleus. In fact, it may be shown that it corresponds to $\frac{1}{2}m\dot{r}^2$, where $\dot{r}$ is the radial velocity $dr/dt$ (see Appendix G). Overall, the equation is very much like the Schrödinger equation in one dimension. Of course, it differs in the rotational energy term, but this is merely a scalar function of $r$. Accordingly, its effect on the radial wave function is much like that of an additional potential energy term.[10] In particular, it would diverge at $r = 0$ if $\ell$ were nonzero, so that all $\ell \neq 0$ wave functions are excluded from the origin [i.e., $R(0) = 0$; see also Figure 6.12].

The final step in solving the Schrödinger equation for the hydrogen atom is inserting the potential energy (6-9) and solving the resulting "$r$-equation."

### The $r$-equation

$$\frac{-\hbar^2}{2m} \frac{1}{r^2} \frac{d}{dr}\left(r^2 \frac{d}{dr}\right) R(r) + \left[\frac{\hbar^2 \ell(\ell+1)}{2mr^2} - \frac{1}{4\pi\epsilon_0} \frac{e^2}{r}\right] R(r) = ER(r)$$

(6-26)

As in the case of the $\Theta(\theta)$ solution, we leave the solution of this intricate differential equation to a higher-level course. The crucial point, as the reader should by now suspect, is that the solutions are all physically unacceptable (divergent) except when the constants take on certain values. The constants $E$ and $\ell$ must obey the following:

$$E = -\frac{me^4}{2(4\pi\epsilon_0)^2 \hbar^2} \frac{1}{n^2} \quad \text{(where } n = 1, 2, 3, \ldots) \tag{6-27}$$

$$\ell = 0, 1, 2, \ldots, n - 1 \tag{6-28}$$

The physical conditions required of the wave function in the third and final dimension have given us a third quantum number, $n$, called the **principal quantum number,** and the property that is quantized according to its value is energy.[11]

---

[10]The sum $\dfrac{\hbar^2 \ell(\ell+1)}{2mr^2} + U(r)$ is often regarded as a single "effective potential energy."

[11]The total energy is negative because the (negative) potential energy is larger than the (positive) kinetic. Only if the electron were unbound—with sufficient kinetic energy to escape its attraction to the proton—would the total energy be positive, a case we don't consider. Energy quantization results if and only if a particle is bound.

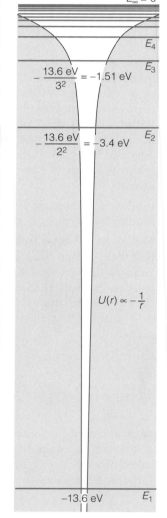

**Figure 6.11** Allowed bound-state energies in the hydrogen atom.

Inserting numerical values in (6-27),

$$E_n = \frac{-13.6 \text{ eV}}{n^2} \qquad (n = 1, 2, 3, \ldots) \qquad \text{(6-29)}$$

Hydrogen atom energy levels

The relationship between the allowed energies is illustrated in the energy-level diagram of Figure 6.11. Energy increases (becomes less negative) with $n$, asymptotically approaching zero. In its ground-state orbit, the electron in a hydrogen atom is 13.6 eV "deep." Since 13.6 eV would be required to remove it from the proton, we say that the electron's **binding energy** is 13.6 eV, or, equivalently, that the **ionization energy** of the hydrogen atom is 13.6 eV. Notably, this is in perfect agreement with the experimental value—quantum mechanics explains the atom! The agreement between theory and experimental evidence is discussed further in Section 6.8.

In passing from the $\phi$-equation to the $\theta$-equation we found that the latter placed restrictions on the quantum number from the former—$m_\ell$ could be no larger than $\ell$. Something similar occurs in passing from the $\theta$-equation to the $r$-equation. Whereas equation (6-18) said that $\ell$ could be any nonnegative integer, according to equation (6-28), it can be no larger than $n - 1$. Again there is a physical reason: The quantum number $n$ specifies the *total* energy, and $\ell$ is related to the rotational energy. Since for a given total energy there is a limit on the rotational energy, $\ell$ should be restricted according to the value of $n$. Taking the restriction into account, we replace equation (6-22):

$$|L| = \sqrt{\ell(\ell + 1)}\,\hbar \qquad (\ell = 0, 1, 2, \ldots, n - 1) \qquad \text{(6-30)}$$

For each set $(n, \ell)$ there is a solution $R_{n,\ell}(r)$ of (6-26), several of which are given in Table 6.5 (which appears later in this section, after normalization is discussed). Collectively, they are known as the associated Laguerre functions. They are expressed in terms of a constant defined as follows:

$$a_0 \equiv \frac{(4\pi\epsilon_0)\hbar^2}{me^2} = 0.0529 \text{ nm} \qquad \text{(6-31)}$$

Bohr radius

This combination of factors, known as the **Bohr radius,** is pervasive in atomic physics, so the definition is very convenient. Equation (6-27), for instance, assumes the somewhat simpler form

$$E = -\frac{\hbar^2}{2ma_0^2}\frac{1}{n^2}$$

Moreover, the Bohr radius is a good gauge of atomic dimensions, as we shall see.

Our solution of the Schrödinger equation for the hydrogen atom is complete. Let us now turn to two important related points—degeneracy and normalization.

## Degeneracy

The hydrogen atom is spherically symmetric. Consequently, there is much degeneracy. Equation (6-27) tells us that the energy of an electron orbiting a proton depends only on $n$. But equation (6-30) tells us that for each value of $n$ there are

**Table 6.4**

**Energy**

| $n$ | 1 | 2 | | | | 3 | | | | | | | | | ... | $n$ |
|---|---|---|---|---|---|---|---|---|---|---|---|---|---|---|---|---|
| $\ell$ | 0 | 0 | 1 | | | 0 | 1 | | | 2 | | | | | ... | $0, \ldots, n-1$ |
| $m_\ell$ | 0 | 0 | −1 | 0 | +1 | 0 | −1 | 0 | +1 | −2 | −1 | 0 | +1 | +2 | ... | $-\ell, \ldots, +\ell$ |

**Degeneracy**

| 1 | 4 | 9 | ... | $n^2$ |
|---|---|---|---|---|

states of different angular momentum $|L|$—different $\ell$. Equation (6-23) in turn tells us that for each $\ell$ there are states of different $L_z$—different $m_\ell$. Thus, since for each $(n, \ell, m_\ell)$ there is a unique wave function $R_{n,\ell}\Theta_{\ell,m_\ell}\Phi_{m_\ell}$, all energy levels except the $n = 1$ ground state are degenerate. Furthermore, the degeneracy increases as $n^2$. Table 6.4 summarizes these ideas.

Given the atom's spherical symmetry, we should not expect $E$ to depend on $m_\ell$. Spatial orientation—a *component* of angular momentum—should not be a factor. But what of $\ell$? It appears in the $r$-equation (6-26); why don't the allowed energies (6-27) depend on it? The condition is known as **accidental degeneracy.** Consider equation (6-25), the $r$-equation *before* a specific $U(r)$ is inserted:

$$\frac{-\hbar^2}{2m}\frac{1}{r^2}\frac{d}{dr}\left(r^2\frac{d}{dr}\right)R(r) + \frac{\hbar^2\ell(\ell+1)}{2mr^2}R(r) + U(r)R(r) = ER(r)$$

In general, the values of $E$ for which this has acceptable solutions *do* depend on $\ell$. It is only in the special case of the simple $1/r$ hydrogen atom potential energy that $E$ "accidentally" does not. *Any* deviation of the potential energy from this simple case would lead to quantum energy levels that depend on $\ell$. In particular, additional electrons orbiting the nucleus would alter the potential energy, thus destroying the accidental degeneracy. Degeneracy and dependence of energy on $\ell$ are fundamental to chemistry. They are discussed further in Chapter 7.

## Normalization

As always, the probability density is the complex square of the wave function.

$$|\Psi(\mathbf{r}, t)|^2 = |\psi(r, \theta, \phi)\phi(t)|^2 = (R(r)\Theta(\theta)\Phi(\phi)\phi(t))^*(R(r)\Theta(\theta)\Phi(\phi)\phi(t))$$

(By common but unfortunate convention, the symbol for the azimuthal *angle* is the same as for the *function* giving the time dependence. Be careful not to confuse them.) In stationary states, the product $\phi(t)^*\phi(t)$ is always unity; the probability density does not vary with time. We know too that $\Phi(\phi)^*\Phi(\phi)$ is unity. Thus, since both $R(r)$ and $\Theta(\theta)$ are real, the probability density becomes

$$|\psi(r, \theta, \phi)|^2 = (R(r)\Theta(\theta))^2 \qquad (6\text{-}32)$$

Probability density

Just as in one-dimensional cases, the total probability of finding the electron some-where in space must be unity. Using the volume element in spherical polar coordinates (Section 6.3), the normalization condition becomes

$$\int_{\text{all space}} |\Psi(\mathbf{r}, t)|^2 \, dV = 1$$

$$\rightarrow \int_{r=0}^{r=\infty} \int_{\theta=0}^{\theta=\pi} \int_{\phi=0}^{\phi=2\pi} (R(r)\Theta(\theta))^2 \, r^2 \sin\theta \, dr \, d\theta \, d\phi = 1$$

Let us separate angular from radial parts.

$$\int_{r=0}^{r=\infty} (R(r))^2 r^2 \, dr \int_{\theta=0}^{\theta=\pi} \int_{\phi=0}^{\phi=2\pi} (\Theta(\theta))^2 \sin\theta \, d\theta \, d\phi = 1$$

Thus far, we have avoided reference to normalization constants. The product of multiplicative constants from all three parts could be considered a single constant. But it turns out to be more convenient to consider the overall constant to be a product of two constants, one that is kept with the angular parts and one with the radial part. The constants are chosen so that each of these integrals is unity.

$$\int_0^\infty (R(r))^2 r^2 \, dr = 1 \quad \text{(6-33)} \qquad \int_0^\pi (\Theta(\theta))^2 \, 2\pi \sin\theta \, d\theta = 1 \quad \text{(6-34)}$$

Note that the trivial integration over $\phi$ has already been carried out.

---

**Example 6.4**

Find the angular normalization constant for $\Theta(\theta)$ for the case $\ell = 1, m_\ell = +1$.

**Solution**

We obtain $\Theta(\theta)$ from (6-19) and Table 6.3. Using $A_{1,+1}$ for the normalization constant, we have

$$\Theta_{1,+1}(\theta) = A_{1,+1} \sin\theta$$

Inserting this in (6-34),

$$(A_{1,+1})^2 \int_0^\pi \sin^2\theta \, 2\pi \sin\theta \, d\theta = 1$$

and carrying out the integral,

$$(A_{1,+1})^2 2\pi \frac{4}{3} = 1 \quad \Rightarrow \quad A_{1,+1} = \sqrt{\frac{3}{8\pi}}$$

---

Table 6.6 lists normalized angular parts of the wave functions, obtained in the same way as in Example 6.4. The radial wave functions $R_{n,\ell}(r)$ given in Table 6.5 also include proper normalization constants; they satisfy (6-33).

**Table 6.5    Radial Solutions of (6-26)**

| $n, \ell$ | $R_{n,\ell}(r)$ |
|---|---|
| 1, 0 | $\dfrac{1}{(1a_0)^{3/2}} 2e^{-r/a_0}$ |
| 2, 0 | $\dfrac{1}{(2a_0)^{3/2}} 2\left(1 - \dfrac{r}{2a_0}\right)e^{-r/2a_0}$ |
| 2, 1 | $\dfrac{1}{(2a_0)^{3/2}} \dfrac{r}{\sqrt{3}\,a_0}e^{-r/2a_0}$ |
| 3, 0 | $\dfrac{1}{(3a_0)^{3/2}}\left(2 - \dfrac{4r}{3a_0} + \dfrac{4r^2}{27a_0^2}\right)e^{-r/3a_0}$ |
| 3, 1 | $\dfrac{1}{(3a_0)^{3/2}} \dfrac{4\sqrt{2}\,r}{9a_0}\left(1 - \dfrac{r}{6a_0}\right)e^{-r/3a_0}$ |
| 3, 2 | $\dfrac{1}{(3a_0)^{3/2}} \dfrac{2\sqrt{2}\,r^2}{27\sqrt{5}\,a_0^2}e^{-r/3a_0}$ |

**Table 6.7**

| Letter | $s$ | $p$ | $d$ | $f$ | $g$ | $h$ |
|---|---|---|---|---|---|---|
| Value of $\ell$ | 0 | 1 | 2 | 3 | 4 | 5 |

**Table 6.6    Angular Solutions (Spherical Harmonics)***

| $\ell, m_\ell$ | $\Theta_{\ell, m_\ell}(\theta)\,\Phi_{m_\ell}(\phi)$ |
|---|---|
| 0, 0 | $\sqrt{\frac{1}{4\pi}}$ |
| 1, 0 | $\sqrt{\frac{3}{4\pi}}\cos\theta$ |
| 1, ±1 | $\sqrt{\frac{3}{8\pi}}\sin\theta\, e^{\pm i\phi}$ |
| 2, 0 | $\sqrt{\frac{5}{16\pi}}(3\cos^2\theta - 1)$ |
| 2, ±1 | $\sqrt{\frac{15}{8\pi}}\cos\theta\sin\theta\, e^{\pm i\phi}$ |
| 2, ±2 | $\sqrt{\frac{15}{32\pi}}\sin^2\theta\, e^{\pm 2i\phi}$ |
| 3, 0 | $\sqrt{\frac{7}{16\pi}}(5\cos^3\theta - 3\cos\theta)$ |
| 3, ±1 | $\sqrt{\frac{21}{64\pi}}(5\cos^2\theta - 1)\sin\theta\, e^{\pm i\phi}$ |
| 3, ±2 | $\sqrt{\frac{105}{32\pi}}\cos\theta\sin^2\theta\, e^{\pm 2i\phi}$ |
| 3, ±3 | $\sqrt{\frac{35}{64\pi}}\sin^3\theta\, e^{\pm 3i\phi}$ |

\* Since the angular parts are always the same, no matter what might be the specific form of the spherically symmetric $U(r)$, they are usually tabulated together. "Spherical harmonic" is just the common name for the product $\Theta(\theta)\Phi(\phi)$.

## Probability Density

Let us now take a look at the whereabouts of hydrogen's electron. Figure 6.12 represents the probability density $(R_{n,\ell}(r)\Theta_{\ell,m_\ell}(\theta))^2$ for various states $(n, \ell, m_\ell)$ as density of shading. (*Note:* Divisions on the axes are units of $a_0$.) They are labeled using the convention of **spectroscopic notation,** in which the number designates the value of $n$ and the letter the value of $\ell$. The scheme for $\ell$ is shown in Table 6.7 (letters advancing alphabetically beyond $f$). Thus, a $3d$ state is one in which $n = 3$ and $\ell = 2$. For each $(n, \ell)$ state, of course, there are $2\ell + 1$ possible $m_\ell$-values.

We note several important features:

1. As always, the probability density is a diffuse "cloud" spread over space.

2. It extends farther from the origin (i.e., the proton) as $n$ increases.

3. In the $s$ states, where $\ell = 0$ and the energy is solely radial, the probability density is spherically symmetric; there is no hint of an orbital plane. In $p$ states, three possible approximate orbits are suggested: one with motion roughly parallel to the z-axis ($m_\ell = 0$), and two roughly in the x-y plane ($m_\ell = \pm 1$). The $d$ states suggest five possible orbits, adding two between the z-axis and x-y plane ($m_\ell = \pm 1$) to the extremes of $m_\ell = 0$ and $m_\ell = \pm 2$ (see also Figure 6.9). On the other hand, the probability should be spheri-

$$|\psi(r, \theta, \phi)|^2 = (R(r)\,\Theta(\theta))^2$$

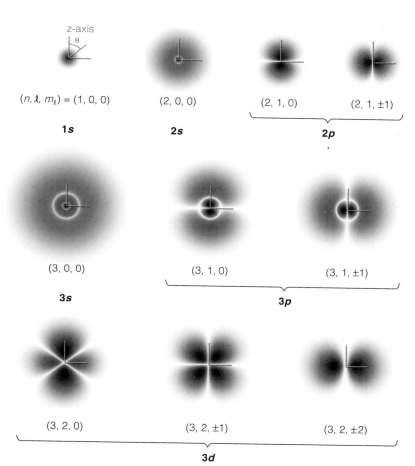

Figure 6.12 Electron probability densities in the hydrogen atom, through $n = 3$.

cally symmetric for any given $n$ and $\ell$ *in an isolated atom*. In this case, as noted in Section 6.4, all $m_\ell$-values would be equally likely. If the reader imagines adding the probability densities for all $m_\ell$-values for any given $n$ and $\ell$, a spherically symmetric average should seem quite plausible.

4. In the $\ell \neq 0$ states, the probability density vanishes at the origin. A particle with angular momentum can never be found there.

5. For fixed $\ell$, as $n$ increases, the number of radial antinodes increases; increasing the total energy but not the rotational energy implies greater radial energy, thus more radial antinodes. By the same token, for fixed $n$, as $\ell$ increases, the number of radial antinodes decreases; increasing the rotational energy at fixed total energy must decrease the radial. Orbits of small $\ell$, with more radial antinodes, may be thought of as somewhat elliptical— there is much motion toward and away from the nucleus. States of the largest $\ell$ (i.e., $\ell = n - 1$) have but one radial antinode. They are more like circular orbits at a fixed radius.

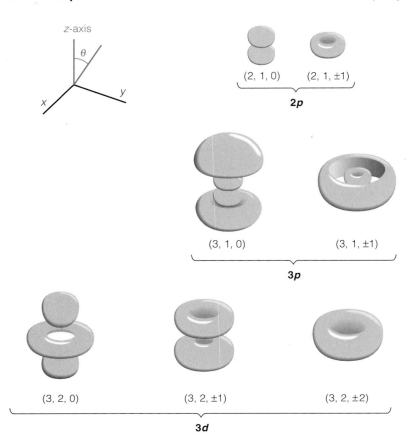

**Figure 6.13** Constant-probability-density surfaces for the hydrogen atom.

The surfaces of constant probability density plotted in Figure 6.13 provide an alternative view. Once again, orientations of approximate orbital planes are apparent. The *s* states are not shown; the surfaces would be simple spheres. (It is important to note that probability densities do not drop to zero abruptly—these surfaces enclose only the vast majority of space where the electron is likely to be found.)

## ◆ 6.6 A Closer Look at $\psi(r, \theta, \phi)$

Having skipped the actual solutions giving $\Theta(\theta)$ and $R(r)$, it would be easy to lose sight of the fact that they do satisfy the Schrödinger equation. In the following example, we demonstrate that at least one representative $R\Theta\Phi$ does, and with proper values of $E$ and $L$. Afterward, we will scrutinize its spatial extent.

If we insert potential energy (6-9) and the $L^2$ operator (6-20) into the Schrödinger equation of (6-10), we begin with a reasonably compact expression.

$$-\frac{\hbar^2}{2m}\frac{1}{r^2}\left[\frac{\partial}{\partial r}\left(r^2\frac{\partial}{\partial r}\right) + \frac{1}{-\hbar^2}\hat{L}^2\right]\psi(r, \theta, \phi)$$

$$+ \left(\frac{-1}{4\pi\epsilon_0}\frac{e^2}{r}\right)\psi(r, \theta, \phi) = E\psi(r, \theta, \phi)$$

Now inserting $R\Theta\Phi$ for $\psi$, and noting that $\hat{L}^2$ contains no derivatives with respect to $r$, we have

$$-(\Theta\Phi)\frac{\hbar^2}{2m}\frac{1}{r^2}\frac{\partial}{\partial r}\left(r^2\frac{\partial R}{\partial r}\right) + \frac{1}{2mr^2}R\hat{L}^2(\Theta\Phi)$$

$$+ \left(\frac{-e^2}{4\pi\epsilon_0 r}\right)(R\Theta\Phi) = E(R\Theta\Phi)$$

$$(6\text{-}35)$$

### Example 6.5

For a hydrogen atom electron in the $(n, \ell, m_\ell) = (2, 1, +1)$ state, (a) write the solution of the time-independent Schrödinger equation, and (b) verify explicitly that it is a solution.

### Solution

(a) Using Tables 6.5 and 6.6,

$$\psi(\mathbf{r}) = \psi(r, \theta, \phi) = R_{2,1}(r)\Theta_{1,+1}(\theta)\Phi_{+1}(\phi)$$

$$= \left[\frac{1}{(2a_0)^{3/2}}\frac{r}{\sqrt{3}\,a_0}e^{-r/2a_0}\right]\left(\sqrt{\frac{3}{8\pi}}\sin\theta\,e^{+i\phi}\right)$$

$$= \frac{1}{8a_0^{5/2}\sqrt{\pi}}re^{-r/2a_0}\sin\theta\,e^{+i\phi}$$

(b) The normalization constants may be ignored because they appear in each term in equation (6-35), and so cancel. Let us attack (6-35) one piece at a time. The first is

$$\frac{\partial}{\partial r}\left(r^2\frac{\partial R}{\partial r}\right) = \frac{\partial}{\partial r}\left[r^2\frac{\partial(re^{-r/2a_0})}{\partial r}\right]$$

$$= \frac{\partial}{\partial r}\left[\left(r^2 - \frac{r^3}{2a_0}\right)e^{-r/2a_0}\right]$$

$$= \left(2r - \frac{2r^2}{a_0} + \frac{r^3}{4a_0^2}\right)e^{-r/2a_0}$$

The second is

$\hat{L}^2(\Theta\Phi)$

$$= \hat{L}^2 \sin\theta\, e^{+i\phi}$$

$$= -\hbar^2 \left[ \csc\theta \frac{\partial}{\partial\theta} \left( \sin\theta \frac{\partial}{\partial\theta} \right) + \csc^2\theta \frac{\partial^2}{\partial\phi^2} \right] \sin\theta\, e^{+i\phi}$$

$$= -\hbar^2 \left[ e^{+i\phi} \csc\theta \frac{\partial}{\partial\theta} \left( \sin\theta \frac{\partial}{\partial\theta} \sin\theta \right) + \sin\theta \csc^2\theta \frac{\partial^2}{\partial\phi^2} e^{+i\phi} \right]$$

$$= -\hbar^2 \left[ e^{+i\phi} \csc\theta \frac{\partial}{\partial\theta} (\sin\theta\cos\theta) + \csc\theta(-1)e^{+i\phi} \right]$$

$$= -\hbar^2 e^{+i\phi} [\csc\theta(\cos^2\theta - \sin^2\theta) - \csc\theta]$$

$$= -\hbar^2 e^{+i\phi} [\csc\theta(1 - 2\sin^2\theta) - \csc\theta]$$

$$= 2\hbar^2 \sin\theta\, e^{+i\phi}$$

or, expressed in an revealing form,

$$\hat{L}^2(\Theta\Phi) = [\sqrt{1(1+1)}\,\hbar]^2(\Theta\Phi)$$

Comparing with (6-21), we see our wave function indeed leads to a 1 where it is supposed to be—the value of $|L|$ is $\sqrt{2}\,\hbar$.

Reinserting the two pieces into (6-35),

$$-\sin\theta\, e^{+i\phi} \frac{\hbar^2}{2m} \frac{1}{r^2} \left[ \left( 2r - \frac{2r^2}{a_0} + \frac{r^3}{4a_0^2} \right) e^{-r/2a_0} \right]$$

$$+ \frac{1}{2mr^2} re^{-r/2a_0} (2\hbar^2 \sin\theta\, e^{+i\phi})$$

$$+ \left( \frac{-e^2}{4\pi\epsilon_0 r} \right) re^{-r/2a_0} \sin\theta\, e^{+i\phi} = Ere^{-r/2a_0} \sin\theta\, e^{+i\phi}$$

The wave function $re^{-r/2a_0} \sin\theta\, e^{+i\phi}$ now cancels in all terms, leaving

$$-\frac{\hbar^2}{2m} \frac{1}{r^2} \left( 2 - \frac{2r}{a_0} + \frac{r^2}{4a_0^2} \right) + \frac{1}{2mr^2} 2\hbar^2 + \frac{-e^2}{4\pi\epsilon_0 r} = E$$

or, after further cancellation,

$$\left( \frac{\hbar^2}{ma_0} - \frac{e^2}{4\pi\epsilon_0} \right) \frac{1}{r} - \frac{\hbar^2}{8ma_0^2} = E$$

But since $a_0$ is by definition $(4\pi\epsilon_0)\hbar^2/me^2$, the term in parentheses is zero, leaving

$$-\frac{me^4}{8(4\pi\epsilon_0)^2\hbar^2} = E$$

According to (6-27), this is indeed what the energy should be in an $n = 2$ state.

Now let us show how the probability densities we obtain from $\psi(r, \theta, \phi)$ agree with the diagrams of Figures 6.12 and 6.13.

Example 6.6

A hydrogen atom electron is in the $(2, 1, +1)$ state. Calculate the probability that it would be found (a) within 30° of the $x$-$y$ plane, irrespective of radius; (b) between $r = 2a_0$ and $r = 6a_0$, irrespective of angle; and (c) within 30° of the $x$-$y$ plane *and* between $r = 2a_0$ and $r = 6a_0$.

Solution

(a) The $x$-$y$ plane is $\theta = 90°$ or $\frac{1}{2}\pi$. We seek the probability of finding an electron in a region of space covering 30° $(\frac{1}{6}\pi)$ either side of $\theta = \frac{1}{2}\pi$. Example 6.4 involved the same state, and there we set the angular normalization integral (6-34) integrated over *all* angles to unity. All we need do to determine a fraction of unity is restrict the angular limits of integration.

$$\text{probability} = \int_{\pi/3}^{2\pi/3} \left( \sqrt{\frac{3}{8\pi}} \sin \theta \right)^2 2\pi \sin \theta \, d\theta$$

$$= \frac{3}{4} \int_{\pi/3}^{2\pi/3} \sin^3 \theta \, d\theta = \frac{3}{4} \frac{11}{12} = \frac{11}{16}$$

A solid angle of $4\pi$ steradians covers all of space. The angle we have considered is $2\pi$ steradians.[12] It might be might said that we have considered half of space. The probability we found is greater than half, so we see that the electron is more likely to be found nearer the $x$-$y$ plane than the $z$-axis. The $\ell = 1$, $m_\ell = \pm 1$ diagrams of Figures 6.9, 6.12, and 6.13 bear this out.

(b) As we see in Figure 6.12, so far as $r$ goes, the probability/shading is greatest near $4a_0$. (We find in Section 6.7 that the most probable radius tends to vary as $n^2 a_0$.) Thus, we calculate the probability that the electron will be in the region where we would most expect to find it. All we need do is restrict the limits of the radial probability integral (6-33). We obtained $R(r)$ in the previous example. Thus,

$$\text{probability} = \int_{2a_0}^{6a_0} (R_{2,1}(r))^2 r^2 \, dr$$

$$= \int_{2a_0}^{6a_0} \left[ \frac{1}{(2a_0)^{3/2}} \frac{r}{\sqrt{3}\, a_0} e^{-r/2a_0} \right]^2 r^2 \, dr$$

$$= \frac{1}{24a_0^5} \int_{2a_0}^{6a_0} r^4 e^{-r/a_0} \, dr$$

After somewhat tedious integration by parts, we find that

probability $= 0.662$

(c) Restricting both radial and angular dependences is equivalent to merely finding the product of the above two probabilities. Thus,

$$\text{probability} = \frac{11}{16} \times 0.662 = 0.455$$

Although we consider only a small region, the probability is nearly 50%.

[12]Integrating over both angular variables gives the solid angle

$$\int_{\theta=0}^{\theta=\pi} \int_{\phi=0}^{\phi=2\pi} \sin \theta \, d\theta \, d\phi$$
$$= 4\pi \text{ steradians}$$

$$\int_{\theta=\pi/3}^{\theta=2\pi/3} \int_{\phi=0}^{\phi=2\pi} \sin \theta \, d\theta \, d\phi$$
$$= 2\pi \text{ steradians}$$

## 6.7 Radial Probability

Often it is useful to know a probability per unit distance in the radial direction alone, irrespective of angle. We find this by first restricting the probability integral (6-33).

$$\text{probability of finding particle between } r_1 \text{ and } r_2 = \int_{r_1}^{r_2} (R(r))^2 r^2 \ dr$$

Now assuming that $r_1$ and $r_2$ differ by only $dr$, the probability $dP$ of finding the electron becomes

$$dP = (R(r))^2 r^2 \ dr$$

Thus, the **radial probability** $P(r)$, a probability *per unit distance,* we define as

$$P(r) = \frac{dP}{dr} = r^2(R(r))^2 \qquad\qquad (6\text{-}36)$$

Radial probability

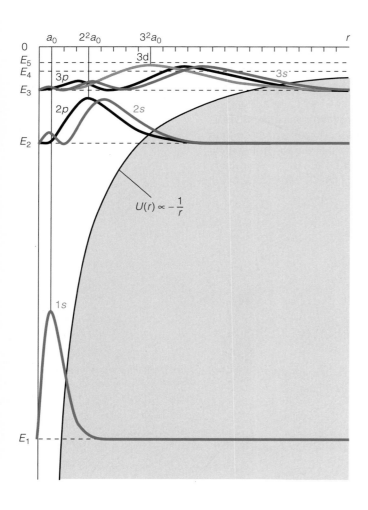

**Figure 6.14** Radial probabilities and energies in the hydrogen atom.

There is a simple argument that (6-36) should include the $r^2$ that it does: Imagine a situation in which the probability *per unit volume* is constant. That is, $R$, $\Theta$, and $\Phi$ are all constant (at least over a large portion of space surrounding the origin). It would be 9 times more likely to find the particle within a 1-nm-thick spherical shell of radius 3 cm than a 1-nm-thick shell of 1-cm radius, simply because the area of the 3-cm-radius shell is 9 times as large. (Remember: The area of a sphere is proportional to $r^2$.) Given a constant probability per unit volume, but a volume 9 times as large, the probability per unit distance (per 1 nm in the radial direction) is naturally 9 times as large. Equation (6-36) would agree; $R$ would be constant and the $r^2$ would give the appropriate "area factor" of 9.

Figure 6.14 shows radial probabilities plotted on axes whose heights correspond to energy for hydrogen atom states through $n = 3$. (Since, as we know, all but the ground state are degenerate, there are multiple states at the higher energies.) Note that all of these standing waves extend into the classically forbidden region. Also, as noted in Section 6.5, at any given $n$, smaller $\ell$ (less rotational energy) implies greater radial energy; there are more antinodes and the radial probability extends farther from the origin. The highest angular momentum states ($\ell = n - 1$) have only one antinode and extend least far from the origin. Again, these states are more nearly circular orbits at a reasonably well-defined radius.

---

### Example 6.7

An electron in a hydrogen atom is in a $3d$ state. What is the most probable radius at which to find the electron?

#### Solution

The most probable radius is where the radial probability is maximum. Looking up $R_{3,2}(r)$ in Table 6.5, we have

$$P(r) = r^2(R_{3,2}(r))^2 = r^2 \left[ \frac{1}{(3a_0)^{3/2}} \frac{2\sqrt{2}\, r^2}{27\sqrt{5}\, a_0^2} e^{-r/3a_0} \right]^2$$

$$= \frac{8}{(3a_0)^3 (27)^2 5 a_0^4} r^6 e^{-2r/3a_0}$$

To find the maximum, we set the derivative to zero. (The normalization constant may be ignored.)

$$\frac{dP(r)}{dr} = \left( 6r^5 + r^6 \frac{-2}{3a_0} \right) e^{-2r/3a_0} = 0 \quad \text{or} \quad r^5 \left( 6 + r \frac{-2}{3a_0} \right) e^{-2r/3a_0} = 0$$

The equation holds at $r = 0$ and $r = \infty$, but these are obviously minima, where $P(r)$ is itself zero [$P(r)$ is nonnegative]. The maximum occurs where the quantity in parentheses is zero.

$$6 + r \frac{-2}{3a_0} = 0 \quad \Rightarrow \quad r_{\text{most probable}} = 9a_0$$

This agrees with the $3d$ plot of Figure 6.14.

In states for which $\ell = n - 1$—that is, those more nearly circular—there is a pattern to the most probable distance from the proton. According to Table 6.5,

$$R_{1,0} \propto r^0 e^{-r/1a_0}$$

$$R_{2,1} \propto r^1 e^{-r/2a_0}$$

$$R_{3,2} \propto r^2 e^{-r/3a_0}$$

In general

$$R_{n,n-1} \propto r^{(n-1)} e^{-r/na_0}$$

Therefore, the radial probability obeys

$$P(r) = r^2 (R(r))^2 \propto r^2 [r^{(n-1)} e^{-r/na_0}]^2 = r^{2n} e^{-2r/na_0}$$

It is left as an exercise to show that this function has a maximum given by

$$r_{\text{most probable}} = n^2 a_0 \qquad (\ell = n - 1) \tag{6-37}$$

Again, the result is in agreement with plots in Figure 6.14. Though $\ell < n - 1$ states are not quite so easily handled, they follow this rule to a fair approximation: *Orbit radius increases approximately as the square of n.*

Expectation values are always found in the same way—by putting an operator between the wave function and its complex conjugate, then integrating over all space. But in the case of the atom, if the quantity of interest is merely a function of the radius [i.e., $f(r)$], the angular parts integrate to unity and the expectation value assumes a simple form.

$$\bar{f} = \int_0^\infty f(r)(R(r))^2 r^2 \; dr = \int_0^\infty f(r) P(r) \; dr \tag{6-38}$$

**Example 6.8**

What is the expectation value of the radius of an electron in a 3d state?

Solution

Here, $f(r)$ is just $r$. With the radial probability found in the previous example, we have

$$\bar{r} = \int_0^\infty r P(r) \; dr = \int_0^\infty r \left[ \frac{8}{(3a_0)^3 (27)^2 5 a_0^4} r^6 e^{-2r/3a_0} \right] dr$$

$$= \frac{8}{(3a_0)^3 (27)^2 5 a_0^4} \int_0^\infty r^7 e^{-2r/3a_0} \; dr$$

The integral is of the form $\int_0^\infty x^m e^{-bx} \; dx = m!/b^{m+1}$, with $b = 2/3a_0$ and $m = 7$. Thus,

$$\bar{r} = \frac{8}{(3a_0)^3 (27)^2 5 a_0^4} 7! \left( \frac{3a_0}{2} \right)^8 = 10.5 a_0$$

In Example 6.7, we found the most probable radius of a $3d$ electron to be $9a_0$. An expectation value, on the other hand, is an average of radii we would find if we repeatedly "looked" for electrons in $n = 3$, $\ell = 2$ states. We have found in Example 6.8 that it is larger than the most probable radius. Although these should be close, there is no reason they should be equal. The plot of $P(r)$ in Figure 6.14 has a single maximum, but while it terminates abruptly at the origin, it extends infinitely far the other way. Thus, the "average" radius *should* be larger than the most probable one.

## 6.8 Evidence of Quantization: Spectral Lines

Decades before the advent of quantum mechanics, simple spectroscopic observations demonstrated that hydrogen does not emit light of all wavelengths, but of only certain values, known as **spectral lines.**[13] Those in the visible portion of the spectrum, 400–700 nm, are

$$\text{experimentally observed} \atop \text{hydrogen wavelengths} \left\{ \begin{array}{l} 656 \text{ nm} \\ 486 \text{ nm} \\ 434 \text{ nm} \\ 410 \text{ nm} \end{array} \right.$$

The prevailing theory failed miserably to predict these observations. According to classical mechanics, the charged electron, constantly accelerating radially inward, should radiate away its energy as electromagnetic waves, emitting a broad range of wavelengths as it spirals into the nucleus.

By trial and error, a formula was found that yielded the experimentally observed wavelengths. Although its form bespoke a tantalizing pattern, there was no underlying theory; it simply worked. The wavelengths obey

$$\frac{1}{\lambda} = R_{\text{H}} \left( \frac{1}{4} - \frac{1}{n^2} \right) \qquad (n = 3, 4, 5, \dots) \tag{6-39}$$

Called the **Rydberg constant**, $R_{\text{H}}$ was merely assigned the value that gave the best agreement with the experimental observations: $1.097 \times 10^7 \text{ m}^{-1}$. Substitution will verify that $n = 3$ yields 656 nm, $n = 6$ yields 410 nm, and so on. Even shorter wavelengths were later found, corresponding to $n = 7$, $n = 8, \dots$.

Quantum mechanics provided the answer. In stationary states, neither the probability density nor charge density of the electron's orbiting standing wave varies with time. Thus, there should be no radiation. Now if the electron is in a state other than the ground state, a downward transition can occur, during which the probability density does oscillate. But it does so only at the frequency just

---

[13]Fortunately, a given spectral line is *not* one well-defined wavelength; rather, it consists of a range of wavelengths around the central value. (Otherwise, the line would be very difficult to resolve.) There are two reasons for this "fuzziness": Doppler broadening, due to thermal motion of the light-emitting atoms, discussed in Exercise 1.35, and the energy-time uncertainty principle, discussed in Exercise 3.50. The two effects are compared in Exercise 52.

right to generate a photon that would carry away the energy difference (see Section 4.11). Since the energy differences are quantized, photons can be emitted at only certain frequencies.

Let us calculate the wavelength of a photon emitted in a transition. As we know, energy depends only on the quantum number $n$. Suppose the electron makes a transition from an $n_i$ state to an $n_f$ state. The energy given the photon is the initial energy of the electron minus the final. Using equation (6-27), we obtain

$$
\begin{aligned}
E &= E_i - E_f \\
&= \left[ -\frac{me^4}{2(4\pi\epsilon_0)^2\hbar^2} \frac{1}{n_i^2} \right] - \left[ -\frac{me^4}{2(4\pi\epsilon_0)^2\hbar^2} \frac{1}{n_f^2} \right] \\
&= \frac{me^4}{2(4\pi\epsilon_0)^2\hbar^2} \left( \frac{1}{n_f^2} - \frac{1}{n_i^2} \right)
\end{aligned}
$$

For a photon, $E = hf = hc/\lambda$. Therefore,

$$
\frac{hc}{\lambda} = \frac{me^4}{2(4\pi\epsilon_0)^2\hbar^2} \left( \frac{1}{n_f^2} - \frac{1}{n_i^2} \right)
$$

or

$$
\begin{aligned}
\frac{1}{\lambda} &= \frac{me^4}{2(4\pi\epsilon_0)^2\hbar^2 hc} \left( \frac{1}{n_f^2} - \frac{1}{n_i^2} \right) \\
&= 1.097 \times 10^7 \text{ m}^{-1} \left( \frac{1}{n_f^2} - \frac{1}{n_i^2} \right)
\end{aligned}
\tag{6-40}
$$

We infer that the wavelengths given by (6-39) result from transitions down to the $n_f = 2$ level from the $n_i = 3, 4, 5, \ldots$ levels. A previously mystifying observation is explained by quantum theory—a great triumph! Logically, there should be wavelengths corresponding to transitions ending at levels other than $n_f = 2$. These too have been observed. For instance, numerous wavelengths in the ultraviolet region of the spectrum are emitted. They result from transitions ending at the $n_f = 1$ level. Involving a greater downward jump, they are of shorter wavelength. Transitions ending at $n_f = 3, 4, \ldots$ represent smaller energy differences and are consequently in the longer-wavelength infrared region of the spectrum and beyond.

## 6.9 Transitions

Figure 6.15 illustrates some of the transitions possible in a hydrogen atom. The arrows signify these as downward transitions, in which the energy lost by the atom might be emitted in the form of a photon. Corresponding upward transitions are certainly possible, but, requiring an energy input, do not occur spontaneously. The transitions are grouped according to the energy level at which they terminate. Each such group has been named for a pioneer in the field of spectroscopy. For example, the **Balmer series,** by definition, comprises all transitions ending at $n_f = 2$. The visible wavelengths given in the previous section are thus members

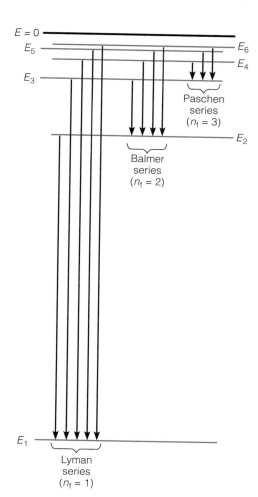

**Figure 6.15** Hydrogen's spectral lines: A photon is emitted when the electron "jumps downward."

of the Balmer series, the first ($n_i = 3$, $n_f = 2$), the second ($n_i = 4$, $n_f = 2$), and so on.

It is natural to wonder what an electron in an $n_i = 3$ state would do. Would it drop to the $n_f = 2$ state, producing a photon corresponding to the first spectral line in the Balmer series ($n_i = 3 \rightarrow n_f = 2$)? Or would it drop fully to $n_f = 1$, contributing to the second line in the **Lyman series** ($n_i = 3 \rightarrow n_f = 1$)? The answer is that it may do either, governed by probabilities (see Section 6.11). But there is an even more basic question: If it is indeed in a "stationary state," why would the electron make any transition at all? Our treatment of the hydrogen atom assumes an isolated atom. However, even in what we might consider a vacuum there are always spontaneously fluctuating electromagnetic fields, known as **vacuum fluctuations.** In effect, these change the all-important potential energy. The theory (Quantum Electrodynamics, or QED) is quite sophisticated. We simply note that vacuum fluctuations perturb the atom in such a way as to cause it continually to assume a lower energy state, if one is available. Consequently, if an electron is *not* in its ground state (and gains no energy from an external source), it will jump down through lower energy levels, producing corresponding photons, and eventually end up there.

Of course, an atom also may be induced to jump to a higher energy level. Indeed, the excess electromagnetic energy emitted by atoms in their otherwise inexorable decay to the ground state must somehow have been absorbed in the first place. Two ways in which an atom may increase its energy are (1) being struck by a massive particle, and (2) being struck by a photon. In the first, the massive particle might be simply another atom in a hot gas. If the colliding particles have sufficient kinetic energy, an atom may absorb a portion of the collision energy and jump to a higher level. The colliding particles of course depart with reduced kinetic energy. In the second, the energy of the photon must be precisely the energy difference from one atomic energy level to another, and the photon is fully absorbed in the process.

---

## Example 6.9

The atoms in a sample of hydrogen are all in their ground states. (The atoms are too cold to bump one another to higher levels.) Now a beam of electrons accelerated through a potential difference of 12.5 V is directed at the sample. What wavelengths will be emitted?

### Solution

All the atoms initially have energy $-13.6$ eV. Each electron in the beam has kinetic energy of 12.5 eV, so in a collision with an accelerated electron an atom could gain at most 12.5 eV, raising its energy to $-13.6$ eV $+$ 12.5 eV $=$ $-1.1$ eV. But $-1.1$ eV is not one of the allowed energies. The allowed energies are

$$-13.6 \text{ eV}, \qquad \frac{-13.6 \text{ eV}}{2^2} = -3.4 \text{ eV}, \qquad \frac{-13.6 \text{ eV}}{3^3} = -1.51 \text{ eV},$$

$$\frac{-13.6 \text{ eV}}{4^2} = -0.85 \text{ eV}, \quad \dots$$

Thus, there is enough energy in a collision to raise the atom no higher than the $n = 3$ level. Thereafter, it will emit characteristic photons as it drops back to the ground state.[14] Three transitions are possible: It may drop directly from the third to the first energy level ($n = 3 \to n = 1$), or it may drop from the third to the second ($n = 3 \to n = 2$), then to the first ($n = 2 \to n = 1$).

In all cases, the energy of the photon emitted in the downward transition is the difference between the atomic energy levels.

$$E_{3 \to 1} = E_3 - E_1 = -1.51 \text{ eV} - (-13.6 \text{ eV}) = 12.1 \text{ eV}$$

$$E_{3 \to 2} = E_3 - E_2 = -1.51 \text{ eV} - (-3.4 \text{ eV}) = 1.9 \text{ eV}$$

$$E_{2 \to 1} = E_2 - E_1 = -3.4 \text{ eV} - (-13.6 \text{ eV}) = 10.2 \text{ eV}$$

[14]We should note that once energy is absorbed, downward transitions occur in a *very* short time, typically $10^{-8}$ s. The probability of a second accelerated electron hitting the excited atom and raising it to an even higher energy is negligible; multiple electrons cannot gang up on one atom.

The wavelengths follow from $E = hc/\lambda$.

$$\lambda_{3 \to 1} = \frac{(6.63 \times 10^{-34} \text{ J·s})(3 \times 10^8 \text{ m/s})}{(12.1 \text{ eV})(1.6 \times 10^{-19} \text{ J/eV})} = 103 \text{ nm}$$

$$\lambda_{3 \to 2} = \frac{(6.63 \times 10^{-34} \text{ J·s})(3 \times 10^8 \text{ m/s})}{(1.9 \text{ eV})(1.6 \times 10^{-19} \text{ J/eV})} = 656 \text{ nm}$$

$$\lambda_{2 \to 1} = \frac{(6.63 \times 10^{-34} \text{ J·s})(3 \times 10^8 \text{ m/s})}{(10.2 \text{ eV})(1.6 \times 10^{-19} \text{ J/eV})} = 122 \text{ nm}$$

The first and third wavelengths are, respectively, the second and first members of the (ultraviolet) Lyman series. The second wavelength is the first in the Balmer series.

Spectroscopic observations are indispensable in many applications, such as astronomy. We shall discuss elements other than hydrogen later. For now, we note only that different elements emit different wavelengths. Each has a unique spectrum that serves as a spectroscopic "fingerprint" of the element. Not only is it possible to identify the element by the wavelengths it emits; the temperature of a sample of the element may also be determined. Cold hydrogen would emit no radiation. All atoms would be in their ground states and would remain there. However, were the sample heated, the interparticle collisions might become sufficiently energetic to cause some atoms to jump to the $n = 2$ level. At such a temperature, the spectrum would be fairly dull. It would consist of essentially a single line, corresponding to transitions from $n = 2$ back down to the ground state. (At any temperature there is, of course, a range of speeds, and some collisions must therefore be energetic enough to allow upward transitions to arbitrarily high levels. But the likelihood is very small until the average kinetic energy of the colliding particles is roughly equal to the energy difference.) Were the temperature raised, the $n = 3$ level would become accessible, allowing a few more wavelengths to be produced in subsequent downward transitions. Were the temperature increased further, the spectrum would become even more rich. Thus, the relative brightness of its characteristic spectral lines is an indication of an element's temperature.

Matter may also reveal its composition through what wavelengths it *absorbs*. For example, many stars, owing to the complex and violent motion of the charged particles within, emit essentially a continuous spectrum of wavelengths. If on its way to Earth this light passes through an intervening cloud of interstellar gas, wavelengths corresponding to differences in the energy levels of the atoms in the gas will be absorbed. As the atoms drop back down to the ground state, they reemit these wavelengths in all directions. These wavelengths are so scattered compared to the rest of the light—continuing unaffected on its way from the star to Earth—that they are essentially missing in the light that arrives. Thus, a missing spectrum positively identifies the element that is doing the absorbing. Figure 6.16 illustrates the phenomenon. The top plot shows hydrogen's spectral lines, while the bottom plot shows what wavelengths of a continuous spectrum would reach Earth after passing through an interstellar cloud of hydrogen. (*Note:* The Balmer series is

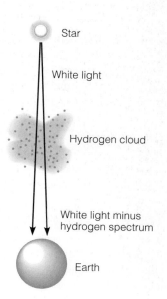

Star

White light

Hydrogen cloud

White light minus
hydrogen spectrum

Earth

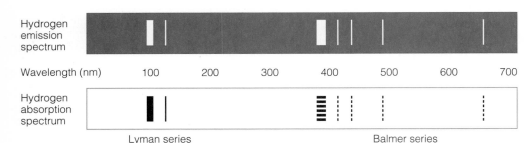

**Figure 6.16** Hydrogen's emission and absorption spectra.

shown dashed in the absorption spectrum because its presence is not guaranteed. Were the gas cold, there would be essentially no atoms *initially* in the $n = 2$ state, so photons whose energies correspond to transitions between the $n = 2$ and higher levels could not be absorbed.)

There are absorption lines in our own Sun's spectrum. Light produced in the interior passes through the Sun's less-hot atmosphere, where hydrogen, sodium, magnesium, iron, and other elements in a gaseous state absorb their characteristic wavelengths.

## ▾ 6.10 Hydrogenlike Atoms

Our solution of the hydrogen atom was based on one all-important piece of information: the potential energy

$$U(r) = -\frac{1}{4\pi\epsilon_0}\frac{e^2}{r}$$

Any application in which the potential energy is of the same form would follow in the same way. "Hydrogenlike" atoms are those with a single electron. Clearly, hydrogen is hydrogenlike. Another example is singly ionized (one electron removed) helium. Normally, helium has two electrons orbiting its two-proton nucleus. If one electron is removed, the remaining one has potential energy:

$$U(r) = -\frac{1}{4\pi\epsilon_0}\frac{(2e)(e)}{r} \qquad \textbf{(Singly ionized helium)}$$

In general, if an atom normally has $Z$ electrons orbiting $Z$ protons, and $(Z - 1)$ of its electrons are removed, the single electron has potential energy:

$$U(r) = -\frac{1}{4\pi\epsilon_0}\frac{Ze^2}{r} \qquad \textbf{(One-electron atom)}$$

Thus, everything that has been said of the hydrogen atom may be said of hydrogenlike atoms by merely replacing $e^2$ by $Ze^2$ wherever it may be found. In particular, the energy levels follow from equation (6-27).

Hydrogen ($Z = 1$):    $E_n = -\dfrac{me^4}{2(4\pi\epsilon_0)^2\hbar^2}\dfrac{1}{n^2} = \dfrac{-13.6\ \text{eV}}{n^2}$

Hydrogenlike:    $E_n = -\dfrac{m(Ze^2)(Ze^2)}{2(4\pi\epsilon_0)^2\hbar^2}\dfrac{1}{n^2} = Z^2\dfrac{-13.6\ \text{eV}}{n^2}$

(6-41)

For $\ell = n - 1$, the approximate radii obey

Hydrogen ($Z = 1$):    $r_n = n^2\dfrac{(4\pi\epsilon_0)\hbar^2}{me^2} = n^2 a_0$

Hydrogenlike:    $r_n = n^2\dfrac{(4\pi\epsilon_0)\hbar^2}{m(Ze^2)} = \dfrac{1}{Z}n^2 a_0$

(6-42)

Bound-state energies are "deeper" (more negative) by $Z^2$, and the orbit radii smaller by $1/Z$.

---

**Example 6.10**

What are the ionization energy and the most probable radius of the electron in the ground state of He$^+$?

**Solution**

The energy in the ground state is given by (6-41), with $Z = 2$.

$$E_1 = (2)^2\frac{-13.6\ \text{eV}}{1^2} = -54.4\ \text{eV}$$

An energy of 54.4 eV would be required to remove this electron.

In the $n = 1$ state, $\ell$ can only be zero (i.e., $n - 1$), so (6-42) is applicable.

$$r_1 = \frac{1}{2}1\ (0.0529\ \text{nm}) = 0.026\ \text{nm}$$

Singly ionized helium is considerably smaller than hydrogen.

---

# ◆ 6.11 Photon Emission: Rules and Rates

In Section 6.9 we noted that photon-producing transitions from one quantum state to another are governed by probabilities. The question of *why* a particle in a "stationary" state would choose to leave it and enter a transitional phase to a lower-energy state has its answer, as noted, in the destabilizing effect of vacuum fluctuations. Our interest here is in *how* a charged particle in transitional state generates a photon, for it is this that makes some transitions highly favored and rapid, while other seemingly likely transitions essentially do not occur at all.

In a transition between two quantum states, a particle occupies neither; its state is somewhat of a combination of the two. We may crudely represent the transitional state as follows:

$$\Psi_{\text{transition}}(\mathbf{r},\ t) = \Psi_i(\mathbf{r},\ t) + \Psi_f(\mathbf{r},\ t) \tag{6-43}$$

$$= \psi_i(\mathbf{r})e^{-i(E_i/\hbar)t} + \psi_f(\mathbf{r})e^{-i(E_f/\hbar)t}$$

where "i" and "f" refer to the initial and final states.[15] Because $E_i$ and $E_f$ are different, this transitional state does not have a well-defined energy. Accordingly, in contrast to a stationary state, the temporal parts of the wave function are very important. The probability density is

$$\Psi_t^*(\mathbf{r},\ t)\Psi_t(\mathbf{r},\ t)$$

$$= [\psi_i(\mathbf{r})e^{-i(E_i/\hbar)t} + \psi_f(\mathbf{r})e^{-i(E_f/\hbar)t}]^* [\psi_i(\mathbf{r})e^{-i(E_i/\hbar)t} + \psi_f(\mathbf{r})e^{-i(E_f/\hbar)t}]$$

$$= |\psi_i(\mathbf{r})|^2 + |\psi_f(\mathbf{r})|^2 + \psi_i^*(\mathbf{r})\psi_f(\mathbf{r})e^{i[(E_i-E_f)/\hbar]t} + \psi_f^*(\mathbf{r})\psi_i(\mathbf{r})e^{i[(E_f-E_i)/\hbar]t}$$

$$\tag{6-44}$$

Bearing in mind that $e^{i\omega t} = \cos(\omega t) + i\sin(\omega t)$, we see that this is an oscillatory function of time, with an angular frequency $\omega = (E_i - E_f)/\hbar$, just right to generate a photon whose energy is $\hbar\omega = E_i - E_f$. This does answer the simple qualitative question of how it is even possible to generate a photon (there is *no* charge density oscillation in a *stationary* state). But to gain a better quantitative understanding, we must investigate more closely the actual process of generating electromagnetic radiation.

The most effective way to generate electromagnetic radiation is via an oscillating electric dipole.[16] Consequently, if an oscillating charge distribution is to be an effective radiator, it should have an electric dipole moment. For point charges $+q$ and $-q$ separated by a distance $d$, the electric dipole moment vector is by definition $\mathbf{p} = qd$, with its direction being from the negative charge toward the positive. Diagram (a) of Figure 6.17 shows such a dipole, while diagram (b) indicates how we might produce an electric dipole radiator/antenna.

In a hydrogen atom, the dipole moment is not quite so simple. It has a compact positive charge (i.e., the proton), but its negative charge is an electron "cloud." Still, $\mathbf{p}$ has an expectation value. With charges of $+e$ at the origin and $-e$ at the point $\mathbf{r}$, the dipole moment would be by definition $-e\mathbf{r}$. Multiplying by the probability of finding the electron at the point $\mathbf{r}$ and integrating over all $\mathbf{r}$ then gives

$$\overline{\mathbf{p}} = \int_{\text{all space}} (-e\mathbf{r})|\Psi(\mathbf{r},\ t)|^2\ dV \tag{6-45}$$

Now, if a function changes sign when $\mathbf{r}$ is replaced by $-\mathbf{r}$, we say that it is "of odd parity." A function "of even parity" is unchanged by the substitution. (These have the same meanings as "odd" and "even" functions, but without the word

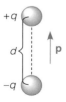

**(a)** Electric dipole

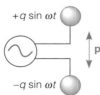

**(b)** Dipole antenna

**Figure 6.17** An oscillating dipole is an antenna emitting radiation.

[15]To be more realistic, allowance should be made for varying proportions of the two states. As it is, the initial and final states are added in equal amounts. However, since the conclusions we draw are not dependent on the proportions, we choose the simplest case.

[16]This assertion is proved in most junior-level texts on electricity and magnetism.

"parity" these terms can sometimes be ambiguous.) If $\Psi$ is of *either* even or odd parity, its square will be of even parity, and the integrand of odd parity. The integral would be zero and there would be no net dipole moment. An important question, then, is whether $\Psi$ has *any* parity. (A function need not be of either parity, e.g., $1 + x$.)

Let us consider the effect on the hydrogen atom wave functions of replacing the position vector $\mathbf{r}$ by $-\mathbf{r}$. In spherical polar coordinates, this involves replacing $\phi$ with $\phi + \pi$ and $\theta$ with $\pi - \theta$. Therefore, the radial wave part $R(r)$ has no bearing on whether a wave function is of even or odd parity. (We do *not* replace $r$ with $-r$, because the scalar radius $r$ is the same at $-\mathbf{r}$ as at $\mathbf{r}$.) For all angular solutions $\Theta(\theta)\Phi(\phi)$, the replacement returns the same function for even values of $\ell$ and the negative of the function for odd values of $\ell$. (It is left as an exercise to verify this for the angular solutions in Table 6.6.) In other words, the wave function $\psi(\mathbf{r}) = R(r)\Theta(\theta)\Phi(\phi)$ is either of even parity or odd parity depending on whether $\ell$ is even or odd. Therefore, *if the electron is in a single stationary state, the probability density $|\Psi|^2$ would always be even, and the atom could have no oscillating electric dipole moment.*

But what of a transitional state? The first two terms of (6-44) are squares of functions of either even or odd parity, and so are even. For the purpose of calculating dipole moment, they may be ignored. The last two terms are complex conjugates of one another. When added, the result is twice the real part of either. Thus,

$$\overline{\mathbf{p}} = -2e \ \text{Re}\left[ e^{i[(E_f - E_i)/\hbar]t} \int_{\text{all space}} \mathbf{r}\, \psi_f^*(\mathbf{r})\psi_i(\mathbf{r}) \ dV \right] \qquad (6\text{-}46)$$

The product $\psi_f^*(\mathbf{r})\psi_i(\mathbf{r})$ will be odd and the dipole moment nonzero only if the initial and final states are of different parity. Thus, since parity depends on $\ell$, we see that $\ell$ must change by an odd number. Actually, even this restriction does not guarantee that an "electric dipole transition" is possible. It turns out that $\Delta\ell$ must be $\pm 1$, and, furthermore, that $\Delta m_\ell$ must be 0 or $\pm 1$.[17] These conditions are known as **selection rules.**

---

### Example 6.11

A hydrogen $3p$ electron is in the state $(n, \ell, m_\ell) = (3, 1, +1)$. Which downward transitions are forbidden by the electric-dipole radiation selection rules?

### Solution

A transition to the $1s$ state is allowed, as is a transition to the $2s$ state; both have $\ell = 0$ and $m_\ell = 0$. But a transition to a $2p$ state, regardless of $m_\ell$, is forbidden because $\Delta\ell$ would be zero.

---

Any time there is great symmetry in the potential energy, selection rules are liable to crop up. For instance, in a diatomic molecule the interparticle force keeps the oscillating atoms at a certain average separation. For large oscillations, the force is stronger for repulsion than for attraction (cf. Figure 4.15). But for small oscillations it behaves essentially as a symmetric simple harmonic oscillator and,

[17]The reason has to do with an intrinsic property of the photon; its nature prevents it from carrying away from an atom more than a certain, fixed amount of angular momentum. The property of intrinsic angular momentum, or "spin," is discussed in Chapter 7.

by arguments very similar to those above, electric dipole transitions can then occur only between states for which $\Delta n = \pm 1$.

## Transition Rate

Classically, the rate at which an oscillating electric dipole radiates energy is proportional to the time average of the square of the dipole moment and to the fourth power of the frequency.[18] In SI units, the relationship is

$$\frac{\text{energy}}{\text{time}} \cong 10^{-16} p_{\text{t-avg}}^2 \omega^4 \quad \text{(SI units)}$$

By relating this energy emission rate to a photon emission rate, we gain a good idea of the factors governing the average lifetime of an excited state.

$$\frac{\text{energy}}{\text{photon}} \frac{\text{photon}}{\text{time}} \cong 10^{-16} p_{\text{t-avg}}^2 \omega^4 \quad \rightarrow \quad \hbar\omega \frac{\text{photon}}{\text{time}} \cong 10^{-16} p_{\text{t-avg}}^2 \omega^4$$

$$\Rightarrow \quad \text{lifetime} = \frac{\text{time}}{\text{photon}} \cong \frac{10^{-18}}{p_{\text{t-avg}}^2 \omega^3} \quad \text{(SI units)} \tag{6-47}$$

The larger the dipole moment and the higher the frequency, the more rapidly the transition would be expected to occur. To quantify this, we must calculate the time average of the square of (6-46). For convenience, we make the following definitions:

$$\int_{\text{all space}} \mathbf{r}\, \psi_f^*(\mathbf{r}) \psi_i(\mathbf{r})\, dV \equiv \overline{\mathbf{r}_{fi}} \qquad \Delta E \equiv E_f - E_i$$

Now using $\text{Re}[(x_1 + iy_1)(x_2 + iy_2)] = x_1 x_2 - y_1 y_2$, equation (6-46) becomes

$$\overline{\mathbf{p}} = -2e\left[\cos\left(\frac{\Delta E}{\hbar}t\right) \text{Re } \overline{\mathbf{r}_{fi}} - \sin\left(\frac{\Delta E}{\hbar}t\right) \text{Im } \overline{\mathbf{r}_{fi}}\right] \tag{6-48}$$

Squaring this vector, we obtain

$$\overline{\mathbf{p}}^2 = 4e^2\left[\cos^2\left(\frac{\Delta E}{\hbar}t\right)(\text{Re } \overline{\mathbf{r}_{fi}})^2 + \sin^2\left(\frac{\Delta E}{\hbar}t\right)(\text{Im } \overline{\mathbf{r}_{fi}})^2\right.$$

$$\left. - 2\cos\left(\frac{\Delta E}{\hbar}t\right)\sin\left(\frac{\Delta E}{\hbar}t\right)(\text{Re } \overline{\mathbf{r}_{fi}}) \cdot (\text{Im } \overline{\mathbf{r}_{fi}})\right]$$

When averaged over a whole period ($T = 2\pi/\omega = 2\pi\hbar/\Delta E$), the last term integrates to zero, while the squares of sine and cosine in the first two terms average to $\frac{1}{2}$. Thus, we obtain the simple result that

$$\overline{\mathbf{p}}_{\text{t-avg}}^2 = 4e^2\frac{1}{2}[(\text{Re } \overline{\mathbf{r}_{fi}})^2 + (\text{Im } \overline{\mathbf{r}_{fi}})^2] = 2e^2|\overline{\mathbf{r}_{fi}}|^2$$

---

### Example 6.12

A hydrogen atom emits a photon as it makes a transition to the ground state from the $(n, \ell, m_\ell) = (2, 1, +1)$ state. (a) What is the approximate dipole separation during the transition? (b) What is the approximate lifetime of the state?

[18]This is invariably proven along with the earlier statement—that an electric dipole is the most efficient radiator.

## Solution

(a) We calculate $\overline{\mathbf{r}_{fi}}$ for the $2p$ to $1s$ transition:

$$\overline{\mathbf{r}_{1s2p}} = \int_{\text{all space}} \mathbf{r}\, \psi^*_{1s}(\mathbf{r})\psi_{2p}(\mathbf{r})\, dV$$

Using $\mathbf{r} = x\hat{\mathbf{x}} + y\hat{\mathbf{y}} + z\hat{\mathbf{z}} = r\sin\theta\cos\phi\hat{\mathbf{x}} + r\sin\theta\sin\phi\hat{\mathbf{y}} + r\cos\theta\hat{\mathbf{z}}$ [19] and the wave functions of Tables 6.5 and 6.6, we have

$$\overline{\mathbf{r}_{1s2p}} = \int_{\text{all space}} (r\sin\theta\cos\phi\hat{\mathbf{x}} + r\sin\theta\sin\phi\hat{\mathbf{y}} + r\cos\theta\hat{\mathbf{z}})$$

$$\times \left[ \frac{1}{(1a_0)^{3/2}} 2e^{-r/a_0} \sqrt{\frac{1}{4\pi}} \right]$$

$$\times \left[ \frac{1}{(2a_0)^{3/2}} \frac{r}{\sqrt{3}\,a_0} e^{-r/2a_0} \sqrt{\frac{3}{8\pi}} \sin\theta\, e^{i\phi} \right] r^2\sin\theta\, dr\, d\theta\, d\phi$$

$$= \frac{1}{8\pi a_0^4} \int_0^\infty r^4 e^{-3r/2a_0}\, dr$$

$$\times \int_{\theta=0}^{\theta=\pi} \int_{\phi=0}^{\phi=2\pi} (\sin^3\theta\cos\phi\, e^{i\phi}\hat{\mathbf{x}} + \sin^3\theta\sin\phi\, e^{i\phi}\hat{\mathbf{y}}$$

$$+ \sin^2\theta\cos\theta\, e^{i\phi}\hat{\mathbf{z}})\, d\theta\, d\phi$$

Evaluation of the integrals is left to the reader. The result is

$$\overline{\mathbf{r}_{1s2p}} = \frac{1}{8\pi a_0^4} \frac{4!}{(3/2a_0)^5} \left( \frac{4}{3}\pi\hat{\mathbf{x}} + i\frac{4}{3}\pi\hat{\mathbf{y}} + 0\hat{\mathbf{z}} \right) = \frac{128}{243} a_0(\hat{\mathbf{x}} + i\hat{\mathbf{y}})$$

Thus,

$$|\overline{\mathbf{r}_{fi}}|^2 = \left( \frac{128}{243} \right)^2 a_0^2 \times 2 = 0.555 a_0^2 \;\Rightarrow\; |\overline{\mathbf{r}_{fi}}| = 0.74 a_0$$

Not surprisingly, the dipole separation is comparable to the Bohr radius.[20]

(b) To use (6-47), we must calculate $p_{t\text{-avg}}^2$ and $\omega$.

$$\overline{\mathbf{p}}_{t\text{-avg}}^2 = 2e^2 |\overline{\mathbf{r}_{fi}}|^2 = 2e^2(0.555 a_0^2)$$

$$= 2(1.6 \times 10^{-19}\text{ C})^2(0.555)(0.053 \times 10^{-9}\text{ m})^2$$

$$= 8.0 \times 10^{-59}\text{ C}^2\cdot\text{m}^2$$

$$\omega = \frac{|E_i - E_f|}{\hbar}$$

$$= \frac{\left| \dfrac{-13.6\text{ eV}}{2^2} - \dfrac{-13.6\text{ eV}}{1^2} \right| \times 1.6 \times 10^{-19}\text{ J/eV}}{1.055 \times 10^{-34}\text{ J}\cdot\text{s}}$$

$$= 1.55 \times 10^{16}\text{ s}^{-1}$$

[19]The reader may wonder why we break the position vector into Cartesian components, rather than using $\mathbf{r} = r\hat{\mathbf{r}}$. The unit vectors in spherical polar coordinates vary in direction (as functions of $\theta$ and $\phi$). Integration is easier using ones that do not.

[20]*Note:* According to (6-48),

$$\overline{\mathbf{p}} = -2e\left[ \frac{128}{243} a_0\hat{\mathbf{x}} \cos\left( \frac{\Delta E}{\hbar}t \right) \right.$$

$$\left. - \frac{128}{243} a_0\hat{\mathbf{y}} \sin\left( \frac{\Delta E}{\hbar}t \right) \right]$$

The dipole moment is real and has both $x$- and $y$-components that oscillate in time.

Inserting in (6-47),

$$\text{lifetime} \cong \frac{10^{-18}}{p_{\text{t-avg}}^2 \omega^3}$$

$$= \frac{10^{-18}}{(8.0 \times 10^{-59}\ \text{C}^2 \cdot \text{m}^2)(1.55 \times 10^{16}\ \text{s}^{-1})^3} = 3.4 \times 10^{-9}\ \text{s}$$

Typical electric dipole transitions are found to occur in about $10^{-8}$ s. This value is in good agreement.

In conclusion, it should be noted that transitions "forbidden" as electric dipole transitions may occur by other processes: The atom may for instance emit radiation as an oscillating *magnetic* dipole. Such an "antenna," however, is much less efficient (even classically) than an electric dipole, so transitions occurring by this process are much slower, typically by a factor of $10^{-4}$. Even more complicated processes are possible (e.g., electric quadrapole), but these are still less efficient and thus slower. Transitions that can occur only by slow processes cause certain states in some atoms to be unusually long lived. These are known as metastable states, and are important to the operation of a laser, discussed in Chapter 8.

## Progress and Applications

Quantum mechanics explains much about the atom. Still, until fairly recently the ability to "hold" atoms and judge the predictions has been very limited. The **ion trap** is doing just this, and a host of important observations is being made. (Other examples are found in the Progress and Applications in Chapter 7.)

■ Recent advances have allowed study of highly ionized and hydrogenlike (i.e., single-electron) atoms over essentially the entire periodic table, up to $U^{91+}$! In the electron-beam ion trap shown in Figure 6.18, an electron beam, with

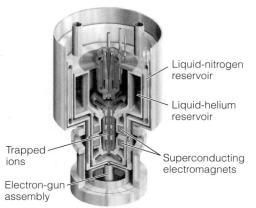

**Figure 6.18** Electron beam ion trap.

Liquid-nitrogen reservoir

Liquid-helium reservoir

Trapped ions

Superconducting electromagnets

Electron-gun assembly

0 1 2 in.

some help from superconducting magnets, confines ions while stripping many of their orbiting electrons via repeated collisions. Highly ionized atoms play a major role in nuclear fusion reactors, energy release in stars, and x-ray lasers. Moreover, stripping away "excess" electrons promises to advance our understanding of relativistic and other "higher-order" effects (see Chapter 7) in "simple" atoms—interactions between the electron, the nucleus it orbits and, owing to vacuum fluctuations, the very space itself.

■ Atomic electrons in excited states usually drop to the ground state within about 10 ns. Recently, an ionized ytterbium atom held in an ion trap was "bumped" up to an excited state whose lifetime was estimated at 10 years! (M. Roberts, et al., *Physical Review Letters,* March 10, 1997, pp. 1876–1879; P. Rigby, *Nature,* March 20, 1997, p. 225.) This state is so long-lived because of the extremely inefficient way in which it must shed its energy, via an "octopole transition." A lifetime this long may be useful in a high-precision atomic clock.

■ The angular dependence of wave functions in the real three-dimensional world is crucial in the study of chemical bonding in molecules and solids, but has gained even more importance in connection with the new "high-$T_C$" superconductors (see Chapter 9). In earlier "low-$T_C$" superconductors, pairs of electrons were bound together in a state of zero angular momentum—a spherically symmetric "s-wave." Recent evidence strongly suggests that in the new superconductors the electrons pair up in a "d-wave." Such a state would have two units of angular momentum and would not be spherically symmetric. Indeed, some of the evidence comes from experimental observations that the superconductivity exhibits anisotropic (nonspherically symmetric) features.

## Basic Equations

**Time-dependent Schrödinger equation**

$$-\frac{\hbar^2}{2m}\nabla^2\Psi(\mathbf{r},\,t)\,+\,U(\mathbf{r})\Psi(\mathbf{r},\,t)\,=\,i\hbar\,\frac{\partial}{\partial t}\,\Psi(\mathbf{r},\,t) \qquad (6\text{-}1)$$

**Normalization**

$$\int_{\text{all space}}|\Psi(\mathbf{r},\,t)|^2\,dV\,=\,1 \qquad (6\text{-}2)$$

**Time-independent Schrödinger equation**

$$-\frac{\hbar^2}{2m}\nabla^2\psi(\mathbf{r})\,+\,U(\mathbf{r})\psi(\mathbf{r})\,=\,E\psi(\mathbf{r}) \qquad (6\text{-}3)$$

**Hydrogen atom potential energy**

$$U(r)\,=\,-\frac{1}{4\pi\epsilon_0}\frac{e^2}{r} \qquad (6\text{-}9)$$

**Hydrogen atom wave function**

$$\psi(r,\,\theta,\,\phi)\,=\,R(r)\Theta(\theta)\Phi(\phi)$$

**Angular momentum quantization**

$$L_z = m_\ell \hbar \qquad (m_\ell = 0, \pm 1, \pm 2, \ldots, \pm\ell) \qquad (6\text{-}23)$$

$$|L| = \sqrt{\ell(\ell + 1)}\,\hbar \qquad (\ell = 0, 1, 2, \ldots, n - 1) \qquad (6\text{-}30)$$

**Radial equation**

$$\frac{-\hbar^2}{2m}\frac{1}{r^2}\frac{d}{dr}\left(r^2\frac{d}{dr}\right)R(r) + \frac{\hbar^2\ell(\ell + 1)}{2mr^2}R(r) + U(r)R(r) = ER(r)$$

$$\underbrace{\qquad\qquad}_{\text{KE}_{\text{radial}}} \qquad \underbrace{\qquad}_{\text{KE}_{\text{rotational}}} \qquad \underbrace{}_{\substack{\text{potential} \\ \text{energy}}} \quad \underbrace{}_{\substack{\text{total} \\ \text{energy}}} \qquad (6\text{-}25)$$

**Energy quantization**

$$E = -\frac{me^4}{2(4\pi\epsilon_0)^2\hbar^2}\frac{1}{n^2} = -\frac{13.6\text{ eV}}{n^2} \qquad (n = 1, 2, 3, \ldots) \qquad (6\text{-}27)$$

**Bohr radius**

$$a_0 \equiv \frac{(4\pi\epsilon_0)\hbar^2}{me^2} = 0.0529\text{ nm} \qquad (6\text{-}31)$$

**Hydrogen atom normalization**

$$\int_0^\infty (R(r))^2 r^2\, dr = 1 \qquad (6\text{-}33) \qquad\qquad \int_0^\pi (\Theta(\theta))^2\, 2\pi \sin\theta\, d\theta = 1 \quad (6\text{-}34)$$

**Radial probability**

$$P(r) = \frac{dP}{dr} = r^2(R(r))^2 \qquad (6\text{-}36)$$

**Expectation value**

$$\overline{f} = \int_0^\infty f(r)P(r)\, dr \qquad (6\text{-}38)$$

**Most probable orbit radius**

$$r_{\text{most probable}} = n^2 a_0 \qquad (\ell = n - 1) \qquad (6\text{-}37)$$

## Summary

In three dimensions, the Schrödinger equation is

$$-\frac{\hbar^2}{2m}\nabla^2\Psi(\mathbf{r}, t) + U(\mathbf{r})\Psi(\mathbf{r}, t) = i\hbar\frac{\partial}{\partial t}\Psi(\mathbf{r}, t)$$

and the complex square of the wave function $|\Psi(\mathbf{r}, t)|^2$ is the probability per unit volume of finding the particle in the vicinity of the position $\mathbf{r}$. In bound states, boundary conditions are imposed on the wave function in all three dimensions, in whatever coordinate system is appropriate. This leads to three quantization conditions and three quantum numbers, which govern the allowed values of the quantized properties. Which physical properties are quantized depends on the system being considered, but in all cases energy is quantized. For each set of three quantum numbers there is a unique wave function. Generally speaking, the more symmetric the situation (the potential energy), the larger will be the number of different wave functions with the same energy—that is, the greater the degeneracy.

In the case of the hydrogen atom, the three quantum numbers are the magnetic quantum number $m_\ell$, which governs allowed values of the $z$-component of angular momentum; the orbital quantum number $\ell$, which governs the magnitude of the angular momentum; and the principal quantum number $n$, which governs energy. These values are given by

$$E = -\frac{me^4}{2(4\pi\epsilon_0)^2\hbar^2}\frac{1}{n^2} \qquad (n = 1, 2, 3, \ldots)$$

$$|L| = \sqrt{\ell(\ell + 1)}\,\hbar \qquad (\ell = 0, 1, 2, \ldots, n - 1)$$

$$L_z = m_\ell\hbar \qquad (m_\ell = 0, \pm 1, \pm 2, \ldots, \pm\ell)$$

There is a unique wave function $\psi(r, \theta, \phi) = R_{n,\ell}(r)\Theta_{\ell,m_\ell}(\theta)\Phi_{m_\ell}(\phi)$ for each set $(n, \ell, m_\ell)$. However, the energy of the hydrogen atom depends only on $n$. Consequently, the degeneracy, the number of unique states with the same energy but different $\ell$- and $m_\ell$-values, grows as $n^2$.

The probability density in the hydrogen atom varies with $r$, the distance from the origin/proton, and $\theta$, the polar angle with whatever "special" axis is established by experiment. The radius at which the electron would most likely be found increases with $n$ approximately according to

$$r \cong n^2 a_0 \quad \text{where} \quad a_0 = \frac{(4\pi\epsilon_0)\hbar^2}{me^2}$$

A quantum-mechanical treatment of the hydrogen atom provides the explanation for observations that cannot be explained classically, including the stability of the atom and the discrete wavelengths of light that it emits.

# EXERCISES

## Section 6.2

1. Show that the function

    $$Ae^{+\sqrt{C_x}\,x} + Be^{-\sqrt{C_x}\,x}$$

    cannot be zero at two points. This makes it an unacceptable solution for the infinite well, since it cannot be continuous with the wave functions outside the walls (which are zero).

2. For the cubic 3D infinite-well wave function

    $$\psi(x, y, z) = A \sin\frac{n_x\pi x}{L} \sin\frac{n_y\pi y}{L} \sin\frac{n_z\pi z}{L}$$

show that the correct normalization constant is $A = (2/L)^{3/2}$.

3. An electron is trapped in a semiconductor "quantum dot," in which it is confined to very small region in all three dimensions. If the lowest-energy transition is to produce a photon of 450-nm wavelength, what should be the well's width (assumed cubic)?

4. An electron is confined to a cubic 3D infinite well 1 nm on a side. (a) What are the three lowest *different* energies possible? (b) To how many different states do these three energies correspond?

5. An electron is trapped in a cubic 3D infinite well. In the states $n_x, n_y, n_z =$ (a) 2, 1, 1, (b) 1, 2, 1, and (c) 1, 1, 2,

what is the probability of finding the electron in the region $(0 \leq x \leq L, \frac{1}{3}L \leq y \leq \frac{2}{3}L, 0 \leq z \leq L)$? Discuss any differences in the three results.

6. Consider a cubic 3D infinite well. (a) How many different wave functions have the same energy as the one for which $(n_x, n_y, n_z) = (5, 1, 1)$? (b) Into how many different energy levels would this level split if the length of one side were increased by 5%? (c) Make a scale diagram, similar to Figure 6.3, illustrating the splitting according to energy of the previously degenerate wave functions. (d) Is there any degeneracy left? If so, how might it be "destroyed"?

7. Consider a cubic 3D infinite well of side length $L$. There are 15 identical particles of mass $m$ in the well, but for whatever reason no more than two particles can have the same wave function. (a) What is the lowest possible *total* energy? (b) In this minimum-total-energy state, at what point(s) would the highest-energy particle most likely be found? (*Note:* Knowing no more than its energy, the highest-energy particle might be in any of multiple wave functions open to it, and with equal probability.)

8. *The 2D Infinite Well:* In two dimensions, the Schrödinger equation is

$$\left(\frac{\partial^2}{\partial x^2} + \frac{\partial^2}{\partial y^2}\right)\psi(x, y) = -\frac{2m(E - U)}{\hbar^2}\psi(x, y)$$

(a) Given that $U$ is a constant, separate variables by trying a solution of the form $\psi(x, y, z) = f(x)g(y)$, then dividing by $f(x)g(y)$. Call the separation constants $C_x$ and $C_y$.

(b) For an infinite well,

$$U = \begin{cases} 0 & 0 < x < L, 0 < y < L \\ \infty & \text{otherwise} \end{cases}$$

What should $f(x)$ and $g(y)$ be outside this well? What functions would be acceptable standing-wave solutions for $f(x)$ and $g(y)$ inside the well? Are $C_x$ and $C_y$ positive, negative, or zero? Imposing appropriate conditions, find the allowed values of $C_x$ and $C_y$.

(c) How many independent quantum numbers are there?

(d) Find the allowed energies $E$.

(e) Are there energies for which there is not a unique corresponding wave function?

## Section 6.4

9. A particle is trapped in a spherical infinite well; the potential energy is zero for $r < a$ and infinite for $r > a$. (a) Does the analysis of Section 6.4 of the angular parts

of $\psi(r, \theta, \phi)$ apply in this case? Explain. (b) Given that the particle has the minimum possible *nonzero* angular momentum, what possible angles might its angular momentum make with the $z$-axis?

10. A particle orbiting due to an attractive central force has angular momentum $1.00 \times 10^{-33}$ kg·m/s. What $z$-components of angular momentum is it possible to detect?

11. (a) Show that the probability density that comes from the complex square of the function

$$\Phi_{m_\ell}(\phi) = Ae^{+im_\ell\phi} + Be^{-im_\ell\phi}$$

is not in general independent of $\phi$. (b) What condition(s) must be met by $A$ and/or $B$ if the probability density is to be spherically symmetric (i.e., independent of $\phi$)?

## Section 6.5

12. (a) For a one-dimensional particle in a box, what is the meaning of $n$? (Specifically, what does knowing it tell us?) (b) What is the meaning of $n$ for a hydrogen atom? (c) For a hydrogen atom, what is the meaning of $\ell$? Of $m_\ell$?

13. An electron is in an $\ell = 3$ state of the hydrogen atom. What possible angles might the angular momentum vector make with the $z$-axis?

14. How many different $3d$ states are there? What physical property (as opposed to quantum number) distinguishes them, and what different values may this property assume?

15. An electron is in an $n = 4$ state of the hydrogen atom. (a) What is its energy? (b) What properties besides energy are quantized, and what values might be found if these properties were to be measured?

16. In Table 6.4, the pattern that develops with increasing $n$ *suggests* that the number of different sets of $(\ell, m_\ell)$ values for a given value of $n$ (i.e., for the same energy) is $n^2$. Prove this mathematically, by summing the allowed values of $m_\ell$ for a given $\ell$ over the allowed values of $\ell$ for a given $n$.

17. Show that the angular normalization constant in Table 6.6 for the case $(\ell, m_\ell) = (1, 0)$ is correct.

18. Show that the normalization constant $\sqrt{15/32\pi}$ given in Table 6.6 for the angular parts of the $\ell = 2, m_\ell = \pm 2$ wave function is correct.

19. Verify the correctness of the normalization constant of the $2p$ radial wave function, given in Table 6.5 as

$$\frac{1}{(2a_0)^{3/2}\sqrt{3}\, a_0}$$

20. A hydrogen atom electron is in a $2p$ state. If no experiment has been done to establish a $z$-component of angu-

lar momentum, it is equally likely to be found with any allowed value of $L_z$. Show that if the probability densities for these different possible states are added (with equal weighting), the result is independent of both $\phi$ and $\theta$.

21. A wave function with a noninfinite wavelength—however approximate it might be—has nonzero momentum and thus nonzero kinetic energy. Even a single "bump" has kinetic energy. In either case, we can say that the function has kinetic energy because it has curvature—a second derivative. Indeed, the kinetic energy operator in any coordinate system involves a second derivative. The only function without kinetic energy would be a straight line. As a special case, this includes a constant, which may be thought of as a function with an infinite wavelength. By looking at the curvature *in the appropriate dimension(s),* answer the following: For a given $n$, is the kinetic energy solely (a) radial in the state of lowest $\ell$, that is, $\ell = 0$, and (b) rotational in the state of highest $\ell$, that is, $\ell = n - 1$?

22. We have noted that for a given energy/$n$, as $\ell$ increases, the motion is more like a circle at a constant radius, the rotational energy increasing as the rotational correspondingly decreases. But is the radial energy zero for the largest $\ell$ values? Calculate the ratio of expectation values, radial energy to rotational energy, for the $(n, \ell, m_\ell) = (2, 1, +1)$ state. Use the operators

$$\hat{KE}_{rad} = \frac{-\hbar^2}{2m} \frac{1}{r^2} \frac{\partial}{\partial r}\left(r^2 \frac{\partial}{\partial r}\right) \quad \text{and}$$

$$\hat{KE}_{rot} = \frac{\hbar^2 1(1+1)}{2mr^2}$$

(*Note:* Strictly speaking, the $KE_{rot}$ operator—discussed in Appendix G—depends on $r$, $\theta$, and $\phi$, but use of equation (6-21) puts it in a form dependent only on $r$.)

## Section 6.6

23. For an electron in the $(n, \ell, m_\ell) = (1, 0, 0)$ state in a hydrogen atom (a) write the solution of the time-independent Schrödinger equation, and (b) verify explicitly that it is a solution with the expected angular momentum and energy.

24. An electron in a hydrogen atom is in the $(n, \ell, m_\ell) = (2, 1, 0)$ state. (a) Calculate the probability that it would be found within 60° of the z-axis, irrespective of radius. (*Note:* This means within 60° of either the $+z$- or $-z$-direction, solid angles totaling $2\pi$ steradians, i.e., "half of space.") (b) Calculate the probability that it would be found between $r = 2a_0$ and $r = 6a_0$, irrespective of angle. (c) What is the probability that it would be found within 60° of the z-axis *and* between $r = 2a_0$ and $r = 6a_0$?

25. Calculate the probability that the electron in a hydrogen atom would be found within 30° of the x-y plane, irrespective of radius, for (a) $\ell = 0$, $m_\ell = 0$; (b) $\ell = 1$, $m_\ell = +1$; and (c) $\ell = 2$, $m_\ell = +2$. (d) As angular momentum increases, what happens to the orbits whose z-components of angular momentum are the maximum allowed?

## Section 6.7

26. What is the most probable distance from the proton of an electron in a 2s state?

27. An electron is in the 3d state of a hydrogen atom. The most probable distance of the electron from the proton is $9a_0$. What is the probability that the electron would be found between $8a_0$ and $10a_0$?

28. Using the functions given in Table 6.5, verify that for the more-circular electron orbits in hydrogen (i.e., $\ell = n - 1$) the radial probability is of the form

$$P(r) \propto r^{2n} e^{-(2r/na_0)}$$

Show that the most probable radius is given by

$$r_{most\ probable} = n^2 a_0$$

29. Consider an electron in the ground state of a hydrogen atom. (a) Sketch plots of $E$ and $U(r)$ on the same axes. (b) Show that, classically, an electron with this energy should not be able to get farther than $2a_0$ from the proton. (c) What is the probability of the electron being found in the classically forbidden region?

30. (a) What is the expectation value of the distance from the proton of a electron in a 3p state? (b) How does this compare with the expectation value in the 3d state, calculated in Example 6.8? Discuss any differences.

31. Imagine two classical charges of $-q$, each bound to a central charge of $+q$. One $-q$ charge is in a circular orbit of radius $R$ about its $+q$ charge. The other oscillates in an extreme ellipse, essentially a straight line from its $+q$ charge out to a maximum distance $r_{max}$. The two orbits have the same energy. (a) Show that $r_{max} = 2R$. (b) Considering the time spent at each orbit radius, in which orbit is the $-q$ charge on average farther from its $+q$ charge?

32. Consider an electron in the ground state of a hydrogen atom. (a) Calculate the expectation value of its potential energy. (b) What is the expectation value of its kinetic energy? (*Hint:* What is the expectation value of the total energy?)

33. Is the potential energy of an electron in a hydrogen atom well defined? Is the kinetic energy well defined? Justify your answers. (You need not actually calculate uncertainties.)

34. Calculate the uncertainties in $r$ for the 2s and 2p states

using the formula

$$\Delta r = \sqrt{\overline{r^2} - \overline{r}^2}$$

What insight does the difference between these two uncertainties convey about the nature of the corresponding orbits?

35. The energy of hydrogen atom wave functions for which $\ell$ is its minimum values of zero is all radial. This is the case for the $1s$ and $2s$ states. The $2p$ state has some rotational energy and some radial. Show that for very large $n$, the states of largest allowed $\ell$ have essentially no radial energy. Do this by comparing the total energy with the rotational energy $L^2/2mr^2$, assuming that $r \cong n^2 a_0$.

36. For the more circular orbits, $\ell = n - 1$ and

$$P(r) \propto r^{2n} e^{-2r/na_0}$$

(a) Show that the coefficient that normalizes this probability is

$$\left(\frac{na_0}{2}\right)^{2n+1} \frac{1}{(2n)!}$$

(b) Show that the expectation value of the radius is given by

$$\overline{r} = n\left(n + \frac{1}{2}\right)a_0$$

and the uncertainty by

$$\Delta r = \sqrt{\frac{n}{2} + \frac{1}{4}}\, na_0$$

(c) What happens to the ratio $\Delta r/\overline{r}$ in the limit of large $n$? Is this large-$n$ limit what would be expected classically?

## Sections 6.8 and 6.9

37. (a) What are the initial and final energy levels for the third (i.e., third-longest wavelength) line in the Paschen series? (See Figure 6.15.) (b) Determine the wavelength of this line.

38. Calculate the "series limit" of the Lyman series of spectral lines. This is defined as the shortest wavelength possible of a photon emitted in a transition from a higher initial energy level to the $n_f = 1$ final level. (Note in Figure 6.15 that the spectral lines of the series "crowd together" at the short-wavelength end of the series.)

39. The only *visible* spectral lines of hydrogen are the four Balmer series lines noted at the beginning of Section 6.8. It is desired to cause hydrogen gas to glow with its characteristic visible colors.

(a) To how high an energy level must the electrons be excited?

(b) Energy is absorbed in collisions with other particles. Assume that after absorbing energy in one collision, an electron jumps down through lower levels so rapidly that it is in the ground state before another collision occurs. If an electron is to be raised to the level found in part (a), how much energy must be available in a single collision?

(c) If such energetic collisions are to be effected simply by heating the gas until the average kinetic energy equals the desired upward energy jump, what temperature would be required? (This explains why heating is an impractical way to observe the hydrogen spectrum. Instead, the atoms are ionized by strong electric fields, as is the air when a static electric spark passes through.)

40. A hydrogen atom absorbs a photon. (a) What wavelength should it be to cause the electron to jump from its ground state to the $n = 4$ state? (b) What wavelength photons might be emitted by the atom following the absorption?

41. To conserve momentum, an atom emitting a photon must recoil, meaning that not all of the energy made available in the downward jump goes to the photon. (a) Find a hydrogen atom's recoil energy when it emits a photon in a $n = 2$ to $n = 1$ transition. (*Note:* The calculation is easiest to carry out if it is assumed that the photon carries essentially all the transition energy, which thus determines its momentum. The result justifies the assumption.) (b) What fraction of the transition energy is the recoil energy?

## Section 6.10

42. Which electron transitions in singly ionized helium yield photons in the 450–500 nm (~blue) portion of the visible range, and what are their wavelengths?

43. Doubly ionized Lithium, $Li^{2+}$, absorbs a photon and jumps from the ground state to its $n = 2$ level. What was the wavelength of the photon?

## Section 6.11

44. Can the transition $2s \rightarrow 1s$ in the hydrogen atom occur by electric dipole radiation? Explain. (Ignore angular momentum conservation.)

45. Verify for the angular solutions $\Theta(\theta)\Phi(\phi)$ of Table 6.6 that replacing $\phi$ with $\phi + \pi$ and $\theta$ with $\pi - \theta$ gives the same function when $\ell$ is even and the negative of the function when $\ell$ is odd.

46. Show that a transition where $\Delta m_\ell = \pm 1$ corresponds to a dipole moment in the $x$-$y$ plane, while $\Delta m_\ell = 0$ corre-

sponds to a moment along the z-axis. (You need consider only the $\phi$-parts of $\psi_i$ and $\psi_f$, which are of the form $e^{im_\ell\phi}$.)

47. Calculate the average time for a hydrogen atom electron in the $(n, \ell, m_\ell) = (3, 2, 0)$ state to make an electric dipole transition to the $(2, 1, 0)$.

**Exercises 48 and 49 refer to a vibrating diatomic molecule that may be treated as a simple harmonic oscillator of mass $10^{-27}$ kg, spring constant $10^3$ N/m, and charge $+e$.**

48. Show that a transition from the $n = 2$ state directly to the $n = 0$ ground state cannot occur by electric dipole radiation.

49. (a) Calculate the lifetime of the first excited state, assuming that it decays by electric dipole radiation to the ground state. (b) What is the wavelength of the photon emitted?

## General Exercises

50. Consider two particles that experience a mutual force but no external forces. The classical equation of motion for particle 1 is $\dot{\mathbf{v}}_1 = \mathbf{F}_{2\text{ on }1}/m_1$ and for particle 2 is $\dot{\mathbf{v}}_2 = \mathbf{F}_{1\text{ on }2}/m_2$, where the dot means a time derivative. Show that these are equivalent to

$$\mathbf{v}_{cm} = \text{constant} \quad \text{and} \quad \dot{\mathbf{v}}_{rel} = \mathbf{F}_{mutual}/\mu$$

where $\mathbf{v}_{cm} \equiv (m_1\mathbf{v}_1 + m_2\mathbf{v}_2)/(m_1 + m_2)$, $\mathbf{F}_{mutual} \equiv \mathbf{F}_{1\text{ on }2} = -\mathbf{F}_{2\text{ on }1}$ and

$$\mu \equiv \frac{m_1 m_2}{m_1 + m_2}$$

In other words, the motion can be analyzed in two pieces: the center of mass motion, at constant velocity; and the relative motion, but in terms of a one-particle equation where that particle experiences the mutual force and has the "reduced mass" $\mu$.

51. The $\psi_{2,1,0}$ state, the 2p state in which $m_\ell = 0$, has most of its probability density along the z-axis, and so is often referred to as a $2p_z$ state. To allow its probability density to "stick out" in other ways, and thus facilitate various kinds of molecular bonding with other atoms, an atomic electron may assume a wave function that is an algebraic combination of multiple wave functions open to it. One such "hybrid state" is the sum $\psi_{2,1,+1}$ + $\psi_{2,1,-1}$. (*Note:* Since the Schrödinger equation is a linear differential equation, a sum of solutions with the same energy is a solution with that energy. Also, normalization constants may be ignored in the following questions.)
    (a) Write this wave function and its probability density in terms of $r$, $\theta$, and $\phi$.

(b) In which of the following ways does this hybrid state differ from its "parts" (i.e., $\psi_{2,1,+1}$ and $\psi_{2,1,-1}$) and from the $2p_z$ state: energy; radial dependence of its probability density; angular dependence of its probability density?
    (c) This state is often referred to as the $2p_x$. Why?
    (d) How might we produce a $2p_y$ state?

52. Spectral lines are fuzzy due to two effects: Doppler broadening and the uncertainty principle. The relative variation in wavelength due to the first effect (see Exercise 1.35) is given by

$$\frac{\Delta\lambda}{\lambda} \cong \frac{\sqrt{3k_BT/m}}{c}$$

where $T$ is the temperature of the sample and $m$ the mass of the particles emitting the light. The variation due to the second effect (see Exercise 3.50) is given by

$$\frac{\Delta\lambda}{\lambda} \cong \frac{\lambda}{4\pi c\ \Delta t}$$

where $\Delta t$ is the typical transition time.
    (a) Suppose the hydrogen in a star has a temperature of $10^5$ K. Compare the broadening of these two effects for the first line in the Balmer series (i.e., $n_i = 3 \to n_f = 2$). Assume a transition time of $10^{-8}$ s. Which effect is more important?
    (b) Under what condition(s) might the other effect predominate?

53. A spherical infinite well has potential energy,

$$U(r) = \begin{cases} 0 & r < a \\ +\infty & r > a \end{cases}$$

Since this is a central force, we may use the Schrödinger equation in the form (6-25)—that is, just before the specific hydrogen atom potential energy is inserted. Show that the following is a solution

$$R(r) = A(\sin\ br)/r$$

Now apply the appropriate boundary conditions and in so doing find the allowed angular momenta and energies for solutions of this form.

54. Residents of Flatworld (a two-dimensional world far, far away) have it easy. Although quantum mechanics (of course) applies in their world, the equations they must solve to understand atomic energy levels involve only two dimensions. In particular, the Schrödinger equation for the one-electron "flatrogen" atom is

$$-\frac{\hbar^2}{2m}\frac{1}{r}\frac{\partial}{\partial r}\left(r\frac{\partial}{\partial r}\right)\psi(r,\theta) - \frac{\hbar^2}{2m}\frac{1}{r^2}\frac{\partial^2}{\partial\theta^2}\psi(r,\theta)$$
$$+ U(r)\psi(r,\theta) = E\psi(r,\theta)$$

(a) Separate variables by trying a solution of the form $\psi(r, \theta) = R(r)\Theta(\theta)$, then dividing by $R(r)\Theta(\theta)$. Show that the "$\theta$-equation" can be written

$$\frac{d^2}{d\theta^2}\Theta(\theta) = C\Theta(\theta)$$

where $C$ is a separation constant.

(b) To be physically acceptable, $\Theta(\theta)$ must be continuous, which, since it involves rotation about an axis, means that it must be periodic. What must be the sign of $C$?

(c) Show that a complex exponential is an acceptable solution for $\Theta(\theta)$.

(d) Imposing the periodicity condition, find the allowed values of $C$.

(e) What property is quantized according to the value of $C$?

(f) Obtain the "$r$-equation."

(g) Given that $U(r) = -b/r$, show that a function of the form $R(r) = e^{-r/a}$ is a solution, but only if $C$ is a certain one of its allowed values.

(h) Determine the value of $a$, and thus find the ground-state energy and wave function of the flatrogen atom.

# Spin and Atomic Physics

I t is often said that in quantum mechanics there are only three bound-state problems solvable "in closed form" (i.e., without numerical methods): the infinite well, the harmonic oscillator, and the simple hydrogen atom. Each is essentially a one-particle problem. (In the hydrogen atom, there could be no potential energy without the proton, but it is ignored otherwise.) We must now confront the "unsolvable." Most real applications are multiple-particle systems, and the logical place to start is an atom with multiple electrons. Still it is true that once multiple electrons orbit the nucleus, the problem is mathematically unsolvable; it is impossible to solve a Schrödinger equation that takes into account the potential energy shared by each electron with the nucleus *and* the potential energy shared between pairs of electrons. The prospect is not so bleak as it might seem. Much may indeed be learned by application of the principles we have studied. However, the first section of this chapter is devoted to a study of an intrinsic property of individual particles: spin. Spin must be understood first because, while of little consequence in a one-particle problem, it is crucial to an understanding of many-particle systems. We introduce this "new" property in the context of a search for evidence of angular momentum quantization in the hydrogen atom.

## 7.1 Evidence of Angular Momentum Quantization: A New Property

The claim that angular momentum is quantized, that a component may take on only certain values along an axis established by experiment, is easily made but perhaps not so easily justified. The hope, of course, is to find a means to either validate or refute the claim once and for all. The first success in this direction sprang from a simple observation: Since any charged particle with an angular momentum is a circulating electric current, it should possess a magnetic dipole moment, and if an atom's angular momentum may take on only certain values, so should its magnetic dipole moment. Therefore, since a magnetic field may be used to exert a force on a magnetic dipole, the magnetic force on the atom should in some sense be quantized.

## Angular Momentum and Magnetic Dipole Moment

Imagine a classical electron orbiting counterclockwise in a circle of radius $r$. Being negatively charged, it represents a clockwise (conventional) current of magnitude $I = e/T$, where $e$ is the fundamental charge and $T$ the period of revolution. Magnetic dipole moment is given by $\mu = IA$, where $A$ is the area encircled by a current loop. Thus,

$$\mu = IA = \frac{e}{T}\pi r^2 = \frac{e}{2\pi r/v}\pi r^2 = \frac{e}{2}vr = \frac{e}{2m_e}(m_e vr) = \frac{e}{2m_e}L$$

where $m_e$ is the electron mass. (In the last step we used the classical formula relating the angular momentum to the mass, speed, and radius of an orbiting point mass.) By the usual right-hand rules, the electron's angular momentum in Figure 7.1 is up and its magnetic dipole moment is down. Therefore, we may write

$$\boldsymbol{\mu_L} = -\frac{e}{2m_e}\mathbf{L} \qquad\qquad (7\text{-}1)$$

(The reason for the subscript on $\mu$ will be discussed shortly.) This derivation is a classical one, based on the assumption that an electron is a particle. It is primarily meant to illustrate simply that a charged orbiting particle has a magnetic dipole moment. But it happens that the result is quantum-mechanically correct; it correctly describes the relationship between the angular momentum of an electron matter wave and the electron's magnetic dipole moment. Let us now see how this relationship may be exploited to reveal angular momentum quantization.

In an external magnetic field $\boldsymbol{B}$, the potential energy of a magnetic dipole is

$$U = -\boldsymbol{\mu}\cdot\mathbf{B}$$

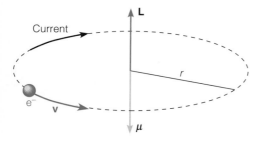

**Figure 7.1**  A charge with angular momentum has a magnetic dipole moment.

Force is the negative gradient of potential energy.

$$\mathbf{F} = -\boldsymbol{\nabla}(-\boldsymbol{\mu}\cdot\mathbf{B}) = \boldsymbol{\nabla}(\mu_x B_x + \mu_y B_y + \mu_z B_z) \tag{7-2}$$

Magnetic dipole moment is an overall property of a current distribution and so is not a function of position. If **B** were also constant (uniform), the potential energy would be constant and the force would be zero; as we learn in electricity and magnetism, the net force on a dipole in a uniform field is zero. To exert a force, the field must be nonuniform.

However, a magnetic dipole in *any* magnetic field experiences a torque $\boldsymbol{\tau} = \boldsymbol{\mu} \times \mathbf{B}$, which causes it to precess about the *B*-field line, as does a gyroscope about a gravitational field line. Suppose **B** points in the *z*-direction. The torque due to this external influence, while not altering $\mu_z$, necessarily forces $\mu_x$ and $\mu_y$ to change continuously, as shown in Figure 7.2. (See the essay entitled "Precession Rate in a Magnetic Field.") Whatever direction **B** points becomes a "special" axis—which we always call the *z*-axis—and it will be impossible to investigate simultaneously any preexisting quantization of $\boldsymbol{\mu}$ along another axis. Since magnetic dipole moment and angular momentum are directly related, it will also be impossible to investigate simultaneously any preexisting quantization of **L** along another axis. We have uncovered an intriguing but satisfying truth: *The theoretical conclusion, from solving the hydrogen atom Schrödinger equation, that only one component of angular momentum is quantized, is in accord with the experimental impossibility of observing quantization of more than one component.* Since all means of probing possible quantization of angular momentum would employ some kind of inherently directional force, there is none that does not, in and of itself, establish a preferred direction—that is, a "*z*-axis."

## Precession Rate in a Magnetic Field

The precessional frequency of the dipole in a magnetic field, called the **Larmor frequency,** is determined as follows: Since $\boldsymbol{\mu}$ and **L** are in opposite directions, the torque $\boldsymbol{\mu} \times \mathbf{B}$ is perpendicular to **B**, $\boldsymbol{\mu}$, and **L**. By the rotational second law $\boldsymbol{\tau} = d\mathbf{L}/dt$, the change in angular momentum $d\mathbf{L}$ is in the direction of the torque. Combining these two observations, $d\mathbf{L}$ is perpendicular to **B** and **L**; that is, the angular momentum vector maintains the same angle with **B**, while its point describes a circle. It precesses in the direction shown in Figure 7.2. Now consider magnitudes. For a precession of $d\phi$, the magnitude $dL$ is $(L\sin\theta)d\phi$, where $L\sin\theta$ is the component of **L** perpendicular to the *z*-axis. Thus,

$$|\boldsymbol{\tau}| = |\boldsymbol{\mu} \times \mathbf{B}| \rightarrow \left|\frac{d\mathbf{L}}{dt}\right| = \left|\left(-\frac{e}{2m_e}\mathbf{L}\right) \times \mathbf{B}\right|$$

$$\frac{dL}{dt} = \frac{e}{2m_e}LB\sin\theta$$

$$\frac{(L\sin\theta)\,d\phi}{dt} = \frac{e}{2m_e}LB\sin\theta$$

$$\frac{d\phi}{dt} = \frac{eB}{2m_e}$$

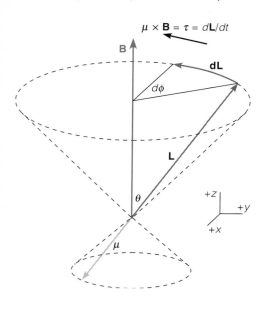

$$\mu \times \mathbf{B} = \tau = d\mathbf{L}/dt$$

**Figure 7.2** Torque, $\mu \times \mathbf{B}$, perpendicular to $\mathbf{L}$ causes $\mathbf{L}$ to precess.

## The Stern–Gerlach Experiment

In the Stern–Gerlach experiment, the poles of a magnet are situated so as to produce a channel in which there is a nonuniform field. In hopes of observing a "quantization of force," a beam of atoms is sent through the channel and deposited at a screen beyond. If force may take on only certain discrete values, then the atoms passing through the channel should be deflected to only certain discrete points on the screen. Classically, orbiting electrons would also possess magnetic dipole moments and so should also experience a force, but since there is no classical reason for quantization along an axis established by the observer, the atoms should be deflected across a *continuous* band.

In Figure 7.3, the atom moves through the center of the apparatus's channel, which lies along the $x$-axis, and $\mathbf{B}$ is strictly in the $y$-$z$ plane. By symmetry, there

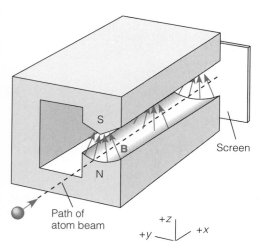

**Figure 7.3** Nonuniform magnetic field exerts a force on a magnetic dipole.

is no $y$-component of force, and since nothing varies with $x$, there is no $x$-component. Thus, only the $z$-component of equation (7-2) remains.

$$\mathbf{F} = \mu_z \frac{\partial B_z}{\partial z}\hat{\mathbf{z}}$$ (7-3)

(*Note:* $\partial B_y/\partial z = 0$ because $B_y = 0$ all along the $z$-axis.) We relate this force to angular momentum via equation (7-1), specifically through the equation's $z$-component.

$$\mathbf{F} = \left(-\frac{e}{2m_e}L_z\right)\frac{\partial B_z}{\partial z}\hat{\mathbf{z}}$$

Again, if $L \neq 0$, the classical expectation is that $L_z$ could be any of the continuum of values from $-L$ to $+L$, leading to a continuous band. But if the $z$-component of angular momentum is quantized, the force exerted on the atom should also be quantized. Using the quantization condition $L_z = m_\ell \hbar$,

$$\mathbf{F} = -\frac{e}{2m_e}(m_\ell\hbar)\frac{\partial B_z}{\partial z}\hat{\mathbf{z}} \quad (m_\ell = -\ell, \ldots, +\ell)$$ (7-4)

Classical expectation
($L \neq 0$)

Suppose hydrogen atoms whose electrons happen to be in $\ell = 1$ states pass through the channel, precessing as they travel. Since the quantum number $m_\ell$ may take on three values, we should expect to see three lines at the screen, corresponding to an upward force ($m_\ell = -1$), a downward force ($m_\ell = +1$), and no force ($m_\ell = 0$), as shown in Figure 7.4.[1] In general, we should expect to see $2\ell + 1$ lines—an odd number.

When a beam of hydrogen atoms was used with the Stern–Gerlach apparatus, however, *two* lines were seen.[2] Were this not baffling enough, ground-state hydrogen atoms have $\ell = 0$, so we should expect the experiment to show but a *single* undeflected line, corresponding to zero force. The reader need not be alarmed. The deductions made thus far are correct; the splitting in the $\ell = 1$ case does occur (see Section 7.8). But splitting even in the $\ell = 0$ case reveals that a magnetic moment not previously considered is coming into play.

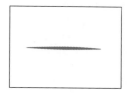
$\ell = 0$ prediction

$\ell = 0$ observation

## Spin

Electrons possess an **intrinsic magnetic dipole moment** and inextricably related **intrinsic angular momentum,** and it is the quantization of this, rather than of the dipole moment related to the *orbital* angular momentum **L,** that is revealed in the Stern–Gerlach experiment. Intrinsic angular momentum is given the name **spin** and the symbol **S.** (Remember: **S** is an angular momentum.) An **intrinsic property** is one that is fundamental to a given particle's nature, an inherent property that cannot be taken away no matter how we might try. Indeed, intrinsic properties are what distinguish one fundamental particle from another (e.g., an electron from

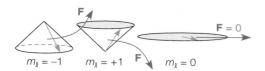
$m_\ell = -1$ $m_\ell = +1$ $m_\ell = 0$

**Figure 7.4** The force on a magnetic dipole depends on its component along the magnetic field, which is quantized.

[1]It is because an atom's behavior in a magnetic field depends on $m_\ell$ that it is known as the "magnetic quantum number."

[2]In the original Stern–Gerlach experiment of 1922, two lines were seen, but neutral *silver* atoms were used. Although the silver atoms, having one $5s$ electron beyond a closed $n = 4$ shell, should behave as atoms of $\ell = 0$, the Stern–Gerlach apparatus was later used with hydrogen (1925, Phipps and Taylor), to rule out any complication multiple electrons might introduce.

a photon). We are already familiar with two intrinsic properties: mass and charge. To these we now add spin.

Although an experimental fact, the very existence of spin is not predicted, nor is its nature explained via a nonrelativistic theory. It does arise naturally, but only from a quantum-mechanical theory that adheres to the principles of special relativity. (In particular, energy, upon which the Schrödinger equation and any relativistically correct generalization must be based, is not given by $E = p^2/2m$ for a free particle, but by $E^2 = p^2c^2 + m^2c^4$.) Relativistic quantum mechanics is beyond the scope of the text. Yet we must have some quantitative grasp of spin, because without it we cannot understand many areas of knowledge of the physical world, including notably most of chemistry. Consequently, we must be content with treating spin in an ad hoc[3] fashion, and the following results of relativistic quantum mechanics must simply be accepted.

1. For an electron, the relationship between intrinsic magnetic dipole moment and intrinsic angular momentum[4] is:

$$\boldsymbol{\mu}_{\mathbf{S}} = -g_e \frac{e}{2m_e} \mathbf{S} \quad (g_e \cong 2) \tag{7-5}$$

This differs from (7-1), the corresponding relationship between $\boldsymbol{\mu}_{\mathbf{L}}$ and $\mathbf{L}$, by $g_e$, the **electron gyromagnetic ratio**.[5] There is no classical explanation of this difference.

2. Somewhat analogous to the relationship $L = \sqrt{\ell(\ell + 1)}\hbar$ between the magnitude of the *orbital* angular momentum and the dimensionless value $\ell$, there is for *all* particles with intrinsic angular momentum $\mathbf{S}$ (not just electrons) a relationship between the magnitude $S$ and a dimensionless value $s$.

$$S = \sqrt{s(s + 1)}\hbar \tag{7-6}$$

In the case of spin, however, the value $s$ is not a quantum number that may take on different values, but instead depends on the kind of particle; it is a given fixed value for a given kind of particle. For an electron it is found that $s = \frac{1}{2}$; for a $W^-$ particle (a fundamental particle discussed in Chapter 11) it is found that $s = 1$; other examples are given in Table 7.2. Indeed, $s$ is the *measure* of this "new" intrinsic property of a given particle. Accordingly, it is often referred to as the "spin of the particle"—we say that the electron is "a spin-$\frac{1}{2}$ particle." This usage is common but somewhat ambiguous. For

---

[3] Expedient; without theoretical foundation.

[4] Because there are two angular momenta and two dipole moments, we adopt symbols to distinguish them. The magnetic dipole moment related to the electron's *orbital* angular momentum $\mathbf{L}$ (i.e., due to orbital motion) is $\boldsymbol{\mu}_{\mathbf{L}}$, while that related to its *intrinsic* angular momentum $\mathbf{S}$ is $\boldsymbol{\mu}_{\mathbf{S}}$.

[5] Introduction of this factor and its nomenclature might seem whimsical; it certainly could be dispensed with merely by putting its numerical value in (7-5). However, it is not exactly 2. Furthermore, there are similar relationships between $\boldsymbol{\mu}$ and $\mathbf{S}$ for other particles, but with different ratios, hence the modifier *electron* gyromagnetic ratio. For instance, $g_{\text{proton}} \cong 5.6$. (Nevertheless, $\boldsymbol{\mu}_{\mathbf{S}, \text{proton}} \ll \boldsymbol{\mu}_{\mathbf{S}, \text{electron}}$ because $m_p \gg m_e$, so the predominant effect in the Stern–Gerlach experiment is due to the electron.) Expression (7-5) becomes general if the "e" subscripts are omitted.

an electron, the magnitude of the intrinsic angular momentum, *also called spin*, is not $\frac{1}{2}$; it is $S = \sqrt{\frac{1}{2}(\frac{1}{2} + 1)}\hbar = \frac{\sqrt{3}}{2}\hbar$.

3. Completely analogous to $L_z = m_\ell \hbar$, the $z$-component of intrinsic angular momentum is inherently quantized; it may take on only those values given by

$$S_z = m_s\hbar \qquad (m_s = -s, -s + 1, \ldots, s - 1, s) \qquad (7\text{-}7)$$

The quantity $m_s$ is a new quantum number—the **spin quantum number.** The values allowed it are those between $-s$ and $+s$, in integral steps. These depend on $s$, but $s$ in turn depends on the kind of particle. Therefore, for a spin-$\frac{1}{2}$ particle such as an electron, $m_s$ may take on two values: $-\frac{1}{2}$ or $+\frac{1}{2}$.[6] For a $W^-$ particle, $m_s$ may take on the values $-1, 0$, or $+1$.

We may now understand the result in the Stern–Gerlach experiment. For $\ell = 0$, the *orbital* magnetic dipole moment $\boldsymbol{\mu}_L$ is zero, so it should not be subject to a force, but there should still be a force on the *intrinsic* magnetic dipole moment. Therefore, in equation (7-3) we should use, rather than $\boldsymbol{\mu}_L = -(e/2m_e)\,\mathbf{L}$ and $L_z = m_\ell \hbar$, the corresponding (7-5) and (7-7): Using 2 for $g_e$,

$$\mathbf{F} = \left(-2\frac{e}{2m_e}S_z\right)\frac{\partial B_z}{\partial z}\hat{\mathbf{z}} = -\frac{e}{m_e}(m_s\hbar)\frac{\partial B_z}{\partial z}\hat{\mathbf{z}} \qquad (m_s = -s, \ldots, +s) \qquad (7\text{-}8)$$

Because $m_s$ takes on two values for an electron, the force should be restricted to two values; there should indeed be two lines, corresponding to $m_s = -\frac{1}{2}$ and $m_s = +\frac{1}{2}$.[7]

---

### Example 7.1

Determine the acceleration of a hydrogen atom in a magnetic field whose rate of change is 10 T/m.

#### Solution

Equation (7-8) applies, where the electron's spin quantum number $m_s$ may be either of the two values $\pm\frac{1}{2}$.

$$F = -\frac{e}{m_e}(m_s\hbar)\frac{\partial B_z}{\partial z}$$

$$= -\frac{1.6 \times 10^{-19}\ \text{C}}{9.11 \times 10^{-31}\ \text{kg}}\left(\pm\frac{1}{2} \times 1.055 \times 10^{-34}\ \text{J·s}\right)10\ \text{T/m}$$

$$= \pm 9.3 \times 10^{-23}\ \text{N}$$

Although the force is on the electron's magnetic moment, the entire atom accelerates.

$$a = \frac{F}{m} = \frac{9.3 \times 10^{-23}\ \text{N}}{1.67 \times 10^{-27}\ \text{kg}} = 5.5 \times 10^4\ \text{m/s}^2$$

[6] For better or for worse, $m_s$ is also often referred to by the name "spin," as are $S$ and $s$! One often hears that "the electron is in a spin-plus-one-half state" or a "spin-minus-one-half state," meaning the value of $m_s$. In such cases, context is usually sufficient to avoid confusion. For example, all electrons by nature have $s = \frac{1}{2}$ and $S = \frac{\sqrt{3}}{2}\hbar$, so to use these terms in reference to an individual electron in a multielectron system is to refer unambiguously to its $m_s$-value.

[7] When $\ell$ is *not* zero, splitting due to the *orbital* magnetic dipole moment, which alone would give an odd number of lines, combines with the splitting due to the spin. This combining can be complicated and is taken up in Section 7.8.

The term "spin" suggests that an electron has angular momentum due to spinning on its axis, in addition to that due to its orbit about the nucleus, as does Earth while orbiting the Sun. However, helpful though the spinning sphere notion may be, the intrinsic angular momentum and magnetic dipole moment of an electron cannot be understood on the basis of a spinning charged object! *Spin is nonclassical to its very roots.* No assumptions of mass and/or charge densities circulating about an axis can explain the observations (see Exercise 1). Indeed, spin has nothing whatsoever to do with spatial motion, something that could presumably be altered; it is truly intrinsic. This would have to be the case in light of the experimental evidence that the electron has no minimum volume; apparently it occupies no space! Even accepting *point* angular momentum—angular momentum without spatial extent—the reader may well wonder how such a non-position-dependent property could ever be incorporated quantitatively into a wave function. In relativistic quantum mechanics, a convenient way to account for spin is via a matrix. This is reasonable: A matrix has a discrete number of elements, so it is well suited to describe a property that is inherently quantized. (See the essay entitled, "An Alternative Approach to Quantum Mechanics.")

In nonrelativistic quantum mechanics, we keep account of spin separately. For an electron in an atom, the quantum numbers $n$, $\ell$, and $m_\ell$ are said to specify the **spatial state** (i.e., wave function) of the electron, while the quantum number $m_s$ specifies the **spin state.** The electron's state is then completely specified by the *four* quantum numbers $n$, $\ell$, $m_\ell$, and $m_s$. It may be represented as follows:

$$m_s = +\tfrac{1}{2}: \; \psi_{n,\ell,m_\ell,+\frac{1}{2}} = \psi_{n,\ell,m_\ell}(r,\,\theta,\,\phi)\,\{\uparrow\} \qquad \text{(Spin-up)}$$

$$m_s = -\tfrac{1}{2}: \; \psi_{n,\ell,m_\ell,-\frac{1}{2}} = \psi_{n,\ell,m_\ell}(r,\,\theta,\,\phi)\,\{\downarrow\} \qquad \text{(Spin-down)} \tag{7-9}$$

The symbols $\{\uparrow\}$ and $\{\downarrow\}$ and terms "spin-up" and "spin-down" are merely different ways of indicating the discrete value of $m_s$. The symbols should not be thought of as multiplying $\psi_{n,\ell,m_\ell}(r,\,\theta,\,\phi)$ in the usual sense, because spatial and spin states are of completely different character. Spin is, however, a necessary part of the overall description of the state, so it is fair to place the symbol alongside the wave function.

## An Alternative Approach to Quantum Mechanics

At about the same time that Erwin Schrödinger was formulating his wave-function approach to quantum mechanics, Werner Heisenberg was formulating an alternative but equivalent approach using matrices instead of functions. In the early years of quantum mechanics, the matrix formulation seemed more clumsy and Schrödinger's approach consequently garnered more attention. The wave-function formulation is also somewhat easier to grasp as an introduction to quantum mechanics. Nevertheless, the matrix formulation is in many ways the more elegant of the two. An arbitrary state is viewed as a combination of a discrete set (matrix elements) of allowed quantum states. And with spin already represented by a matrix, replacing the wave function by a matrix makes for a consistent way of incorporating all aspects of the state.

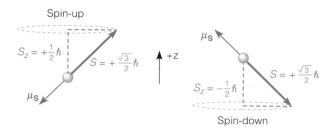

**Figure 7.5** The two possible spin states of an electron.

Figure 7.5 illustrates the electron's two possible spin states. Although crudely depicting the electron as a simple round object, it clarifies the relationship between the magnitude of the intrinsic angular momentum and its two possible $z$-components. As $\mathbf{L}$ may be found to make only certain angles with the $z$-axis in the simple hydrogen atom, an experiment to determine the $z$-component of the intrinsic angular momentum $\mathbf{S}$ has only two possible outcomes.

### Spin and Degeneracy

Owing to its two possible states, spin should increase the degeneracy in the hydrogen atom by a factor of 2. Were the energy of the atom independent of $m_s$, for each set of $n$, $\ell$, $m_\ell$, there would be two possible values of $m_s$ and thus two different states with the same energy. Doubling the number of states with the same $n$ would increase the degeneracy from $n^2$ to $2n^2$. However, with spin comes the possibility of new interactions in the atom, which may destroy some of the symmetry and so *reduce* the degeneracy. Several sections later in the chapter, beginning with Section 7.6, are devoted to studying these more complex interactions.

### The Spin of the Photon

Although in this chapter we concentrate mostly on the behavior of spin-$\frac{1}{2}$ electrons, it is important to mention the intrinsic angular momentum of the photon. The photon is of spin 1 (i.e., $s = 1$), so its intrinsic angular momentum $S$ is $\sqrt{1(1 + 1)}\hbar$. A consequence is that if an atom undergoes a transition in which a photon is produced, there is a limit on how much its angular momentum may change; the atom's angular momentum quantum number may change by no more than unity. This restriction manifests itself in so-called **selection rules,** which crop up often in discussions of spectral emissions.

## 7.2 Identical Particles: Symmetrization

As we now know quite well, quantum mechanics limits the knowledge we may have about a one-particle system. However, it places further distinct limitations on our knowledge of multiparticle systems. In Section 7.3, we shall see why these limitations are inextricably linked to spin. But their foundation can be understood independent of the notion of spin, and it is here that we begin.

Imagine that two particles share the same space. Whether we view them as two particles, two waves, or a single combined wave, it is still reasonable to speak of two "particles," in that we may isolate a system that has twice the *intrinsic properties* of a known fundamental particle. For instance, a system of two electrons has a charge of $2 \times (-1.6 \times 10^{-19}\,\text{C})$ no matter how it is viewed. By "share the same space," we mean that the particles are not physically isolated; there is some region of space in which they "overlap," a region where either particle may be found. By this definition, two electrons rattling around in a box "share the same space."

We first seek an appropriate two-particle Schrödinger equation. A likely candidate is one with an operator for the sum of the kinetic energies, an operator for potential energy and, on the other side, the energy.

$$(\hat{KE}_1 + \hat{KE}_2)\psi + \hat{U}\psi = \hat{E}\psi$$

For the sake of clarity, let us restrict our attention to one dimension. Position $x$ appears in a one-particle wave equation. If we are to specify kinetic and potential energies in a two-particle wave equation in an analogous way, we have no choice but to use an $x_1$ and an $x_2$. In effect, each particle has a kinetic energy, and the potential energy should depend in some general way upon the positions of the particles. Nevertheless, we are not simply adding two one-particle wave equations; we are formulating a true two-particle wave equation in which the particles are not treated completely independently, but are described by a single, overall wave function $\psi(x_1, x_2)$ (just as a function describing an electromagnetic wave may represent many photons). Thus, we have

$$\left(-\frac{\hbar^2}{2m}\frac{\partial^2}{\partial x_1^2} - \frac{\hbar^2}{2m}\frac{\partial^2}{\partial x_2^2}\right)\psi(x_1, x_2) + U(x_1, x_2)\psi(x_1, x_2) = E\psi(x_1, x_2)$$

$$(7\text{-}10)$$

where $E$ represents the energy of the whole, two-particle system.

Let us now attempt to solve this equation via separation of variables, by assuming that

$$\psi(x_1, x_2) = \psi_a(x_1)\,\psi_b(x_2) \qquad (7\text{-}11)$$

where $\psi_a$ and $\psi_b$ are in general different functions. As usual, we insert this product into (7-10), then divide by it.

$$\frac{\left(-\dfrac{\hbar^2}{2m}\psi_b(x_2)\dfrac{\partial^2 \psi_a(x_1)}{\partial x_1^2} - \dfrac{\hbar^2}{2m}\psi_a(x_1)\dfrac{\partial^2 \psi_b(x_2)}{\partial x_2^2}\right) + U(x_1, x_2)\psi_a(x_1)\psi_b(x_2)}{\psi_a(x_1)\psi_b(x_2)}$$

$$= \frac{E\psi_a(x_1)\psi_b(x_2)}{\psi_a(x_1)\psi_b(x_2)}$$

Now canceling and rearranging a bit,

$$\frac{-\hbar^2}{2m}\frac{1}{\psi_a(x_1)}\frac{\partial^2 \psi_a(x_1)}{\partial x_1^2} + \frac{-\hbar^2}{2m}\frac{1}{\psi_b(x_2)}\frac{\partial^2 \psi_b(x_2)}{\partial x_2^2} + U(x_1, x_2) = E \qquad (7\text{-}12)$$

As in the one-particle case, the functional form of the potential energy dictates the solution of the equation, but it is solvable for only a few special cases. In particular, (7-12) is not as it stands "variables-separate" (which was our goal), because $U(x_1, x_2)$ need not be a sum of separate functions of $x_1$ and $x_2$. Were the particles subject only to *external* forces, then since the effect of an external force on one particle in a system is independent of the position of another particle in that system, the total potential energy would indeed be given by

$$U(x_1, x_2) = U_1(x_1) + U_2(x_2) \qquad \text{(External forces only)}$$

Dependencies on $x_1$ and $x_2$ in (7-12) could then be algebraically separated. However, such a restriction precludes any consideration of forces *between* particles in the system. The potential energy associated with an *internal* force between two particles in a system is generally a function of the absolute value of their separation, $|x_1 - x_2|$, and this cannot be expressed as a simple sum. For example, the potential energy shared by two electrons is

$$U(x_1, x_2) = \frac{e^2}{(4\pi\epsilon_0)|x_1 - x_2|} \neq U_1(x_1) + U_2(x_2) \qquad \text{(Interparticle force)}$$

Since any realistic application involving multiple charged particles must take into account the interparticle Coulomb interactions, instances of truly separable potential energy are rare indeed. Even so, we may understand several important features of multiple-particle systems by considering an "unrealistic" simple case.

## Particles in a Box

Suppose two identical particles that do *not* exert forces on one another are confined in an infinite well; each particle is subject only to the external forces exerted by the walls, and its potential energy is thus independent of the other particle's position. (Clearly these cannot be charged particles.) Equation (7-12) becomes

$$\frac{-\hbar^2}{2m}\frac{1}{\psi_a(x_1)}\frac{\partial^2 \psi_a(x_1)}{\partial x_1^2} + \frac{-\hbar^2}{2m}\frac{1}{\psi_b(x_2)}\frac{\partial^2 \psi_b(x_2)}{\partial x_2^2} + U(x_1) + U(x_2) = E$$

or

$$\left( \frac{-\hbar^2}{2m}\frac{1}{\psi_a(x_1)}\frac{\partial^2 \psi_a(x_1)}{\partial x_1^2} + U(x_1) \right) + \left( \frac{-\hbar^2}{2m}\frac{1}{\psi_b(x_2)}\frac{\partial^2 \psi_b(x_2)}{\partial x_2^2} + U(x_2) \right) = E$$

$$\underbrace{\phantom{xxxxxxxxxx}}_{C_1} \qquad\qquad\qquad \underbrace{\phantom{xxxxxxxxxx}}_{C_2} \qquad \text{(7-13)}$$

where $U$ is the infinite-well potential energy. By the usual separation-of-variables arguments, each of the two terms in parentheses must be a constant. Designating them $C_1$ and $C_2$, we have

$$\frac{-\hbar^2}{2m}\frac{1}{\psi_a(x_1)}\frac{\partial^2 \psi_a(x_1)}{\partial x_1^2} + U(x_1) = C_1$$

and

$$\frac{-\hbar^2}{2m}\frac{1}{\psi_b(x_2)}\frac{\partial^2 \psi_b(x_2)}{\partial x_2^2} + U(x_2) = C_2$$

Each is identical to the one-particle Schrödinger equation for a particle in a box, except that $E$ is replaced by $C$. Therefore, using equations (4-16),

$$\psi_n(x_1) = \sqrt{\frac{2}{L}} \sin\frac{n\pi x_1}{L} \qquad C_1 = \frac{n^2\pi^2\hbar^2}{2mL^2}$$

and

$$\psi_{n'}(x_2) = \sqrt{\frac{2}{L}} \sin\frac{n'\pi x_2}{L} \qquad C_2 = \frac{n'^2\pi^2\hbar^2}{2mL^2}$$

We now have quantum numbers to distinguish the particular wave functions of each particle, so we can replace subscripts $a$ and $b$ with $n$ and $n'$. Correspondingly, wave function (7-11) becomes

$$\psi(x_1, x_2) = \psi_n(x_1)\psi_{n'}(x_2) \tag{7-14}$$

Now inserting the values of $C_1$ and $C_2$ into (7-13), we obtain $E$.

$$\frac{n^2\pi^2\hbar^2}{2mL^2} + \frac{n'^2\pi^2\hbar^2}{2mL^2} = E \tag{7-15}$$

Not surprisingly, the total energy of the two-particle system is the sum of the individual infinite-well energies of states $n$ and $n'$.

### A Not-So-Small Problem

Let us examine (7-14) more closely. Suppose $n = 4$ and $n' = 3$.

$$\psi_4(x_1) = \sqrt{\frac{2}{L}} \sin\frac{4\pi x_1}{L} \qquad \psi_3(x_2) = \sqrt{\frac{2}{L}} \sin\frac{3\pi x_2}{L}$$

The total wave function (7-14) would be

$$\psi(x_1, x_2) = \psi_4(x_1)\psi_3(x_2) = \frac{2}{L} \sin\frac{4\pi x_1}{L} \sin\frac{3\pi x_2}{L}$$

We would expect that the square of the wave function would give us the probability density $P(x_1, x_2)$ for this two-particle system:

$$P(x_1, x_2) = (\psi(x_1, x_2))^2 = \left(\frac{2}{L}\right)^2 \sin^2\frac{4\pi x_1}{L} \sin^2\frac{3\pi x_2}{L} \tag{7-16}$$

As a logical generalization of its meaning in the one-particle case, this would be the probability of simultaneously finding particle 1 within a unit distance at $x_1$ and particle 2 within a unit distance at $x_2$—that is, probability/(distance$_1$ · distance$_2$). As the reader may verify, if (7-16) is integrated over both $x_1$ and $x_2$ between limits 0 and $L$, the result is unity. Sensibly, the probability of finding both particles *somewhere* within the box should be unity.

Figure 7.6 shows the *individual* wave functions $\psi_4$ and $\psi_3$ and corresponding probability densities for the fourth and third particle-in-a-box energy levels. Figure 7.7 shows the *total* (product) probability density (7-16), a function of the two

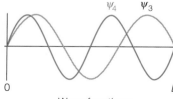

Wave functions

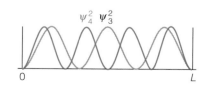

Probability densities

**Figure 7.6** Wave functions and probability densities of the third and fourth infinite-well states.

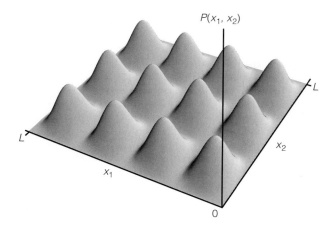

$P(x_1, x_2)$

**Figure 7.7** Asymmetric probability density.

variables $x_1$ and $x_2$.[8] Note how the behaviors with respect to $x_1$ and $x_2$ differ; there are four bumps in the probability density as a function of $x_1$ but only three as a function of $x_2$. In particular, we see that, independent of the value of $x_2$, the probability (density) of finding particle 1 at the center of the box is zero. Equation (7-16) agrees—inserting $x_1 = \frac{1}{2}L$,

$$P(\tfrac{1}{2}L, x_2) = \left(\frac{2}{L}\right)^2 \sin^2 \frac{4\pi(\frac{1}{2}L)}{L} \sin^2 \frac{3\pi x_2}{L} = 0$$

On the other hand, the probability of finding particle 2 at the center of the box is nonzero in general, depending on the value of $x_1$.

$$P(x_1, \tfrac{1}{2}L) = \left(\frac{2}{L}\right)^2 \sin^2 \frac{4\pi x_1}{L} \sin^2 \frac{3\pi(\frac{1}{2}L)}{L} = \left(\frac{2}{L}\right)^2 \sin^2 \frac{4\pi x_1}{L} \neq 0$$

This cannot be! To claim that particle 2 may be found at a given location but that particle 1 may never be found there is to profess what cannot be known! Particles 1 and 2 are *identical;* they bear no distinctive label and cannot be watched in an attempt to keep track of them separately, and their wave functions share the same space. Identical particles sharing the same space are said to be **indistinguishable.** If an experiment is performed and a particle found, we simply cannot know which particle it is. Because it is impossible to *verify* any asymmetry in the behaviors of the two particles, we require that the probability density obey the following restriction:

Probability density must be unchanged if particle labels are switched.

The reader may be uneasy with this restriction. It is one thing to alter a theory because it renders predictions that can be experimentally verified as *false*. But it is quite another to alter a theory simply because it renders predictions that *cannot be verified* experimentally (because we cannot know which particle has been found). However, it may be said that quantum mechanics itself is a theory adapted to what it is possible to verify. Were we not led to interpret the wave function as "probability amplitude" by the observation that in a double-slit experiment a

[8]The box is not two-dimensional. The diagram plots $x_1$ and $x_2$ along different axes merely for clarity; they are, of course, really parallel.

particle's position cannot be predicted with certainty? In any event, the restriction is necessary; the far-reaching and verifiable consequences we will soon discuss cannot be explained otherwise.

How might $P(x_1, x_2)$ be altered so that switching particle labels 1 and 2 leaves it unchanged? We might try merely adding the same function with the 1 and 2 switched:

$$P(x_1, x_2) \stackrel{?}{=} \left(\frac{2}{L}\right)^2 \sin^2 \frac{4\pi x_1}{L} \sin^2 \frac{3\pi x_2}{L} + \left(\frac{2}{L}\right)^2 \sin^2 \frac{4\pi x_2}{L} \sin^2 \frac{3\pi x_1}{L}$$

This is indeed symmetric (i.e., the same function) under interchange of the particle labels. But it is unacceptable because, quite simply, it is not the square of a solution of the Schrödinger equation. To produce a symmetric probability density, we must start with the *wave function*. There are two ways of modifying solution (7-14) to meet the requirement:

$$\psi_S(x_1, x_2) \equiv \psi_n(x_1)\psi_{n'}(x_2) + \psi_{n'}(x_1)\psi_n(x_2) \quad \text{(Symmetric)}$$

$$\psi_A(x_1, x_2) \equiv \psi_n(x_1)\psi_{n'}(x_2) - \psi_{n'}(x_1)\psi_n(x_2) \quad \text{(Antisymmetric)}$$

(7-17)

The Schrödinger equation is a *linear* differential equation—for which sums and differences of solutions are also solutions—so both of these are as valid mathematically as (7-14). But in solutions (7-17) the notion that one of the particles occupies state $n$ and the other state $n'$ is abandoned; both are *combinations* of particle 1 in state $n$, particle 2 in state $n'$ and the reverse, particle 1 in state $n'$, particle 2 in state $n$. To accept either is to accept simply that "states $n$ and $n'$ are occupied." Our immediate concern is whether wave functions (7-17) lead to a symmetric probability density. A simple argument shows that they do: If the particle labels are switched, the "symmetric" wave function is unchanged and the "antisymmetric" wave function switches sign (hence their names); in neither case would the *square* change. (Only the probability density is required to be symmetric, because the wave functions themselves cannot be observed.) Writing out the probability densities further clarifies the point.

$$P_S(x_1, x_2) = (\psi_S(x_1, x_2))^2 = (\psi_n(x_1)\psi_{n'}(x_2))^2 + (\psi_{n'}(x_1)\psi_n(x_2))^2$$
$$+ 2\psi_n(x_1)\psi_{n'}(x_2)\psi_{n'}(x_1)\psi_n(x_2)$$

$$P_A(x_1, x_2) = (\psi_A(x_1, x_2))^2 = (\psi_n(x_1)\psi_{n'}(x_2))^2 + (\psi_{n'}(x_1)\psi_n(x_2))^2$$
$$- 2\psi_n(x_1)\psi_{n'}(x_2)\psi_{n'}(x_1)\psi_n(x_2)$$

In either case, interchanging particle labels 1 and 2 would leave the expression unchanged. For convenience, we refer to these probability densities as "symmetric" or "antisymmetric,"—that is, according to the corresponding *wave function*—but both *probability densities* are symmetric.

Both the symmetric and antisymmetric probability densities preserve certain features of the *un*symmetrized probability density (7-16): When properly normalized (see Example 7.2), they yield unity when integrated over all values of $x_1$ and $x_2$, and they may be interpreted as a probability of finding particle 1 within a unit distance at $x_1$ and particle 2 within a unit distance at $x_2$. Now, however, the probability is the same if 1 and 2 are switched.

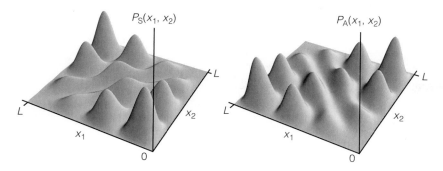

To illustrate their similarities and differences, the probability densities $P_S(x_1, x_2)$ and $P_A(x_1, x_2)$ for the two types of wave function are shown in Figure 7.8 for our special case $n = 4$, $n' = 3$. In common with the unsymmetrized probability density of Figure 7.7, both have 12 bumps. But while they are distributed asymmetrically between $x_1$ and $x_2$ in the unsymmetrized case, they are divided symmetrically in $P_S(x_1, x_2)$ and $P_A(x_1, x_2)$. Even so, there is a crucial distinction between $P_S(x_1, x_2)$ and $P_A(x_1, x_2)$: $P_S(x_1, x_2)$ is largest where the particles are close together, where $x_1$ and $x_2$ are near the same value, along the line $x_1 = x_2$. On the other hand, $P_A(x_1, x_2)$ is largest where the particles are far apart; its largest bumps occur when $x_1$ is near $L$ and $x_2$ near 0 and when $x_1$ is near 0 and $x_2$ near $L$. In fact, $P_A(x_1, x_2)$ is *identically zero* when $x_1 = x_2$. It may be said that the particles are on average farther apart in an antisymmetric spatial state than they are in a symmetric one.

### Example 7.2

Two particles in a box occupy the $n = 4$ and $n' = 3$ states. (a) Write the symmetric and antisymmetric wave functions. (b) Find the proper normalization constants. (c) For each, calculate the probability that both particles would be found in the left-hand side of the box (i.e., between 0 and $\frac{1}{2}L$). Compare these with what the probability would be if the symmetrization requirement were ignored.

### Solution

(a) With $\psi_n$ being the $n = 4$ state and $\psi_{n'}$ the $n' = 3$, equations (7-17) become

$$\psi_S(x_1, x_2) = \sin\frac{4\pi x_1}{L} \sin\frac{3\pi x_2}{L} + \sin\frac{3\pi x_1}{L} \sin\frac{4\pi x_2}{L}$$

$$\psi_A(x_1, x_2) = \sin\frac{4\pi x_1}{L} \sin\frac{3\pi x_2}{L} - \sin\frac{3\pi x_1}{L} \sin\frac{4\pi x_2}{L}$$

(b) Using the symbol $A$ for the normalization constant and a "$\pm$" to treat both cases, we set the integral over all possible locations of both particles equal to unity:

$$\int_0^L \left[ A\left( \sin\frac{4\pi x_1}{L} \sin\frac{3\pi x_2}{L} \pm \sin\frac{3\pi x_1}{L} \sin\frac{4\pi x_2}{L} \right) \right]^2 dx_1 \, dx_2 = 1$$

Dividing both sides by $A^2$ and rearranging a bit, we obtain

$$\frac{1}{A^2} = \int_0^L \sin^2 \frac{4\pi x_1}{L}\, dx_1 \int_0^L \sin^2 \frac{3\pi x_2}{L}\, dx_2$$

$$+ \int_0^L \sin^2 \frac{3\pi x_1}{L}\, dx_1 \int_0^L \sin^2 \frac{4\pi x_2}{L}\, dx_2$$

$$\pm\, 2 \int_0^L \sin \frac{4\pi x_1}{L} \sin \frac{3\pi x_1}{L}\, dx_1 \int_0^L \sin \frac{3\pi x_2}{L} \sin \frac{4\pi x_2}{L}\, dx_2$$

The first four integrals are standard particle-in-a-box normalization integrals (cf. Section 4.5), each equal to $\frac{1}{2}L$. The last two are identical to each other. Using trigonometric identities,[9] the integrands are $\frac{1}{2}[\cos(\pi x/L) - \cos(7\pi x/L)]$, which when integrated give $(L/2\pi) \cdot [\sin(\pi x/L) - \frac{1}{7}\sin(7\pi x/L)]$. Evaluated at their limits, these are zero. The symmetric and antisymmetric states thus have the same normalization constant.

$$\frac{1}{A^2} = \frac{1}{2}L\frac{1}{2}L + \frac{1}{2}L\frac{1}{2}L + 0 \quad \Rightarrow \quad A = \frac{\sqrt{2}}{L}$$

(c)  Here we integrate only over values of $x_1$ and $x_2$ between 0 and $\frac{1}{2}L$.

probability $(0 \le x_1 \le \frac{1}{2}L,\, 0 \le x_2 \le \frac{1}{2}L)$

$$= \int_0^{L/2} \left[ \frac{\sqrt{2}}{L}\left( \sin \frac{4\pi x_1}{L} \sin \frac{3\pi x_2}{L} \pm \sin \frac{3\pi x_1}{L} \sin \frac{4\pi x_2}{L} \right) \right]^2 dx_1\, dx_2$$

$$= \frac{2}{L^2} \int_0^{L/2} \sin^2 \frac{4\pi x_1}{L}\, dx_1 \int_0^{L/2} \sin^2 \frac{3\pi x_2}{L}\, dx_2$$

$$+ \frac{2}{L^2} \int_0^{L/2} \sin^2 \frac{3\pi x_1}{L}\, dx_1 \int_0^{L/2} \sin^2 \frac{4\pi x_2}{L}\, dx_2$$

$$\pm\, 2\frac{2}{L^2} \int_0^{L/2} \sin \frac{4\pi x_1}{L} \sin \frac{3\pi x_1}{L}\, dx_1 \int_0^{L/2} \sin \frac{3\pi x_2}{L} \sin \frac{4\pi x_2}{L}\, dx_2$$

Due to their symmetry about the box's center (cf. Figure 7.6), the first four integrals are half what they were before: $\frac{1}{4}L$. The last two are

$$\frac{L}{2\pi}\left( \sin \frac{\pi x}{L} - \frac{1}{7}\sin \frac{7\pi x}{L} \right)\Bigg|_0^{L/2} = \frac{4L}{7\pi}$$

Thus,

$$\text{probability} = \frac{2}{L^2}\frac{L}{4}\frac{L}{4} + \frac{2}{L^2}\frac{L}{4}\frac{L}{4} \pm \frac{4}{L^2}\frac{4L}{7\pi}\frac{4L}{7\pi}$$

$$= \frac{1}{4} \pm 0.132 = \begin{cases} 0.382 & \textbf{(Symmetric)} \\ 0.118 & \textbf{(Antisymmetric)} \end{cases}$$

Were we to ignore the symmetrization requirement, the probability that either particle individually would be found on the left-hand side would be one-half, so that the probability of both being on the left would be one-fourth. The symmetrization requirement yields two states: the symmetric,

[9]That is,

$$\frac{1}{2}[\cos(4a - 3a) - \cos(4a + 3a)]$$

$$= \frac{1}{2}[\cos(4a)\cos(3a) + \sin(4a)\sin(3a)]$$

$$- \frac{1}{2}[\cos(4a)\cos(3a) - \sin(4a)\sin(3a)]$$

$$= \sin(4a)\sin(3a)$$

in which the probability of both being on a given side is greater than one-fourth, and the antisymmetric, in which it is less than one-fourth. (The reader should study Figure 7.8 to see that it and the probabilities we have found are in qualitative agreement.)

## An Important Observation

Before moving on, we note that $\psi_S$ and $\psi_A$ have the same energy (the same as the unsymmetrized wave function; see Exercise 6). Accordingly, if the particles do not interact with one another in any way, it is equally likely that the system would be in the antisymmetric state as in the symmetric state. But this would not be the case if an interparticle force were present. Were the two particles electrons instead of noninteracting particles, the antisymmetric state would be of lower energy. The simple reason is that, as we have just observed, in the antisymmetric state the electrons are less likely to be found close together—the repulsive energy would be reduced. This argument holds even though introduction of the interparticle potential energy renders the Schrödinger equation unsolvable except by numerical methods.

## Spin-State Symmetry

To ensure a symmetric probability density, we must be concerned not only with the *spatial* state of a multiparticle system, but also with its spin state. It is not enough to consider merely the probability of finding particle 1 at $x_1$ and particle 2 at $x_2$; we must consider the probability of finding particle 1 at $x_1$ *in a given spin state* and particle 2 at $x_2$ *in a given spin state*. The spin state and spatial state (wave function) are both essential parts of the overall description of a system. We will put the two together in Section 7.3. In preparation, however, having already done so for spatial states alone, let us briefly look at symmetrization of spin states alone, restricting our attention to spin-$\frac{1}{2}$ particles.

We begin by generalizing our spin notation to two particles. In the symbol for the spin state, the first arrow represents the value of $m_s$ for particle 1 and the second arrow the value of $m_s$ for particle 2. This is analogous to the arguments $(x_1, x_2)$ of a function $\psi(x_1, x_2)$. For two particles, there are four possibilities.

$$
\begin{array}{cccc}
\mathbf{m}_{s1}\ \mathbf{m}_{s2} & \mathbf{m}_{s1}\ \mathbf{m}_{s2} & \mathbf{m}_{s1}\ \mathbf{m}_{s2} & \mathbf{m}_{s1}\ \mathbf{m}_{s2} \\
\{\ \uparrow\quad \uparrow\ \} & \{\ \uparrow\quad \downarrow\ \} & \{\ \downarrow\quad \uparrow\ \} & \{\ \downarrow\quad \downarrow\ \} \\
+\tfrac{1}{2}\ +\tfrac{1}{2} & +\tfrac{1}{2}\ -\tfrac{1}{2} & -\tfrac{1}{2}\ +\tfrac{1}{2} & -\tfrac{1}{2}\ -\tfrac{1}{2}
\end{array}
$$

Of these four, there are three symmetric combinations and one antisymmetric combination. The former are known collectively as a "triplet of spin states" and the latter as a "singlet spin state."

$$
\left.
\begin{array}{c}
\{\uparrow\uparrow\} \\
\{\uparrow\downarrow\} + \{\downarrow\uparrow\} \\
\{\downarrow\downarrow\}
\end{array}
\right\}
\begin{array}{l}
\textbf{Triplet} \\
\textbf{(symmetric)}
\end{array}
\qquad
\{\uparrow\downarrow\} - \{\downarrow\uparrow\}
\begin{array}{l}
\textbf{Singlet} \\
\textbf{(antisymmetric)}
\end{array}
$$

The symmetries are easily seen: If the left (particle 1) and right (particle 2) arrows are switched, the triplet states are unchanged, while the singlet state switches sign. The singlet state and the middle triplet state bear similarities to the antisymmetric and symmetric spatial states. They should, simply because adding and subtracting states with particle labels switched are ways of ensuring a combined state with some kind of symmetry. There are two "extra" states here because, while we have no argument against the particles' spins being both up or both down, for a *well-defined energy* of $E_n + E_{n'}$ we cannot have both in spatial state $n$ or both in spatial state $n'$ (since these would not be of the proper energy).

# 7.3 Exchange Symmetry and the Exclusion Principle

Let us now combine spatial and spin symmetrizations to obtain states whose overall (spatial plus spin) character, called **exchange symmetry,** is either symmetric or antisymmetric under exchange of particle labels. For individual-particle spatial states [10] $\psi_n$ and $\psi_{n'}$, where $n$ and $n'$ stand for all quantum numbers necessary to specify the *spatial* states (e.g., $n$, $\ell$, $m_\ell$), the symmetric and antisymmetric combinations are given in Table 7.1. The symmetric ones are those in which spatial and spin states are both symmetric or both antisymmetric; switching $x_1$ and $x_2$ *and* the particles' spin arrows has no effect when both are symmetric, and no *net* effect when both are antisymmetric, since both spatial and spin states would switch sign. Correspondingly, the antisymmetric combinations are those in which, of the spatial and spin states, one is symmetric and the other antisymmetric; switching $x_1$ and $x_2$ and the spin arrows would switch the overall sign.

## The Connection Between Symmetrization and Spin

We are now prepared to understand one of the central ideas of the chapter—why spin is so crucial in explaining the behavior of multiple-particle systems. *It is fundamental to nature that a system of multiple indistinguishable particles will be in an overall state of definite exchange symmetry, and whether it is symmetric or antisymmetric depends on the spin of the particles involved.* Theory predicts and experiment verifies that integral-spin particles—that is, those for which $s$ is 0, 1, 2, . . .—invariably manifest a symmetric overall state. The generic name for integral-spin particles is **boson** (bō′ zän). Particles of half-integral spin (e.g., $s =$

[10]The term "individual particle" will appear often in the following discussion. It is important to distinguish the individual states that each particle may occupy from the overall state of a multiparticle system.

**Table 7.1   Combined Spatial- and Spin-State Symmetries (two spin-$\frac{1}{2}$ particles)**

| Symmetric | | Antisymmetric | |
|---|---|---|---|
| $(\psi_n(x_1)\psi_{n'}(x_2) + \psi_{n'}(x_1)\psi_n(x_2))$ $\begin{cases} \{\uparrow\uparrow\} \\ \{\uparrow\downarrow\} \ + \ \{\downarrow\uparrow\} \\ \{\downarrow\downarrow\} \end{cases}$ | | $(\psi_n(x_1)\psi_{n'}(x_2) - \psi_{n'}(x_1)\psi_n(x_2))$ $\begin{cases} \{\uparrow\uparrow\} \\ \{\uparrow\downarrow\} \ + \ \{\downarrow\uparrow\} \\ \{\downarrow\downarrow\} \end{cases}$ | |
| $(\psi_n(x_1)\psi_{n'}(x_2) - \psi_{n'}(x_1)\psi_n(x_2))\, (\{\uparrow\downarrow\} - \{\downarrow\uparrow\})$ | | $(\psi_n(x_1)\psi_{n'}(x_2) + \psi_{n'}(x_1)\psi_n(x_2))\, (\{\uparrow\downarrow\} - \{\downarrow\uparrow\})$ | |

**Table 7.2**

| Fermions<br>Antisymmetric State | | Bosons<br>Symmetric State | |
|---|---|---|---|
| **Particle** | $s$ | **Particle** | $s$ |
| Electron, $e^-$ | $\frac{1}{2}$ | Pion, $\pi°$ | 0 |
| Proton, p | $\frac{1}{2}$ | Alpha-particle, $\alpha$<br>(helium nucleus) | 0 |
| Neutron, n | $\frac{1}{2}$ | Photon, $\gamma$ | 1 |
| Neutrino, $\nu$ | $\frac{1}{2}$ | Deuteron, d<br>(bound n-p) | 1 |
| Omega, $\Omega^-$ | $\frac{3}{2}$ | Graviton | 2 |

$\frac{1}{2}, \frac{3}{2})$ manifest an antisymmetric overall state and are called by the generic name **fermion** (fer′ mē än). An electron, for example, is a spin-$\frac{1}{2}$ particle and is thus a fermion. Table 7.2 lists several examples of bosons and fermions.

The principle we have introduced here might seem innocuous, but it has resounding ramifications. Let us investigate perhaps the most far-reaching.

## Fermions: The Exclusion Principle

The most common massive particles are, of course, electrons, protons, and neutrons. Since all are fermions, a multiparticle system of electrons, protons, or neutrons will be in an antisymmetric overall state. Consider the simple case of two fermions occupying spatial states $n$ and $n'$ (where $n$ and $n'$ again stand for all quantum numbers necessary to specify the spatial state). If $n$ and $n'$ are equal, the antisymmetric spatial state is identically zero!

$$\psi_n(x_1)\psi_n(x_2) - \psi_n(x_1)\psi_n(x_2) = 0$$

Necessarily, then, if the wave function is indeed to describe two *existing* particles, the spatial state must be symmetric (which is not zero). But from Table 7.1 we see that this in turn requires the spin state to be antisymmetric, with one particle spin up ($m_s = +\frac{1}{2}$) and the other down ($m_s = -\frac{1}{2}$). Equal spatial quantum numbers implies different spin quantum numbers. The fermions cannot have the same *full* set of quantum numbers—they cannot be in the same individual-particle state! This is what we know as the **exclusion principle.**

> No two indistinguishable fermions may occupy
> the same individual-particle state.

Exclusion principle

Since to specify its full set of quantum numbers is to fully specify a particle's state, an equivalent way of stating this principle is that *no two indistinguishable fermions may have the same set of quantum numbers.* Although we have considered only the simple case of two fermions, the exclusion principle holds for any number.[11] It was discovered by Wolfgang Pauli in 1924, and the achievement won him the Nobel prize (1945).

The exclusion principle applies only to *indistinguishable* fermions. As noted earlier, this means that the fermions must (1) be of the same kind, and (2) share

[11] There is a methodical way to express an antisymmetric overall state for *any number* of particles. It is called the **Slater determinant.** In this form, it is easy to see that switching the spatial and spin states of any two particles switches the sign of the overall state *and* that there is no acceptable (nonzero) state unless all particles have different sets of quantum numbers. The Slater determinant is discussed in several of the exercises.

the same space. Of course, all electrons are identical and all protons are identical, but it is quite possible to tell electrons from protons; their charge and mass serve as labels. There are no symmetry requirements involving two particles that can be distinguished, so the exclusion principle does not apply to a system of different kinds of fermions (e.g., the electron and proton of the hydrogen atom).

The second condition is a more delicate issue. Inasmuch as wave functions in real situations do not end abruptly in space, but tend to fall off exponentially, it may be argued that *everything in the universe* shares the same space, that the wave functions of all things overlap. This being the case, there should be an overall state for all the electrons in the universe, antisymmetric under exchange of any two particle labels and with no two electrons occupying the same state! However, the condition is of no significant consequence for electrons far apart, those whose wave functions overlap only negligibly. A good argument involves a logical extension of a simple special case.

Consider a system of two electrons, each in a different infinite well, in which wave functions *do* end abruptly. In this case, the electrons *would be distinguishable:* Suppose particle 1 is in the left well and particle 2 in the right. Because there is absolutely no mingling of the wave functions, it is certain that the probability of finding particle 1 somewhere in the left well at a later time is unity and the probability of finding particle 2 there is zero. No requirement of symmetry in the two-particle state could sensibly or justifiably be imposed. Logically, then, to the extent that the electron wave functions of two isolated atoms do not *significantly* overlap, the electrons of one atom are distinguishable from those of another. We are therefore justified in broadening the exclusion principle as follows: No two fermions of the same kind may occupy the same state *in an individual atom or other nearly isolated system.*

The importance of the exclusion principle cannot be overstated. As we shall soon see, were there no exclusion principle, nature would be profoundly different. It is the very foundation of chemistry.

### Bosons

The exclusion principle is a direct consequence of the inherent features of an antisymmetric overall state. Perhaps the best way to appreciate this is to look at what could occur if fermions manifested a symmetric overall state, as do bosons. Consider the symmetric two-particle states of Table 7.1. If $n = n'$, the symmetric spatial state $\psi_n(x_1)\psi_n(x_2) + \psi_n(x_1)\psi_n(x_2)$ is nonzero and so any of the triplet spin states would be perfectly acceptable, including the two-spins-up or two-spins-down states. Indistinguishable bosons may indeed possess the same set of quantum numbers; the exclusion principle does *not* apply to bosons.[12]

## 7.4 Multielectron Atoms and the Periodic Table

To gain a grasp of basic quantum mechanics in three dimensions, the hydrogen atom is a splendid object of study. But isolated hydrogen atoms are comparatively rare on Earth. To shed meaningful light on our surroundings, we must understand multielectron atoms. The exclusion principle is of prime importance: No two elec-

[12]Interestingly, bosons, always the counterpoint to fermions, actually "prefer" to be in the same state. This leads to the effect of "Bose-Einstein condensation," responsible for the unusual properties of liquid helium. The topic is discussed in Section 8.7.

trons may be in the same individual-particle state. Thus, the lowest energy state of an atom with $Z$ protons in its nucleus—of "atomic number" $Z$—would have its $Z$ electrons fill up the $Z$ lowest-energy quantum states, one electron per state. It may fairly be said that the whole of chemistry is a consequence. Before we explore this assertion, however, we must touch upon some related points.

First, we note that since multiple electrons alter the potential energy, the spatial states are not the same functions as in the hydrogen atom. Still, they have much the same character and are again distinguished according to values of the quantum numbers $n$, $\ell$, and $m_\ell$. In Chapter 6, we found that the energy in the hydrogen atom depends only on the principal quantum number $n$. We noted that this "accidental degeneracy" is due to the potential energy in the Schrödinger equation having a simple $1/r$ dependence, and that any deviation would cause the energy to depend also on $\ell$. Interelectron forces are a deviation, and the consequent dependence of energy on $\ell$ we may understand as follows: In Section 6.5, we noted that a smaller $\ell$ may be thought of as a more elliptic orbit. For instance, a $3s$ ($\ell = 0$) electron is likely to be found both far from and close to the origin, while in a $3d$ ($\ell = 2$) state its radius is more nearly uniform, more like a circle, as depicted in the margin (see also Figure 6.14). In a circular orbit, an electron orbits not only the nucleus, but all electrons closer to the nucleus than itself. We say that the inner electrons **screen** the outer from the nucleus. In effect, electrons in circular/large-$\ell$ orbits orbit a smaller positive charge and so have a smaller *negative* potential energy than they would without the other electrons; they are in a relatively high energy state. On the other hand, an elliptical/small-$\ell$ orbit of the same average radius (i.e., the same $n$) brings the electron closer to the nucleus, which is therefore not screened as effectively, resulting in a larger negative potential energy. Of course, an elliptic orbit also extends *farther* from the origin than a circular one. But as it turns out, the energy-lowering effect at small radii predominates: *Energy is lower for lower-$\ell$/more-elliptical orbits.* Although the multielectron atom is, strictly speaking, "unsolvable," approximation techniques have been very successful (one of the best is discussed in Section 7.5) and provide the quantitative support for this claim.

Approximation techniques also provide the theoretical support for an important rule-of-thumb: As electrons fill states in an atom, they do so according to the lowest value of $n + \ell$, with "preference" given to $n$; that is, in cases of equal $n + \ell$, the state of lowest $n$ is of lower energy. We refer to those electrons with a given value of $n$ as being "in the $n$th **shell,**" and the subset of these with a given value of $\ell$ as being "in the $n$-$\ell$ **subshell.**" Thus, the $1s$ shell fills first, and the $2s$ subshell ($n + \ell = 2$) fills before the $2p$ subshell ($n + \ell = 3$), which in turn fills before the $3s$ subshell ($n + \ell$ is also 3, but $n$ is lower in the $2p$). The $3p$ is next, but the $4s$ subshell ($n + \ell = 4$) fills before the $3d$ ($n + \ell = 5$). Table 7.3 illustrates this

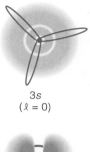

$3s$
($\ell = 0$)

$3d$
($\ell = 2$)

**Table 7.3**

| Subshell $n_\ell$ | 1s | 2s | 2p | 3s | 3p | 4s | 3d | 4p | 5s | 4d | 5p | 6s | 4f | 5d | 6p | 7s | 5f | 6d |
|---|---|---|---|---|---|---|---|---|---|---|---|---|---|---|---|---|---|---|
| $n + \ell$ | 1 | 2 | 3 | 3 | 4 | 4 | 5 | 5 | 5 | 6 | 6 | 6 | 7 | 7 | 7 | 7 | 8 | 8 |
| Number of electrons $2(2\ell + 1)$ | 2 | 2 | 6 | 2 | 6 | 2 | 10 | 6 | 2 | 10 | 6 | 2 | 14 | 10 | 6 | 2 | 14 | 10 |

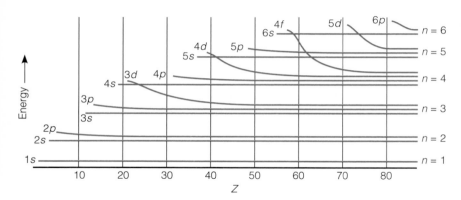

**Figure 7.9** The energy ordering of electron states varies with Z.

"schedule" as well as the number of electrons that can fit into each subshell (the product of the number of allowed $m_s$-values and $m_\ell$-values: $2(2\ell + 1)$). However, due to the complex interactions among the electrons, as outer subshells fill, the ordering of *inner* subshells changes. For instance, if only the 1$s$, 2$s$, and 2$p$ subshells are full, the 4$s$ is of lower energy than the 3$d$, and so is the next subshell to fill. But if subshells are filled out to $n = 4$ or more, the 3$d$ is of lower energy than the 4$s$. In fact, as more and more electrons orbit the nucleus, the tendency is for inner-shell energies *not* to intermingle; a lower $n$ is a lower energy (while a lower $\ell$ is still a lower energy *within* a shell). This progressive change in subshell ordering is depicted schematically in Figure 7.9.

With the exclusion principle and the dependence of energy on $\ell$ in mind, we may now understand the different chemical behaviors of the elements.

## Chemical Behavior and the Periodic Table

Nothing is so fundamental to chemistry as the periodic table, and the basis of it is quantum mechanics. The periodicity is in the chemical behaviors of the elements, which depend almost entirely on the number and binding energies of each element's **valence electrons**—weakly bound electrons that dangle at the periphery of the electron cloud. The behaviors can be understood only via a quantum-mechanical treatment where the exclusion principle plays a central role. Without the exclusion principle, the ground states of *all* atoms would have all electrons in the 1$s$ state, bound very tightly. They would have spherically symmetric probability densities, could possess no external electric fields (as would an asymmetric probability/charge density) to interact with other atoms, and would thus be chemically inert. On the contrary, the exclusion principle tells us that there is room for only two electrons in the $n = 1$ level (and eight in the $n = 2$, etc.), so most elements unavoidably have some electrons "sticking out."

A chemical reaction involves electrons being in some way exchanged or rearranged between atoms. Hydrogen has one electron, which is naturally all it has to offer in a chemical reaction. It has one valence electron. Having an energy of $-13.6$ eV, the energy required to free this electron from its nucleus is 13.6 eV. Let us now consider the other elements. As we do, however, remember that each time we imagine adding another electron we also add a proton to the nucleus. We study the likely chemical behaviors of *neutral* atoms (not weird cases of many electrons orbiting one proton).

   Helium has a spins-opposite pair of $1s$ electrons. Its valence would be 2 but for the fact that these two complete the $n = 1$ shell. Being in the same *spatial* state (differing only in their spin state), they are not significantly screened from the nucleus by one another; they feel an attractive force to a positive nuclear charge twice that of the hydrogen atom. Overlooking the interelectron repulsion, each would behave as an electron orbiting a nucleus of $Z = 2$, with an energy of $-54.4$ eV (a value obtained in Example 6.10). The interelectron repulsion raises this only to about $-40$ eV. In comparison to hydrogen, then, the electrons in helium are very tightly bound and thus extremely reluctant to engage in chemical reactions. Helium is accordingly known as a **noble gas** and its valence is said to be *zero*. By convention, we indicate the occupation of an element's atomic levels via its ground-state **electronic configuration.** These are shown in the margin for hydrogen and helium. The superscripts indicate the number of electrons in each subshell.

   Beyond helium, additional electrons must go into the next-lowest energy levels. For $Z = 3$ and $Z = 4$, the $2s$ fills; then for $Z = 5$ to $Z = 10$, the $2p$ fills. As the $n = 2$ shell begins to fill, the electrons are considerably less tightly bound than the $n = 1$ electrons. Not only are they farther from the nucleus than the $n = 1$ electrons, they are also screened by these same electrons from the full nuclear charge of $+Ze$. Consequently, it is the $n = 2$ electrons that participate in chemical reactions; the $n = 1$ electrons are *not* part of the valence. The valence of these elements thus progresses from 1, the single $2s$ electron of $Z = 3$ lithium, to 8 in $Z = 10$ neon.

   With the filling of the $p$ states a question arises: Now that there is freedom to occupy different *spatial* states (i.e., different $m_\ell$-values) within the subshell, how indeed do they fill? Because energy is independent of $m_\ell$, one $m_\ell$ state is as likely as another to be "chosen" by the first $2p$ electron (i.e., for $Z = 5$ boron). But there is a correlation once other electrons are added. Consider $Z = 6$ carbon, which has two $2p$ electrons: As noted in Section 7.2, the repulsive energy between electrons is lowered if the *spatial* state is antisymmetric; but to be in such a state, *the $m_\ell$-values must not be the same*. Furthermore, the antisymmetric spatial state requires a symmetric (triplet) spin state—that is, one with spins aligned. The trend continues with $Z = 7$ nitrogen. Each of its three $2p$ electrons has a different $m_\ell$-value and their spins are all aligned.[13]

   As the $n = 2$ shell nears completion, the outer electrons, being all at about the same radius, are not screened much by one another; meanwhile they "see" an ever larger nuclear charge $Z$. Thus, they become increasingly tightly bound to the nucleus. (Note the relative lowering of the $2p$ subshell in Figure 7.9.) By $Z = 10$, the $n = 2$ shell is complete.

   The abrupt change in chemical behavior near a complete shell is worthy of close scrutiny. Sodium has $Z = 11$. Its last electron cannot fit into the $n = 2$ shell; it must go in to the $3s$. Rarely do tightly bound, inner-shell electrons participate in chemical reactions, so sodium has only this lone valence electron. It is far from the nucleus and weakly bound, and this accounts for sodium's enthusiastic chemical activity. Fluorine ($Z = 9$), on the other hand, lacking one electron to complete its $n = 2$ shell, prefers to capture a weakly bound electron of *another* atom rather

H:   $1s^1$

He:   $1s^2$

C:   $1s^2 2s^2 2p^2$

[13]This tendency for spins to be aligned in unfilled subshells, a direct consequence of the electrons' "desire" to spread out among as many $m_\ell$-values as possible, is known as "Hund's rule."

than give up any of its many, relatively tightly bound outer electrons. Like sodium, it is quite chemically reactive, but in somewhat the opposite sense; it is the taker rather than the giver and is thus said to be of valence $-1$.[14] If a sodium atom and fluorine atom are brought close together, the fluorine atom will seize sodium's dangling, weakly bound valence electron, completing its more tightly bound $n = 2$ shell. Although transferring charge to turn neutral atoms into positive and negative ions requires energy, the "hole" in fluorine's $n = 2$ shell is deep and the net result is a lower-energy state. An **ionic bond** is formed, in which the ions are held together by electrostatic attraction. To complete the comparison, $Z = 10$ neon differs by just one electron from both sodium and fluorine, but its behavior is vastly different from both. Its $n = 2$ shell is full; it has no valence electrons to offer, as does sodium, and no holes to fill, as does fluorine. It thus joins helium as a zero-valence noble gas, only grudgingly participating in chemical reactions with other elements.

F:   $1s^2 2s^2 2p^5$

Ne:   $1s^2 2s^2 2p^6$

Na:   $1s^2 2s^2 2p^6 3s^1$

As $Z$ increases beyond 10, the $3s$ then the $3p$ states fill, and the valence (now excluding both the full $n = 1$ and $n = 2$ shells) increases just as it does in filling the $n = 2$ levels. Therefore, the chemical behaviors of $Z = 11$ sodium, one electron beyond the filled $n = 2$ shell, and $Z = 3$ lithium, one beyond the filled $n = 1$ shell, are very similar; so too are the behaviors of $Z = 12$ magnesium and $Z = 4$ beryllium, both with two valence electrons beyond a filled shell; and so on. *This is the "periodicity" of the periodic table!* Shown in Figure 7.10, its first column is valence 1, containing $Z = 1$, $Z = 3$, and $Z = 11$; the second column is valence 2, containing $Z = 4$, $Z = 12$, and so on. The eighth and last column does *not* correspond to a filled $n = 3$ shell; the $3p$ states are full, but the $3d$ states remain. Furthermore, as Table 7.3 indicates, the $3d$ states fill only after the $4s$. However, the energy jump from the $3p$ to the $4s$ level is large, so that although $Z = 18$ argon does not have a full $n = 3$ shell, it is relatively tightly bound. Thus, in common with $Z = 10$ and $Z = 2$, it behaves as a noble gas of valence 0. By these same arguments, $Z = 17$ chlorine, in the seventh column, is chemically similar to $Z = 9$ fluorine; both are valence $-1$.

After the filling of the $3p$ subshell, the $4s$ fills. Because of the large energy jump involved, this begins a new sequence with valences 1 ($Z = 19$) and 2 ($Z = 20$). There is a break, however, between valences 2 and 3, between the completion of the $4s$ subshells and the filling of the $4p$ subshells. In this break are found the so-called **transition elements,** which involve filling of $d$ subshells after the shell of *higher n* has already begun to fill. As the $3d$ states (of which there are ten) fill, the electronic configuration does not always vary consistently with $Z$. The first deviation is $Z = 24$ chromium, in which the $3d$ level increases by 2 over $Z = 23$, while the $4s$ level loses an electron. The energy levels are relatively close, and such "exceptional" behavior is not uncommon. Although there are many arguments and "rules" meant to explain deviations from consistent filling, the simple truth is that the forces in multielectron atoms are so complicated as to make theoretical prediction of the ground-state electronic configuration difficult. In the final analysis, the lowest-energy configuration is determined experimentally.

By $Z = 30$, the $3d$ level has filled completely, beyond which the $4p$ fills in a consistent way, adding a third period of elements with valences $3-8$. As does the completion of the $3p$, the completion of the $4p$ level (but neither the $4d$ nor $4f$) results in a noble gas; again the complete $4p$ subshell is fairly tightly bound and the jump to the next level, the $5s$, is large.

[14]Likewise, $Z = 8$ oxygen, two electrons short of a complete shell, is said to be "valence $-2$." There is no clear distinction between a valence of $-2$ and of $+6$, nor between $-1$ and $+7$. Valences are usually specified according to the most common behavior—that is, whether the atom tends to receive or donate electrons in a chemical reaction.

# Periodic Table of the Elements

Symbol —— **He** —— 2 —— Atomic number Z
4.0026 —— Atomic mass*
Electronic configuration —— 1s² 
(if different from pattern)

| Subshell | ns¹ | ns² | (n−2)f | (n−1)d¹ | (n−1)d² | (n−1)d³ | (n−1)d⁴ | (n−1)d⁵ | (n−1)d⁶ | (n−1)d⁷ | (n−1)d⁸ | (n−1)d⁹ | (n−1)d¹⁰ | np¹ | np² | np³ | np⁴ | np⁵ | np⁶ |
|---|---|---|---|---|---|---|---|---|---|---|---|---|---|---|---|---|---|---|---|
| Typical valence | +1 | +2 | | | | | | | | | | | | +3 | +4 | +5 | −2 | −1 | 0 |
| Shell n 1 | **H** 1 1.00794 | | | | | | | | | | | | | | | | | | **He** 2 4.0026 1s² |
| 2 | **Li** 3 6.941 | **Be** 4 9.01218 | | | | | | | | | | | | **B** 5 10.811 | **C** 6 12.011 | **N** 7 14.0067 | **O** 8 15.9994 | **F** 9 18.9984 | **Ne** 10 20.1797 |
| 3 | **Na** 11 22.9898 | **Mg** 12 24.3050 | | | | | | | | | | | | **Al** 13 26.9815 | **Si** 14 28.0855 | **P** 15 30.9738 | **S** 16 32.066 | **Cl** 17 35.4527 | **Ar** 18 39.948 |
| 4 | **K** 19 39.0983 | **Ca** 20 40.078 | | **Sc** 21 44.9559 | **Ti** 22 47.88 | **V** 23 50.9415 | **Cr** 24 51.9961 3d⁵4s¹ | **Mn** 25 54.9381 | **Fe** 26 55.847 | **Co** 27 58.9332 | **Ni** 28 58.6934 | **Cu** 29 63.546 3d¹⁰4s¹ | **Zn** 30 65.39 | **Ga** 31 69.723 | **Ge** 32 72.61 | **As** 33 74.9216 | **Se** 34 78.96 | **Br** 35 79.904 | **Kr** 36 83.80 |
| 5 | **Rb** 37 85.4678 | **Sr** 38 87.62 | | **Y** 39 88.9059 | **Zr** 40 91.224 | **Nb** 41 92.9064 4d⁴5s¹ | **Mo** 42 95.94 4d⁵5s¹ | **Tc** 43 (98) | **Ru** 44 101.07 4d⁷5s¹ | **Rh** 45 102.906 4d⁸5s¹ | **Pd** 46 106.42 4d¹⁰5s⁰ | **Ag** 47 107.868 4d¹⁰5s¹ | **Cd** 48 112.411 | **In** 49 114.82 | **Sn** 50 118.710 | **Sb** 51 121.757 | **Te** 52 127.60 | **I** 53 126.904 | **Xe** 54 131.29 |
| 6 | **Cs** 55 132.905 | **Ba** 56 137.327 | La–Yb | **Lu** 71 174.967 | **Hf** 72 178.49 | **Ta** 73 180.948 | **W** 74 183.85 | **Re** 75 186.207 | **Os** 76 190.2 | **Ir** 77 192.2 | **Pt** 78 195.08 5d⁹6s¹ | **Au** 79 196.967 5d¹⁰6s¹ | **Hg** 80 200.59 | **Tl** 81 204.383 | **Pb** 82 207.2 | **Bi** 83 208.980 | **Po** 84 (209) | **At** 85 (210) | **Rn** 86 (222) |
| 7 | **Fr** 87 (223) | **Ra** 88 (226) | Ac–No | **Lr** 103 (260) | **Rf** 104 (261) | **Db** 105 (262) | **Sg** 106 (263) | **Bh** 107 (262) | **Hs** 108 (265) | **Mt** 109 (266) | | | | | | | | | |

**Lanthanides**

| f¹ | f² | f³ | f⁴ | f⁵ | f⁶ | f⁷ | f⁸ | f⁹ | f¹⁰ | f¹¹ | f¹² | f¹³ | f¹⁴ |
|---|---|---|---|---|---|---|---|---|---|---|---|---|---|
| **La** 57 138.906 5d¹6s² | **Ce** 58 140.115 4f¹5d¹6s² | **Pr** 59 140.908 | **Nd** 60 144.24 | **Pm** 61 (145) | **Sm** 62 150.38 | **Eu** 63 151.965 | **Gd** 64 157.25 4f⁷5d¹6s² | **Tb** 65 158.925 4f⁸5d¹6s² | **Dy** 66 162.50 | **Ho** 67 164.930 | **Er** 68 167.26 | **Tm** 69 168.934 | **Yb** 70 173.04 |

**Actinides**

| f¹ | f² | f³ | f⁴ | f⁵ | f⁶ | f⁷ | f⁸ | f⁹ | f¹⁰ | f¹¹ | f¹² | f¹³ | f¹⁴ |
|---|---|---|---|---|---|---|---|---|---|---|---|---|---|
| **Ac** 89 (227) 6d¹7s² | **Th** 90 232.038 6d²7s² | **Pa** 91 (231) 5f²6d¹7s² | **U** 92 238.029 5f³6d¹7s² | **Np** 93 (237) 5f⁴6d¹7s² | **Pu** 94 (244) | **Am** 95 (243) | **Cm** 96 (247) 5f⁷6d¹7s² | **Bk** 97 (247) | **Cf** 98 (251) | **Es** 99 (252) | **Fm** 100 (257) | **Md** 101 (258) | **No** 102 (259) |

* In atomic mass units, u. Averaged over naturally occurring isotopes. Values in parentheses are mass numbers of most stable known isotopes.

**Figure 7.10** Periodic table of the elements.

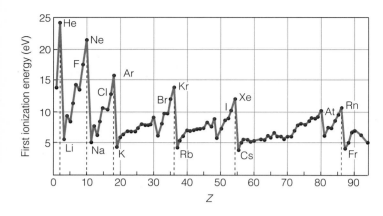

**Figure 7.11** First ionization energies of the elements.

The next row in the periodic table, beginning with $Z = 37$ rubidium, is very similar to the previous. The $5s$ fills, then a lower-$n$ subshell, the $4d$, then the $5p$. The row beginning with $Z = 55$ cesium reveals an expansion similar to that accommodating the transition elements, caused by an increase in the number of lower-$n$ subshells that fill "late." The $6s$ fills, then in turn the $4f$, $5d$, and $6p$. The elements corresponding to the filling of the $4f$ subshell are the 14 **lanthanides**—after lanthanum, the first in this $Z = 57$–70 series—also known as the **rare earths** (a misnomer, as many are quite abundant in nature). The row beginning with $Z = 87$ francium mirrors the previous row. The $7s$ fills, then the $5f$, comprising the 14-member $Z = 89$–102 **actinides,** then the $6d$. The $7p$ would fill next, adding yet another valence 3 to valence 8 period, but elements have not yet been found of such high $Z$.

The periodicity of electron energies is also apparent in the experimentally determined first ionization energies shown in Figure 7.11. (The *first* ionization energy is the energy needed to remove the first electron from the atom. The situation changes, of course, after the first is removed.) Local maxima are found at the noble gases (He, Ne, Ar, Kr, Xe, and Rn), in which electrons are tightly bound. In the elements immediately after the noble gases (Li, Na, K, Rb, Cs, and Fr), the energy needed to remove the single valence electron drops precipitously; this electron is easily stolen by an electron-hungry atom. In the elements immediately before (F, Cl, Br, I, and At), the electrons are relatively tightly bound—and that they have room for one more electron explains their appetite.

# ◆ **7.5** An Approximate Solution

Although related in many ways, the wave functions and energies of the states ($1s$, $2s$, etc.) in multielectron atoms differ from those of a one-electron atom like hydrogen. In this section, after briefly examining the difficulty of solving the problem, we discuss one of the simplest and most successful techniques of solving it approximately.

Generalizing the two-particle case (7-10), the Schrödinger equation for $N$ particles in three dimensions is

$$-\frac{\hbar^2}{2m}\sum_i \nabla_i^2\, \psi(\mathbf{r}_1, \mathbf{r}_2, \mathbf{r}_3, \ldots) + \sum_{i<j} U(|\mathbf{r}_i - \mathbf{r}_j|)\, \psi(\mathbf{r}_1, \mathbf{r}_2, \mathbf{r}_3, \ldots)$$

$$= E\psi(\mathbf{r}_1, \mathbf{r}_2, \mathbf{r}_3, \ldots) \tag{7-18}$$

The subscript on the $\nabla^2$ operator indicates that it involves partial derivatives with respect to coordinates of the $i$th particle, and the potential energy term sums interparticle potential energies between all pairs of particles. (Since potential energy is always shared, to include terms for $i > j$ would be superfluous—in fact, overcounting by a factor of 2.) To solve (7-18) for the multielectron atom, the nucleus and its orbiting electrons would be the particles, and $U(|\mathbf{r}_i - \mathbf{r}_j|)$ would be the familiar $(1/4\pi\epsilon_0)(q_i q_j/|\mathbf{r}_i - \mathbf{r}_j|)$. Our task would be greatly simplified if we could, as in Section 7.2, break the Schrödinger equation into separate equations for each particle: one equation for electron 1 involving $U(x_1)$, another for electron 2 involving $U(x_2)$, and so on. The difficulty is that interparticle potential energies like $U(|\mathbf{r}_i - \mathbf{r}_j|)$ simply do not separate into a $U(\mathbf{r}_i)$ and a $U(\mathbf{r}_j)$. Having no alternative, we resort to simplifying approximations. We now outline a well-used and reasonably accurate method: the Hartree approximation, or **Hartree model.**

## The Hartree Model

We seek to digest the $N$-particle Schrödinger equation into identical one-particle equations for each electron, yet somehow taking into account interelectron forces. We do this by assuming that each electron orbits in an **effective potential energy,** the net effect of the positive nucleus and a spherically symmetric cloud of negative charge associated with the other electrons' probability densities. This is an approximation in that the repulsive forces between electrons are not treated as true interparticle forces, dependent on interparticle separation, but rather as average forces independent of the actual locations of other electrons. However, though simplified, the interparticle forces are not ignored. Indeed, to a good approximation the electron cloud distribution *is* nearly spherically symmetric, and it is quite reasonable that a given electron would be affected by the others in some average kind of way.

But how do we know what the effective potential energy would be? An iterative process is followed. It begins with an initial assumption of $U(r)$, based upon reasonable arguments about the electric field experienced by the electrons: According to Gauss's law, any spherically symmetric charge distribution surrounding a cavity produces no net electric field *inside* the cavity and a point-charge field *outside* the entire distribution. Therefore, at the location of the hypothetical innermost electron (Figure 7.12), inside the cavity in the charge cloud of the other electrons, the electric fields of the surrounding electrons cancel, leaving only the field due to the nucleus, a positive point charge at the origin.

$$E\text{-field} = \frac{Ze}{(4\pi\epsilon_0)r^2} \qquad \text{(Small } r\text{)}$$

At the other extreme, the outermost electron orbits not only the positive nucleus but also the $Z - 1$ other electrons, which produce a point-charge field appropriate to a charge of $(Z - 1) \times (-e)$.

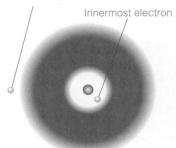

Outermost electron

Innermost electron

**Figure 7.12** How the atom looks to the innermost and outermost electrons, in the Hartree model.

$$E\text{-field} = \frac{Ze}{(4\pi\epsilon_0)r^2} + \frac{-(Z-1)e}{(4\pi\epsilon_0)r^2} = \frac{e}{(4\pi\epsilon_0)r^2} \qquad (\textbf{Large } r)$$

The outermost electron experiences the field of a single positive charge. (As noted in the previous section, inner electrons screen outer electrons, to varying degrees, from experiencing the full attraction of the $Z$ protons in the nucleus.) A reasonable approximation to the electric field for all values of $r$ would join these two extremes with some smooth function. One way is to assume a simple function $Z(r)$ for effective charge "seen" by an electron, a function that is unity as $r \to \infty$ and $Z$ as $r \to 0$. For instance,

$$E\text{-field} = \frac{Z(r)e}{(4\pi\epsilon_0)r^2} \quad \text{where } Z(r) \equiv 1 + (Z-1)e^{-br} \qquad (\textbf{All } r)$$

From the electric field felt by a typical electron, we obtain its potential energy via integration:

$$U(r) = (-e)V(r) \quad \text{where } V(r) = \int_r^\infty E(r)\, dr$$

Having obtained a plausible effective potential energy, the next step in the Hartree approximation is to solve the resulting one-particle Schrödinger equation for the allowed wave functions and corresponding energies of a representative electron. (This is a task for numerical methods on a computer, as any deviation of the potential energy from a simple $1/r$ dependence makes solution a far more difficult undertaking.) The ground state of a given element would then have its $Z$ electrons occupying the $Z$ lowest-energy states consistent with the exclusion principle.

It might seem that the method is complete at this point. On the contrary, the wave functions obtained will not agree with the potential energy assumed responsible for their form! The underlying assumption is that the effective potential is due to the distribution of electron charge. But with the wave functions now in hand, we may actually *calculate* the charge density and thus the potential energy, and it is *not* the one we initially assumed. The procedure is to sum $\rho(r) = -e \times |\psi(\mathbf{r})|^2$ for each of the occupied states, giving the net charge density, use Gauss's law to find the field, then integrate this to find the potential energy. Our initial assumption merely got the ball rolling; it was not based on any specific knowledge of the electrons' locations, so it should not be surprising that the new *calculated* potential energy differs.

The Hartree approximation (Figure 7.13) proceeds by using the new potential energy to solve for a new generation of wave functions and energies. But these are based on the charge density associated with the *now-old* wave functions. Again the wave functions do not correspond to the potential energy from which they are obtained. The approximation is at this point not self-consistent, but it does approach consistency. Thus, we repeat the process, calculating a new charge density and potential energy and in turn yet another generation of wave functions and energies. At some point, a new generation does not differ significantly from the previous; that is, the potential energy associated with the charge density obtained from a given advanced generation of wave functions, when put back into the Schrödinger equation, returns the same wave functions. The iterative process has converged. The model is not a complete description, but is at least self-consistent.

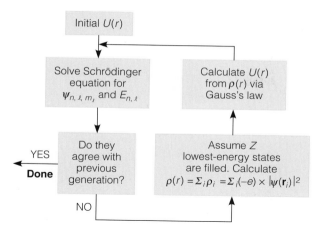

**Figure 7.13** Hartree model's interative process.

The Hartree model is often referred to as a **self-consistent field** (meaning electric field) approach.

We will not present the quantitative side of the Hartree model. We merely note that it is a very good first step toward understanding wave functions and energies in multielectron atoms. Although more refined and advanced models/approximations are needed to understand some behaviors, the Hartree model alone explains quite well much of what has been revealed by experiment. Perhaps its most important feature is its ability to show how energy varies with $\ell$.

## ◆ 7.6 The Spin-Orbit Interaction

In this section and the two that follow, we begin to study how the atom's various angular momenta/dipole moments may interact. We start with the simplest case, by returning to the one-electron hydrogen atom. Section 7.9 applies the ideas to multielectron atoms.

In an atom, the electron, with its intrinsic magnetic dipole moment, is essentially an orbiting magnet, and so should interact with any magnetic fields present. Our simple solution to the hydrogen atom assumed that the only potential energy in the atom is that due to the Coulomb attraction between nucleus and electron. It is certainly oversimplified if there is any magnetic field present.

We need look no further than the atom itself; moving charges produce magnetic fields. Because it orbits, the electron produces a magnetic field. But from the electron's perspective (Figure 7.14) it is the nucleus that is in *orbit,* producing a magnetic field in which the electron's intrinsic magnetic dipole moment, related to its *spin,* has an orientation energy—hence the name "spin-orbit" interaction.

A detailed study of the interaction is too involved to present here. But with fairly simple considerations we may gain a good qualitative understanding of it, as well as a reasonable idea of how much it might alter the atomic energy levels. Strictly speaking, this "new" interaction invalidates the naive solution of the hydrogen atom in Chapter 6, but we will see that the energy involved is relatively small, so the effect may be considered a minor perturbation on an essentially correct description.

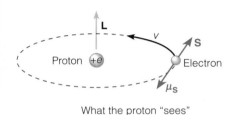

What the proton "sees"

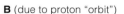

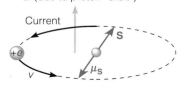

What the electron "sees"

**Figure 7.14** The electron feels a magnetic field, due the "orbiting" proton, in the same direction as **L**.

From introductory electricity and magnetism, we know that the orientation energy of a dipole $\boldsymbol{\mu}$ in a magnetic field **B** is $U = -\boldsymbol{\mu} \cdot \mathbf{B}$. Here the dipole moment is the electron's intrinsic $\boldsymbol{\mu}_{\mathbf{S}}$ and the field is due to orbital motion. To keep these relationships straight, we write

$$U = -\boldsymbol{\mu}_{\mathbf{S}} \cdot \mathbf{B}_{\text{due to L}} \qquad (7\text{-}19)$$

To determine **B,** as a crude approximation we assume that the hydrogen atom's electron "sees" the proton orbiting in a circle about itself, as in Figure 7.14. The magnetic field at the center of a current loop is given by $B = \mu_0 I/2r$. We can use this to relate $B$ to $L$ by relating $I$ to its cause—orbital motion:

$$I = \frac{e}{T} = \frac{e}{2\pi r/v} = \frac{e}{2\pi m_e r^2} m_e vr = \frac{e}{2\pi m_e r^2} L$$

Here we have assumed that whether the motion is viewed as proton orbiting electron or vice versa, the orbit radius, speed, and direction are the same. Therefore, by multiplying and dividing by the *electron* mass, we relate $I$ to the electron's orbital angular momentum. Noting that, by the usual right-hand rules, **L** is in the same direction as the **B** felt by the electron, we then have

$$B = \frac{\mu_0 I}{2r} = \frac{\mu_0}{2r} \frac{e}{2\pi m_e r^2} L \quad \rightarrow \quad \mathbf{B} = \frac{\mu_0 e}{4\pi m_e r^3} \mathbf{L} \qquad (7\text{-}20)$$

Finally, inserting this and equation (7-5) into (7-19), the orientation energy is

$$\begin{aligned} U &= -\boldsymbol{\mu}_{\mathbf{S}} \cdot \mathbf{B}_{\text{due to L}} \\ &= -\left(-g_e \frac{e}{2m_e} \mathbf{S}\right) \cdot \left(\frac{\mu_0 e}{4\pi m_e r^3} \mathbf{L}\right) = g_e \frac{\mu_0 e^2}{8\pi m_e^2 r^3} \mathbf{S} \cdot \mathbf{L} \end{aligned} \qquad (7\text{-}21)$$

The energy is high (positive) when the two angular momenta are more nearly aligned and low (negative) when more nearly opposite.[15] This makes sense. If **S**

[15]Qualifications like "more nearly" are necessary because **L** and **S** cannot in general be *precisely* parallel or opposite. Angular momentum vectors add at only certain quantized angles. Therefore, terms like "aligned," "parallel," "anti-aligned," and "opposite" should not be taken literally. Addition of angular momenta is discussed in Section 7.7.

**Figure 7.15** Electron's two possible orientation energies.

were *aligned* with **L**, $\mu_S$ would be *opposite* **L** and therefore **B**; as we know from electricity and magnetism, the highest-energy state occurs when a magnetic dipole is opposite the magnetic field. Figure 7.15 illustrates the point.

To give us an order-of-magnitude idea of the energy involved in the spin-orbit interaction, the product of the two vectors' magnitudes $|\mu_S||B|$ will do. The first magnitude is straightforward:

$$|\mu_S| = g_e \frac{e}{2m_e}|S| \cong 2\frac{e}{2m_e}\left(\frac{\sqrt{3}}{2}\hbar\right)$$

$$= \frac{1.6 \times 10^{-19} \text{ C}}{9.11 \times 10^{-31} \text{ kg}}\frac{\sqrt{3}}{2}(1.055 \times 10^{-34} \text{ J}) = 1.6 \times 10^{-23} \text{ A·m}^2$$

To determine $B$ via (7-20), itself a rough approximation, we must make reasonable assumptions about $r$ and $\ell$. First we note that in a state in which $\mathbf{L} = 0$, there should be no orientation energy at all; if no charge has any consistent orbital motion, there is no magnetic field. Therefore, the interaction should come into play only when $\ell$ is 1 or larger, which in turn requires $n$ to be 2 or larger. Choosing the simplest case, a $2p$ state, $L = \sqrt{\ell(\ell + 1)}\hbar = \sqrt{2}\hbar$ and $r = n^2a_0 = 4a_0$.

$$B = \frac{\mu_0 e}{4\pi m_e(4a_0)^3}\sqrt{2}\hbar$$

$$= \frac{(4\pi \times 10^{-7} \text{ T·m/A})(1.6 \times 10^{-19} \text{ C})}{4\pi(9.11 \times 10^{-31} \text{ kg})(4 \times 0.0529 \times 10^{-9} \text{ m})^3}\sqrt{2}(1.055 \times 10^{-34} \text{ J·s})$$

$$= 0.28 \text{ T}$$

Considering the size of the particles involved, this is a strong field, thousands of times Earth's field at its surface. Putting the two together,

$$U \cong |\mu_S||B| = (1.6 \times 10^{-23} \text{ A·m}^2)(0.28 \text{ T}) \cong 10^{-23} \text{ J} \sim 10^{-4} \text{ eV}$$

Electron energies in the hydrogen atom are typically a matter of electronvolts ($\sim 10^{-19}$ J), so we see that the orientation energy of the electron's intrinsic magnetic dipole moment in the atom's internal magnetic field is only about $10^{-4}$ of the overall energy. Small though it is, the spin-orbit interaction is clearly discernible in the so-called **fine structure** of hydrogen's spectral lines. Energy levels predicted in ignorance of spin to be degenerate are split according to the orientation of **L** and **S**. The states in which **L** and **S** are aligned are of slightly higher energy than those in which **L** and **S** are antialigned. Photons produced in transitions to or from these states can thus be of slightly different energies.

# ◆ **7.7** Adding Angular Momenta

Angular momentum vectors do not add in an arbitrary way, but by strict quantum-mechanical rules. In this section, we discuss these rules mostly in the context of the adding **S** and **L.** But *the same addition rules apply in all cases in which various angular momenta interact within a system*—spin angular momentum with orbital angular momentum, spin with spin, orbital with orbital, and so on.

The angular momenta **S** and **L** in an atom are not independent of one another. The whole idea of the "spin-orbit" interaction is that spin and orbital angular momenta affect each other through a magnetic field internal to the atom. One result of this coupling of **S** and **L** is that if quantization is probed with a weak *external B*-field (much weaker than the internal, so as not to overwhelm the internal interaction) we do not observe independent quantization of each, but rather of the vector sum: the **total angular momentum J.**

$$\mathbf{J} = \mathbf{L} + \mathbf{S} \tag{7-22}$$

The individual magnitudes $L$ and $S$ are not altered; both follow the usual form: $L = \sqrt{\ell(\ell + 1)}\,\hbar$ and $S = \sqrt{s(s + 1)}\,\hbar$. But the vectors **L** and **S** may add in only certain ways, at certain angles. We have some reason to expect this. Recall from the discussion of the Stern–Gerlach experiment that if spin could somehow be ignored, **L** could make only certain angles—corresponding to $2\ell + 1$ possible $m_\ell$-values—with the $z$-axis of the *external B*-field. We now know that orbital motion produces an *internal B*-field parallel to **L,** an "internal $z$-axis." It is certainly plausible, then, that the intrinsic angular momentum **S** should make only certain angles with **L**. Indeed, we might expect there to be $2s + 1$ such allowed angles, corresponding to the $2s + 1$ allowed values of $m_s$.

This is in fact the case. It is found that the magnitude $J$ and the $z$-component $J_z$ of the total angular momentum are restricted to the following values:

$$J = \sqrt{j(j + 1)}\,\hbar \quad (j = |\ell - s|, |\ell - s| + 1, ..., \ell + s - 1, \ell + s) \tag{7-23}$$

$$J_z = m_j\hbar \qquad (m_j = -j, -j + 1, ..., j - 1, j) \tag{7-24}$$

To each allowed magnitude $J$ would correspond a different angle $\theta$ between **L** and **S,** and the number of $j$-values in integral steps from $|\ell - s|$ to $\ell + s$ is indeed $2s + 1$ (the same as from $-s$ to $+s$). Of course, in applying this rule to hydrogen, with its lone electron, $s$ will always be $\frac{1}{2}$ and $j$ accordingly may take on only the two values $\ell + \frac{1}{2}$ and $\ell - \frac{1}{2}$, which correspond respectively to **L** and **S** "aligned" and "antialigned." (In multielectron atoms, total spin can be greater than $\frac{1}{2}$ and $j$ may take on more values. See Section 7.9.)

In one sense, equation (7-23) resembles standard rules for vector addition: Given vectors **A** and **B**, of magnitudes $A$ and $B$, the magnitude $C$ of the vector sum **C** = **A** + **B** cannot be greater than $A + B$ nor less than $|A - B|$. However, the angular momentum magnitudes $L$, $S$, and $J$ are not simply proportional to the values $\ell$, $s$, and $j$. Rather, each is proportional to $\sqrt{i(i + 1)}$, where $i$ stands for $\ell$, $s$, or $j$. Consequently, $J$ is strictly less than $L + S$ and (excepting $\ell = s$) greater than $|L - S|$, which in turn implies that **L** and **S** are never exactly parallel nor

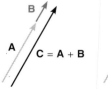

opposite. The terms "aligned" and "antialigned" must therefore not be taken literally.

Let us prove that **L** and **S** cannot be parallel. The maximum value of $j$ allowed by (7-23) is $\ell + s$. The corresponding total angular momentum magnitude is

$$J_{max} = \sqrt{(\ell + s)(\ell + s + 1)}\,\hbar$$

Were **L** and **S** parallel, the magnitude of the vector sum would be

$$L + S = \sqrt{\ell(\ell + 1)}\,\hbar + \sqrt{s(s + 1)}\,\hbar$$

We can see most easily how these compare by squaring them:

$$J_{max}^2 = (\ell^2 + \ell + s^2 + s + 2\ell s)\hbar^2$$

$$(L + S)^2 = (\ell^2 + \ell + s^2 + s + 2\sqrt{\ell^2 s^2 + \ell^2 s + s^2 \ell + \ell s}\,)\hbar^2$$

By inspection,

$$J_{max} < L + S$$

Since $J$ is strictly less than the sum of the magnitudes of **L** and **S,** we conclude that rule (7-23) forbids **L** and **S** to be exactly parallel. It is left as an exercise to show that at its minimum value, for which $j = |\ell - s|$, the magnitude of the total angular momentum (again excepting $\ell = s$) obeys

$$J_{min} > |L - S|$$

Since $|L - S|$ would be the magnitude of the vector sum if **L** and **S** were opposite, we conclude that (7-23) also does not allow **L** and **S** to be exactly opposite.

Summarizing the significance of equations (7-23) and (7-24): The atom's internal $B$-field causes **L** and **S** to be "locked together" at one of certain allowed angles, corresponding to the $2s + 1$ values of $J$ allowed by (7-23). Were we to investigate angular momentum quantization by applying a weak external $B$-field, as in the Stern–Gerlach experiment, it is quantization of $J_z$ that we would observe. For atoms with a given allowed value of $J$, we would find $J_z$ to be one of the $2j + 1$ values allowed by equation (7-24).[16]

The relationships between **L, S,** and **J** are illustrated in Figure 7.16 for the case $\ell = 2$, $s = \frac{1}{2}$. Given $\ell$ and $s$, the magnitudes $L$ and $S$ are fixed, but there are two possible angles between **L** and **S**: one aligned $(j = \ell + s = \frac{5}{2})$, a relatively high energy state (cf. Figure 7.15), and one antialigned $(j = \ell - s = \frac{3}{2})$, a low-energy state. Each of the two corresponding total angular momenta may make any of $2j + 1$ possible angles with the $z$-axis of an external field. In the $j = \frac{5}{2}$ case, there are six values, and in the $j = \frac{3}{2}$ case, there are four. But as in Figure 6.8 for **L** alone, although a **J** vector maintains its angle *with* the $z$-axis, it may point in any of an infinite number of directions *about* the $z$-axis, leaving $J_x$ and $J_y$ uncertain. Vectors **L** and **S** have similar freedom to revolve as a unit about the **J** vector.

Since an external field reveals quantization of $J_z$, it may seem that $m_\ell$ and $m_s$ are now quantum numbers of no relevance. Indeed, with the spin-orbit interaction coupling **L** and **S** into a total **J**, we regard $n$, $\ell$, $j$, and $m_j$ as the relevant—usually referred to as "good"—quantum numbers.[17] However, the total number of possible states is the same whether $m_\ell$ and $m_s$ or $j$ and $m_j$ are truly relevant. Consider

[16]The splitting into two lines in Section 7.1 pertained to the special case $\ell = 0$, and if there is no **L**, the *total* angular momentum should simply be **S.** Equation (7-23) agrees: $j$ may take on values between $|0 - \frac{1}{2}|$ and $|0 + \frac{1}{2}|$, that is, the single value $\frac{1}{2}$. Equation (7-24) then tells us that $m_j$ can be $-\frac{1}{2}$ or $+\frac{1}{2}$. Clearly, **J** and **S** are interchangeable in this case. Although the experiment does demonstrate quantization of $J_z$, in the $\ell = 0$ case it is trivially equivalent to demonstrating quantization of $S_z$ alone.

[17]We might call $s$ "good," but for hydrogen's single electron, it is not a quantum number that may take on different values.

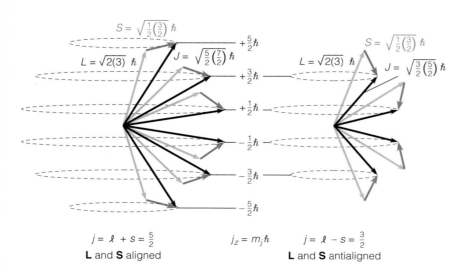

$$j = \ell + s = \tfrac{5}{2}$$
**L** and **S** aligned

$$j_z = m_j \hbar$$

$$j = \ell - s = \tfrac{3}{2}$$
**L** and **S** antialigned

the $\ell = 2$, $s = \tfrac{1}{2}$ case in Figure 7.16: *Ignoring* any coupling of **S** and **L**, $L_z = m_\ell \hbar$ may take on five $(2\ell + 1)$ values, and for each of these, there would be two possible values of $m_s$—a total of ten different states. On the other hand, $j$ may be $\tfrac{5}{2}$ with six possible $z$-components, or $\tfrac{3}{2}$ with four possible $z$-components, also a total of ten states. If there is no strong external $B$-field, the latter view is correct; the states do form groups according to $j$. But a sufficiently strong external field can overwhelm the coupling of intrinsic angular momentum to the atom's internal $B$-field. In this case, both **L** and **S** are *independently* quantized—each precesses with one fixed component along the $z$-axis of the external field—and $m_\ell$ and $m_s$ again become relevant indicators of the system's state—that is, "good" quantum numbers. These two different behaviors in an external field are discussed further in Section 7.8.

---

### Example 7.3

Identify the different total angular momentum states possible for the case $\ell = 3$.

### Solution

If we ignore any coupling of **L** and **S,** the number of states is simply the number of allowed values of $m_\ell$ times 2 $(m_s = \pm\tfrac{1}{2}$ for each): $(2\ell + 1) \times 2 = 14$. As previously noted, the number should be the same whether or not the spin-orbit interaction is in force. According to (7-23), the total angular momentum quantum number $j$ is allowed only two values. These and the corresponding allowed values of $m_j$ are given by

$$j_{max} = \ell + s = \tfrac{7}{2} \qquad m_j = -\tfrac{7}{2}, -\tfrac{5}{2}, -\tfrac{3}{2}, -\tfrac{1}{2}, +\tfrac{1}{2}, +\tfrac{3}{2}, +\tfrac{5}{2}, +\tfrac{7}{2}$$

$$j_{min} = \ell - s = \tfrac{5}{2} \qquad m_j = -\tfrac{5}{2}, -\tfrac{3}{2}, -\tfrac{1}{2}, +\tfrac{1}{2}, +\tfrac{3}{2}, +\tfrac{5}{2}$$

There are 14 states. The corresponding values of the actual physical properties are then

$$J_{max} = \sqrt{\tfrac{7}{2}(\tfrac{7}{2} + 1)}\,\hbar = \tfrac{\sqrt{63}}{2}\hbar \qquad J_z = \pm\tfrac{7}{2}\hbar,\ \pm\tfrac{5}{2}\hbar,\ \pm\tfrac{3}{2}\hbar,\ \pm\tfrac{1}{2}\hbar$$

$$J_{min} = \sqrt{\tfrac{5}{2}(\tfrac{5}{2} + 1)}\,\hbar = \tfrac{\sqrt{35}}{2}\hbar \qquad J_z = \pm\tfrac{5}{2}\hbar,\ \pm\tfrac{3}{2}\hbar,\ \pm\tfrac{1}{2}\hbar$$

## Relativity and Hydrogen Atom Energy Levels

Because spin is relativistic in origin, the spin-orbit interaction may be viewed as a relativistic effect. As noted earlier, the interaction destroys some of the symmetry in the hydrogen atom and so reduces the degeneracy; if **L** and **S** are aligned $(j = \ell + \tfrac{1}{2})$, the energy is higher than if **L** and **S** are antialigned $(j = \ell - \tfrac{1}{2})$—*all other things being equal*. However, the spin-orbit interaction is not the only relativistic cause of discrepancy between the naive energy levels of Chapter 6 and those actually observed. Another involves the ellipticity of orbits. Recall that for a given $n$, a wave function of *smaller* $\ell$ (i.e., less rotational energy), is one of greater radial energy. A wave of small rotational but large radial energy may be thought of as a wave of greater ellipticity. Also in Chapter 6 we noted that energy being independent of $\ell$ is "accidental," a consequence of the simplicity of the Coulomb potential energy, and that any mathematical alteration of the problem destroys this degeneracy. Although the details are too complex to present here, the more complicated relativistic dependence of energy on speed is such an alteration. It is found that a more elliptic (small $\ell$) orbit is of lower energy than a more circular (large $\ell$) orbit of the same $n$. In contrast to the spin-orbit interaction, this relativistic factor affects even the $n = 1$, $\ell = 0$ state; it too is of lower energy than we would otherwise expect. Of course, the two effects—the spin-orbit interaction and the dependence of energy on ellipticity—coexist. Taking both into account, theory predicts and experiment agrees that (1) the energies in the hydrogen atom are all slightly lower than expected nonrelativistically, and (2) the deviation depends on the total angular momentum. That is, the smaller the value of $J$, the lower is the energy. Consequently, to specify states of different energy, we incorporate into the spectroscopic notation of Chapter 6 a subscript indicating the value of the total angular momentum quantum number $j$. Thus, a $3d$ state in which **L** and **S** are antialigned (i.e., $j = \ell - s = 2 - \tfrac{1}{2} = \tfrac{3}{2}$) is a $3d_{3/2}$ state, and it is of slightly lower energy than the $3d$ state in which **L** and **S** are aligned $(j = 2 + \tfrac{1}{2} = \tfrac{5}{2})$, designated $3d_{5/2}$.

Figure 7.17 illustrates the relativistic splitting of hydrogen atom energy levels. States that by nonrelativistic theory should be degenerate are shifted downward and split according to the total angular momentum. Note that the $3d_{3/2}$ and $3p_{3/2}$ states—of the same $j$—are of the same energy. Again, this is a combined result of two effects. The ellipticity-dependence argument suggests that the $3p$ state, of lower angular momentum and so greater ellipticity than the $3d$ state, should be of *lower* energy. But the $3p_{3/2}$ state has **L** and **S** aligned $(j = \tfrac{3}{2} = 1 + \tfrac{1}{2} = \ell + s)$, while in the $3d_{3/2}$ they are antialigned $(j = \tfrac{3}{2} = 2 - \tfrac{1}{2} = \ell - s)$. This spin-orbit interaction argument suggests that the $3p_{3/2}$ state should be *higher* energy. As it

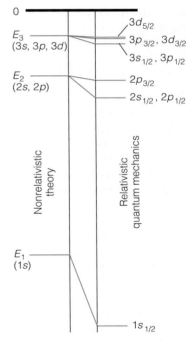

**Figure 7.17** Relativistic splitting of hydrogen atom energies (greatly exaggerated).

turns out, the two effects balance. Still, the relative heights of all levels in the figure harmonize with both arguments: for **LS**-aligned states, larger $\ell$ corresponds to higher energy; and for a given $\ell$, **LS**-aligned states are higher energy than **LS**-antialigned states.

The splitting of the energy levels in Figure 7.17 is greatly exaggerated. The spacing of the nonrelativistic energies levels (according to $n$) is correct. But the slopes showing how the true states deviate from these simple levels are all steeper than they should be by about $10^4$(!), specifically by the constant value $1/\alpha^2$. Called the **fine-structure constant**, $\alpha$ is defined as

$$\alpha \equiv \frac{e^2}{(4\pi\epsilon_0)\hbar c} \cong \frac{1}{137} \tag{7-25}$$

Fine-structure constant

This combination of factors appears often in a relativistic treatment of the hydrogen atom, and its square $\sim 10^{-4}$ is a fair measure of the deviations of the energies from nonrelativistic expectation. (Exercise 25 suggests how to find this factor in the spin-orbit interaction energy (7-21); it is no coincidence that this energy is about $10^{-4}$ times typical hydrogen atom energies.) Clearly, the relativistic splitting of the hydrogen atom energy levels is small, but it does not escape our notice. Precise spectroscopic equipment is able to discern the slight wavelength differences of photons produced in transitions expected *non*relativistically to be of the same energy. For instance, a transition from the $3d_{3/2}$ state to the $2p_{1/2}$ state yields a photon of slightly shorter wavelength (higher energy) than does a transition from the $3s_{1/2}$ state to the $2p_{3/2}$ state. Explanation of the fine structure of hydrogen's spectral lines was one of the great early triumphs of relativistic quantum mechanics.[18]

# ◆ 7.8 External Magnetic Fields and the Z-axis

In the simple hydrogen atom of Chapter 6, energy depended only on the principle quantum number $n$. In Sections 7.6 and 7.7, we learned that spin and other relativistic factors complicate matters; energy also depends on total angular momentum, the quantum number $j$. Still, without an external influence such as an experimental observation, there is no preferred direction in space and so no dependence of energy upon $m_\ell$, $m_s$, or $m_j$; $z$-components are meaningless if there is no $z$-axis. But what happens if we introduce an external influence?

The most obvious influence to consider, given the atom's magnetic dipole moments (spin and orbital), is an external magnetic field. States degenerate in the absence of an external $B$-field should split into different energies depending on the orientation of magnetic dipole moments in that field. How they split depends on the strength of the external field. As it turns out, it is easiest to analyze cases somewhat at opposite extremes. If the external field is weak compared to the atom's internal field, the spin-orbit interaction "locks" **S** and **L** into a total angular momentum **J**, and the quantization of this **J** along the external $z$-axis gives a quantized interaction energy between the external field and $\mu_J$ (the locked-together $\mu_S$ and $\mu_L$). This is the Zeeman effect. If the external field is strong, the spin-orbit coupling is broken, and $\mu_S$ and $\mu_L$ have independent orientation energies, quantized according to the independent quantizations of **S** and **L** along the external field. This is the Paschen–Back effect.

[18]Even "fine structure" does not tell the whole story of hydrogen's spectral lines. It neglects other, smaller effects. There is the "Lamb shift," an interaction between the electron and electromagnetic vacuum fluctuations, which splits states of a given $n$ and $j$ according to the value of $\ell$. Then there is interaction between the electron and the relatively small magnetic moment of the proton, responsible for the so-called "hyperfine structure," which splits each $(n, j, \ell)$ level into a "doublet," higher energy when electron and proton spins are aligned, lower when antialigned.

## The Zeeman Effect (weak field)

In a weak external $B$-field, $\mathbf{S}$ and $\mathbf{L}$ are coupled by the atom's internal magnetic field into a total angular momentum $\mathbf{J}$ that is quantized. A consequence is that $\boldsymbol{\mu}_\mathbf{S}$ and $\boldsymbol{\mu}_\mathbf{L}$ are also coupled into a total magnetic dipole moment $\boldsymbol{\mu}_\mathbf{J}$, and different orientations of $\boldsymbol{\mu}_\mathbf{J}$ with $\mathbf{B}_{ext}$ have different energies. Were $\boldsymbol{\mu}_\mathbf{J}$ opposite $\mathbf{J}$ (as $\boldsymbol{\mu}_\mathbf{S}$ is opposite $\mathbf{S}$ and $\boldsymbol{\mu}_\mathbf{L}$ is opposite $\mathbf{L}$), the allowed angles between $\boldsymbol{\mu}_\mathbf{J}$ and $\mathbf{B}_{ext}$ would be the same as between $\mathbf{J}$ and $\mathbf{B}_{ext}$: $\theta = \cos^{-1}(J_z/J)$ (cf. Example 6.2). However, $\boldsymbol{\mu}_\mathbf{J}$ and $\mathbf{J}$ are *not* collinear, for the simple reason that the proportionality constant between $\boldsymbol{\mu}_\mathbf{S}$ and $\mathbf{S}$ is not the same (is essentially twice as large) as that between $\boldsymbol{\mu}_\mathbf{L}$ and $\mathbf{L}$. From equations (7-1) and (7-5),

$$\boldsymbol{\mu}_\mathbf{L} = -\frac{e}{2m_e}\mathbf{L} \qquad \boldsymbol{\mu}_\mathbf{S} \cong -2\frac{e}{2m_e}\mathbf{S} \qquad\qquad \text{(7-26)}$$

(Henceforth, we will use the approximate value 2 for the electron gyromagnetic ratio.) There is no similar fundamental relationship between $\boldsymbol{\mu}_\mathbf{J}$ and $\mathbf{J}$; the total magnetic dipole moment is simply the vector sum of $\boldsymbol{\mu}_\mathbf{L}$ and $\boldsymbol{\mu}_\mathbf{S}$.

$$\boldsymbol{\mu}_\mathbf{J} = \boldsymbol{\mu}_\mathbf{L} + \boldsymbol{\mu}_\mathbf{S} = -\frac{e}{2m_e}(\mathbf{L} + 2\mathbf{S}) \qquad\qquad \text{(7-27)}$$

The relationships are shown in Figure 7.18. Since $\mathbf{L} + 2\mathbf{S}$ (the direction of $\boldsymbol{\mu}_\mathbf{J}$) is obviously not $\mathbf{L} + \mathbf{S}$ (the direction of $\mathbf{J}$), $\boldsymbol{\mu}_\mathbf{J}$ and $\mathbf{J}$ are not collinear. Thus, while $\mathbf{J}$ precesses simply at a fixed angle about the external field/$z$-axis, $\boldsymbol{\mu}_\mathbf{J}$ need only do so on average; it might also spin *about an axis through* $\mathbf{J}$. In fact, it does; it must precess right along with $\mathbf{S}$ and $\mathbf{L}$ about $\mathbf{J}$. There are two precessions because there are two magnetic fields: $\mathbf{J}$ precesses about the *external* field/$z$-axis, and the atom's *internal* field locks $\mathbf{S}$ into a fixed angle with $\mathbf{L}$, while still allowing the two to precess as a unit about $\mathbf{J}$. The internal field being by assumption stronger allows

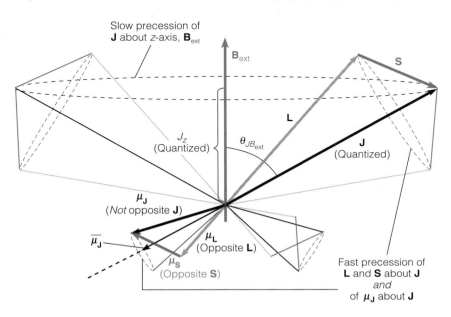

Slow precession of **J** about $z$-axis, $\mathbf{B}_{ext}$

$\mathbf{B}_{ext}$

**S**

$J_z$ (Quantized)

$\theta_{JB_{ext}}$

**L**

**J** (Quantized)

$\boldsymbol{\mu}_\mathbf{J}$ (*Not* opposite **J**)

$\overline{\boldsymbol{\mu}_\mathbf{J}}$

$\boldsymbol{\mu}_\mathbf{L}$ (Opposite **L**)

$\boldsymbol{\mu}_\mathbf{S}$ (Opposite **S**)

Fast precession of **L** and **S** about **J** *and* of $\boldsymbol{\mu}_\mathbf{J}$ about **J**

**Figure 7.18** In a weak external field, **J** precesses slowly about **B** with $J_z$ fixed (quantized), while **L** and **S** precess rapidly about **J**.

us to simplify what would otherwise be a very difficult analysis. Precessional frequency is directly related to the strength of the field (see the essay entitled "Precession Rate in a Magnetic Field"). Accordingly, the spinning of **S**, **L**, and $\boldsymbol{\mu_J}$ about **J** is much faster than the precession of **J** about the external field. Therefore, to calculate the orientation energy of $\boldsymbol{\mu_J}$ in the external field, we may simply use an average $\overline{\boldsymbol{\mu_J}}$ vector: the component of $\boldsymbol{\mu_J}$ opposite **J**. The orientation energy may then be written

$$U = -\overline{\boldsymbol{\mu_J}} \cdot \mathbf{B}_{ext} = -\overline{\mu_J}\, B_{ext}\, \cos \theta_{\overline{\mu_J} B_{ext}}$$

Since $\overline{\boldsymbol{\mu_J}}$ is opposite **J**, $\theta_{\overline{\mu_J} B_{ext}} = \pi - \theta_{JB_{ext}}$, so that

$$U = \overline{\mu_J}\, B_{ext}\, \cos \theta_{JB_{ext}}$$

Now, as noted earlier, the angle $\theta_{JB_{ext}}$ between **J** and **B** is given by

$$\cos \theta_{JB_{ext}} = \frac{J_z}{J}$$

so that

$$U = \overline{\mu_J} \frac{J_z}{J} B_{ext} \qquad (7\text{-}28)$$

As we might have expected, we see that in the presence of an external field, energies depend on the orientation of **J** relative to that field (the $z$-axis).

We know that $J_z$ depends on $m_j$, but to see fully what (7-28) tells us we must address $\overline{\mu_J}$. Since $\overline{\boldsymbol{\mu_J}}$ is the component of $\boldsymbol{\mu_J}$ opposite **J**, its magnitude is given by $\overline{\mu_J} = \mu_J |\cos \theta_{\mu_J J}|$, and using $\boldsymbol{\mu_J} \cdot \mathbf{J} = \mu_J J \cos \theta_{\mu_J J}$, this becomes $\overline{\mu_J} = |\boldsymbol{\mu_J} \cdot \mathbf{J}|/J$. Equation (7-27) and $\mathbf{J} = \mathbf{L} + \mathbf{S}$ may then be used to express this in terms of $L$, $S$, and $J$, and hence $\ell$, $s$, and $j$. The details of these steps are left to Exercise 31. The result is

$$\overline{\mu_J} = \frac{e}{2m_e} \frac{3j(j+1) - \ell(\ell+1) + s(s+1)}{2\sqrt{j(j+1)}} \hbar$$

so that (7-28) becomes

$$U = \left( g_{Lande} \frac{e}{2m_e} J_z \right) B_{ext} \qquad (7\text{-}29a)$$

where

$$g_{Lande} \equiv \frac{3j(j+1) - \ell(\ell+1) + s(s+1)}{2j(j+1)} \qquad (7\text{-}30)$$

Comparing with our initial expression $U = -\overline{\boldsymbol{\mu_J}} \cdot \mathbf{B}_{ext}$, we see that the quantity in parentheses in (7-29a) must be $-\overline{\mu_{Jz}}$. Note that it resembles $-\overline{\mu_{Lz}} = (e/2m)\, L_z$ and $-\mu_{Sz} = 2(e/2m)\, S_z$. But the relationship here between $\overline{\mu_{Jz}}$ and $J_z$ involves a more complicated proportionality constant, known as the **Lande g-factor.**[19]

Finally, using $J_z = m_j \hbar$, (7-29a) may be rewritten

$$U = g_{Lande} \frac{e}{2m_e} m_j \hbar B_{ext} \qquad (7\text{-}29b)$$

[19]*Note:* it is correct in the special cases; that is, $g_{Lande} = 1$ in the $L$-only case, $s = 0$, $j = \ell$, and $g_{Lande} = 2$ in the $S$-only case, $\ell = 0$, $j = s$.

It is worthwhile to reiterate that it is practically impossible to solve the hydrogen atom problem without approximation while taking into account the effects of the spin-orbit interaction and external fields. Nevertheless, when these effects are small, the correct energies are given merely by adding the energies associated with these effects to the simple hydrogen atom energies.

$$E_{\text{weak external field}} = E_{\text{zero field}} + \frac{e}{2m_e} g_{\text{Lande}}\, m_j \hbar B_{\text{ext}} \qquad (7\text{-}31)$$

In a weak external magnetic field, energy depends on $n$, $\ell$, $j$, and $m_j$

We may thus conclude: A weak external field splits otherwise degenerate atomic states according to $m_j$ into $2j + 1$ states above and below the zero-field level, and the size of the splitting depends, through $g_{\text{Lande}}$, on the quantum numbers $j$ and $\ell$. One result is that the force exerted by a *nonuniform* external field ($\partial B/\partial z \neq 0 \Rightarrow \partial U/\partial z \neq 0 \Rightarrow F_z \neq 0$) would be different for states of different $m_j$. In other words, a Stern–Gerlach-type experiment would break a single beam of atoms into $2j + 1$ beams.[20] Another result is that in the presence of an external field, a spectral line corresponding to a transition from one $(n, j)$ state to another will be split into several lines.

---

**Example 7.4**

A sample of hydrogen atoms is placed in a 0.05-T magnetic field. (a) Into how many levels is the $1s_{1/2}$-level split, and into how many is the $2p_{3/2}$ split? (b) In each case, how large is the energy splitting? (c) Because the intrinsic angular momentum of the photon is unity, in a photon-producing transition from one state to another, $m_j$ cannot change by more than 1. Into how many lines is the $2p_{3/2}$ to $1s_{1/2}$ spectral line split by the field? (d) What is the energy spacing between the lines? (e) By about how much is the wavelength of a spectral line shifted?

(*Note:* The field is weaker than the estimated 0.3-T internal field of Section 7.6.)

**Solution**

(a) Since $j = \frac{1}{2}$ in the $1s_{1/2}$ state, $m_j$ can be either of the values: $\pm\frac{1}{2}$. Thus, the level is split due to the different dipole-field orientations into two different energies, one of slightly higher energy and one slightly lower than the zero-field level. Similarly, $j = \frac{3}{2}$ implies that $m_j$ can be any of the four values: $\pm\frac{3}{2}, \pm\frac{1}{2}$. Thus, four energies result.

(b) First we calculate $g_{\text{Lande}}$ for the two states.

$$g_{\text{Lande},\,1s_{1/2}} = \frac{3\frac{1}{2}(\frac{1}{2} + 1) - 0 + \frac{1}{2}(\frac{1}{2} + 1)}{2\frac{1}{2}(\frac{1}{2} + 1)} = 2$$

$$g_{\text{Lande},\,2p_{3/2}} = \frac{3\frac{3}{2}(\frac{3}{2} + 1) - 1(1 + 1) + \frac{1}{2}(\frac{1}{2} + 1)}{2\frac{3}{2}(\frac{3}{2} + 1)} = \frac{4}{3}$$

[20]In the $\ell = 0$, spin-only case considered in Section 7.1, $j = s = \frac{1}{2}$ and $m_j = m_s = \pm\frac{1}{2}$. The beam would split into two.

We see that the splitting of the $2p_{3/2}$ states is only $\frac{2}{3}$ that of the $1s_{1/2}$. Figure 7.19 illustrates the relative spacing. Now using (7-29b),

$$U_{1s_{1/2}} = 2\frac{1.6 \times 10^{-19}\ C}{2(9.11 \times 10^{-31}\ kg)}\left\{\pm\tfrac{1}{2}\right\}(1.055 \times 10^{-34}\ \text{J·s})(0.05\ \text{T})$$

$$= \left\{\pm\tfrac{1}{2}\right\}9.3 \times 10^{-25}\ \text{J} = \left\{\pm\tfrac{1}{2}\right\}5.8 \times 10^{-6}\ \text{eV}$$

and

$$U_{2p_{3/2}} = \frac{4}{3}\frac{1.6 \times 10^{-19}\ C}{2(9.11 \times 10^{-31}\ kg)}\begin{Bmatrix}\pm\tfrac{3}{2}\\ \pm\tfrac{1}{2}\end{Bmatrix}(1.055 \times 10^{-34}\ \text{J·s})(0.05\ \text{T})$$

$$= \begin{Bmatrix}\pm\tfrac{3}{2}\\ \pm\tfrac{1}{2}\end{Bmatrix}6.2 \times 10^{-25}\ \text{J} = \begin{Bmatrix}\pm\tfrac{3}{2}\\ \pm\tfrac{1}{2}\end{Bmatrix}3.9 \times 10^{-6}\ \text{eV}$$

The $m_j$-values are, as always, spaced in integral steps, so we see that the energy splitting in the $1s_{1/2}$ is $5.8 \times 10^{-6}$ eV and in the $2p_{3/2}$ is $3.9 \times 10^{-6}$ eV. Given that the external field is weak compared to the internal, it is not surprising that these energy shifts are even smaller than the spin-orbit energy of Section 7.6.

(c) Using (7-31), with $E_{1s_{1/2}}$ and $E_{2p_{3/2}}$ representing the zero-field energies, a transition from a $2p_{3/2}$ state to a $1s_{1/2}$ state would involve an energy difference of

$$\Delta E = \left(E_{2p_{3/2}} + \frac{e}{2m_e}\frac{4}{3}\begin{Bmatrix}\pm\tfrac{3}{2}\\ \pm\tfrac{1}{2}\end{Bmatrix}\hbar B_{ext}\right) - \left(E_{1s_{1/2}} + \frac{e}{2m_e}2\left\{\pm\tfrac{1}{2}\right\}\hbar B_{ext}\right)$$

Transitions in which $m_j$ changes from $+\frac{3}{2}$ in the $2p_{3/2}$ to $-\frac{1}{2}$ in the $1s_{1/2}$ or from $-\frac{3}{2}$ in the $2p_{3/2}$ to $+\frac{1}{2}$ in the $1s_{1/2}$ are forbidden by the given selection rule. This leaves only six: $+\frac{3}{2}$ to $+\frac{1}{2}$, $-\frac{3}{2}$ to $-\frac{1}{2}$, and the four possible transitions between $m_j = \pm\frac{1}{2}$ states. As the reader may easily verify, they may be written together as

$$\Delta E = (E_{2p_{3/2}} - E_{1s_{1/2}}) + \frac{e}{2m_e}\hbar B_{ext}\begin{Bmatrix}\pm\tfrac{5}{3}\\ \pm\tfrac{3}{3}\\ \pm\tfrac{1}{3}\end{Bmatrix}$$

There are six equally spaced photon energies. This may also be deduced from Figure 7.19.

(d) From the previous equation we see that the spacing is

$$\frac{e\hbar}{2m_e}B_{ext}\frac{2}{3} = \frac{1.6 \times 10^{-19}\ C}{2(9.11 \times 10^{-31}\ kg)}(1.055 \times 10^{-34}\ \text{J·s})(0.05\ \text{T})\frac{2}{3}$$

$$= 3.1 \times 10^{-25}\ \text{J} = 1.9 \times 10^{-6}\ \text{eV}$$

(e) Since the splittings we have found are so small, the transition energy is still essentially

$$E_2 - E_1 = \frac{-13.6\ \text{eV}}{2^2} - \frac{-13.6\ \text{eV}}{1^2} = 10.2\ \text{eV} = 1.6 \times 10^{-18}\ \text{J}$$

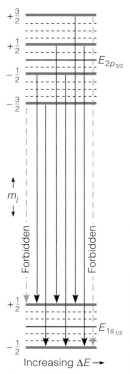

**Figure 7.19** In a weak magnetic field, there are six allowed $2p$ to $1s$ transitions.

Thus, the energy difference between spectral lines is about $1.9 \times 10^{-6}$ eV/10.2 eV $\cong 2 \times 10^{-7}$ times the energy of the given line. This causes a similar shift in the wavelength. The wavelength of this first line in the Lyman series is about 122 nm (see Section 6.9), so the external field would split it into six spectral lines spaced at about $2.5 \times 10^{-5}$ nm.

## The Paschen–Back Effect (strong field)

Analysis of the splitting of degenerate levels is much easier if the external $B$-field is relatively strong—perhaps several tesla. So strong a field overwhelms the internal field that would otherwise couple **L** and **S** together. Orbital and spin angular momenta then exhibit independent quantizations along the external field/$z$-axis, as depicted in Figure 7.20. Using (7-26), the orientation energies are given by

$$U_L = -\boldsymbol{\mu_L} \cdot \mathbf{B}_{ext} = -\mu_{L_z} B_{ext} = -\left(-\frac{e}{2m_e} L_z\right) B_{ext} = \frac{e}{2m_e} m_\ell \hbar B_{ext}$$

and

$$U_S = -\boldsymbol{\mu_S} \cdot \mathbf{B}_{ext} = -\mu_{S_z} B_{ext} = -\left(-2\frac{e}{2m_e} S_z\right) B_{ext} = 2\frac{e}{2m_e} m_s \hbar B_{ext}$$

Note the similarities to (7-29b). Although a "strong" field is able to break the spin-orbit coupling, its effect is still usually small enough that the overall energies are once again obtained by merely adding these perturbations to the zero-field energies.

$$E_{\text{strong external field}} = E_{\text{zero field}} + \frac{e}{2m_e}(m_\ell + 2m_s)\hbar B_{ext} \qquad (7\text{-}32)$$

In a strong external magnetic field, energy depends on $n$, $\ell$, $m_\ell$, and $m_s$

The reader is encouraged to compare this result with the corresponding weak-field result (7-31), and also to compare Figures 7.18 and 20. This should help in understanding why the "good" quantum numbers (besides $n$ and $\ell$) are $j$ and $m_j$ in the weak-field case, but $m_\ell$ and $m_s$ in a strong field.

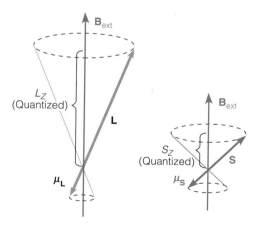

**Figure 7.20** A strong external field forces both **L** and **S** to precess independently about **B**.

# ◆ **7.9** Multielectron-Atom Excitation Spectra

Photon emissions from the elements provide important information about the structure of the electronic energy levels and also provide an excellent means of distinguishing one element from another. In this section, we take a brief look at photon emission due to electron excitations above the ground state, usually in the visible and near-visible range. Another form of spectral emission, **characteristic x rays,** we take up in the Section 7.10.

To understand the richness of spectral lines in excited atoms, it is necessary to understand how the various angular momenta interact, because this determines how otherwise degenerate levels split. In a one-electron atom such as hydrogen, there is a spin-orbit interaction between **S** and **L**. In multielectron atoms, there are many spin and orbital angular momenta to interact. Naturally the interactions can be complex. Let us solidify our grasp of the quantum rules by considering **total spin.**

## Total Spin

As noted in Section 7.7, the quantum rules are the same when adding any angular momenta. As **L** and **S** may add to a total angular momentum of magnitude $J = \sqrt{j(j+1)}\hbar$, where $j$ may take on any integral value between $\ell + s$ and $|\ell - s|$, two spin angular momenta may add to a total spin angular momentum $\mathbf{S}_T$ of magnitude $S_T = \sqrt{s_T(s_T + 1)}\hbar$, where $s_T$ may take on any integral value between $s + s$ and $|s - s|$. For two spin-$\frac{1}{2}$ particles such as electrons, $s_T$ may take on only the values 1 and 0, corresponding to $S_T = \sqrt{2}\hbar$ and $S_T = 0$. The $z$-component $S_{Tz}$ also obeys the usual quantization rules, so we should expect that for $s_T = 1$ there will be states with three possible $z$-components: $+\hbar, 0, -\hbar$. These are the triplet states of Section 7.2. The singlet state is the one with total spin angular momentum zero, and zero $z$-component. Table 7.4 summarizes these points.

Figure 7.21 depicts how the two particles' spins add to form the triplet states. In all three states, the spins are "aligned" (as nearly as the quantum addition rules allow), adding to a total spin $\mathbf{S}_T$ of the same magnitude, $\sqrt{2}\hbar$.[21] As usual, while this angular momentum makes a fixed angle with the $z$-axis, it may point in any direction in the $x$-$y$ plane. The $S_{Tz} = +\hbar$ state has both spins up, more or less, as suggested by the symbolic representation of Table 7.4. The $S_{Tz} = -\hbar$ state is similar but with both spins down. In the $S_{Tz} = 0$ state, one particle is spin-up and the other spin-down, though it cannot be said which is which. Again, this fits with the symbolic representation: the $S_{Tz} = 0$ state is a *combination* of $S_{1z}$ up, $S_{2z}$ down and $S_{1z}$ down, $S_{2z}$ up. Note that even though their $z$-*components* add to zero, the

**Table 7.4    Total Spin States (two spin-$\frac{1}{2}$ particles)**

| Symmetric | | | Antisymmetric |
|---|---|---|---|
| Triplet: $s_T = 1, S_T = \sqrt{2}\hbar$ | | | Singlet: $s_T = 0, S_T = 0$ |
| {↑↑} | {↑↓} + {↓↑} | {↓↓} | {↑↓} − {↓↑} |
| $S_{Tz} = +\hbar$ | $S_{Tz} = 0$ | $S_{Tz} = -\hbar$ | $S_{Tz} = 0$ |

[21] As it should be, this magnitude is greater than the $\frac{\sqrt{3}}{2}\hbar$ of a single spin-$\frac{1}{2}$ particle.

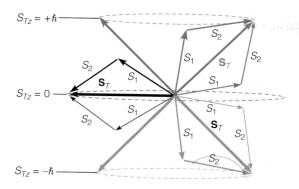

**Figure 7.21** Formation of triplet states.

*spins* are "aligned"; they add to a total spin greater than either. In the singlet state (not depicted), the particles' *spins*—not just $z$-components—are truly opposite, perfectly antialigned, giving $S_T = 0$.

## LS Coupling

For lighter atoms, the usual interaction among the various angular momenta is as follows: The valence electrons' orbital angular momenta interact to form a quantized total orbital angular momentum $\mathbf{L}_T$; the spins interact to form a quantized total spin angular momentum $\mathbf{S}_T$; "finally," there is an overall spin-orbit interaction between $\mathbf{S}_T$ and $\mathbf{L}_T$, giving a quantized "grand total" angular momentum $\mathbf{J}_T$. Each angular momentum addition follows the standard rules we have just used to find total spin. This kind of interaction is known as *LS* coupling—the "final step" is coupling of $L$ and $S$. At its root is the tendency for the interelectron repulsion to produce some correlation in the spatial states, including of course $\ell$-values, which in turn, because the *overall* state must be antisymmetric, demands some correlation between the spin states. (The argument in Section 7.4 that electrons in unfilled shells should spread out among $m_\ell$-values is similar.) Only "after" these two correlations does a relatively weak overall spin-orbit interaction come into play.[22]

The character of excited energy levels in monovalent atoms is much the same as in hydrogen, because the inert inner-shell electrons effectively screen the valence electron from all but one positive charge of the nucleus. To sort out the effects of true multielectron interactions, let us examine the simplest case of interaction between two valence electrons, one in an excited state and the other "left behind." Consider a helium atom where one electron is in an excited state while the other remains in the $1s$ state. (This is not an unrealistic special case. Energy sufficient to raise *both* electrons out of the $1s$ state would be far more than sufficient to liberate *one* of them *completely,* simply ionizing the atom.) Suppose the excited level is the $2p$. By the angular momentum addition rules, the quantum number $\ell_T$ for the total orbital angular momentum may be anywhere from $|\ell_2 - \ell_1|$ to $(\ell_2 + \ell_1)$ in integral steps. Since $\ell$ is 0 for the $1s$ and 1 for the $2p$, $\ell_T$ can be only 1. The spins couple to form a total spin $s_T$ anywhere from $|s_2 - s_1|$ to $(s_2 + s_1)$ in integral steps, which, for spin-$\frac{1}{2}$ electrons, means 0 or 1. As already noted, these are the singlet (antialigned) and triplet (aligned) states. Finally, $\mathbf{L}_T$

---

[22]The other kind of interaction involves a spin-orbit coupling between *each* electron's $\mathbf{S}$ and $\mathbf{L}$, yielding a quantized J for each, "followed by" a coupling of all these individual $\mathbf{J}$ vectors into a quantized "grand total" angular momentum $\mathbf{J_T}$. As the "final step" is coupling of individual $\mathbf{J}$ vectors, rather than $\mathbf{S}_T$ and $\mathbf{L}_T$, this kind of interaction is known as *jj*-coupling. It predominates in high-$Z$ atoms, where the intense magnetic field of the highly charged "orbiting" nucleus favors an "immediate" spin-orbit interaction for each electron. Left until last is a coupling of $J$'s that reduces the relatively weak interelectron repulsion.

and $\mathbf{S}_T$ couple via a spin-orbit interaction to form a total angular momentum $\mathbf{J}_T$, where $j_T$ may take on any value between $|\ell_T - s_T|$ and $(\ell_T + s_T)$ in integral steps.[23] For $s_T = 0$, this is 1 (i.e., $\ell_T$) while for $s_T = 1$, it may be 0, 1, or 2. We see then that the excited electron has open to it many different $2p$ states. We designate these using the notation shown in the margin. Thus, the $(\ell_T, s_T, j_T) = (1, 0, 1)$ state is the singlet $2\,^1P_1$ (meaning *spin* singlet) state, the $(1, 1, 1)$ state is the triplet $2\,^3P_1$ state, etc.[24] In this notation the ground state is the singlet $1\,^1S_0$. (There is no triplet $1\,^3S_0$ state, as the exclusion principle requires spins to be antialigned when two electrons are in the same spatial state.)

Table 7.5 summarizes how *LS* coupling governs one-electron excited states in helium, through $n = 3$. In general, different *LS* coupling states are of different

Multielectron spectroscopic notation:

$n^{\,2s_T+1}\ell_{T\,j_T}$

**Table 7.5** *LS* Coupling in Helium
(one electron excited)

| $n_1\ell_1 n_2\ell_2$ | $\ell_T$ $\begin{array}{c}\lvert\ell_2 - \ell_1\rvert\\ \cdots\\ (\ell_2 + \ell_1)\end{array}$ | $s_T$ $\begin{array}{c}\lvert s_2 - s_1\rvert\\ \cdots\\ (s_2 + s_1)\end{array}$ | $j_T$ $\begin{array}{c}\lvert\ell_T - s_T\rvert\\ \cdots\\ (\ell_T + s_T)\end{array}$ | State |
|---|---|---|---|---|
| $1s1s$ | 0 | 0 | 0 | $1\,^1S_0$ |
| $1s2s$ | 0 | 0 | 0 | $2\,^1S_0$ |
|  |  | 1 | 1 | $2\,^3S_1$ |
| $1s2p$ | 1 | 0 | 1 | $2\,^1P_1$ |
|  |  | 1 | 0 | $2\,^3P_0$ |
|  |  |  | 1 | $2\,^3P_1$ |
|  |  |  | 2 | $2\,^3P_2$ |
| $1s3s$ | 0 | 0 | 0 | $3\,^1S_0$ |
|  |  | 1 | 1 | $3\,^3S_1$ |
| $1s3p$ | 1 | 0 | 1 | $3\,^1P_1$ |
|  |  | 1 | 0 | $3\,^3P_0$ |
|  |  |  | 1 | $3\,^3P_1$ |
|  |  |  | 2 | $3\,^3P_2$ |
| $1s3d$ | 2 | 0 | 2 | $3\,^1D_2$ |
|  |  | 1 | 1 | $3\,^3D_1$ |
|  |  |  | 2 | $3\,^3D_2$ |
|  |  |  | 3 | $3\,^3D_3$ |

[23]To follow the angular momentum addition rules, the quantum *numbers* are most important, but the actual angular momenta may always be found from these in the usual way:

$$L_T = \sqrt{\ell_T(\ell_T + 1)}\hbar,$$
$$S_T = \sqrt{s_T(s_T + 1)}\hbar,$$

and

$$J_T = \sqrt{j_T(j_T + 1)}\hbar.$$

[24]It is conventional to use uppercase letters (S, P, D, F, etc.) for the total orbital angular momentum quantum number $\ell_T$, rather than the lowercase letters used for an individual electron $\ell$.

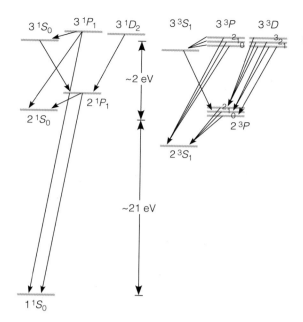

**Figure 7.22** Helium's rich spectrum.

energy,[25] so that rather than simple $n = 3 \to 1$, $3 \to 2$, and $2 \to 1$ transitions, there are many related spectral lines among these levels. Even so, the number of transitions among the states is restricted by selection rules:

$$\Delta \ell_T = \pm 1 \qquad \Delta s_T = 0 \qquad \Delta j_T = 0, \pm 1 \quad (j_T = 0 \nrightarrow j_T = 0)$$

As usual, these are based on two considerations: angular momentum addition rules (angular momentum is carried away by a *unit* spin photon) and the ability of the atom in transition to oscillate as an electric dipole (see Section 6.11). Figure 7.22 (not to scale) shows the wealth of spectral lines possible among just the excited states through $n = 3$. Since $\Delta s_T$ must be zero, the lines naturally divide into those among the $s_T = 0$ singlet states and those among the $s_T = 1$ triplet states. Interestingly, from the triplet $2S$ state a transition down to the ground state is not "allowed." This is a good example of a **metastable state,** upon which the operation of lasers depends. Transitions to the ground state can occur only by means other than electric-dipole photon generation (e.g., interatomic collisions), means that are invariably slow.

We will not delve deeper into multielectron-atom excitation spectra, except to note that if the element is subjected to a weak external magnetic field, its spectral

---

[25]Many factors, often competing, influence which states will be of lower energy. In order of prominence, the major ones are as follows: (1) The preference for an antisymmetric spatial state favors (usually results in a lower energy for) a symmetric spin state—a *larger* $s_T$. (2) The reduction in electrostatic repulsion effected by having multiple $\ell \neq 0$ electrons orbit *in the same direction,* "passing one another" less often than if counterrevolving, favors a *larger* $\ell_T$. (3) The spin-orbit interaction favors a *smaller* $j_T$. For a complete analysis, we refer the interested reader to a dedicated atomic physics text.

lines are further enriched—split according to $m_{j_T}$ by the Zeeman effect. In fact, equations (7-29) and (7-30) may be used directly, with $\ell_T$, $s_T$, and $j_T$ replacing $\ell$, $s$, and $j$.[26]

# 7.10 Multielectron-Atom x-Ray Spectra

Multielectron atoms may do something a one-electron atom cannot—emit electromagnetic radiation due to transitions "below" the ground state. In the ground state of a multielectron atom, electrons completely fill the lowest-energy states up to some maximum energy. If one of these electrons is somehow knocked out of an inner shell, a "hole" is left behind. An electron in a higher energy state may then jump down into the hole, leaving a higher-energy hole; a yet-higher-energy electron may then jump down into this hole, and so on. In each of these transitions, a photon may be produced. While excitations above the ground state's maximum occupied energy involve no more than tens of electronvolts (note the ionization energies in Figure 7.11), energies deep within the inner shells of most elements are typically thousands of electronvolts. Transitions within these deep levels are therefore in the energy range of x rays. (This range of wavelengths, from 10 nm to $10^{-2}$ nm, corresponds to photon energies of $10^3$–$10^5$ eV.)

Since energy levels are quantized, only certain x rays can be emitted by a given element, and since the electronic structure of each element is different, these **characteristic x rays** of an element serve as a "fingerprint" by which it may be identified. Even so, because inner-shell electronic structure is largely independent of the behavior of the valence electrons, with their temperamental periodicities, the variation of characteristic x-ray wavelengths with $Z$ is smooth (see Exercise 35). Of course, to observe x-ray spectra, we must eject the inner-shell electron in the first place. A common way is to hit the sample with an external electron beam. The result is a continuous spectrum due to bremsstrahlung (see Section 2.2), having nothing to do with the specific element, upon which are superimposed the element's characteristic lines, as depicted in Figure 7.23. Special notation is used to refer to the x-ray photons generated in inner-shell transitions. The $n = 1$ shell is referred to as the "$K$-shell," the $n = 2$ as the "$L$-shell," the $n = 3$ as the "$M$-shell," and so on, and these letters are used to indicate the shell at which a transition *terminates*. A subscript, advancing from $\alpha$ on through the Greek alphabet, designates how many shells higher the transition began: $\alpha$ begins one shell higher, $\beta$ two shells higher, and so forth. Thus, a $K_\alpha$ line is produced in a transition ending

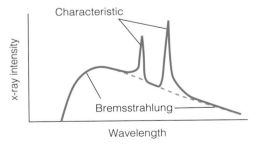

**Figure 7.23** An x-ray spectrum.

[26] In the early days of atomic physics, a distinction was made between a "normal" Zeeman effect and an "anomalous" Zeeman effect. In the former, levels are split according to a $z$-component of *orbital* angular momentum only, which would always lead to an odd number of levels ($2\ell + 1$), as noted in Section 7.1. Cases in which levels split into an even number (as in Example 7.4) could not be understood, because spin was as yet unknown. These were the "anomalous" cases. In retrospect, we see that the "normal" behaviors were merely those cases of multivalent atoms in which the total spin is *zero*.

at the $K$-shell ($n = 1$) and beginning at the $L$-shell ($n = 2$).[27] A $L_\beta$ line is from the $N$-shell ($n = 4$) down to the $L$-shell. The characteristic x rays of the elements are tabulated in many references and provide an important resource for determining elemental composition.

## Progress and Applications

■ Quantum computing requires a **quantum logic gate.** In conventional digital computing, data bits—zeros and ones—must be stored. Any "two-state device," such as a simple on-off wall switch, could accomplish the task. Quantum computing, however, uses qubits (quantum bits), in which the state is really a *combination* of two states. One of the ways being studied to produce a quantum logic gate uses a beryllium ion in an rf ion trap, the spin of the lone valence electron being the pair of states, either up or down (C. Monroe, et al., *Physical Review Letters,* December 18, 1995, pp. 4714–4717). While it is true that an experiment would measure one or the other, this does not mean that spin *is* either one or the other before (or without) such a measurement. For example, if an electron's spin is measured as up along the $z$-axis, a subsequent measurement *along a perpendicular axis*—for instance, the $x$—might find it either up or down, with equal probability. In effect, the first measurement left the electron in a state that was an equal combination of spin-up and spin-down along the $x$-axis. In general, a spin state may be an arbitrarily weighted combination of spin-up and spin-down. Thus, rather than being only two states, the number of different states is infinite. This increase in information makes quantum computing intriguing but also brings with it a problem: Noise is much more likely to obscure a continuously variable combination than a simple "on or off."

■ Interactions of the atom's electrons with magnetic fields, of both internal and external origin, can be very complicated, but are crucial in many areas of modern physics and are thus the subject of considerable scrutiny. In an interesting application, "forbidden" magnetic dipole transitions (see Section 6.11) between states differing in their spin-orbit orientation energy are studied using an electron beam ion trap (C. A. Morgan, et al., *Physical Review Letters*, March 6, 1995, pp. 1716–1719). Analysis of x rays from transitions deep within occupied electron energy levels is common, but these transitions are between "deep" states laid bare by stripping away most of the orbiting electrons. Moreover, the states' energies differ by a relatively small amount, so the photons produced in the transitions are in the visible and near-visible range. Such radiation may one day be useful in monitoring the temperature in nuclear fusion reactors.

■ The operation of a laser is discussed in Chapter 8, but the simple fact that it is a source of very tightly controlled light, of precise wavelength and direction, makes it an indispensable tool in atomic physics. **Laser cooling** may sound like a contradiction in terms (don't lasers burn things up?), but it is really a very clever technique to slow down atoms, one that is often used in ion traps. Suppose that the lowest-energy photon an atom may emit as it drops to its ground state is of wavelength $\lambda_0$. Now lasers producing photons whose wavelength is slightly longer than $\lambda_0$, with corresponding lower energy, are aimed from

[27]Transitions *within* a given shell are not energetic enough to be part of an element's x-ray spectrum.

opposite directions at the atom (ordinarily in its ground state). Were the atom stationary, the lasers' photons could not be absorbed; their energy is too small. However, if the atom moves toward laser 1, the wavelength of the photon approaching from the laser is Doppler-shifted toward shorter wavelength. It is now able to be absorbed by the atom, exciting an upward jump. However, to conserve momentum, the atom will be slowed down. Meanwhile, it could not have absorbed photons from laser 2, since these, coming from a *receding* source, are Doppler-shifted toward longer wavelength. Thus, whichever direction the atom heads, it will absorb only photons from ahead, not behind. It must slow down! This technique cannot slow the atom to zero speed; the atom reemits the photon in an arbitrary direction, and so will, on average, end up with the momentum an atom would have when recoiling from such an emission, essentially that of the photon itself. Still, this single-photon recoil speed is small. Refinements involving the interference of polarized laser beams have lowered even this barrier. Temperatures below a microkelvin are now routine, with the present limit being in the nanokelvin range. As perusal of Progress and Applications throughout the text will show, low temperatures allow study of a host of phenomena.

## Basic Equations

**Orbital magnetic dipole moment**

$$\boldsymbol{\mu}_{\mathbf{L}} = -\frac{e}{2m_e}\mathbf{L} \tag{7-1}$$

**Spin (intrinsic) magnetic dipole moment**

$$\boldsymbol{\mu}_{\mathbf{S}} = -g_e\frac{e}{2m_e}\mathbf{S} \ (g_e \cong 2) \tag{7-5}$$

**Intrinsic angular momentum—spin**

$$S = \sqrt{s(s+1)}\hbar \tag{7-6}$$

**Z-component of spin**

$$S_z = m_s\hbar \ (m_s = -s, -s+1, \ldots, s-1, s) \tag{7-7}$$

**Symmetric and antisymmetric wave functions**

$$\psi_S(x_1, x_2) \equiv \psi_n(x_1)\psi_{n'}(x_2) + \psi_{n'}(x_1)\psi_n(x_2) \quad \text{(Symmetric)} \tag{7-17}$$

$$\psi_A(x_1, x_2) \equiv \psi_n(x_1)\psi_{n'}(x_2) - \psi_{n'}(x_1)\psi_n(x_2) \quad \text{(Antisymmetric)}$$

**Angular momentum addition rules**

$$\mathbf{J} = \mathbf{L} + \mathbf{S} \tag{7-22}$$

$$J = \sqrt{j(j+1)}\hbar \ (j = |\ell - s|, |\ell - s| + 1, \ldots, \ell + s - 1, \ell + s) \tag{7-23}$$

$$J_z = m_j\hbar \qquad (m_j = -j, -j+1, \ldots, j-1, j) \tag{7-24}$$

# Summary

Electrons and other fundamental particles possess intrinsic angular momentum that is inherently quantized. It is called spin. The measure of this property is the value of $s$, a characteristic of a given kind of particle. The relationship between this numerical value and the actual angular momentum $S$ of the particle is

$$S = \sqrt{s(s + 1)}\hbar$$

The electron is a "spin-$\frac{1}{2}$" particle, whose intrinsic angular momentum is thus $S = \frac{\sqrt{3}}{2}\hbar$. Inextricably related to spin is an intrinsic magnetic dipole moment.

$$\boldsymbol{\mu}_S = -g_e \frac{e}{2m_e}\mathbf{S} \qquad (g_e \cong 2)$$

Prediction of these properties comes from a relativistically correct treatment of quantum mechanics. An important result is that besides the three "spatial" quantum numbers $n$, $\ell$, and $m_\ell$ that come from a three-dimensional nonrelativistic treatment of the atom, a fourth quantum number, the spin quantum number $m_s$, is necessary to specify the state of an electron in an atom. The quantum number $m_s$ is restricted to the values

$$m_s = -s, -s + 1, \ldots, s - 1, s$$

Directly related to these are the allowed values of the particle's $z$-component of intrinsic angular momentum.

$$S_z = m_s\hbar$$

Spin quantization can be revealed in a nonuniform magnetic field: Because the $z$-component of intrinsic angular momentum $\mathbf{S}$ is quantized, the $z$-component of magnetic dipole moment $\boldsymbol{\mu}_S$ is quantized, and the force on the dipole in a nonuniform field is therefore quantized.

Determining the state of a many-particle system requires solution of the $N$-particle Schrödinger equation. Simple solutions are products of individual-particle functions:

$$\psi(x_1, x_2, x_3, \ldots) = \psi_a(x_1)\psi_b(x_2)\psi_c(x_3) \cdots$$

However, such a product solution implies knowledge beyond what is possible to know. In particular, it corresponds to an asymmetric probability density. This is not allowed because asymmetry in the locations of indistinguishable particles cannot be verified. To be acceptable, the probability density of a system of $N$ indistinguishable particles must be unchanged if any two particle labels are switched/exchanged.

Solutions of the Schrödinger equation with the same total energy but not violating this requirement are symmetric or antisymmetric sums of product solutions. For two particles, these are

$$\psi_S(x_1, x_2) = \psi_a(x_1)\psi_b(x_2) + \psi_b(x_1)\psi_a(x_2) \qquad \textbf{(Symmetric)}$$

$$\psi_A(x_1, x_2) = \psi_a(x_1)\psi_b(x_2) - \psi_b(x_1)\psi_a(x_2) \qquad \textbf{(Antisymmetric)}$$

The symmetric state is unchanged if particle labels 1 and 2 are exchanged; the antisymmetric switches sign. Each corresponds to a probability density unchanged under exchange of particle labels, but the two are of quite different character. The particles are more likely to be found close together in the symmetric, and far apart in the antisymmetric.

Whether a given state is symmetric or antisymmetric depends not only on the spatial state, specified by $n$, $\ell$, and $m_\ell$ in spherical polar coordinates, but also the spin state, specified by $m_s$. Two electrons might, for instance, be in a symmetric spatial state but antisymmetric spin state. Since switching particle labels would switch the sign of the overall state, switching the spin but not the spatial state, the overall state is antisymmetric.

The importance of the exchange symmetry, symmetric versus antisymmetric, of a kind of particle is that a system of indistinguishable particles will always be found to be in a state that is either symmetric or antisymmetric, depending on the kind of particle, in particular on its spin. Particles whose spin $s$ is an integer manifest symmetric overall states and are collectively known as bosons. Particles whose spin $s$ is half-integral manifest antisymmetric overall states and are collectively known as fermions.

Systems of electrons will always manifest an antisymmetric overall state, since electrons are $s = \frac{1}{2}$ fermions. However, because the antisymmetric combination of any two states with the same quantum numbers, (i.e., identical individual-particle states) is identically zero, no two indistinguishable fermions may occupy the same individual-particle state. Therefore, no two electrons in the same atom may possess the same set of quantum numbers. This is the exclusion principle.

The exclusion principle is the foundation of chemistry. Electrons orbiting a nucleus cannot all occupy the lowest-energy individual-particle state. In particular, only two may occupy the lowest-energy spatial state $n$, $\ell$, $m_\ell = 1, 0, 0$: one with $m_s = +\frac{1}{2}$ and one with $m_s = -\frac{1}{2}$. The greater the number of electrons, the higher must be the value of $n$ of the highest-energy electrons. Consequently, many electrons may be far from the nucleus and screened from its attractive force by intervening lower-energy electrons. The number of valence electrons—weakly bound outer electrons—available to participate in chemical reactions with other atoms depends intimately on the number of electrons allowed in lower-energy shells and subshells ($n$- and $\ell$-values).

Complete characterization of multielectron-atom behavior is problematic, because the Schrödinger equation cannot be solved easily when multiple interparticle forces are involved. Approximation techniques of varying degrees of sophistication are required. Nevertheless, understanding of much atomic behavior follows from simple one-particle quantum mechanics and application of the exclusion principle.

## EXERCISES

### Section 7.1

1. The electron is known to have a radius no larger than $10^{-16}$ m. (a) Treating it as a classical spinning sphere of this radius, what is the largest possible value of its moment of inertia? (b) Given the magnitude of its intrinsic angular momentum, what is the smallest possible value of its angular velocity and corresponding equatorial speed?

2. In the Stern–Gerlach experiment, how much would a hydrogen atom emanating from a 500 K oven (KE = $\frac{3}{2}k_B T$) be deflected in traveling one meter through a magnetic field whose rate of change is 10 T/m?

3. Show that the frequency at which the electron's *intrinsic* magnetic dipole moment would precess in a magnetic field is given by $\omega \cong eB/m_e$. Calculate this frequency for a field of 1.0 T.

4. A hydrogen atom in its ground state is subjected to an external magnetic field of 1.0 T. What is the energy difference between the spin-up and spin-down states?

5. In the phenomenon of **electron spin resonance,** a photon of the proper frequency may induce a "spin-flip" in the single valence electron of a sodium atom in an external magnetic field. At a particular frequency, photons are readily absorbed because their energy is exactly equal to the energy difference between the spin-up and spin-down orientations of the electron in the magnetic field. (a) Suppose the magnetic field is given by **B** = (0.40 T)$\hat{\mathbf{z}}$. What must be the frequency of an incoming photon to induce a "spin-flip"? (b) Is the electron initially spin-up, or spin-down?

## Section 7.2

6. Show that the symmetric and antisymmetric combinations of equation (7-17) are solutions of the two-particle Schrödinger equation (7-10) of the same energy as $\psi_n(x_1)\psi_{n'}(x_2)$, the unsymmetrized product (7-14).

7. Two particles in a box occupy the $n = 1$ and $n' = 2$ states. (a) Show that the normalization constants for both the symmetric and antisymmetric states are the same as that given in Example 7.2. (b) Calculate for both states the probability that both particles would be found in the left-hand side of the box (i.e., between 0 and $\frac{1}{2}L$).

8. Two particles in a box have a total energy $(5\pi^2\hbar^2/2mL^2)$. (a) Which states are occupied? (b) Make a sketch of $P_S(x_1, x_2)$ versus $x_1$ for points along the line $x_2 = x_1$. (c) Make a similar sketch of $P_A(x_1, x_2)$. (d) Repeat parts (b) and (c) but for points on the line $x_2 = L - x_1$. *Note:* $\sin[m\pi(L-x)/L] = (-1)^{m+1}\sin(m\pi x/L)$.

## Section 7.3

9. What is the minimum possible energy for five (noninteracting) spin-$\frac{1}{2}$ particles of mass $m$ in a one-dimensional box of length $L?$ What if the particles were spin-1? What if the particles were spin-$\frac{3}{2}$?

10. **Slater Determinant.** Expressing states of overall antisymmetric character for many fermions can be tedious. A compact way is the Slater determinant:

$$\begin{vmatrix} \psi_{n_1}(x_1)m_{s1} & \psi_{n_2}(x_1)m_{s2} & \psi_{n_3}(x_1)m_{s3} & \cdots & \psi_{n_N}(x_1)m_{sN} \\ \psi_{n_1}(x_2)m_{s1} & \psi_{n_2}(x_2)m_{s2} & \psi_{n_3}(x_2)m_{s3} & \cdots & \psi_{n_N}(x_2)m_{sN} \\ \psi_{n_1}(x_3)m_{s1} & \psi_{n_2}(x_3)m_{s2} & \psi_{n_3}(x_3)m_{s3} & \cdots & \psi_{n_N}(x_3)m_{sN} \\ \cdots & \cdots & \cdots & \cdots & \cdots \\ \psi_{n_1}(x_N)m_{s1} & \psi_{n_2}(x_N)m_{s2} & \psi_{n_3}(x_N)m_{s3} & \cdots & \psi_{n_N}(x_N)m_{sN} \end{vmatrix}$$

It is based on the fact that for $N$ fermions there must be $N$ different individual-particle states, or sets of quantum numbers. The $i$th state has *spatial* quantum numbers (which might be $n_i$, $\ell_i$, and $m_{\ell_i}$) represented simply by $n_i$, and spin quantum number $m_{si}$. Were it occupied by the $j$th particle, the state would be $\psi_{n_i}(x_j)m_{si}$. A column corresponds to a given state, and a row to a given particle. For instance, the first column corresponds to individual-particle state $\psi_{n_1}(x_j)m_{s1}$, where $j$ progresses (through the rows) from particle 1 to particle $N$. The first row corresponds to particle 1, which successively occupies all individual-particle states (progressing through the columns).

(a) What property of determinants ensures that the overall state is zero if any two individual-particle states are identical?

(b) What property of determinants ensures that switching the labels on any two particles switches the sign of the overall state?

11. The Slater determinant is introduced in Exercise 10. Show that if states $n$ and $n'$ of the infinite well are occupied, and both spins are up, the Slater determinant yields the antisymmetric overall state:

$$(\psi_n(x_1)\psi_{n'}(x_2) - \psi_{n'}(x_1)\psi_n(x_2))\left\{+\tfrac{1}{2}, +\tfrac{1}{2}\right\}$$

12. The Slater determinant is introduced in Exercise 10. Show that if states $n$ and $n'$ of the infinite well are occupied, the particle in state $n$ having spin-up and the particle in $n'$ having spin-down, the Slater determinant yields the antisymmetric overall state:

$$\psi_n(x_1)\psi_{n'}(x_2)\left\{+\tfrac{1}{2}, -\tfrac{1}{2}\right\} - \psi_{n'}(x_1)\psi_n(x_2)\left\{-\tfrac{1}{2}, +\tfrac{1}{2}\right\}$$

13. The overall state found in Exercise 12 for two particles in a box is neither of the two one-spin-up, one-spin-down antisymmetric states given in Table 7.1. (a) Show that it is a linear combination of these two states, and in so doing that its spatial state alone has no definite exchange symmetry, nor does its spin state alone. (b) Were the two particles to repel one another, would you expect the overall state to be that given in Exercise 12, or one of the two given in Table 7.1? Explain.

14. A lithium atom has three electrons. These occupy individual-particle states corresponding to the sets of four quantum numbers given by $(n, \ell, m_\ell, m_s) =$

$(1, 0, 0, +\frac{1}{2})$, $(1, 0, 0, -\frac{1}{2})$, and $(2, 0, 0, +\frac{1}{2})$. Using $\psi_{1,0,0}(x_j)\{+\frac{1}{2}\}$, $\psi_{1,0,0}(x_j)\{-\frac{1}{2}\}$, and $\psi_{2,0,0}(x_j)\{+\frac{1}{2}\}$ to represent the individual-particle states when occupied by particle $j$, use the Slater determinant (see Exercise 10) to find an expression for an antisymmetric overall state. Your answer should be *sums* of terms like

$$\psi_{1,0,0}(x_1)\psi_{1,0,0}(x_2)\psi_{2,0,0}(x_3)\ \{+\tfrac{1}{2}, -\tfrac{1}{2}, +\tfrac{1}{2}\}$$

## Section 7.4

15.  Write the electronic configurations for phosphorus, germanium, and cesium.
16.  Element 114 has never been found, but what would we expect its valence to be?
17.  Were it to follow the standard pattern, what would be the electronic configuration of element 119?
18.  Consider rows 2 and 5 in the periodic table. Why should fluorine, in row 2, be more reactive than iodine, in row 5, while lithium, in row 2, is less reactive than rubidium, in row 5?
19.  Consider row 4 of the periodic table. The trend is that the $4s$ subshell fills, then the $3d$, then the $4p$. (a) Judging by adherence to and deviation from this trend, what might be said of the energy difference between the $4s$ and $3d$ relative to that between the $3d$ and $4p$? (b) Is this also true of row 5? (c) Are these observations in qualitative agreement with Figure 7.9? Explain.

## Section 7.7

20.  Identify the different total angular momentum states $(j, m_j)$ allowed a $3d$ electron in a hydrogen.
21.  Show that, unless $\ell = s$, $\mathbf{L}$ and $\mathbf{S}$ cannot be exactly opposite; that is, show that at its minimum possible value, for which $j = \ell - s$, the magnitude $J$ of the total angular momentum is strictly greater than the difference $|L - S|$ between the magnitudes of the orbital and intrinsic angular momentum vectors.
22.  Assuming that the spin-orbit interaction is not overwhelmed by an external magnetic field, what is the minimum angle the *total* angular momentum vector $\mathbf{J}$ may make with the $z$-axis in a $3d$ state of hydrogen?
23.  What is the angle between $\mathbf{L}$ and $\mathbf{S}$ in a (a) $2p_{3/2}$ (b) $2p_{1/2}$ state of hydrogen?
24.  The well-known **sodium doublet,** two yellow spectral lines of very close wavelength, 589.0 nm and 589.6 nm, is caused by splitting of the $3p$ energy level, due to the spin-orbit interaction.

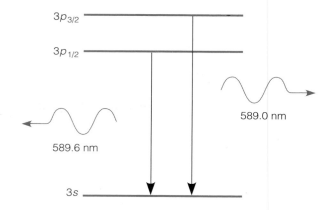

$3p_{3/2}$

$3p_{1/2}$

589.0 nm

589.6 nm

$3s$

In its ground state, sodium's single valence electron is in the $3s$ level. It may be excited to the next higher level, the $3p$, then emit a photon as it drops back to the $3s$. However, the $3p$ is actually two levels: the $3p_{3/2}$ and the $3p_{1/2}$. In the $3p_{3/2}$, $\mathbf{L}$ and $\mathbf{S}$ are aligned, and in the $3p_{1/2}$ they are antialigned. Since the transitions start from slightly different initial energies yet have identical final energies (the $3s$ being $\ell = 0$ and thus having no "orbit" to lead to spin-orbit interaction), there are two different wavelengths possible for the emitted photon. Calculate the difference in energy between the two photons, and from this obtain a rough value of the average strength of the internal magnetic field experienced by the valence electron.

25.  The hydrogen spin-orbit interaction energy given in equation (7-21) is $g_e(\mu_0 e^2/8\pi m_e^2 r^3)\ \mathbf{S}\cdot\mathbf{L}$. Using a reasonable value for $r$ in terms of $a_0$, and the relationships $S = \frac{\sqrt{3}}{2}\hbar$ and $L = \sqrt{\ell(\ell+1)}\hbar$, show that this energy is proportional to a typical hydrogen atom energy by the factor $\alpha^2$, where $\alpha$ is the fine structure constant.
26.  Whether adding spins to get total spin, spin and orbit to get total angular momentum, or total angular momenta to get a "grand total" angular momentum, addition rules are always the same: Given $J_1 = \hbar\sqrt{j_1(j_1+1)}$ and $J_2 = \hbar\sqrt{j_2(j_2+1)}$, where $J$ is an angular momentum (orbital, spin, or total) and $j$ a quantum number, the total is $J_T = \hbar\sqrt{j_T(j_T+1)}$, where $j_T$ may take on any value between $|j_1 - j_2|$ and $j_1 + j_2$ in integral steps; and for each value of $j_T$, $J_{T_z} = m_{j_T}\hbar$, where $m_{j_T}$ may take on any of $2j_T + 1$ possible values in integral steps from $-j_T$ to $+j_T$. Since separately there would be $2j_1 + 1$ possible values for $m_{j1}$ and $2j_2 + 1$ for $m_{j2}$, the total number of states should be $(2j_1 + 1)(2j_2 + 1)$. Prove this; that is, show that the sum of the $2j_T + 1$ values for $m_{j_T}$ over all the allowed values for $j_T$ is $(2j_1 + 1) \cdot (2j_2 + 1)$. [*Note:* Here we are proving in general what

we verified in Example 7.3 for the special case $j_1 = 3$ (orbital angular momentum), $j_2 = \frac{1}{2}$ (spin).]

27. The general rule for adding angular momenta is given in Exercise 26. When adding angular momenta with $j_1 = 2$ and $j_2 = \frac{3}{2}$, (a) what are the possible values of the quantum number $j_T$ and the total angular momentum $J_T$, (b) how many different states are possible, and (c) what are the $(j_T, m_{jT})$ values for each of these states?

## Section 7.8

28. The spin-orbit interaction splits the hydrogen $4f$ state into many. (a) Identify these states and rank them in order of increasing energy. (b) If a weak external magnetic field were now introduced (weak enough that it does not disturb the spin-orbit coupling), into how many different energies would each of these states be split?

29. The angles between $\mathbf{S}$ and $\boldsymbol{\mu}_\mathbf{S}$ and between $\mathbf{L}$ and $\boldsymbol{\mu}_\mathbf{L}$ are 180°. What is the angle between $\mathbf{J}$ and $\boldsymbol{\mu}_\mathbf{J}$ in a $2p_{3/2}$ state of hydrogen?

30. The Zeeman effect occurs in sodium just as in hydrogen; sodium's lone $3s$ valence electron behaves much as hydrogen's $1s$. Suppose sodium atoms are immersed in a 0.1-T magnetic field.
    (a) Into how many levels is the $3p_{1/2}$ level split?
    (b) Determine the energy spacing between these states.
    (c) Into how many lines is the $3p_{1/2}$ to $3s_{1/2}$ spectral line split by the field?
    (d) Describe quantitatively the spacing of these lines.
    (e) The "sodium doublet" (589.0 nm and 589.6 nm) is two spectral lines, $3p_{3/2} \rightarrow 3s_{1/2}$ and $3p_{1/2} \rightarrow 3s_{1/2}$, which are split according to the two different possible spin-orbit energies in the $3p$ state (see Exercise 24). Determine the splitting of the sodium doublet (the energy difference between the two photons). How does it compare to the line splitting of part (d), and why?

31. The component of $\boldsymbol{\mu}_\mathbf{J}$ along $\mathbf{J}$ is $\overline{\mu_J}$ and is given by

$$\overline{\mu_J} = \frac{|\boldsymbol{\mu}_\mathbf{J} \cdot \mathbf{J}|}{J}$$

Starting with $\boldsymbol{\mu}_\mathbf{J} \cdot \mathbf{J}$, and using $\boldsymbol{\mu}_\mathbf{J} = -(e/2m_e)\,(\mathbf{L} + 2\mathbf{S})$, $\mathbf{J} = \mathbf{L} + \mathbf{S}$ and the usual relationships $L = \sqrt{\ell(\ell + 1)}\,\hbar$, $S = \sqrt{s(s + 1)}\,\hbar$, and $J = \sqrt{j(j + 1)}\,\hbar$, show that

$$\overline{\mu_J} = \frac{e}{2m_e}\,\frac{3j(j + 1) - \ell(\ell + 1) + s(s + 1)}{2\sqrt{j(j + 1)}}\,\hbar$$

(*Note:* $J^2 = L^2 + S^2 + 2\mathbf{L} \cdot \mathbf{S}$.)

## Section 7.9

32. What is the angle between the spins in a triplet state?

33. (a) Show that, taking into account the possible $z$-components of $J$, there are a total of 12 $LS$-coupled states corresponding to $1s2p$ in Table 7.5. (b) Show that this is the same number of states there would be for two electrons occupying the $1s$ and $2p$ if $LS$ coupling were ignored.

34. As is done for helium in Table 7.5, determine for a carbon atom the various states allowed according to $LS$ coupling. The coupling is between carbon's two $2p$ electrons (its filled $2s$ subshell not participating), one of which always remains in the $2p$ state. Consider cases in which the other is as high as the $3d$ level. (*Note:* When both electrons are in the $2p$, the exclusion principle restricts the number of states. The only allowed states are those in which $s_T$ and $\ell_T$ are both even or both odd.)

## Section 7.10

35. Estimate $K_\alpha$ characteristic x-ray wavelengths: A hole has already been produced in the $n = 1$ shell, and an $n = 2$ electron is poised to jump in. Assume that the electron behaves as though in a "hydrogenlike" atom (see Section 6.10), with energy given by $Z_{eff}^2(-13.6 \text{ eV}/n^2)$. Before the jump, it orbits $Z$ protons, one remaining $n = 1$ electron, and (on average) half its seven fellow $n = 2$ electrons, for a $Z_{eff}$ of $Z - 4.5$. After the jump, it orbits $Z$ protons and half of its fellow $n = 1$ electron, for a $Z_{eff}$ of $Z - 0.5$. Obtain a formula for $1/\lambda$ versus $Z$. Compare the predictions of this model with the experimental data given in the accompanying figure and Table 7.6.

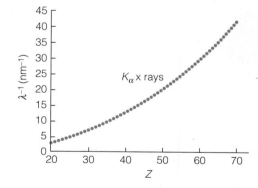

**Table 7.6** $K_\alpha$ x Rays

| Z | $\lambda^{-1}(\text{nm}^{-1})$ |
|---|---|
| 20 (Ca) | 3.0 |
| 30 (Zn) | 7.0 |
| 40 (Zr) | 12.7 |
| 50 (Sn) | 20.3 |
| 60 (Nd) | 29.9 |
| 70 (Yb) | 41.8 |

**General Exercises**

36. (a) Estimate the repulsive energy between helium's electrons. Do this by comparing the energy required to remove the first electron (see Figure 7.11), then the second, from the remaining "hydrogenlike" atom, to the energy required to remove both if there were no repulsion—that is, if both electrons behaved simply as though in a "hydrogenlike" atom. (b) How far apart would two electrons have to be to produce such a repulsive energy? (c) How does this distance compare to the approximate orbit radius in the $Z = 2$ hydrogenlike atom?

# 8

# Statistical Mechanics

Think of it! We are surrounded by countless air molecules, whose velocities we cannot know, much less control. Were they all to move in the same direction we would be in big trouble. Why don't they? Whenever there are countless particles free to move about, the answer is always the same—probabilities.

In statistical mechanics, our concern is making predictions about properties and behaviors in systems where the number of particles is huge—typically Avogadro's number. We cannot be absolutely certain about things, simply because it is impossible to know what each of the particles is doing. Faced with incomplete knowledge, we make *statistical* predictions based on the applicable laws of *mechanics*—classical or quantum. Necessarily, our prediction of what will be observed is the average of all the things that might possibly occur. Air pressure, for instance, a smooth effect according to human senses, we find to be perfectly explained as an average of forces exerted in isolated microscopic collisions.

It may seem ironic, since increasing the number of particles would appear to increase the amount of information we *don't* know, but statistical mechanics is able to make successful predictions indeed *because* the number of particles is large. Averages over very large numbers are much more precise than averages over small numbers. The average force per unit area exerted on our skin by the countless air molecules in a room is so extraordinarily predictable as to seem constant. If rooms contained "only" thousands of particles, but exerting the same average force per area by colliding more energetically, the effect would be a terrifying series of sporadic blows. Consider another example—an experiment in which a lone volunteer is placed in the center of a long hallway. What will be the location of the "particle" later? Obviously, a prediction is almost certain to be wrong. There are many things the subject might choose to do: Walk one way, run the other, stop to think. But if $10^{23}$ volunteers were available (there are fewer than $10^{10}$ people on Earth) and released in the hallway, a prediction of the *average* location of the "particles" at a later time would be accurate. If they move independently—they do not decide to stick together—their average location is precisely the location where they were released; as many will go one way down the hall at a given speed as the other way at the same speed. Even if they do not move independently, there might be some way to model the correlation, and an accurate prediction yet result.

A system in which the number of particles is large enough that predictions become very precise is known as a **thermodynamic system.** It is only in such macroscopic systems that the average properties of pressure, temperature, and density (or concentration) have real meaning. The molecules of air in a room, the electrons in a conducting wire, the photons within a bed of glowing coals; each constitutes a thermodynamic system. Two colliding electrons do not.

**Thermodynamic system:**
Countless particles; precise averages

Statistical mechanics is not nonclassical physics in the same sense as special relativity and quantum mechanics. Rather, it is a distinct area of physics that applies to many others, classical as well as quantum. The answer to the obvious question "Why study it now?" is that we are soon to encounter branches of modern physics that require it. A gas laser is a thermodynamic system of gas molecules, and a silicon microchip is a thermodynamic system of atoms bound in a solid lattice.

In this chapter, we will be concerned mostly with understanding energy distributions, which underlie many important behaviors of thermodynamic systems. An energy distribution specifies the fraction of a system's particles that have a given allowed energy. Generally speaking, the higher the energy of a state, the smaller will be the number of particles in that state. To show how large numbers lead to reliable predictions, however, let us begin with another kind of distribution in one of the simplest thermodynamic systems imaginable.

## 8.1  A Simple Thermodynamic System

The simplest thermodynamic system in which a distribution might be specified is $N$ point particles held in a container divided in half by an imaginary line. (In statistical mechanics, the symbol invariably chosen for the total number of particles in a system is $N$.) An example would be the air molecules in a room with right and left halves, which we refer to as the "two-sided room." The distribution we consider here is not an energy distribution, but a much simpler one: the number of particles on a given side. The reader may wonder why we bother with so "trivial" a case; the result is "obvious"—half the particles should be found on each side. It is not so much this simple answer that is important here, but the ease with which we can show that deviations from the expected distribution are negligible when $N$ is large. The basic idea is this: There are only two possible "states" of a given air molecule, that is, being on the right side and being on the left—and the probability of being in either of these states is one-half. Nevertheless, if the room contained only four molecules, it would not be safe to say that half are on each side. (The probability of having *all* on the same side is, in fact, 12.5%.) *But*

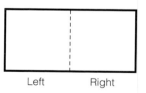

Left          Right

$N = \begin{array}{l}\text{Total number}\\ \text{of particles}\end{array}$

*it would be safe to say that half are on each side if the room contained a truly macroscopic number of molecules.*

Being statistical predictions, distributions rest upon probabilities, and probabilities rest upon "numbers of ways" of doing something (e.g., arranging 20 apples in 5 baskets). For our two-sided room example, let us consider the number of ways $W^N_{N_R}$ of arranging $N$ air molecules so that $N_R$ are on the right side. This number of ways is given by the binomial coefficient:

$$W^N_{N_R} = \binom{N}{N_R} \equiv \frac{N!}{N_R!(N - N_R)!} \tag{8-1}$$

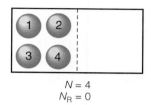

$N = 4$
$N_R = 0$

For instance, were $N$ equal to 4, there is only one way of obtaining no molecules on the right side ($N_R = 0$)—that is, all four on the left. Equation (8-1) agrees:

$$W^4_0 = \binom{4}{0} \equiv \frac{4!}{0!4!} = 1$$

(Logically, the number of ways of obtaining four on the right side must be the same; the binomial coefficient is unchanged if $N_R$ is replaced by $N - N_R$.) The number of ways of obtaining one molecule on the right ($N_R = 1$) should be four, since any of the particles might be on that side. Again, (8-1) agrees:

$$W^4_1 = \binom{4}{1} \equiv \frac{4!}{1!3!} = 4$$

The number of ways of obtaining two on one side is also correctly given:

$$W^4_2 = \binom{4}{2} \equiv \frac{4!}{2!2!} = 6$$

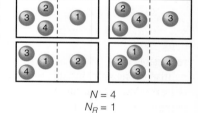

$N = 4$
$N_R = 1$

Now a point fundamental to statistical mechanics: All ways of arranging *specific* particles on given sides are equally probable.[1] But a state in which the distribution of particles is uniform (irrespective of which specific particles are on which side) is the most probable, simply because there is a greater *number* of ways of obtaining it.

To see how large numbers of particles lead to a precise average, let us consider cases of increasing $N$. For the case $N = 4$, the numbers of ways of obtaining the five distributions with different $N_R$ are plotted in Figure 8.1(a). The total number of ways is $1 + 4 + 6 + 4 + 1 = 16$ and is approximately the area under the blue curve. Figure 8.1(b) plots numbers of ways of arranging 40 molecules. Of the 41 possible distributions, the maximum still occurs at $N_R = \frac{1}{2}N$, but it is *much* larger, $10^{11}$ rather than 6. Even more important, the distributions in which $N_R$ is far from $\frac{1}{2}N$ constitute a smaller *fraction* of the total number of ways of distributing molecules (again, the area under the curve). In other words, the curve is peaked more sharply at $N_R = \frac{1}{2}N$ than in the $N = 4$ case. The trend continues for $N = 400$. Only a very small fraction of the total number of ways is far from the $N_R = \frac{1}{2}N$ distribution. Consequently, the probability of finding a significantly nonuniform distribution is very small. With a realistic macroscopic number of $10^{23}$, Figure 8.1(d), the total number of ways of distributing molecules is contained within a minute

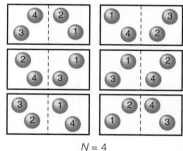

$N = 4$
$N_R = 2$

[1] Each way occurs with a probability of $\frac{1}{2} \times \frac{1}{2} \times \ldots = \frac{1}{2}^N$, because the probability of a given particle being on a given side is $\frac{1}{2}$ and the total probability is the product, since particles are assumed to move independently.

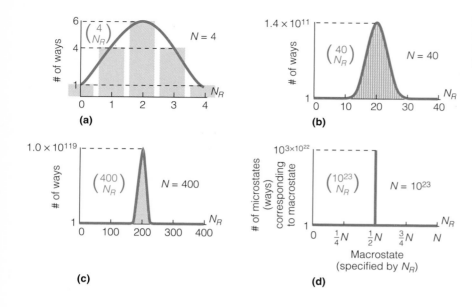

**Figure 8.1** Number of distributing particles on two sides of a room—variation as total number of particles increases from 4 to $10^{23}$.

region of $N_R = \frac{1}{2}N$. True, there is a *possibility* of the air in a room being predominantly on one side, but it is so extraordinarily unlikely as to be dismissed. We have shown what we set out to show: Each *individual particle* has a probability of $\frac{1}{2}$ of being in either of its two possible "states" (i.e., left or right), and the *distribution* of particles will indeed have half in one state and half in the other if it is a truly thermodynamic system.

## Microstates and Macrostates

The two-sided room provides a good framework for learning some useful terminology. It is crucial to bear in mind the distinction between the microscopic and the macroscopic. As shown in Figure 8.1(d), in the $N = 10^{23}$ case there are $10^{3\times 10^{22}}$ ways of obtaining the distribution in which $N_R = \frac{1}{2}N$. Each "way" is a different **microstate.** The microstate is the <u>state</u> of the system given complete <u>micro</u>scopic knowledge of the states of the individual particles. In the two-sided room, the individual-particle states are simply "left" and "right."[2]

Realistically, however, such microscopic knowledge is unobtainable. No real observer could keep track of molecules crossing the room and know which ones are on which side. The $10^{3\times 10^{22}}$ microstates corresponding to $N_R = \frac{1}{2}N$ would be utterly indistinguishable. Of course, we assume that it is plausible to know at least the *number* (or concentration) of molecules on the two sides, but this constitutes knowledge of only the **macrostate.** To know the macrostate of a system is to know the properties that do not depend on the exact microscopic states of every individual particle: the overall properties of number, energy, and volume, and the average local properties of pressure, temperature, and particle concentration. In the two-sided room, the $N_R = \frac{1}{4}N$ macrostate is that in which the particle concentration is three times as large on the left side as on the right (i.e., a ratio of $\frac{3}{4}$ to $\frac{1}{4}$). In the $N = 4$ case, four microstates correspond to this macrostate. The $N_R = 0$ macrostate has a vacuum on the right. There is only one way to do this; only one micro-

[2] In *quantum* statistical mechanics, the individual-particle state is, of course, a quantum state. In *classical* statistical mechanics, it is the particle's position and velocity (see Section 8.5, on classical averages). Our simple classical example is one in which the position is merely either "left" or "right," and the velocity is not even considered.

state corresponds to this macrostate. The $N_R = \frac{1}{2}N$ macrostate has equal particle concentrations on the two sides and is the one obtainable in the greatest number of microscopic ways (e.g., 6 for $N = 4$, and $10^{3 \times 10^{22}}$ for $N = 10^{23}$). Because it corresponds to the greatest *number* of microstates, all of which are assumed equally probable, it is the most probable macrostate. The most probable macrostate is what we know as the **equilibrium state.**[3]

## Equilibrium

If left alone, thermodynamic systems eventually reach **equilibrium.** Equilibrium is a macrostate whose macroscopic properties—pressure, temperature, and particle concentration—*do not change with time.* Begun in a *nonequilibrium* macrostate, a thermodynamic system will pass through other macrostates, its macroscopic properties varying in space and time, while inexorably approaching its unique unchanging equilibrium state. Again, it is important to distinguish between the microscopic and the macroscopic. If air were released on one side of a room and allowed to undergo free expansion, both the microstate and the macrostate would change; individual molecules move from one side to the other and even a casual observer would detect the macroscopic wind. But the air will eventually attain an unchanging uniform distribution. When this equilibrium point is reached, the microstate still changes—molecules still move about the room—but the macrostate, as an observer can tell, does not.

The laws of *dynamics* do not explain the approach to equilibrium; the laws of *statistics* do. Given an initial nonuniform air distribution, it may well seem from the observer's macroscopic viewpoint that the system is "driven" toward equilibrium by a "force": a difference in particle concentration between the sides of the room. Under different conditions, a temperature difference might seem to a macroscopic observer the "force" that drives a system initially of nonuniform temperature to an equilibrium state of uniform temperature. However, such "forces" originate neither in Newton's laws of motion nor in the laws of quantum mechanics. Microscopically, *nothing* drives the system to equilibrium. It approaches equilibrium simply because a thermodynamic system always has a vastly "most probable macrostate," to which *essentially all* of the microscopic ways correspond.

The $N_R = \frac{1}{2}N$ macrostate in Figure 8.1(d) is a perfect example. The system might begin in a highly improbable macrostate, one for which the number of ways/ microstates is small. But given time to allow significant microscopic rearrangement, it is extraordinarily unlikely that it would not later *and for all time thereafter* be found in one of the huge preponderance of microstates corresponding to the "most probable macrostate," and if the system's macrostate is the same forever, its macroscopic properties do not change with time. This unchanging, overwhelmingly most probable macrostate defines the equilibrium state.

Generalizing from our simple two-state system, in a system where individual particles have available to them many different *energy* states, we might initially distribute the total energy in an arbitrary way—half the particles in their ground states and half in their 37th quantum levels, for instance. But regardless of its initial state, the system would in time assume its most probable state, with a predictable fraction of the particles in each individual-particle state. It is such "equilibrium energy distributions" that we study in this chapter.

[3]A rule of thumb: The word "way" almost always refers to a microstate. Expressions such as "microscopic ways" are thus redundant, but will often be used to strengthen the association in the reader's mind. The ambiguous term "state," when applied to a system, usually refers to the macrostate. (When applied to a particle, it refers to the individual-particle state.)

# ▼ 8.2 Entropy and Temperature

Everyone has some idea of what temperature means, but to the student of science it may be somewhat disturbing when it suddenly appears *in an equation*. Since it appears in all the energy distributions we will consider, it is helpful to see how the idea becomes a quantitative property. It begins with entropy.

The second law of thermodynamics says that the disorder—that is, the entropy—of an isolated system will not decrease. In contrast to other basic laws of physics, this law has a "reason": *A more disordered state is a more probable state,* and thermodynamic systems invariably move toward more probable states. In the $N = 4$ case of the two-sided room, for instance, the $N_R = 2$ macrostate is more probable than the $N_R = 0$, and it is more disordered; a state with two molecules on each side is less ordered than one in which all are on one side. Given that disorder increases with probability, and probability with number of microscopic ways, we may give the concept of disorder a logical quantitative definition by relating it to the number of ways of obtaining the macrostate. For several reasons we choose to define entropy $S$ as proportional to the *natural logarithm* of the number of ways:

$$S \equiv k_B \ln W \qquad\qquad (8\text{-}2)$$

where $k_B$ is the Boltzmann constant. Because the equilibrium macrostate corresponds to the greatest number of microscopic ways, (8-2) says that *entropy is maximum at equilibrium.*

Why the choice of natural logarithm? First, since numbers of ways tend to be proportional to numerical factors to the $N$th power (e.g., $\frac{1}{2}^N$), this definition gives an entropy roughly proportional to the size of (the number of particles in) the system. Consider the two-sided room with $N = 10^{23}$: In the $N_R = \frac{1}{2}N$ macrostate, $W = 10^{3 \times 10^{22}}$ and $S = k_B \ln W = k_B 7 \times 10^{22}$. It is no coincidence that the factor multiplying the Boltzmann constant is of the same order of magnitude as $N$. Secondly, it makes entropy "additive." The total number of ways of arranging two independent systems is the product of the separate numbers of ways ($W_a \times W_b$), and the log of a product is the sum of the logs, so the total entropy is then the sum. Lastly, by this definition, a state so completely ordered that it can be obtained in but one way has zero entropy, as in the totally ordered $N_R = 0$ macrostate of the two-sided room.

Relating entropy to numbers of ways is fairly easy in the two-sided room. In more complex realistic cases it is, of course, more difficult. Nevertheless, if $W$ could somehow be calculated, the change in entropy according to (8-2) would be

$$\Delta S = S_f - S_i = k_B(\ln W_f - \ln W_i) = k_B \ln \frac{W_f}{W_i}$$

Fortunately, there is an alternative way to calculate entropy changes that is in most cases easier to use (and with which the reader is probably already familiar). Before we see how it fits in, let us investigate the link between entropy and temperature.

## Temperature

In thermodynamics, temperature is given a clear logical definition. Identified by layman and scientist alike as the measure of the ability to transfer thermal energy, its definition naturally involves the phenomenon of heat transfer.

Imagine two thermodynamic systems of fixed volume, perhaps two rigid containers of gas. The two systems cannot exchange *mechanical* energy by doing work on one another, but are in "thermal contact," so that *thermal* energy can flow via microscopic collisions. Suppose heat is flowing slowly from system 1 to system 2. (We might say that system 1 is of higher temperature, but this is not yet part of our vocabulary.) Slow heat flow between the two allows each individual system time to assume a new equilibrium state and entropy at each step of the process. Now consider a short time interval during which an infinitesimal amount of heat flows from system 1 to system 2. The total change in entropy is the sum of the changes.

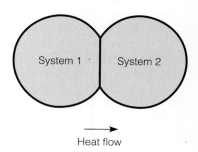

Heat flow

$$dS_{total} = dS_1 + dS_2$$

Energy exchange alone causes the entropy change, so $dS$ may be expressed as $(\partial S/\partial E)\, dE$. The partial derivative keeps constant other things, such as volume, that might affect the entropy.[4]

$$dS_{total} = \left(\frac{\partial S}{\partial E}\right)_2 dE_2 + \left(\frac{\partial S}{\partial E}\right)_1 dE_1$$

System 2 gains the energy lost by system 1, so we must have $dE_1 = -dE_2$.

$$dS_{total} = \left[\left(\frac{\partial S}{\partial E}\right)_2 - \left(\frac{\partial S}{\partial E}\right)_1\right] dE_2$$

Now if systems 1 and 2 are far from being in equilibrium with each other, the entropy change of the total two-container system must be nonzero (as it moves toward its maximum). Thus, $\partial S/\partial E$ would have to be be *different* for system 1 than for system 2. But suppose that only an infinitesimal amount of heat remains to be transferred from 1 to 2 for the total system to attain equilibrium. Since the total entropy is maximum at equilibrium, $dS$ would have to be zero.[5] This would require that $\partial S/\partial E$ be the *same* for both. In other words, systems 1 and 2 will be in equilibrium with each other if and only if $\partial S/\partial E$ is the same for both. *This is exactly the property we ascribe to temperature.* For several reasons,[6] temperature is defined as inversely proportional to $\partial S/\partial E$.

---

[4]In general, a system's entropy depends not only on its internal energy, but also on its volume and the numbers of different kinds of particles (e.g., chemical species). It might be increased by heating at constant volume ($\Delta E > 0$, $\Delta V = 0$), by free expansion without energy change ($\Delta E = 0$, $\Delta V > 0$), or in a spontaneous chemical reaction in which neither volume nor internal energy changes. For the purpose of defining temperature, it is simplest to keep volume and particle numbers fixed.

[5]Think of the entropy of the total system as a function of the energy transferred. If energy transferred in one direction causes the entropy to *increase* to its maximum value, further energy transfer in the *same* direction must cause it to *decrease*. The function's rate of change must be zero at equilibrium. A process carried out infinitesimally close to equilibrium is known as a "reversible" process. Entropy does not change in reversible processes.

[6]In spontaneous heat transfer from system 1 to 2, the disorder of system 2 must increase by more than the disorder of system 1 decreases; that is, $\partial S_2/\partial E > |\partial S_1/\partial E|$. A definition involving a reciprocal properly assigns to system 1 the larger value of $T$; the definition also agrees with the ideal gas law.

$$\frac{\partial S}{\partial E} \equiv \frac{1}{T} \tag{8-3}$$

A quick comparison of (8-2) and (8-3) raises an obvious question: How on Earth do we take a derivative with respect to $E$ of a number of ways? The answer, unfortunately, is much too detailed to present here, but we can clarify the main point by turning the question around. Rather than taking a derivative of (8-2) to find temperature, may we integrate (8-3) to find entropy? Equation (8-3), with its partial derivative, relates temperature to a change in entropy at constant volume. In such a zero-work case, it would be valid to write $dS = (1/T)\, dE$. But since $dE$ would then equal the heat added to the system, we might just as well write $dS = (1/T)\, dQ$. An extension of the approach used earlier shows that the latter equation is correct even when two systems are allowed to exchange energy via work (see Exercise 4). Integrating this, we obtain a commonly used result:

$$\Delta S = S_f - S_i = \int_i^f \frac{dQ}{T} \qquad \left( \begin{array}{c} \text{Over reversible} \\ \text{path from} \\ \text{initial to final state} \end{array} \right) \tag{8-4}$$

We now state without proof that this equation, arising from (8-3), is equivalent to $\Delta S = k_B \ln(W_f/W_i)$! There is indeed a way of relating numbers of ways to the more-familiar quantities of internal energy, heat and work. To calculate entropy changes, equation (8-4) is usually easier, but it must not be forgotten that the definition in (8-2) is equivalent and puts the concept of disorder on the solid, quantitative foundation of probabilities.

## Example 8.1

An ideal gas of $10^{23}$ air (nitrogen) molecules is initially confined to one side of a room, then expands to fill the whole room. Its internal energy $E$ does not change. (a) Calculate its entropy change using equation (8-2). (*Note:* Stirling's approximation $M! \cong \sqrt{2\pi M}\, M^M e^{-M}$ will be helpful. It is correct within $\frac{1}{10}\%$ for $M > 10^3$, becoming even more accurate as $M$ increases.) (b) Entropy/disorder is a state function, and so should change by the same amount no matter how the constant-energy expansion is carried out. Assume that the air is kept always close to equilibrium by doing work against a slow-moving piston, while heat is constantly added to maintain the internal energy. Calculate its entropy change using equation (8-4).

### Solution

(a) Equation (8-2) tells us that $\Delta S = k_B \ln(W_f/W_i)$. In our simple two-sided room, there is only *one* way of having all particles on one side, and the number of ways of having $\frac{1}{2}N$ on each side is $N!/(\frac{1}{2}N!)(\frac{1}{2}N!)$. Thus,[7]

$$\frac{W_f}{W_i} = \frac{N!/(\frac{1}{2}N!)(\frac{1}{2}N!)}{1} = \frac{N!}{(\frac{1}{2}N!)^2}$$

[7]With our sole criterion being whether a particle is on one side or the other, we are indeed *defining* the initial entropy to be zero. We would not ordinarily think of the disorder of air confined to half a room as zero, but the *change* in entropy/disorder due to a simple constant-internal-energy expansion is the same no matter what state is defined to be zero.

Since $N$ is so large, the approximation is valid:

$$\frac{W_f}{W_i} = \frac{\sqrt{2\pi N}\, N^N e^{-N}}{\left(\sqrt{2\pi \frac{1}{2}N}\,(\frac{1}{2}N)^{N/2}e^{-N/2}\right)^2} = \frac{\sqrt{2\pi N}}{\pi N}\frac{N^N}{(\frac{1}{2}N)^N}\frac{e^{-N}}{e^{-N}} = \sqrt{\frac{2}{\pi N}}\, 2^N$$

Now using a property of logarithms,

$$\Delta S = k_B \ln\left(\sqrt{\frac{2}{\pi N}}\, 2^N\right) = k_B\left[\ln 2^N + \ln\sqrt{\frac{2}{\pi N}}\right]$$

For $N = 10^{23}$, the second term in the brackets is absolutely negligible compared to the first. The ratio of the numbers of ways is dominated by the factor $2^N$. Thus,

$$\Delta S = k_B \ln 2^N = N k_B \ln 2$$

Now inserting numerical values,

$$\Delta S = 10^{23}(1.38 \times 10^{-23}\text{ J/K})\ln 2 = 0.96\text{ J/K}$$

(b) Since the heat added equals the work done, we may replace $dQ$ with $dW = P\,dV$, and using $PV = Nk_BT$, this becomes $dW = (Nk_BT/V)\,dV$. Thus,

$$\Delta S = \int_i^f \frac{(Nk_BT/V)\,dV}{T} = Nk_B\int_i^f \frac{dV}{V} = Nk_B \ln\frac{V_f}{V_i} = Nk_B \ln 2$$

The result is the same! We haven't proven that (8-2) and (8-4) are equivalent, but we have at least verified that they are equivalent in this special case. Notably, *any time* we use (8-4) to find a change in entropy we are at the same time, through its equivalence to (8-2), determining a ratio of numbers of ways, final to initial.

It is worth reiterating that unconfined air will expand, its entropy increase, because the state in which the air is evenly distributed is a vastly more probable state than one in which it is unevenly distributed; the ratio of numbers of ways goes as $2^N$! Thus, it would be extraordinarily unlikely for the system to move away from equilibrium. The second law boldly makes the proclamation official: The system *cannot* move away from equilibrium.

# ▼ 8.3 Time Reversal and the Second Law

Among the basic laws of physics, the second law of thermodynamics is unique in several ways. As we discuss these differences, we will gain valuable insight into the important distinction between microscopic and macroscopic knowledge.

Almost all basic laws of physics apply to both macroscopic and microscopic phenomena. For instance, when two objects collide, we accept that momentum and (total) energy are conserved whether the objects are microscopic fundamental particles or macroscopic balls of clay; the "macroscopic law" is accepted as a logical consequence of the "microscopic law." This cannot be said of the second law of thermodynamics, however, because *entropy has no microscopic meaning.*

In statistical mechanics, the properties we expect to observe in a given macrostate are averages over all possible microscopic ways of obtaining the macrostate, and entropy is the logarithm of the total number of the microstates. Naturally, were we to have complete *microscopic* knowledge (within quantum-mechanical limits), there would not be *multiple* microscopic *possibilities* and thus no reason to speak of a total number thereof. *Entropy has meaning only so long as we admit that we cannot have microscopic knowledge of a macroscopic system.*

Besides having only macroscopic meaning, the second law of thermodynamics is unique in another way. Almost all basic laws of physics are **time-reversal invariant.** Were we to watch a motion picture of some evolving physical system, the system would obey these laws equally well whether the picture is shown forward or backward. For instance, a ball thrown upward on the Moon (i.e., with no air resistance) and returning groundward would appear to obey the second law of motion and the universal law of gravitation the same way forward as backward in time. However, were we to consider instead a process in which *entropy* obviously changes, things would be different. While momentum and energy conservation would appear to hold for every microscopic collision in either time direction, and could thus provide no clue, we would definitely know that a movie in which air enters a puncture and reinflates a tire is being run backward. *Irreversible* (one-way, entropy-increasing) processes are the only true test, for all *microscopic* laws are indeed time-reversal invariant. Because entropy would increase in a motion picture run forward but decrease in one run backward, we say that the second law of thermodynamics exhibits **time-reversal asymmetry.**

Time-reversal asymmetry and the fact that entropy has only macroscopic meaning are inextricably related. To see the connection, we return to our trusty two-sided room. Suppose we begin with a macrostate in which one-third of the particles are on the left and two-thirds are on the right. The corresponding number of microscopic ways of arranging particles is

$$W_i = \frac{N!}{(\frac{1}{3}N)!(\frac{2}{3}N)!}$$

The system's final macrostate has particles equally distributed between the sides and the corresponding number of microstates is

$$W_f = \frac{N!}{(\frac{1}{2}N)!(\frac{1}{2}N)!}$$

As we know, $\Delta S = k_B \ln(W_f/W_i)$. In Exercise 6, it is shown that

$$\frac{W_f}{W_i} \cong \frac{\sqrt{8}}{3}\left(\frac{32}{27}\right)^{N/3}$$

This quotient is greater than unity, indicating that entropy increases, but more important for our purposes is that for large $N$ it is *huge.* We conclude that the number of microstates corresponding to the initial state is an *insignificant* fraction of those corresponding to the final equilibrium state.

Now we may understand why we should not expect entropy to behave symmetrically in both time directions. Forward in time, the system starts in one of the comparatively few microstates corresponding to the improbable macrostate $N_R = \frac{2}{3}N$, $N_L = \frac{1}{3}N$, and presumably moves (labeled *a* in Figure 8.2) toward the equilibrium macrostate $N_R = N_L = \frac{1}{2}N$. Is this guaranteed? Since we could not know the

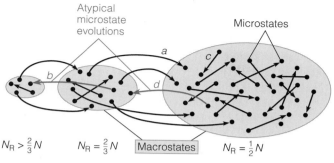

Figure 8.2 Very few microstates evolve toward a less probable macrostate, further from equilibrium.

Atypical microstate evolutions

Microstates

Macrostates

$N_R > \frac{2}{3}N$    $N_R = \frac{2}{3}N$    $N_R = \frac{1}{2}N$

Toward equilibrium →

initial *micro*state, the system might have begun in some extraordinarily atypical microstate in which as time passed the number on the right actually *increased* (labeled *b*). Most probably, though, no such trick would have been played on us. (See the essay entitled "The Probability of Entropy Increase.")

## The Probability of Entropy Increase

That the number of particles on two sides of a room should approach equality is easier said than proved. Initially, there are twice as many particles on the right. The probability that the first particle crossing is from the right side to the left is thus twice as large as the probability that it is from left to right, that is, $\frac{2}{3}$ to $\frac{1}{3}$. Were there only three particles, there would be a 1-in-3 chance that the lone particle on the left would be the next to cross over, and we might be tempted to say that there is a one-third probability that entropy will decrease. But this would not be a violation of the second law of thermodynamics, for the simple reason that entropy is defined only for macroscopic systems. Even with $10^{23}$ particles, there is still a one-third probability that the first particle crossing is from left to right, but neither would this be a violation of the second law—a change of one particle is not a *macro*-

scopic change! For a true macroscopic change, the probability of a change away from equilibrium—that is, toward a lower-entropy state—is negligible.

We can begin to see why by imagining waiting just long enough for *three* particles to cross over. As shown in the figure below, there are eight possible sequences of crossings. The total probability of a decrease in number on the more crowded right side would no longer be $\frac{2}{3}$, as it is for one crossing, but rather $\frac{20}{27}$, while the probability of a move toward a greater imbalance in numbers would now be only $\frac{7}{27}$ rather than $\frac{1}{3}$. The trend continues, and for a truly macroscopic number of particles crossing over, the probability of the number on the right increasing is absolutely negligible. The change in entropy, now something with meaning, will definitely be positive.

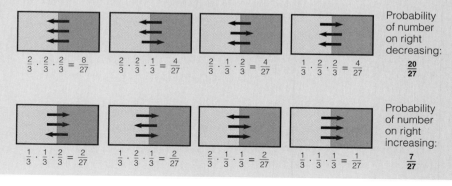

$$\frac{2}{3}\cdot\frac{2}{3}\cdot\frac{2}{3}=\frac{8}{27}\qquad \frac{2}{3}\cdot\frac{2}{3}\cdot\frac{1}{3}=\frac{4}{27}\qquad \frac{2}{3}\cdot\frac{1}{3}\cdot\frac{2}{3}=\frac{4}{27}\qquad \frac{1}{3}\cdot\frac{2}{3}\cdot\frac{2}{3}=\frac{4}{27}$$

Probability of number on right decreasing: $\frac{20}{27}$

$$\frac{1}{3}\cdot\frac{1}{3}\cdot\frac{2}{3}=\frac{2}{27}\qquad \frac{1}{3}\cdot\frac{2}{3}\cdot\frac{1}{3}=\frac{2}{27}\qquad \frac{2}{3}\cdot\frac{1}{3}\cdot\frac{1}{3}=\frac{2}{27}\qquad \frac{1}{3}\cdot\frac{1}{3}\cdot\frac{1}{3}=\frac{1}{27}$$

Probability of number on right increasing: $\frac{7}{27}$

If the picture (the behavior labeled *a*) is run backward, however, the game is rigged! We "start" in a *macro*state of apparent equilibrium, but unbeknownst to us the *micro*state is one of a minute fraction of the microstates for which the number on the right increases significantly as time passes (backward). The number of equilibrium microstates is huge, and the overwhelming majority do not behave in such a weird way. Rather, they preserve the equilibrium macrostate (as does *c*) in either time direction indefinitely. Thus, running a movie backward is not fair, for while we ordinarily assume that all microstates are equally probable, in a backward movie of an irreversible process we are deceitfully focusing on an extremely atypical microstate that happens to move away from equilibrium as time goes backward. Indeed, there are such rogues in the *forward* time direction (labeled *d*), but without similarly rigging the game, we cannot single out these atypical microstates either. On the contrary, given the real constraints on our knowledge, we will virtually always observe the behavior of a *typical* microstate, moving inexorably toward the unchanging, most probable, and disordered macrostate: equilibrium.

## 8.4 The Boltzmann Distribution

One of the most important and useful predictions of statistical mechanics is how energy will be distributed among a large number of particles in equilibrium. In the following sections, we study several such **equilibrium energy distributions** and their applications. Each distribution yields the number of particles in a state of a given energy (e.g., how many will be in their $n = 1$ states, how many in their $n = 2$, and so on). Although the word "equilibrium" may suggest uniformity, it is the most disordered state microscopically. Energy is constantly being exchanged randomly between particles; some will have less than the average energy, others more. It is one of the wonders of physics that knowing only whether the particles are bosons, fermions, or "classically distinguishable" particles we can say how many will have a given energy. Once again, the simple reason is that the likely state of a system becomes quite predictable when $N$ approaches Avogadro's number.

We begin with the **Boltzmann distribution,** a special case that applies when particles are so spread out that the indistinguishability of identical bosons or fermions may be ignored. A good example is a gas under ordinary conditions. The gas molecules, though identical, are so far apart as to not share the same space. Their wave functions do not significantly overlap; they can be distinguished (see Section 7.2).

A rigorous derivation of the Boltzmann distribution from principles of statistics is fairly involved. In the interest of showing most clearly how and why energy is distributed as it is, let us begin with a simple model. Figure 8.3 represents in a fanciful way a system of $N$ identical but separately identifiable (i.e., distinguishable) harmonic oscillators, which exchange energy in some unspecified way; one oscillator jumps to a higher level as another drops to a lower. We choose the harmonic oscillator because its energy depends in a simple way on a single quantum number: $E = (n + \frac{1}{2})\hbar\omega_0$ (cf. Section 4.8). We simplify things further by suppos-

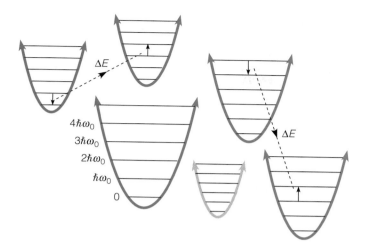

Figure 8.3 Harmonic oscillators exchanging energy.

ing that the potential energy is shifted (as is always allowed) by $-\frac{1}{2}\hbar\omega_0$, so that the ground-state energy is zero. The energy of the $i$th oscillator in its $n_i$th energy level is thus

$$E_{n_i} = n_i\hbar\omega_0 \qquad (n_i = 0, 1, 2, \ldots) \tag{8-5}$$

and the total energy of the $N$ oscillators is

$$E = \sum_{i=1}^{N} n_i\hbar\omega_0 = M\hbar\omega_0 \qquad \left(\text{where } M \equiv \sum_{i=1}^{N} n_i\right) \tag{8-6}$$

Note that the integer $M$, the sum of the quantum numbers of all oscillators, is directly proportional to the total energy.

To find the probability of a given particle/oscillator being in a given energy state, we must use some elementary probability and statistics. An axiom of statistics is that, of numerous equally likely possibilities, the probability that a given subset will occur is the number of ways the subset might occur divided by the total number of ways of obtaining all possibilities. The probability of obtaining subset $a$ is thus

$$P_a = \frac{\text{number of ways of obtaining subset } a}{\text{total number of ways of obtaining all possibilities}}$$

Applying this to our case, the probability that particle $i$ will possess energy $E_{n_i}$ (that its quantum number will be $n_i$) is

$$P_{n_i} = \frac{\text{number of ways energy can be distributed with } n_i \text{ fixed}}{\text{total number of ways energy can be distributed}} \tag{8-7}$$

Here we are invoking the same postulate of statistical mechanics used in the two-sided room of Section 8.1: All microscopic ways of distributing a given total energy are assumed *equally likely*. The probability thus depends merely on relative *numbers* of ways/microstates.

Now, elementary statistics tells us that the number of ways $N$ integers (i.e., the particles' quantum numbers) can be added to give the integer $M$ is $(M + N - 1)!/M!(N - 1)!$. This would be the total number of ways the energy could be distributed among the $N$ quantum numbers. The number of ways energy can be distributed with $n_i$ fixed is the number of ways the $N - 1$ *other* quantum numbers can be added to give $M - n_i$ (proportional to the *remainder* of the energy). This is $[(M - n_i) + (N - 1) - 1]!/(M - n_i)![(N - 1) - 1]!$. Therefore,

$$P_{n_i} = \frac{[(M - n_i) + (N - 1) - 1]!/(M - n_i)![(N - 1) - 1]!}{(M + N - 1)!/M!(N - 1)!} \qquad (8\text{-}8)$$

To gain some feel for numerical values, suppose that $N = 10$ and $M = 50$. That is, there are 10 particles/oscillators and the sum of their quantum numbers is 50. The average energy per particle is

$$\overline{E} = \frac{E}{N} = \frac{M\hbar\omega_0}{N} = 5\hbar\omega_0 \qquad (8\text{-}9)$$

An average oscillator occupies its fifth energy level. The probability that particle $i$ has the average energy, that $n_i = 5$ and the quantum numbers of the remaining 9 particles add to 45, is

$$P_{n_i=5} = \frac{(45 + 9 - 1)!/45!(9 - 1)!}{(50 + 10 - 1)!/50!(10 - 1)!} = \frac{8.86 \times 10^8}{1.26 \times 10^{10}} = .0705$$

The probabilities given by (8-8) for all values of $n_i$ are shown in Figure 8.4. The curve drops sharply as $n_i$ increases. This leads us to an important conclusion:

> Varying the energy of just one of the many particles causes a sharp change in the number of ways of distributing the remaining energy among the others; the greatest freedom to distribute the remaining energy occurs when that one particle has the least energy. Therefore, the more probable state for a given particle, the state in which the number of ways of distributing the energy among all particles is greatest, is one of lower energy.[8]

Equation (8-8) is cumbersome and limited to the special case of harmonic oscillators. In a system of infinite wells, for instance, $E$ is proportional to $n$ *squared*

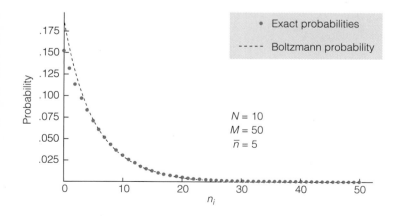

Probability

$N = 10$
$M = 50$
$\overline{n} = 5$

- Exact probabilities
- - - - - Boltzmann probability

$n_i$

**Figure 8.4** Probability of a given oscillator being in its $n_i$ state, and Boltzmann probability.

[8]*Note*: Although the $n_i = 0$ state is the most probable, higher energies are possible, so the average energy must be greater than that of the lowest-energy state. Indeed, we know that it is $5\hbar\omega_0$.

and the expression would be completely different. However, it may be shown that in the limit of large systems of distinguishable particles, *all* cases converge to the Boltzmann probability.[9]

$$P(E_n) = Ae^{-E_n/k_BT} \qquad (8\text{-}10)$$

Boltzmann probability

This is the probability that in a large system at temperature $T$ an individual particle will be in state $n$ of energy $E_n$, where $n$ stands for the set of all quantum numbers necessary to specify the individual-particle state (e.g., $n$, $\ell$, $m_\ell$, $m_s$). Probability drops exponentially with energy.

Importantly, the Boltzmann probability (and all the distributions yet to come) can be applied to subsystems of larger systems. In small subsystems, there might be significant fluctuations, but the prediction still holds on average, becoming more precise as the subsystem grows. The large remainder of the overall system is what in thermodynamics is called a **reservoir.** The temperature, appearing in all the distributions we discuss, is a well-defined property of the overall macroscopic system. (See Section 8.2 and Appendix H.)

The proportionality constant $A$ in equation (8-10) follows from the requirement that the sum of the probabilities of a given particle occupying all possible states must be unity. It is left as an exercise to show that for our system of harmonic oscillators[10]

$$1 = \sum P(E_{n_i}) = \sum_{n_i=0}^{\infty} Ae^{-E_{n_i}/k_BT} = \sum_{n_i=0}^{\infty} Ae^{-n_i\hbar\omega_0/k_BT} \qquad (8\text{-}11)$$

$$\Rightarrow \quad A = (1 - e^{-\hbar\omega_0/k_BT})$$

The Boltzmann probability for harmonic oscillators thus becomes

$$P(E_{n_i}) = (1 - e^{-\hbar\omega_0/k_BT})e^{-n_i\hbar\omega_0/k_BT} \qquad (8\text{-}12)$$

Boltzmann probability for harmonic oscillators

The exact formula (8-8) involves factorial functions of $N$ and $M$, whereas (8-12) is an exponential function of $T$. How can they be compared? The simplest way is to ensure that equation (8-12) corresponds to the same average energy, given in (8-9): $M\hbar\omega_0/N$. We find the average energy from equation (8-12) by multiplying the energy of a given state of oscillator $i$, $n_i\hbar\omega_0$, by the probability that that state will occur, then summing over all allowed states.

$$\overline{E} = \sum E_{n_i}P(E_{n_i}) = \sum_{n_i=0}^{\infty} n_i\hbar\omega_0[(1 - e^{-\hbar\omega_0/k_BT})e^{-n_i\hbar\omega_0/k_BT}]$$

It is left as an exercise to show that the result is

$$\overline{E} = \frac{\hbar\omega_0}{e^{\hbar\omega_0/k_BT} - 1} \qquad (8\text{-}13)$$

Setting this equal to $M\hbar\omega_0/N$ then gives us a relationship between $T$ and $M/N$. It is also left as an exercise to show that equation (8-12) may then be written

$$P(E_{n_i}) = \frac{1}{(M/N) + 1}e^{-n_i\ln(1 + N/M)} \qquad (8\text{-}14)$$

Plotted in Figure 8.4, the exponential Boltzmann distribution agrees quite well with the exact probabilities of (8-8). The slight deviation is due only to the fact that the $N$ we have considered is so small. In a realistic macroscopic limit, the

[9]The proof is found in Appendix H, following a derivation of the "quantum distributions" discussed later in the chapter. It is deferred because (1) it is probably better for the reader first to gain an understanding of what information the Boltzmann distribution conveys, without being distracted by a rather tedious derivation, and (2) it cannot be fully appreciated apart from the quantum distributions.

[10]The value of $n_i$ may, of course, be no larger than $M$, in which case this one particle possesses all the energy of the $N$-particle system. However, since the probability drops to negligible values long before $E_i$ approaches the total energy, extending the sum to infinity introduces no significant error.

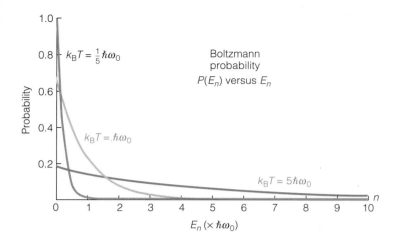

messy detailed probabilities and the Boltzmann distribution coincide precisely. [Equation (8-8) becomes (8-14); see Exercise 13.]

Figure 8.5 shows the Boltzmann probability (8-12) plotted versus energy at three different temperatures: $k_B T/\hbar\omega_0 = \frac{1}{5}$, 1, and 5. Temperature is related to average particle energy, and $1\hbar\omega_0$ is the first energy above the zero-energy ground state. Therefore, the curves represent cases in which the average particle energy is, respectively, less than, comparable to, and greater than the first-excited-state energy. The plots make sense: At the low temperature, it is quite unlikely that a given particle would be in any energy level higher than the ground state; at the high temperature, on the other hand, occupation of the $n = 0$ ground state is not much more likely than occupation of the $n = 1, n = 2, n = 3, \ldots$ states.

Usually we are interested in the *number* of particles expected in a given state. Accordingly we reexpress the Boltzmann distribution by multiplying the probability (8-10) by the total number of particles:

$$N(E_n)_{\text{Boltz}} = NAe^{-E_n/k_B T} \qquad (8\text{-}15)$$

Boltzmann distribution

(Again, $n$ represents the set of all quantum numbers necessary to specify the state.) In our special case of harmonic oscillators, we found the constant $A$ by using the fact that the total probability must be unity. Equivalently, summing (8-15) must give $N$. Because they give a number of particles occupying a given state, energy distributions are often referred to as **occupation number** distributions. Although we cannot devote the space to justifying it, the principle of Section 8.1 still applies. As the numbers occupying the individual-particle states become large, they accordingly become precise; a thermodynamic system started with any odd distribution of energy would progress inexorably toward a vastly most probable equilibrium macrostate in which occupation number drops exponentially with energy.

**Example 8.2**

Consider a sample of hydrogen atoms at a temperature of 300 K. (Ordinarily, hydrogen exists as diatomic molecules. We consider individual atoms for the sake of simplicity.) (a) What is the ratio of the number of atoms in their $n = 2$ states to the number in their ground states? (b) At what temperature would this ratio be one-tenth?

## Solution

(a) Equation (8-15) gives the number *in a given state* of energy $E_n$. For instance, the number in the state $(n, \ell, m_\ell, m_s) = (2, 1, +1, +\frac{1}{2})$ is

$$N(E_{2, 1, +1, +\frac{1}{2}}) = NA \exp\left(-\frac{E_{2, 1, +1, +\frac{1}{2}}}{k_B T}\right)$$

In the simplest analysis, energy depends only on $n$, so that inclusion of the other quantum numbers might seem superfluous. However, it reminds us that since there are eight $n = 2$ states (i.e., a degeneracy of $2n^2$) all with the same energy, the number of atoms with energy $E_2$ is actually 8 times larger.[11] (There will, for instance, be as many in 2, 0, 0, $-\frac{1}{2}$ states as in 2, 1, +1, +$\frac{1}{2}$-states.) Similarly, there are two states for which $n = 1$. Thus,

$$\frac{\text{number with energy } E_2}{\text{number with energy } E_1} = \frac{8 \times N(E_{2, 1, +1, +\frac{1}{2}})}{2 \times N(E_{1, 0, 0, +\frac{1}{2}})}$$

(8-16)

$$= 4\frac{NAe^{-E_2/k_B T}}{NAe^{-E_1/k_B T}} = 4e^{-(E_2 - E_1)/k_B T}$$

Note that in finding *ratios* the total number and multiplicative constant cancel.

The hydrogen atom energies are $-13.6$ eV/$n^2$, so that

$$E_2 - E_1 = \frac{-13.6\ eV}{2^2} - \frac{-13.6\ eV}{1^2} = 10.2\ eV$$

and at 300 K,

$$k_B T = (1.38 \times 10^{-23}\ \text{J/K})(300\ \text{K}) = 4.14 \times 10^{-21}\ \text{J} = 0.0259\ \text{eV}$$

Therefore,

$$\frac{\text{number with energy } E_2}{\text{number with energy } E_1} = 4e^{-10.2/0.0259} \cong 10^{-171}$$

At room temperature, the probability of finding even a single hydrogen atom in an excited state is practically zero.[12]

(b) Inserting in (8-16),

$$0.1 = 4e^{-10.2\ eV/k_B T} \quad \Rightarrow \quad T \cong 32{,}000\ \text{K}$$

The surface of our Sun is approximately 6000 K, giving a ratio of about $10^{-8}$. We see then that except at very high temperatures the overwhelming majority of hydrogen atoms are in their ground states.

[11] Multiplying a number of particles *in a particular state* of a given energy by the *number of states* with that energy to obtain the number of particles with that energy is something we will encounter often.

[12] The attentive reader may perceive a difficulty associated with increasing $n$: Degeneracy increases, but the energy difference is limited by 13.6eV. See Exercise 14 for an interesting discussion of this point.

## Averages

Perhaps the most common task to which distributions are put is calculating averages. In general, the average of a quantity $Q$ is its value $Q_n$ for a particle in state $n$, multiplied by the number of particles in that state, summed over all possible states, then divided by the total number of particles.

$$\overline{Q} = \frac{\sum\limits_{n} Q_n N(E_n)}{N} = \frac{\sum\limits_{n} Q_n N(E_n)}{\sum\limits_{n} N(E_n)} \tag{8-17}$$

Of the two forms shown, the latter is often more convenient; the normalization constant cancels, so it need not be calculated. Note that we have already used this method of averaging: In the steps leading to $\overline{E}$ in (8-13), the sums $\sum P(E_n)$ and $\sum E_n P(E_n)$ were calculated (and, in effect, divided). Simply multiplying each by $N$ and replacing an $E_n$ by a $Q_n$ in the second sum gives the denominator and numerator of (8-17).

## From a Sum to an Integral

In statistical mechanics, the spacing of the quantum levels is often much smaller than typical particle energies. This is certainly the case classically, where discrete quantum levels are unresolvable, but is true even in some quantum thermodynamic systems. If quantum levels are indeed "closely spaced," a sum over states may be replaced by an integral. As partial justification let us consider the condition of unit probability (8-11):

$$\sum_{n} A e^{-E_n/k_{\mathrm{B}}T}\,\Delta n = 1 \tag{8-18}$$

where $\Delta n$ is unity. The ratio of successive terms in this sum is

$$\frac{e^{-E_{n+1}/k_{\mathrm{B}}T}}{e^{-E_n/k_{\mathrm{B}}T}} = e^{-(E_{n+1}-E_n)/k_{\mathrm{B}}T}$$

If $E_{n+1} - E_n$ is much smaller than $k_{\mathrm{B}}T$, the ratio is nearly unity. Thus, the summand in (8-18) would change very slowly as the sum progresses through the allowed quantum numbers. This is the usual justification for replacing a sum by an integral.

A quantum number is not the most convenient variable of integration (i.e., $\Delta n \to dn$). Energy is a logical alternative. However, to replace an integral over values of $n$ (or $n$, $\ell$, $m_\ell$, $m_s$, or another such set) to one over values of $E$, a ratio giving the number of allowed quantum states dn per energy interval dE is required. This ratio—states per energy—is given the symbol $D(E)$ and the descriptive name **density of states.** We may represent the substitution as follows:

Density of states, defined

$$\Delta n \;\to\; dn = \frac{dn}{dE}\,dE \equiv D(E)\,dE \quad \text{where } D(E) \equiv \frac{dn}{dE} \tag{8-19}$$

In contrast to the occupation number distribution, the form of the density of states depends on the particular system of interest. Let us consider a system of harmonic oscillators, ignoring spin. Again, this is the simplest system because energy is directly proportional to a single quantum number. Since the energy levels are equally spaced, the number of states per unit change in energy is a constant.

$$E = n\hbar\omega_0 \quad \rightarrow \quad dE = \hbar\omega_0 \, dn \quad \Rightarrow \quad dn = \frac{1}{\hbar\omega_0} dE$$

Thus,

$$D(E) = \frac{dn}{dE} = \frac{1}{\hbar\omega_0}$$

Density of states: Oscillators (spin ignored)

To allow for different possible spin states, we note that for each energy there would be $(2s + 1)$ such states, so the number of states per unit energy would simply be multiplied by this factor. (In Section 8.6, where spin cannot be ignored, we will use $D(E) = (2s + 1)/\hbar\omega_0$.) States associated with degenerate quantum numbers—those upon which energy does *not* depend—may always be accounted for this way. (In effect we did this for the $\ell$, $m_\ell$, and $m_s$ quantum numbers in Example 8.2.) It is more difficult to see how sense can be made of (8-19) if energy actually depends on multiple quantum numbers. This issue we take up in Section 8.7.

To confirm the validity of our switch from sum to integral, let us repeat some of our earlier harmonic oscillator calculations. Consider first the normalization constant. Although different when obtained via integration, it is related in a very sensible way.

$$1 = \sum_n Ae^{-E_n/k_BT}(\Delta n) \quad \rightarrow \quad 1 = \int_0^\infty Ae^{-E/k_BT}\left(\frac{1}{\hbar\omega_0} dE\right) \tag{8-20}$$

Carrying out the integration,

$$1 = A\frac{k_BT}{\hbar\omega_0} \quad \Rightarrow \quad A = \frac{\hbar\omega_0}{k_BT}$$

This is precisely what (8-11) becomes in the limit $k_BT \gg \hbar\omega_0$—that is, when $k_BT$ is much greater than the energy-level spacing. Equation (8-15) now becomes

$$N(E)_{Boltz} = \frac{N\hbar\omega_0}{k_BT}e^{-E/k_BT} \tag{8-21}$$

$N(E)_{Boltz}$ for oscillators ($k_BT \gg \hbar\omega_0$)

If our switch to integration is self-consistent, integrating the number of particles per energy over all energies should yield $N$. However, as are all the distributions we consider, (8-21) is a number of particles per *state*. *To obtain the number of particles per energy, we must multiply by the number of states per energy—the density of states.* This idea is so pervasive in statistical mechanics as to merit committing to memory.

| $dN$ | $=$ | $N(E)$ | $\times$ | $D(E) \, dE$ |
|:---:|:---:|:---:|:---:|:---:|
| Number of particles within $dE$ of $E$ | | Number of particles in a given state of energy $E$ | | Number of states within $dE$ of $E$ |

With this in mind, we should have

$$N = \int_0^\infty N(E)D(E)\,dE = \int_0^\infty NAe^{-E/k_B T}\frac{1}{\hbar\omega_0}\,dE \tag{8-22}$$

We can easily see that this is correct; it is just (8-20) multiplied by $N$. Finally, the average energy should follow by multiplying a given $E$ by the number of particles per energy, $N(E)D(E)$, summing (integrating) over all energies, then dividing by $N$.

$$\overline{E} = \frac{\int E\,N(E)D(E)\,dE}{N} = \frac{\int_0^\infty E\,N(E)\frac{1}{\hbar\omega_0}\,dE}{N} \tag{8-23}$$

It is left as an exercise to verify that the result is

$$\overline{E} = k_B T \tag{8-24}$$

As can easily be shown, this is indeed the $k_B T \gg \hbar\omega_0$ limit of (8-13).

Equation (8-23) serves as a model for averaging via integration. It is the same as (8-17), but with (the implicitly assumed) $\Delta n$ replaced by $D(E)dE$.

Average

$$\overline{Q} = \frac{\sum_n Q_n N(E_n)}{N} = \frac{\sum_n Q_n N(E_n)}{\sum_n N(E_n)} \tag{8-17}$$

$$\overline{Q} = \frac{\int Q(E)N(E)D(E)\,dE}{N} = \frac{\int Q(E)N(E)D(E)\,dE}{\int N(E)D(E)\,dE} \tag{8-25}$$

## ▾ 8.5  Classical Averages

The Boltzmann distribution is applicable only when indistinguishability of identical particles may be ignored. As previously noted, there are *quantum* thermodynamic systems in which the particles are distinguishable. Nevertheless, the Boltzmann distribution is often said to be a classical distribution,[13] because it is the correct one in all systems in the limit where quantum-mechanical behaviors converge to the classical.

Classical averages naturally involve not summation over allowed quantum states, but integration over the continuum of "classical states"—that is, particle positions and velocities. The energy $E_n$ becomes $E(\mathbf{r}, \mathbf{v})$—potential energies depend on position $\mathbf{r}$ and kinetic on velocity $\mathbf{v}$. For the volume element in three dimensions, we use the symbol $d^3r$, where the superscript signifies three spatial variables. For the differential element associated with the three velocity variables, $dv_x\,dv_y\,dv_z$ in rectangular coordinates, we choose the analogous symbol $d^3v$. With (8-17) as our model, we find classical averages as follows:

---

[13] It may, in fact, be derived completely classically—that is, without recourse to sums over quantum states.

$$\overline{Q} = \frac{\int Q(\mathbf{r}, \mathbf{v}) N(E(\mathbf{r}, \mathbf{v}))_{\text{Boltz}} \, d^3r \, d^3v}{\int N(E(\mathbf{r}, \mathbf{v}))_{\text{Boltz}} \, d^3r \, d^3v} \qquad (8\text{-}26)$$

Classical average

Before using (8-26), we must address a subtle point: In going from a sum over states in (8-17) to an integral over energies in (8-25) we replaced $\Delta n$ by $D(E) \, dE$, but in (8-26) it is replaced directly by $d^3r \, d^3v$. Is a "density of classical states" needed, and if so, aren't there infinitely many position-velocity states per $d^3r \, d^3v$? Perhaps surprisingly, the uncertainty principle provides the answer. Position and velocity/momentum simply cannot be known simultaneously with absolute precision, so it really makes no sense to speak of an infinite number of such states within an interval. There is an inescapable granularity, a spacing between $(\mathbf{r}, \mathbf{v})$ values that can be declared truly distinct. It can be shown that there is a one-to-one correspondence between quantum-mechanical states and "distinguishable" position-velocity states. Thus, the replacement involves a proportionality constant that cancels, top and bottom, in (8-26). In the classical limit, of course, the granularity is unresolvable and continuous integration is valid.

### Example 8.3

By approximating the atmosphere as a column of classical particles of mass $m$ at uniform temperature $T$ in a uniform gravitational field $g$, calculate the average height of an air molecule above Earth's surface.

### Solution

Probability depends on energy. In this example, we have

$$E(\mathbf{r}, \mathbf{v}) = \text{KE} + U_{\text{grav}} = \frac{1}{2}m(v_x^2 + v_y^2 + v_z^2) + mgy \equiv \frac{1}{2}mv^2 + mgy$$

With $Q = y$, we may now use (8-26).

$$\overline{y} = \frac{\int y \, NA \exp\left[-(\frac{1}{2}mv^2 + mgy)/k_B T\right] dx \, dy \, dz \, dv_x \, dv_y \, dv_z}{\int NA \exp\left[-(\frac{1}{2}mv^2 + mgy)/k_B T\right] dx \, dy \, dz \, dv_x \, dv_y \, dv_z}$$

The constant $NA$ cancels top and bottom, leaving the all-important exponential Boltzmann factor. Furthermore, integrations over all variables but $y$ separate and cancel top and bottom. Carrying out the remaining integrations, we obtain

$$\overline{y} = \frac{\displaystyle\int_0^\infty y e^{-(mgy/k_B T)} \, dy}{\displaystyle\int_0^\infty e^{-(mgy/k_B T)} \, dy} = \frac{k_B T}{mg}$$

The result makes sense qualitatively. At a higher temperature, particles are more energetic and so should rise to greater heights. Conversely, if either $m$ or $g$ were larger, we would not expect particles to rise to heights as great. Inserting the mass of an $N_2$ molecule, $m = 4.7 \times 10^{-26}$ kg, and assuming a uniform temperature of 273 K (although the air is *not* of uniform temperature), the result is 8.2 km, a fair approximate value. (It would be somewhat lower for the atmosphere's $O_2$.)

## The Maxwell Speed Distribution

The distribution of speeds in an ideal gas is a standard topic in introductory classical thermodynamics. Known as the Maxwell speed distribution, it is just one of the many faces of the Boltzmann distribution.

Suppose we are interested in determining the average of some property that depends only on the gas particle's speed $v$. Examples are the speed itself, the square of the speed, and kinetic energy. Let us refer to this property as $f(v)$. In the Boltzmann factor, we insert the translational kinetic energy, which depends on $v$.[14]

$$E(\mathbf{r}, \mathbf{v}) = \frac{1}{2}m(v_x^2 + v_y^2 + v_z^2) \equiv \frac{1}{2}mv^2$$

We then insert this and $f(v)$ into (8-26), whereupon spatial integrations cancel.

$$\overline{f(v)} = \frac{\int f(v)e^{-\frac{1}{2}mv^2/k_BT}\, d^3r\, d^3v}{\int e^{-\frac{1}{2}mv^2/k_BT}\, d^3r\, d^3v} = \frac{\int f(v)e^{-\frac{1}{2}mv^2/k_BT}\, d^3v}{\int e^{-\frac{1}{2}mv^2/k_BT}\, d^3v}$$

Since $f(v)$ depends only on speed rather than velocity, we may further simplify things by using not rectangular velocity components but spherical polar coordinates: $v_x, v_y, v_z \rightarrow v, \theta, \phi$. By analogy with $dV = r^2 \sin\theta\, dr\, d\theta\, d\phi$ in spatial coordinates (see Section 6.3), the "volume" element in velocity coordinates is $dv_x\, dv_y\, dv_z \rightarrow v^2 \sin\theta\, dv\, d\theta\, d\phi$. With this in place of $d^3v$, the angular integrals cancel and the remaining integral is over the single variable of speed.

$$\overline{f(v)} = \frac{\int f(v)e^{-\frac{1}{2}mv^2/k_BT} v^2 \sin\theta\, dv\, d\theta\, d\phi}{\int e^{-\frac{1}{2}mv^2/k_BT} v^2 \sin\theta\, dv\, d\theta\, d\phi} = \frac{\int_0^\infty f(v)e^{-\frac{1}{2}mv^2/k_BT} v^2\, dv}{\int_0^\infty e^{-\frac{1}{2}mv^2/k_BT} v^2\, dv}$$

The denominator is a standard gaussian integral, whose value is $\sqrt{\pi}/2(k_BT/m)^{3/2}$. Rearranging a bit, we arrive at

$$\overline{f(v)} = \int_0^\infty f(v)\left[\sqrt{\frac{2}{\pi}}\left(\frac{m}{k_BT}\right)^{3/2} v^2 e^{-\frac{1}{2}mv^2/k_BT}\right] dv \qquad (8\text{-}27)$$

An often-used average is the **rms speed,** defined $v_{rms} \equiv \sqrt{\overline{v^2}}$. It is left as an exercise to show that (8-27) gives $v_{rms} = \sqrt{3k_BT/m}$.

Viewing (8-27) as $\overline{f(v)} = \int f(v)\, dP = \int f(v)(dP/dv)dv$, where $P$ is probability, we see that the quantity in brackets is a probability per unit speed.

$$P(v)_{\text{Maxwell}} = \frac{dP}{dv} = \sqrt{\frac{2}{\pi}}\left(\frac{m}{k_BT}\right)^{3/2} v^2 e^{-\frac{1}{2}mv^2/k_BT}$$

Often, the *number* of particles *per unit volume* per unit speed is of greater interest. To obtain this, we merely multiply the probability per unit speed by the total number of particles and divide by the volume. The result is known as the **Maxwell speed distribution:**

[14]Polyatomic ideal gases have internal degrees of freedom: rotational and vibrational energies. These depend on *relative* velocities between atoms in the molecule and on interatomic separations. We need not include them because $f(v)$ is a function only of the molecule's overall translational speed. The integrations over other variables would separate and cancel top and bottom.

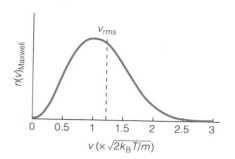

**Figure 8.6** Maxwell speed distribution.

$$n(v)_{\text{Maxwell}} = \frac{N}{V}\sqrt{\frac{2}{\pi}}\left(\frac{m}{k_{\text{B}}T}\right)^{3/2} v^2 e^{-\frac{1}{2}mv^2/k_{\text{B}}T} \qquad (8\text{-}28)$$

Maxwell speed distribution

The Maxwell speed distribution is shown in Figure 8.6. Since the plot increases from zero to a maximum before decreasing, it might be tempting to conclude that probability does not strictly decrease with energy. On the contrary, the probability of occupation of a *given state/velocity* does decrease exponentially with energy, but the *number of states* (velocity vectors) at any given speed increases as the square of the speed. Consider an analogy: The set of position vectors of a fixed magnitude $r$ describes a sphere of radius $r$ centered at the origin. Doubling $r$ would quadruple the area of the sphere. In effect, the "number of position vectors" of magnitude $r$, given arbitrary but fixed solid-angle spacing, increases as the area of a sphere, proportional to $r^2$. The same may be said of velocity. The "number of velocity vectors" of speed $v$ increases as $v^2$. It is the three-dimensional velocity element $dv_x\, dv_y\, dv_z = v^2 \sin\theta\, dv\, d\theta\, d\phi$ that introduces this factor. This is sensible, since we know that the element $d^3r\, d^3v$ is the classical analog that replaces $\Delta n$ and thus accounts for "number of states." Nevertheless, even though the number of states/velocities of a given speed increases with speed, causing the Maxwell distribution to increase initially, the exponentially decaying probability eventually wins out and causes the distribution to fall to zero.

## 8.6 Quantum Distributions

In the ten-oscillator example of Section 8.4, we said that the number of ways of distributing energy with $E_{n_i}$ fixed was the number of ways the other particles' quantum numbers could add up to account for the system's remaining energy. We implicitly assumed that switching the labels of any two particles yielded a *different* way of distributing energy. For instance, if $n_j = 5$ and $n_k = 2$ for particles $j$ and $k$ was one way, then $n_j = 2$ and $n_k = 5$ (and no other changes) was a different way. However, if the particles are indistinguishable, these are *not* different ways, and if numbers of ways are different quantum-mechanically than classically, probabilities will also be different. From these ideas, perhaps seeming of little practical interest, arise many startling and far-reaching consequences.

Recall from Chapter 7 that there are two very different types of indistinguishable particles: bosons (integral spin) and fermions (half-integral spin). We now investigate how and why the distributions of energy in systems of these two should differ from the Boltzmann distribution and from each other. To see the basic distinctions most easily, we consider a simple example: four particles, $a$, $b$, $c$, and $d$, bound by a harmonic oscillator potential energy. It is convenient to define the ground-state energy as zero and the spacing between levels as $\delta E$.

$$E_i = n_i \hbar \omega_0 \equiv n_i\, \delta E \qquad (n_i = 0, 1, 2, \ldots)$$

Now suppose that the system has total energy $2\,\delta E$. That is,

$$n_a + n_b + n_c + n_d = 2$$

Given this constraint, how may the energy be distributed among the particles? Table 8.1 shows all possible ways/microstates for distinguishable particles. We see that the smallest quantum number is most common, the largest least; as always, *the probable number of particles in a given state decreases with increasing energy.*

If the particles are indistinguishable bosons, however, *there are no labels,* and truly different states are distinguished only by having different *numbers* of particles at different levels.

Table 8.2 shows how the energy may be distributed in this case. Again, the probable number of particles drops with increasing energy, *but in a different way.*

**Table 8.1    Ways of distributing energy $2\delta E$ among four distinguishable particles: $a$, $b$, $c$, $d$**

| $n$ | | | Ways | | | | | | | | Number of times $n$ appears | Probability $P$ (#/40) | Probable number of particles ($P \times 4$) |
|---|---|---|---|---|---|---|---|---|---|---|---|---|---|
| 2 | $a$ | $b$ | $c$ | $d$ | | | | | | | 4 | 0.1 | 0.4 |
| 1 | | | | | $ab$ | $ac$ | $ad$ | $bc$ | $bd$ | $cd$ | 12 | 0.3 | 1.2 |
| 0 | $bcd$ | $acd$ | $abd$ | $abc$ | $cd$ | $bd$ | $bc$ | $ad$ | $ac$ | $ab$ | 24 | 0.6 | 2.4 |
| | | | | | | | | | Totals | | 40 | 1.0 | 4.0 |

**Table 8.2    Ways of distributing energy $2\delta E$ among four indistinguishable bosons**

| $n$ | Ways | | Number of times $n$ appears | Probability $P$ (#/8) | Probable number of particles ($P \times 4$) |
|---|---|---|---|---|---|
| 2 | X | | 1 | 0.125 | 0.5 |
| 1 | | XX | 2 | 0.25 | 1.0 |
| 0 | XXX | XX | 5 | 0.625 | 2.5 |
| | | Totals | 8 | 1.0 | 4.0 |

**Table 8.3** Ways of distributing energy $2\delta E$ among four indistinguishable fermions $\left(s = \frac{1}{2}\right)$

| $n$ | Ways | Number of times $n$ appears | Probability $P$ (#/4) | Probable number of particles $(P \times 4)$ |
|---|---|---|---|---|
| 2 | | 0 | 0.0 | 0 |
| 1 | XX | 2 | 0.5 | 2 |
| 0 | XX | 2 | 0.5 | 2 |
| | *Totals* | 4 | 1.0 | 4 |

The basic reason is that there are fewer label permutations as more particles have the same $n$, and multiply occupied $n$ tend be lower-than-average values. The distinguishable case thus "undercounts" lower $n$ values, so that bosons have a higher probability of being in the $n = 0$ state than would distinguishable particles. For indistinguishable $s = \frac{1}{2}$ fermions, on the other hand, as shown in Table 8.3, states in which more than two particles have the same $n$ are strictly forbidden by the exclusion principle. The distinguishable accounting clearly "overcounts" these, and fermions thus have a lower probability of occupying the $n = 0$ state than would distinguishable particles. Figure 8.7 summarizes the data given in the tables.

As in Chapter 7, where we demanded that the probability density be symmetric, the reader may be reluctant to accept that *physical consequences* arise from the mere fact of our inability to distinguish identical particles. (Why should the particles abide by these probabilities? How do *they* know that we can't distinguish them?) Here as there, however, the conclusions we draw agree with experimental evidence. As we shall soon see, there is a sensible limit in which both bosons and fermions indeed behave just as we would expect distinguishable particles to behave. But, in general, they behave in their own special ways *by nature*.

Let us now move on from our simple four-particle example to the distributions' quantitative forms for large numbers of particles. For distinguishable particles, detailed case-specific probabilities converge to the Boltzmann distribution. For quantum-mechanically indistinguishable particles, probabilities converge to two different distributions: the **Bose–Einstein** for bosons, and the **Fermi–Dirac** for fermions. The derivation of these **quantum distributions** is found in Appendix H. Although no new principles are introduced, it is fairly involved. Nevertheless,

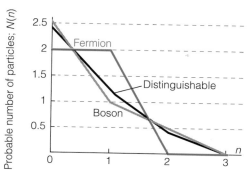

**Figure 8.7** The probable number of particles at the allowed energies depends on whether the particles are bosons, fermions, or distinguishable.

the student who wishes to understand how they can be derived in complete ignorance of the specific system to which they might be applied (knowing only
whether the particles are bosons, fermions, or classically distinguishable) should
expend the effort. Here we merely present the distributions and study how they
compare.

$$
\text{case-specific} \atop \text{probabilities} \quad \underset{\text{large } N}{\longrightarrow} \quad N(E) = \begin{cases} \dfrac{1}{Be^{+E/k_B T}} & \begin{array}{l} \text{Boltzmann} \\ \text{distribution} \\ \text{(distinguishable)} \end{array} & \text{(8-29)} \\[2em] \dfrac{1}{Be^{+E/k_B T} - 1} & \begin{array}{l} \text{Bose–Einstein} \\ \text{distribution} \\ \text{(bosons)} \end{array} & \text{(8-30)} \\[2em] \dfrac{1}{Be^{+E/k_B T} + 1} & \begin{array}{l} \text{Fermi–Dirac} \\ \text{distribution} \\ \text{(fermions)} \end{array} & \text{(8-31)} \end{cases}
$$

All three are decreasing functions of $E$; the higher the energy of a state, the
fewer will be the particles occupying that state. In all cases, the constant $B$ is related to the total number of particles. They differ most conspicuously as to whether
$-1$, $0$, or $+1$ is added to the exponential in the denominator. The consequences
can be drastic. But to see this, we must consider the constants $B$, for these depend
on temperature in ways that do depend on the specific system.

## Oscillators Revisited

The simplest case to consider is, as always, a system of harmonic oscillators,[15] but
with bosons and fermions in the picture we must now include spin. We assume
that $N$ is large and integration over energies valid. To find $B$ for distinguishable
particles, we note that condition (8-22), for oscillators with spin ignored, determines the multiplicative constant $A$ to be $\hbar\omega_0/k_B T$. Including spin, the density of
states changes from $1/\hbar\omega_0$ to $(2s + 1)/\hbar\omega_0$, so that this condition becomes

$$
N = \int_0^\infty N(E)D(E)dE = \int_0^\infty NAe^{-E/k_B T}\frac{2s + 1}{\hbar\omega_0}dE
$$

Accordingly, $A$ must become $(\hbar\omega_0/k_B T)/(2s + 1)$—that is, its previous value divided by $2s + 1$. Comparing $N(E)$ here with its form in (8-29), we see that the
multiplicative constants are simply related.

$$
B = \frac{1}{NA} = k_B T\frac{2s + 1}{N\hbar\omega_0}
$$

---

[15]This is not the most realistic of applications; the oscillators would move in only one dimension and must be assumed to all share the same well, so that they are quantum-mechanically indistinguishable. However, it reveals with a minimum of complexity important behaviors common to all
applications.

The procedure to find $B$ for the quantum distributions is similar and is left as an exercise. The results are

$$
\text{Harmonic oscillator distributions} \begin{cases} N(E)_{\text{Boltz}} = \dfrac{\mathcal{E}}{k_{\mathrm{B}} T} \dfrac{1}{e^{E/k_{\mathrm{B}} T}} & \text{Distinguishable} & (8\text{-}32) \\[3ex] N(E)_{\text{B-E}} = \dfrac{1}{\dfrac{e^{E/k_{\mathrm{B}} T}}{1 - e^{-\mathcal{E}/k_{\mathrm{B}} T}} - 1} & \text{Bosons} & (8\text{-}33) \\[4ex] N(E)_{\text{F-D}} = \dfrac{1}{\dfrac{e^{E/k_{\mathrm{B}} T}}{e^{+\mathcal{E}/k_{\mathrm{B}} T} - 1} + 1} & \text{Fermions} & (8\text{-}34) \end{cases}
$$

All are expressed in terms of a constant $\mathcal{E}$ defined for convenience by

$$
\mathcal{E} \equiv \frac{N \hbar \omega_0}{2s + 1} \tag{8-35}
$$

The significance of $\mathcal{E}$ is that if exactly one particle occupied each allowed *state*—one in each of the $2s + 1$ spin states at each energy level, or quantum number—the topmost energy level occupied would be $N\hbar\omega_0/(2s + 1)$.

Let us first investigate the quantum distributions' behaviors at high temperature. We shall see that this is the classical limit. If the temperature/average energy is much larger than $\mathcal{E}$, the probability of any two particles occupying the same state should be small. Rather than being "packed" into lower-energy states, they would be spread out sparsely over individual-particle states up to very high energies. And if every microstate does have all particles in different individual-particle states, the differences between the distributions should disappear. (Since the number of permutations in the distinguishable case would be the same for essentially every microstate, $N!$, it would give the same *relative* weighting to each as the boson case, which has no permutations. The fermion case should also agree, because no microstates would have to be "thrown out" to obey the exclusion principle.) All cases might as well involve classical distinguishable particles. Indeed, it is easily shown (see Exercise 23) that

$$
k_{\mathrm{B}} T \gg \mathcal{E} \quad \Rightarrow \quad \left.\begin{array}{c} N(E)_{\text{B-E}} \\ N(E)_{\text{F-D}} \end{array}\right\} \;\rightarrow\; N(E)_{\text{Boltz}}
$$

The high-temperature $k_{\mathrm{B}} T = 5\mathcal{E}$ plots of Figure 8.8 illustrate the point; the occupation number is much less than unity at all energies and the three distributions nearly coincide.

The state of affairs is vastly different at the low temperature extreme, in which $k_{\mathrm{B}} T \ll \mathcal{E}$ and we might expect multiple particles to inhabit the lower energy levels. In the $k_{\mathrm{B}} T = \frac{1}{5}\mathcal{E}$ plots, the Bose–Einstein distribution lies below the classical Boltzmann at intermediate energy levels, but is larger at the very smallest values of $E$; there is a greater tendency for bosons to crowd together into the lowest energy levels. The Fermi–Dirac, on the other hand, lies above the classical at intermediate energies and below at small values of $E$. *Classically,* we would expect a

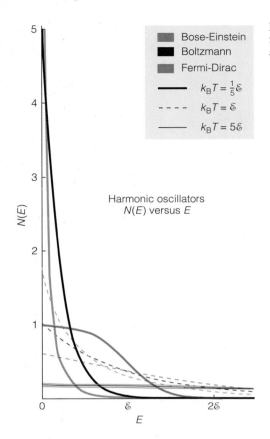

**Figure 8.8** The three distributions for oscillators at low, intermediate, and high temperature.

cold system to have multiple particles in the lowest-energy states, but the number of fermions can never exceed unity in any state. Thus, the lower the temperature, the more pronounced will be the fermion curve's low-energy bulge, suggesting a packing of particles as "close" as the exclusion principle allows. Note the qualitative agreement between the $k_B T = \frac{1}{5}\mathcal{E}$ plots of Figure 8.8 and the plots in Figure 8.7. Both are low temperature. (In fact, $2\,\delta E$ is the lowest energy possible for fermions in the four-particle example.) [16]

Table 8.4 further illustrates the differences between the distributions, via their $E = 0$ values in the low-temperature limit $k_B T \ll \mathcal{E}$ (obtained in Exercise 24). Given their relationships to the classical case, $N(0)_{\text{B-E}} \gg N(0)_{\text{Boltz}} \gg N(0)_{\text{F-D}}$, we might say that while fermions refuse to occupy the same low-energy state,

**Table 8.4**

| Oscillator | Classical | Boson | Fermion |
|---|---|---|---|
| $N(0)$ $(k_B T \ll \mathcal{E})$ | $\dfrac{\mathcal{E}}{k_B T}$ | $e^{\mathcal{E}/k_B T}$ | 1 |

[16]*Note:* In Figure 8.7, the "state" was the *spatial* state only, allowing two spin-$\frac{1}{2}$ fermions per state. In all subsequent discussions, each "state" is the total spatial-plus-spin state, allowing only one fermion per state.

bosons actually "prefer" it. Summarizing our findings in the simple harmonic oscillators case,

At high temperatures, particles are spread out sparsely over individual-particle states. With occupation numbers much less than unity, it is largely irrelevant whether the particles are bosons, fermions, or classically distinguishable, and the quantum distributions thus agree with the classical.

At low temperatures, bosons tend to congregate in the lowest-energy individual-particle state, while fermions tend to fill states, one particle per state, up to some maximum energy.

### Example 8.4

Write an expression for the average energy of a system of indistinguishable harmonic oscillators. Allow for either the boson or fermion case. (Assume that $k_B T \gg \hbar\omega_0$, so that integration is valid.)

Solution

We find averages using the general form (8-25).

$$\overline{E} = \frac{\int E\,N(E)D(E)\,dE}{N}$$

We must use $N(E)_{B\text{-}E}$ for bosons and $N(E)_{F\text{-}D}$ for fermions. A "density of states," on the other hand, has nothing to do with how many particles might fill those states. It is the same as in Section 8.4 for distinguishable oscillators (including spin): $(2s + 1)/\hbar\omega_0$. (The result of $D(E) = dn/dE$ is the same because $E = n\hbar\omega_0$ still holds, and, regardless of the *kind* of particle, the $2s + 1$ allowed spin states would simply multiply the number of states at a given $n$.) Now using $\pm$ and $\mp$ as needed (tops signs bosons, bottom signs fermions) to distinguish (8-33) and (8-34), we have

$$\overline{E} = \frac{1}{N} \int_0^\infty E\left[\frac{\pm 1}{(e^{E/k_B T})/(1 - e^{\mp \mathscr{E}/k_B T}) - 1}\right]\frac{(2s + 1)}{\hbar\omega_0}\,dE$$

The integral in Example 8.4 is not easily executed. The results, $\overline{E}(T)$, are plotted in Figure 8.9. At high temperatures ($k_B T \gg \mathscr{E}$), both boson and fermion cases are essentially plots of a linear relationship $\overline{E} = k_B T$, the same as the classical. The particles are spread so diffusely among the states that quantum indistinguishability is unimportant. Even so, the energy is slightly lower for bosons, which tend to congregate at lower levels more than would be expected classically, and slightly higher for fermions, which refuse to settle into already-occupied lower-energy states. At lower temperatures, the average energy for bosons approaches zero, while that for fermions approaches $\frac{1}{2}\mathscr{E}$, the logical average when levels are filled uniformly from zero to a maximum of $\mathscr{E}$. It should be noted that these smooth curves rely upon the assumption that integration is valid. This is the case because

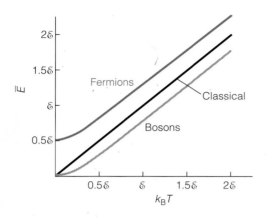

**Figure 8.9** Average oscillator energy versus temperature for the three types of particles.

we have considered temperatures such that $k_BT$ is comparable to $\mathcal{E}$, which is proportional to $\hbar\omega_0$ *times N*. At *very* low temperatures, where $k_BT$ is comparable to the spacing $\hbar\omega_0$ between levels, *summation* would be required. The boson case is particularly sensitive because the minus sign in the denominator of (8-30) causes a pronounced increase in $N(E)$ near $E = 0$. One result is that at low temperature the vast majority of particles in a boson gas may drop into the ground state. The amazing "superfluid" behavior of liquid helium is a manifestation, as we see in Section 8.7.

## The Fermi Energy

The Fermi–Dirac is a most curious distribution. As Figure 8.8 shows, at low temperature the occupation number is nearly unity out to an energy where it quickly drops to nearly zero. The energy where it drops is known as the **Fermi energy** $E_F$. Strictly speaking, the Fermi energy is defined as the energy at which the occupation number is one-half. The convenience of this definition is that the constant $B$ in (8-31) may be written in terms of $E_F$ for any system. Given that $N(E)_{\text{F-D}} = \frac{1}{2}$ at $E = E_F$, we have

$$N(E_F)_{\text{F-D}} = \frac{1}{Be^{+E_F/k_BT} + 1} = \frac{1}{2} \quad \Rightarrow \quad B = e^{-E_F/k_BT}$$

Thus

$$N(E)_{\text{F-D}} = \frac{1}{e^{(E-E_F)/k_BT} + 1} \tag{8-36}$$

In this form, it is easily seen that in the limit $T \to 0$ the Fermi–Dirac distribution is a "step function": unity for $E < E_F$ and zero for $E > E_F$.

The appearance of $E_F$ in (8-36) is somewhat deceptive, as we might conclude that it is independent of temperature. But how could it be? As shown in the $k_BT = 5\mathcal{E}$ plot of Figure 8.8, at high temperatures, particles are so spread out that the occupation number is less than $\frac{1}{2}$ at *all* energy levels. If $N(E)$ is never $\frac{1}{2}$, there *is* no "Fermi energy." The variation of $E_F$ with temperature is dependent on the system considered. It is left as an exercise to find $E_F(T)$ in the harmonic oscillators case. Most important is its value at $T = 0$: It is $\mathcal{E}$—and it should be! At $T = 0$,

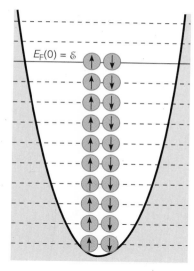

$E_F(0) = \mathcal{E}$

Figure 8.10 Fermion oscillators in the lowest-energy state possible.

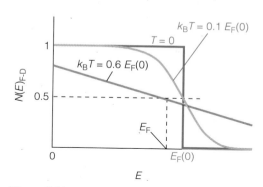

Figure 8.11 The Fermi–Dirac distribution at $T = 0$ and above. For low temperatures, the Fermi energy $E_F$ differs little from its $T = 0$ value.

spin-$\frac{1}{2}$ fermions would be packed one pair per level into lowest-energy states up to a maximum level, and this was exactly how $\mathcal{E}$ was defined. Figure 8.10 illustrates schematically the filling of oscillator states at $T = 0$.

Some idea of the variation of $E_F$ with $T$ is provided in Figure 8.11, which shows $N(E)_{\text{F-D}}$ for oscillators at three different low temperatures. The $T = 0$ plot drops from unity to zero abruptly at $E_F(0)$. In practical applications, this energy level can be quite high. For electrons in a typical metal, for instance, it is hundreds of times the value of $k_B T$ at room temperature (see Example 8.5). The exclusion principle forces the electrons to occupy levels much higher than would classically distinguishable particles. The plot of $k_B T = 0.1 E_F(0)$ shows that as long as $k_B T$ is considerably less than $E_F(0)$ the occupation number passes through $\frac{1}{2}$ at nearly the same energy—the Fermi energy is essentially constant. Furthermore, the distribution deviates from a true step function nearly symmetrically about $E_F$. It is only when $k_B T$ becomes comparable to $E_F(0)$ that the bulge is significantly depleted and the occupation number passes through $\frac{1}{2}$ at significantly lower energy. When we speak of a Fermi energy, we are usually referring to a system of fermions at low temperature, $k_B T \ll E_F(0)$. Therefore, since the Fermi energy doesn't vary much at such low temperatures, it is justifiable that we refer to "the" Fermi energy as if it were a constant.

Although many of our observations here have been based on a system of oscillators, they are borne out qualitatively in all systems of fermions.

## 8.7 The Quantum Gas

We now apply the quantum distributions to a situation more realistic than a system of ideal oscillators. In so doing, we will begin to understand behaviors that a "classical physicist" would find completely baffling.

The system we now consider is one in which $N$ massive particles are free to move in three dimensions, as are molecules in a classical gas. (The massless photon is treated in the next section.) However, we are not considering ordinary classical gases, but systems in which particle concentration is high enough that quantum indistinguishability must be taken into account. It may be a "gas" of bosons, such as helium atoms (spin-0), or it may be a "gas" of fermions, such as conduction electrons (spin-$\frac{1}{2}$) in a metal. The particles are assumed to be moving freely, yet still bound to some region, so that they occupy discrete quantum states. The logical model is the infinite well, in which particles experience no force except at the confining walls. Of course, to allow for motion in three dimensions, we must use the 3D well.

Primarily, we wish to be able to calculate averages. But just as in the oscillators case, there are two things we must determine first: the density of states and the constant $B$ in distributions (8-30) and (8-31). (Again, although the distributions apply to all systems in the forms given, the constant $B$ is system-dependent, as is the density of states). Let us start with $D(E)$ for the three-dimensional infinite well.

In Chapter 6, it is shown that the energy in a cubic three-dimensional infinite well is given by (6-8):

$$E_{n_x, n_y, n_z} = (n_x^2 + n_y^2 + n_z^2) \frac{\pi^2 \hbar^2}{2mL^2} \qquad (8\text{-}37)$$

The density of states is nearly trivial for one-dimensional harmonic oscillators because $E$ is directly proportional to the lone quantum number $n$. We see that it is now not quite so easy to determine the number of quantum states in an energy interval $dE$. To do so, we resort to a standard "trick": Picture the set $(n_x, n_y, n_z)$ as coordinates in three-dimensional "quantum-number space," each discrete point representing a discrete quantum state, as shown in Figure 8.12. The radial "distance" from the origin in this space would be given by

$$r_n = \sqrt{n_x^2 + n_y^2 + n_z^2}$$

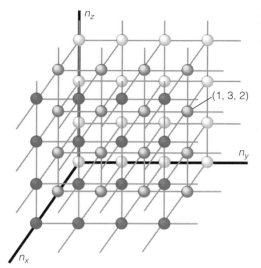

**Figure 8.12** A graphical representation of the allowed states in a 3D infinite well.

Thus,

$$E = \frac{\pi^2 \hbar^2}{2mL^2} r_n^2 \quad \text{or} \quad r_n = \sqrt{\frac{2mL^2 E}{\pi^2 \hbar^2}} \tag{8-38}$$

Sensibly, the "radius" increases with energy. Now, if the energy increases by $dE$, the radius increases by $dr_n$, so an energy *interval* of $dE$ encloses a volume in "quantum-number space" of $\frac{1}{8} 4\pi r_n^2 \, dr_n$, i.e., a one-eighth section of a spherical shell of thickness $dr_n$. But a volume in quantum-number space is a set of quantum states, so we may write

$$\text{\# states in interval } dE = \frac{1}{8} 4\pi r_n^2 \, dr_n \tag{8-39}$$

Let us express this in terms of $E$. Differentiating both sides of (8-38), we have

$$dr_n = \sqrt{\frac{mL^2}{2\pi^2 \hbar^2 E}} \, dE$$

Now inserting this and (8-38) itself into (8-39),

$$\text{\# states in interval } dE = \frac{1}{8} 4\pi \left( \frac{2mL^2 E}{\pi^2 \hbar^2} \right) \sqrt{\frac{mL^2}{2\pi^2 \hbar^2 E}} \, dE$$

$$= \frac{\pi}{\sqrt{2}} \left( \frac{mL^2}{\pi^2 \hbar^2} \right)^{3/2} E^{1/2} \, dE$$

or

$$dn = \frac{m^{3/2} V}{\pi^2 \hbar^3 \sqrt{2}} E^{1/2} \, dE$$

where $V$ is the volume $L^3$ of the confining well. Thus, the number of (spatial) quantum states per energy is

$$D(E) = \frac{dn}{dE} = \frac{m^{3/2} V}{\pi^2 \hbar^3 \sqrt{2}} E^{1/2}$$

Finally, to include spin we merely multiply this by the number of spin states, $2s + 1$.

$$D(E) = \frac{(2s + 1)m^{3/2} V}{\pi^2 \hbar^3 \sqrt{2}} E^{1/2} \tag{8-40}$$

Density of states:
Gas in 3D well (massive)

(A reminder: A density of states exists no matter how many particles may occupy those states. It is the same for bosons, fermions or distinguishable particles.)

Let us now turn to the constant $B$ in equations (8-30) and (8-31). It is related to the number of particles and so may be found as follows: Using a $\mp$ to distinguish (8-30) for bosons from (8-31) for fermions,

$$N = \int N(E)D(E) \, dE = \int_0^\infty \frac{1}{Be^{+E/k_B T} \mp 1} \left[ \frac{(2s + 1)m^{3/2} V}{\pi^2 \hbar^3 \sqrt{2}} \right] E^{1/2} \, dE \tag{8-41}$$

Unfortunately, here we hit a roadblock: The integral in (8-41) cannot be solved in closed form, and calculating averages via (8-25) is thus no simple matter. Even so, we can learn much from approximations. Perhaps the average of greatest interest is the average particle energy. Exercise 37 outlines the steps that give $B$ as a power series of diminishing terms, which can then be used in (8-25) to find the average energy, also as a series. The result is

$$\overline{E} = \frac{\int E\, N(E)D(E)\, dE}{\int N(E)D(E)\, dE} \cong \frac{3}{2} k_B T \left[ 1 \mp \frac{\pi^3 \hbar^3 \sqrt{2}}{(2s+1)(2\pi m k_B T)^{3/2}} \left( \frac{N}{V} \right) + \cdots \right]$$

$$(8\text{-}42)$$

In contrast to our one-dimensional oscillators system, in which volume was not even a consideration, the average energy in a quantum gas depends both on temperature *and* particle density $N/V$. Notably, expression (8-42) is only useful to show the *beginnings* of deviation from classical behavior; terms become rapidly small only when the temperature is not too low and the particle density not too high.

The first term in (8-42) is sensible; it is the average energy of a classical gas particle. That the effects of quantum indistinguishability arise in the second term may be seen in two ways. First, classical particle behavior should prevail if the particles have *short* wavelengths and *large* separations (i.e., they do not share the same space). Wavelength depends on speed, and the rms speed of a gas particle is $\sqrt{3k_B T/m}$ (cf. Section 8.5). Thus,

$$\lambda = \frac{h}{p} = \frac{h}{mv} \propto \frac{h}{m\sqrt{k_B T/m}} \quad \text{or} \quad \lambda \propto \frac{h}{\sqrt{m k_B T}}$$

The linear distance $d$ between particles is roughly the cube root of the volume per particle—that is, $d \propto (V/N)^{1/3}$. Putting these together, we see that the second term in equation (8-42) is proportional to $(\lambda/d)^3$. It will indeed be small when particles' wavelengths are much smaller than interparticle separations.

The second way is based upon how diffusely the particles are spread out among the energy levels. To determine this, let us calculate the Fermi energy at $T = 0$. The Fermi energy is a limiting energy only in a fermion gas. But corresponding to a one-particle-per-state situation, it serves as a good measure of how spread out the particles are among the states in all cases. We start with the expression for the total number.

$$N = \int_0^\infty N(E)_{\text{F-D}} D(E)\, dE$$

Since in the limit $T \to 0$, $N(E)_{\text{F-D}}$ is unity for $E < E_F$ and zero for $E > E_F$, we may simply integrate $D(E)$ from zero to $E_F$ and solve for $E_F$.

$$N = \int_0^{E_F} \frac{(2s+1)m^{3/2}V}{\pi^2 \hbar^3 \sqrt{2}} E^{1/2}\, dE = \frac{(2s+1)m^{3/2}V}{\pi^2 \hbar^3 \sqrt{2}} \left( \frac{2}{3} E_F^{3/2} \right)$$

$$E_F = \frac{\pi^2 \hbar^2}{m} \left[ \frac{3}{(2s+1)\pi\sqrt{2}} \left( \frac{N}{V} \right) \right]^{2/3}$$

$$(8\text{-}43)$$

(This expression is important in its own right. *Even at zero temperature,* the greater the density of fermions, the higher will be their energies.) Rearranging,

$$\frac{\hbar^3}{(2s + 1)m^{3/2}}\left(\frac{N}{V}\right) \propto E_F^{3/2}$$

Thus, the second term of (8-42) is proportional to $(E_F/k_B T)^{3/2}$. Sensibly, if $k_B T$ is much larger than $E_F$, the particles will be spread so diffusely among high-energy states that the effects of quantum indistinguishability should be negligible.

In summary: A gas will behave classically when the "quantum term" in (8-42) is small. Two equivalent ways of expressing this are

$$\left(\frac{\lambda}{d}\right)^3 \ll 1 \quad \text{or} \quad \frac{N}{V}\frac{\hbar^3}{(mk_B T)^{3/2}} \ll 1 \quad \textbf{(Classical)} \tag{8-44}$$

From the quantum term, we also see that at a given temperature the average particle energy in a boson gas is less than in a classical one (as was true for oscillators, cf. Figure 8.9). Being related to average energy, the pressure is also lower. Similarly, the average energy and pressure in a fermion gas are higher than in a classical one.

As noted earlier, the air in a room need not be treated as indistinguishable particles—criteria (8-44) are met. But in the many important applications where a bound system of particles must be so treated, the 3D infinite-well approximation often yields excellent results. One example arises in cosmology; by treating them as systems of gravitationally bound fermions, we may predict the radii of so-called white dwarf stars and neutron stars (see Exercises 53 and 54). A more common application is that of the electrostatically bound conduction electrons in a metal.

## Conduction-Electron Energy Levels

In metals, some of each atom's valence electrons are shared among all atoms, moving essentially freely in three dimensions through the solid. They are known as **conduction electrons** and are a good example of a fermion gas. This "gas" is confined in a potential well whose origin is illustrated in Figure 8.13. It is the net effect of the potential energies due to the positive ions spaced regularly throughout the solid. The top diagram shows the potential energy of an electron in the presence of a single positive ion—a simple Coulomb potential energy. Three ions would, of course, make the potential energy more complex. But as more ions are added to a given region of space, the net potential energy begins to assume the form of a finite well. In a crystalline solid, the density of positive ions is so great that a finite well becomes an excellent approximation. Often, an even simpler approximation is acceptable: The conduction electrons are treated as though confined in an *infinite* well, in which case our analysis of the quantum gas applies. Let us investigate whether the electron gas in a typical metal must indeed be treated as a quantum gas.

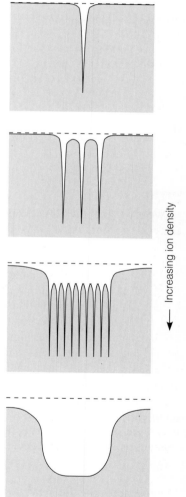

Increasing ion density

**Figure 8.13** Packing in more positive ions takes the single-atom potential energy to that of a finite well.

### Example 8.5

Assume that the conduction electrons in a piece of silver behave as a fermion gas, each atom contributing one electron. Calculate the Fermi energy. (The density of silver is $10.5 \times 10^3 \, \text{kg/m}^3$.)

### Solution

Equation (8-43) gives the Fermi energy in terms of properties easily determined. We know the mass and spin of electrons, but we must also determine the number of conduction electrons per unit volume. At one conduction electron per atom, this is equivalently the number of silver atoms per unit volume.

$$\frac{N}{V} = \frac{\text{number of atoms}}{\text{volume}} = \frac{\text{mass/volume}}{\text{mass/atom}}$$

$$= \frac{10.5 \times 10^3 \, \text{kg/m}^3}{107.9 \, \text{u} \times 1.66 \times 10^{-27} \, \text{kg/u}} = 5.86 \times 10^{28} \, \text{m}^{-3}$$

Thus,

$$E_F = \frac{\pi^2 \hbar^2}{9.11 \times 10^{-31} \, \text{kg}} \left[ \frac{3}{(2)\pi \sqrt{2}} 5.86 \times 10^{28} \, \text{m}^{-3} \right]^{2/3}$$

$$= 8.8 \times 10^{-19} \, \text{J} = 5.5 \, \text{eV}$$

Were we to apply classical physics to the conduction electron gas, the average energy would be $\frac{3}{2} k_B T$. At a room temperature of 300 K, this is only about 0.04 eV. The example shows that even at *zero* temperature electrons in silver must fill states up to 5.5 eV! Here we have an excellent example of $k_B T \ll E_F$, where (8-42) would be an awful approximation, its "quantum term" much larger than its classical first term. Since $k_B T$ at room temperature is only about $\frac{1}{100} E_F$, the occupation number distribution would resemble the extreme nonclassical $T = 0$ plot of Figure 8.11. To electrons in a metal, room temperature is very cold. Temperatures hundreds of times higher would be needed before they could spread out among higher energy levels, and since such temperatures would vaporize most metals, we conclude that conduction electron behavior is always a decidedly quantum-mechanical affair.

We may now understand other properties of metals. Work function, the minimum energy required to remove an electron from a metal, is simply the energy difference between the Fermi energy and the top of the finite well in which the conduction electrons are bound (Figure 8.14).

$$\phi = U_0 - E_F$$

Moreover, it is the difference in the work functions of different metals that is responsible for the phenomenon of **contact potential**—the potential difference that develops when different kinds of metal come into contact. Suppose two kinds of metal are electrically isolated and at the same electric potential, arbitrarily designated $V = 0$, and that the work function of metal 1 is greater than that of metal 2,

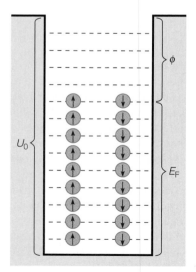

**Figure 8.14** Electron energies in a "cold" metal.

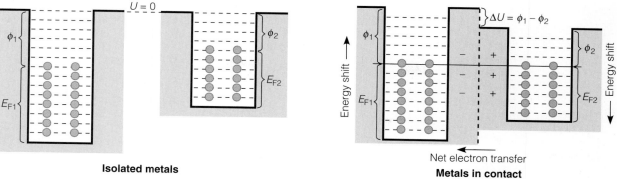

**Isolated metals**  **Metals in contact**

**Figure 8.15** Contact potential arises from unequal work functions.

as shown in Figure 8.15. When they are put in contact, the most energetic electrons in metal 2—that is, those $\phi_2$ below the top of their finite well—are of greater energy than the most energetic in metal 1. Seeking a lower energy state, electrons from metal 2 enter metal 1. This effectively shifts the energy levels of metal 1 upward, through an increase in electrostatic repulsion, while correspondingly shifting the energy levels in metal 2 downward. The transfer of electrons stops when the two Fermi energies are at the same level.[17] At this point, the electrostatic potential energy is higher in metal 1 than in metal 2 by $U_1 - U_2 = \phi_1 - \phi_2$. Thus, the potential difference is $V_1 - V_2 = (\phi_1 - \phi_2)/(-e) = (\phi_2 - \phi_1)/e$. Metal 1, negatively charged, is at lower potential.

## Bose–Einstein Condensation and Liquid Helium

At very low temperatures, the minus sign in the denominator makes the Bose–Einstein distribution (8-30) very large near $E = 0$. The physical consequence in a boson gas is a decidedly nonsmooth dropping of particles into the lowest-energy quantum state—and amazing behavior! The best-known example is helium-4. As we have noted, under normal conditions gases are so diffuse that it is irrelevant whether the particles are bosons or fermions. Nevertheless, the helium-4 atom is a spin-0 boson; the total spin of the nucleus is zero and the two $1s$ electrons must be in a spin-0 singlet state. (By contrast, the helium-3 atom, with an odd number of spin-$\frac{1}{2}$ fermions in its nucleus, is a fermion. Henceforth, by "helium" we mean helium-4, the vastly more abundant form.) Reluctant to solidify, helium atoms move as a fluid down to very low temperatures. At some point, of course, interparticle forces become significant and the fluid is no longer a simple gas, as we have assumed—it becomes a liquid. Even so, at very low temperatures, a considerable fraction of its atoms "condense" into the same lowest-energy individual-particle state. If there are cases where quantum effects should be perceptible to humans, this certainly qualifies.[18]

Indeed, liquid helium exhibits so-called **superfluid** characteristics: negligible viscosity and near perfect thermal conductivity. Both are consequences of the order necessarily accompanying a macroscopic system in which a substantial fraction of the particles move essentially as one. Where motion is disordered, viscosity and resistance to heat flow follow, but ordered motion implies streamlined

[17]Only a very small fraction of the total number of conduction electrons need transfer to shift energy levels by the few electronvolts needed. See Exercise 35 for an order-of-magnitude estimate.

[18]About 10% of the atoms are in the ground state. Recently, essentially pure "Bose–Einstein condensates" have been produced in samples of column-I elements dilute enough to be considered true gases. See Progress and Applications.

**Figure 8.16** Superfluid liquid helium creeping out of its quarters. (Courtesy of A. Leitner, Rensselaer Polytechnic Institute.)

collective behaviors. In liquid helium, the transition to the largely ordered state occurs at about 2.2 K. An order-of-magnitude prediction of this temperature follows from the criteria of (8-44) (see Exercise 29).

Superfluid properties are responsible for a remarkable phenomenon shown in Figure 8.16: If at very low temperature a bowl is lowered into liquid helium, then raised above the surface, liquid helium will crawl up the sides of the bowl and spill over the edge until the bowl is empty, seemingly defying gravity. The mechanism is capillarity combined with negligible viscosity and perfect thermal conductivity. Helium adheres to the inside surface of the bowl, creeping in a film up the side as does ordinary water in a capillary tube. A height limit is reached in an ordinary liquid due to temperature variations between liquid and container: When the container is hotter, the film evaporates; when colder, vapor condenses back to the liquid below. This, however, cannot occur in liquid helium, because of its near perfect thermal conductivity. Additional helium moves upward in the thin film, unimpeded by viscosity, to grasp the bowl at ever greater heights until the top edge is reached.

## 8.8 Massless Bosons: The Photon Gas

Photons, being spin-1, are bosons, and to study a gas of photons, we need a new density of states. The quantum gas density of states (8-40) is not applicable, because it is based on the energy quantization condition (8-37) for *massive* particles.

The correct density of states is derived in Appendix C.[19] It is expressed however as a number of states per unit frequency, rather than per unit energy.

$$dn = \frac{8\pi V}{c^3} f^2 \, df \qquad\qquad \text{(C-8)}$$

Using $E = hf$, we may convert this to the usual number per unit energy.

$$dn = \frac{8\pi V}{c^3} \frac{E^2 \, dE}{h^3} \quad \text{or} \quad D(E) \equiv \frac{dn}{dE} = \frac{8\pi V}{h^3 c^3} E^2 \qquad\qquad \text{(8-45)}$$

Density of states: Photon gas

The photon application differs in another important way. While situations abound where a fixed number of massive objects is confined to some volume, never is this the case for photons. A bowl containing liquid helium does not emit and absorb helium atoms (we might say it "reflects" them), but it is constantly emitting and absorbing photons. The point is that in a photon gas the number is not "chosen" beforehand; it is determined by whatever conditions lead to equilibrium between the photons and the surrounding matter. It may seem strange to think of a gas of photons having a temperature and being in equilibrium with matter. But we can certainly imagine a nonequilibrium overall state where one is hotter than the other, where energy on average is transferred from photons to matter or vice versa, and if they exchange energy, but are otherwise isolated, an overall most probable/disordered state of one uniform temperature must inevitably result. Although the photons do obey the Bose–Einstein distribution once this equilibrium is reached, a question remains: We have said that the constant $B$ in $N(E)_{\text{B-E}}$ is related to the total number of bosons, but what if that number is not fixed? It may be shown that when the number of photons is determined by the conditions of equilibrium energy exchange with matter, $B = 1$.

Because the number of photons is not predetermined in the photon gas, the average particle energy is generally of less interest than the *total* energy, that is, the average energy times $N$. A gas of photons is a thermodynamic system, so it is appropriate to refer to the total energy as "internal energy." Following (8-25),

$$E_{\text{internal}} = \overline{E}N = \int E \, N(E) D(E) \, dE$$

Inserting $N(E)_{\text{B-E}}$, with $B = 1$, and the photon gas density of states,

$$E_{\text{internal}} = \int_0^\infty E \frac{1}{e^{E/k_B T} - 1} \frac{8\pi V}{h^3 c^3} E^2 \, dE \qquad\qquad \text{(8-46)}$$

With the temporary definition $x \equiv E/k_B T$, this may be rewritten

$$E_{\text{internal}} = \frac{8\pi V}{h^3 c^3} (k_B T)^4 \int_0^\infty \frac{x^3}{e^x - 1} \, dx$$

The value of the integral is $\pi^4/15$. Thus,

$$E_{\text{internal}} = \frac{8\pi^5 k_B^4}{15 h^3 c^3} V T^4 \qquad\qquad \text{(8-47)}$$

The energy per unit volume in a gas of photons grows as temperature to the *fourth* power. Usually we expect *average* particle energies to be proportional to $T$ to the

[19]The derivations are similar, however, because both view the states as standing waves in a 3D well.

first power. The total energy grows so quickly here because the *number* of particles/photons grows as $T$ to the third power (see Exercise 40).

---

**Example 8.6**

Internal energy in the form of electromagnetic radiation permeates all matter, since all matter contains oscillating charged particles. Assume that an ideal monatomic gas at STP (273 K, 1 atm) is in equilibrium with electromagnetic radiation. Find the ratio of the photon internal energy to the internal kinetic energy of the gas atoms.

**Solution**

For the electromagnetic radiation,

$$\frac{E_{\text{photon}}}{V} = \frac{8\pi^5 k_B^4}{15h^3 c^3}(300 \text{ K})^4 = 6.1 \times 10^{-6} \text{ J/m}^3$$

The kinetic energy of the atoms is $N$ times the average energy of $\frac{3}{2}k_B T$. Thus, using the ideal gas law, we may write

$$E = N\overline{E} = \frac{3}{2}Nk_B T = \frac{3}{2}PV$$

so that

$$\frac{E}{V} = \frac{3}{2}(1.013 \times 10^5 \text{ Pa}) = 1.5 \times 10^5 \text{ J/m}^3$$

The ratio is

$$\frac{6.1 \times 10^{-6}}{1.5 \times 10^5} = 4 \times 10^{-11}$$

We conclude that under ordinary conditions the fraction of a thermodynamic system's total energy in the form of electromagnetic radiation is rather small.

---

## Thermodynamics and Electromagnetic Radiation

Quantum statistical mechanics provides the important quantitative link between temperature and electromagnetic radiation. Previously, the nature of the electromagnetic radiation coming from an object of a given temperature was mostly a matter of observation. For example, the Stefan–Boltzmann law originated as merely a statement of the experimental observation that the radiation's intensity depends on $T^4$. The law is very important because it is what governs heat transfer via electromagnetic radiation, yet it lacked theoretical justification. Accordingly, equation (8-47), from which it follows (see Exercise 42), was one of statistical mechanics' more celebrated early achievements.

We gain further insight by examining the integrand in (8-46). It is the contribution to the total electromagnetic energy of the photons whose energies lie in a small range $dE$ around $E$.

$$dE_{internal} = E \frac{1}{e^{E/k_B T} - 1} \frac{8\pi V}{h^3 c^3} E^2 \, dE$$

Let us reexpress this in terms of frequency. Using $E = hf$, we have the contribution due to the photons in the frequency range $df$ around $f$.

$$dE_{internal} = \frac{hf^3}{e^{hf/k_B T} - 1} \frac{8\pi V}{c^3} \, df \quad \text{or} \quad \frac{dE}{df} = \frac{hf^3}{e^{hf/k_B T} - 1} \frac{8\pi V}{c^3} \quad \text{(8-48)}$$

Plotted in Figure 8.17, this is Planck's **spectral energy density,** also derived in Appendix C. Although an individual photon's energy is, of course, proportional to $f$, $dE/df$ initially increases more rapidly because the density of states increases as energy (frequency) *squared;* for a given number-per-*state* $N(E)$, more states $D(E)$ *available* in a given range $dE$ (or $df$) means that more photons will be occupying states in that range. Ultimately, the decreasing Bose–Einstein distribution wins out; while very high energy states exist, essentially no photons will be found with such high frequencies. Clearly, the total photon energy has a maximum, and it can be shown that the frequency at which it occurs is directly proportional to the temperature. (This is the essence of Wien's law, which, like the Stefan–Boltzmann "law," was originally put forth without theoretical justification. For its derivation see Exercise 43.) We see this proportionality in the figure. The 2300 K curve has a maximum at a frequency of 238 THz, and the 4600 K curve at 476 THz. The $T^4$ dependence is also shown: The total energy in the 4600 K case (the area under the curve) is 16 ($2^4$) times that in the 2300 K. Extrapolating to still lower temperatures, we can easily see why something at near room temperature (such as a human body) emits relatively little electromagnetic radiation, and essentially all of it at wavelengths longer than the visible range of 400–700 nm. "Red-hot" is very hot!

It is noteworthy that we obtained equation (8-48) by treating electromagnetic radiation as indistinguishable bosons. But Planck's own derivation (in Appendix C) succeeds without observing indistinguishability requirements! This is possible because Planck's derivation treats the radiation *solely as standing waves,* which by their very differences in numbers of antinodes *are distinguishable.* The Boltzmann distribution, which Planck used, is then the correct one. There are two different but equivalent routes to the same physical truth.

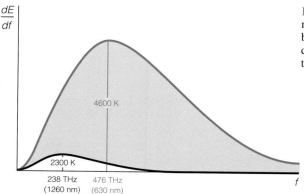

**Figure 8.17** Electromagnetic energy radiated by an object versus frequency at two different temperatures.

## 8.9 The Laser

Few modern scientific discoveries have attracted more attention than the laser. Surgery, communication, nuclear fusion, information storage and retrieval, reading product bar codes, precision alignment, and holography are but a few of its applications. The heart of its utility is its production of **coherent** light. Light is coherent when it travels in only one direction, is of a single wavelength, and is in phase, meaning that there are definite wave fronts where the oscillating electromagnetic field is at a maximum at a given instant. In contrast, light from an incandescent light bulb diverges in all directions, contains a broad spectrum of wavelengths, and comprises disjoint pieces of waves produced by countless *independent* oscillating sources in the hot filament. While ordinary light can be focused into a near unidirectional beam with lenses or mirrors, and gas vapor lights may produce reasonably well-defined wavelengths, a laser does these things automatically. Above all, the "in phase" nature of its light cannot be duplicated with ordinary sources. Part of laser light's exceptional power is the supreme constructive interference among its photons. How is this accomplished?

Light from ordinary sources is produced chiefly by **spontaneous emission** (Figure 8.18). Suppose the light-producing material is a gas, such as in a fluorescent bulb or sodium streetlight. An electron excited to a higher atomic energy level drops to a lower level in a short but unpredictable time, producing a photon. Transitions between the same levels in the many gas atoms produce photons of the same wavelength, but uncoordinated, moving in arbitrary directions and not in phase with each other. As we learned in Chapter 6, electrons may be promoted to higher atomic energy levels by absorption of a photon of precisely the correct

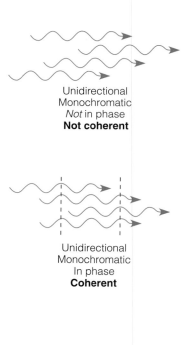

**Figure 8.18** Three different photon-atom interactions.

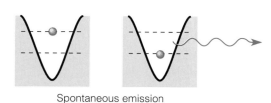

Spontaneous emission

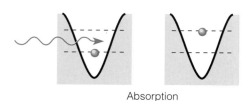

Absorption

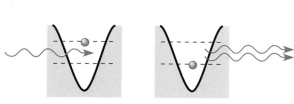

Stimulated emission

wavelength. Absorption and spontaneous emission are two electron-photon interactions. There is a third: **stimulated emission.** An electron in one energy level, if influenced by a passing photon whose energy is precisely the difference between that level and a lower-energy one, may be induced to drop to the lower energy level. In essence, the first photon stimulates the emission of a second of precisely the same wavelength. Moreover, the second photon moves in the same direction as and in phase with the first; they are coherent. The word "laser" stands for light amplification by the stimulated emission of radiation.

To understand how stimulated emission may be exploited to produce a coherent light source of *many* photons, we must consider the exchange and balance of energy between a material and the electromagnetic radiation with which it interacts.

Let us restrict our attention to transitions between representative energy levels 1 and 2 of the material's atoms. Upward transitions $(1 \rightarrow 2)$ are caused by photon absorption (also known as *stimulated* absorption). The rate at which they occur should be proportional to the number of atoms $N_1$ already occupying level 1 and to the number of photons of the required energy $E_2 - E_1 \equiv \Delta E$. In a photon gas, the number of photons of energy $\Delta E$ is, as always, the product of the occupation number and the density of states. Thus, the absorption rate is given by

$$R_{abs} = B_{abs} \underset{\substack{\text{Atoms in} \\ \text{level 1}}}{N_1} \times \underset{\substack{\text{Photons of} \\ \text{energy } \Delta E}}{N(\Delta E)D(\Delta E)}$$

where $B_{abs}$ is the proportionality constant. There are two mechanisms for downward transition: spontaneous and stimulated emission. The rates of both should be proportional to the number of atoms already occupying level 2. The rate of stimulated emission, requiring a photon to initiate the process, should also be proportional to the number of photons of energy $\Delta E$. Spontaneous emission on the other hand occurs whether or not there is a preexisting photon gas and so must be independent of the number of photons present.[20]

$$R_{emit} = R_{spon} + R_{stim}$$

$$= A_{spon} \underset{\substack{\text{Atoms in} \\ \text{level 2}}}{N_2} + B_{stim} \underset{\substack{\text{Atoms in} \\ \text{level 2}}}{N_2} \times \underset{\substack{\text{Photons of} \\ \text{energy } \Delta E}}{N(\Delta E)D(\Delta E)}$$

A different symbol is used for the spontaneous emission proportionality constant, in accordance with a convention. It was Einstein who first charted the course we now follow, and the coefficients are known universally as the **Einstein $A$ and $B$ coefficients.** The ratio of emission to absorption rate is of central importance:

$$\frac{R_{spon} + R_{stim}}{R_{abs}} = \underset{\text{Spontaneous}}{\frac{A_{spon}N_2}{B_{abs}N_1 N(\Delta E)D(\Delta E)}} + \underset{\text{Stimulated}}{\frac{B_{stim}N_2 N(\Delta E)D(\Delta E)}{B_{abs}N_1 N(\Delta E)D(\Delta E)}}$$

or

$$\frac{R_{spon} + R_{stim}}{R_{abs}} = \underset{\text{Spontaneous}}{\frac{N_2}{N_1}\frac{A_{spon}}{B_{abs}N(\Delta E)D(\Delta E)}} + \underset{\text{Stimulated}}{\frac{N_2}{N_1}\frac{B_{stim}}{B_{abs}}} \qquad \text{(8-49)}$$

[20]As discussed in Section 6.9, although an electron in a "stationary state" might seem to have no cause to make a downward transition, even if the atom is isolated it is still affected by electromagnetic vacuum fluctuations. These triggers for spontaneous emission are completely independent of whatever else might happen to occupy the space.

Consider for the time being a state of *equilibrium.* Although no macroscopic change can be seen, atoms are continually jumping back and forth between states 1 and 2, emitting and absorbing photons of energy $\Delta E$. Even so, the atoms obey the Boltzmann distribution (to a good approximation, especially in a gas), so that

$$\frac{N_2}{N_1} = e^{-(E_2 - E_1)/k_B T} = e^{-\Delta E/k_B T}$$

The photons behave as a boson gas at temperature $T$. Using equation (8-45),

$$N(\Delta E)D(\Delta E) = \frac{1}{e^{\Delta E/k_B T} - 1} \frac{8\pi V}{h^3 c^3}(\Delta E)^2$$

Thus, *in equilibrium* the emission-to-absorption ratio (8-49) is

$$\frac{R_{spon} + R_{stim}}{R_{abs}} = e^{-\Delta E/k_B T}\frac{A_{spon}}{B_{abs}}\frac{h^3 c^3}{8\pi V(\Delta E)^2}(e^{\Delta E/k_B T} - 1) + e^{-\Delta E/k_B T}\frac{B_{stim}}{B_{abs}}$$

or

$$\frac{R_{spon} + R_{stim}}{R_{abs}} = \frac{A_{spon}}{B_{abs}}\frac{h^3 c^3}{8\pi V(\Delta E)^2}(1 - e^{-\Delta E/k_B T}) + e^{-\Delta E/k_B T}\frac{B_{stim}}{B_{abs}}$$

In equilibrium, however, the rate at which atoms make upward transitions from level 1 to level 2 must equal the downward transition rate. This is known as the "principle of detailed balance."[21] Simply put, the right-hand side of the previous equation must be unity. Since the temperature is arbitrary, this is possible only if both coefficients of $e^{-\Delta E/k_B T}$ are unity (so that these terms cancel and leave unity). Thus,

$$\frac{A_{spon}}{B_{abs}}\frac{h^3 c^3}{8\pi V(\Delta E)^2} = 1 \quad \text{and} \quad \frac{B_{stim}}{B_{abs}} = 1$$

We have reached an intriguing conclusion: $B_{stim} = B_{abs}$. This means that if the numbers of atoms occupying the two levels happened to be equal, a stray photon of the proper frequency is just as likely to induce absorption as stimulated emission. (In equilibrium, lower energies are more highly populated, so that absorption is more probable than stimulated emission, with spontaneous emission making up the difference.) We wish to make stimulated emission the more probable!

Considering equilibrium has allowed us to find ratios of the $A$ and $B$ coefficients. Since they *are constants,* they are the same even if the system is not in equilibrium. But the transition *rates* depend also on numbers of atoms in the states, which will be different if the system is not in equilibrium. Is it possible then to enhance stimulated emission by establishing the proper nonequilibrium state? To answer this, we return to equation (8-49), the emission-to-absorption ratio *before* equilibrium was assumed. Now knowing the ratios of the $A$ and $B$ coefficients, we see that it becomes simpler.

$$\frac{R_{spon} + R_{stim}}{R_{abs}} = \underbrace{\frac{N_2}{N_1}\frac{8\pi V(\Delta E)^2}{h^3 c^3 N(\Delta E)D(\Delta E)}}_{\text{Spontaneous}} + \underbrace{\frac{N_2}{N_1}}_{\text{Stimulated}} \qquad (8\text{-}50)$$

[21]Equilibrium demands only that the numbers in the two levels not change on average. This could still hold if direct $(2 \to 1)$ transitions were balanced by *indirect* $(1 \to 2)$ transitions via a third energy level, in which case the direct $(1 \to 2)$ and $(2 \to 1)$ transitions rates would not be equal. Nevertheless, experimental evidence indicates that they are equal.

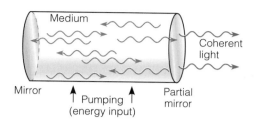

**Figure 8.19** Basic components of a laser.

   As we might have guessed, to establish a condition where stimulated emission far exceeds absorption, we need only cause $N_2$ to far exceed $N_1$; a disproportionate number of electrons must somehow be gotten to the higher energy level. Such an overpopulation of higher levels is known as a **population inversion** and is, of course, contrary to the natural exponential decrease with energy. It must be established by some external means. Nevertheless, so long as it exists, one photon of the proper frequency, perhaps the result of spontaneous emission, may become two, which may become four, and so on, very quickly resulting in a large number of coherent photons.

   Before investigating how a population inversion might be established, let us take a look at the basic elements of a laser, shown in Figure 8.19. Because various spontaneous emissions could initiate separate "outbreaks" of coherence in different directions, the medium is tuned to amplify the coherent light in one direction. Parallel mirrors are placed at the medium's two ends. Photons not parallel to the axis have relatively few opportunities to induce stimulated emission, while those parallel to the axis reflect back and forth, providing many opportunities. The length along the axis is tuned to a standing-wave condition $L = n\lambda/2$, so that waves moving in opposite directions constructively interfere. In the resulting **resonant cavity,** the radiation parallel to the axis is greatly amplified, and off-axis radiation is diminished. By making the mirror at one end only partially reflecting, a fraction of coherent light is continuously allowed to exit the medium.

   There are several ways of adding the external energy needed to promote electrons to the higher levels. One is **optical pumping,** used most often in solid-state lasers, where the laser medium is a solid. The medium is subjected to a very intense light of broad spectrum, of which at least a fraction is of the required frequency. Although much of the input energy is dissipated as heat, a significant amount becomes coherent light. Usually, optical pumping in a solid-medium laser is pulsed to allow time for cooling. Figures 8.20 and 8.21 show two examples of optical pumping. The first is the original ruby laser. The optical pumping is done by a flashtube, which is wound around a ruby crystal, the "lasing" material. Figure 8.21 shows a one-pass laser amplifier, in which a bank of high-intensity lamps "charges up" neodymium glass plates just prior to a laser beam's entry at one end; when the beam enters, atoms in the glass deexcite and a much more powerful beam emerges. (The beam is polarized, and light is passed without reflection by orienting the plates at "Brewster's angle.")

   Another means of energy input is **electric discharge pumping,** used in gas lasers. Atoms are raised to higher energy levels as the discharge ionizes some atoms and produces generally violent motion. Because excess heat dissipation is

**Figure 8.20** The original laser, with spiral flashtube surrounding resonant cavity of ruby.

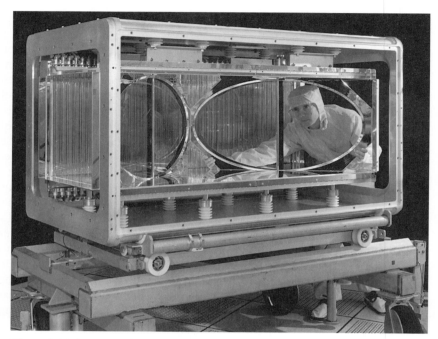

**Figure 8.21** A one-pass laser amplifier.

less of a problem in a gas, it is usually possible to sustain continuous (as opposed to pulsed) operation.

A vexing problem remains: Optical pumping can increase the population of a higher energy level only when the number of electrons at the lower level is greater. No matter how intense, it cannot increase a higher level's population beyond the point where the numbers in the two levels are equal, for at this point absorption would be no more probable than stimulated emission. (Remember: $B_{stim} = B_{abs}$.) Nor would this stimulated emission be useful, because it would be induced by the incoherent photons from the optical pumping source. Electric discharge alone is also incapable of establishing a population inversion. Then how can it be done? The problem is solved by using a medium with a **metastable state.** Most downward atomic transitions occur spontaneously within about $10^{-8}$ s. A few (occurring via relatively ineffective mechanisms; cf. Section 6.11) take much longer,

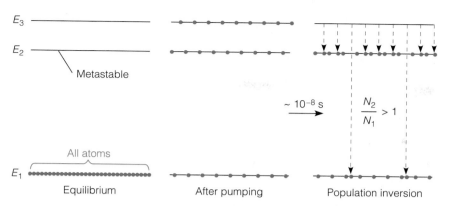

**Figure 8.22** Energy transitions in the three-level laser.

typically $10^{-3}$ s. The upper level in such a transition is known as a metastable (meaning nearly stable) state.

To see how this solves the problem, let us consider a simplified model of a laser—an "atom" with only three energy levels: the ground state $E_1$, an excited metastable state $E_2$, and a yet higher (non-metastable) state $E_3$, from which electrons drop preferentially to $E_2$ rather than fully down to $E_1$ (Figure 8.22). The temperature is low ($k_BT < E_2 - E_1$), so that essentially all atoms are in the ground state. Now, very intense optical pumping is applied, so intense that a point of equal absorption and stimulated emission is reached between all levels: $N_1 \cong N_2 \cong N_3$. Then the pumping is stopped. After about $10^{-8}$ s, all electrons will have dropped from the "fast" $E_3$ state, but more to $E_2$ than to $E_1$. For the next $\sim10^{-3}$ s, $N_2$ is greater than $N_1$, and the situation is ripe for a cascade of stimulated emissions between $E_2$ and $E_1$. (In fact, it is not necessary that pumping be stopped. Continuous output is possible as long as pumping is able to keep ahead of the stimulated emission, which would otherwise quickly deplete $E_2$.)

What we have just described is the **three-level laser.** A way of increasing efficiency is the **four-level laser** (Figure 8.23), in which another "fast" level is employed, above the ground state but below the metastable state. Let us call this the "near-ground" state. Its energy is $E_1$ and we redefine the ground state as $E_0$. The

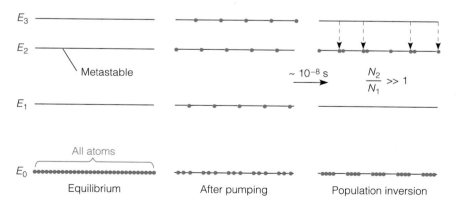

**Figure 8.23** Four-level laser.

benefit of this arrangement is that in the first $\sim 10^{-8}$ s after pumping, all excited states but the metastable one depopulate. Thus, rather than dropping to a non-empty ground state, electrons in the metastable state may drop to the essentially empty near-ground state. The four-level laser is more efficient simply because the ground state is not the terminal state for stimulated emission; a virtually complete population inversion between $E_2$ and $E_1$ is obtained without having to pump the ground state so near empty as in the three-level. In practice, lasers often employ even more than four energy levels, but the basis of the four-level still applies. (The semiconductor diode laser, discussed in Section 9.9, is different. A population inversion is still needed, and it is effectively a four-level, but it has some interesting twists, including a completely different pumping mechanism.)

No matter how many levels are employed, only certain substances meet the criteria required for laser operation. As noted earlier, the first laser used a ruby crystal as the medium. Three-level and optically pumped, ruby lasers produce 694.3-nm light. Four-level neodymium glass lasers, used in laser fusion experiments (see Section 10.7), produce 1060-nm infrared light. The most common gas laser is the helium–neon, producing red-orange 632.8-nm light. It employs a roundabout path to establish a population inversion in neon: Energy passes first from electric discharge to metastable helium states, then via helium–neon collisions to neon states at essentially the same energies as those in the helium, followed by stimulated emission to lower neon states. Driven by rapidly expanding applications, exciting advances are being made regularly in finding new laser materials with improved capabilities.

## ◆ 8.10 Specific Heats

The ability of a substance to store energy internally is of great concern in many applications. To be truly understood, it must be viewed as a macroscopic average of microscopic energy storage. Thus, the topic is a good example of the utility of statistical mechanics.

A basis for understanding specific heats is the **equipartition theorem.** When derived using *classical* statistical mechanics, it may be expressed as follows:

> A particle in a thermodynamic system at equilibrium will manifest $\frac{1}{2}k_B T$ of energy on average for each independent variable (degree of freedom) on which its energy depends quadratically.

Equipartition theorem

Examples of independent variables and quadratic dependence are a one-dimensional oscillator, $E = \frac{1}{2}mv^2 + \frac{1}{2}\kappa x^2$, which has two degrees of freedom; an atom in a monatomic ideal gas, $E = \frac{1}{2}mv_x^2 + \frac{1}{2}mv_y^2 + \frac{1}{2}mv_z^2$, which has three; and an object rotating about a single fixed axis, $E = \frac{1}{2}I\omega^2$, which has one. We will not prove the theorem, but will instead investigate the circumstances in which quantum-mechanical considerations demand that it be applied with care. As we shall see, low temperatures are the problem. First, though, let us see why the theorem is so important.

Specific heat is defined as the amount of heat required to warm a unit mass by 1 degree of temperature, and is measured in J/kg·K. We will find it more convenient to determine the related **molar heat capacity:** energy per *mole* per degree of temperature, in J/mol·K. According to the equipartition theorem, for each degree of freedom,

$$\overline{E}_{\text{one particle}} = \frac{1}{2}k_{\text{B}}T \quad \text{(Per degree of freedom)}$$

Since the total internal energy is just the average energy times the number, the internal energy of 1 mole of particles (Avogadro's number $N_A$) would be

$$E_{\text{internal per mole}} = N_A\frac{1}{2}k_{\text{B}}T$$

Using the definition of the ideal gas constant, $R \equiv N_A k_{\text{B}}$, we obtain

$$R \equiv N_A k_{\text{B}}$$
8.315 J/mol·K

$$E_{\text{internal per mole}} = \frac{1}{2}RT \quad \text{(Per degree of freedom)}$$

Now molar heat capacity at constant volume (i.e., when no energy is exchanged via mechanical work) is the rate at which the internal energy per mole changes with temperature.

$$C_v = \frac{\partial E_{\text{internal per mole}}}{\partial T}$$

Thus, we see the importance of the equipartition theorem: Yielding the coefficient relating $E$ and $T$, it is the route to heat capacity.

## Gases

According to the equipartition theorem, the internal energy per mole of a monatomic gas is $\frac{3}{2}RT$ (degrees of freedom: $v_x$, $v_y$, $v_z$) and the molar specific heat at constant volume is thus $\frac{3}{2}R$. A diatomic gas molecule, on the other hand, may also possess energy associated with rotation about the two axes perpendicular to the axis through the atoms. Calling the latter the $x$-axis, this adds 2 degrees of freedom: $\omega_y$ and $\omega_z$, the angular velocities about $y$ and $z$. (There is no rotational energy about the $x$-axis because all of the mass lies along that axis.) A diatomic molecule may also vibrate along its $x$-axis, adding the 2 degrees of freedom of a one-dimensional oscillator: $v_{\text{relative}}$ and $x_{\text{relative}}$. According to classical statistical mechanics, then, a gas of diatomic molecules, with 7 degrees of freedom, should have $E_{\text{internal per mole}} = \frac{7}{2}RT$ and $C_v = \frac{7}{2}R$.[22]

Diatomic hydrogen's molar heat capacity is about $\frac{7}{2}R$—but only above several thousand degrees Kelvin! At room temperature, its heat capacity is $\frac{5}{2}R$ (Figure 8.24); it does not appear to store energy in vibrational motion. The reason is that, so far as quantum oscillator states are concerned, room temperature is very cold; $k_{\text{B}}T$ is much less than $\hbar\omega_0$, the energy jump to the first excited state.[23] Consequently, if we wish to find the average oscillator energy, though the Boltzmann

[22]Our goal here is not to identify all possible ways gas molecules might store energy, but rather to show how quantum statistical mechanics governs energy storage among *representative* ways. We go no further than two-atom molecules.

[23]Since both rotation and vibration involve two degrees of freedom, knowing only that the heat capacity is low by $\frac{2}{2}R$ is not enough to distinguish them. But a diatomic molecule's quantum vibrational and rotational levels have distinctive natures (see Section 9.3), and theory and experiment agree that it is hydrogen's *vibrational* excited states that are "out of reach" at room temperature.

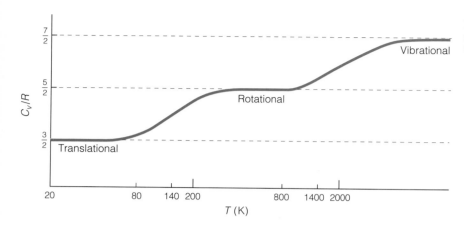

**Figure 8.24** Variation of hydrogen's molar heat capacity with temperature.

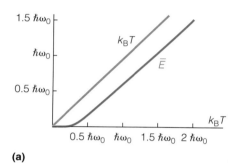

(a)

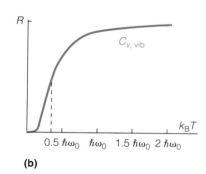

(b)

**Figure 8.25** Average energy and molar heat capacity of a system of oscillators at temperatures where $k_B T \sim \hbar\omega_0$.

distribution is correct for a diffuse gas of molecules, we may not use *integration*. We integrated to obtain (8-24), $\overline{E} = k_B T$, and it clearly won't do: $E_{per\ mole}$ would be $RT$, and $C_v$ would be the *constant R;* it would not suddenly "kick in" at 2000 K. Equation (8-13), obtained via summation, is the correct route to $C_v$.

$$\overline{E} = \frac{\hbar\omega_0}{e^{\hbar\omega_0/k_B T} - 1}$$

As shown in Figure 8.25(a) (and noted in Section 8.4), at high temperature, $k_B T \gg \hbar\omega_0$, this vibrational energy approaches the expected linear increase with temperature—$k_B T$. This agrees with the classical equipartition theorem: 2 degrees of freedom gives $2 \times \frac{1}{2}k_B T$. But at low temperature, $k_B T \ll \hbar\omega_0$, the average energy increases only very slowly, because few of the molecules would be energetic enough to occupy any state other than the ground state.

Let us now see how the heat capacity is affected by vibrational quantization. For a mole of oscillators, the internal energy due solely to vibrational motion would be

$$E_{internal\ per\ mole,\ vib.} = N_A \frac{\hbar\omega_0}{e^{\hbar\omega_0/k_B T} - 1}$$

Differentiation with respect to $T$ then gives the portion of the heat capacity due to vibrational motion.

$$C_{v,\text{vib.}} = N_A k_B \frac{(\hbar\omega_0)^2 e^{\hbar\omega_0/k_B T}}{(k_B T)^2 (e^{\hbar\omega_0/k_B T} - 1)^2} = R \frac{(\hbar\omega_0)^2 e^{\hbar\omega_0/k_B T}}{(k_B T)^2 (e^{\hbar\omega_0/k_B T} - 1)^2}$$

Figure 8.25(b), essentially the slope of Figure 8.25(a), plots $C_{v,\text{vib.}}$. Its resemblance to the vibrational jump in Figure 8.24 is conspicuous. From zero, it jumps to practically $R$ at $k_B T \cong 0.3\hbar\omega_0$. At around this temperature, translational kinetic energies should be great enough to begin an interchange with vibrational energies. Given that $\omega_0$ is $8.28 \times 10^{14}$ rad/s for the hydrogen molecule (see Example 4.5), the corresponding temperature works out to be 1900 K, in good agreement with Figure 8.24.

The "earlier" jump in hydrogen's heat capacity is due to the beginnings of energy exchange with *rotational* quantum states. The basic idea is the same: Unless the temperature is at least comparable to the jump to the first excited *rotational* state, rotational states cannot participate in energy storage. Diatomic molecule rotational energy levels are discussed in Chapter 9. Here we simply note that it is because hydrogen's rotational level spacing is much smaller than its vibrational spacing that the rotational jump occurs at a much lower temperature.

In summary, we see that the spacing between quantum energy levels must be taken into account when applying the equipartion theorem to predict heat capacity. At low temperatures, certain degrees of freedom effectively store no energy.

## Solids

In a solid, atoms cannot move freely. They oscillate about fixed locations and do not have rotational degrees of freedom. We may view them as being held in place by "springs" (interatomic forces) in the three spatial dimensions. Since there are 2 degrees of freedom for a one-dimensional oscillator, we should expect an atom oscillating in three dimensions to have 6 degrees of freedom: $x$, $y$, $z$, $v_x$, $v_y$, and $v_z$. Accordingly, the internal energy per mole should be $\frac{6}{2}RT$ and the molar heat capacity $3R$. In a metal, however, the conduction electrons do not oscillate with the individual atoms, but move freely about the material. They constitute a gas co-existing with the solid. With no vibrational or rotational energies (being "monatomic"), they should add a factor of $\frac{3}{2}R$ to the molar heat capacity of the solid. Classically, then, the molar heat capacities of solids should obey:

Insulator:  $C_v = 3R$

(Classical)

Conductor:  $C_v = 4.5R$

By now, we should not be surprised that classical arguments break down at low temperatures. In particular, even at room temperature the molar heat capacities of conductors do not differ greatly from those of insulators—and we already have the reason. At $T = 0$, electrons in a metal fill the lowest-energy states possible up to a maximum $E_F$. But at ordinary temperatures, $k_B T$ is so much less than $E_F$ that it might as well be zero (cf. Example 8.5). Any temperature *change* within the ordinary range would still have the electrons packed into essentially the same energy levels, and if the electrons' average energy changes very little with temperature,

**Table 8.5**

| Solid | $c_v$ (J/g·K) | $m$ (g/mol) | $C_v$ (J/mol·K) |
|-------|---------------|-------------|-----------------|
| Li | 3.51 | 6.9 | 24.2 |
| Al | 0.90 | 27.0 | 24.3 |
| P | 0.77 | 31.0 | 23.8 |
| S | 0.71 | 32.1 | 22.6 |
| Fe | 0.45 | 55.8 | 25.1 |
| Ag | 0.23 | 107.9 | 24.8 |
| | | | $3R = 24.9$ |

their heat capacity is very small.[24] Accepting, then, that the electrons should not be much of a factor, Table 8.5 shows just how well classical statistical mechanics does predict the heat capacity *due to motion of the atoms* for certain crystalline solids at near room temperature (298 K). For both insulators and conductors, the heat capacities are very nearly $3R$. Nevertheless, even this prediction fails for many solids—diamond, for instance, as we shall see in an example—if the *atoms in the lattice* also find room temperature too cold.

To see how the heat capacity of the atoms in a solid behaves at low temperature, we might be tempted to apply the same analysis as for vibrating gas molecules, though without allowance for free translation or rotation. But the situation is not so simple. Gas molecules may usually be treated as "noninteracting." Although they do of course exchange energy, they do so only one collision at a time. They are essentially free and separate particles whose vibrational motions are independent of one another. In a solid, however, atomic vibrations are strongly *interdependent*. The motion of one affects all those around it, the motions of which affect all those around them, and so on, in a wavelike fashion. How may we analyze the *collective* motions of the atoms?

There are two different but equivalent approaches, the same two used to analyze electromagnetic radiation bound in some region: the "standing-wave approach," used to derive Planck's spectral energy density in Appendix C, and the "boson gas approach" of Section 8.8. There are great similarities between vibrations bound in a solid object and electromagnetic "vibrations" bound to a cavity. In either case, the standing-wave approach treats the oscillations as standing waves of different energies. Standing waves are distinguishable by their differences in numbers of antinodes, and they therefore obey the Boltzmann distribution. The boson gas approach, on the other hand, treats the energy as residing in a gas of indistinguishable boson particles. In the case of electromagnetic radiation, the particles are, of course, photons. We now assert that mechanical waves in a solid may also be treated as a gas of particles, known as **phonons.** A phonon—a quantum of vibration—may travel from one point in a solid to another, carrying an energy *hf*.

Phonon (boson): The quantum of vibrational energy; $E = hf$

---

[24]It increases from zero linearly with $T$, though. See Exercise 50.

It does not carry spin; the phonon is a spinless boson. The reader may be suspicious about phonons; are they *real* particles? The answer is no, not in the same sense as photons or electrons. They have no existence apart from the medium of propagation. However, the complete agreement between the two views, wave and particle, is but one of many evidences that *vibrations in a solid may be treated as comprising an integral number of discrete phonons rather than diffuse waves.*

## The Debye Model

Whichever approach is taken, the steps to finding the heat capacity of a crystalline solid are at many points the same and are generally referred to as the **Debye model.** Because it most closely parallels the treatment of photons in Section 8.8, let us use the "boson gas approach." We begin by finding the internal energy of a phonon gas, as we did for a photon gas in equation (8-46). To adapt this equation to phonons, it need only be modified to account for three principal differences.

*Maximum Particle Energy:* While the photons in a cavity may be of arbitrarily short wavelength, there is an absolute lower limit on the wavelength of vibrations in a crystal. Figure 8.26 illustrates the reason. In a vibrational wave, adjacent atoms oscillate at the same frequency but are at different points in their cycles. A long wavelength is the result of a small phase difference between adjacent atoms, as shown in Figure 8.26(a). Figure 8.26(b) shows a short wavelength, in which adjacent atoms are one-half cycle and therefore one-half wavelength apart. Let us refer to this as $\lambda_{min}$. As suggested in Figure 8.26(c), it may be shown that for any

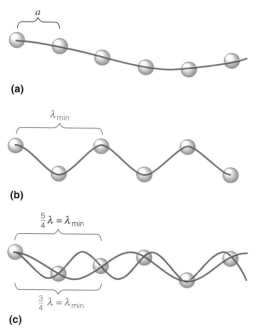

(a)

(b)

(c)

**Figure 8.26** In a crystal, only vibrational wavelengths greater than $2a$ make sense.

wavelength shorter than $\lambda_{min}$ there is a wavelength longer than $\lambda_{min}$ for which the atoms would be at the same points in their cycles. (The proof is left as an exercise.) Thus, it makes no sense to speak of a wave for which $\frac{1}{2}\lambda$ is less than the spacing between adjacent atoms, since it is physically identical to one for which $\frac{1}{2}\lambda$ is greater than the spacing; the sine waves coincide *at all discrete atoms*.[25] Accordingly, the minimum wavelength is 2*a*, twice the atomic spacing. But a minimum wavelength implies a maximum frequency and so a maximum phonon energy $E_{max}$. The integration in (8-46) must be terminated at $E_{max}$.

*Propagation Speed:* All photons move at *c* in a vacuum. Vibrational waves in a solid, however, not only do not travel at *c;* they do not all travel at the *same* speed. (The variation in $v = f\lambda = \omega/k$ is described by a dispersion relation, characteristic of the medium, that gives $\omega$ as a function of *k*. See Section 5.3.) As an approximation, we assume that they *do* move at the same speed, the speed of sound $v_s$ characteristic of the material. We must replace *c* by $v_s$.

*Longitudinal Oscillation:* The density of states (8-45), obtained in Appendix C, includes a factor of 2 for light's two independent (transverse) polarizations. In a solid, there is energy storage in *longitudinal* vibrations as well as two polarizations of transverse vibration. Assuming equal speeds for transverse and longitudinal waves, the net effect is simply to multiply (8-45) by $\frac{3}{2}$.

Taking the three factors into account, equation (8-46) becomes

$$E_{internal} = \int_0^{E_{max}} E \frac{1}{e^{E/k_B T} - 1} \frac{12\pi V}{h^3 v_s^3} E^2 \, dE \tag{8-51}$$

Still we have yet to explicitly relate $E_{max}$ to the minimum wavelength 2*a*.

In one dimension, atomic spacing *a* is length divided by number, *L/N*, and the minimum phonon wavelength would thus be 2*L/N*. In three dimensions, determining the minimum wavelength is somewhat more involved. Let us consider a crystalline solid in which atoms are found at the corners of cubes, as in Figure 8.27. The number of atoms along one side is not *N* but $N^{1/3}$. Therefore, the spacing between atoms is $L/N^{1/3}$ and the minimum wavelength "along one axis" would be $2L/N^{1/3}$. Wavelength is not a vector, however; it doesn't have components. But intimately related to it is something that *is* a vector: wave number. (Remember: $\mathbf{p} = \hbar\mathbf{k}$ is a vector form of $p = h/\lambda$.) Using $k = 2\pi/\lambda$, the corresponding *maximum* wave-number component $k_x$ would be $\pi N^{1/3}/L$. In three dimensions, then, the wave-number vector may point anywhere within a cube in "wave-number space" with one corner at zero and of side length $\pi N^{1/3}/L$ (Figure 8.28). Its maximum magnitude would obviously depend on its direction, greatest when pointing toward the far corner. However, by finding a spherical quadrant that encloses the same "volume," we obtain a reasonable average maximum magnitude: The "radius" of the sphere would be $k_{max}$. Equating the "volumes" of the cube and spherical quadrant,

$$\left(\frac{\pi N^{1/3}}{L}\right)^3 = \frac{1}{8}\left(\frac{4}{3}\pi k_{max}^3\right) \quad \Rightarrow \quad k_{max} = \left(\frac{6\pi^2 N}{L^3}\right)^{1/3} = \left(\frac{6\pi^2 N}{V}\right)^{1/3}$$

[25] Importantly, since the sine waves do not coincide at all points *in space*, the argument applies only in cases where discrete points alone are "real." In particular, since electromagnetic waves are not mechanical and do not propagate via the motion of a discrete set of atoms, there is no similar restriction for electromagnetic radiation in a vacuum. We might say that the spacing of discrete points is zero, and so is the minimum wavelength.

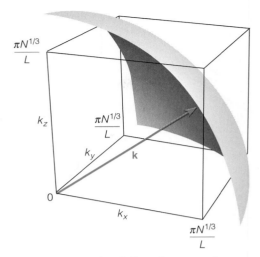

Figure 8.27 Atoms in a crystalline solid.

Figure 8.28 A cube of allowed wave numbers and a spherical quadrant of equal "volume."

We now need only relate $k$ to $E$. For a simple plane wave, $k = 2\pi/\lambda = 2\pi f/v$, so that $E = hf = hvk/2\pi$. Thus, with our assumption that all waves move at the same speed $v_s$, we have

$$E_{max} = \frac{hv_s k_{max}}{2\pi} = hv_s \left(\frac{3N}{4\pi V}\right)^{1/3}$$

The following definition will significantly streamline our work:

$$T_D \equiv \frac{E_{max}}{k_B} = \frac{hv_s}{k_B}\left(\frac{3N}{4\pi V}\right)^{1/3} \tag{8-52}$$

Being an energy divided by the Boltzmann constant (in J/K), $T_D$ has units of temperature. It is known as the **Debye temperature,** and it depends on properties of the given material (i.e., $N/V$ and $v_s$). Equation (8-51) now becomes

$$E_{internal} = \int_0^{k_B T_D} E \frac{1}{e^{E/k_B T} - 1} \frac{9N}{(k_B T_D)^3} E^2 \, dE \tag{8-53}$$

We simplify this further by redefining the dummy variable of integration:

$$E/k_B T \equiv x$$

Thus,

$$E_{internal} = 9k_B N \frac{T^4}{T_D^3} \int_0^{T_D/T} \frac{x^3}{e^x - 1} \, dx$$

Finally, given that $N = N_A$ and using $k_B N_A \equiv R$, we obtain

$$E_{internal\ per\ mole} = 9R \frac{T^4}{T_D^3} \int_0^{T_D/T} \frac{x^3}{e^x - 1} \, dx \tag{8-54}$$

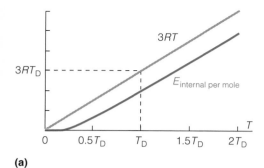

**(a)**

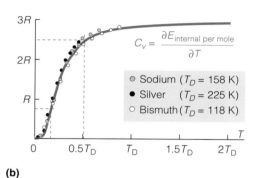

**(b)**

**Figure 8.29** Debye model average energy and molar heat capacity, and actual heat capacities of solids at low temperature.

Unfortunately, the integral cannot be carried out in closed form. Figure 8.29(a) shows the result of a numerical calculation. As expected, the molar internal energy approaches the classical $3RT$ at high temperature, but we also see that "high" temperature for a solid is judged relative to its Debye temperature. This makes sense, because $T_D$ is proportional to $v_s$, and speeds of sound increase with the strength of interatomic forces; stronger "springs" imply larger jumps between quantum levels. Figure 8.29(b) shows the molar heat capacity. At high temperature, it approaches the classical $3R$, while at low temperature, it falls to zero (as $T^3$—see Exercise 49). Plotted with the theoretically predicted heat capacity are measured heat capacities of several crystalline solids of different Debye temperatures. Even given the approximations made (e.g., a uniform speed for longitudinal and transverse waves, geometry-dependent averaging), agreement is excellent and greatly strengthens the case for treating the vibrational energy in a crystalline solid as a phonon gas.[26]

**Example 8.7**

The molar heat capacity of diamond is 6.0 J/mol·K $\cong 0.73R$ at 293 K. (a) According to the Debye model, at what temperature would the molar heat capacity be $2.5R$? (b) What is the maximum ("cutoff") phonon frequency of diamond?

[26]The theory fails for "amorphous" solids such as glasses, in which the atoms are *not* arranged in a regular pattern. Liquids are similarly difficult to analyze.

Solution

(a) From Figure 8.29(b), we see that a heat capacity of $0.73R$ occurs at around $T = 0.16T_D$. Therefore,

$$T_D \cong \frac{293 \text{ K}}{0.16} = 1830 \text{ K}$$

A heat capacity of $2.5R$ occurs around $0.51T_D = 0.51(1830) \cong 930$ K, in good agreement with experiment.

(b) Using (8-52), the maximum phonon energy is

$$E_{max} = k_B T_D = (1.38 \times 10^{-23} \text{ J/K})(1830 \text{ K}) = 2.5 \times 10^{-20} \text{ J}$$

Thus,

$$f_{max} = \frac{E_{max}}{h} = 3.8 \times 10^{13} \text{ Hz}$$

## Progress and Applications

■ Liquid helium and its superfluid properties have been studied for decades. There are many examples of even numbers of fermions being stuck together well enough that, so far as their interactions with each other go, they behave as bosons; with an even number of spin-$\frac{1}{2}$ particles, the total angular momentum quantum number would have to be an integer. Helium-4 (two electrons, two protons, two neutrons) is a good example. Being a liquid, however, with strong interparticle forces, it at best only approximates a "condensed" Bose–Einstein gas—nearly all particles in the ground state. Only fairly recently has a true **Bose–Einstein condensate** (BEC) been achieved. The first success was in a dilute gas of several thousand rubidium-87 atoms (rubidium-87 has 37 electrons, 37 protons, and 50 neutrons, an even number) in which laser cooling and evaporation combined to lower the temperature to less than 0.1 $\mu$K (M. H. Anderson, et al., *Science*, July 14, 1995, pp. 198–201). The achievement is illustrated in Figure 8.30 by a computer-generated plot of particle number versus velocity; the unusually prominent bump near zero gives clear evidence of

**Figure 8.30** Number of rubidium atoms, plotted vertically, versus velocity in two dimensions before (left), near (middle), and at achievement of an essentially pure BEC.

the occurrence. Bose–Einstein condensation has since been achieved in lithium and sodium. In all the experiments, the atoms were held in a magnetic trap, although each had its own clever variation.

■ Long a matter of curiosity, helium-3 is the subject of much study nowadays, not only for its own unique properties, but also because it may shed light on other areas of modern physics not fully understood. As we have noted, helium-4 is a boson and acts as we might expect of a boson: At low temperature, it tends toward Bose–Einstein condensation and superfluid behavior. Helium-3, with two electrons, two protons, and *only one* neutron, is a fermion—but it also becomes a superfluid! The question is how. Much evidence now suggests that in helium-3 at low temperature *pairs of atoms* may interact in such a way as to behave as a collection of bosons. If true, they should exhibit Bose–Einstein condensation, indeed becoming a superfluid at low temperature. Still, while it is easy to claim that atom pairing is how helium-3 becomes a superfluid, justifying the claim is quite another matter. The work continues. An important reason for the interest is that superconductivity, which shares many remarkable traits with superfluidity, occurs by *electron* pairing. The pairing is well understood in some superconductors, but not in the newer and potentially more useful "high-$T_c$" superconductors (see Chapter 9). Unraveling the pairing mechanism in helium-3 may help.

■ A host of alternatives to the standard laser discussed in the chapter have been investigated over the years, and many have been put into use. Most lasers operate in gaseous or solid media containing countless excited atoms. As they jump between energy levels, these atoms add their photons to the beam. One drawback is intensity fluctuations; a system of numerous interacting atoms inevitably results in a fluctuating number of photons. An experimental device featuring greater control is the **single-atom laser,** in which a beam of excited barium atoms passes through a resonant cavity on average *one atom at a time* (K. An, et al., *Physical Review Letters,* December 19, 1994, p. 3375). When an atom enters (Figure 8.31), the light trapped between the mirrors induces the atom's excited electron to drop to a lower level, adding a photon to the beam "on schedule." The single-atom laser promises to add much to our understanding of the fundamental interaction of atoms with light. Another type of laser

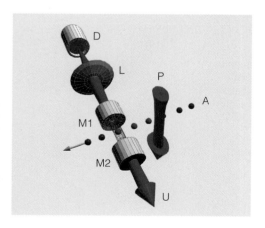

**Figure 8.31** Schematic diagram of the single-atom laser. Excitation field P excites barium atoms A, which deexcite in a resonant cavity composed of mirrors M1 and M2. Lens L focuses the light into detector D.

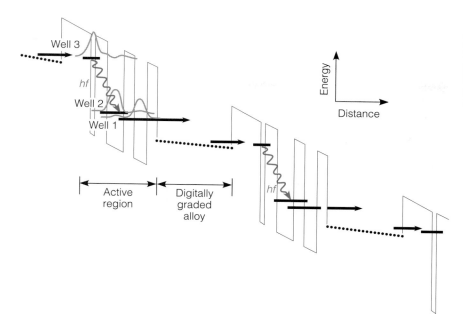

**Figure 8.32** The quantum cascade laser's staircase of wells. An electron tunnels into well 3 and emits a photon as it drops to well 2, repeating the process many times. An external electric field gives the energy levels their overall slope.

being developed is the **quantum cascade laser.** Here the radiation is produced by electrons "cascading" through, and tunneling between, a "downward staircase" of quantum wells embedded in a semiconductor, as depicted in Figure 8.32. Since the energy levels involved are not those of the atoms in the bulk material, but rather of the embedded quantum wells, the wavelength can be chosen simply by making the wells of the proper size. Besides covering a wide range of wavelengths, its potential strengths include portability and relatively high power. A different kind of laser actually in wide use is the **free-electron laser** (FEL), in which the "lasing medium" consists of free electrons sent through a series of magnets. The magnets cause the electrons to "wiggle" along their paths. All accelerating (i.e., wiggling) charges emit electromagnetic radiation, and radiation from the FEL's wiggled electrons is coherent and very powerful. Free-electron lasers are generally not "table-top" machines, but large devices relying on the beam from a high-energy electron accelerator; several universities and other research organizations have FEL facilities. Their high power and easily tunable frequency make them attractive for applications in biology, medicine, nuclear fusion, materials science, and many other areas.

■ While in this chapter we have concentrated on thermodynamic systems where the behavior is predictable, the scientific community has only comparatively recently recognized that there are many systems exhibiting a special kind of unpredictability: **chaos.** It arises in nearly every physical and biological science, in many seemingly unrelated situations. The unifying thread is that these situations all involve a phenomenon governed by a differential equation that is *nonlinear.* The differential equations we encounter in this text are linear; they depend on a function and its derivatives to only the first power. Any dependence on the function or its derivative to any other power makes a differential equation nonlinear, and quite often there is no mathematical technique able to solve it. Consequently, we must fall back on numerical approximations. But then we run into another quirk of nonlinear differential equations: Extremely small changes in the conditions at one point may result in extremely large differences at a later point. (The atmosphere's behavior is governed by nonlinear fluid dynamics, and it is often mused that the weather at one place may hinge on the earlier flap of a butterfly's wing on the other side of the world.) Thus, what actually happens may appear completely unpredictable.

A common example is a nonlinear (non-Hooke's law) oscillator driven by a periodic driving force: For certain values of the force's amplitude, the oscillator's period is that of the driving force. However, if the amplitude is changed ever so slightly, although the driving *period* has not changed, the *system* oscillates at a period larger by a factor of 2; further change in the amplitude may later reveal another abrupt doubling of the system's period. At some point, a small change may produce *aperiodic* behavior! It may seem random; it may seem like noise; but it isn't, for it is governed by an underlying mathematical relationship (however intractable).

Such deterministic but seemingly unpredictable behavior is what we call chaos, and is found in mechanical oscillators, electrical circuits, fluid flow, optics, chemistry, population systems, biological development, heart fibrillation, the weather, and on and on. Much work is now being done on controlling the onset of chaos in cases where it is unwanted, or extracting some kind of sense from it in cases where it is either uncontrollable or a natural characteristic of the system under study. Chaos is a huge and expanding field. We cannot hope to do it justice here. The interested reader is encouraged to consult the literature: It is a common topic in scientific journals, and many new books dedicated to chaos are published each year.

## Basic Equations

| Entropy | | Temperature | |
|---|---|---|---|
| $S \equiv k_{\mathrm{B}} \ln W$ | (8-2) | $\dfrac{\partial S}{\partial E} \equiv \dfrac{1}{T}$ | (8-3) |

**Boltzmann probability**

$$P(E_n) = A e^{-E_n/k_{\mathrm{B}}T} \qquad (8\text{-}10)$$

**Distributions**

$$N(E) = \begin{cases} \dfrac{1}{Be^{+E/k_{\mathrm{B}}T}} & \text{Boltzmann distribution} \quad \text{(Distinguishable)} & (8\text{-}29) \\[3ex] \dfrac{1}{Be^{+E/k_{\mathrm{B}}T} - 1} & \text{Bose–Einstein distribution} \quad \text{(Bosons)} & (8\text{-}30) \\[3ex] \dfrac{1}{Be^{+E/k_{\mathrm{B}}T} + 1} & \text{Fermi–Dirac distribution} \quad \text{(Fermions)} & (8\text{-}31) \end{cases}$$

Density of states, defined

$$D(E) \equiv \frac{dn}{dE} \tag{8-19}$$

Density of states, examples

Oscillators:           Gas (massive):                Gas (photon):

$$\frac{2s + 1}{\hbar\omega_0}$$        $$\frac{(2s + 1)m^{3/2}V}{\pi^2\hbar^3\sqrt{2}}E^{1/2}$$        $$\frac{8\pi V}{h^3 c^3}E^2$$

$$\tag{8-40}$$           $$\tag{8-45}$$

Number of particles in energy range $dE$

$$dN = N(E)D(E)\, dE$$

Average

$$\overline{Q} = \frac{\sum\limits_n Q_n N(E_n)}{N} = \frac{\sum\limits_n Q_n N(E_n)}{\sum\limits_n N(E_n)} \tag{8-17}$$

$$\overline{Q} = \frac{\int Q(E)N(E)D(E)\, dE}{N} = \frac{\int Q(E)N(E)D(E)\, dE}{\int N(E)D(E)\, dE} \tag{8-25}$$

Classical average

$$\overline{Q} = \frac{\int Q(\mathbf{r},\ \mathbf{v})N(E(\mathbf{r},\ \mathbf{v}))_{\text{Boltz}}\, d^3r\, d^3v}{\int N(E(\mathbf{r},\ \mathbf{v}))_{\text{Boltz}}\, d^3r\, d^3v} \tag{8-26}$$

# Summary

Statistical mechanics is the area of physics in which statistical considerations are used to predict the average properties and behaviors of thermodynamic systems. A thermodynamic system is one in which the number of particles is so large that deviations from statistically predicted average behavior are very small.

In a thermodynamic system in equilibrium, the number of particles in a given individual-particle quantum state—the occupation number $N(E)$—obeys a distribution that is independent of the system (e.g., harmonic oscillator versus gas) but does depend on the kind of particle. Distinguishable particles obey the Boltzmann distribution, bosons the Bose–Einstein, and fermions the Fermi–Dirac.

$$N(E)_{\text{Boltz}} = \frac{1}{Be^{+E/k_{\text{B}}T}} \quad N(E)_{\text{B-E}} = \frac{1}{Be^{+E/k_{\text{B}}T} - 1} \quad N(E)_{\text{F-D}} = \frac{1}{Be^{+E/k_{\text{B}}T} + 1}$$

Distinguishable                 Bosons                          Fermions

Occupation number decreases with increasing energy in all three cases, and the quantum distributions converge to the "classical" Boltzmann distribution whenever quantum indistinguishability may be ignored, whenever the occupation number is much less than unity. The quantum distributions differ most at low temperature and/or high density. The Bose–Einstein tends more toward having all particles congregating in the lowest-energy quantum state, while the Fermi–Dirac tends toward an occupation number of unity (one particle per state) up to the so-called Fermi energy, at which the occupation number drops abruptly to zero.

The average of a property $Q$ (such as energy) is calculated via

$$\overline{Q} = \frac{\sum_n Q_n N(E_n)}{\sum_n N(E_n)}$$

where $n$ represents all quantum numbers necessary to specify the individual-particle state. If the spacing between these states is small, the sum may be replaced by an integral:

$$\overline{Q} = \frac{\int Q(E)N(E)D(E)\, dE}{\int N(E)D(E)\, dE}$$

The quantity $D(E)$ is known as the density of states and is the number of quantum states in the energy range $dE$. Its form depends on the system to which it is applied.

Among the things quantum statistical mechanics explains are the energies of conduction electrons in a metal, a fermion gas; the unusual properties of liquid helium, a boson gas; and the relationship between temperature, energy, and frequency in the electromagnetic gas of photons in equilibrium with matter.

The laser is a device producing coherent light via stimulated emission. A non-equilibrium population inversion is established in the light-producing medium, in which the number of electrons in a higher-energy state is greater than that in a lower-energy state. If the higher state is metastable, electrons promoted to it by external optical pumping or electric discharge are prevented from quickly returning and reestablishing equilibrium. A passing photon whose energy equals the difference between the higher and lower states then precipitates deexcitation (stimulated emission) and the newly produced photon is coherent with the first. With a population inversion, the process quickly multiplies—one producing two, two producing four, and so on—and a powerful coherent beam results.

## EXERCISES

### Section 8.1

1. Consider the two-sided room. (a) Which is more likely to have an imbalance of five particles (i.e., $N_R = \frac{1}{2}N + 5$): a room with $N = 20$, or with $N = 60$? (*Note:* The total number of ways of distributing particles, the sum of $W_{N_R}^N$ from 0 to $N$, is $2^N$.) (b) Which is more likely to have an imbalance of 5% (i.e., $N_R = \frac{1}{2}N + 0.05N$)? (c) An average-size room is quite likely to have a trillion more air molecules on one side than the other. Why may we say that precisely half will be on each side?

2. A two-sided room contains six particles, $a$, $b$, $c$, $d$, $e$ and $f$, with two on the left, four on the right. (a) Describe the macrostate. (b) Identify the possible microstates.

(*Note:* With only six particles, this is not a thermody-namic system, but the general idea still applies, and the number of combinations is tractable.)

3.  Consider a room divided by imaginary lines into three equal parts. Sketch a two-axis plot of the number of ways versus $N_{left}$ and $N_{right}$ for the case $N = 10^{23}$. ($N_{middle}$ is not independent, being of course $N - N_{right} - N_{left}$.) Represent the number of ways by density of shading.

## Section 8.2

4.  The diagram shows two systems that may exchange both thermal and mechanical energy, via a movable heat-conducting partition. Since both $E$ and $V$ may change, we consider the entropy of each system to be a function of both: $S(E, V)$. Considering the exchange of thermal energy only, we argued in Section 8.2 that it was reason-able to equate $1/T$ to $\partial S/\partial E$. In the more general case, $P/T$ is also equated to something. (a) Why should pres-sure come into play, and to what might $P/T$ be equated? (*Note:* Check to see if the units make sense.) (b) Given this relationship, show that $dS = dQ/T$. (Remember the first law of thermodynamics.)

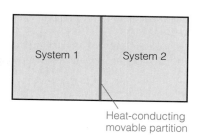

Heat-conducting
movable partition

5.  A "cold" object, $T_1 = 300$ K, is briefly put in contact with a "hot" object, $T_2 = 400$ K, and 60 J of heat flows from the hot object to the cold one. The objects are then separated, their temperatures having changed negli-gibly, due to their large sizes. (a) What are the changes in entropy of each object and the system as a whole? (b) Knowing only that these objects are in contact and at the given temperatures, what is the ratio of the proba-bilities of their being found in the second (final) state to that of their being found in the first (initial) state?

## Section 8.3

6.  For large numbers, the factorial function can be re-placed using Stirling's approximation:

$$N! \cong \sqrt{2\pi N}\, N^N e^{-N}$$

    (a) What is the percent error in Sterling's

approximation for $N = 10$ and for $N = 60$ (neither of which is really "large")?

(b)  The number of ways of obtaining $\frac{1}{3}N$ particles on one side of a room and $\frac{2}{3}N$ on the other is $N!/(\frac{1}{3}N)!\,(\frac{2}{3}N)!$. The number of ways of obtain-ing an equal distribution is $N!/(\frac{1}{2}N)!\,(\frac{1}{2}N)!$. Using Stirling's approximation, show that the ratio, equal to unequal, is approximately

$$\frac{\sqrt{8}}{3}\left(\frac{32}{27}\right)^{N/3}$$

## Section 8.4

7.  Four distinguishable harmonic oscillators $a$, $b$, $c$, and $d$ may exchange energy. The energies allowed particle $a$ are $E_a = n_a \hbar \omega_0$, particle $b$ are $E_b = n_b \hbar \omega_0$, and so on. Consider an overall state (macrostate) in which the to-tal energy is $3\hbar\omega_0$. One possible microstate would have particles $a$, $b$, and $c$ in their $n = 0$ states and particle $d$ in its $n = 3$ state; that is, $(n_a, n_b, n_c, n_d) = (0, 0, 0, 3)$. (a) List *all* possible microstates. (b) What is the probabil-ity that a given particle will be in its $n = 0$ state? (c) An-swer part (b) for all other possible values of $n$. (d) Plot the probability versus $n$.

8.  In a large system of distinguishable harmonic oscilla-tors, how high does the temperature have to be for the probability of occupying the ground state to be less than one-half?

9.  In a large system of distinguishable harmonic oscilla-tors, how high does the temperature have to be for the probable number of particles occupying the ground state to be less than one?

10. Show that for the Boltzmann distribution applied to a system of simple harmonic oscillators, the condition of unit probability

$$1 = \sum P(E_{n_1}) = \sum_{n_1=0}^{\infty} A e^{-n_1 \hbar \omega_0/k_B T}$$

    implies that $A = 1 - e^{-\hbar \omega_0/k_B T}$. (*Hint:* With $e^{-\hbar\omega_0/k_B T} \equiv x$, the sum is $\sum_{n=0}^{\infty} x^n$.)

11. Derive equation (8-13). First make the definition $\beta \equiv 1/k_B T$ then $x \equiv e^{-\beta\hbar\omega_0}$. Thus,

$$\frac{dx^n}{d\beta} = \frac{d}{d\beta} e^{-\beta n\hbar\omega_0} = -n\hbar\omega_0 x^n$$

    The sum preceding (8-13) then becomes

$$\overline{E} = (1 - x)\sum_{n=0}^{\infty} -\frac{dx^n}{d\beta} = -(1 - x)\frac{d}{d\beta}\sum_{n=0}^{\infty} x^n$$

12. By setting the Boltzmann average energy

$$\overline{E} = \frac{\hbar\omega_0}{e^{\hbar\omega_0/k_BT} - 1}$$

equal to the exact oscillator average energy $M\hbar\omega_0/N$, find the relationship between $T$ and $M/N$. Use this to obtain equation (8-14) from (8-12).

13. Show that in the limit of large numbers the exact probability of equation (8-8) becomes the Boltzmann probability of (8-14). (Use $K!/(K - k)! \cong K^k$, which holds when $k \ll K$.)

14. In Example 8.2, a ratio is obtained of number of particles expected in the $n = 2$ state to that in the ground state. Rather than the $n = 2$ state, consider arbitrary $n$.
    (a) Show that the ratio is

    $$\frac{\text{number with energy } E_n}{\text{number with energy } E_1} = n^2 e^{-13.6 \text{ eV}(1 - n^{-2})/k_BT}$$

    (Hydrogen atom energies are $E_n = -13.6 \text{ eV}/n^2$.)
    (b) What is the limit of this ratio as $n$ becomes very large? Can it exceed unity? If so, under what condition(s)?
    (c) In Example 8.2, we found that even at the temperature of the Sun's surface (6000 K) the ratio for $n = 2$ is only $10^{-8}$. For what value of $n$ would the ratio be 0.01?
    (d) Is it realistic that the number of atoms with high $n$ could be greater than the number with low $n$?

15. Consider a system of one-dimensional spinless particles in a box, somehow exchanging energy.
    (a) Show that

    $$D(E) = \frac{m^{1/2}L}{\hbar\pi\sqrt{2}} \frac{1}{E^{1/2}}$$

    (b) What is it about the one-dimensional infinite well energy levels that causes the number of states per unit energy to *decrease* with energy?

16. Obtain equation (8-24) from equations (8-21) and (8-23).

## Section 8.5

17. Show that the rms speed of a gas molecule, defined as $v_{rms} \equiv \sqrt{\overline{v^2}}$, is given by $\sqrt{3k_BT/m}$.

18. (a) Calculate the average speed of a gas molecule in a classical ideal gas. (b) What is the average velocity of a gas molecule?

19. (a) Using the Maxwell speed distribution, determine the most probable speed of a particle of mass $m$ in a gas at temperature $T$. (b) How does this compare to $v_{rms}$? Explain.

20. Determine the relative probability of a gas molecule being within a small range of speeds around $2v_{rms}$ to being in the same range of speeds around $v_{rms}$.

21. To obtain the Maxwell speed distribution, we assumed a uniform temperature, a kinetic-only energy of $E(\mathbf{r}, \mathbf{v}) = \frac{1}{2}m(v_x^2 + v_y^2 + v_z^2)$, and that we wished to find the average of an arbitrary function of $v$. Along the way we obtained a probability per unit speed, $P(v)$. (a) Assuming a uniform temperature, an energy of $E(\mathbf{r}, \mathbf{v}) = \frac{1}{2}m(v_x^2 + v_y^2 + v_z^2) + mgy$, and that we wish to find the average of an arbitrary function of $y$, obtain a probability per unit height, $P(y)$. (b) Assuming a temperature of 300 K, how much less is the density of the atmosphere's $N_2$ at an altitude of 800 m (about 3000 ft) than at sea level? (c) What of the $O_2$ in the atmosphere?

## Section 8.6

22. There is a simple argument, practically by inspection, that distributions (8-29), (8-30), and (8-31) should agree whenever occupation number is much less than unity. Provide the argument.

23. In the case of simple harmonic oscillators, with $\mathscr{E} \equiv N\hbar\omega_0/(2s + 1)$, show that the Bose–Einstein and Fermi–Dirac distributions converge to the Boltzmann in the limit $k_BT \gg \mathscr{E}$.

24. Obtain the expressions in Table 8.4 for the three distributions' $E = 0$ occupation numbers in the low-temperature limit $k_BT \ll \mathscr{E}$.

25. Using $N\hbar\omega_0/(2s + 1) \equiv \mathscr{E}$, $D(E) = (2s + 1)/\hbar\omega_0$ (both correct for all systems of harmonic oscillators), and

    $$N = \int_0^\infty N(E)D(E) \, dE$$

    obtain equations (8-33) and (8-34) from equations (8-30) and (8-31). [*Note:* $\int_0^\infty (Be^z \pm 1)^{-1} dz = \pm \ln(1 \pm 1/B)$.]

26. (a) From equations (8-34) and (8-36), obtain an expression for $E_F(T)$, the Fermi temperature for a collection of fermion oscillators. (b) Show that $E_F(0) = \mathscr{E}$. (c) Plot $E_F(T)$ versus $k_BT/\mathscr{E}$ from zero to $k_BT/\mathscr{E} = 1.5$. (d) By what percent does the Fermi energy drop from its maximum $T = 0$ value when $k_BT$ rises to 25% of $\mathscr{E}$?

## Section 8.7

27. Calculate the Fermi energy for copper, which has a density of $8.9 \times 10^3 \text{ kg/m}^3$ and one conduction electron per atom. Is room temperature "cold"?

28. Copper has one conduction electron per atom and a density of $8.9 \times 10^3 \text{ kg/m}^3$. By the criteria of equation (8-44), show that at room temperature (300 K) the conduction electron gas must be treated as a quantum gas of indistinguishable particles.

29. Obtain an order-of-magnitude value for the temperature at which helium might begin to exhibit quantum (superfluid) behavior. See (8-44). (Helium's specific gravity is about 0.12.)

30. Find the density of states $D(E)$ for a two-dimensional infinite well (ignoring spin), in which

$$E_{n_x, n_y} = (n_x^2 + n_y^2) \frac{\pi^2 \hbar^2}{2mL^2}$$

31. We investigate here whether quantum indistinguishability must be taken into account for gas molecules in the air. Assume a temperature of 300 K and a pressure of 1 atm. (a) Calculate the average separation between nitrogen molecules in the air (assume air is 80% nitrogen). (b) Calculate the wavelength of a typical air (nitrogen) molecule. (c) Should air behave classically, or as a quantum gas?

32. The **Fermi velocity** $v_F$ is defined by $E_F = \frac{1}{2}mv_F^2$, where $E_F$ is the Fermi energy. The Fermi energy for conduction electrons in sodium is 3.1 eV. (a) Calculate the Fermi velocity. (b) What would be the wavelength of an electron with this velocity? (c) If each sodium atom contributes one conduction electron to the electron gas and sodium atoms are spaced roughly 0.37 nm apart, is it necessary, by the criteria of equation (8-44), to treat the conduction electron gas as a *quantum* gas?

33. To obtain equation (8-43), we calculated $N$ as a function of $E_F$, assuming $T = 0$. Calculate $E$ as a function of $E_F$, and use this to show that the minimum energy of a gas of spin-$\frac{1}{2}$ fermions may be written

$$E = \frac{3}{10} \left( \frac{3\pi^2 \hbar^3}{m^{3/2}V} \right)^{2/3} N^{5/3}$$

34. Using a procedure similar to that used to obtain equation (8-43), show that at very low temperature, $T \cong 0$, the average energy of a conduction electron in a metal is $\frac{3}{5}E_F$.

35. This problem investigates what fraction of the available charge must be transferred from one conductor to another to produce a typical contact potential. (a) As a rough approximation, treat the conductors as 10 cm × 10 cm square plates 2 cm apart, a parallel-plate capacitor, so that $q = CV$ where $C = \epsilon_0 (0.01 \text{ m}^2/0.02 \text{ m})$. How much charge must be transferred from one plate to the other to produce a potential difference of 2 V? (b) Approximately what fraction would this be of the total number of conduction electrons in a 100-g piece of copper, which has one conduction electron per atom?

36. The maximum wavelength light that will eject electrons from metal 1 via the photoelectric effect is 410 nm. For metal 2, it is 280 nm. What will be the potential difference if these two metals are put in contact?

37. Derive equation (8-42). Start with the expression for $N$:

$$N = \int_0^\infty N(E)D(E)\, dE$$

Insert the quantum gas density of states and an expression for the distribution, using $\pm$ to distinguish the Bose–Einstein from the Fermi–Dirac. Then change variables: $E = y^2$. Factor $Be^{+y^2/k_B T}$ out of the denominator. In the integrand will be a factor

$$\left( 1 \mp \frac{1}{B} e^{-y^2/k_B T} \right)^{-1}$$

Using $(1 \mp \epsilon)^{-1} \cong 1 \pm \epsilon$, a sum of two integrals results, each of gaussian form. The integral thus becomes two terms in powers of $1/B$. Repeat the process, but instead finding an expression for $E_{total}$ in terms of $1/B$, using

$$E_{total} = \int_0^\infty E\, N(E)D(E)\, dE$$

Divide the expression for $E_{total}$ by the expression for $N$, both in terms of $1/B$. Now $1/B$ can be safely eliminated by using the "lowest-order" expression for $N$ in terms of $1/B$.

## Section 8.8

38. The temperature of our Sun's surface is about 6000 K. (a) At what wavelength is the spectral emission of the Sun a maximum? (Refer to Exercise 43.) (b) Is there something conspicuous about this wavelength?

39. At what wavelength does the human body emit the maximum electromagnetic radiation? Use Wien's law from Exercise 43 and assume a skin temperature of 70°F.

40. (a) Show that the number of photons per unit volume in a photon gas of temperature $T$ is approximately ($2 \times 10^7 \text{ K}^{-3}\text{m}^{-3})T^3$. (*Note:* $\int_0^\infty x^2(e^x - 1)^{-1}\, dx \cong 2.40$). (b) Show that the average photon energy in a cavity at temperature $T$ is given by: $\bar{E} \cong 2.7k_B T$.

41. (a) Obtain an expression for the heat capacity per unit volume, in J/K·m$^3$, for electromagnetic radiation. (b) What is its value at 300 K?

42. The electromagnetic intensity thermally radiated by a body of temperature $T$ is given by

$$I = \sigma T^4 \quad \text{where } \sigma = 5.67 \times 10^{-8} \text{ W/m}^2\cdot\text{K}^4$$

This is known as the **Stefan–Boltzmann law.** Show that this law follows from the energy per volume given in equation (8-47). (*Note:* Intensity, power per unit area, is the product of the energy per unit volume and distance per unit time. But since intensity is a flow in a given direction away from the blackbody, the correct speed is not $c$. For radiation moving uniformly in all directions, the average *component* of velocity in a given direction is $\frac{1}{4}c$.)

43. According to **Wien's law,** the wavelength $\lambda_{max}$ at which the thermal emission of electromagnetic energy from a body of temperature $T$ is maximum obeys

$$\lambda_{max} T = 2.898 \times 10^{-3} \text{ m·K}$$

Show that this law follows from equation (8-48). To do this, use $f = c/\lambda$ to reexpress (8-48) in terms of $\lambda$ rather than $f$, then obtain an expression that when solved would yield the wavelength at which this function is maximum. The transcendental equation cannot be solved exactly, so it is enough to show that $\lambda = (2.898 \times 10^{-3} \text{ m·K})/T$ solves it to a reasonable degree of precision.

### Section 8.10

44. At high temperature, the average energy of a classical one-dimensional oscillator is $k_B T$, and for an atom in a monatomic ideal gas it is $\frac{3}{2} k_B T$. Explain the difference using the equipartition theorem.

45. Classically, what would be the average energy of a particle in a system of particles free to move in the $x$-$y$ plane while rotating about the $z$-axis?

46. Classically, what would be the average energy per mole of a vibrating membrane: a two-dimensional sheet of particles, each bound to its four closest neighbors by four springs oriented along the two perpendicular axes?

47. Prove that for any sine function $\sin(kx + \phi)$ of wavelength shorter than $2a$, where $a$ is the atomic spacing, there is a sine function with a wavelength longer than $2a$ that has the same values at the points $x = a, 2a, 3a,$ and so on. (*Note:* It is probably easier to work with wave number rather than wavelength. We seek to show that for every wave number *greater* than $\pi/a$ there is an equivalent one less than $\pi/a$.)

48. The Debye temperature of copper is 345 K. (a) Estimate its molar heat capacity at 100 K using the plot in Figure 8.29(b). (b) Determine its corresponding specific heat and compare it with the experimental value of 0.254 J/g·K.

49. From equation (8-54), show that the specific heat (per mole) of a crystalline solid varies as $T^3$ for $T \ll T_D$.

50. The contribution of the *electrons* to the heat capacity of a metal is small but measurable. To determine it, we might try to calculate the total energy via the integral $\int E N(E)D(E) \, dE$. As we know, the integral is problematic for a 3D gas. But to get a rough idea, let us break it into pieces, so to speak—treating $E$ and $N(E)D(E) \, dE$ separately. Only those electrons near the top of the distribution gain energy, and to a good approximation they

gain $E \cong k_B T$. The other part, the product $N(E)_{F-D}D(E)$, is shown in the diagram for $T = 0$ and $T \neq 0$. Since the integral of $N(E)_{F-D}D(E) \, dE$ is both the area under the graph and the number, the *fraction* of electrons excited is the ratio of the area of the blue-shaded triangle to the total area under the graph. It is convenient to find the total area using the $T = 0$ graph, where $N(E)_{F-D}$ is simply unity up to $E_F$ and the density of states, proportional to $E^{1/2}$, may be written $D(E_F)E^{1/2}/E_F^{1/2}$.

(a) Show that the fraction excited is $\sim 3Nk_B T/8E_F$.

(b) Show that $C_v$ is a linear function of $T$ and obtain the proportionality constant in terms of $E_F$.

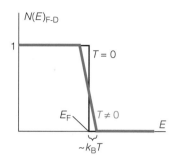

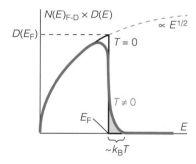

### General Exercises

51. Nuclear density is approximately $10^{17}$ kg/m$^3$. (a) Treating them as a gas of fermions bound together by the (nonelectrostatic) "strong internucleon attraction," calculate $E_F$ for the neutrons in lead-206 (82 protons and 124 neutrons). (b) Treating them the same way, what would $E_F$ be for the protons? (c) In fact, the energies of the most energetic neutrons and protons, those that should be at the Fermi energy, are essentially equal in lead-206. What has been left out of parts (a) or (b) that might account for this?

52. From elementary electrostatics, the total electrostatic

potential energy in a sphere of uniform charge $Q$ and radius $R$ is given by

$$U = \frac{3}{5} \frac{1}{4\pi\epsilon_0} \frac{Q^2}{R}$$

(a)  What would be the energy per charge in a lead nucleus if it could be treated as 82 protons distributed uniformly throughout a sphere of radius $7 \times 10^{-15}$ m?

(b)  How does this result fit with Exercise 51?

53.  When a star has nearly burned up its internal fuel, it may become a **white dwarf.** It is crushed under its own enormous gravitational forces to the point where the exclusion principle for the electrons becomes a factor. A smaller size would decrease the gravitational potential energy, but assuming the electrons to be packed into the lowest-energy states consistent with the exclusion principle, "squeezing" the potential well necessarily increases the energies of all the electrons (by shortening their wavelengths). If gravitation and the electron exclusion principle are the only factors, there is a minimum total energy and corresponding equilibrium radius.

(a)  Treat the electrons in a white dwarf as a quantum gas. The minimum energy allowed by the exclusion principle (see Exercise 33) is:

$$E_{electrons} = \frac{3}{10} \left( \frac{3\pi^2 \hbar^3}{m_e^{3/2} V} \right)^{2/3} N^{5/3}$$

Note that as the volume $V$ is decreased the energy does increase. For a neutral star, the number of electrons, $N$, equals the number of protons. Assuming that protons account for half the white dwarf's mass $M$ (neutrons accounting for the other half), show that the minimum electron energy may be written

$$E_{electrons} = \frac{9\hbar^2}{80 \, m_e} \left( \frac{3\pi^2 M^5}{m_p^5} \right)^{1/3} \frac{1}{R^2}$$

where $R$ is the star's radius.

(b)  The gravitational potential energy of a sphere of mass $M$ and radius $R$ is given by

$$E_{grav} = -\frac{3}{5} \frac{GM^2}{R}$$

Taking both factors into account, show that the minimum total energy occurs when

$$R = \frac{3\hbar^2}{8G} \left( \frac{3\pi^2}{m_e^3 m_p^5 M} \right)^{1/3}$$

(c)  Evaluate this radius for a star whose mass is equal to that of our Sun, $\sim 2 \times 10^{30}$ kg.

(d)  White dwarfs are comparable to the size of Earth. Does the value in part (c) agree?

54.  Exercise 53 discusses the energy balance in a white dwarf. The tendency to contract due to gravitational attraction is balanced by a kind of incompressibility of the electrons due to the exclusion principle.

(a)  Matter contains protons and neutrons, which are also fermions. Why do the electrons become a hindrance to compression before the protons and neutrons?

(b)  Stars several times our Sun's mass have sufficient gravitational potential energy to collapse further than a white dwarf; they are able to force essentially all their matter to become neutrons (the electrons and protons combine in the process). When they cool off, an energy balance is reached similar to that in the white dwarf, but with the neutrons filling the role of the incompressible fermions. The result is a **neutron star.** Repeat the process of Exercise 53, but assuming a body consisting solely of neutrons. Show that the equilibrium radius is given by

$$R = \frac{3\hbar^2}{2G} \left( \frac{3\pi^2}{2m_n^8 M} \right)^{1/3}$$

(c)  Show that the radius of a neutron star whose mass is twice that of our Sun is only about 10 km.

# Bonding: Molecules and Solids

I n this chapter, we study how atoms bond together to form molecules and solids, and what characteristics and behaviors arise when they do. Among the topics covered are molecular spectra, crystalline structure, electrical conductivity, and semiconductor theory. As in the case of multielectron atoms, it is impossible to analyze multiatom bonding solely from "first principles"—to solve the Schrödinger equation. On the other hand, physics would not have advanced far had physicists tackled only those problems that could be solved "with pencil and paper." Thus, it is of great importance that we learn to combine first principles with acceptance of various models and plausible qualitative arguments, if they are found to agree with the experimental evidence.

## 9.1 When Atoms Come Together

Underlying this chapter is the idea that when isolated atoms approach one another, electrons are affected by more than "their own" atom. Their wave functions are no longer isolated-atom wave functions, but are solutions of the Schrödinger equation for a more complex multiatom potential energy. Even so, the resulting *molecular* states are often recognizable as closely related to or combinations of the atomic states (wave functions) of the isolated atoms.

Fortunately, many important features of this combining of isolated-atom states may be demonstrated in a much simpler, one-dimensional system. Let us consider

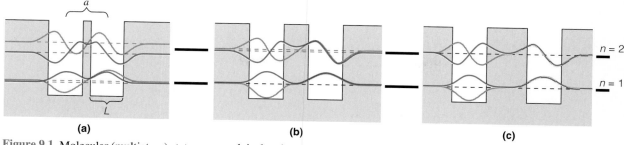

**Figure 9.1** Molecular (multiatom) states approach isolated-atom states as atomic separation increases.

the possible quantum states of a single electron in the presence of two "atoms," represented by one-dimensional finite wells of width $L$ and separation $a$. We choose the finite well as a model because it is capable of holding an electron bound, as is an atom, and allows wave functions to wander into the classically forbidden region (as the even simpler *infinite* well would not). We must allow for mingling of wave functions among "atoms."

Figure 9.1(a) shows the wave functions and energy levels of the four lowest-energy states in wells/atoms whose separation is comparatively small. Although they somewhat resemble the four lowest-energy wave functions of a *single* finite well (cf. Figure 4.14), these states form closely spaced pairs. At a larger separation, Figure 9.1(b), the energies of the lower pair are essentially equal and their wave functions very similar, virtually coincident in the "right atom" and opposite in the "left atom." The energies of the upper pair have also become closer. At a still larger separation, Figure 9.1(c), each pair has converged to a single energy.

It is no coincidence that *pairs* approach equal energy. The lowest-energy pair converges at great separation to two states that if algebraically added and subtracted are the $n = 1$ states of the *two* isolated "atoms." In other words, the two "molecular" states are merely different linear combinations of the two $n = 1$ atomic states. Similarly, the upper pair converges to states that are equivalent to (mere recombinations of) the two $n = 2$ isolated-atom states. Were there three wells, there would be three $n = 1$ "molecular" states, which at large separation would converge to the three isolated-atom $n = 1$ states. (Figure 9.27 shows four wells.) In general, when $N$ atoms come together, their $n = 1$ isolated-atom states combine to form a band of $N$ related molecular $n = 1$ states; their $n = 2$ isolated-atom states combine to form an $N$-state $n = 2$ band, and so on. A given band's energies cluster around the energy of the isolated-atom state and spread apart as the atomic separation decreases. Figure 9.2 illustrates schematically the combining of atomic states of $N$ wells/atoms to produce $N$-state molecular bands.

As Figure 9.1 shows, truly molecular states do not "belong" (asymmetrically) to one atom or another; rather, they mingle indiscriminately among them. This raises a troubling question: If the two atoms/wells are moved very far apart, it would seem that the lone electron would still be "shared" between them, but shouldn't it belong strictly to one or the other? The answer is yes—but only if the atomic separation is infinite! Only when $a$ is infinite do the two $n = 1$ molecular

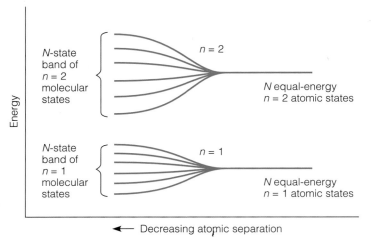

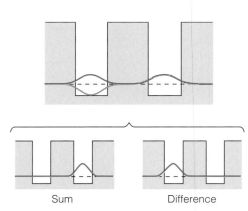

Figure 9.2 At small separations, each atomic state becomes an $N$-state band.

Figure 9.3 Atoms far apart can have "their own" electrons.

states have precisely the same energy. If their energies are different, the electron will occupy the lower energy one; if the same, the electron may occupy either, *or equivalently the sum or difference*. As we see in Figure 9.3, the sum and difference do genuinely belong to one atom or the other. Realistically, atoms are *effectively* infinitely far apart and the molecular states are of equal energy whenever $a$ is much larger than the electron wave's penetration depth. In such a case, an electron would orbit strictly one atom or the other.

## 9.2 Molecules

Atoms form molecules when the molecular state is of lower energy than the separated atoms. The behavior of the atoms' valence electrons is all-important. As they rearrange, relative to their separated-atom orbits, a lower energy may result. But how? Let us begin our study of molecules by considering the simplest possible case: the $H_2^+$ molecule, two protons and one electron. The total energy may be divided into two parts: (1) the electron's energy—its kinetic energy plus the negative, attractive potential energy it shares with the protons—and (2) the positive, repulsive potential energy shared by the two protons.[1] Consider first the electron's energy. Were the protons separated by a great distance, the lowest possible energy would have the electron simply bound to one or the other, its energy $-13.6$ eV. (Putting it somewhere between the two distantly separated protons would yield essentially zero, i.e., a higher, energy.) Were the protons fused, we would essentially have helium, but with only one electron. This is also fairly simply analyzed; as shown in Example 6.10, the electron's energy would be $-54.4$ eV. Sensibly, then, we expect the electron's lowest possible energy to decrease with decreasing proton separation. However, the positive electrostatic potential energy shared by the protons *increases* as their separation decreases. At large separations it is zero, and at negligible separation it is arbitrarily large and positive. Thus, the *total* en-

[1] The protons are treated as classical particles. Compared to the electron they are essentially fixed, due to their much larger mass. The motion of the nuclei is taken up in Section 9.3.

ergy is $-13.6$ eV at large separations, and arbitrarily large and positive at negligible separation.[2] Now the important question: Is there a minimum, less than $-13.6$ eV? If the protons' repulsive energy were to increase faster than the electron's energy decreases, there would be no advantage to forming a molecular bond. Since it is an experimental fact that $H_2^+$ does form a stable molecule, the answer must be "yes."

Of course, we might be interested in proving this via the Schrödinger equation. As always, we would begin with the potential energy. Given protons fixed at $x = -\frac{1}{2}a$ and $x = +\frac{1}{2}a$, the electron at position $\mathbf{r}$ would be a distance $|\mathbf{r} - (-\frac{1}{2}a\hat{\mathbf{x}})|$ from one and $|\mathbf{r} - (+\frac{1}{2}a\hat{\mathbf{x}})|$ from the other. The corresponding potential energy is given in Figure 9.4. Unfortunately, although it is the simplest imaginable for a molecule, it renders the Schrödinger equation solvable only by numerical approximation techniques. When this is done, however, it is found that the electron's energy does vary from $-13.6$ eV for large $a$ to $-54.4$ eV for $a = 0$ and that when the protons' repulsive energy is added there is a minimum total energy. Its value is $-16$ eV, and it occurs at a proton separation of $a \cong 0.1$ nm—in agreement with experiment.

But what of the wave function? A separation of 0.1 nm is only about twice the Bohr radius of the hydrogen atom, so the electron in $H_2^+$ should be strongly influenced by both protons. Indeed, we find that the ground-state wave function is symmetric about the midpoint between the protons—an even function of $x$—and has the vast majority of its probability/charge density in the region between them. The attraction both protons share with the intervening electron cloud is the root of the molecular bond's lower energy.

Adding a second electron would, of course, yield the neutral $H_2$ molecule. Just as for multielectron *atoms*, however, this drastically complicates the Schrödinger equation. As we might guess, though, the ground state has both electrons (spins opposite) in the same lowest-energy spatial state as in $H_2^+$. The energy-minimizing sharing of a spins-opposite pair of valence electrons is known as a **covalent bond.** Its strength derives from the electron pair's spatial state being centrally located and thus attracted to both positive ions; loosely speaking, both atoms in $H_2$ lay claim to two electrons. It is not surprising, then, to see behavior somewhat like helium, very stable and less chemically reactive than would be the separated atoms. Taking the idea further, valence $-1$ elements should also attain greater stability (lowered energy) by sharing a pair of electrons in a covalently bonded diatomic molecule. Fluorine ($Z = 9$), with seven electrons in the $n = 2$ shell, becomes somewhat neonlike ($Z = 10$) by forming diatomic $F_2$, each atom donating an electron to an equally shared pair. The covalent molecules $Cl_2$, $Br_2$,

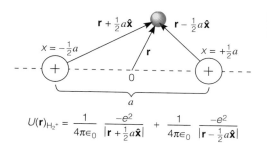

$$U(\mathbf{r})_{H_2^+} = \frac{1}{4\pi\epsilon_0}\frac{-e^2}{|\mathbf{r}+\frac{1}{2}a\hat{\mathbf{x}}|} + \frac{1}{4\pi\epsilon_0}\frac{-e^2}{|\mathbf{r}-\frac{1}{2}a\hat{\mathbf{x}}|}$$

**Figure 9.4** The $H_2^+$ molecule and its potential energy.

[2]Helium does not have infinite energy. Protons actually begin to attract each other (via the "strong" force) when in the same nucleus. We may ignore this point, however, because a minimum energy arises long before the protons are close enough to be considered in the same nucleus, and we are seeking the minimum energy for a true *diatomic* molecule.

and $I_2$ form in a corresponding way. With only six electrons in the $n = 2$ shell, oxygen atoms must share two electron pairs to form the stable molecule $O_2$. Similarly, three pairs are shared in $N_2$. This triple sharing makes nitrogen unusually inert for a "non-noble" gas.

## Bonding States

Useful though it is, the view of covalent bonding as atoms acquiring noble gas character by sharing specific pairs of valence electrons is oversimplified. First, the quantum states of a *molecule* are inherently different than those of isolated noble gas *atoms*. Secondly, "nonbonding" electrons may also be shared. A closer look at the electron wave functions for the simple $H_2^+$ molecular potential energy provides good insight.

Consider Figure 9.5. As in the finite-well "atoms" case of Figure 9.1, if the protons are distantly separated, there are two equal-energy molecular $n = 1$ states spread equally between the atoms, and their sum and difference are none other than the isolated-atom $1s$ states. But at small separations the electron's wave function is affected by both protons; the *atomic* states no longer exist! Rather, there are two molecular $n = 1$ states, with different energies, distributed equally between the atoms. One is an odd function of position and the other an even [$\psi(x) = -\psi(-x)$ and $\psi(x) = \psi(-x)$, respectively]. The even one is given the symbol $\sigma 1s$ and is known as a **sigma bonding orbital.**[3] It is the lower energy of the two because both protons are close to the whole electron cloud. The odd one, given the symbol $\sigma^* 1s$, is called a **sigma antibonding orbital.** It is of higher energy, higher even than the atomic states, because the electron avoids the shared region between the two atoms,[4] and the close proximity of the "bare" protons then leads to an actual net repulsion.

Figure 9.6 shows a way of visualizing how the two $1s$ molecular wave functions might arise from two $1s$ atomic wave functions, without actually solving the

**Figure 9.5** Small separation yields high and low energy molecular $1s$ states.

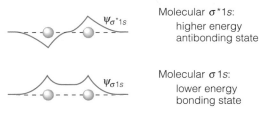

**Large proton separation**

Molecular $\sigma^* 1s$:
higher energy
antibonding state

Molecular $\sigma 1s$:
lower energy
bonding state

**Small proton separation**

[3] The qualifier "sigma" is used to distinguish this type of bond from another that can occur in $p$ orbitals but not $s$ orbitals. The nature of the distinction is discussed later in the section.

[4] Whereas $\psi(x) = \psi(-x)$ places no limitation on $\psi(0)$, $\psi(x) = -\psi(-x)$ requires that $\psi(0) = 0$.

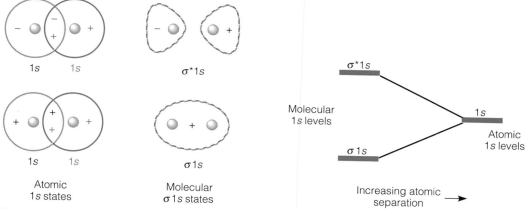

**Figure 9.6** Formation of 1s bonding and anti-bonding states.

**Figure 9.7** Molecular and atomic 1s energy levels.

Schrödinger equation. The plus and minus signs do *not* represent the sign of a charge (electrons being of course strictly negative); they represent the sign of the wave function. Remember that all wave functions contain an $e^{-i\omega t}$ and so change sign periodically. The even $\sigma 1s$ state results from the isolated-atom states combining in phase, while the odd $\sigma^* 1s$ results from their combining a half-cycle out of phase. The former is thus large between the atoms and the latter is small. The plus and minus are, of course, irrelevant when considering the probability/charge density, the complex *square* of the wave function.

In the spirit of Figure 9.2, the energy-level diagram of Figure 9.7 shows the separation-dependent link between the 1s atomic states and the corresponding molecular states. It is the $\sigma 1s$ molecular state that is occupied by the lone electron in $H_2^+$ and by the spins-opposite pair in neutral $H_2$, as shown in Figure 9.8. A molecular bond forms because it is of lower energy than an isolated-atom 1s state. The $\sigma^* 1s$ molecular state is referred to as "antibonding" because its occupation tends to discourage bonding, since its energy is considerably higher than the atomic 1s state. Even so, it is possible for an "antibonding" state to be occupied in a *bound* molecule. Although helium doesn't form an $He_2$ molecule, since its third and fourth electrons would have to occupy its $\sigma^* 1s$ level, giving a higher average energy than for isolated atoms, $He_2^+$ *does* exist. Two electrons in the $\sigma 1s$ state and one in the $\sigma^* 1s$ state give a lower average energy than three electrons in atomic 1s states (i.e., isolated He and $He^+$). Strictly speaking, then, since both bonding and antibonding states are spread equally between the atoms, *all* of a molecule's electrons must be treated as shared.

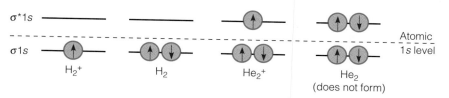

**Figure 9.8** Filling of the 1s levels.

In atoms where the $n = 2$ levels are occupied, there is richer detail in the molecular states. Atomic $2s$ states combine to form a high energy $\sigma^*2s$ and low energy $\sigma 2s$ in essentially that same way as for the $1s$. To understand how $2p$ atomic states might combine, we must digress briefly.

The angular dependences of the three equal-energy hydrogen $2p$ spatial states are given in Table 6.6 as

$$\psi_{2,1,0} \propto \cos\theta \qquad \psi_{2,1,\pm 1} \propto \sin\theta e^{\pm i\phi}$$

They may be combined to yield three equivalent atomic states with probability densities oriented along three independent axes (Figure 9.9). The $m_\ell = \pm 1$ states are combined by addition and subtraction.

$$\psi_{2p_z} = \psi_{2,1,0} \propto \cos\theta$$

$$\psi_{2p_x} = \psi_{2,1,+1} + \psi_{2,1,-1} \propto \sin\theta\cos\phi \tag{9-1}$$

$$\psi_{2p_y} = \psi_{2,1,+1} - \psi_{2,1,-1} \propto \sin\theta\sin\phi$$

The states $\psi_{2p_x}$ and $\psi_{2p_y}$ bear the same relationships to the $x$- and $y$-axes as the $\psi_{2,1,0}$ does to the $z$-axis. (Remember: $x = r\sin\theta\cos\phi$, $y = r\sin\theta\sin\phi$, $z = r\cos\theta$.) It is only due to the influence of nearby atoms that the $2p$ states might prefer to arrange themselves this way—to produce a bond of lower energy.[5]

Now, for *two* atoms there are *six* $2p$ spatial states, and when identical atoms bond, they arrange themselves into six molecular levels of different energies. Choosing the molecular axis as the $x$-axis, they are depicted in Figure 9.10. (Once again, the plus and minus refer to the sign of the wave function. For instance, $\psi_{p_x}$ is $+\sin\theta$ when $\phi = 0$ and $-\sin\theta$ when $\phi = 180°$). The $2p_x$ orbitals combine as do the $1s$ and $2s$ states to form a pair of bonding and antibonding states in which *charge density is largest along the molecular axis*. This is the criterion for the label $\sigma$-bond. The $2p_y$ and $2p_z$ also form bonding (centrally shared) and antibonding (divided) states, but the charge density is largest off-axis. These are known as $\pi$-bonds. By symmetry, the energies of the $\pi 2p_y$ and $\pi 2p_z$ are equal, as are those of the $\pi^*2p_y$ and $\pi^*2p_z$.

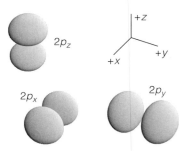

**Figure 9.9** Three perpendicular $2p$ atomic states.

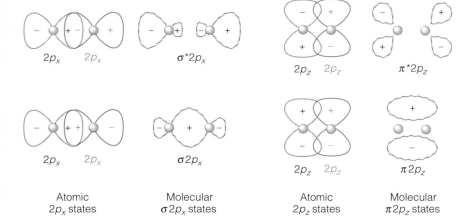

| Atomic $2p_x$ states | Molecular $\sigma 2p_x$ states | Atomic $2p_z$ states | Molecular $\pi 2p_z$ states |

**Figure 9.10** Formation of $2p$ bonding and antibonding states.

[5]In Section 6.4, the two-term solution $\Phi(\phi) = C\sin(m_\ell\phi) + D\cos(m_\ell\phi)$ was discarded as not spherically symmetric. But when multiple atoms come into play, "asymmetric" sines and cosines may well be appropriate.

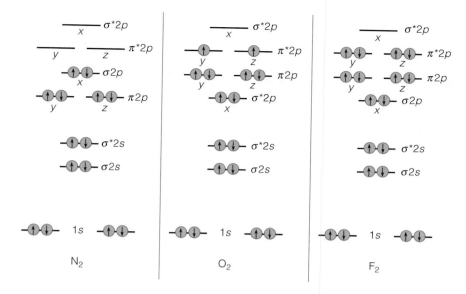

**Figure 9.11** Molecular energy levels in nitrogen, oxygen, and fluorine.

Figure 9.11 shows how the molecular levels fill in the purely covalent $N_2$, $O_2$, and $F_2$. We note several features:

1. At these values of $Z$, the $1s$ levels are not part of the valence. They are so closely confined to their respective atoms as to be considered approximate *atomic* $1s$ levels.

2. As occurs in atomic energy levels (cf. Figure 7.9), the ordering of levels varies from one element to another. The $\sigma 2p$, for instance, is higher than the $\pi 2p$ in $N_2$, but lower in $O_2$ and $F_2$.

3. Oxygen's two $\pi^* 2p$ electrons occupy different states, for the same reason that electrons in an individual atom's unfilled subshell tend to spread out among the $m_\ell$-values—so that they may assume a lower-energy antisymmetric spatial state. This spins-aligned condition gives $O_2$ a permanent magnetic dipole moment.

4. In contrast to the simpler shared-pairs-of-electrons view, we see here that *all* $n = 2$ electrons are shared. Even so, there is a connection: We might say that only the electrons in attractive bonding states *in excess* of those in repulsive antibonding states result in a *net* lowering of energy. For instance, $O_2$ has 6 bonding and 2 antibonding $2p$ electrons, a difference of 4—that is, a *net* two pairs of bonding electrons. Nitrogen is particularly inert chemically because none of its $2p$ electrons is required to occupy an antibonding state.

Being symmetric, covalent molecules comprising two *identical* atoms have no electric dipole moment. In bonds between atoms of different elements, however, atomic states are distorted asymmetrically; electrons are not shared equally. The result is a so-called **polar covalent** bond (Figure 9.12). For instance, when hydrogen and fluorine bond to form HF, the sharing of hydrogen's $n = 1$ and fluorine's

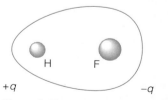

**Figure 9.12** A polar covalent bond.

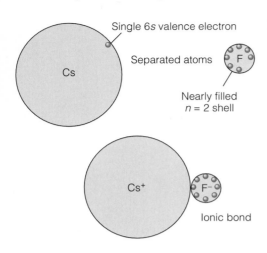

Single 6s valence electron

Separated atoms

Nearly filled
n = 2 shell

Ionic bond

**Figure 9.13** An ionic bond.

n = 2 electrons is not symmetric. The fluorine end has a significant excess of negative charge, leaving an equal positive charge excess at the hydrogen end.

Asymmetric charge distribution is most pronounced when a small, compact atom that is one electron short of noble gas structure meets a large, spongy atom with a lone valence electron far from its nucleus. In CsF (Figure 9.13), the most extreme case, the fluorine does not *share* cesium's valence electron; it wholly appropriates it. The fluorine ion assumes a stable, spherically symmetric neon structure, and the cesium ion is left with a similarly stable xenon structure. Such a bond is not *co*valent; it is known as an **ionic bond.** It is true that energy would have to be expended to produce the positive ion by detaching its valence electron. But the final energy is lower because even more energy is gotten out in adding this electron to the negative ion—a complete shell is a minimum-energy state. In contrast to covalent bonds, where sharing occurs along a specific axis, ionic bonds have no directionality. They are maintained simply by the electrostatic attraction between the closed-shell, spherically symmetric ions. In general, molecular bonds exhibit characteristics somewhere between purely covalent and purely ionic.

Thus far, we have considered only diatomic molecules. In complex molecules, to achieve the lowest possible energy, atomic states may "hybridize" in a great number of ways. Let us consider a particularly important example, which leads to the tetrahedral arrangement common in carbon compounds. When surrounded by electron-hungry atoms, carbon often shares all four of its $n = 2$ electrons. Just as the three 2p states $\psi_{2,1,0}, \psi_{2,1,+1}, \psi_{2,1,-1}$ may hybridize to form the $\psi_{2p_x}, \psi_{2p_y}$, and $\psi_{2p_z}$, all four of carbon's $n = 2$ states may combine to form four equal-energy atomic states.[6] In these states, known as **hybrid $sp^3$** states, the electrons are most likely to be found at the vertices of a tetrahedron, as shown in Figure 9.14, and

---

[6]To this point we have discussed combining only states of *equal* energy, but from Chapter 7 we know that an s state should be of lower energy than a p state. Nevertheless, the impetus for hybridization is invariably to produce lower-energy bonds with other atoms, and assuming there are such external factors, there may well be good reason (i.e., a lower overall energy) for an electron to occupy a combination of s and p states. See also Exercise 7.

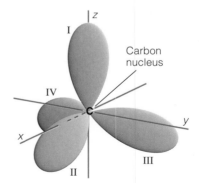

Carbon nucleus

**Figure 9.14** Carbon's $sp^3$ tetrahedral lobes.

**Table 9.1   Hybrid $sp^3$ states**

| Lobe | Wave function (not normalized) |
|------|--------------------------------|
| I | $-\psi_{2s} + \psi_{2p_z}$ |
| II | $-\psi_{2s} - \frac{1}{3}\psi_{2p_z} + \frac{\sqrt{8}}{3}\psi_{2p_x}$ |
| III | $-\psi_{2s} - \frac{1}{3}\psi_{2p_z} - \frac{\sqrt{2}}{3}\psi_{2p_x} + \frac{\sqrt{6}}{3}\psi_{2p_y}$ |
| IV | $-\psi_{2s} - \frac{1}{3}\psi_{2p_z} - \frac{\sqrt{2}}{3}\psi_{2p_x} - \frac{\sqrt{6}}{3}\psi_{2p_y}$ |

are thus poised to bond with four other atoms. (To obtain four lobes, identical except in orientation, $n = 2$ states may be combined as shown in Table 9.1. See also Exercise 6.)

The simplest example of tetrahedral bonding is diamond, a solid form of pure elemental carbon, in which each carbon atom bonds covalently to four others (Figure 9.15). Each bond is a pair of electrons sharing the same molecular spatial state, the "bonding" combination of atomic $sp^3$ lobes from both participating atoms, as shown in Figure 9.16. The presence of four covalent bonds per atom makes diamond a very strong crystalline structure. A typical molecular *compound* in the tetrahedral configuration is methane, $CH_4$, in which a hydrogen atom electron and one of the carbon's four are shared at each lobe. Although the sharing is not equal, the hydrogen being slightly positive, the arrangement of the lobes at the 109.5° tetrahedral angle (see Exercise 5) is symmetric and leaves methane with zero electric dipole moment.

Elements near carbon in the periodic table often assume a similar geometry, as shown in Figure 9.17. In ammonia, $NH_3$, nitrogen's "extra" $n = 2$ electron (it has five) takes the place of the "missing" hydrogen electron; but with the proton also missing, that lobe has a negative charge. Ammonia thus possesses a significant electric dipole moment, accounting for many of its properties as a solvent. One

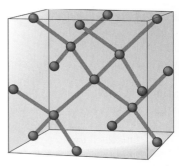

**Figure 9.15** The crystalline structure of diamond.

**Figure 9.16** The carbon–carbon bond.

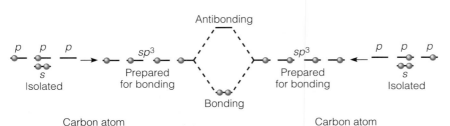

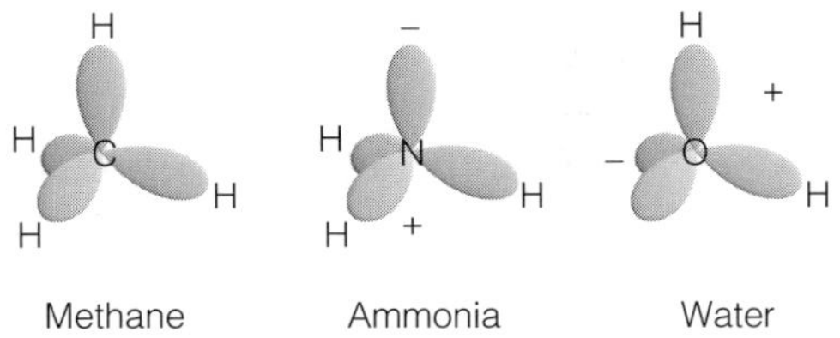

**Figure 9.17** Similar electron orbits.

step further removed, water, $H_2O$, has oxygen's two extra $n = 2$ electrons replacing hydrogen atom electrons at two lobes. Naturally, this configuration also has an electric dipole moment. The angle between the carbon–hydrogen bonds is found to be 107.3° in ammonia and 104.5° in water, very close to the tetrahedral angle.[7]

## 9.3 Rotation and Vibration

In addition to electronic energies, a molecule may store energy in rotational motion and the vibrational motion of its atoms. Not only are these forms of energy storage important to the heat capacity of the material, but the fact that they are quantized manifests itself in the spectral emissions of the particular molecular species. These spectral "fingerprints" aid in determining identity and concentration. We may grasp the basic ideas in the context of a simple diatomic molecule.

As we noted in the previous section, the total energy in a diatomic system initially decreases as the distant atoms are brought closer together, because both nuclei are attracted to the electron cloud that becomes ever more concentrated in the region between them. It reaches a minimum, then becomes quite large at very small separations, due to the repulsion of the nuclei. Qualitatively, then, the atoms (nuclei plus nonvalence electrons) behave as though sharing a potential energy of the form depicted in Figure 9.18, where $x_1$ and $x_2$ are their positions and $x = x_1 - x_2$ thus the atomic separation. Yet this raises a concern: In discussing *electron* states, we ignored the motion of the atoms. Being bound to each other, however, the atoms, in common with all bound particles, *must* move! Surely this affects the electronic states—but how?

The situation is not as complicated as it might seem. It turns out that atomic motion usually involves relatively little energy. The atoms can jump around among their vibrational and rotational states without bumping the electrons to other electron states, where their charge cloud, and thus their role in the interatomic potential energy, would be different. (If the electron state *does* change, there is a corresponding change in the interatomic potential energy; see Figure 9.24.) Therefore, we may treat the molecule as two point masses sharing a smooth potential energy of the form in Figure 9.18. Although it seems we still know very little about this potential energy, if the atomic motion is indeed small, there are really only two things we need to know: its equilibrium separation $a$ and its "spring constant" $\kappa$.

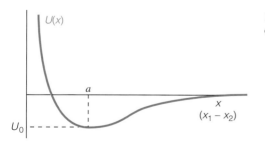

**Figure 9.18** Potential energy in a diatomic molecule.

[7]In these molecules, which lack the four-direction symmetry of methane, the hybridization of the $n = 2$ states is not simply the tetrahedral $sp^3$. Combined with it is a certain amount of the 90° $p_x$, $p_y$, $p_z$ hybridization, accounting for the deviations from the tetrahedral angle.

Near a local minimum, any smooth potential energy "looks" like an ideal spring—a parabola. To justify this, let us expand our unknown potential energy in a power series about its local minimum, the equilibrium atomic separation $a$.

$$U(x) \cong U(a) + \frac{1}{1!}\frac{dU(x)}{dx}\bigg|_a (x - a) + \frac{1}{2!}\frac{d^2U(x)}{dx^2}\bigg|_a (x - a)^2 + \cdots$$

The first term is the potential energy minimum, labeled $U_0$ in Figure 9.18. The second term is zero because $dU/dx$ is zero at the minimum of $U(x)$—that is, at $a$. Dropping higher-order terms, we have

$$U \cong U_0 + \frac{1}{2}\frac{d^2U(x)}{dx^2}\bigg|_a (x - a)^2$$

This is a parabola of minimum of $U_0$ and centered a distance $a$ from the origin. Now making the definition:

$$\frac{d^2U(x)}{dx^2}\bigg|_a \equiv \kappa \qquad (9\text{-}2)$$

we obtain

$$U \cong U_0 + \frac{1}{2}\kappa(x - a)^2$$

We also define a new "separation" $x_r$, the amount by which the actual atomic separation $x$ exceeds $a$:

$$x_r \equiv x - a$$

This separation is positive if $x$ is larger than $a$, negative if smaller. Finally, $U_0$ merely shifts all possible energies by a constant value. As in classical mechanics, potential energy may be redefined (shifted down or up) so that this constant value is zero. Thus,

$$U(x_r) \cong \frac{1}{2}\kappa x_r^2$$

If the actual atomic separation $x$ deviates only slightly from $a$ (i.e., the vibrational energy is not too large), then $x_r$ will be small and the power series expansion accurate. Accordingly, we model a diatomic molecule as in Figure 9.19: two point masses, an average distance of $a$ apart, that may rotate as a whole and may oscillate along their axis as though connected by an ideal Hooke's law spring. According to (9-2), the force constant $\kappa$ of the interatomic spring depends on the curvature of the potential energy at its minimum.

The quantum mechanical solution of the problem proceeds with an argument that there is an equivalent one-particle system, shown in Figure 9.20, in which the potential energy is still $\frac{1}{2}\kappa x_r^2$. The resulting one-particle Schrödinger equation is very similar to others we have encountered, specifically the harmonic oscillator and the angular parts of the hydrogen atom. The arguments and solution are discussed in Appendix I. The result is that the molecular energy levels depend

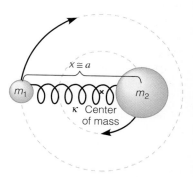

**Figure 9.19** A simplified diatomic molecule.

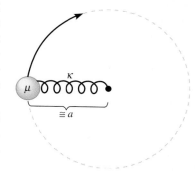

**Figure 9.20** One-particle equivalent of diatomic molecule.

on two quantum numbers $n$ and $\ell$, related to vibrational and rotational energy, respectively.[8]

$$E_{n,\ell} = E_{\text{vib}} + E_{\text{rot}} = \left(n + \frac{1}{2}\right)\hbar\sqrt{\frac{\kappa}{\mu} + \frac{\hbar^2\ell(\ell+1)}{2\mu a^2}} \qquad \begin{matrix} n = 0, 1, 2, \ldots \\ \ell = 0, 1, 2, \ldots \end{matrix}$$

(9-3)

While not affecting the energy, different rotational orientations are allowed, governed (as in the hydrogen atom) by a quantum number restricted to the values $m_\ell = 0, \pm 1, \ldots, \pm\ell$.

The mass of the equivalent single particle is given by

$$\mu \equiv \frac{m_1 m_2}{m_1 + m_2}$$

(9-4)

and is known as the **reduced mass.** Were one of the particles of overwhelming mass, the reduced mass would simply be that of the other ($m_2 \gg m_1 \Rightarrow \mu = m_1$). This makes sense: With an essentially infinite mass at one end of a spring, the only thing moving, vibrationally or rotationally, would be the small particle at the other end. In general, however, the reduced mass is less than both $m_1$ and $m_2$.

### Example 9.1

The diatomic HD molecule has an ordinary hydrogen atom and a deuterium atom. (In deuterium, a neutron joins the proton in the nucleus.) Their masses are 1.007 u and 2.013 u, respectively, and the bond length is 0.074 nm. At what temperature would the ratio of molecules in $\ell = 1$ rotational states to those with no rotational energy be one-tenth? (Assume that all are in their ground vibrational states.)

### Solution

In many ways this is similar to Example 8.2. Since there are three different rotational orientations (values of $m_\ell$) with $\ell = 1$, but only one for $\ell = 0$, the ratio of the numbers of molecules with these $\ell$-values is

$$\frac{\text{number with energy } E_{0,1}}{\text{number with energy } E_{0,0}} = 3 \times \frac{e^{-E_{0,1}/k_B T}}{e^{-E_{0,0}/k_B T}} = 3e^{-(E_{0,1} - E_{0,0})/k_B T}$$

To calculate energies, we need the reduced mass. Using (9-4),

$$\mu = \frac{(1.007\text{ u})(2.013\text{ u})}{1.007\text{ u} + 2.013\text{ u}} = 0.671\text{ u}$$

As noted, the reduced mass is less than either individual mass. If the vibrational state does not change, the energy difference given by (9-3) is

[8]Do not confuse this $\ell$ with that for *electronic* energy levels. Although they originate in solutions of identical differential equations, one has to do with angular momentum quantization of orbiting electrons and the other with that of an entire molecule.

$$E_{0,1} - E_{0,0} = \frac{\hbar^2 1(1 + 1)}{2\mu a^2} - \frac{\hbar^2 0(0 + 1)}{2\mu a^2} = \frac{\hbar^2}{\mu a^2}$$

$$= \frac{(1.055 \times 10^{-34} \text{ J·s})^2}{(0.671 \times 1.66 \times 10^{-27} \text{ kg})(0.074 \times 10^{-9} \text{ m})^2}$$

$$= 1.82 \times 10^{-21} \text{ J} = 0.011 \text{ eV}$$

We seek a ratio of one-tenth.

$$\frac{1}{10} = 3 \exp\left[\frac{-1.82 \times 10^{-21} \text{ J}}{(1.38 \times 10^{-23} \text{ J/K})T}\right] \Rightarrow T \cong 40 \text{ K}$$

Odd though it may seem, at temperatures much below 40 K, the vast majority of the molecules are simply incapable of rotating! There is insufficient translational kinetic energy to bump them to nonzero quantum rotational levels.

The example provides the basis for understanding hydrogen's rotational heat capacity "jump" (Figure 9.21). Rotation should begin to contribute to energy storage as soon as levels *above* the ground state become accessible, and the choice of a one-tenth probability is a fair measure. Although the fact that ordinary $H_2$ has *identical* atoms makes its analysis a bit more complicated, certain conclusions are still valid. The $H_2$ molecule has the same bond length as HD but a somewhat smaller reduced mass (~0.5 u). Thus, we should expect the jump to begin somewhat above 40 K—as Figure 9.21 clearly shows. (The vibrational jump is discussed in Section 8.10.)

## Spectra

With vibrational/rotational energies restricted to certain discrete values, molecules should emit and absorb only certain wavelengths of light. Spectral emissions and absorptions should therefore be a good test of our diatomic molecule model.

Clearly, if some of the molecule's vibrational/rotational energy is given to a photon, $n$ or $\ell$ or both must decrease. But the possible transitions are restricted by selection rules (as are electronic transitions in atoms; see Section 6.11). First, since the photon is spin-1, the system's total angular momentum before emission can equal that after emission only if $\ell$ for the molecule changes by unity, leading

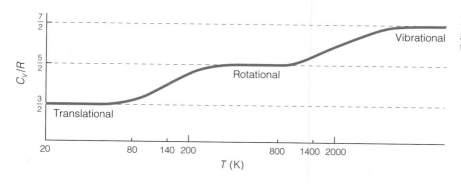

**Figure 9.21** Variation of hydrogen's molar heat capacity with temperature.

to the selection rule $\Delta\ell = \pm 1$. Secondly, electromagnetic radiation is emitted or absorbed most effectively by a charge distribution when it oscillates as an electric dipole. Although charge density in a *stationary* state produces no electromagnetic radiation, in a *transition* between vibrational states it does oscillate. It may be shown that a quantum-mechanical simple harmonic oscillator oscillates as an electric dipole only in transitions between energy states whose quantum numbers differ by unity. Thus, to the extent that the potential energy of the diatomic molecule may be treated as parabolic,[9] the selection rule for vibrational levels is $\Delta n = \pm 1$.

To lose vibrational energy, the molecule must, of course, jump to the next lower $n$. But it is generally the case that vibrational energy spacing is much larger than rotational spacing, so energy will decrease whether $\ell$ increases or decreases. Thus, we obtain the possible energies of an emitted photon by subtracting the molecule's final energy from its initial for both cases: $\Delta\ell = \pm 1$. Using equation (9-3),

$$E_{\text{photon}} = \begin{cases} E_{n_i,\ell_i} - E_{n_i-1,\ell_i+1} \\ E_{n_i,\ell_i} - E_{n_i-1,\ell_i-1} \end{cases}$$

$$= \begin{cases} \hbar\sqrt{\dfrac{\kappa}{\mu}} - (\ell_i + 1)\dfrac{\hbar^2}{\mu a^2} & (\ell_i = 0, 1, 2, \ldots) \\ \hbar\sqrt{\dfrac{\kappa}{\mu}} + \ell_i\dfrac{\hbar^2}{\mu a^2} & (\ell_i = 1, 2, 3, \ldots) \end{cases}$$

(Since $\ell$ is strictly nonnegative, $\ell_i$ cannot be zero when $\Delta\ell = -1$.) Let us rewrite this in a way that better illustrates the character of the spectrum,

$$E_{\text{photon}} = \hbar\sqrt{\dfrac{\kappa}{\mu}} \mp I\dfrac{\hbar^2}{\mu a^2} \qquad (I = 1, 2, 3, \ldots) \tag{9-5}$$

Photons will be observed whose energies are spaced equally on either side of $\hbar\sqrt{\kappa/\mu}$. The energy $\hbar\sqrt{\kappa/\mu}$ itself is conspicuously absent because angular momentum conservation requires that any transition involve a (unit) change in $\ell$. By similar arguments, the energies of photons *absorbed* by the molecule are given by the same formula.

Figure 9.22 shows molecular rotational/vibrational levels for $n = 0$ and 1 and $\ell = 0, 1, 2,$ and 3, along with the transitions allowed in photon emission and absorption. The photon energies constitute a **vibration-rotation band,** with a "hole" in the middle due to the forbidden $\Delta\ell = 0$ transition. Figure 9.23 shows how well the theory agrees with experiment. It plots intensity of light *absorbed* by diatomic HCl versus frequency and hence photon energy. Each spike in this vibration-rotation absorption band represents missing transmitted light—missing because photons of that frequency are readily absorbed. Note the regular spacing and the hole in the middle.[10] (Each spike is split in two because chlorine in nature has two isotopes of slightly different mass. There are thus two different reduced masses, so the allowed energies should indeed split into two slightly different sets.)

[9]At higher $n$-values, the true anharmonic nature of the interatomic potential energy is revealed, invalidating the selection rule; the change in $n$ need not be unity.

[10]That the spacing is not quite equal is due to effects we have ignored; "centrifugal force" causes $a$ to increase somewhat with $\ell$, reducing the rotational energy. Thus, $\Delta\ell = +1$ absorptions are somewhat less energetic than the formula suggests, and $\Delta\ell = -1$ absorptions are more energetic.

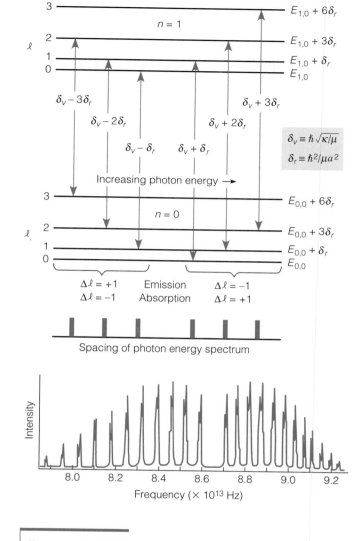

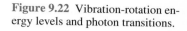

**Figure 9.22** Vibration-rotation energy levels and photon transitions.

**Figure 9.23** Absorption spectrum of HCl.

## Example 9.2

From the data given in Figure 9.23, determine (a) the approximate bond length, and (b) the effective force constant in an HCl molecule.

### Solution

(a) From the tenth line to the left of the "hole" to the tenth to the right (20 steps) is a frequency range of $1.2 \times 10^{13}$ Hz—an average spacing of $6.0 \times 10^{11}$ Hz. Thus, the photons represented by these lines differ in energy by

$$(6.63 \times 10^{-34} \text{ J·s})(6.0 \times 10^{11} \text{ Hz}) = 4.0 \times 10^{-22} \text{ J} = 0.0025 \text{ eV}$$

This must equal the energy spacing indicated in equation (9-5), $\hbar^2/\mu a^2$. We must find $\mu$. Using data from the periodic table (Figure 7.10), averaged over naturally occurring isotopes,

$$m_H = 1.01 \text{ u} \qquad m_{Cl} = 35.5 \text{ u}$$

$$\mu = \frac{(1.01 \text{ u})(35.5 \text{ u})}{1.01 \text{ u} + 35.5 \text{ u}} = 0.982 \text{ u} = 1.63 \times 10^{-27} \text{ kg}$$

Thus,

$$4.0 \times 10^{-22} \text{ J} = \frac{(1.055 \times 10^{34} \text{ J·s})^2}{(1.63 \times 10^{-27} \text{ kg})a^2} \quad \Rightarrow \quad a = 1.3 \times 10^{-10} \text{ m}$$

(b) The hole falls at $8.65 \times 10^{13}$ Hz, or a photon energy $hf$ of

$$(6.63 \times 10^{-34} \text{ J·s})(8.65 \times 10^{13} \text{ Hz}) = 5.73 \times 10^{-20} \text{ J} = 0.358 \text{ eV}$$

According to equation (9-5), the hole corresponds to the vibrational energy spacing of $\hbar\sqrt{\kappa/\mu}$. (Note how much larger it is than the rotational spacing.) Thus,

$$5.73 \times 10^{-20} \text{ J} = (1.055 \times 10^{-34} \text{ J·s})\sqrt{\frac{\kappa}{1.63 \times 10^{-27} \text{ kg}}}$$

$$\Rightarrow \quad \kappa = 482 \text{ N/m}$$

Generally speaking, transitions where *electrons* change states (e.g., the Balmer series in hydrogen) involve much greater energy differences (>1 eV) than those involving molecular vibrational transitions (>0.1 eV), which are in turn more energetic than molecular rotational energy differences (<0.1 eV). Although not to proper scale, Figure 9.24 illustrates the relationships. The curves represent two

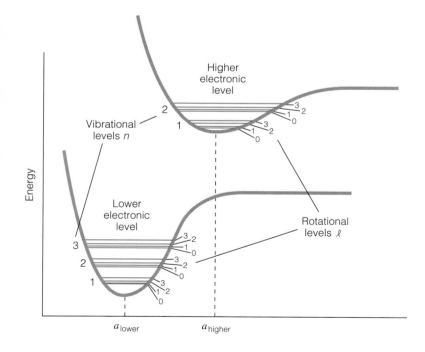

**Figure 9.24** Molecular vibrational and rotational levels for two different electron levels.

different electron states, perhaps the ground state and the first excited state. When an electron is in an excited state, the atoms are still bound to each other, but are farther apart on average and the interatomic potential energy well is wider and shallower. Even so, for each electronic state there are closely spaced vibrational levels and more closely spaced rotational levels. Thus, while we might expect to see simple spectral lines in hydrogen (molecules, not atoms) due to electron jumps, closer scrutiny reveals a fine structure due to molecular vibrational and rotational energy changes, causing one line to be many closely spaced lines.[11]

## 9.4  Crystalline Solids

Let us now turn our attention from molecules, the bonding of a small number of atoms to form somewhat larger particles, to crystalline solids, the bonding of a large number of atoms to form regular structures of macroscopic size. All elements and compounds form solids at sufficiently low temperature and/or high pressure. Most often, the atoms form a **crystal lattice,** in which certain atoms are found at certain locations in a microscopic unit, which is itself repeated countless times in all dimensions.[12] They do this because it results in a state of lowest energy. Typical atomic spacing is five to ten times the Bohr radius $a_0$, or a few angstroms (1 Å ≡ $10^{-10}$ m).

Geometrical considerations show that there is a limited number of **lattice types,** ways in which atoms can be arranged in a regular geometrical pattern.[13] Figure 9.25 shows several of the most common. The **body-centered cubic** lattice is a repetition of cubes with atoms at the corners and center. It is found in the solid forms of the periodic table's first-column elements, as well as in a number of transition metals, such as iron, chromium and tungsten. The **face-centered cubic** is also a repetitive cubic structure, but with atoms at the centers of the cubic *faces* instead of at the center of the cube. The noble gases (except helium) solidify in this structure, as do many transition elements—for example, copper, silver, and gold. The **hexagonal closest packed** structure has a six-sided symmetry, and shares with the face-centered cubic the distinction of resulting in the smallest volume per atom for a lattice of identical spheres. Elements with this structure are found throughout the periodic table—helium, magnesium, zinc, titanium, osmium, and many "rare earths." Compounds form the same structures as elements.

---

[11]Lacking a permanent electric dipole moment, *symmetric* diatomic molecules (e.g., $H_2$, $N_2$) do not have spectra due to pure vibration-rotation transitions. However, molecular rotational and vibrational quantum states may change when the orbiting *electrons* make transitions between electronic states, and the existence of rotational and vibrational levels is confirmed in the superimposition of their characteristic spacings upon the electronic spectra.

[12]Another possibility is that the atoms form an *amorphous solid,* in which there is no long-range order; angles and lengths of "bonds" are so irregular that the locations of atoms separated by more than a few intervening atoms are almost completely unrelated. Familiar examples of amorphous solids are glass and rubber.

[13]In three dimensions, there are 14 possible lattice types. To give some idea of what these "geometrical considerations" are, *two*-dimensional space can be completely filled up with equilateral triangles, squares or hexagons, *but not pentagons.* There is no two-dimensional lattice with a five-sided symmetry.

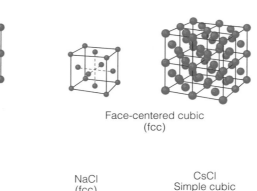

**Figure 9.25** Crystal lattices.

For instance, sodium chloride is a face-centered cubic structure in which a sodium chlorine pair replaces the individual atom. In fact, the sodium and chorine ions each independently form a face-centered cubic arrangement, since the atoms of one are a fixed displacement from those of the other. Cesium chloride, though resembling the body-centered cubic, is an example of a **simple cubic** structure—formed by each ionic type independently.

Besides their lattice geometry, crystalline solids are often categorized according to how the valence electrons are bound in the solid. Four categories are generally recognized.

## Covalent Solid

In a **covalent solid,** such as diamond, each atom shares covalent bonds with those surrounding it, resulting in an unbroken network of strong bonds. Such solids are comparatively hard, due to the inherent strength of the covalent bond, and have high melting points. They are poor electrical conductors because all valence electrons are "locked" into bonds between adjacent atoms. The crystal lattice assumes a geometry determined by the directionality of the interatomic covalent bonds. In the case of diamond, it is face-centered cubic.

## Ionic Solid

When atoms with nearly filled shells meet atoms with weakly bound valence electrons, the former may seize electrons from the latter, producing an **ionic solid.** The solid is held together by the strong electrostatic attraction between the ions. Thus, ionic solids are comparatively hard, with high melting points. Because the transfer of electrons leaves both positive and negative ions with noble gas electronic structure, electrons are not free to respond to electric fields—ionic solids are poor electrical conductors. The ionic bond lacking any sort of directionality, the geometry of the lattice is determined by whatever arrangement leads to the lowest electrostatic energy. This varies according to the relative sizes of the ions. In any case, lowest energy results when charges alternate—that is positive-negative-positive, and so on. Since ionic solids depend upon asymmetry between atoms, *elements* do not form ionic solids. For instance, NaCl is an ionic solid because each sodium atom may transfer its lone valence electron to a chlorine atom, giving each atom a noble gas electronic configuration. But noble gas structure cannot be attained in the same way in a solid of sodium *only*.

## Metallic Solid

Except for noble gases, all elements have valence electrons, but most elements simply do not have the proper *number* of valence electrons to bond with surrounding atoms into covalent solids. Elements and compounds with "leftover" valence electrons form a **metallic solid.** One of the reasons there are relatively few *non-metals* among the elemental solids is that, in most cases, no arrangement of covalent bonds can possibly accommodate all the valence electrons; some are necessarily excess. When bound to their isolated atoms, valence electrons have relatively high energy. In a metallic solid, however, the atomic valence states mix, much as in a diatomic covalent bond, but encompassing all the atoms in the crystal. This mixing produces a virtual continuum of states whose energies are lower than those of isolated atoms. Shared by all atoms, the valence electrons occupying these states move rather unfettered about the crystal lattice, forming an electron gas (Section 8.7), a distinguishing feature of the metallic solid.

The attraction that holds a metallic solid together is between the positive ions and the pervasive electron gas, and the ions assume whatever spacing leads to the lowest energy of the whole system. Since the cohesive forces are not in the form of covalent bonds *between the ions* forming the lattice, metals are often malleable and possess lower melting points than covalent solids. The valence electrons that are free to move among the continuum of states, the so-called **conduction electrons,** readily respond to an external electric field. Accordingly, metallic solids are generally excellent conductors of electricity.

## Molecular Solid

It would seem that noble gases, if forming crystalline solids at all, should not do so by any of the foregoing means. Their complete electronic structure provides no incentive to share electrons, and their electrons are so tightly bound that no energy reduction is possible by allowing them to move freely throughout the solid. But this should also be true of *molecules* whose bonding is so tight as to result in near noble gas stability, such as the covalent molecules $H_2$, $N_2$, $O_2$, $F_2$, $CH_4$, $NH_3$, and $H_2O$. Nevertheless, "noble molecules" still have cause to form a crystalline solid, known fittingly as a **molecular solid.** When atoms are close together, even when disinclined to share electrons, the charge distributions as a whole still interact. For noble gas atoms (e.g., Ne, Ar) and symmetric covalent molecules (e.g., $H_2$, $N_2$), with no *permanent* electric dipole moment, a given atom/molecule may have the effect of an electric dipole upon its neighbor *at any instant*. This in turn causes its neighbor's charge distribution also to assume an instantaneous *induced* electric dipole moment. Although the molecules' dipole moments fluctuate about zero, the fluctuations are always correlated so as to produce a net attraction—a lower energy. (The attraction is crudely depicted in Figure 9.26, blue arrows representing dipole moments at one instant, gray at another.) The *induced* dipole-dipole attraction is known as the **London force.** Compared to the electron sharing in covalent solids, and the monopole-monopole electrostatic attraction in ionic solids, dipole-dipole attractions are very weak. Accordingly, the temperature must be very low for atoms/molecules to form a solid lattice by this means, and even so the solid is fairly soft. The geometry of the lattice depends on the size and shape of the atoms/molecules. Composed of "noble" particles, molecular solids lack free electrons; they are poor electrical conductors.

If the molecules in a molecular solid have permanent electric dipole moments (e.g., $NH_3$ and $H_2O$), an additional, somewhat stronger dipole-dipole attraction exists. It is particularly strong in the case of water, and accounts for the unusually high melting point of this molecular solid.

Just as for individual molecules, bonding in solids can be very complex and is often a combination of the various types. There are, for instance, always London forces between atoms, and bonds in compounds are often partly covalent, partly ionic. A good indicator of the complexity of bonding in solids is the progressive change in character among the tetravalent elements: carbon ($Z = 6$), silicon ($Z = 14$), germanium ($Z = 32$) and tin ($Z = 50$). Although at $0°C$, all four solids exhibit the covalent diamond structure, electrical conductivity increases progressively from diamond, an insulator, to tin, a fair conductor. Apparently, electrons are not free to move about the solid in diamond, but are free to do so in tin.[14] How can this be explained? The somewhat unsatisfying answer is that even "covalent" and "ionic" solids must often be treated as though all the valence electrons are shared by all atoms, just as in a metallic solid. Different conductivities are then based on the fact that the electronic states do not form an unbroken continuum of energies from which the electrons may choose—they separate into energy bands.

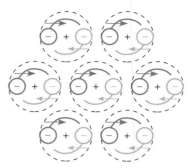

**Figure 9.26** Molecular solids are held together by alignment of instantaneous electric dipole moments.

## 9.5 Energy Bands

If all of a metal's valence electrons are released to the solid as a whole, why do we find that the number of *conduction* electrons per atom—those actually free to respond to an electric field—often does not equal the number of *valence* electrons per atom? And if even nonmetals must be considered as freely sharing electrons throughout the solid, how can they be electrical insulators? The answers rest upon the fact that the well into which the valence electrons are entrusted is not a *simple* well; electrons may not possess arbitrary energy, but are restricted to **energy bands.**

In reference to Figure 9.1 we noted that the four wave functions resemble the four lowest-energy states of a single well, but at the same time are grouped into pairs centered on the energies of the individual wells—they exhibit features of both the individual atoms and of one large well. Figure 9.27, showing wave

[14]At around $23°C$, tin undergoes a transformation, abandoning the diamond structure for a more truly metallic one.

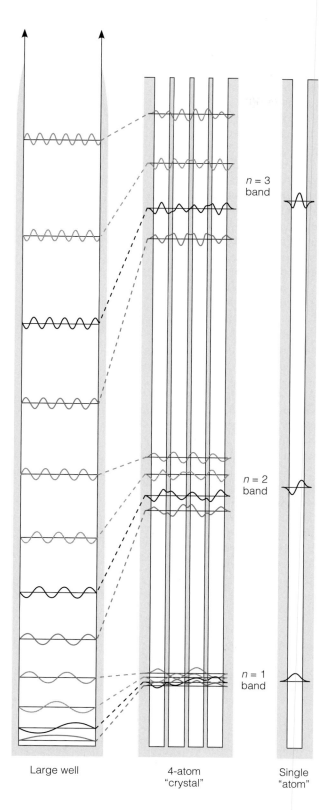

**Figure 9.27** A crystal shares features of both the individual atoms and one large well.

Large well

4-atom "crystal"

Single "atom"

functions and energy levels for four finite wells, further illustrates these points: Though somewhat distorted, the wave functions in the "crystal" (center diagram) have about the same wavelengths, the same numbers of antinodes, as those of one large well (left diagram). Since the electrons in the crystal may thus possess any of the wave numbers they might have in one large well, we see that the individual ions have little effect on the electrons' allowed *kinetic* energies ($\hbar^2 k^2/2m$). On the other hand, the *potential* energy due to the positive ions causes the electrons' *total* energies to be grouped into bands centered on the single-atom states (right diagram).

Consider, for example, the $n = 2$ band: The top-of-the-band wave function is of just the right wavelength to be zero at points halfway between the "atoms," where the periodic potential energy is at its maximum. Its potential energy is thus low and its total energy scarcely higher than the corresponding large well wave function. The next-lower wave function is of smaller wave number/kinetic energy, but being thus ever so slightly "out of synch," it is not zero between all atoms. Its potential energy is somewhat higher and its total not much lower than the top-of-the-band state. The next-lower state is lower still in kinetic but higher again in potential, and so again is only slightly lower in total energy. The bottom-of-the-band state is nearly at its *maximum* at all interatomic points. Thus, while of still lower kinetic energy, it is of very high potential, again leaving only a slight decrease in the total. Note that all the $n = 2$ band states resemble the $n = 2$ single-atom state: two antinodes per "atom."

The situation changes markedly at the next-lower energy state. This $n = 1$ top-of-the-band state more closely resembles an $n = 1$ single-atom state, with one antinode per atom. Moreover, returning to the same nodes-between-atoms condition as at the top of the $n = 2$ band, not only is its *kinetic* energy lower that the $n = 2$ bottom-of-the-band; its *potential* energy is much lower. Thus, from the bottom of the $n = 2$ to the top of the $n = 1$ band, the total energy drops greatly—there is a "band gap."

Figure 9.28, plotting energy versus wave number, further illustrates the effect of the ionic potential. In one large well of width $L$, the allowed wave numbers $k$ would be multiples of $\pi/L$ and the energy would be purely kinetic: $\hbar^2 k^2/2m$. The dashed parabola shows how energy would vary with $k$. In the four-atom "crystal," the allowed $k$-values, and hence kinetic energies, are essentially the same (as noted), so plotting $E$ versus $k$ shows clearly the grouping caused by the ionic potential; bottom-of-the-band states have unusually high (potential) energy.

## As N Becomes Large

With our four-atom "crystal" of Figure 9.27 as a guide, we can easily understand important features of the $N$-atom case, represented in Figure 9.29:

1. A top-of-the-band state of unusually low potential energy occurs when the wave function is zero at points midway between the atoms.[15] For an atomic separation $a$, the condition required for the $n$th band is $a = n\lambda/2$, or $k = n\pi/a$. At $n$ antinodes per atom, this state would have $N \times n$ antinodes over the whole crystal. The state of next-higher $k$, having just one more antinode overall, would be of *essentially the same wave number*. But with an extra

[15] Because the positive ions attract the electrons, points *midway between* atoms will always be of *high* potential energy.

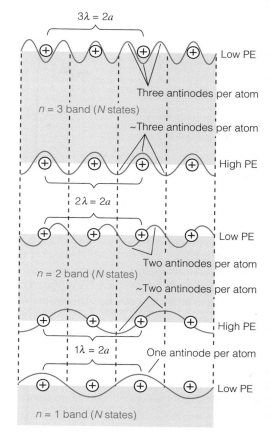

**Figure 9.29** Band gaps occur $\lambda = 2a/n$.

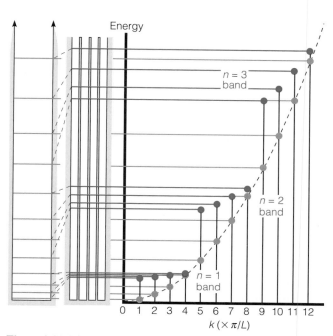

**Figure 9.28** The periodic ionic potential groups the large-well states into bands.

wavelength "squeezed in," it would be nearly maximum at a great many interatomic points. It is a state of unusually high potential energy—the bottom of the next band. *The jump between bands n and n + 1 occurs at $k = n\pi/a$.*

2. The condition of $n$ antinodes per atom, $a = n\lambda/2$, is the same as the quantization condition for a *single* atom/well of width $a$. Thus, *there is one jump—one band—for each single-atom state.*

3. With $N \times n$ antinodes overall in the state atop band $n$ and $N \times (n + 1)$ in the state atop band $n + 1$, they are separated by $N$ states: *Each band consists of N states.*

Because $N$ is so large, a band is essentially a continuum of energies. When $a$ is not too small, they spread above and below the energy of the single-atom state, as

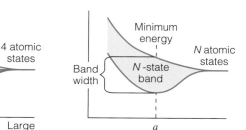

**Figure 9.30** Behavior of states as atomic separation varies.

in Figure 9.30. *How much* above and below depends on how much overlap there would be between *adjacent* atoms' wave functions. Consequently, the **band width** depends not on the *number* of atoms, but on the *atomic spacing,* increasing as *a* decreases (see Figure 9.1). At very small separations, the repulsion of the positive ions, ignored in our four-atom model, leads to a rapid rise in overall energy. The crystal forms with a spacing at which the overall energy is minimum.

We will not attempt to solve the Schrödinger equation for a real crystal, but the solutions have a few features worthy of note. It may be proven that they are of the following general form, known as **Bloch functions:**

$$\psi = e^{ikx}w_k(x) \quad \text{where } w_k(x + a) = w_k(x) \tag{9-6}$$

As do those illustrated in Figure 9.27, these combine characteristics of a large infinite well and an individual atom. The part related to the individual positive ions is the $w_k(x)$ term. It is of the same period $a$ as the microscopic ionic potential energy, and might crudely be viewed as a periodic repetition of individual-atom wave functions.

The "large-well part" is the $e^{ikx}$. Note however that, rather than the sines and cosines typical of standing waves in an infinite well, this complex exponential is a traveling plane wave, characteristic of a free particle. For a single *macroscopic* well, the allowed energies are nearly as unrestricted as for truly free particles. Moreover, in a real crystal a great many boundary conditions are imposed on the wave functions by the great many atoms, appearing within the larger well as microscopic wells embedded periodically. Thus, it may be said that the periodicity of $w_k(x)$ is the most important condition imposed on the wave functions, and the relative few conditions at the physical edges of the crystal may simply be ignored. It is reasonable, then, that the free-particle $e^{ikx}$ is the appropriate large-well part of the wave function. Now it is true that because a plane wave has constant amplitude, it would "feel" only an average potential energy; it would be concentrated nowhere. But while most $\psi$ are essentially free traveling waves, those at band edges are modified by $w_k(x)$ in such a way that the total solution $\psi$ becomes a true standing wave. Indeed it becomes two: a lower-energy one with nodes midway between the positive ions, and a higher-energy one with antinodes at these points.

Summarizing: From the *macroscopic* crystal as a whole arises a continuum of allowed energies and free-particle, traveling-wave solutions; while from the *microscopic* ionic potential energy arises periodic variations in these solutions and grouping of allowed energies into bands related to the individual-atom states. Figure 9.31 plots allowed energies versus wave number in a one-dimensional crys-

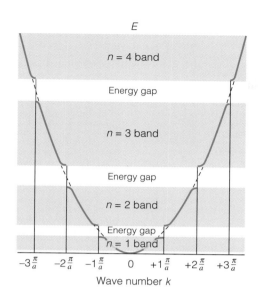

E

n = 4 band

Energy gap

n = 3 band

Energy gap

n = 2 band

Energy gap
n = 1 band

$-3\frac{\pi}{a}$   $-2\frac{\pi}{a}$   $-1\frac{\pi}{a}$   $0$   $+1\frac{\pi}{a}$   $+2\frac{\pi}{a}$   $+3\frac{\pi}{a}$

Wave number $k$

**Figure 9.31** Bands and gaps in a one-dimensional crystal.

tal where $N$ is not small and $k$ may be of either sign (an electron may move either direction). The allowed energies are essentially the parabolic continuum of kinetic energies available to an electron roaming freely in a large infinite well, except near $k = n\pi/a$, where the ionic potential leads to a jump from an unusually low energy at the top of one band to an unusually high energy at the bottom of the next.[16] The intervening region of forbidden energies is known as an **energy gap.**

Importantly, we have thus far discussed only the nature of states *available* to an electron in a crystal. We have not addressed the question of how many electrons might be present to *fill* those states, and it is this very issue that governs whether a material will be a conductor, an insulator, or a semiconductor.

## Electrical Conduction

Before studying the relationship between energy bands and conductivity, it is necessary that we review aspects of electrical conduction that may be understood from a *classical* point of view, and identify those that require a quantum-mechanical approach.

In the classical picture, electrons are free to roam about in a conductor. They may suffer collisions with the positive ions, however, and this is a necessary ingredient in understanding why the electrical current is proportional to the applied field (Ohm's law), rather than being a linearly increasing function of time. Were there no collisions, electrons subject to a constant electric field would constantly accelerate, giving a constantly increasing current. Collisions represent a retarding force. In essence, in a practically instantaneous time interval, the electrons attain terminal speed, in which a balance is reached between the accelerating field and the decelerating effect of collisions.

In the classical picture a typical electron moves randomly with an average speed $\bar{v}$ characteristic of the temperature. After a collision with a positive ion, it is just as likely to be moving one direction as another. With no external electric field,

[16]In Figure 9.28, all the energies are above the large-well parabola because the individual wells necessarily raise the *average* potential energy of all states. In Figure 9.31, this systematic shift has been chosen to be zero.

the average velocity of all the electrons is zero. But when an electric field is present, the electrons gain in the time between collisions a component of velocity opposite the field. Let us define $\tau$ as the average time between collisions, or **collision time.** At any instant in time, a given electron may have been freely accelerating for an arbitrary span of time, but an average electron will have been accelerating for a time $\tau$. It will thus have acquired an added velocity component of magnitude $(eE/m_e)\tau$, where $e$ is the fundamental charge, $E$ the electric field strength, and $m_e$ the electron mass. This slight shift in the more-rapid random thermal motion toward velocities opposite the field is known as "drift velocity." The resulting current density—current per unit area—is found as follows:

$$ j \equiv \frac{\text{charge}}{\text{time·area}} = \frac{\text{charge}}{\text{distance·area}}\frac{\text{distance}}{\text{time}} = \frac{e \times \text{number}}{\text{volume}}\frac{\text{distance}}{\text{time}} $$

Thus,

$$ j = (e\rho)v_{\text{drift}} = (e\rho)\left(\frac{eE}{m_e}\right)\tau = \left(\frac{e^2\rho\tau}{m_e}\right)E \tag{9-7} $$

where $\rho$ is the number of free charge carriers per unit volume. The effect, a current density, is proportional to the cause, an electric field—Ohm's law. The proportionality constant is known as the **conductivity** of the material and given the symbol $\sigma$.

$$ j = \sigma E \quad \text{where } \sigma = \frac{e^2\rho\tau}{m_e} $$

Classically, conductivity decreases as the collision time $\tau$ decreases, as collisions become more frequent. One way to decrease the collision time is to increase the temperature; faster-moving particles collide more frequently with obstructions around them. Thus, the resistance—inversely proportional to conductivity—of conductors increases with temperature.

Much of the classical view of electrical conductivity is valid quantum-mechanically. Collisions are still viewed as the origin of electrical resistance. However, electrons are not disposed to collide with any and all positive ions. Quite the contrary, we find that in metals—though Ohm's law still holds—the collision time is much larger than can be explained by a classical theory in which electrons may collide with all positive ions. The quantum-mechanical explanation is that, except near edges of bands, the states allowed an electron in a periodic crystal are essentially those of a free particle; the electron is a *wave,* upon which the positive ions have little effect! It is not the regular array of positive ions that upsets the otherwise free electrons. Rather, *deviations from regularity* perturb the electron wave and thus determine resistance. At room temperature, the vibrational motion of the positive ions is the most important deviation. Here the relevant collisions may be viewed as between electrons and phonons (the quanta of vibration—see Section 8.10). As the temperature increases, so does the number of phonons present, contributing to an increase of resistance. At very low temperatures ($\lesssim 10$ K), vibrational motion is considerably diminished and the predominant sources of collisions are microscopic **lattice imperfections.** These may be either "point defects"—vacancies (missing atoms) or impurities (atoms of a different element)—or more widespread disruptions of crystal regularity. But in any case, the quantum

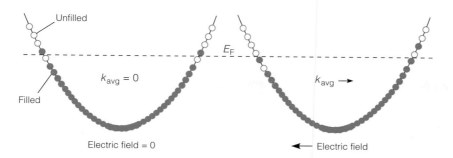

**Figure 9.32** The effect of an external electric field on the occupation of electron states.

wells at certain locations are altered and the electron wave perturbed. In contrast to phonons, the abundance of lattice imperfections is largely independent of temperature, and so is the resistance they cause.

Figure 9.32 illustrates important features of conduction in a large one-dimensional well—with no apparent band gaps—at low temperature.[17] Electrons occupy all states up to the Fermi energy, whereupon the occupation number drops rapidly from unity to zero. As in the classical view, the effect of an external electric field is to shift the electrons' momenta toward those in which $k$ ($= p/\hbar$) is opposite the field. But due to the exclusion principle, the quantum-mechanical response is a *net* shift only among the highest-energy states: from high energy states where the momentum is in the direction of the field to even higher energy states where it is opposite the field. Classically or quantum-mechanically, however, electrons do not continue to climb to higher energy levels (do not accelerate indefinitely). The energy-reducing effect of collisions invariably produces a balance, so that a given electric field results in a given net momentum and corresponding current density. Of course, were there no states *available* at higher energy, because of a band gap, there could be no response to the external field at all. As we see in the next section, this is the fundamental difference between a conductor and an insulator.

## 9.6 Conductors, Insulators, and Semiconductors

In three dimensions, atomic spatial states depend on three quantum numbers, and in a crystalline solid each becomes a band. By the simplest argument, lithium ($Z = 3$) should be a conductor and beryllium ($Z = 4$) an insulator.[18] As we know, each energy band comprises $N$ *spatial* states, where $N$ is the number of atoms in the crystal. Therefore, including spin, each band is able to hold $2N$ electrons. However, in monovalent lithium, there are only $N$ valence electrons in a solid of $N$ atoms; because the $2s$ spatial state in the lithium *atom* is only half full, the $2s$ *band* in *solid* lithium is only half full. Thus, as shown in Figure 9.33, there is still half a band of unfilled energy levels available to the $2s$ electrons. When an external electric field is applied, they are free to enter these states and thus participate

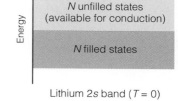

Lithium $2s$ band ($T = 0$)

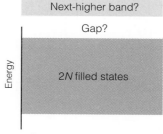

Beryllium $2s$ band ($T = 0$)

**Figure 9.33** Lithium's partially filled band—and beryllium's full one?

---

[17] As shown in Example 8.5, even room temperature may be considered cold for a normal metal.

[18] These are the simplest cases of interest, since hydrogen and helium form molecular solids, the latter only at pressures above about 25 atm.

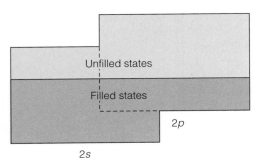

Unfilled states

Filled states

2p

2s

**Figure 9.34** Overlap may change full and empty bands into one partially filled band.

in electrical conduction (cf. Figure 9.32). A beryllium atom, on the other hand, has a full 2s level. Accordingly, the solid, with 2N valence electrons, should have a full 2s band. Assuming a significant energy gap to the next-higher band, electrons in beryllium should be unable to freely absorb energy and enter different states—that is, to respond to an external field. Beryllium should be an insulator.

Beryllium is a conductor. It is the last assumption that is incorrect. Given the degeneracies or near degeneracies (levels split by a very small amount) in atomic energy levels, it should not be surprising that some bands overlap—that is, they are not separated by a gap. As depicted schematically in Figure 9.34, we find that the 2s and 2p bands overlap significantly in beryllium. This being the case, the system's ground state will have electrons that would otherwise have to occupy states to the top of the 2s band instead filling the lower energy states of the 2p, until the highest occupied state is at the same level in both. Consequently, the electrons in beryllium also have a continuum of unfilled higher-energy states available just above them, making beryllium a conductor.

Relating atomic $(n, \ell, m_\ell)$ states to the bands in a solid can be rather complicated. We will be able to grasp the essentials of conductivity without the complication, by simply regarding overlapping bands as one band and considering how electrons fill these combined bands. The highest-energy band that would be completely full at zero temperature is defined as the **valence band,** and the band just above it as the **conduction band.** Whereas in the conduction band there would be freedom to absorb energy, in the valence band there is not; the electrons are packed full into all the states in the band. Figure 9.35 illustrates the features that distinguish insulators, conductors, and semiconductors. (Plotted alongside the

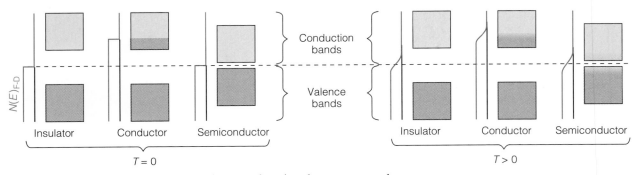

**Figure 9.35** Band filling of insulator, conductor, and semiconductor at zero and nonzero temperature.

bands in each case is the Fermi–Dirac distribution. As discussed in Chapter 8, at $T = 0$ the occupation number drops from unity to zero abruptly at the Fermi energy. At low nonzero temperatures, it drops less precipitously.) A material will be a conductor if there is significant occupation of the conduction band and a significant range of unfilled energy levels immediately above. Even at zero temperature, electrons would be free to absorb any amount of energy from an external field, so there will always be a response. A material will be an insulator if the conduction band is empty and the gap between it and the valence band is large.[19] Since the minimum amount of energy an electron in the valence band could absorb is large, there is essentially no response to an external electric field. Even at low nonzero temperatures, the number of electrons energetic enough to occupy a state in the conduction band is negligible.

The term "semiconductor" is applied to materials that are insulators at zero temperature, but in which the gap between the valence and conduction bands is small. By somewhat arbitrary definition, an insulator with a band gap smaller than about 2 eV is considered a semiconductor. With such a small gap, the temperature does not have to be very high for there to be significant occupation of the conduction band, and freedom to respond to a field. Thus, while at zero temperature semiconductors are insulators, at ordinary temperatures their conductivity is significant (though not as high as that of a true conductor). Notably, while the conductivity of a "conductor" decreases with temperature, that of a semiconductor increases, because the *number* of electrons in the conduction band increases markedly with temperature.[20]

Semiconducting elements commonly used in electronic devices are silicon and (to a lesser extent) germanium. Both are "covalent" solids with the same lattice geometry as diamond. Nevertheless, they are properly treated via band theory, as though all atoms share all valence electrons. The four valence electrons per atom completely fill a band—the valence band—formed from the four tetrahedral molecular bonding states, the $3sp^3$ in silicon and $4sp^3$ in germanium. The four unoccupied antibonding states form the empty higher-energy conduction band (see Fig. 9.16).[21] The bands are separated by an energy gap: 1.1 eV in silicon and 0.7 eV in germanium. (A similar band structure is found in the $n = 2$ states of diamond. But with a band gap of 5.4 eV, diamond qualifies as an insulator.)

## ◆ 9.7  The Conductivity Gap

To better appreciate the differences between conductors, insulators, and semiconductors, let us apply statistical mechanics to determine the number of thermally excited electrons in each of these materials.

---

[19]Insulators are not incredible accidents; cases in which one band is full and the next empty are common. Not only do we expect an atom with no partially filled subshells to form a solid with no partially filled bands, but, just as band *overlap* can turn expected insulators into conductors (e.g., beryllium), unfilled atomic subshells can *split* into isolated full and empty bands (cf. Figure 9.16), turning expected conductors into insulators (e.g., boron).

[20]This factor more than compensates for the reduction of conductivity due to increased collision frequency.

[21]Each atom alone has four *spatial* and, including spin, eight total $s$ and $p$ states, but it is not spin that splits each of the four $sp^3$ spatial states into two different energies; it is the presence at each lobe of a lobe from another atom.

## Example 9.3

Since there are no allowed states in a band gap, the density of states drops to zero at both edges of a band. Assume that between the edges it is a fairly smooth bump, as shown in Figure 9.36. In a normal conductor, the occupation number drops off quickly somewhere near the middle of the conduction band, at an energy $E_F$, measured from the band's bottom. Show that the fraction of conduction electrons occupying excited states—that is above $E_F$—is of the order of $k_B T/E_F$.

### Solution

To find a fraction, we need the number excited and the total number. The number above $E_F$ we find by integrating the product of $N(E)_{F-D}$ and the density of states from $E_F$ to the top of the band.

$$N_{excited} = \int_{E_F}^{E_{top}} N(E)_{F-D} D(E) \; dE$$

Since the Fermi–Dirac distribution falls to zero so quickly, the integrand is nonzero only within a very small range beyond $E_F$. In this range, the density of states is essentially constant, its value at $E_F$. Removing it from the integral, we have

$$N_{excited} \cong D(E_F) \int_{E_F}^{E_{top}} N(E)_{F-D} \; dE$$

We now insert $N(E)_{F-D}$ from (8-36), and since it drops quickly to zero, we may continue the integration to infinity without affecting the result.

$$N_{excited} \cong D(E_F) \int_{E_F}^{\infty} \frac{1}{e^{(E-E_F)/k_B T} + 1} \; dE$$

The value of the integral is $k_B T \ln 2$. Thus,

$$N_{excited} \cong D(E_F) k_B T \ln 2$$

The total number of electrons in the conduction band we find by integrating $N(E)_{F-D} D(E)$ over all energies in the band.

$$N = \int N(E)_{F-D} D(E) \; dE$$

Since the total number is independent of the temperature, we may carry out this calculation assuming that $T = 0$. This greatly simplifies the integral because $N(E)_{F-D}$ is unity up to $E_F$ and zero thereafter.

$$N = \int_0^{E_F} 1 D(E) \; dE$$

Were $D(E)$ a linearly increasing function, the integral would be simply $\frac{1}{2} D(E_F) E_F$. In fact, $D(E)$ is somewhat like a convex bump, as shown in Figure 9.36. Although this would lead to a coefficient a bit larger than $\frac{1}{2}$, the important factor is the product $D(E_F) E_F$.

$$N \cong D(E_F) E_F$$

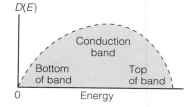

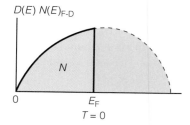

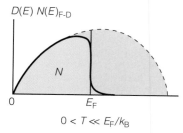

**Figure 9.36** Conductor band filling.

Similarly discarding the earlier ln 2, our quotient is

$$\frac{N_{\text{excited}}}{N} \cong \frac{D(E_F)k_B T}{D(E_F)E_F} = \frac{k_B T}{E_F}$$

We found in Example 8.5 that for silver $E_F = 5.5$ eV. Since $k_B T = 0.026$ eV at 300 K, we see that the fraction of conduction electrons occupying excited states in a typical conductor is approximately $\frac{1}{200}$, or about $\frac{1}{2}$%. Small though it is, with approximately $10^{23}$ electrons occupying the band, the number of thermally excited charge carriers is huge. Now let us gain some idea of just how scarce free charge carriers should be in an insulator or a semiconductor.

### Example 9.4

As a crude approximation, assume that the density of states for electrons in an insulator or semiconductor is a constant $D$, except in the gap between valence and conduction bands, where it is zero, as shown in Figure 9.37. Show that the fraction of valence electrons excited to states in the conduction band is governed by the factor $e^{-E_{\text{gap}}/2k_B T}$, where $E_{\text{gap}}$ is the width of the valence-conduction gap. Assume that $k_B T \ll E_{\text{gap}}$ and that $E_F$ is in the exact middle of the gap.[22]

#### Solution

As in Example 9.3, the total number of electrons in the valence band is most easily found at $T = 0$, where $N(E)_{\text{F-D}}$ is unity across the entire band. Since the density of states is also constant, the integration is trivial:

$$N = \int_{\substack{\text{val. band} \\ \text{bottom}}}^{\substack{\text{val. band} \\ \text{top}}} N(E)_{\text{F-D}} D \, dE = DE_v$$

At $T \neq 0$, the number excited to the conduction band is

$$N_{\text{excited}} = \int_{\substack{\text{cond. band} \\ \text{bottom}}}^{\substack{\text{cond. band} \\ \text{top}}} N(E)_{\text{F-D}} D \, dE$$

Also as in Example 9.3, we may extend the integral's upper limit to infinity with no effect; the occupation number is essentially zero well before the top. Noting that the bottom of the band is at $E = E_F + \frac{1}{2}E_{\text{gap}}$, we thus have

$$N_{\text{excited}} = \int_{E_F + \frac{1}{2}E_{\text{gap}}}^{\infty} N(E)_{\text{F-D}} D \, dE$$

Now removing the constant density of states from the integral and inserting equation (8-36), we obtain

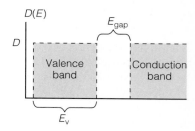

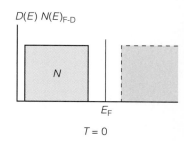

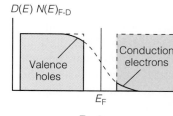

**Figure 9.37** Insulator/semiconductor band filling.

---

[22]The number of electrons excited *from* the valence band must equal the number excited *to* the conduction band; and the deviation of $N(E)_{\text{F-D}}$ from a true step function is symmetric about $E_F$ (cf. Figure 8.11). For both statements to hold, given our simple density of states, $E_F$ must be right in the middle of the gap.

$$N_{\text{excited}} = D \int_{E_F + \frac{1}{2}E_{\text{gap}}}^{\infty} \frac{1}{e^{(E-E_F)/k_B T} + 1} \, dE = Dk_B T \ln(1 + e^{-E_{\text{gap}}/2k_B T})$$

Given that $k_B T \ll E_{\text{gap}}$, the exponential will be very small. We may therefore use the approximation $\ln(1 + \epsilon) \cong \epsilon$ for $\epsilon \ll 1$, leaving

$$N_{\text{excited}} = Dk_B T \, e^{-E_{\text{gap}}/2k_B T}$$

Finally, division by $N = DE_v$ yields

$$\frac{N_{\text{excited}}}{N} = \frac{k_B T}{E_v} e^{-E_{\text{gap}}/2k_B T}$$

Bands tend to be of order 10 eV in width (see Exercise 18), and $k_B T$ at room temperature is of order $10^{-2}$ eV, so the multiplicative factor is approximately of order $10^{-3}$. This quotient is roughly analogous to the one in the previous example for a metal. Here, however, the exponential factor leads to a great suppression of electrons in excited states. (It should be noted that a more realistic density of states would change only the less-important multiplicative factor.) Given a band gap of 5 eV, typical for an insulator, the exponential factor at room temperature would be $\sim 10^{-42}$! Since $N$ is "only" of order $10^{23}$, we should expect no electrons whatsoever in the conduction band of an insulator, and thus no current in response to an external electric field. The difference in conductivity between conductors and insulators is one of the most stark contrasts in nature, and was baffling to early physicists. We see that it is rooted in the existence of an energy gap. For a semiconductor, with a gap of $\sim 1$ eV, the exponential factor is of order $10^{-8}$. Its extreme sensitivity to modest changes in $E_{\text{gap}}$ is shown by the fact that a decrease in $E_{\text{gap}}$ by only a factor of 5 increases the fraction excited by over 30 orders of magnitude! Grown to $10^{-8}$ of the total (or $10^{-11}$, taking the multiplicative factor into account), the number of conduction electrons in a semiconductor is significant.

## 9.8  Semiconductor Theory

The simple electrical circuits of resistors, capacitors, and inductors studied in introductory electricity and magnetism are of limited utility; without a means of *amplification*, electronic communication—public address systems, telephone, radio, television—would be impossible. Related to amplification is electronic switching—that is, the ability to change current from off (a zero) to on (a one), the basis of digital computing. (A switch is merely an amplifier that knows no middle ground.) Without this, the microchip control so pervasive in today's civilization could not exist. A central question is, why are modern electronics and semiconductors so inextricably related? As we shall see, two basic devices of electronic control are the **diode** (Figure 9.38), through which current may flow but one way (distinguishing it utterly from resistors, capacitors, and inductors),

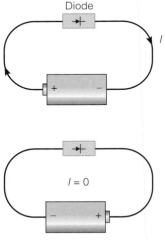

**Figure 9.38** A diode allows current to flow only one way.

and the **transistor** (Figure 9.39), by which a small current in one circuit may control a large current in another—the essence of amplification. Semiconductors are the most convenient way to accomplish these and other tasks because they can possess charge carriers of *both* positive and negative sign, whose concentrations can be easily varied. Now it is true that the positive ions in a solid do not move about freely, but in semiconductors the net response of the (negative!) electrons to an external field can be precisely the same as freely moving positive charges. These effective positive charge carriers are known as **holes**.

## Holes

The identification of effective electron behavior with the motion of positive charges is much deeper than the simple notion of "conventional current." In introductory electric circuits, a current of negative charges moving opposite an electric field is often treated as a conventional current of positive charges moving in the direction of the field. For most purposes, the two are indistinguishable. Convenient though this view may be, however, it is easily demonstrated that the charge carriers in an ordinary conductor, such as copper, are *not* positive. The simplest way involves the Hall effect, in which positive-carrier and negative-carrier currents behave differently in an external *magnetic* field. Holes, on the other hand, behave precisely as positive charges whether the field is electric or magnetic. For all practical purposes, positive charge carriers do exist. Let us now characterize holes, and see why they are to be found in semiconductors but not ordinary conductors.

A hole is the absence of an electron in the valence band. In a semiconductor at zero temperature, the conduction band is completely empty and the valence band completely full. The valence electrons have no freedom to respond to an electric field because there are no empty states easily accessible into which they might move. But at *nonzero* temperature, if an electron is promoted via thermal excitation from the top of the valence band to the bottom of the conduction band, a "hole" remains (Figure 9.40). Not only is there a new electron free to move in the conduction band, but the empty state it leaves behind represents a certain freedom in the *valence* band. Suppose the promoted electron had occupied state $n$ in the valence band—the hole is in state $n$. Now an electron in state $n'$ in the valence band may jump into state $n$, creating a vacancy in state $n'$—the hole has changed states. This process is repeated continuously. Thus, while the valence electrons do not move freely, always constrained to move only into the one unoccupied state, that unoccupied state changes arbitrarily. The hole is as free to move among states as an *electron* in an *empty* band would be.

Now imagine dividing a full valence band into a single electron free to move among states and the balance of the electrons in the band—that is, the corresponding hole. Assuming an electrically neutral material, since the electron has charge $-e$, the balance of the band, the hole, has a net charge $+e$. Furthermore, since there is no response to an external field in a *full* band—in particular, there is no current—a hole in a given state must have the same velocity as would the single electron if occupying that state. We conclude then that *a hole behaves as a charge $+e$ and moves as would an electron free in the valence band.*

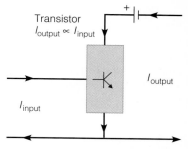

Figure 9.39 In a transistor, a small current controls a large one.

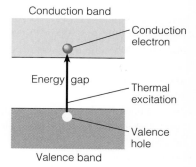

Figure 9.40 Thermal excitation creating a pair of charge carriers.

Still a thorny problem remains: If moving as would a free electron, it seems that a hole should move opposite an external electric field, and if so, it should tend to cancel the current due to the electron "up" in the conduction band. There should be no current when both positive (valence band hole) and negative (conduction band electron) move *in the same direction.* In fact, an external field does produce a net current; the electron in the conduction band moves opposite the field, but *the hole moves in the direction of the field.* The surprising reason is that a free electron near the top of the valence band would move in the direction of the field—in the direction a positive charge would move! At the heart of this peculiarity is **effective mass,** which for top-of-the-valence-band electrons is negative!

By the classical second law of motion, mass is $F_{net}/a$. Mass being positive goes along with the acceleration always being in the direction of $F_{net}$. For electrons in a solid, $F_{net}$ would include external forces, such as external electromagnetic fields, and forces internal to the solid, due to the positive ions. However, since we are usually interested in the net response (i.e., $a$) to an applied external force, it is sensible to refer to an effective mass defined as $F_{ext}/a$. Nevertheless, there *are* internal forces, so it is possible for the *net* force to be opposite the *external* force, meaning that $a$ would be opposite $F_{ext}$. Thus may effective mass be negative. Taking into account forces internal to the medium through which it moves, it is shown at the end of the section that the effective mass of an electron in a crystalline solid is given by

$$m_{eff} = \hbar \left( \frac{d^2\omega}{dk^2} \right)^{-1} \tag{9-8}$$

As usual, $\omega$ is $E/\hbar$.

It is helpful to see what (9-8) predicts for a free particle—that is, in the presence of no medium at all. In such a kinetic-only case, $E = p^2/2m = \hbar^2 k^2/2m$, so that $\omega = \hbar k^2/2m$. Thus,

$$m_{eff} = \hbar \left( \frac{d^2}{dk^2} \frac{\hbar k^2}{2m} \right)^{-1} = \hbar \left( \frac{\hbar}{m} \right)^{-1} = m \quad \text{(Free particle)}$$

This is certainly reassuring; with no internal forces to interfere, we should expect the ratio of external force to acceleration to be the usual mass.

To see how effective mass fits in with hole motion in a semiconductor, let us return to the plot of electron energy versus wave number in a crystalline solid (Figure 9.31). As shown in Figure 9.41, for $k$-values near the middle of a band, energy follows the same trend it would in one large infinite well—the parabola $\omega = \hbar k^2/2m$; the electrons are unaffected by internal forces and the effective mass is the true mass. The most energetic electrons in an ordinary conductor occupy such middle-of-the-band states and so respond to an external field as free particles. At $k$-values near the edges of a band, however, the ionic potential becomes significant—internal forces affect the electrons markedly. An electron near the bottom of a band, such as one thermally excited to the conduction band in a semiconductor, occupies a state where the curvature $d^2\omega/dk^2$ is greater than for a free particle. Thus, it behaves as though less massive than usual. On the other hand, an electron near the top of a band, where thermal excitation would leave a hole, occupies a state in which $d^2\omega/dk^2$ is negative, and so has negative effective mass. Accord-

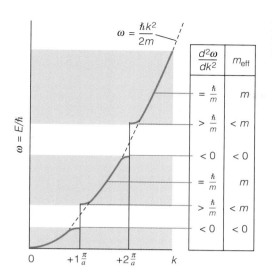

$$\omega = \frac{\hbar k^2}{2m}$$

| $\dfrac{d^2\omega}{dk^2}$ | $m_{\text{eff}}$ |
|---|---|
| $= \dfrac{\hbar}{m}$ | $m$ |
| $> \dfrac{\hbar}{m}$ | $< m$ |
| $< 0$ | $< 0$ |
| $= \dfrac{\hbar}{m}$ | $m$ |
| $> \dfrac{\hbar}{m}$ | $< m$ |
| $< 0$ | $< 0$ |

**Figure 9.41** An electron's effective mass depends on what state it occupies.

ingly, it accelerates *opposite* the external force, and so in the direction of an external electric field. Moving as would such an electron, a hole near the top of the valence band accelerates in the direction a positive charge would accelerate.[23] Figure 9.42 illustrates how conduction electrons and valence holes respond to an external field.

Clearly, the criteria defining holes do not apply to an ordinary conductor. The active band is the conduction band, and while far from empty, it is far from full (cf. Figure 9.35). Rather than having only a few free spaces at the negative-mass top of a band, it has many normal-mass electrons with plenty of "room" to respond to an electric field.

## Hole Energy

It is helpful to contrast hole energy with electron energy. A lower-energy state results when an electron jumps down from a higher-energy state to a hole in a lower-energy state, but in the process the hole jumps upward (Figure 9.43). Viewing the hole as the bearer of energy, to attain a lower-energy state a hole must move upward—opposite the direction an electron would move. A useful analogy is an air bubble rising through water: The system's energy decreases whether we view it as the sinking of the water (sea of electrons) or the rising of a bubble (hole). As an electron tends to sink to the lowest-energy state open, a hole tends to "float" as high as possible. In particular, an electron may jump from one band down to a lower one, provided there is room (a hole) for it, and a hole may jump to a higher band, provided that at least one electron occupies the band (that it isn't full of holes).

---

[23] It is interesting to note that there are inflection points in Figure 9.41, where $d^2\omega/dk^2$ is zero. The effective mass of an electron in such a state is infinite! The explanation of this surprising conclusion is that, owing to particularly strong internal forces between the solid and the electron in such a state, the external force is communicated to the solid as a whole. Clearly, the electron would seem very massive.

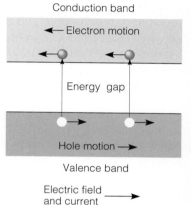

**Figure 9.42** Conduction at $T > 0$ in a semiconductor.

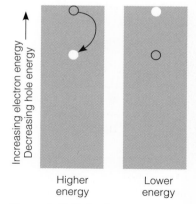

**Figure 9.43** Holes float.

## Doping

The semiconductors we have considered thus far are referred to as **intrinsic semiconductors,** pure materials in which the number of electrons thermally excited to the conduction band necessarily equals the number of holes left behind in the valence band. By **doping**—addition of small amounts of impurity elements—we produce **extrinsic** semiconductors. These have a preponderance of either conduction electrons or valence holes and, in either case, a much larger concentration of charge carriers than the pure intrinsic semiconductor. In the following discussion, we assume the intrinsic semiconductor to be of the most common type: a "covalent" lattice of tetravalent (valence 4) atoms, such as silicon or germanium.

An **n-type** extrinsic semiconductor is produced by interspersing in an intrinsic semiconductor a small fraction of **impurity** atoms of a pentavalent (valence 5) element, such as phosphorus or arsenic. Typical doping would have one of every $10^5$ atoms of the intrinsic material replaced in the covalent lattice by an impurity atom. In the intrinsic semiconductor, the valence band is completely full. Each pentavalent impurity atom adds an extra electron for which there is no room, since only four of its electrons may share in the solid's tetrahedral structure. This does not automatically put the extra electron in the conduction band, however, for each impurity atom also has an extra *positive* charge, and this complicates things. The effect of these extra positive charges is to create new electronic states, known as **donor states,** one for each impurity atom. They may be thought of as states in which the extra electrons are bound to the extra positive charges at the impurity sites, and the energy needed to free the electrons from this bond is the amount by which the donor states lie below the conduction band. Due to the dielectric polarizability of the intrinsic material, however, the attractiveness of the extra positive charges is dulled; they bind the extra electrons very weakly. As shown in Figure 9.44, donor states are typically only 0.01–0.07 eV below the conduction band, so close that even fairly low temperatures readily excite electrons from them into the conduction band. (Exercise 25 outlines a simple, hydrogenlike approximation that gives a reasonably accurate value for the binding energy.) Because the impurity atoms are so sparse as to be essentially isolated from one another, donor states do not form a band. Therefore, the earlier arguments about holes do not apply. Vacated donor states do not behave as free positive charge carriers, but the electrons vacating these states do become free negative charge carriers in the conduc-

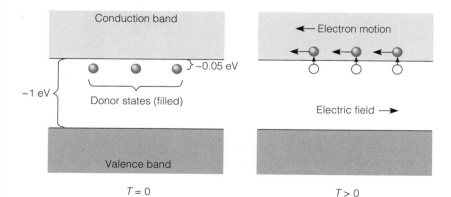

**Figure 9.44** Conduction in an n-type semiconductor.

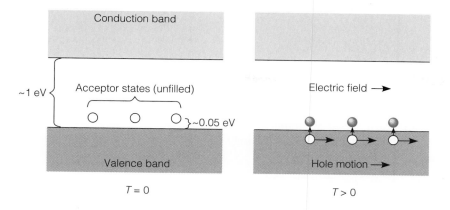

**Figure 9.45** Conduction in a p-type semiconductor.

tion band. Thus, in an n-type (negative) semiconductor, the predominant charge carriers are conduction band electrons.

By complementary arguments, a **p-type** (positive) extrinsic semiconductor is produced using impurity atoms of a trivalent (valence 3) element, such as aluminum or gallium (Figure 9.45). Since each impurity atom lacks one electron to complete the tetrahedral covalent bonding, we might expect the valence band not to be full. But because each impurity atom also has one less positive charge, electrons added to complete the tetrahedral bonding would not be bound so tightly as those in the pure intrinsic solid. Thus, the unfilled states are slightly above the valence band. They are known as **acceptor states**. Alternatively, as *filled* donor states in an n-type semiconductor may be viewed as electrons bound just below the conduction band, the *unfilled* states in a p-type semiconductor may be viewed as holes bound just above the valence band. Relative to the surrounding lattice of tetravalent positive ions, the trivalent ion behaves as a point of negative charge (though again dulled by dielectric polarization), able to bind the hole in place. Bearing in mind that holes are free to move about once they reach the *valence* band, and that hole energy increases in the downward direction, the hole/acceptor states should indeed be bound just above the valence band. As Figure 9.45 depicts, little thermal excitation is needed for holes to jump down (for electrons in the valence band to jump to acceptor states). Like donor states, acceptor states do not form a band, so electrons jumping to these states from the valence band are not free to respond to an external field, but the holes left behind are certainly free to do so. In a p-type semiconductor, the predominant charge carriers are valence band holes.

In both n-type and p-type semiconductors, conductivity is much greater than in the intrinsic material. The simple reason is that the jumps from the valence band to the acceptor states and from the donor states to the conduction band are much smaller than the intrinsic material's valence–conduction gap: ~0.05 eV versus ~1.0 eV. As a rough measure, let us use a result obtained in Example 9.4. Of potential charge carriers in the *intrinsic* material, only the fraction $e^{-1\,\text{eV}/2(0.03\,\text{eV})} \cong 10^{-8}$ would actually be available for conduction; but $e^{-0.05\,\text{eV}/2(0.03\,\text{eV})} \cong 10^0$ of the potential extrinsic charge carriers would be available—that is, valence holes in p-type and conduction electrons in n-type. Therefore, even with only a one in $10^5$ level of doping, the number of charge carriers in a doped semiconductor is much larger than in the pure intrinsic material. It is true that there is always some thermal excitation across the *entire* valence-to-conduction gap, creating electron-hole pairs in both n-type and p-type semiconductors. Both will therefore always harbor charge carriers of the "wrong" sign (i.e., valence holes in n-type and conduction electrons in p-type), known as **minority carriers.** But these are small in number. The important charge carriers are the **majority carriers:** valence holes in p-type and conduction electrons in n-type.

What makes doped semiconductors unique is the ease with which the sign and abundance of the charge carriers may be arbitrarily varied, via impurity type and concentration. As we see in the next section, these qualities are indispensable to the fabrication of the simple devices basic to modern electronics.[24]

---

### ◆ Effective Mass

In Section 5.3, it is shown that the velocity of a classical particle, called group velocity, is related to its matter wave properties $\omega$ and $k$ by

$$v_{particle} = v_{group} = \frac{d\omega}{dk}$$

The internal forces between the particle and the medium through which it moves are taken into account in the expression of $\omega$ as a function of $k$ (cf. Example 5.3). Thus, the acceleration of a particle in any medium, including a crystalline solid, may be written

$$a = \frac{d}{dt}v_{particle} = \frac{d}{dt}\left(\frac{d\omega}{dk}\right) = \frac{d^2\omega}{dk^2}\frac{dk}{dt}$$

To find *effective* mass, we divide the *external* force (only) by this acceleration, expressed in terms of $\omega$ and $k$. To make sense of the quotient, we must express the external force in the same terms. This we do by noting that (ignoring losses due to frictional forces) it is the external force that does the net work on a particle moving through a medium. The force times the particle's velocity is the work per unit time (power):

$$F_{ext}v_{particle} = \frac{dE}{dt}$$

Using $E = \hbar\omega$, we then have

$$F_{ext}v_{particle} = \hbar\frac{d\omega}{dt} = \hbar\frac{d\omega}{dk}\frac{dk}{dt} = \hbar v_{particle}\frac{dk}{dt}$$

or

$$F_{ext} = \hbar\frac{dk}{dt}$$

Thus,

$$m_{eff} \equiv \frac{F_{ext}}{a} = \frac{\hbar\dfrac{dk}{dt}}{\dfrac{d^2\omega}{dk^2}\dfrac{dk}{dt}} = \frac{\hbar}{\dfrac{d^2\omega}{dk^2}}$$

[24] In some elements, called "semimetals" (e.g., bismuth, antimony), electrons fill states to a point at which there is a *slight* overlap between bands. With some electrons occupying the low-lying states in the conduction band and leaving empty states at top of the valence band, there is conduction due to *both* conduction electrons and valence holes. Nevertheless, semimetals cannot fill the role of semiconductors. As we soon see, the energy *gap* is crucial to the operation of important circuit devices.

The most common use of a diode is in a "recti-fier." The standard electrical power supplied by power companies is ac (alternating current), in which the current's direction oscillates back and forth at 60 Hz. This is preferable to the alternative, dc (direct current), because the range of voltages needed by home and business users is huge. The only easy way of stepping-up or stepping-down voltage is with a transformer (an integral part of the user's electrical device), and transformers, depending on electromagnetic induction, require time-varying current. Nevertheless, once of the proper voltage, the electricity must often be rectified—converted to a constant dc. A diode allows current to flow only during parts of the cycle when it flows the proper way, and a "filter capacitor" smoothes out the otherwise choppy output.

## 9.9   Semiconductor Devices

Let us now study how doped semiconductors may be exploited to produce two of the most important circuit elements.

### The Diode

There are many applications in which it is necessary to allow current to flow only one way. But to actually build a device that does so automatically, known as a **diode,** is another matter. (See the essay entitled "Why Make a Diode?") We now have a solution. Asymmetric current flow can be achieved through a simple *physical* asymmetry: We produce a diode by putting an n-type semiconductor in contact with a p-type semiconductor. The area of contact is known as a **p-n junction.** As shown in Figure 9.46, if we apply a potential difference, with the p-type side at higher potential, known as a **forward-biased** condition, free holes in the p-type and free electrons in the n-type flow toward the junction. When they meet, they **recombine,** electrons dropping down from the conduction band across the energy gap to fill the holes in the valence band (or holes jumping *up* to the conduction band). Current flows continuously as electrons are added at the low-potential side, holes are added (valence electrons removed) at the high potential side, and the two recombine at the junction. (The energy released as the conduction electrons drop

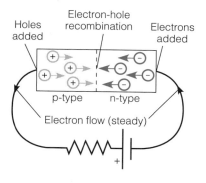

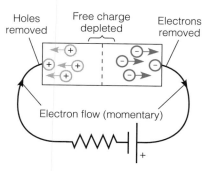

**Figure 9.46** The behavior of the charge carriers in forward- and reverse-biased p-n junctions.

**Forward bias**

**Reverse bias**

down to the valence band is the source of light in an LED, or light-emitting diode.) On the other hand, if the n-type is at the higher potential, known as **reverse-biased,** both holes and electrons move *away* from the junction, electrons exiting the n-type and holes the p-type (filled by electrons from the external circuit). While in the forward-biased condition, energy is *released* as free electrons and holes recombine, in the reverse-biased condition, nothing is available to pull apart electron-hole pairs, to promote electrons from the valence to the conduction band.[25] Thus, a region devoid of free charge carriers quickly forms, and current flow is almost instantly zero.

The energy-level diagrams of Figure 9.47 clarify the goings-on. For convenience, we make the plausible assumption that if the p-type and n-type are separate, both p-type acceptor levels and n-type donor levels are half full (i.e., they define the Fermi energy in their respective types). Now suppose they are put in contact, but without an externally applied potential difference—an **unbiased** condition. Electrons in the n-type conduction band are at higher energy than any in the p-type, and holes in the p-type valence band are at higher energy than any vacant states in the n-type. Thus, conduction electrons and valence holes, freely crossing the junction due to random diffusion, may combine and release energy. But this migration cannot persist; adding electrons into the p-type shifts all its electron energy levels upward (due to the electrons' mutual repulsion), while the corresponding loss of electrons in the n-type shifts its energy levels downward. Only a very small fraction need cross to produce an internal electric field strong enough to discourage further crossing. (The Fermi energies become equal.) This **built-in electric field** is typically $10^6$ V/m and occupies the region surrounding

[25]While this is usually the case, an important exception is the solar cell, in which the electrons are "forced upward" to the higher-energy side by the photons in light deliberately aimed at the junction. The process is essentially opposite to that in an LED.

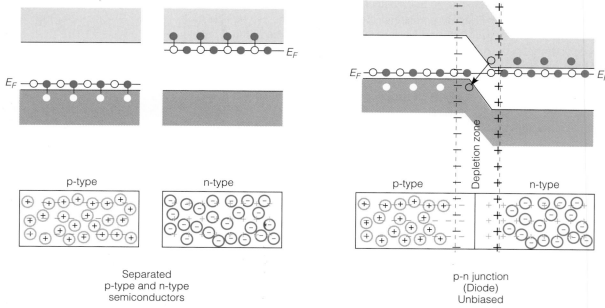

Separated
p-type and n-type
semiconductors

p-n junction
(Diode)
Unbiased

**Figure 9.47** When p-type and n-type are joined, charges transfer as electron energy levels equalize, producing a dogleg in the bands.

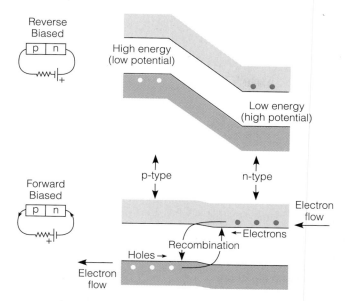

**Figure 9.48** Bands in a p-n junction under reverse and forward bias.

the junction, known as the **depletion zone,** where free charge carriers are no longer to be found.

Applying a reverse bias—lower potential on the p-type side—exaggerates the dogleg in the energy bands (Figure 9.48), shifting the p-type energies still higher relative to the n-type. (Remember: Low potential is high potential *energy* for a negative electron.) Again no current flows: Always restricted to moving through the allowed states in the distorted conduction band, the n-type conduction electrons are even less able to "climb the hill" to the p-type side than before.[26] Holes cannot make the "dive."

A point important to the coming discussion of the transistor is that any minority carriers, *were they present,* would very much like to cross the junction in a *reverse-biased* condition. Conduction electrons in the p-type would readily slide downward to the n-type conduction band, while valence holes in the n-type would float upward to the p-type valence band. However, minority carriers in the diode arise only from thermal excitation of electron-hole pairs across the entire gap. These being few, the resulting "thermal current" is small.[27]

If the diode is forward biased, the n-type conduction band is shifted upward, toward that of the p-type. Conduction electrons near the junction will readily wander (the term is "diffuse") across as soon as the unevenness is eliminated, and once in the same physical region, they and valence holes engage in recombination, releasing an energy per event of about $E_{gap}$. For a band gap of 1 eV, we would need a forward bias of only about 1 V. Actually, the diode begins to turn on slightly before the conduction bands are even, because the conduction electrons are spread across an energy range of roughly $k_B T$. But since this is only about $\frac{1}{40}$ eV at room temperature, the turn-on is rather sharp. Indeed, in practice the voltage across a forward-biased diode is always very close to $E_{gap}/e$.[28] The conduction band unevenness being thus very small, the electrons are not "pushed" (electrostatically) across the junction; they simply wander.

[26]Electrons might circumvent the climb up the conduction band potential barrier by tunneling through to the p-type valence. The "tunnel diode" is designed to exploit this (see Exercise 32), but it is not a significant effect in a normal diode.

[27]One of the reasons silicon is preferable to germanium in most applications is that it has a larger band gap; thermal excitation will thus produce fewer minority carriers and so a smaller minority carrier current.

[28]Any bias voltage *in excess* of $E_{gap}/e$ would effectively be a potential difference across a fairly good conductor. Thus, in a circuit a forward-biased diode behaves more or less as an opposing battery of fixed potential drop $E_{gap}/e$, about a volt, in series with a simple wire; the potential drop never differs much from $E_{gap}/e$.

## The Transistor

The transistor is a straightforward application of ideas we have already discussed. If we join in succession three alternating extrinsic semiconductors (with certain refinements), we produce a transistor. A transistor can be either a pnp or npn, depending on which type is found in the center. Since the operation of the pnp is fully complementary, we restrict our attention to the npn.

The p-type center region of an npn transistor is called the **base** and the n-type regions the **emitter** and **collector.**[29] In operation (Figure 9.49), the emitter-base junction is forward biased, and conduction electrons therefore flow from the emitter into the base. These electrons are discouraged from recombining with holes in the base by making the base of small size and very light doping. The beauty of the transistor is that these conduction electrons are *minority* carriers in the base and the collector-base junction is reverse-biased. Consequently, while a small fraction do recombine and pass out the base, the rest are quite happy to "slide down" to the collector's conduction band. This is the essence of amplification: While the emitter-base bias (the input) controls the amount of current that gets into the base from the emitter, most of it goes out the collector (the output); the emitter-collector current is *proportional* to the emitter-base current *and typically a hundred times greater!* For their 1948 discovery of the transistor, John Bardeen, Walter Brattain, and William Shockley were awarded the 1956 Nobel prize in physics.

Figure 9.50 shows the rudiments of a transistor amplifier. The input, a microphone, introduces small-amplitude voltage, and thus current, oscillations in the

[29] Although the emitter and collector might seem interchangeable, the two junctions are biased differently—one forward, one reverse. The n-type region of the reverse-biased junction heats up as the conduction electrons plummet to its conduction band from the higher levels. This "collector" is thus usually made relatively large and often mounted on a heat sink. Doping level also differs between emitter and collector, to optimize each to its role.

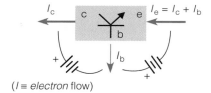

$I_e = I_c + I_b$

($I \equiv$ *electron* flow)

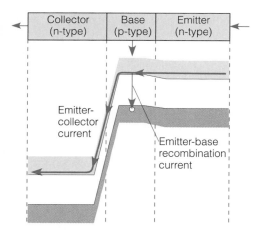

**Figure 9.49** Bands and charge flow in an npn transistor.

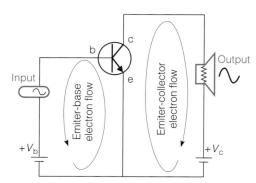

**Figure 9.50** The elements of a transistor amplifier.

emitter-base circuit. Corresponding large-amplitude current oscillations in the emitter-collector circuit then produce a much more powerful signal at the output/speaker. (*Note:* The collector-base junction *is* reverse-biased; $V_c$ is generally much larger than $V_b$ and keeps the collector at higher potential than the base.)

The so-called "bipolar" transistor discussed here is only one of many semiconductor amplification devices. Another is the MOSFET, discussed in Exercise 31. Its electrical properties coupled with its ease of fabrication make it one of the most common switching devices in integrated circuits (Figure 9.51).

## Integrated Circuits

Doped semiconductors provide a very convenient route to microelectronic circuits. A silicon chip can be fabricated in which the doping is arbitrarily n-type or p-type in successive minuscule regions. Consequently, many transistors can be fit into a very small space. Moreover, single n-type or p-type regions may be doped heavily to serve as highly conducting "wires" or lightly to serve as resistors. In this way, a single microscopic silicon chip may contain an entire complex circuit (Figure 9.52), hence the name **integrated circuit.** A discussion of production techniques and circuits may be found in most modern electronics texts.

## The Diode Laser

Certainly one of the today's more important semiconductor devices is the diode laser. (The basics of laser operation are discussed in Section 8.9.) As noted earlier, the energy given up when electrons and holes recombine across the gap of a forward-biased diode is the source of light in an LED; essentially, $E_{gap}$ becomes the photon energy. It is also the light source in the diode laser, but for this purpose the usual sparse doping won't do. Were there only a relative few electrons in the conduction band and holes in the valence band, the upper state of the light-producing transition would be far from overpopulated and the lower one far from empty. There would be no population inversion. As shown in Figure 9.53, the key is heavy doping. High impurity-atom density causes the occupied donor and unoccupied acceptor levels to spread into **impurity bands,** which overlap, respectively, the n-type conduction and p-type valence bands. If at least one of these impurity bands is fairly wide, a photon energy considerably greater than $E_{gap}$ is required to induce *absorption;* the energy jump from the highest occupied valence state to the lowest unoccupied conduction state, from the bottom of empty acceptor band to the top of full donor band, is $E_{Fc} - E_{Fv}$ and is greater than $E_{gap}$. However, the

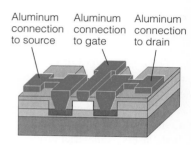

Aluminum connection to source    Aluminum connection to gate    Aluminum connection to drain

**Figure 9.51** The basic design in an integrated circuit of a single MOSFET, whose width is typically only about a micrometer.

**Figure 9.52** A microprocessor integrated circuit chip containing millions of transistors, perched on a stack of pennies.

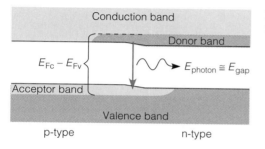

Conduction band

Donor band

$E_{Fc} - E_{Fv}$    $E_{photon} \cong E_{gap}$

Acceptor band

Valence band

p-type          n-type

**Figure 9.53** A population inversion in a forward-biased laser diode.

minimum-energy *emission* transition is from the bottom of the donor band to the top of the acceptor, roughly $E_{gap}$. Thus, for photons whose energy is at least $E_{gap}$ but less than $E_{Fc} - E_{Fv}$, an essentially complete population inversion exists.[30]

As in other types of lasers, a resonant cavity is needed to amplify the desired wavelength. In the diode laser, it is formed simply by cleaving the ends of the crystal parallel to each other and at the proper separation. (Semiconductors' high refractive index gives sufficiently complete reflection.) Because it too often gives up its recombination energy in forms other than light, silicon is a poor choice for diode lasers. A material producing photons with much higher efficiency is gallium arsenide, a compound lattice formed of trivalent (column-III gallium) and penta-valent (column V arsenic) atoms. Other combinations symmetric about column IV (both III–V and II–VI) are also being used. Diode lasers typically have effi-ciencies—electrical power to light power—greater then 50%, and these "opto-electronic" devices are playing an ever more important role in modern electronics.

## ▾ 9.10 Superconductivity

Some materials at low temperature have essentially *zero* electrical resistance! Figure 9.54 shows resistivity ($1/\sigma$) versus temperature for tin and copper. While copper retains resistance down to the lowest measurable temperatures, tin's resis-tivity plummets to zero at its **critical temperature** $T_c$, a characteristic of the ma-terial. (This is unrelated to the "critical temperature" of a real gas.) In such a state, the material is known as a **superconductor.** About 40% of the natural elemental metals have been found to become superconductors at low temperature. In this section, we discuss features common to superconducting materials and the mecha-nisms by which superconductivity occurs.

With no resistance, an electric current established in a superconducting mate-rial should persist indefinitely—*without an applied voltage.* Indeed, many experi-ments have been done testing this very prediction, one indicating a *minimum* time for significant current decay of 100,000 years!

Another feature of superconductors is that they exclude magnetic field lines, depicted in Figure 9.55. If a material is below its critical temperature (i.e., in a

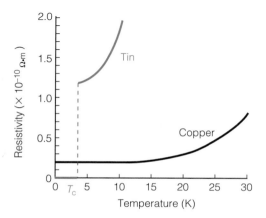

**Figure 9.54** Copper always has electrical resistance; tin becomes a superconductor.

[30]The tops and bottoms of the donor and acceptor bands correspond roughly to the four levels of a four-level laser. Since the electrons hovering above the p-type acceptor band are "injected" from the n-type donor band (or, equivalently, holes are injected from the p-type acceptor to hover below the n-type donor band), the pumping mechanism is known as "cur-rent injection."

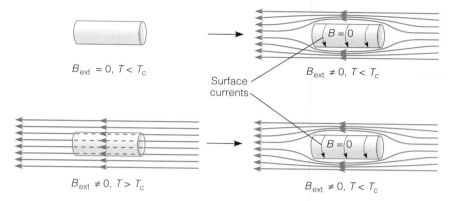

$B_{ext} = 0, \, T < T_c$

$B_{ext} \neq 0, \, T < T_c$

Surface currents

$B_{ext} \neq 0, \, T > T_c$

$B_{ext} \neq 0, \, T < T_c$

**Figure 9.55** When it drops below its critical temperature, a superconductor expels magnetic field lines.

superconducting state) in a region of zero magnetic field, then an external magnetic field is introduced, the field lines will curve around the object, never passing through. This might be expected on the basis of Faraday's law: Any change in magnetic flux through a cross section would induce an emf, and if there is no resistance, an arbitrarily large surface current is possible, able to produce an induced internal field that cancels the external. However, it can be shown that even perfect conductivity would allow an external field to creep slowly from the surface into the material. Moreover, there is a closely related magnetic behavior that cannot be attributed merely to a complete lack of resistance: If a superconducting material *above* its critical temperature is *initially* permeated by a magnetic field, *then* cooled, the magnetic field lines will be expelled as it drops below its critical temperature. The result is the same—a superconducting state that excludes magnetic field lines. Since the external field is not varied, we might expect no currents to be induced that might cancel the external field. But surface currents do arise—and they persist indefinitely. To explain magnetic effects, then, we must attribute to superconductors not only perfect conductivity, but perfect diamagnetism. This being the case, a superconductor will always repel a magnet. A vivid example is shown in Figure 9.56, in which a small magnet hovers above a disk of

**Figure 9.56** To exclude the field lines of the permanent magnet (above), a superconducting disk (below) becomes an opposing magnet.

superconducting material. Starting at room temperature with the magnet resting on the disk, liquid nitrogen (77 K) is added to the dish surrounding the disk; the disk cools to below its critical temperature and the magnet is pushed upward, its field lines expelled from the disk. Exclusion of magnetic field lines by a superconductor is known as the **Meissner effect.**

For all superconducting materials, there is a limit on the strength of the external field they can exclude. If the field is too strong, superconductivity is destroyed and the field penetrates the material in the normal way. The field strength at which this occurs is known as the **critical field** $B_c$. It is temperature dependent—highest at $T = 0$ and dropping to zero at the critical temperature. It is natural to wonder what use might be made of superconductivity. Such thoughts inevitably lead to the reason so much is made of superconductors' magnetic properties: According to Ampere's law, electrical current is always accompanied by magnetic fields, and if there is a limit on the magnetic field allowed in a material, then there will be a limit on its ability to carry current. Let us now study two types of superconductor and their important magnetic properties.

## Type-I and Type-II Superconductors

The earliest known superconductors were what are now referred to as Type-I superconductors. Figure 9.57(a) shows the dependence of critical field $B_c$ on temperature in the case of lead. Type-I superconductors are characterized by a sharp transition as the external magnetic field increases—from field-excluding superconducting state to field-penetrating non-superconducting state. Figure 9.58(a) shows how the net and induced fields vary with the external at a given $T < T_c$. Up to $B_c$ the net field is zero. The induced field, produced by currents at the sample's surface, perfectly cancels the external. Beyond $B_c$ the material is "normal," the external field passing through freely. Superconducting *elements* tend to be Type-I. Their critical field values and critical temperatures are relatively low (typically $0.01–0.1T$ and $1–9$ K, respectively), and their practical usefulness thus limited.

Type-II superconductors are distinguished by their inclination to spontaneously form microscopic **vortices,** resulting in two critical field values. A vortex is a superconducting region of circulating current, surrounding a "tube" in which the material is in the non-superconducting state and through which field lines pass, known as a **normal core** (Figure 9.59). At external field strengths below a characteristic value $B_{c1}$, a Type-II superconductor behaves just like a Type-I: Surface

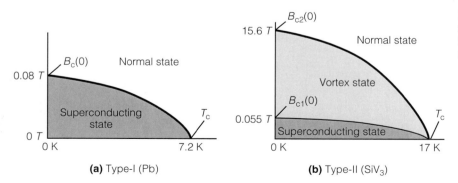

**(a)** Type-I (Pb)

**(b)** Type-II (SiV$_3$)

**Figure 9.57** Variation of critical magnetic fields with temperature for Type-I and Type-II superconductors.

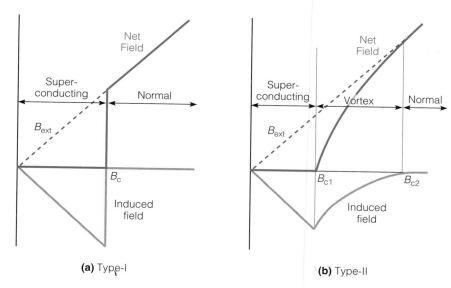

(a) Type-I

(b) Type-II

**Figure 9.58** External, induced, and net fields in Type-I and Type-II superconductors.

currents completely exclude the external field. Above $B_{c1}$ the **vortex state** begins. With the formation of vortices, some field lines pass through the material via normal cores while the surrounding regions remain electrically superconducting. As the field is increased, vortices become more dense (Figure 9.60), allowing more field lines to pass through. The vortex state ends—the entire sample becomes a normal conductor—when the external field exceeds the characteristic value $B_{c2}$. Partial flux penetration in the vortex state is shown clearly by the net-field curve of Figure 9.58(b). By the time $B_{c2}$ is reached, all of the external fields lines pass through—that is, $B_{ind} = 0$ and $B_{net} = B_{ext}$. As shown in Figure 9.57(b), both $B_{c1}$ and $B_{c2}$ decrease to zero as $T$ approaches $T_c$.

Metallic compounds and alloys tend to be Type-II superconductors. They typically have critical temperatures more than twice as high and critical fields two to three orders of magnitude higher than Type-I. Since one of the uses of superconducting materials is the wiring of electromagnets carrying large currents and producing very strong fields, Type-II superconductors have wider practical application.

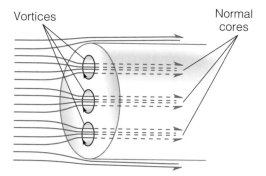

**Figure 9.59** Magnetic field lines passing through vortices in a Type-II superconductor.

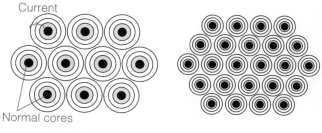

**Figure 9.60** Vortices become more dense as field strength increases. (Both fields shown are between $B_{c1}$ and $B_{c2}$.)

## BCS Theory

Although there are obvious differences between Type-I and Type-II superconductors, the mechanism by which superconductivity occurs is the same for both: the ordered motion of the electrons in pairs, known as **Cooper pairs.** The motions of all pairs are so strongly correlated and ordered that they respond to an external electric field as one, and so are unable to scatter individually in ways that ordinarily result in electrical resistance.

Whatever the specific mechanism might be, there are two important clues that it depends intimately on interactions between the conduction electrons and the lattice of positive ions:

1. The critical temperature of a given material varies with the average mass $M$ of the positive ions, according to $T_c \propto M^{-1/2}$. This is known as the **isotope effect.** Thus, samples of the same element but with different proportions of that element's various isotopes[31] have slightly different critical temperatures. That their mass affects the temperature at which superconductivity occurs clearly indicates that the positive ions are somehow involved in the process.

2. As a rule, elements that superconduct at low temperatures are poor conductors at room temperature and good ordinary conductors do not become superconductors at low temperature. This would certainly suggest that strong interactions/collisions between electrons and the lattice, the bane of conductivity at room temperature, are a *requirement* for superconductivity at low temperature.

The next question, then, is how the electron-lattice interaction can form a Cooper pair.

At low temperatures, two electrons in a solid can experience a net attraction, mediated by a phonon.[32] As shown in Figure 9.61, an electron moving through a

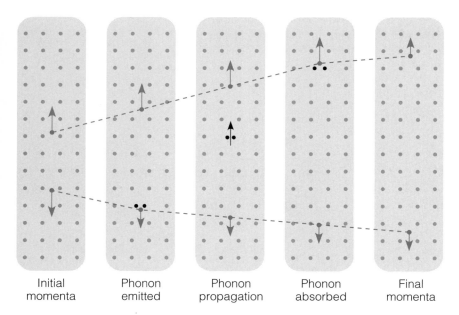

**Figure 9.61** A phonon-mediated attraction between electrons.

| Initial momenta | Phonon emitted | Phonon propagation | Phonon absorbed | Final momenta |

[31]The number of protons in the nucleus determines the element, but the *isotope* of the element is determined by the number of protons *plus neutrons.* For instance, lithium exists in nature in two isotopes: Li-6 (7.5%), with 3 protons and 3 neutrons, and Li-7 (92.5%), with 3 protons and 4 neutrons. Although the same element, Li-7 atoms are more massive than those of Li-6. Isotopes are discussed further in Section 10.1.

[32]Oscillations in a solid lattice may be treated as a collection of particles called phonons. The phonon, the quantum of vibration, is discussed in Section 8.10.

solid causes a local distortion of the lattice, attracting the positive ions and thus creating a region of unusually high positive charge concentration. This distortion pulls on the electron as the two go their separate ways through the lattice. We say that the electron has emitted a phonon. As it approaches another electron, the positive concentration also pulls on it; the second electron absorbs the phonon. The net effect is that the electrons experience equal momentum changes toward one another, just as would particles sharing an attractive force. Under the proper conditions, the attraction is stronger than the electrons' Coulomb repulsion—they will be bound. Even so, the binding energy is small, typically $10^{-3}$ eV, so the electron separation is fairly large, on the order of 1 $\mu$m. Since this is much larger than atomic spacings in a lattice, a great many Cooper pairs may overlap one another.

The explanation of superconductivity involves not merely a possible binding attraction between pairs of electrons, but the system as a whole. It is known as **BCS theory.**[33] It encompasses essentially all behaviors common to Type-I and Type-II superconductors: perfect conductivity, the Meissner effect, the existence of a critical temperature, and so on. Let us look at a few of its more important points.

According to BCS theory, a very low energy state occurs when the most energetic conduction electrons, those near $E_F$, occupy available states in spins-opposite Cooper pairs, with *all pairs having the same net momentum.* In the state of zero current, each pair has zero net momentum, and when current flows, the pair momentum common to all is nonzero (Figure 9.62). The result in any case is a very ordered state. Although members of a pair experience momentum changes, the centers of mass of all pairs have precisely the same momentum.

It is helpful to describe how the electron states fill at $T = 0$, shown in Figure 9.63. In contrast to our earlier cases, where no particlelike electron-phonon encounters were considered, the occupation number does not drop to zero abruptly at $E_F$. Consider a pair of electrons whose energies are near $E_F$. If not engaging in a phonon-mediated attraction, their total energy would be approximately $2E_F$, but the attraction gives the pair a lower total energy. On the other hand, the pair cannot engage in the attraction, cannot emit and absorb phonons, unless they are indeed free to enter unoccupied states of different momenta in response. So it happens that electrons near the Fermi level form pairs in the free range of states slightly

[33]Named after its discoverers Bardeen, Cooper, and Schrieffer, who received the 1972 Nobel prize in physics for their work of 15 years earlier.

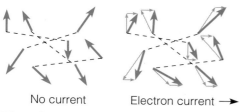

No current      Electron current →

**Figure 9.62** All Cooper pairs have the same momentum, zero when current is zero, nonzero otherwise.

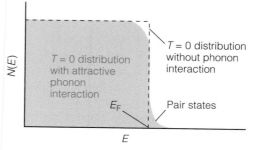

**Figure 9.63** Although electrons in a superconductor occupy energies above $E_F$, the overall state is of lower energy, due to the phonon-mediated pair attraction.

above $E_F$, yet with energies lower than if filling states to $E_F$ without the attraction. Only electrons near $E_F$ form pairs and thus participate in superconductivity, as those far below would require more energy to promote to free levels near $E_F$ than could be compensated for by the attraction.

Electrical resistance is zero in a superconducting state because, while lattice vibrations (i.e., phonons) are actually an integral part of the state, the other cause of resistance, lattice imperfections, effectively cannot scatter electrons. The smallest disruption of the collective, ordered motion of the pairs would involve the breaking apart of at least one Cooper pair. But the pair-binding energy, small though it is, is still larger than the $k_BT$ available in a collision with a lattice imperfection at low temperature. Thus, such collisions effectively cannot occur! The lowest-energy, all-pairs-the-same state of current flow will persist. Because of the finite minimum disruptive energy, there is a finite energy gap between a material's low-temperature superconducting state and the lowest-energy excited state in which electrical resistance—random scattering of individual electrons—is manifest.[34]

### Type-I or Type-II?

The distinction between Type-I and Type-II superconductors lies in the material's **mean free path**—the average distance its electrons travel before undergoing a collision. A material in which the mean free path is large will have Cooper pairs interlocked in a coherent, orderly way over a vast region. Thus, they tend to assume states homogeneous throughout the specimen, completely excluding magnetic fields. Type-I superconductors fit this description. Type-II superconductors, on the other hand, have a relatively short mean free path. Accordingly, order can be much more short-ranged, allowing for a microscopic network of flux-carrying normal cores within an otherwise superconducting region.

## High-$T_C$ Superconductors

In 1986 began a field-day for experimental physicists, and a new challenge for theoretical physicists. A new variety of superconductor was discovered: the **high-$T_c$ superconductor**. These are compounds of copper and oxygen with other elements, and qualify as ceramics. The discovery was astonishing because ceramics are excellent *insulators* at ordinary temperatures. The first one found was $La_{2-x}Ba_xCuO_4$, for which Bednorz and Müller won the 1987 Nobel prize in physics. (The $x$ means that the compound is basically $La_2CuO_4$, but with barium atoms replacing lanthanum atoms at random locations.) The most startling thing was that the critical temperature, 30 K, was significantly higher than that of any other material then known. A feverish hunt soon led to ceramics with yet higher critical temperatures. A milestone was reached when $YBa_2Cu_3O_7$ was found to have a critical temperature of 92 K, exceeding the 77 K boiling point of liquid nitrogen. Previous work on superconductivity required the materials to be cooled with liquid helium, a precious commodity. The new material could be cooled using relatively plentiful liquid nitrogen. Of course, a huge milestone will be reached when and if a material is found whose critical temperature is above room temperature. The highest critical temperature to date is that of $HgBa_2Ca_2Cu_3O_8$, at 133 K.

[34] Although they share the trait that there are no available states within, the energy gap in superconductivity and band gaps in ordinary solids are different effects.

The high-$T_c$ superconductors possess the Type-II traits of forming vortices and having two critical field values. However, the mechanism of superconductivity differs in significant ways from the standard BCS behavior of the Type-I and Type-II. For one thing, high-$T_c$ superconductors do not as a rule exhibit the isotope effect; electron pairs still play a role, but the interaction with the bulk material is different. The fact that all high-$T_c$ superconductors have alternating planes of copper and oxygen atoms has led to an understanding of the basic electron/lattice interaction, but a comprehensive theory has yet to emerge.

High-$T_c$ superconductors are important because of their high critical temperatures and high critical fields. However, although these ceramics are easily fabricated, they are not so easily worked; they cannot be formed into wires as conveniently as can metals. Most methods now being implemented to surmount this drawback involve using many strands or ribbons of the material bound together to form a cable (Figure 9.64). Continuing development offers promise that the high-$T_c$ superconductors will soon gain the widespread use that their electrical and magnetic properties make attractive.

**Figure 9.64** A many-filament high-$T_c$ cable, able to carry 3.3 kA at liquid nitrogen temperature.

## Uses

One of the present uses of superconducting materials is in superconducting magnets for particle accelerators. Because current flows without resistance, little heat is generated and little power wasted. Strong magnetic fields can be sustained while incurring only the expense of maintaining the liquid helium coolant, necessary simply to prevent the material from rising above its critical temperature. Obviously, with a room-temperature superconductor, even this expense could be eliminated. Other proposed uses, particularly for the hoped-for room temperature superconductors, are computers, where the ever-decreasing size of microchip circuitry has led to an increasing heat dissipation problem; power lines, where resistance heating wastes power and mandates high-voltage/low-current transmission, with its unwieldy step-up/step-down transformers; and rail transportation, in which friction can be practically eliminated by exploiting the repulsion between magnets and superconductors to suspend the vehicle above the rail without physical contact. Certainly, if workable, very high temperature superconductors can be found, uses will multiply tremendously. The hunt is at present one of the most active areas of physics research. Exciting discoveries lie ahead, discoveries that will delight and challenge experimentalist and theorist alike, and revolutionize society.

## Progress and Applications

The study of solids and liquids, altogether known as **condensed matter,** is now the most active area of research in physics. Great discoveries and advances have been made in recent decades, and the end is not in sight.

■ From the discovery of the **buckyball** to the present, **fullerenes** have been feverishly studied for applications ranging from lubricants to high-strength fibers to superconductors. Fullerenes are complex structures composed entirely of

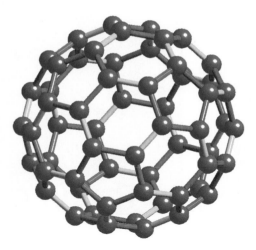

**Figure 9.65** The buckyball—a single molecule of 60 carbon atoms. The hexagons (but not the pentagons) have double bonds, three per ring, shown in lighter gray.

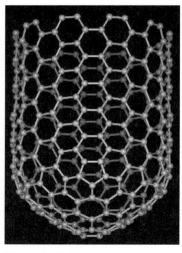

**Figure 9.66** The closed end of a nanotube, with its 120-atom endcap.

carbon atoms. The buckyball, $C_{60}$, the first discovered (H. W. Kroto, et al., *Nature,* November 14, 1985, p. 162–163) consists of 60 carbon atoms bound into a single molecule in a soccerball-type configuration (Figure 9.65). Mixed with the proper doping, solids made of buckyballs can be metallic or even superconductors. Larger configurations may contain hundreds of carbon atoms in a single molecule. Many are closed ellipsoidal structures, but **nanotubes** (Figure 9.66) are also being studied for both mechanical and electrical properties—capable of being insulators, conductors, or semiconductors. A typical nanotube is closed at one end by half of a $C_{240}$ sphere, giving it a diameter on the order of a nanometer, and has a length measured in micrometers. The fullerenes and similar many-atom structures now being fabricated from other elements represent a new and largely unexplored field of technology. Not surprisingly, applications unimagined only a few years ago continue to arise.

■ That hydrogen should become a metal at high pressure was first suggested by Eugene Wigner over a half-century ago but was not confirmed until the 1990s (S. T. Weir, et al., *Physical Review Letters,* March 11, 1996, pp. 1860–1863). Applying enormous pressure to a fluid state of hydrogen, decreasing the atomic separation, causes its bands to widen. The valence-conduction band gap shrinks from 15 eV, an insulator, to about 0.3 eV, small enough for thermal agitation to produce many conduction electrons and qualify the sample as metallic. The success certainly adds to our knowledge of transitions between insulator and metal, and may aid nuclear fusion research, where the behavior of hydrogen at high pressures is of central importance. Furthermore, Jupiter and Saturn have long been thought to contain hydrogen in a metallic state, so the achievement is also important to our understanding of these mysterious giants.

■ *Crystalline* solids, while common and the most easily analyzed, do not encompass all solids. A familiar exception is glass, which forms an amorphous solid. Recent decades have seen great advances in understanding the behavior of amorphous solids. Amorphous silicon is now a common material in "thin-film transistors" and has found wide use in image sensors, active-matrix displays and solar cells. Another "exceptional" solid is the quasicrystal, in which "forbidden" symmetries, such as ten-sided ones, are allowed, but at the expense of lacking some of an ordinary crystal's translational periodicity. Since its discovery in the mid-1980s (D. Shechtman, et al., *Physical Review Letters,* November 12, 1984, p. 1951–1953), both physicists and mathematicians have been at work unraveling the quasicrystal's mysteries.

- The quest for ever faster long-range communication has focused attention on the diode laser. Light, having vastly more "bumps" per second (i.e., frequency) to carry bits of data than ordinary electronic signals or radio waves, is an obvious choice of information carrier, with the optical fiber as its flexible medium. Still this leaves the question of long-range conveyance, which means amplification at regular intervals. An excellent solution has been found in the EDFA, erbium-doped fiber amplifier, depicted in Figure 9.67. At an amplifying station, light from a laser diode is added just ahead of a special erbium-doped section of fiber, "charging up" the erbium atoms. The signal then causes stimulated emission in the erbium and emerges amplified. As a source of amplification, the diode laser, with its built-in pumping mechanism, is both compact and efficient.

- In the excitement surrounding the discovery and exploration of the high-$T_c$ superconductors, the "cuprates," other new types of superconductors have arrived on the scene with less fanfare. **Fullerides,** fullerenes doped with monovalent impurity atoms, such as potassium and rubidium, have been found to superconduct at temperatures as high as 30 K. While this is below the critical temperatures of the cuprates, it is higher than all earlier Type-I and Type-II superconductors. Recent evidence suggests that their superconductivity mechanism is much like the standard BCS of the earlier types, rather than that of the cuprates. Accordingly, research into why their critical temperatures are as high as they are promises to add to our understanding of superconductivity.

Another new kind of superconductor is the **heavy-fermion superconductor.** A common factor in these materials is electrons in $f$ states (i.e., $\ell = 3$), for they all contain either lanthanides or actinides, the elements in which the $f$ states are filling. The name "heavy-fermion" comes from the fact that the electron's effective mass is 100–1000 times the free-electron mass. Although their

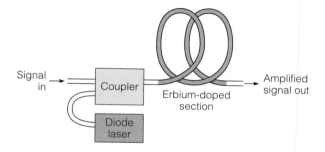

Signal in → Coupler → Erbium-doped section → Amplified signal out

Diode laser

**Figure 9.67** The elements of an erbium-doped fiber amplifier used in fiber optic communication.

critical temperatures are quite low, typically less than 1 K, heavy-fermion superconductors are garnering attention due to their unique superconductivity mechanism, different from both the standard BCS and the cuprates.

## Summary

Bonding between atoms involves rearrangement of the atoms' valence electrons. Molecules and solids form whenever the states the electrons occupy result in a lower energy than for isolated atoms. These many-atom states may be thought of as combinations of single-atom states. When $N$ atoms bond, the $n$th single-atom state becomes an $N$-state band.

In molecules, a covalent bond occurs when two spins-opposite electrons, one from each of two identical atoms, occupy a molecular spatial state centered between the atoms. The energy is low because both atoms attract both electrons. If the atoms are not identical, the electrons will not be shared equally. The extreme case, where a $-1$ valence atom wholly appropriates the valence electron of a $+1$ valence atom, is known as an ionic bond. The attraction is then simply between negative and positive ions. The character of most molecular bonds is somewhere between purely covalent and purely ionic.

Pure solids usually form in a regular geometric arrangement of atoms known as a crystal lattice, of which there are many geometrical types. Regardless of the lattice type, bonding is usually categorized as one of four different kinds: covalent, ionic, metallic, and molecular. The bonding in the first two is as in individual molecules. A metallic solid's valence electrons are shared not by pairs of atoms, but by all atoms. They move freely through the solid, giving metals high electrical conductivity. Noble gases atoms and tightly bound molecules do not share electrons, but form molecular solids via relatively weak dipole-dipole attractions.

In no kind of solid can the valence electrons always be treated as belonging to one atom or one pair of atoms. Rather, they must often be treated as bound in the solid as a whole. The potential energy is essentially that of one large potential well, but with a superimposed periodic variation due to regularly spaced, positive ions. Because $N$ is so large, each $N$-state energy band, formed of single-atom states, is a virtual continuum of allowed energies. The range of forbidden energies between bands is known as a band gap.

A solid's electrical conductivity is related to filling of its bands. The topmost band that is ordinarily full (at $T = 0$) is known as the valence band. If electrons fill part of the next-higher band, the conduction band, the material will be an electrical conductor; the highest-energy electrons have many immediately higher states into which they may enter and thus respond to an external electric field. If electrons fill states only to the top of the valence band, the material will be an insulator. Having no easily accessible states, the electrons have no freedom to respond.

An insulator with a small valence-conduction band gap is called a **semiconductor.** Common examples are the "covalent" solids silicon and germanium, with band gaps of 1.1 eV and 0.7 eV, respectively. With so small a gap, simple thermal excitation can promote a significant number of electrons from the valence band to the otherwise empty conduction band. These respond to an external field just as do electrons in a conductor. Furthermore, the resulting valence band vacancies also contribute to electrical conductivity. They are known as holes, and respond to an external field as positive charge carriers.

The numbers of conduction band electrons and valence band holes can be varied freely by doping: introducing into the lattice a small fraction of impurity atoms. Trivalent impurity atoms give a p-type semiconductor, in which the number of valence holes is unusually large, while an n-type semiconductor, where conduction electrons predominate, is produced using pentavalent impurity atoms. Important devices are made by assembling various combinations of n-type and p-type semiconductors. Putting together one of each type in a p-n junction results in a diode, which allows electrical current to pass only one way. Three put together, either npn or pnp, form a transistor, a basic element of amplification.

## EXERCISES

### Section 9.1

1. Formulate an argument explaining why the even wave functions in Figure 9.1 should be lower in energy than their odd partners.

### Section 9.2

2. From the diagrams in Figure 6.12 and the qualitative behavior of the wave functions they represent, argue that a combination of the $2p_z$ (i.e., $m_\ell = 0$) and the negative of $2s$ would produce a function that sticks out preferentially in the *positive* z-direction. This is known as a **hybrid** *sp* state.

3. **Electron affinity** is a property specifying the "appetite" of an element for gaining electrons. Elements, such as fluorine and oxygen, that lack only one or two electrons to complete shells can achieve a lower energy state by absorbing an external electron. For instance, in uniting an electron with a neutral chlorine atom, completing its $n = 3$ shell and forming a $Cl^-$ ion, 3.61 eV of energy is liberated. Suppose an electron is detached from a sodium atom, whose ionization energy is 5.14 eV, then transferred to a (faraway) chlorine atom. (a) Must energy on balance be put in by an external agent, or is some energy actually liberated, and how much? (b) The transfer leaves the sodium with a positive charge, and the chlorine with a negative. Energy can now be gotten out by allowing these ions to draw close, forming a molecule. How close must they approach to recover the energy expended in part (a)? (c) The actual separation of the atoms in a NaCl molecule is 0.24 nm. How much lower in energy is the molecule than the separated neutral atoms?

4. Exercise 3 outlines how energy may be gotten out by transferring an electron from an atom that easily loses an electron to one with a large appetite for electrons, then allowing the two to approach, forming an ionic bond.
   (a) Consider separately the cases of hydrogen bonding with fluorine and sodium bonding with fluorine. In each case, how close must the ions approach to reach "break even," where the energy needed to transfer the electron between the separated atoms is balanced by the electrostatic potential energy of attraction? (The ionization energy of hydrogen is 13.6 eV, while that of sodium is 5.1 eV, and the electron affinity of fluorine is 3.40 eV.)
   (b) Of HF and NaF, one is considered to be an ionic bond and the other a covalent bond. Which is which, and why?

5. The vertices of a tetrahedron are four vertices of a cube symmetrically chosen so that no two are adjacent. Show that the angle between the vertices of a tetrahedron is 109.5°.

6. Referring to Table 9.1, lobe I of the hybrid $sp^3$ states combines the spherically symmetric $s$ state with the $p$ state that is oriented along the z-axis, and thus sticks out in the $+z$-direction (see Exercises 2 and 7). If Figure 9.14 is a true picture, then in a coordinate system rotated counterclockwise about the y-axis by the tetrahedral angle, lobe II should become lobe I. In the new frame, while y-values are unaffected, what were values in the z-x plane become values in the $z'$-$x'$ frame, according to $x = x' \cos \alpha + z' \sin \alpha$ and $z = z' \cos \alpha - x' \sin \alpha$, where $\alpha$ is 109.5°, or $\cos^{-1}(-\frac{1}{3})$.
   (a) Show that lobe II becomes lobe I. Note that since neither the $2s$ state nor the *radial* part of the $p$

states is affected by a rotation, only the angular parts given in equations (9-1) need be considered.

(b) Show that if lobe II is instead rotated about the $z$-axis, by simply shifting $\phi$ by $\pm 120°$, the result is lobes III and IV.

(c) We have shown that all lobes have the same shape and that lobe I (i.e., the $z$-axis) makes the tetrahedral angle with the other three. Now if lobe I is thought of as a vector with components $(0, 0, 1)$, lobes II, III, and IV would have components $(\sqrt{\frac{8}{9}}, 0, -\frac{1}{3}), (-\sqrt{\frac{2}{9}}, \sqrt{\frac{6}{9}}, -\frac{1}{3})$ and $(-\sqrt{\frac{2}{9}}, -\sqrt{\frac{6}{9}}, -\frac{1}{3})$. Show that the angles between all pairs of "vectors" II, III, and IV are 109.5°. (Remember: $\cos \beta = \mathbf{A} \cdot \mathbf{B}/AB$.)

7. In Section 9.2, we discussed $p_x$, $p_y$, and $p_z$ states—linear combinations of the $m_\ell = (-1, 0, +1)$ isolated-atom $p$ states—and the hybrid $sp^3$ states, combinations of all four $s$ and $p$ spatial states. Another kind of hybrid state, which sticks out in just one direction, is the $sp$, a linear combination of a single $p$ state and an $s$ state. Consider an arbitrary combination of the $2s$ state with the $2p_z$ state, which is oriented along the $z$-axis. Let us represent this by $\cos \tau \, \psi_{2,0,0} + \sin \tau \, \psi_{2,1,0}$. (The factors involving $\tau$ ensure that the total is normalized. Were normalization carried out, cross terms would integrate to zero, leaving $\cos^2 \tau \int |\psi_{2,0,0}|^2 + \sin^2 \tau \int |\psi_{2,1,0}|$, which is unity.)

(a) Calculate the probability that an electron in such a state would be in the $+z$-hemisphere. (*Note:* Here the cross terms do *not* integrate to zero.)

(b) What value of $\tau$ leads to the maximum probability, and what is the corresponding ratio of $\psi_{2,0,0}$ to $\psi_{2,1,0}$?

(c) What is the value of the maximum probability in the $+z$-direction?

(d) (Optional) Using a computer, make a density (shading) plot of the probability density—density versus $r$ and $\theta$—for the $\tau$-value found in part (b).

## Section 9.3

8. The interatomic potential energy in a diatomic molecule (Figure 9.18) has many features: a minimum energy, an equilibrium separation, a curvature, and so on. (a) Upon what features do the rotational energy levels depend? (b) Upon what features do the vibrational levels depend?

9. The bond length of the $N_2$ molecule is 0.11 nm and its effective spring constant is $2.3 \times 10^3$ N/m. At room temperature, (a) what would be the ratio of molecules with rotational quantum number $\ell = 1$ to those with $\ell = 0$ (at the same vibrational level), and (b) what would be the ratio of molecules with vibrational quantum

number $n = 1$ to those with $n = 0$ (with the same rotational energy)?

10. The bond length of the $N_2$ molecule is 0.11 nm and its effective spring constant is $2.3 \times 10^3$ N/m. (a) From the size of the energy jumps for rotation and vibration, determine whether either of these modes of energy storage should be active at 300 K. (b) According to the equipartition theorem, the heat capacity of a diatomic molecule storing energy in rotations but not vibrations should be $\frac{5}{2}R$ (3 translational + 2 rotational degrees of freedom), while if also storing energy in vibrations, it should be $\frac{7}{2}R$ (adding 2 vibrational degrees). Nitrogen's molar heat capacity is 20.8 J/mol·K at 300 K. Does this agree with your findings in part (a)?

11. The effective force constant of the molecular "spring" in HCl is 480 N/m and the bond length is 0.13 nm.

(a) Determine the energies of the two lowest-energy vibrational states.

(b) For these energies, determine the amplitude of vibration if the atoms could be treated as oscillating *classical* particles.

(c) For these energies, by what percent does the atomic separation fluctuate?

(d) Calculate the classical vibrational frequency $\omega_{vib} = \sqrt{\kappa/\mu}$ and rotational frequency $\omega_{rot} = L/I$. For the rotational frequency, assume that $L$ is its lowest nonzero value, $\sqrt{1(1+1)}\hbar$, and that the moment of inertia $I$ is $\mu a^2$.

(e) Is it valid to treat the atomic separation as fixed for rotational motion while changing for vibrational?

12. The carbon monoxide molecule CO has an effective spring constant of 1860 N/m and a bond length of 0.113 nm. Determine four wavelengths of light that carbon monoxide might absorb in vibration-rotation transitions.

13. Vibrational-rotation spectra are rich! For the CO molecule (data are given in Exercise 12), how many rotational levels are there between the ground vibrational state and the first excited vibrational state?

14. As noted in Example 9.1, the HD molecule differs from $H_2$ in that a deuterium atom replaces a hydrogen atom. (a) What effect, if any, does the replacement have upon the bond length and force constant? Explain. (b) What effect, if any, does it have upon the rotational energy levels? (c) What effect, if any, does it have upon the vibrational energy levels?

15. From the qualitative shapes of the interatomic potential energies in Figure 9.24, would you expect the vibrational levels in the excited electronic state to be spaced the same, farther apart, or closer together than those in the lower-energy electronic state? Explain. What about the rotational levels?

## Section 9.4

16. For the four kinds of crystal binding—covalent, ionic, metallic, and molecular—how would the density of valence electrons vary throughout the solid? Would it be constant, centered on the atoms, largest between the atoms, or would it alternate, with a net charge density positive at one atom and negative at the next?

17. The energy necessary to break the ionic bond between a sodium ion and fluorine ion is 4.99 eV. The energy necessary to separate the sodium and fluorine ions that form the ionic NaF crystal is 9.30 eV per ion pair. Explain the difference.

## Section 9.5

18. Assuming an interatomic spacing of 0.15 nm, obtain a rough value for the width (in eV) of the $n = 2$ band in a one-dimensional crystal.

19. The density of silver is $10.5 \times 10^3$ kg/m$^3$. (a) What is the approximate atomic separation? (b) Silver's conductivity is $6.3 \times 10^7$ $\Omega^{-1} \cdot$m$^{-1}$, and it has one conduction electron per atom. How far would a conduction electron travel in one collision time $\tau$? Assume that it has a kinetic energy of about 5.5 eV—that is, the Fermi energy. (c) Do silver's conduction electrons collide readily with all its positive ions?

## Section 9.6

20. Carbon (diamond) and silicon have the same covalent crystal structure, yet diamond is transparent while silicon is opaque to visible light. Argue that this should be the case based only on the difference in band gaps; that of the insulator diamond is roughly 5 eV and that of the semiconductor silicon is about 1 eV.

21. In diamond, carbon's four full (bonding) $s$ and $p$ spatial states become a band and the four empty (antibonding) ones become a higher-energy band. Considering the trend in the band gaps of the tetravalent elements at $T = 0°$C, explain why it might not be surprising that "covalent" tin behaves as a conducting metallic solid.

## Section 9.7

22. The resistivity of silver is $1.6 \times 10^{-8}$ $\Omega \cdot$m at room temperature (300 K), while that of silicon is about 10 $\Omega \cdot$m. (a) Show that this disparity follows, at least to order of magnitude, from the approximate 1-eV band gap in silicon. (b) What would you expect for the room temperature resistivity of diamond, which has a band gap of about 5 eV?

23. Show that for a room temperature semiconductor with a band gap of 1 eV, a temperature rise of 4 K would raise the conductivity by about 30%.

24. The Fermi velocity $v_F$ is defined by $E_F = \frac{1}{2}mv_F^2$, where $E_F$ is the Fermi energy. The Fermi energy for silver is 5.5 eV. (a) Calculate the Fermi velocity. (b) What would be the wavelength of an electron with this velocity? (c) How does this compare to the lattice spacing of 0.41 nm? Does the order of magnitude make sense?

## Section 9.8

25. As a crude approximation, an impurity pentavalent atom in a (tetravalent) silicon lattice can be treated as a one-electron atom—the extra electron orbits a net positive charge of unity. Because this "atom" is not in free space, however, $\epsilon_0$, the permitivity of free space, must be replaced by $\kappa\epsilon_0$, where $\kappa$ is the dielectric constant of the surrounding material. The hydrogen atom ground-state energies would thus become

$$E = -\frac{me^4}{2(4\pi\kappa\epsilon_0)^2\hbar^2}\frac{1}{n^2} = \frac{-13.6 \text{ eV}}{\kappa^2 n^2}$$

Given $\kappa = 12$ for silicon, how much energy must be put in to free a donor electron in its ground state? (Actually, the effective mass of the donor electron is less than $m_e$, so this prediction is somewhat high.)

## Section 9.9

26. The photons emitted by an LED (light-emitting diode) arise from the energy given up in electron-hole recombinations across the energy gap. How large should the energy gap be to give photons at the red end of the visible spectrum (700 nm)?

27. A typical built-in electric field across the depletion zone of an unbiased diode is $10^6$ V/m. If the band gap is 1.0 eV, how wide is the depletion zone?

28. In a diode laser, electrons dropping from conduction band, across the gap, and into the valence band produce the photons that add to the coherent light. The ZnTe laser has a band gap of 2.25 eV. About what wavelength laser light would you expect it to produce?

29. It is often said that the transistor is the basic element of amplification, yet it supplies no energy of its own. Exactly what is its role in amplification?

30. Sketch an energy versus position diagram, complementary to Figure 9.49, showing valence hole motion and conduction electron participation in an operating pnp transistor.

**31.** In many kinds of integrated circuit, the preferred element of amplification/switching is not the "bipolar" transistor discussed in the chapter, but the somewhat similar MOSFET: <u>m</u>etal <u>o</u>xide <u>s</u>emiconductor <u>f</u>ield-<u>e</u>ffect <u>t</u>ransistor. The accompanying diagram shows one in its "normally off" state: conduction electrons cannot flow from the n-type **source,** analogous to the emitter, "over the bump" in the p-type region, to the n-type **drain,** analogous to the collector. (An npn arrangement is shown, but just as for the bipolar transistor, a pnp would yield the complementary device.) The important difference is that rather than a direct electrical contact to the p-type region, as in the base of the bipolar, the center lead, the **gate,** is a metal contact bonded to the p-type region but electrically insulated from it by a thin layer of metal oxide.

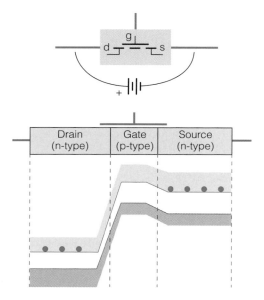

(a) Explain how applying a bias to the gate can cause this device to turn on: Should the gate bias voltage be positive or negative (relative to the source)? Why is the control mechanism referred to as "field effect"?

(b) The MOSFET is often said to be a "unipolar" device because valence holes (conduction electrons in the pnp device) do not play the important role that they do in the bipolar. Explain: Would you expect a significant current through the gate due to electron-hole recombination in the p-type region? Why or why not?

(c) A low input-impedance device is one in which there are large oscillations in input current for small oscillations in the input voltage. Correspond-

ingly, a high input-impedance device has a small input current for a large input voltage. Bearing in mind that the voltage across the forward-biased base-emitter diode of a bipolar transistor is always about $E_{gap}/e$, while the input current is proportional to the output current, would you say that the bipolar transistor has low or high input impedance? What about the MOSFET?

**32.** The diagram shows the energy bands of a **tunnel diode** as the potential difference is increased.

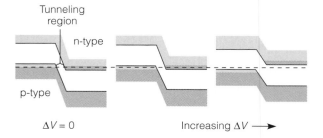

As in the diode laser, heavy doping produces a donor band at the bottom of the n-type conduction band and an acceptor band at the top of the p-type valence. In all unbiased diodes, the depletion zone between the n-type and p-type bands constitutes a potential barrier (see Section 5.1), but in the tunnel diode it is so thin that significant tunneling occurs. The current versus voltage plot follows. In contrast to a normal diode, significant current begins to flow as soon as there is an applied voltage (i.e., before the bias voltage is $E_{gap}/e$.) It then *decreases* (so-called negative resistance), before again increasing in the normal way. Explain this behavior.

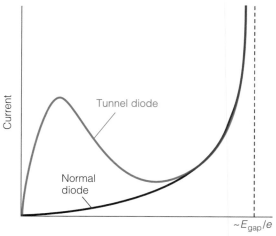

## Section 9.10

33. The diagram shows an idealization of the "floating magnet trick" of Figure 9.56. Before it is cooled, the superconducting disk on the bottom supports the small permanent magnet simply by contact. After cooling, the magnet floats. Make a crude sketch showing what new magnetic fields arise. Where are the currents that produce them?

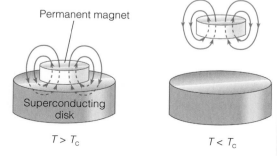

Permanent magnet

Superconducting disk

$T > T_c$

$T < T_c$

34. The "floating magnet trick" is depicted in Exercise 33. Why must the disk on the bottom be a superconductor? Why wouldn't a permanent magnet with a similar magnetic field pattern work? (*Hint:* What happens when you hold two ordinary magnets so that they repel, then release one of them?)

35. The magnetic field at the surface of a long wire of radius $R$ and carrying a current $I$ is $\mu_0 I / 2\pi R$. How large a current could a 0.1-mm-diameter niobium wire carry without exceeding its 0.2 T critical field?

# 10

# Nuclear Physics

In the preceding chapters, we have tended toward the study of ever larger objects. We began with simple particles—mostly electrons and photons—moved to one-electron atoms, then multielectron atoms, then on to molecules and solids. In some sense, to continue this trend would lead to other disciplines—chemistry, engineering, and so on. There is much physics to be studied in the other "direction," however, and the remainder of the text is aimed that way.

Although all atoms contain nuclei, a "study of the atom," or of chemistry (how atoms interact), is concerned exclusively with the behavior of the orbiting electrons. The nuclei are assumed essentially inert. But nuclei are *not* always inert. For one thing, they can fragment, spontaneously or through external influence, and such "nuclear reactions" release enormous amounts of energy compared to chemical reactions. In nuclear physics, it is the orbiting electrons that are of little concern. It is true they are bound to the nuclei, but compared to the energies within the nucleus itself, electron binding energies are negligible. The nucleus is the focus. On the other hand, the nucleus is an even more difficult problem than the atom. Much has been learned, but much has yet to be learned, and nuclear physics will thus be an active area of research for some time to come.

The importance of nuclear physics cannot be overstated. It explains the operation of nuclear reactors and nuclear weapons; nuclear magnetic resonance is used routinely to produce images of the body's interior; radioactivity—energetic particles emanating from nuclei—is both a useful diagnostic tool, via radioactive tagging and dating, and, as we continue to unearth and artificially produce more radioactive materials, a mounting disposal problem. Just as important as these reasons, however, the study of nuclear physics is an essential step in our quest to unravel the fundamental structure of the physical universe.

# 10.1  Basic Structure

All nuclei consist of protons and neutrons, known collectively as **nucleons.** (Important properties are given in Table 10.1.) Figure 10.1 very crudely depicts a helium atom. Two electrons orbit the nucleus, which comprises four nucleons: two protons and two neutrons. We use the symbol $N$ for the number of neutrons and $Z$, the **atomic number,** for the number of protons (equal, of course, to the number of orbiting electrons in the neutral atom). The symbol $A$, called the **mass number,** we use for the total number of nucleons, $Z + N$. For helium: $Z = 2$, $N = 2$ and the mass number $A$ is 4.

As we found in Chapter 7, an element's chemical behavior depends only on the number of electrons orbiting its nucleus—the value of $Z$. But nuclei of the same element rarely behave alike if they have different numbers of *neutrons.* For instance, while the nucleus of ordinary hydrogen is simply a proton, about 15 of every 100,000 hydrogen atoms on Earth have nuclei consisting of one proton *and one neutron.* Known as deuterium, or "heavy hydrogen," this atom is roughly twice as massive as ordinary hydrogen. *Chemically,* deuterium is hydrogen. It participates in chemical reactions—rearrangements of orbiting electrons—precisely as does ordinary hydrogen, and for this reason, it is impossible to separate the two chemically. But deuterium's nucleus is entirely different. It participates in *nuclear* reactions, rearrangements of nucleons, in vastly different ways.

Nuclei with the same number of protons but different numbers of neutrons are said to be **isotopes** of the same element. Thus, we say that there are two naturally occurring isotopes of hydrogen, ordinary hydrogen and deuterium (Figure 10.2). Both are **stable;** if isolated from external influences, they remain unchanged indefinitely. However, hydrogen has a third isotope, "tritium," with one proton and *two* neutrons in the nucleus. Although it may be produced artificially, it is essentially absent in nature, primarily because it is **unstable;** it is subject to radioactive decay, spontaneously undergoing a nuclear reaction and becoming a different element. Radioactive decay, the fate of unstable nuclei, is discussed in Section 10.5.

We use the convention "element-$A$" to refer to different isotopes of the same element. Thus, we may refer to the three isotopes of hydrogen as hydrogen-1, hydrogen-2, and hydrogen-3. The term **percent natural abundance** refers to the

**Table 10.1**

|  | Charge | Mass | |
|---|---|---|---|
| **proton** | $+e$ | $1.6726231 \times 10^{-27}$ kg | $= 1.007276$ u |
| **neutron** | $0$ | $1.6749286 \times 10^{-27}$ kg | $= 1.008665$ u |
| **electron** | $-e$ | $9.109 \times 10^{-31}$ kg | $= 5.486 \times 10^{-4}$ u |

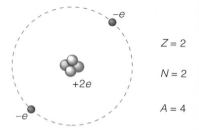

**Figure 10.1**  A crude picture of helium.

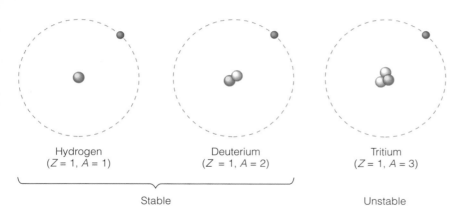

Hydrogen
($Z$ = 1, $A$ = 1)

Deuterium
($Z$ = 1, $A$ = 2)

Tritium
($Z$ = 1, $A$ = 3)

Stable

Unstable

**Figure 10.2** Isotopes of hydrogen.

percentage that a particular isotope constitutes of the total amount of a given element in nature. The percent natural abundance of hydrogen-1 is 99.985% and of hydrogen-2 (deuterium) is 0.015%.[1]

Helium also has two naturally occurring isotopes: helium-3 (two protons, one neutron) and the more abundant and familiar helium-4 (two protons, two neutrons). Carbon-12 (six protons, six neutrons), the most abundant of carbon's naturally occurring isotopes, has been adopted as the basis of a standard measure of mass: the **atomic mass unit** u is defined to be one-twelfth the mass of the carbon-12 atom (including electrons). Some elements have a host of naturally occurring isotopes; tin has ten. Many, such as aluminum, have only one. Some have none; they are simply not found in nature. For all elements, however, isotopes that do *not* occur naturally may either be artificially produced, or they may result from the radioactive decay of other nuclei. All these are unstable; they do not survive indefinitely but undergo radioactive decay after some span of time. On the other hand, while no isotopes *absent* in nature are stable (at least, none has yet been artificially produced), some isotopes *present* in nature are unstable. For instance, uranium-238 and thorium-232, both more abundant in nature than silver, are radioactive. Given their long "half-lives" (discussed in Section 10.6), billions of years must pass before the amounts of these isotopes significantly decrease by radioactive decay.

$$1 \text{ u} = 1.660559 \times 10^{-27} \text{ kg}$$

## Size

The first experiments to probe the nucleus involved directing small particles toward a thin foil of gold. It was hoped that some insight could be gained by analyzing the paths along which the particles would be scattered by collisions with individual gold atoms. The particles of choice, being readily produced and detected (after scattering) were $\alpha$-particles, the positively charged helium nuclei that remain after the electrons are removed from the atoms. Some $\alpha$-particles passed through essentially undeflected. (They missed!) A few rebounded backward, in the direction from which they came. Evidently these had encountered a compact,

Scattered
$\alpha$-particles

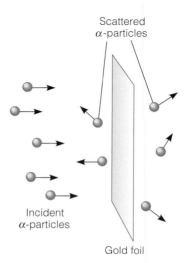

Incident
$\alpha$-particles

Gold foil

---

[1]Henceforth, "hydrogen" is understood to mean exclusively "ordinary" hydrogen, or hydrogen-1. Hydrogen-2 and hydrogen-3 will usually be referred to explicitly by their own names: "deuterium" and "tritium," respectively.

heavy, positively charged object: the nucleus. The size of the nucleus was found by increasing the $\alpha$-particle's speed. As an $\alpha$-particle approaches a nucleus, its kinetic energy is converted to electrostatic potential energy. Its point of closest approach is reached when all its initial kinetic energy is converted. Increasing the initial kinetic energy would allow it to approach closer, and so at a high enough initial energy the $\alpha$-particle is able to breach the nuclear surface. When it does, there is an obvious change in the scattering, since the force is no longer simply electrostatic repulsion. By setting the minimum initial kinetic energy for which this occurs equal to the final electrostatic potential energy, we solve for the corresponding distance of closest approach—the radius of the nucleus.

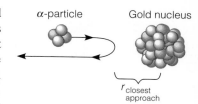
α-particle           Gold nucleus

$r$ closest approach

Experiments on many different elements indicate that nuclei are roughly spherical, with radii obeying the approximate relation:

$$r = A^{1/3} \times R_0 \qquad (10\text{-}1)$$

where $A$ is the mass number and $R_0$ is $1.2 \times 10^{-15}$ m. A femtometer, $10^{-15}$ m, is thus a good measure of nuclear size. Typical *atomic* dimensions (electron orbits) are five orders of magnitude larger—of order $10^{-10}$ m. In other words, if the whole atom with its orbiting electron cloud were the size of Earth, its nucleus would cover as much ground as a good-size baseball field.

Scale of nuclear dimensions:
1 fm = $10^{-15}$ m

The dependence of radius on $A$ in equation (10-1) is conspicuous. The total volume is

$$V_{\text{total}} = \frac{4}{3}\pi r^3 = \frac{4}{3}\pi(A^{1/3}R_0)^3 = A \times \left(\frac{4}{3}\pi R_0^3\right) \qquad (10\text{-}2)$$

It is proportional to the total number of nucleons, which, since neutrons and protons are of nearly equal mass, makes it proportional to total mass: All nuclei have approximately the same density.

$$\rho = \frac{\text{mass}}{\text{volume}} = \frac{A \times m_{\text{nucleon}}}{A \times \left(\frac{4}{3}\pi R_0^3\right)} \qquad (10\text{-}3)$$

$$\cong \frac{1.67 \times 10^{-27} \text{ kg}}{\frac{4}{3}\pi(1.2 \times 10^{-15} \text{ m})^3} \cong 10^{17} \text{ kg/m}^3$$

This is about 14 orders of magnitude denser than typical "solids"!

Constant density suggests that, as depicted in the various diagrams, nucleons in the nucleus behave as incompressible objects packed as close together as possible. Although this view is oversimplified, experimental evidence confirms that nucleons have a repulsive "hard core." Furthermore, the volume allowed each nucleon in such a close-packed arrangement would be the total volume divided by the total number $A$—that is, simply $\frac{4}{3}\pi R_0^3$. Thus, the effective radius of an individual nucleon is approximately $R_0$, or about 1 femtometer.

## 10.2  Binding

Upon learning that the nucleus contains many positively charged protons very close together, and knowing that the electrostatic force causes like charges to repel, many students of science have wondered what prevents the nucleus from

## The Force Name Game

The "strong force" is truly fundamental, engaged in by quarks, fundamental particles (see Chapter 11). The force we *refer to in this chapter* as the strong force is engaged in by *nucleons,* which are not fundamental (being combinations of quarks). It is more properly called the "*residual* strong force" and is a coarse manifestation of the true strong force. As it is conveniently descriptive, we will often refer to the residual strong force as the "internucleon attraction."

The explanation of the term "electromagnetic/weak" is that the familiar electromagnetic force ($\mathbf{F} = q\mathbf{E} + q\mathbf{v} \times \mathbf{B}$) is only a part of a more complex fundamental force whose other part is the so-called weak force. They are really just two aspects of the same force, the electroweak force, but under ordinary conditions they manifest themselves so differently that it is convenient to regard them as two different forces. The "weak part" is the weaker of the two (sensibly) and has the shortest effective range of any force. Accordingly, unlike the electromagnetic force, it is not easily revealed.

exploding. In one sense, the answer is very simple: There is another force at play—an attractive force that holds the nucleons together. Surely it isn't electromagnetic, but neither could it be gravitational; simple calculations show that this relatively weak attractive force is woefully inadequate to hold protons so close together against their electrostatic repulsion. Although these two familiar forces are certainly in effect in the nucleus, they compete with a much stronger force, aptly named the **strong force.** As mass is the property responsible for gravitational interactions, and charge for electromagnetic interactions, *all nucleons possess a property that causes them to attract one another when close together via the strong force.*

This "new" force is one of the fundamental forces of nature: gravitational, electromagnetic/weak, and strong. (See the essay entitled "The Force Name Game.") All forces—weight, tension, forces of contact, friction—appear to be manifestations of one or another of these basic forces. A rough comparison of their relative strengths is given in Table 10.2. (They are discussed together more fully in Chapter 11.) The values are based on certain assumptions of charge, mass, separation, and so on. The strong force is not so easily characterized as the more familiar forces. For one thing, it does not diminish gradually with particle separation, as do the familiar $1/r^2$ forces; it is "strong" only for nucleon separations less than about 2 fm—that is, only when nucleons are not much farther apart than the effective nucleon diameter! The relative strengths in Table 10.2 assumes such

**Table 10.2**

| Force | Relative Strength | Range |
|---|---|---|
| **Strong (Residual)** | 1 | $\sim 1$ fm |
| **Electromagnetic** | $\sim 10^{-2}$ | Long: $\propto 1/r^2$ |
| **Weak** | $\sim 10^{-6}$ | $\sim 10^{-3}$ fm |
| **Gravitational** | $\sim 10^{-39}$ | Long: $\propto 1/r^2$ |

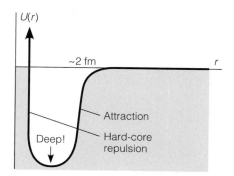

**Figure 10.3** Elements of the internucleon (strong force) potential energy.

small separations. At greater separation, the strong force is negligible, weaker than both the gravitational and electromagnetic. As we shall see, this plays an important role in the stability of the nucleus. Electrostatic repulsion, although weaker than the strong force at close range, becomes a significant destabilizing effect in larger nuclei, where protons are often too far apart to attract each other via the strong force. (In no ordinary circumstance is gravitation a significant effect. Weird unearthly circumstances, such as black holes, are discussed in Section 1.13.)

It is one thing to *claim* that because the nucleus would otherwise explode there must be another force, hidden from our senses, but where is the proof? Much of it comes from scattering experiments. One nucleon is put on a collision course with another, in all possible combinations (proton–proton, proton–neutron, neutron–neutron), and the nature of the internucleon force is then inferred by analyzing the details of the scattering. Figure 10.3 illustrates some of what has been learned. It depicts the strong-force potential energy shared by a nucleon at the origin and another at an arbitrary distance *r*. As we have noted, the strong force is attractive, though with an incompressible "hard core"; short-ranged; and strong. We also find that it is charge-independent—the same between two protons as between two neutrons as between a proton and a neutron—but spin-dependent; that is, it depends on the orientation of the nucleons' spins. Not surprisingly, there is no simple formula for it. In fact, it has yet to be fully characterized. Fortunately, we may grasp much nuclear physics through an understanding of only the few properties we have just enumerated.

## Stability—A Theoretical Model

With all nucleons engaging in an extremely strong attraction, it is tempting to imagine that we could bring together any combination of protons and neutrons and have a stable nucleus. But this is not the case. The vast majority would either not stick together at all or would be unstable, disintegrating in radioactive decay in a time span ranging from an instant to an eon. A few would be stable.

We now investigate the effects of various factors on the stability of nuclei. In this qualitative discussion, we will not as a rule even say whether a given combination of nucleons should stick together or not, nor if it does stick together, whether it would stay that way (i.e., be stable). We consider only what factors *tend* to make it more or less stable. A nucleus is more stable when its constituents are bound in a state of lower energy, requiring an external agent to expend a greater amount of energy to pull out a representative nucleon. Energy is all-important.

At points, our discussion may seem rather speculative, but this is dictated by the great complexity of the strong force. When a phenomenon is so complex that it defies formulation of a comprehensive theory, or relationships are not fully known, we construct a model based on the evidence at hand. Essentially, the model is well-informed but simplified theoretical guesswork. It should, of course, agree with the current experimental evidence, but quite likely it also predicts things currently beyond experimental observation. If these predictions agree with later experimental findings, the model is strengthened and may be claimed to have advanced the understanding of the phenomenon. If they do not agree, the model must be modified.

We begin our model-building with the simplest conceivable nuclei, those with only two nucleons, then move on to nuclei of arbitrary size. In a later section, we discuss well-established models that introduce refinements of the one we begin here.

## Two-Nucleon Nuclei

The simplest possible combinations of nucleons are two protons (p–p), two neutrons (n–n), and a proton–neutron pair (p–n). Which, if any, should be more stable? Let us look at the factors: The short-range strong attraction should create something like a narrow but deep potential energy well, and if the force doesn't "care" whether the particles are protons or neutrons, the well should be the same for all three combinations. This is no guarantee that any would be bound—some wells can be too shallow to have any bound states (cf. Exercise 4.18)—but at least they should all be the same. In these small nuclei, the nucleons would be close enough that the strong nuclear attraction should be overwhelming. Nevertheless, the p–p combination would have some Coulomb repulsion—raising the energy—and this argues that it should be slightly less stable than the other two.[2]

All these arguments are valid, but the truth is that only the p–n forms a bound nucleus. Known as the **deuteron,** it is the nucleus of the stable deuterium atom. The n–n and p–p don't stick together at all. Our arguments have overlooked one of the strong force's properties. The internucleon attraction is spin-dependent, and it is stronger when spins are aligned. Thus, the lowest-energy state possible for two nucleons in a well should have both in the ground spatial state and spins aligned. This state is forbidden to the p–p and n–n by the exclusion principle, because each contains two identical fermions; but it *is* available to the p–n, since it is composed of different fermions. Is this indeed why the p–n forms but the p–p and n–n do not? The experimental evidence indicates that the deuteron is only barely bound; its potential well has no excited states, only the one ground state—and the spins *are* aligned (i.e., its total spin $s_T$ is 1). If the p–n cannot bind any other way—with spins opposite or with one nucleon in a higher spatial state—the others should not bind at all. The energy in the deuteron is crudely

---

[2]Generally speaking, an attraction represents a low energy; an energy input is needed to pull things apart. A repulsion is a high energy; energy can, in fact, be extracted because the particles would fly apart on their own. As a guide, it may help to keep in mind $U_{elec} \propto q_1 q_2/r$. This is positive for like charges (repulsion) and negative for opposite charges (attraction).

depicted in Figure 10.4. (The repulsive nucleon cores are ignored.) Two nucleons occupy the lone bound state in a well produced by their mutual attraction.

### Nuclei of Arbitrary Nucleon Number

We now consider the more general case, but we have already encountered the predominant factors governing stability: the strong force, Coulomb repulsion, and the exclusion principle. Let us take a look at them one at a time.

*Strong Internucleon Attraction* Since a nucleon attracts all others nearby, adding a nucleon to a small nucleus should cause all the nucleons to be more tightly bound. A two-nucleon nucleus has only one bond, but a three-nucleon nucleus has three. Thus, whereas the former has one-half bond *per nucleon,* the latter would have one bond per nucleon—twice as many. A four-nucleon nucleus would have six bonds (pairs), or one-and-a-half bonds per nucleon. A representative nucleon would be harder to extract than in a three-nucleon nucleus. This increase in bonds-per-nucleon cannot continue indefinitely, however. Were there no "hard core" to nucleons, each *could* be close enough to attract all others; all nuclei would collapse to nearly the same small size, and nuclear density would increase linearly with nucleon number. But the nucleons' incompressibility causes volume to increase in proportion to nucleon number, and as nuclear size grows, a given nucleon cannot bind to others more than about 2 fm away. Thus, once completely surrounded, nucleons should have a fixed number of bonds per nucleon, independent of nuclear size.[3] Of course, nucleons at the *surface* are *not* completely surrounded, but as the nuclear sphere grows, the number at the surface, proportional to the surface area, does not keep up with the total number, proportional to the volume (area/volume = $4\pi r^2 / \frac{4}{3}\pi r^3 \propto 1/r$). The *fraction* of surface nucleons decreases. Altogether, then, as the nucleus grows, the average number of bonds per nucleon should initially increase fairly rapidly, then slowly approach a constant, as most nucleons become nearly surrounded. Figure 10.5 shows the trend. Rather than "bonds per nucleon," however, it is labeled as "binding energy per nucleon." It must be remembered that we are considering the *energy* that *binds* a representative nucleon, the energy that would have to be supplied to extract it from the whole.

*Coulomb Repulsion* There is electrostatic repulsion between all pairs of protons in the nucleus. In effect, this positive potential energy shifts all proton energies closer to the top of the finite well in which all nucleons are bound. This shifting is

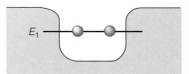

**Figure 10.4** The deuteron's neutron and proton bound by their attractive potential energy.

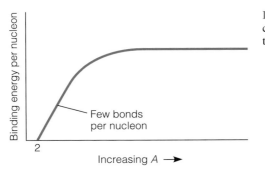

**Figure 10.5** Binding energy per nucleon due to strong internucleon attraction only.

[3]We say that the strong force "saturates": A nucleon can attract no more than the fixed number of others that can fit into its immediate vicinity.

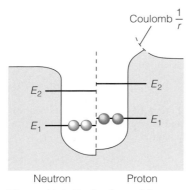

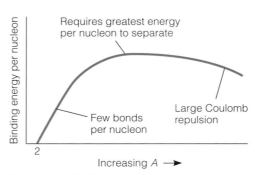

**Figure 10.6** Coulomb repulsion raises proton energies slightly.

**Figure 10.7** Binding energy per nucleon due to both strong internucleon attraction and Coulomb repulsion.

depicted for the case of helium ($Z = 2$, $N = 2$) in Figure 10.6. In small nuclei, all nucleons are close together, so that each repulsion is more than compensated for by a strong internucleon attraction; there is a strong net attraction between all nucleons. But in large nuclei, pairs of protons can be too far apart to attract via the strong force, and so add to the nucleus an uncompensated net repulsion. Thus, while under control in small nuclei, Coulomb repulsion should be an increasingly destabilizing factor in larger nuclei. The energy needed to extract an "average" nucleon becomes progressively smaller. Combining the Coulomb repulsion effect with our previous arguments, we should expect the binding energy per nucleon to vary as in Figure 10.7. Accordingly, *somewhere between the extremes there should be a mass number more stable than all others, a mass number at which the nucleons are the most tightly bound possible.*

*The Exclusion Principle*    At this point, it might seem that the best way to make a stable nucleus would be to build it of essentially all neutrons, since only protons contribute to Coulomb repulsion. There is some validity in the argument, but it overlooks an important factor. Protons and neutrons sharing the same nucleus must obey the exclusion principle—*each independently.* A consequence is that (*ignoring* Coulomb repulsion), for a given $A$, the state in which the number of protons equals the number of neutrons is of lowest energy.

With Figure 10.8 as a guide, we show this as follows: Suppose that $Z$ does equal $N$. The lowest-energy state consistent with the exclusion principle would have all the neutron energy levels filled, each with a spins-opposite pair, up to some maximum energy. Considering only the charge-independent strong force, protons occupy an identical well, and so would fill levels the same way up to the same maximum energy. Now, for a given total number of nucleons, a decrease by 1 in the proton number would mean an increase by 1 in neutron number. Since all lower energy levels are filled, the added neutron would have to occupy the next-higher energy level, which is higher than the previously occupied proton level. The total energy would increase. The argument applies equally well to an increase

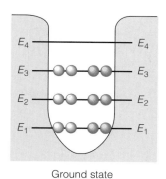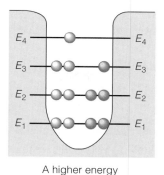

Ground state    A higher energy

**Figure 10.8** Other factors aside, the exclusion principle argues that $N = Z$ is of lower energy.

in the proton number and corresponding decrease in neutron number. Thus, the state $Z = N$ is of lowest energy. The argument breaks down if either top level initially has only *one* neutron or proton, rather than a spins-opposite pair, but is resurrected if we consider exchanging two or more neutrons for protons, or vice versa. The point is that energy is lowest when $Z$ and $N$ are nearly equal.

Now let us factor in Coulomb repulsion: In small nuclei, all proton-pair repulsions are overwhelmed by the strong attraction, so $Z \cong N$ should be the more stable state. In large nuclei, the uncompensated Coulomb repulsions may shift the proton energies upward enough to alter the order of corresponding neutron and proton energy levels, as depicted in Figure 10.9. The lowest-energy state—that is, with the tops of the proton and neutron levels roughly equal—then should have more neutrons than protons. A balance is struck between the neutrons serving as a nonrepulsive "glue," and being forced by increasing numbers into higher-energy states.

Overall, then, our model indicates that the energy should be lowest, the binding energy per nucleon greatest, at some intermediate mass number $A$, but it is not a function of $A$ alone. For small $A$, it should have a *relative* maximum at $Z \cong N$, and for large $A$, the relative maximum should occur at $N > Z$.

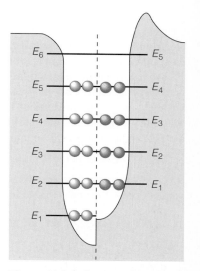

**Figure 10.9** In large nuclei, when Coulomb repulsion becomes significant, the lowest energy should have $N > Z$.

## Stability—The Experimental Truth

Although theories have been quite successful, the truth ultimately comes through experimental evidence. If the internucleon force is indeed a very strong attraction, a large amount of energy should have to be added to separate a nucleus into individual nucleons. The total energy of the separated parts should be significantly greater than that of the bound nucleus. With $E_{internal} = mc^2$, it follows that the total *mass* of the parts would have to be greater than that of the bound nucleus! This is true. Consider the deuteron:

$$\text{deuteron mass} = 2.013553 \text{ u}$$

$$\text{proton mass} + \text{neutron mass} = 1.007276 \text{ u} + 1.008665 \text{ u} = 2.015941 \text{ u}$$

$$\text{mass of parts} - \text{mass of whole} = 2.015941 \text{ u} - 2.013553 \text{ u} = 0.002388 \text{ u}$$

2.013553 u

1.007276 u       1.008665 u

The mass difference is what tells us the binding energy BE. The energy that must be supplied to pull the nucleons apart is the difference between the final and initial energies:

$$BE = m_f c^2 - m_i c^2 = (\text{mass of parts} - \text{mass of whole})c^2 = \Delta m c^2$$

$$= 0.002388 \text{ u} \times 1.661 \times 10^{-27} \text{ kg/u} \times (3 \times 10^8 \text{ m/s})^2$$

$$= 3.57 \times 10^{-13} \text{ J} = 2.22 \text{ MeV}$$

The same arguments hold for the bound electron–proton system we know as the hydrogen atom. Given that 13.6 eV must be added to pull the atom apart, the preceding formula shows that the atom should be less massive than the sum of the separate electron and proton masses by $2.4 \times 10^{-35}$ kg. The important difference is in the *fractional* mass change. It is very small for the hydrogen atom: $(2.4 \times 10^{-35} \text{ kg})/(1.67 \times 10^{-27} \text{ kg}) \cong 10^{-8}$. For the deuteron, it is $(0.002388 \text{ u})/(2.013553 \text{ u}) \cong 10^{-3}$. Although $\frac{1}{10}\%$ may still not sound like much, as a mass-energy conversion it is huge! Yet the five orders of magnitude difference between electronic and nuclear binding energies is typical. The simple reason is that nuclear binding involves much stronger forces.

The true measure of how tightly bound a representative nucleon is to its nucleus is the binding energy *per nucleon,* and by calculating it from experimental values, we may judge our theoretical model. For the deuteron, it is 2.22 MeV/2, or 1.11 MeV per nucleon. For arbitrary nuclei, we find total binding energy just as we did for the deuteron: To completely separate all nucleons requires an energy of

$$BE = (\text{mass of parts} - \text{mass of whole})c^2$$

Since tables invariably list atomic rather than nuclear masses, we will find it convenient to use the approximate formula:

$$BE = (Zm_H + Nm_n - M_{^A_Z X})c^2 \qquad (10\text{-}4)$$

Here, the mass of the parts has the neutron number $N$ multiplying the neutron mass $m_n$, as it should. The proton number $Z$, on the other hand, multiplies not the proton mass $m_p$, but rather the hydrogen *atomic* mass $m_H$. Therefore, this term is too large, by the mass of $Z$ electrons. However, $M_{^A_Z X}$ stands for the *atomic* mass of the isotope of element $X$ with $Z$ protons (and electrons) and $A$ nucleons, and thus subtracts the proper number of electron masses. Still the formula is not exact. As we noted in the case of hydrogen, the mass of a whole atom is not simply the mass of its nucleus plus the mass of $Z$ electrons; the mass of the whole is less than the sum of its parts by the *electronic* binding energy. The formula ignores electron binding energies, but they are usually negligible. (*Note:* $m_H$ is given in the margin, along with a useful conversion factor.)

$$m_H = 1.007825 \text{ u}$$
$$1 \text{ u} \times c^2 = 931.5 \text{ MeV}$$

## Example 10.1

Calculate the binding energy per nucleon of an iron-56 nucleus.

### Solution

From the atomic mass data of Appendix J, we find the mass of iron-56:

$$^{56}_{26}\text{Fe} = 55.934939 \text{ u}$$

The neutron number $N$ is $A - Z = 56 - 26 = 30$. Now using (10-4),

$$\text{BE} = (26 \times 1.007825 \text{ u} + 30 \times 1.008665 \text{ u} - 55.934939 \text{ u})c^2$$

$$= (0.528461 \text{ u})c^2$$

The iron nucleus is less massive than the sum of its parts by more than half a nucleon mass! The given conversion factor may be used to express $c^2$ in MeV/u.

$$(0.528461 \text{ u}) \times 931.5 \text{ MeV/u} = 492.3 \text{ MeV}$$

There are 56 nucleons, so the binding energy per nucleon is

$$\frac{\text{BE}}{\text{nucleon}} = \frac{492.3 \text{ MeV}}{56} = 8.79 \text{ MeV}$$

Figure 10.10 plots BE/nucleon for the naturally occurring isotopes. That it validates our theoretical model may be seen by considering different views. Figure 10.11 is a top view. It shows the locations in the $Z$-$N$ plane of the naturally occurring nuclei. They describe a path known as the **curve of stability.** As our model predicted, for small $A$ (distance from the origin), the greatest stability occurs at $N = Z$, but tends toward $N > Z$ as $A$ increases. Figure 10.12 is a cross section of Figure 10.10, the "slice" taken roughly along the curve of stability. Since BE/nucleon varies with both $N$ and $Z$, no two-axes graph can represent it for all isotopes. But Figure 10.12 shows the general trend with nuclear size—and is conspicuously similar to Figure 10.7! From the deuteron's 1.11 MeV, it rises to a maximum at about $A = 60$ of 8.79 MeV, shared by iron-56, iron-58, and nickel-62. Of all elements, these have the most stable, tightly bound nuclei (as opposed to *electronic configuration,* of which helium is the most stable). It then falls slowly to 7.57 MeV for uranium-238.

So much is made of binding energy *per nucleon* because of its importance to releasing nuclear energy. While it is true that *binding energy* tends to increase

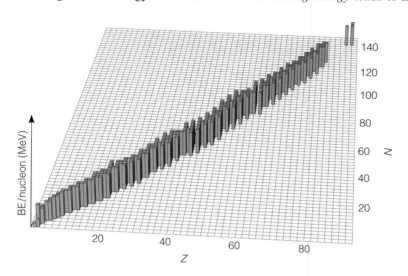

**Figure 10.10** Binding energy per nucleon versus $Z$ and $N$.

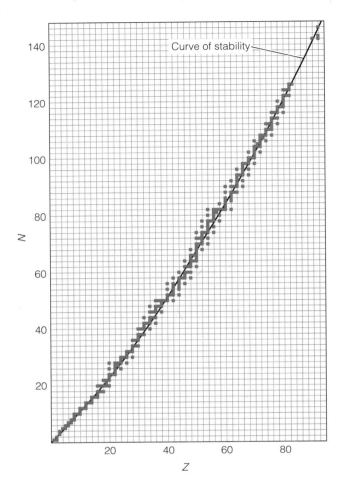

**Figure 10.11** Stable nuclei cluster along the curve of stability.

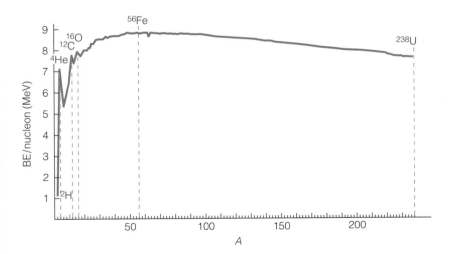

**Figure 10.12** Binding energy per nucleon versus $A$.

with $A$ (e.g., 2.22 MeV for the deuteron, 492 MeV for iron-56, and 1802 MeV for uranium-238), this is due mostly to the simple fact that the *number* of nucleons increases and each is bound by some net attraction to its nucleus. But if we were interested in finding a minimum-energy state for a given macroscopic amount of matter, we would not group nucleons into nuclei of the largest $A$. Given $I$ total nucleons, divided into groups (i.e., nuclei) of $A$ nucleons, the total binding energy would be the number of groups $I/A$ times the binding energy BE of each nucleus: $(I/A) \cdot$ BE or $I \cdot$ (BE$/A$). Clearly, nuclei with the highest binding energy *per nucleon* would give the highest overall binding energy, the lowest-energy state, allowing us to extract the greatest amount of energy for other purposes. We return to the idea of releasing nuclear energy in Section 10.7.

Another feature of Figure 10.12 is that helium-4, carbon-12, and oxygen-16 are unusually tightly bound for their "location" among nuclei. All have $Z = N$, but there are other reasons for their great stability: $N$ and $Z$ are even numbers and, in the case of helium and oxygen, they are "magic" numbers. Explanation of these effects requires a more quantitative nuclear model.

# ▼ 10.3 Nuclear Models

We now discuss two of the most useful theoretical models of the nucleus. Each has its emphasis; they agree in some ways, but each rests upon unique assumptions and simplifications, and makes its own important predictions. There is no single comprehensive theory. Until and unless the nucleus is "completely understood," there will always be room for complementary models.

## The Liquid Drop Model

The force between the molecules in a drop of water is strongly repulsive at very small separations, but attractive at greater separations. In fact, the potential energy qualitatively resembles Figure 10.3. This accounts for the fact that while liquid water is essentially incompressible it also exhibits a reluctance to be pulled apart, a cohesiveness that is responsible for surface tension. To move a molecule to the surface, it must be pulled away from some of the water molecules that would otherwise attract it on all sides. This takes energy. Accordingly, the lowest-energy state of a water droplet is one in which the surface area is minimum—that is, a sphere.

Similarities between the energies in a liquid drop and in the nucleus give the **liquid drop model** its name. In both cases, interior particles are attracted to all particles in the immediate vicinity—a certain number of bonds per particle adding to the total energy holding the constituents together. Were all particles interior, the total binding energy would be directly proportional to the number of particles. Particles at the surface, however, share fewer bonds with others. This reduces the total binding energy, but more so for small "drops," where almost all particles are exterior.

The centerpiece of the liquid drop model is an expression for the binding energies of the nuclei. Since to know binding energy is to know mass, it may also be said that the model yields nuclear masses. We have already discussed qualitatively the separate factors on which it is built. The quantitative expression we now derive is a sum of terms based upon them. Although the terms are obviously approximations, in each case we will justify the form.

*Volume Term*  If we ignore the fact that not all nucleons are surrounded in the interior of the nuclear volume, the total binding energy should be directly proportional to the number of nucleons. Thus, we begin our sum with the term shown in the margin. While the proportionality constant $c_1$ has been theoretically predicted, the terms yet to come include similar constants, and all the constants are usually chosen so as to give the best overall agreement with experimentally determined values. The question might then arise: Is this really a theory, or simply a way of presenting experimental findings? The answer is that it is a simplified theoretical *model* of a very complex system, *guided* by experimental knowledge. The known binding energies per nucleon describe a certain shape as a function of $Z$ and $N$, and, as Figure 10.10 shows, it is not a simple shape. For many functions with arbitrary constants (e.g., $ax^2 + bx + c$), it would simply be impossible to fit the known shape. That the functional form of the liquid drop model can be made to do so as well as it does is worthy evidence that the assumptions upon which it is based are valid. Furthermore, as should be the case for any good model, it may be used to predict binding energies of isotopes yet unknown.

$$BE = c_1 A + \cdots$$

*Surface Term*  The binding energy of surface nucleons is lower; they are not completely surrounded by attractive particles. Accordingly, the total binding energy should be *reduced* by a factor proportional to the surface area, which is in turn proportional to volume to the two-thirds power. We know that the volume is proportional to the number of nucleons, so we obtain the term shown in the margin.

$$-c_2 A^{2/3}$$

*Coulomb Term*  Coulomb repulsion raises the overall potential energy, *reducing* the binding energy. The electrostatic potential energy shared by two protons separated by $r$ is proportional to $e^2/r$. In a nucleus of $Z$ protons, there are $Z(Z-1)$ pairs, and typical proton separation is of the order of the nuclear radius, which we know, from equation (10-1), is proportional to $A^{1/3}$.

$$-c_3 \frac{Z(Z-1)}{A^{1/3}}$$

*Asymmetry Term*  Coulomb repulsion aside, a lower-energy state occurs when $N \cong Z$, due to the exclusion principle. To quantify the effect, we first note that neutron and proton energy levels due to the strong attraction alone should be identical. Let us assume that they are filled to equal heights, $E_F$, and that the levels are roughly equally spaced, $\delta E$ apart, as shown in Figure 10.13. If $j$ neutrons are now changed to protons (or vice versa), the top neutrons must go from neutron level $E_F$ to proton level $E_F + (j/2)\,\delta E$. The others must make an equal jump. If the energy of each of $j$ particles increases by $(j/2)\,\delta E$, the total energy increases by $(j^2/2)\,\delta E$. We can easily express $j$ in terms of $Z$ and $N$. If we start with $Z = N$, then gain $j$ protons and lose $j$ neutrons, $Z - N$ must now be $2j$, giving $j = \frac{1}{2}(Z - N)$. Thus, an asymmetry between $Z$ and $N$ should increase the overall energy (decrease the binding energy) by a factor proportional to $(N - Z)^2\,\delta E$. It remains to quantify the level spacing $\delta E$. Since nucleons are incompressible, adding them to a nucleus increases the physical dimensions of nuclear well. Increasing the size of a potential well decreases all its energy levels (cf. $E_{\text{infinite well}} = n^2\pi^2\hbar^2/2mL^2$), and thus the spacing between levels. But how much? With closer spacing between levels but more particles to fill them, it is quite possible that the height of the topmost occupied level, $E_F$, would be the same no matter what might be the mass number $A$. In fact, there are good arguments that it should be the same for all.[4] If so, the average spacing would be $\delta E = E_F/(\frac{1}{4}A)$, so that $\delta E \propto A^{-1}$. Factoring this in, we obtain the term shown in the margin. It is zero if $Z = N$ and reduces the binding energy as the asymmetry between $N$ and $Z$ increases.

$$-c_4\frac{(N-Z)^2}{A}$$

Altogether, we have

$$\text{BE} = c_1 A - c_2 A^{2/3} - c_3\frac{Z(Z-1)}{A^{1/3}} - c_4\frac{(N-Z)^2}{A} \tag{10-5}$$

Semi-empirical binding energy formula

Trial and error has shown that the sum best fits experimentally determined binding energies with the following values, all in MeV:

$$c_1 = 15.8 \qquad c_2 = 17.8 \qquad c_3 = 0.71 \qquad c_4 = 23.7$$

Because the forms of the terms are based upon theoretical arguments but the coefficients are chosen to fit experimental data, equation (10-5) is known as the **_semi-empirical_ binding energy formula.**

**Example 10.2**

What binding energy per nucleon is predicted by the liquid drop model for the iron-56 nucleus?

**Solution**

Using (10-5) and the coefficients provided,

$$\text{BE} = 15.8(56) - 17.8(56)^{2/3} - 0.71\frac{26(25)}{(56)^{1/3}} - 23.7\frac{(30-26)^2}{56}$$

$$= 496.9 \text{ MeV}$$

$$\frac{\text{BE}}{\text{nucleon}} = \frac{496.9 \text{ MeV}}{56} = 8.87 \text{ MeV}$$

This differs from the true experimental value of Example 10.1 by less than 1%.

Figure 10.14 shows the predictions of the semi-empirical binding energy formula plotted against the known values of BE/nucleon (Figure 10.12). The curves are quite close. Again, that it is possible to fit the experimental curve even this well is indicative of the merit of the liquid drop model's assumptions. Predicted less well are binding energies of very light nuclei, in which we might expect the average behaviors assumed in the model to be poor approximations. Helium in particular is much more tightly bound than the model predicts. As noted earlier, the reason is that both $N$ and $Z$ are even numbers and "magic" numbers. Terms to account for these two effects are often appended to the semi-empirical binding

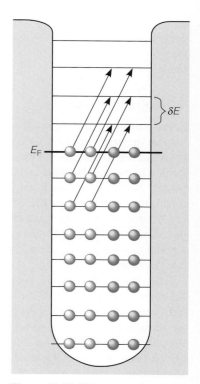

**Figure 10.13** If $j$ neutrons become protons, the energy increases by $\frac{1}{2}j^2\,\delta E$.

[4]According to equation (8-43), for fermions confined in a volume $V$, $E_F$ is proportional to the density $N/V$, which is about the same for all nuclei.

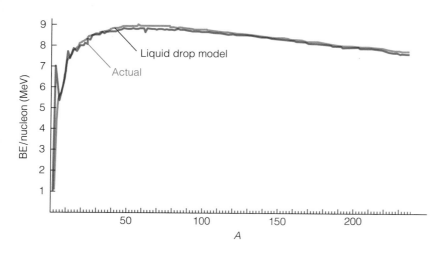

**Figure 10.14** Liquid drop model and actual BE/nucleon compared.

energy formula, but in most cases (helium the most notable exception) introduce only small corrections. Explanation of these effects really belongs to the shell model.

## The Shell Model

Some indication of the complexities of the internucleon strong force is given by the fact that the two most useful nuclear models are so vastly different. Except for the asymmetry term's use of the exclusion principle, the liquid drop model is classical, while the shell model is thoroughly quantum-mechanical. In the liquid drop model, essentially classical particles interact via strong interparticle forces, and total binding energy is predicted. In the shell model, detailed interparticle forces are ignored; the particles are treated as occupying quantum states in some average potential well. The unique successes of the shell model are predictions of nuclear angular momenta and tendencies for both $N$ and $Z$ to be magic numbers or at least even numbers.

The shell model is to the nucleus what the treatment of Chapters 6 and 7 is to the atom. However, because the strong force is so complex, the form of the potential energy is much more speculative than for the atom. In the simplest analysis, the nucleons are assumed to orbit independently of one another in a spherically symmetric potential well. A rounded finite well has been most widely used, but certain basic aspects of nuclear behavior may be understood with an infinite-well assumption.

**Example 10.3**
Nuclei in excited states can lower their energy by "gamma emission," in which a photon carries away the excess energy. As a crude approximation, assume that the nucleons behave as through trapped in a one-dimensional (!) infinite well. Using an appropriate order-of-magnitude width $L$, determine a reasonable energy for the photon emitted as a nucleon makes a transition from one energy level to another.

↑ Solution

The energy levels in the infinite well are

$$E = \frac{\hbar^2 \pi^2 n^2}{2mL^2}$$

An appropriate width would be a typical nuclear diameter. Suppose $A = 100$. Using equation (10-1),

$$2r = 2(100)^{1/3} \times 1.2 \times 10^{-15} \text{ m} \cong 10^{-14} \text{ m}$$

The mass of a nucleon is about $1.67 \times 10^{-27}$ kg. Thus,

$$E = \frac{(1.055 \times 10^{-34} \text{ J·s})^2 \pi^2 n^2}{2(1.67 \times 10^{-27} \text{ kg})(10^{-14} \text{ m})^2} = (3.3 \times 10^{-13} \text{ J})n^2 \cong (2 \text{ MeV})n^2$$

We might expect energy jumps to be of several MeV. Although this is certainly a crude approximation, it is no coincidence that many of the photons emitted by nuclei have energies of the order of magnitude of 1 MeV.

The shell model was proposed when it was found that nuclear stability exhibited some periodicity. Careful inspection of Figure 10.11 shows a disproportionate number of nuclei with either $N$ or $Z$ or both equal to one of the so-called magic numbers: 2, 8, 20, 28, 50, 82, 126. Apparently, great stability accompanies these numbers. Since it was known that atoms of atomic number 2, 10, 18, 36, 54, ... are unusually stable because they possess complete shells or subshells, it was natural to attribute magic numbers to similar behavior in nuclei. After allowance for a strong nuclear spin-orbit interaction, this hypothesis was confirmed.

Another success of the shell model is that when expanded slightly to allow for specific interparticle forces (besides the average central force), it explains the **pairing effect**—the tendency for $Z$ or $N$ to be simply even numbers. This is also apparent in Figure 10.11, in the clustering of stable nuclei along lines of even $N$ and $Z$. Finally, the shell model predicts nuclear angular momenta, which agree well with experimental observation. This should not be too surprising, since we know that solution of the Schrödinger equation for a spherically symmetric potential energy (central force) yields quantized angular momentum states.

## ▾ 10.4 Nuclear Magnetic Resonance and MRI

Since angular momentum is quantized in the nucleus, so is the nuclear magnetic dipole moment. The technique of **Nuclear Magnetic Resonance,** or NMR, exploits this quantized magnetic moment to produce images of the body's interior.

The most important requirements in producing such an image are getting whatever one uses for a probe into the body; interacting in different ways with different features once inside; and detecting the interaction from the outside—all this without damage to the tissues. (The most convenient candidate—light—fails because its penetration of the body is very short-ranged.) These requirements are fulfilled to some extent by x rays: They pass easily into the body; they are absorbed at different rates by materials of different density as they interact with electron clouds in various regions; and the intensity that passes through the body unabsorbed is easily detected. In a simple x ray, the transmitted intensity registers on photographic film and yields an image of the average density through which the x rays passed. It is different, for instance, where it passed through skin, muscle, and bone than where it passed through skin and muscle only. However, x-ray photons typically have energies measured in keV, easily able to tear ions apart, and thus carry with them a degree of biological hazard.

To produce an image using nuclear magnetic resonance, electromagnetic radiation is sent into the body to interact not with the electron charge densities, but directly with the nuclei. The body is immersed in an external magnetic field, so that the magnetic moments of the nuclei have quantized orientation energies along that axis. To cause a jump to a different orientation energy, only photons of a certain frequency may be absorbed; only these meet the necessary resonance condition. As the nuclei jump back down, the absorbed energy is re-radiated in all directions and easily detected. The spacing of the jumps depends on the strength of the external magnetic field.

In **magnetic resonance imaging** (MRI) of the human body (Figure 10.15), the field strength is chosen so that photons in the radio frequency range have the proper energy to excite jumps in hydrogen (ubiquitous in the body). Such radiation passes easily into the body, but, with photon energies far less than an electronvolt, poses no biological threat. However, MRI has advantages beyond the relative low energy of the photons:

1. Using an external magnetic field that is *nonuniform,* MRI is better able to "look" at specific depths within the body. Since magnetic orientation energies are directly proportional to the strength of the field, the resonant frequency will be different in a plane where the field has one value than in a plane where it has another. Thus, resonant frequencies vary predictably with position. In any case, the image is constructed by computer from data taken from various directions.

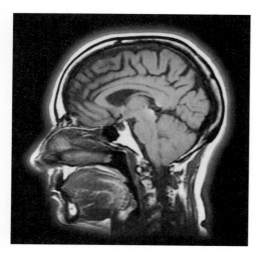

**Figure 10.15** A healthy, living human brain, as seen by MRI.

2. The absorption process is more detailed. The stimulation to higher orientation energies is time-dependent in a way that is in turn dependent on the nature of the surrounding nuclei and atoms (since the nuclei are not truly free). This time-dependence then serves as additional information distinguishing one material or tissue from another.

3. While hydrogen is a common choice, the frequency of the incoming radiation can be tuned to resonant frequencies of different elements, which may provide an important alternative "view."

Although its medical applications are best known, NMR is being used increasingly in basic scientific research. Among other things, it has been used to further our understanding of the all-carbon geometric structures known as fullerenes and the still mysterious high-$T_c$ superconductors (see Section 9.10).

## 10.5  Radioactivity

A nucleus may be bound, yet still in a state of relatively high energy, and if it is possible for it to attain a lower-energy, more tightly bound state, it will eventually do so. Nuclei fitting this description are termed unstable, or radioactive, and the process is known as **radioactive decay.** Radioactive decay usually takes the form of the spontaneous emission of a small energetic particle of one kind or another, as in the photon emission of the preceding example. Of course, energy is conserved. The kinetic energy increases, but it is just the natural result of the decrease in the internal/mass energy. The symbol $Q$ is used for the kinetic energy released.

$$\Delta KE = -\Delta mc^2 = -(m_f - m_i)c^2$$

$$Q = (m_i - m_f)c^2 \tag{10-6}$$

Let us take a look at the various forms of radioactive decay.

### Alpha Decay

As shown in Figure 10.12, the helium nucleus ($\alpha$-particle) is unusually tightly bound for its size; that is, its mass is less than the sum of its parts by an unusually large amount. It is four nucleons in a state of low (internal) energy. The figure also shows that the heaviest of nuclei (largest $A$) tend to be less tightly bound than those of somewhat smaller mass number. It should not be too surprising, then, that a large nucleus, called a **parent nucleus,** might split into a lighter, more tightly bound nucleus, called a **daughter nucleus,** plus a tightly bound $\alpha$-particle (Figure 10.16). Since the $\alpha$-particle takes away two neutrons and two protons, the daughter nucleus has two fewer of each than the parent.

Parent

Daughter          $\alpha$-particle

**Figure 10.16** $\alpha$ Decay.

**E x a m p l e   1 0 . 4**

Alpha decay is one of the ways uranium-238 attains a lower-energy state. What is the daughter nucleus and how much (kinetic) energy is released?

### Solution

Uranium-238 has 92 protons and $238 - 92 = 146$ neutrons, so the daughter has 90 protons and 144 neutrons, identifying it as thorium-234. The reaction may be represented as follows:

$$^{238}_{92}\text{U} \rightarrow {}^{234}_{90}\text{Th} + {}^{4}_{2}\text{He}$$

Obtaining the masses from Appendix J, the change in kinetic energy is

$$Q = (m_i - m_f)c^2$$

$$= [238.050784 \text{ u} - (234.043593 \text{ u} + 4.002603 \text{ u})] \times 931.5 \text{ MeV/u}$$

$$= 4.27 \text{ MeV}$$

*Note:* As we did in calculating binding energies, we have used atomic masses rather than true nuclear masses. Again, though, the "extra" electrons included in the initial mass are subtracted in the final $(92 = 90 + 2)$, and the small electron binding energies may be ignored.

Although the $\alpha$-particle and daughter share the energy released in $\alpha$-decay, momentum conservation requires that in cases where something splits into "large" and "small" pieces, the small piece—in this case, the $\alpha$-particle—possess nearly all the kinetic energy.[5] The daughter's recoil energy is very small.

## Beta Decay

The uncertainty principle may be used to show that electrons cannot be confined in the nucleus (see Exercise 3.51). Yet we find that they are *emitted* by some nuclei. We conclude that they are spontaneously created, then immediately depart. When created and emitted in this way, an electron is called a $\beta^-$-**particle,** and the process is known as $\beta^-$ **decay.**

The phenomenon of $\beta^-$ decay is certainly curious. If charge is to be conserved, what must be happening in the nucleus? It is found that after $\beta^-$ decay, the remaining nucleus, again known as a daughter, has one more proton and one less neutron than the parent had. *In effect, a neutron in the parent spontaneously changes into a proton and an electron.* Charge is conserved.

But this is not the whole story. Were the parent to decay into *only* a daughter and a $\beta$-particle, the $\beta$-particle would possess essentially all the available kinetic energy (even more so than in $\alpha$ decay, since an electron is very small). This is not what we find. Figure 10.17 shows $\beta$-particle energies found in a large number of $\beta$ decays. They vary widely and are usually much less than what *should* be available given the known mass decrease: $Q = [m_{\text{parent}} - (m_{\text{daughter}} + m_\beta)]c^2$. A $\beta$ decay with only two product particles would violate energy conservation!

The problem is even bigger, though. The simplest example of $\beta$ decay is neutron decay. While neutrons are stable when bound in stable nuclei, a *free* neutron is not. Because a free proton is significantly less massive than a free neutron, a neutron may attain a lower energy/mass via $\beta^-$ decay into a proton. However, if a spin-$\frac{1}{2}$ neutron were to become *only* a spin-$\frac{1}{2}$ proton and spin-$\frac{1}{2}$ electron, the final total spin (see Section 7.9) would be either 1 or 0. Angular momentum could not

[5] $\text{KE}_{\text{small}} = p^2/2m_{\text{small}}$ and $\text{KE}_{\text{large}} = p^2/2m_{\text{large}}$, where $p$ is the same for both, yields $\text{KE}_{\text{small}} = (m_{\text{large}}/m_{\text{small}})\text{KE}_{\text{large}}$.

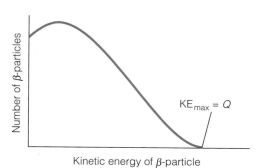

Figure 10.17 The "mysterious" variation in $\beta$-particle energies.

be conserved! Allowance for the creation of photons (spin-1) would not solve the problem, since the total spin would still be an integer.

This seeming violation of fundamental laws of physics caused much consternation among early nuclear physicists. Wolfgang Pauli postulated a solution: The decay produces an unseen additional particle, which carries away angular momentum and variable portions of the available kinetic energy. Charge and angular momentum could be conserved if it were uncharged and spin-$\frac{1}{2}$. (Three final $\frac{1}{2}$ spins—proton, electron, neutrino—can add to the neutron's initial $\frac{1}{2}$.) Moreover, since the $\beta$-particle does sometimes acquire a kinetic energy essentially equal to $Q$, calculated for a final mass of *only* daughter and $\beta$, it would seem that the mystery particle is virtually massless. After considerable effort, Pauli's unseen particle was found and the fundamental laws saved.

We call the "extra" particle created in $\beta^-$ decay an **antineutrino,** symbol $\bar{\nu}$ (Figure 10.18). Its discovery lagged its prediction because it interacts very feebly with matter of any kind (usually passing right through!), so it is very difficult to detect. Being virtually massless, it does not come into play in calculating $Q$. Its energy is essentially all kinetic, and is simply part of the energy released.

How a neutron actually changes into other particles is a question we leave for Chapter 11. Our primary interest here is how $\beta$ decay allows a nucleus to decrease its energy. Consider, for example, boron-12, $^{12}_{5}B$, which has five protons and seven neutrons. If a neutron changes into a proton and an electron, the daughter nucleus has six each protons and neutrons. Since lighter nuclei tend to be more tightly bound when $N = Z$, and furthermore when both are even numbers, it is reasonable that boron-12 might attain a lower energy via $\beta^-$ decay.

Simply to account for nucleon number and charge, the $\beta^-$-particle is often represented as $_{-1}^{0}\beta^-$; that is, it has no nucleons but carries a charge opposite that of a proton. Thus, we represent the $\beta^-$ decay of boron-12 as

$$^{12}_{5}B \rightarrow {}^{12}_{6}C + {}_{-1}^{0}\beta^- + \bar{\nu}$$

The energy released in this reaction is again $-\Delta mc^2$. However, since there are five electrons in boron-12 and six in carbon-12, using *atomic* masses would include an extra electron in the final mass. We circumvent the difficulty simply by allowing this extra electron mass to account for the mass of the $\beta^-$-particle.[6] Therefore, using masses from Appendix J,

$$Q = (12.014352 \text{ u} - 12 \text{ u}) \times 931.5 \text{ MeV/u} = 13.4 \text{ MeV}$$

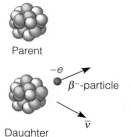

Parent

$-e$

$\beta^-$-particle

Daughter

$\bar{\nu}$

Figure 10.18 $\beta^-$ Decay.

$^6 m_i - m_f = $ (mass of $^{12}_{5}B$ nucleus) $-$ (mass of $^{12}_{6}C$ nucleus $+ m_{\beta-}$)
$= ([\text{mass of } {}^{12}_{5}B \text{ atom} - 5m_e])$
$- ([\text{mass of } {}^{12}_{6}C \text{ atom} - 6m_e] + m_e)$
$= \text{mass of } {}^{12}_{5}B \text{ atom} - \text{mass of } {}^{12}_{6}C \text{ atom}$

We represent neutron decay similarly:

$$\,_0^1 n \rightarrow \,_1^1 p + \,_{-1}^0 \beta^- + \bar{\nu}$$

Using the atomic mass of hydrogen to account for both the proton mass and electron mass, the energy released is

$$Q = (1.008665\ u - 1.007825\ u) \times 931.5\ \text{MeV/u} = 0.78\ \text{MeV}$$

Free neutrons survive approximately 15 minutes on average.

Complementary to $\beta^-$ decay is another form of $\beta$ decay: $\beta^+$ decay. Consider $\,_7^{12}N$, which has five neutrons and seven protons. It could become $\,_6^{12}C$, with six each, if only it could change a proton to a neutron. This is effectively what occurs in $\beta^+$ decay (Figure 10.19). But to conserve charge, a *positively* charged electron must be produced. The **positron,** the antiparticle of the electron, is this particle. Similarly, the "extra" particle created is not an antineutrino, but its antiparticle the **neutrino,** $\nu$. Antiparticles and the rules governing their creation are discussed in Chapter 11. For now, we simply note that the positron has the same mass as the electron, but opposite charge, while the neutrino, like the antineutrino, is uncharged, spin-$\frac{1}{2}$ and of negligible mass. Nitrogen-12 $\beta^+$-decays according to

$$\,_7^{12}N \rightarrow \,_6^{12}C + \,_{+1}^0 \beta^+ + \nu$$

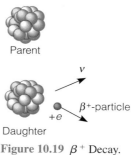

Parent

Daughter

**Figure 10.19** $\beta^+$ Decay.

The atomic mass of the daughter would be one electron mass too small to cancel the number of electrons included in the parent atomic mass. Therefore, to find the energy released, we must include *two* electron masses with the final mass (one to cancel the extra parent electron, one for the $\beta^+$).[7]

$$Q = (12.018613\ u - (12\ u + 2 \times 5.486 \times 10^{-4}\ u)) \times 931.5\ \text{MeV/u}$$

$$= 16.3\ \text{MeV}$$

As in all forms of radioactive decay, the final mass is less than the initial.

Interestingly, while a free neutron is unstable, liable to $\beta^-$-decay to a proton, a free proton cannot $\beta^+$-decay to a neutron, since mass would *increase*. Only in the nucleus is $\beta^+$ decay possible. There, a proton *could* decay to a "more massive" neutron, but because of the many interrelated effects (e.g., decreased Coulomb repulsion) still result in a lower *total* mass/internal energy. Although free proton decay is predicted in many "unified theories" (see Section 11.7), it cannot occur by neutron–positron creation.

There is a third form of $\beta$ decay, **electron capture** (Figure 10.20). A nucleus with too many protons, the type that might tend to $\beta^+$-decay, may approach $N = Z$ by another means: It may seize an orbiting electron and effectively turn this and a proton into a neutron in the nucleus, meanwhile emitting a neutrino. Thus, as in $\beta^+$ decay, $Z$ decreases by 1 and $N$ increases by 1. Many nuclei that $\beta^+$-decay also engage in electron capture. Often, electron capture is "energetically preferred." In $\beta^+$ decay, the mass of the parent nucleus must be at least one electron mass *greater* than the daughter, since a positron is created. In electron capture, however, the electron already exists, so the parent's mass can be as small as one electron

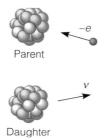

Parent

Daughter

**Figure 10.20** Electron capture.

---

[7]$m_i - m_f = $ (mass of $\,_7^{12}N$ nucleus) $-$ (mass of $\,_6^{12}C$ nucleus $+ m_{\beta^+}$) $= $ ([mass of $\,_7^{12}N$ atom $- 7m_e$]) $-$ ([mass of $\,_6^{12}C$ atom $- 6m_e$] $+ m_e$) $= $ mass of $\,_7^{12}N$ atom $-$ (mass of $\,_6^{12}C$ atom $+ 2m_e$)

mass *less* than the daughter. Carbon-11, though engaging in both, "prefers" electron capture approximately 100 to 1, in either case becoming boron-11.

$$^{11}_{6}C + {}^{0}_{-1}\beta^{-} \rightarrow {}^{11}_{5}B + \nu$$

The energy released in electron capture depends simply on the difference in the *atomic* masses of parent and daughter; the "extra" electron in the parent atomic mass accounts for the mass of the captured electron.

Figure 10.21 shows the effect on $Z$ and $N$ of the decays we have discussed. By altering $Z$ and $N$ differently, they are different ways of approaching the curve of stability. Heavy nuclei with too many protons often $\alpha$-decay ($Z$ and $N$ both reduced by 2), light nuclei with too many protons $\beta^{+}$-decay, or engage in electron capture ($Z$ reduced by 1, $N$ increased by 1), and nuclei with too many neutrons $\beta^{-}$-decay ($Z$ increased by 1, $N$ reduced by 1).

Also shown in Figure 10.21 are the **decay series** of thorium-232 and uranium-238, two radioactive but reasonably abundant naturally occurring isotopes. A radioactive decay need not result in a stable state; many daughter nuclei are themselves radioactive. Thus, one decay may be followed by a series of $\alpha$ and $\beta$ decays. Moreover, many unstable isotopes may decay in more than one way. Thus, at certain points a decay series may **branch,** with a certain probability of going either way. Even so, a stable nucleus will eventually be reached. Thorium-232 ends up as lead-208, and uranium-238 as lead-206. It is true that lead is not one of the most stable nuclei. As we know, the most tightly bound are around $A = 60$. But these lead isotopes are still stable simply because there is no form of radioactive decay available that can take them to a state any lower in energy/mass. Conversely, there are isotopes that appear to be on the curve of stability but are unstable because there *is* a form of decay that leads to an even lower mass (e.g., potassium-40; see Exercise 15).

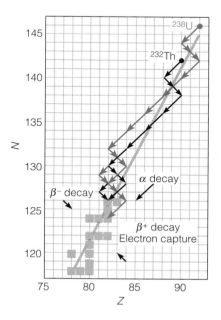

**Figure 10.21** The "directions" of $\alpha$ and $\beta$ decays, and the decay series of uranium-238 and thorium-232.

## Gamma Decay

In $\alpha$ and $\beta$ decay, one element becomes another; the charge of the daughter differs from that of the parent. This is not the case, though, if the particle the nucleus emits is a photon; $Z$ and $N$ do not change. A photon emitted from the nucleus is called a $\gamma$-particle and the decay a $\gamma$ decay (Figure 10.22).

$\gamma$-particle

**Figure 10.22** $\gamma$ Decay.

    Gamma decay often accompanies or follows other forms of decay. In many decays, the daughter nucleus is *not* left in its lowest-energy state. It may take a step in that direction by emitting a photon. In other reactions, such as spontaneous fission, $\gamma$-particles are simply a natural product of the process.

## Spontaneous Fission

An important form of radioactive decay *not* involving the emission of small particles is **spontaneous fission,** in which a nucleus splits (fissions) into two nuclei (Figure 10.23). It is much like $\alpha$ decay, but rather than breaking into small and large parts, the pieces tend to be both of intermediate mass number. The driving force behind spontaneous fission is the reduction of the destabilizing effect of Coulomb repulsion. It is true that fission increases the total surface area. This *alone* would *raise* the overall energy, requiring more nucleons to be pulled to the surface against their mutual attraction. But breaking into two smaller nuclei also decreases the number of protons too far apart to attract each other, so the number exhibiting a net repulsion (see Section 10.2) is reduced. In spontaneous fission, the reduction in Coulomb repulsion more than compensates for the increased "surface energy."

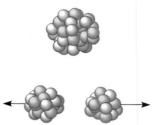

**Figure 10.23** Spontaneous fission.

## Radioactive Identification

Mixing a radioactive isotope with a nonradioactive material gives the material an unmistakable "signature"—recognizable by the energy of the particle emitted. Scientific research exploits this in numerous ways. To determine how some non-radioactive materials move about or how they stick to one another, radioactive atoms are introduced, either as a loose mixture or by being bonded to specific molecular groups. The subsequent behavior of the material is then easily followed by looking for the telltale decays of the radioactive tracer or tag. In one of many biological research applications, a complex molecule known as a steroid is tagged (covalently bonded) with tritium, a $\beta$-emitter, then mixed with a protein. The *steroid's* ability to bind to the protein is clearly indicated by how much *tritium* ends up stuck to the protein. Let us now look at another characteristic of radioactive decay that is also often exploited in scientific investigation.

## 10.6  The Law of Radioactive Decay

All forms of radioactive decay fundamentally change the parent nucleus, and the nucleus may decay in a particular way only once. A useful analogy is a light bulb, which may burn out but once, and then it is fundamentally different from a working light bulb. Furthermore, decays are governed by probabilities (cf. Section 5.2

for $\alpha$ decay); it is impossible to know exactly when a given nucleus will decay. In a large sample of a radioactive isotope, one nucleus may decay right away and another after a very long time. Still there is a predictable *average* time, characteristic of the particular decay; it may be long for the $\alpha$ decay of isotope $X$ and short for the $\beta^+$ decay of isotope $Y$. The light bulb analogy is again helpful. One never knows when a light bulb will burn out. But in a huge office building with thousands of light bulb $X$, there would be a predictable number burning out per day; an average lifetime would be apparent. In a similar building with thousands of light bulb $Y$, a competitor's lower-quality unit, a predictable average lifetime would also be apparent, though it might be much shorter.

No matter what the lifetime, though, if we increase the size of the sample (building), the number decaying (burning out) per unit time should increase proportionally. In other words, the *change* per unit of time in the number of nuclei present should be directly proportional to the number present.

$$\frac{dN}{dt} \propto N \tag{10-7}$$

We make this an equation by using the symbol $\lambda$ for the proportionality constant. Since the change in the number of radioactive nuclei is always negative—the number present is constantly decreasing—the equation requires a minus sign.[8]

$$\frac{dN}{dt} = -\lambda N \tag{10-8}$$

Differential equations of this form are common in physics. For instance, a capacitor discharging through a resistor obeys the same relationship (see Figure 10.24). In each case, the rate of decrease of a quantity that is not being replenished ($N$ or $Q$) is proportional to the quantity itself. As we shall see, the proportionality constant ($\lambda$ or $1/RC$) is inversely related to the average time to decay. We solve the equation by separating variables,

$$\frac{dN}{N} = -\lambda\,dt$$

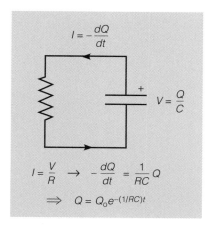

**Figure 10.24** A capacitor's charge decays exponentially with time.

[8]Here, $N$ stands for the number of nuclei present, not the number of neutrons in a nucleus. Unfortunately, the use of $N$ for both quantities is nearly universal, as is the use for the decay constant of $\lambda$, which is not a wavelength.

Integrating time from zero to $t$ as the number of nuclei goes from $N_0$ to $N$, we have

$$\int_{N_0}^{N} \frac{dN'}{N'} = -\lambda \int_0^t dt' \quad \text{or} \quad \ln\left(\frac{N}{N_0}\right) = -\lambda t$$

or

$$N = N_0 e^{-\lambda t} \tag{10-9}$$

The number of nuclei falls exponentially with time. It is worth reiterating that equation (10-7), upon which (10-9) is based, is valid only if the number of parent nuclei yet to decay is large. Only then will the exponential decay be smooth. If the number is small (the office building is small, or large but with most light bulbs already burnt out), decays will be more sporadic and the number remaining will fluctuate considerably about an exponentially decreasing trend.

The constant $\lambda$ is called the **decay constant.** Equation (10-9) says that the larger the value of $\lambda$, the more rapidly the number $N$ decreases. Equation (10-8) agrees, of course; a large $\lambda$ corresponds to a large decay rate. When we speak of "decay rate," though, we usually mean an absolute value—$R$, in decays per second. Thus, equation (10-8) is often written as

$$R = \lambda N \tag{10-10}$$

Naturally, since the number of nuclei decreases exponentially, so does the rate.

Let us see how $\lambda$ is related to the average lifetime of the parent nuclei. A general property of exponential functions is that they decrease by a certain fraction of their value in equal successive intervals of time. If in time interval 1 the number drops to half its initial value, in interval 2 it will drop to half of this value, one-quarter of the initial value; in interval 3 it will drop to half this value, one-eighth the initial one, and so on. The time required for the number to fall to half its initial value is, in fact, the characteristic by which decay rates are usually categorized. It is called the **half-life,** and is given the symbol $T_{1/2}$. Figure 10.25 shows the decrease of $N$ in terms of half-lives. The tie-in to the decay constant is as follows: By definition, $t$ is $T_{1/2}$ when $N$ is $\frac{1}{2}N_0$. Thus,

$$\frac{1}{2}N_0 = N_0 e^{-\lambda T_{1/2}}$$

Canceling $N_0$ and taking natural logarithms,

$$\ln \frac{1}{2} = -\lambda T_{1/2}$$

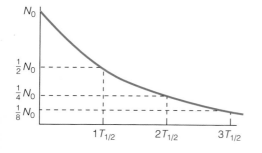

**Figure 10.25** Exponential decay.

$$\lambda = \frac{\ln 2}{T_{1/2}}$$

(10-11)

The decay constant is large when the half-life is small, and vice versa. (Strictly speaking, the half-life is not the average lifetime, but they are related by a simple proportionality constant; see Exercise 22.)

Given the complexities of nuclear binding and the possibility of multiple decay modes, it is not a simple matter to predict half-lives. Many have been measured, however, covering a range from about $10^{-22}$ s to $10^{17}$ years. Table 10.3 gives a few examples. Many more are given in Appendix J.

**Table 10.3**

| Isotope | Decay Mode | Half-life |
|---|---|---|
| $^{35}_{20}Ca$ | $\beta^+$ | 50 ms |
| $^{3}_{1}H$ | $\beta^-$ | 12.3 yr |
| $^{238}_{92}U$ | $\alpha$ | $4.5 \times 10^9$ yr |

**Example 10.5**

A vessel holds 2 $\mu$g of tritium. (a) What is its initial decay rate? (b) How much time will elapse before the amount of tritium falls to 1% of its initial value?

**Solution**

(a) Decay rate is proportional to the number of radioactive nuclei present, which we find in the usual way. Using the atomic mass from Appendix J,

$$N = \frac{2 \times 10^{-9} \text{ kg}}{3.02 \text{ u} \times 1.66 \times 10^{-27} \text{ kg/u}} = 4.0 \times 10^{17}$$

The decay constant we find from equation (10-11) and the half-life in Table 10.3.

$$\lambda = \frac{\ln 2}{12.3 \text{ yr}} \times \frac{1 \text{ yr}}{3.16 \times 10^7 \text{ s}} = 1.78 \times 10^{-9} \text{ s}^{-1}$$

Now using (10-10),

$$R = \lambda N = (1.78 \times 10^{-9} \text{ s}^{-1})(4 \times 10^{17}) = 7.1 \times 10^8 \text{ decays/s}$$

Although the number decaying each second is certainly large, it is still a very small fraction of the total number.

(b) Inserting $N = \frac{1}{100}N_0$ in equation (10-9),

$$\frac{1}{100}N_0 = N_0 e^{-\lambda t} \quad \rightarrow \quad \ln \frac{1}{100} = -\lambda t$$

$$t = \frac{-\ln \frac{1}{100}}{1.78 \times 10^{-9} \text{ s}^{-1}} = 2.6 \times 10^9 \text{ s} = 81.7 \text{ yr}$$

Since the number drops to 1%, the decay rate also will have dropped to 1%. [*Note:* Dividing the time by the half-life gives 81.7/12.3 = 6.64. Thus, the time is 6.64 half-lives. Correspondingly, $\left(\frac{1}{2}\right)^{6.64}$ is $\frac{1}{100}$.]

## Radioactive Dating

An important use of the radioactive decay law is radioactive dating, in which we attempt to determine the age of something old, such as a fossil, a rock, or an artifact. There are several methods, using different radioactive isotopes. All rely on assumptions about the abundance of certain isotopes in the past. If the initial number of nuclei in a sample is known, and the final number, a calculation of its age is straightforward.

In the most common method, carbon-14 dating, the radioactive isotope is carbon-14, which $\beta^-$-decays with half-life of 5730 years. The assumption is that the abundance of carbon-14 in the biosphere[9] was the same in the past as it is now. The rationale is that, although carbon-14 does of course decay, it is continuously replenished by the action of cosmic rays, high-energy photons from space. When these strike nitrogen in the upper atmosphere, they initiate a nuclear reaction that produces the isotope, which then spreads throughout the biosphere, maintaining the ratio of carbon-14 to stable carbon nuclei at about $1.3 \times 10^{-12}$. Carbon-14 dating works only for (formerly) living organisms; while a plant or animal lives, it exchanges carbon with the environment, through its food supply and the atmosphere. Since isotopes of the same element are chemically identical, a living organism should take in a fraction of carbon-14 atoms equal to the fraction in the environment. When it dies, it ceases to exchange carbon with the environment, and from that point onward the amount of carbon-14 decays exponentially.

## Example 10.6

What is the age of a fossil that contains 6 g of carbon and has a decay rate of 27 decays per minute?

### Solution

At 12 g/mol, the fossil contains $\frac{1}{2}$ mol of carbon, or $3.01 \times 10^{23}$ carbon atoms, nearly all of which would be the stable carbon-12. The fraction that would initially have been carbon-14 is

$$N_0 = (1.3 \times 10^{-12})(3.01 \times 10^{23}) = 3.9 \times 10^{11}$$

Knowing the decay constant

$$\lambda = \frac{\ln 2}{T_{1/2}} = \frac{\ln 2}{5730 \times 3.16 \times 10^7 \text{ s}} = 3.83 \times 10^{-12} \text{ s}^{-1}$$

we may find the present number via (10-10).

$$\frac{27 \text{ decays}}{60 \text{ s}} = 3.83 \times 10^{-12} \text{ s}^{-1} \times N \implies N = 1.18 \times 10^{11}$$

Now using (10-9),

$$1.18 \times 10^{11} = (3.9 \times 10^{11})e^{-(3.83 \times 10^{-12} \text{ s}^{-1})t}$$

or

$$t = 3.14 \times 10^{11} \text{ s} \cong 9900 \text{ yr}$$

To determine the ages of rocks and other things that do not exchange carbon with the biosphere, other methods of radioactive dating must be used. The "age" of a rock is the time that has passed since it formed a solid. One technique for

[9]The part of our world inhabited by and interacting with biological organisms.

determining ages of rocks uses the decay of potassium-40 to argon-40. The assumption here is that when the rock formed, it contained no argon at all; since argon is an inert gas, it does not form compounds and it easily passes out of a molten mass. Thus, any Ar-40 atoms present now should be the result of radioactive decay since the solid formed. A slight difficulty arises in determining the number of K-40 decays: It is not simply the number of Ar-40 atoms present, for K-40 also decays to calcium-40, and the rock might well have been "contaminated" with Ca-40 when it formed. However, decays to Ca-40 and Ar-40 occur in fixed ratio, so the total number of K-40 decays is *directly proportional* to the number of Ar-40 atoms. Adding the number of K-40 atoms decayed to the number present gives the initial number, and with the initial number, present number, and half-life, the age can be calculated. Other radioactive isotopes found in various types of rock and commonly used for dating are uranium-238 (decaying to lead-206), thorium-232 (decaying to lead-208), and rubidium-86 (decaying to strontium-86).

## 10.7 Nuclear Reactions

The term **nuclear reaction** refers to any occurrence in which nucleons are changed or exchanged between nuclei, much as electrons are exchanged between atoms in a chemical reaction. Radioactivity is a form of nuclear reaction, but a *spontaneous* one. We now study nuclear reactions that are *induced* by striking a nucleus with another particle. An example:

$$^{10}_{5}B + ^{1}_{0}n \rightarrow ^{7}_{3}Li + ^{4}_{2}He$$

Boron-10 is stable, but introduction of a neutron destabilizes the nucleus, which then breaks into two pieces. The change in kinetic energy is again given by equation (10-6). Using Appendix J,

$$(10.012937\ u + 1.008665\ u - 7.016003\ u - 4.002603\ u) \times 931.5\ MeV/u$$

$$= 2.79\ MeV$$

Mass decreases and kinetic energy increases. In chemistry, this would be called an "exothermic" reaction. Of course, "endothermic" reactions are possible; if the reactants have kinetic energy before the reaction, it is quite possible for the products to have less kinetic energy and therefore more mass. Mass increases in the following reaction, used to produce monoenergetic neutrons:

$$^{7}_{3}Li + ^{1}_{1}H \rightarrow ^{7}_{4}Be + ^{1}_{0}n$$

It is left as an exercise to show that $Q = -1.64$ MeV.

Exothermic reactions are of particular interest. To release kinetic energy, mass must decrease, which means that the products must be more tightly bound than the reactants. There are two ways to do this. Since binding energy per nucleon peaks at around $A = 60$, a given number of nucleons would be in a lowest-energy state if grouped into sets of 60. In nature, nucleons are not all grouped this way. Therefore, energy can be released by putting lighter nuclei together or breaking heavier nuclei apart—the processes of fusion and fission.

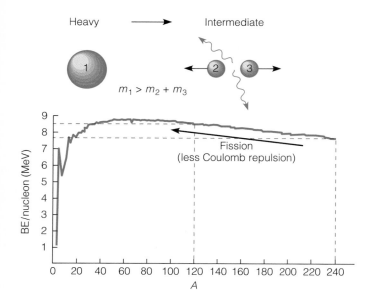

Heavy ⟶ Intermediate

$m_1 > m_2 + m_3$

Fission
(less Coulomb repulsion)

**Figure 10.26** Decreasing BE/
nucleon via fission.

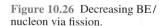

## Fission

Heavy nuclei have less binding energy per nucleon than do those of intermediate mass number (Figure 10.26). Thus, if a heavy nucleus breaks into intermediate-size pieces, the nucleons end up on average in a lower-energy state—a deeper "energy hole." Lower internal energy means smaller mass and an increase in kinetic energy. As noted in connection with spontaneous fission, the driving force is the reduction of Coulomb repulsion.

Isotopes that fission spontaneously are rare. Of greater practical use is **induced fission,** by which we may cause less rare isotopes to release energy at a time of our choosing. Examples of induced U-235 fission are

$$^{235}_{92}U + ^{1}_{0}n \rightarrow ^{141}_{56}Ba + ^{92}_{36}Kr + 3\,^{1}_{0}n$$

$$^{235}_{92}U + ^{1}_{0}n \rightarrow ^{140}_{54}Xe + ^{94}_{38}Sr + 2\,^{1}_{0}n \qquad (10\text{-}12)$$

$$^{235}_{92}U + ^{1}_{0}n \rightarrow ^{132}_{50}Sn + ^{101}_{42}Mo + 3\,^{1}_{0}n$$

As depicted in Figure 10.27, a uranium-235 nucleus absorbs a neutron, becoming briefly an excited state of uranium-236, then fissions in one of many possible ways. (We use uranium-235 as an example, but there are other nuclei that behave similarly, notably plutonium-239.) The kinetic energy released is shared by the "fission fragments" (barium and krypton in the first reaction), the freed neutrons, and $\gamma$-particles. Furthermore, the fission fragments tend to be radioactive and release more energy in the form of $\beta$-particles and neutrinos. Because the energy release is rather complicated, a simple calculation of $Q$ is not possible. But we can estimate its value: According to Figure 10.26, from mass numbers near 235 to those near 120 (a rough average of the product nuclei), binding energy per nucleon increases by about 1 MeV. Thus, in a reaction in which about 200 nucleons of uranium end up near $A = 120$, binding energy should increase by about 200 MeV.

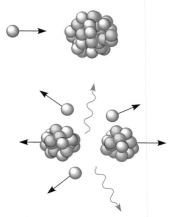

**Figure 10.27** Neutron-induced fission, freeing more neutrons.

All the reactions in (10-12) have neutrons among the products. The tendency for the ratio of $N$ to $Z$ to increase with mass number is the cause. Heavy nuclei have many more neutrons than protons; but intermediate-mass nuclei are most tightly bound when $Z$ and $N$ are more nearly equal. Thus, when a heavy nucleus breaks up, the fragments would have too many neutrons. A lower energy results if the excess neutrons are simply freed.

The presence of *multiple* neutrons among the products of an induced fission reaction is fateful—a **chain reaction** is possible. Suppose that each reaction liberates $m$ product neutrons. If a neutron is required to initiate a reaction, and each reaction produces $m$ neutrons, then each reaction may initiate $m$ other reactions; each of these may in turn initiate $m$ others, and so on, with each successive generation releasing $m$ times the energy of the previous. Using $E_0$ for the energy released in each fission, the energy released in the $j$th generation would be

$$E_j = E_0 m^j \tag{10-13}$$

The energy released would increase exponentially with time!

Since uranium-235 nuclei do exist on Earth, and may certainly absorb neutrons, it is natural to wonder why people on Earth don't stumble into chain reactions. It is one thing for multiple neutrons to be *produced,* but quite another for all of them to be *absorbed* by other uranium-235 nuclei thereafter. Since uranium-235 nuclei are extremely rarefied in nature, the chance of any reaction setting off another is remote. To ensure that a chain reaction will be self-sustaining—each reaction setting off at least one other—the "fissionable material" must be arranged in the proper size and shape, into a so-called **critical assembly.** The reader may have heard the term "critical mass." It is true that there is a minimum mass for a chain reaction to be self-sustaining, but it is the minimum only for a completely pure sample arranged in a sphere. Geometry and impurities are factors. One of the ways of "losing" neutrons is through the surface area. A mass of uranium-235 that would be critical if arranged in a sphere would not be critical if formed into a flat sheet. There would be too much area from which to lose neutrons. A sphere has the minimum surface area to volume ratio. But even for a sphere there is a *minimum* size, because the ratio increases as the size of the sphere decreases. Impurities come into play because they may *absorb* free neutrons (without fissioning)—another loss. To compensate, the assembly would have to be larger.

Whether an assembly is critical or not is measured by the **multiplication constant** $k$, the *average* number of reactions set off by a given reaction. Were three neutrons liberated in each reaction and each to induce another reaction, the number of reactions in each generation would be 3 times the number in the previous generation; $k$ would be its maximum possible value of 3. However, the reactions in (10-12) show that some fissions liberate fewer neutrons. Losses at the surface and absorptions by impurities would further reduce the multiplication constant. Thus, to give the true energy release for a given generation, we replace $m$ in equation (10-13) by $k$.

$$E_j = E_0 k^j$$

A critical assembly requires a multiplication constant of at least unity. If $k > 1$, the energy released increases exponentially with time. If $k = 1$, successive generations would all release the same amount of energy—a steady chain reaction. If

$k < 1$ (i.e., if each reaction induces on average less than one reaction), the energy per generation will *decrease* exponentially with time; a chain reaction would not be self-sustaining, and whatever reaction may have been started would die out.

An **uncontrolled chain reaction** is the basis of the so-called "atomic bomb." [10] Subcritical ($k < 1$) parts are brought quickly together to form a critical assembly ($k > 1$), and the rapid exponential increase in energy results in an explosion, releasing five to six orders of magnitude more energy per kilogram than explosives that rely upon chemical reactions.

### The Fission Reactor

A steady *unchanging* generation of energy requires a multiplication factor of unity. The goal of a fission reactor is, given an assembly in which $k$ may *exceed* unity, to apply controls so as to keep the multiplication factor *at* unity. **Control rods** are made of materials that readily absorb neutrons, remaining otherwise inert. Boron and cadmium are commonly used. If introduced into a region where a chain reaction is occurring, control rods reduce the multiplication factor. Since they can only *decrease k*, and a value of unity must be maintained, a fission reactor must have a *potential* multiplication constant *greater* than unity.

Uranium-235 is the most common fuel in today's fission reactors. But it is very difficult to concentrate. Natural uranium is 99.3% U-238, which absorbs neutrons but rarely fissions. Because they are the same element, the two cannot be separated by chemical methods. The alternatives are tedious, as they rely only on the small mass difference between U-235 and U-238. Methods that have been used include (1) deflection of the ions at different radii in a magnetic field, the "cyclotron method"; and (2) different diffusion rates in a gaseous state through a porous barrier, the "gaseous barrier diffusion method." In any event, reactors do not use pure U-235. Some are able to operate with only the small fraction of U-235 present in natural uranium, while others use uranium in which the component of U-235 has been enriched. How much the uranium must be enriched is related to the effectiveness of the **moderator.** Neutrons produced in fission—"fast" neutrons—tend to have kinetic energy of ~1 MeV. Generally speaking, U-235 is more likely to absorb "slow," or thermal, neutrons, whose kinetic energy is characteristic of the ambient temperature: $\overline{KE} = \frac{3}{2}k_B T \ll 0.1$ MeV. It is the moderator's task to slow the neutrons down, thus increasing the likelihood that they will induce a U-235 fission. Simple mechanics shows that the best way to slow down a particle is a head-on collision with an equal-mass particle. Since protons and neutrons have about the same mass, ordinary water, with single-proton hydrogen nuclei, would seem a good choice of moderator. Unfortunately, hydrogen readily absorbs neutrons—becoming stable deuterium. Thus, the requirements for a moderator are small nuclear size *and* no appetite for absorbing neutrons. They are met adequately by graphite (carbon) and "heavy water," in which the deuterium atoms already contain one neutron and do not readily absorb another.[11] Even so, most reactors in the United States get by with ordinary water; their fuel is sufficiently enriched in U-235 as to more than compensate for the parasitic effect of neutron absorption in hydrogen.

Plutonium-239 behaves very much like U-235, and so is also useful as a fission fuel. On the other hand, it is not to be found in nature. It may be "bred" in a **breeder reactor,** in which the presence of U-238 is actually desirable. When U-

[10]Historically, breaking the *nucleus* apart was referred to as "splitting the atom," when in fact stripping electrons from an atom in a chemical reaction splits the atom. Nevertheless, early fission weapons were termed "atomic bombs," and the name stuck. Later weapons that involved fusion of hydrogen isotopes were termed "hydrogen bombs." Both are "nuclear bombs."

[11]Helium-4 is light and has no appetite for neutrons, but being in all practical conditions a gas, its density is too low to be of use.

238 absorbs a neutron, it becomes for a brief instant uranium-239, then $\beta^-$-decays twice, becoming Pu-239, which then serves as fuel for the reactor.

All nuclear reactors currently producing commercial power are fission (as opposed to fusion) reactors. The energy released is harnessed in the same way as in fossil fuel power plants; it is used to boil water and drive steam turbines. We address the safety concerns connected with fission reactors after discussing fusion, so that the two may be compared.

## Fusion

Light nuclei, like heavy ones, are less tightly bound than those of intermediate mass number (Figure 10.28). Thus, in a process complementary to fission, if light nuclei are thrust together to form somewhat heavier ones, the total mass should decrease and kinetic energy increase. This is fusion. A typical reaction is

$$\underset{2.0141\ u}{{}^{2}_{1}\text{H}} + \underset{1.0078\ u}{{}^{1}_{1}\text{H}} \rightarrow \underset{3.0160\ u}{{}^{3}_{2}\text{He}} + \gamma \qquad (Q = 5.48 \text{ MeV})$$

As we see, helium-3 is less massive than hydrogen and deuterium combined. Coulomb repulsion is not much of a factor in these small nuclei; the lower internal energy is due to the fact that the number of attractive (negative energy) strong bonds among the nucleons goes from one (in the deuterium) to three.

Fusion is the means by which stars generate energy. The most abundant element in stars is hydrogen, and a series of fusion reactions, known as the **proton–proton cycle,** leads it inexorably toward the tightly bound helium-4. Shown in Figure 10.29, it begins with two protons uniting to form a deuteron.

**1:** $\qquad {}^{1}_{1}\text{H} + {}^{1}_{1}\text{H} \rightarrow {}^{2}_{1}\text{H} + {}^{0}_{+1}\beta + \nu \qquad (Q = 0.42 \text{ MeV})$

As we know, there is no bound state of two protons. For the fusion to occur, a proton must first be induced to become a neutron (plus a positron and a neutrino).

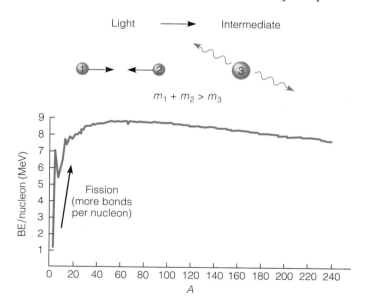

Figure 10.28 Decreasing BE/nucleon via fusion.

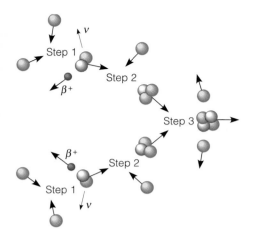

Step 1
Step 2
Step 3
$\beta^+$
$\nu$
$\beta^+$
Step 2
Step 1
$\nu$

**Figure 10.29** The proton–proton fusion cycle.

This involves the "weak force" (see Section 11.3), and as a result the process is relatively slow. In the next fusion, the deuteron combines with another proton to form helium-3.

**2:** $\quad {}^2_1H + {}^1_1H \rightarrow {}^3_2He \quad (Q = 5.48 \text{ MeV})$

In the final step, two helium-3 nuclei fuse to form helium-4.

**3:** $\quad {}^3_2He + {}^3_2He \rightarrow {}^4_2He + {}^1_1H + {}^1_1H \quad (Q = 12.9 \text{ MeV})$

This step must be preceded by two of steps 1 and 2. Altogether, the process results in four protons (six from steps 1 and 2 minus two from step 3) becoming helium-4, plus positrons and neutrinos. Meanwhile, 24.7 MeV of energy is released. (The total is increased further by pair annihilation of the positrons.)

While the proton–proton cycle is the primary energy source in our Sun, in stars where the temperature and helium concentration are high enough, another process occurs. Two ${}^4_2He$ nuclei fuse to form ${}^8_4Be$, which is unstable and would naturally decay back to two helium-4 nuclei (see Example 1.11). But if enough fast-moving helium-4 nuclei are around, another will fuse before the decay can occur, resulting in tightly bound ${}^{12}_6C$. With this nucleus begins the **carbon cycle,** a much more rapid way of "burning" hydrogen.

**1:** ${}^{12}_6C + {}^1_1H \rightarrow {}^{13}_7N$    **2:** ${}^{13}_7N \rightarrow {}^{13}_6C + {}^{\ 0}_{+1}\beta + \nu$    **3:** ${}^{13}_6C + {}^1_1H \rightarrow {}^{14}_7N$

**4:** ${}^{14}_7N + {}^1_1H \rightarrow {}^{15}_8O$    **5:** ${}^{15}_8O \rightarrow {}^{15}_7N + {}^{\ 0}_{+1}\beta + \nu$    **6:** ${}^{15}_7N + {}^1_1H \rightarrow {}^{12}_6C + {}^4_2He$

In this cycle, carbon-12 is simply a catalyst, appearing at the end as the beginning. The net effect is the same as in the proton–proton cycle—four protons becoming helium-4 plus two positrons and two neutrinos. But it proceeds much faster, for at no point does it require a proton to become a neutron *before* fusion can occur. At high enough concentrations and temperatures, elements even heavier than carbon may be formed. However, since beyond about $A = 60$, nuclei become *less* tightly bound, such heavy elements are much less likely to be formed in the random fusion within a star.[12]

Although fission may occur spontaneously, there is no such thing as spontaneous fusion. Nuclei are all positively charged and must be *forced* together. They

[12]Capture of neutrons freed in other reactions does occur, however, and when followed by $\beta^-$ decay, leads to the production of elements of higher Z.

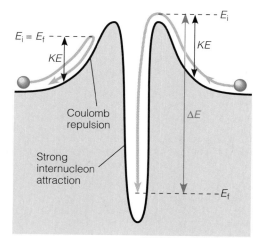

**Figure 10.30** Nuclear fusion: over the Coulomb hurdle, then into the strong-force well.

may well be capable of fusing via the short-range internucleon attraction, but only if their long-range Coulomb repulsion can be surmounted. An analogy: Imagine trying to roll a boulder up the side of a volcano so that it falls into a deep hole at the top. If the boulder (one nucleus) is not headed at sufficient speed directly toward the center of the volcano (another nucleus), it will be deflected to the side or simply roll back down the volcano (the left-hand "boulder" in Figure 10.30). Only if rolled fast enough in the proper direction (the right-hand "boulder") can it settle into the low-energy hole—whose bottom is *very* far below "ground level." Only then can the large available energy $\Delta E$ be released.

By these arguments, high density and temperature should favor fusion. A greater density of nuclei implies more frequent collisions, and a higher temperature means that a greater fraction will have enough energy to surmount the Coulomb barrier. The conditions are easily met in a star, but not so easily on Earth. The first success was the so-called "hydrogen bomb"; the high density and temperature were produced by setting off a fission device (atomic bomb). If properly arranged near the fissioning materials, hydrogen isotopes will fuse in abundance. Pound for pound, fusion explosions are more energetic than fission explosions for the simple reason that the binding energy per nucleon increases more steeply from light toward intermediate than from heavy toward intermediate (cf. Fig. 10.12).

## The Fusion Reactor

The greater their charge, the greater becomes the difficulty in forcing two nuclei close enough to fuse. Therefore, to achieve significant fusion under conditions not so impractical as detonation of a fission device, the following reactions of $Z = 1$ nuclei are most promising (D stands for deuterium and T for tritium):

$$_1^2\text{D} + {}_1^2\text{D} \rightarrow \begin{cases} {}_1^3\text{T} + {}_1^1\text{H} & (Q = 4.0 \text{ MeV}) \\ {}_2^3\text{He} + {}_0^1\text{n} & (Q = 3.3 \text{ MeV}) \end{cases}$$

$$_1^2\text{D} + {}_1^3\text{T} \rightarrow {}_2^4\text{He} + {}_0^1\text{n} \qquad (Q = 17.6 \text{ MeV})$$

The two D–D reactions occur with about equal probability.

Even these best hopes, however, require conditions of extreme density and temperature. To gain some idea, let us set the average thermal kinetic energy equal to the electrostatic potential energy two $+e$ charges would have when separated by the rough nucleon radius of 1 fm:

$$\frac{1}{4\pi\epsilon_0}\frac{q_1 q_2}{r} = \frac{3}{2}k_B T$$

$$\left(9 \times 10^9 \frac{\text{N·m}^2}{\text{C}^2}\right)\frac{(1.6 \times 10^{-19}\text{ C})^2}{10^{-15}\text{ m}} = \frac{3}{2}(1.38 \times 10^{-23}\text{ J/K})T \quad \Rightarrow \quad T \cong 10^{10}\text{ K}$$

This overestimates the temperature required. As we know, at any temperature there is a range of speeds. Moreover, high density also enhances the probability of fusion. The density is extremely high in our Sun's interior, allowing it to continue burning its fuel at "only" about $10^7$ K. Still, even $10^7$ K would vaporize any part of a solid structure it touches, so to avoid contamination, the fuel in a fusion reactor has to be confined without solid structures! Two means are being studied most vigorously: magnetic confinement and inertial confinement.

*Magnetic Confinement* In a magnetic field, a charged particle is compelled to describe a spiral path along and about the field lines. This may be exploited to confine nuclei. The most common and thus far most promising means of magnetic confinement employs a toroidal magnetic field (Figure 10.31). A gas of deuterium and tritium is heated sufficiently to produce an ionized **plasma** of positive nuclei and negative electrons, and these ions spiral about and along the circular field lines inside the toroid. Of course, this merely confines the plasma away from the walls of the toroid; it does not ensure pressure or temperature high enough to produce significant fusion. Several heating methods have been tried. One squeezes the plasma with additional magnetic fields; another injects high-energy neutral particles, which are unaffected by the fields and so can be directed easily to a location most effective in heating the fuel. Nevertheless, *confinement* is still a problem. Not only do random collisions between ions in the plasma allow fuel to escape the magnetic field, all methods of heating introduce other forces that tend to destabilize or scatter the fuel. High density and temperature may be achieved at one point, but the fuel escapes at another. Attempting to hold a plasma in a magnetic field while squeezing it is much like squeezing water between one's hands; the harder one squeezes, the quicker it squirts out the cracks.

*Inertial Confinement* Inertial confinement involves simply delivering simultaneous "blows" to a quantity of fuel from many sides so quickly that it has no opportunity to slip away (Figure 10.32). In one application, the blow is struck by heavy ions accelerated to high speed; another uses laser pulses. At the Omega Laser Facility at the University of Rochester, a target containing deuterium and tritium is struck simultaneously from all sides by 60 beams of 351-nm-wavelength laser light. In an instant, the target is vaporized and the fuel transformed into a high-pressure, high-temperature plasma in which significant fusion should occur.

The goal of all confinement methods is to produce more energy *from* fusion than is needed to establish the conditions necessary to *initiate* fusion. Producing magnetic fields, accelerating particles, running lasers, and so on; all these things

**Figure 10.31** Confinement in a toroidal magnetic field.

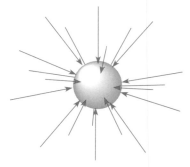

**Figure 10.32** Inertial confinement —struck from all sides.

take energy. If no significant fusion occurs, this energy is simply lost, radiated away in one form or another. The need to increase output relative to input is why the D–T is the preferred fuel in today's fusion experiments; it yields more energy per reaction than D–D. Even so, current experiments have not achieved the efficiency needed for commercial use—a truly practical break-even. When efficiencies are improved, however, the D–D will be the more attractive reaction. Tritium must be produced artificially, but we have oceans of deuterium—15 of every 100,000 hydrogen atoms on Earth. As a fuel source, it is practically inexhaustible.

## Fission Versus Fusion

As practical sources of energy, fission and fusion share certain characteristics, good and bad. Since both exploit the energy associated with the strong force, rather than the relatively weak electromagnetic forces of "chemical reactors," they do not use much fuel; neither requires (or would require) the huge volumes needed by conventional fossil fuel power plants. Both, however, cause the structures that confine the fuel to become radioactive. The useful lives of these structures are limited, and disposal of such "hot" materials is a thorny problem. Also—though this is not a problem unique to *nuclear* reactors—the second law of thermodynamics guarantees that for a given amount of useful energy generated there will be a given amount of waste heat dumped into the environment. The greater the energy generated, the greater will be the thermal pollution problem.

In other important aspects, fission and fusion differ significantly:

*Fuel* While believed to be more abundant than silver in Earth's crust, uranium, the fuel of most of today's fission reactors, is not as plentiful as future energy demands would require. Furthermore, before it can be mined, a potentially destructive operation in itself, rich deposits must be located. Plutonium can be "bred," but it is highly toxic chemically and must therefore be handled with extreme care. Deuterium, the future fusion fuel, is chemically harmless hydrogen. It is fairly easily distilled from natural water and is quite abundant.

*Stability* In a fission reactor, *all* of the fuel must be present, since it all contributes to sustaining the chain reaction. If the control mechanisms that prevent exponential growth break down, a serious runaway may result. In a fusion reactor, there is no chain reaction and thus no after-the-fact control apparatus required; the fuel is fed in only as needed and no runaway consumption of all the fuel is possible. An analogy: In a fusion reactor, the fuel is fed from the fuel tank to the heater as needed; in a fission reactor, a match is set to the fuel tank and controls are applied to keep the heat output at a steady level.

*Combustion Products* Fission products tend to be fairly heavy elements that are themselves radioactive, often with very long half-lives. They may endanger the environment for ages. Disposal of this waste is therefore a serious issue. Fusion products, mostly helium isotopes and tritium, are much less hazardous. Helium is harmless. Tritium *is* radioactive, but its half-life is relatively short and it is not chemically poisonous. It is far less "dirty" than typical fission products.

All things considered, the advantages offered by fusion certainly merit further pursuit to make it a commercially viable energy source.

# Progress and Applications

■ Different views of the nucleus, represented by the liquid drop and shell models, continue to coexist, due to the inherent complexity of the strong force and the difficulty of piecing together evidence from collisions in different energy ranges. Recent experimental results, however, have challenged some previous assumptions based on a liquid drop model. The liquid drop model successfully explains why some nuclei fission if rotating fast enough and deformed sufficiently: The "surface tension," a result of strong attraction between all nucleons, is broken (Figure 10.33). Although the simplest shell model calculations assume independent particles moving in a spherical well, there are techniques that have proved able to handle highly deformed nuclei. These had suggested the possible stability of certain unusually oblong "hyperdeformed" nuclei with huge rotational angular momenta (approaching $100\hbar$). According to the liquid drop model, such nuclei would have been expected to fission immediately, but the evidence now is that they need not (D. R. LaFosse, *Physical Review Letters*, June 26, 1995, p. 5186–5189). Even so, this is just one piece of the puzzle, and there are still numerous behaviors that can be explained only by a model including strong correlations in the motion of all nucleons.

■ As if the unsolved physics were not enough, the confusing world of science funding makes it difficult to predict the progress of fusion research. Nevertheless, several promising projects are now either underway or seem likely to be supported. In the realm of magnetic confinement is a collaborative effort involving several laboratories and universities in the United States: the NSTX, National Spherical Torus Experiment, to be located at the Princeton Plasma Physics Laboratory. Named after the shape of the magnetic field that confines the plasma (Figure 10.34), this reactor has its field lines not only along "lati-

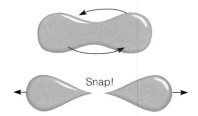

**Figure 10.33** A nuclear "drop" spun to pieces.

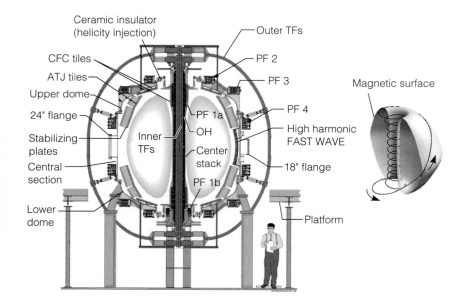

**Figure 10.34** The National Spherical Torus Experiment and its interesting field pattern.

tudes," as in a conventional toroidal field pattern, but also along longitudes. Many studies have indicated that this shape should have several advantages in securing high pressures with fewer instabilities. A much larger undertaking is the ITER, International Thermonuclear Experimental Reactor. As the name suggests, this project has the pledged support of many countries worldwide. It is expressly dedicated to producing, in essence, a first-generation fusion power plant. Of conventional "tokamak" toroidal design, this immense machine, depicted in Figure 10.35, is intended to sustain "burns" of perhaps an hour at a temperature of around $10^8$ K and to produce net power measured in gigawatts. Some physicists have raised concerns that the project may be too ambitious, given the present state of our knowledge and resources.

■ While suffering from the same uncertainties, inertial confinement experiments have also managed to march onward. One of the largest is the proposed NIF, National Ignition Facility (Figure 10.36), to be constructed at Lawrence Livermore National Laboratory. In a technique known as "indirect-drive," a promising departure from the earlier "direct-drive," a heavy-metal (gold or lead) vessel containing the deuterium–tritium fuel capsule would be struck simultaneously by 192 laser beams. The metal vessel serves as an intermediary, converting the laser energy first to x rays, which then compress the fuel. Although its primary goal is to demonstrate net energy gain in inertial confine-

**Figure 10.35** The ITER—the first fusion power plant?

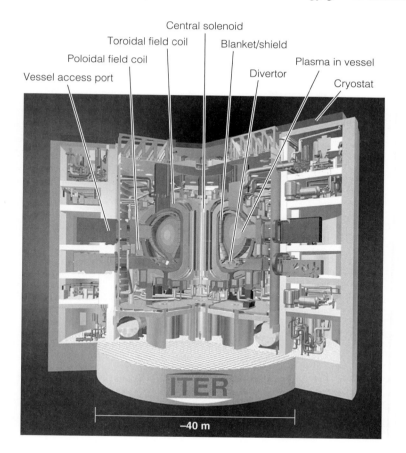

Central solenoid
Toroidal field coil
Blanket/shield
Poloidal field coil
Plasma in vessel
Vessel access port
Divertor
Cryostat

ITER

−40 m

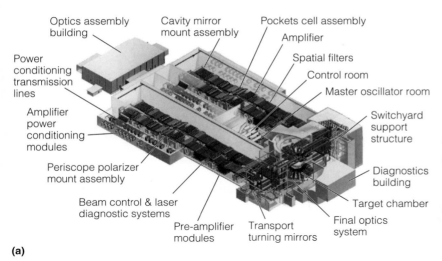

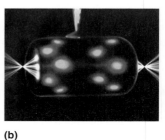

**(a)**

**(b)**

**Figure 10.36** a) The proposed National Ignition Facility. b) Laser beams incident at the ends of a metal vessel that contains the D–T fuel capsule.

ment, the NIF is intended to have several applications in the field of plasmas and fusion. Similar projects have been proposed in France, Japan, and other countries.

## Basic Equations

**Nuclear radius**

$$r = A^{1/3} \times R_0 \quad (R_0 = 1.2 \times 10^{-15} \text{ m})$$   (10-1)

**Binding energy**

$$BE = (Zm_H + Nm_n - M_{Z}^{A}X)c^2$$   (10-4)

**Semi-empirical binding energy formula**

$$BE = c_1A - c_2A^{2/3} - c_3\frac{Z(Z-1)}{A^{1/3}} - c_4\frac{(N-Z)^2}{A}$$   (10-5)

$$c_1 = 15.8 \quad c_2 = 17.8 \quad c_3 = 0.71 \quad c_4 = 23.7$$

**Energy released**

$$Q = (m_i - m_f)c^2$$   (10-6)

**Radioactive decay law**

$$N = N_0e^{-\lambda t}$$   (10-9)      $$R = \lambda N$$   (10-10)

**Half-life**

$$\lambda = \frac{\ln 2}{T_{1/2}}$$

(10-11)

## Summary

The nucleus is composed of protons and neutrons, known collectively as nucleons. The mass number $A$ is the number of nucleons and the atomic number $Z$, which identifies the element, is the number of protons. Nuclei of the same atomic number but different numbers of neutrons are called isotopes of the given element. Nuclear radii are measured in femtometers (1 fm $= 10^{-15}$ m), about five orders of magnitude smaller than radii typical of electron orbits in atoms. All nuclei have about the same density, suggesting a closest-packed arrangement of incompressible nucleons.

The predominant forces in the nucleus are the strong force, a strong but short-range attraction between all nucleons, and the electrostatic force, a weaker but more far-reaching repulsion between all protons. Along with the exclusion principle, they cause nuclei of intermediate mass number to be most stable; for nuclei around $A = 60$, the greatest energy per nucleon would have to be expended to separate the nucleons. Furthermore, light nuclei are most stable when the numbers of protons and neutrons are about equal, while in heavy nuclei, the most tightly bound arrangement has more neutrons than protons, reducing the destabilizing effect of Coulomb repulsion.

The energies binding nucleons to their nuclei are about five orders of magnitude larger than those binding electrons to their atoms. Thus, much more so than for the atom, the mass/energy of a nucleus is significantly less than that of its separated parts.

Because of the complex forces in the nucleus, and our incomplete knowledge of the strong force, the nuclear "problem" cannot be solved exactly. Models must be used. The liquid drop model and shell model have been particularly successful.

Unstable nuclei attain a lower-energy state via radioactive decay. Forms include $\alpha$ decay, in which the nucleus expels a bound two-proton–two-neutron particle; $\beta$ decay, in which a neutron becomes a proton, or vice versa, and an electron or positron is created; electron capture, in which an orbiting electron is captured and combined with a proton, producing a neutron; and $\gamma$ decay, in which a photon carries away excess nuclear energy.

The rate of radioactive decay is probabilistic. Consequently, the number of radioactive nuclei present in a sample drops exponentially with time. The most common measure of the decay rate is half-life, the time required for the number of nuclei present to decay to half its initial value.

A nuclear reaction occurs when nucleons are changed or exchanged among nuclei. Reactions may be either endothermic or exothermic. Particularly important exothermic reactions are fission and fusion. These exploit the very tight binding of intermediate-mass nuclei. In fission, heavy nuclei are split, resulting in a more tightly bound, lower-mass final state. In fusion, light nuclei are brought together, also resulting in a lower-mass final state. In either case, the "lost" mass energy becomes the kinetic energy of the products.

## E X E R C I S E S

### Section 10.1

1. What fraction of space is actually occupied by iron nuclei in a "solid" piece of iron? (The density of iron is $7.87 \times 10^3 \text{ kg/m}^3$.)
2. At what speed would $\alpha$-particles have to be directed at gold foil if some are to contact gold nuclei?
3. From the natural abundances and atomic masses given in Appendix J of the two naturally occurring isotopes of boron, determine the average atomic mass of natural boron, and compare this with the value given in the periodic table of Chapter 7.
4. Determine the approximate ratio of the diameter of a uranium nucleus ($A = 238$) to that of a beryllium nucleus ($A = 9$).

### Section 10.2

5. Calculate the binding energy per nucleon of carbon-12.
6. Calculate the binding energy per nucleon of technetium-98.

### Section 10.3

7. According to the semi-empirical binding energy formula, what should be the binding energy per nucleon of technetium-98?
8. Nuclei of the same mass number but different $Z$ are known as **isobars.** Oxygen-15 and nitrogen-15 are isobars. (a) In which of the factors considered in nuclear binding (represented by terms in the semi-empirical binding energy formula) do these two isobars differ? (b) Which of the isobars should be more tightly bound? (c) Is the conclusion drawn in parts (a) and (b) supported by the decay mode information of Appendix J? Explain. (d) Calculate the binding energies of oxygen-15 and nitrogen-15. By how much do they differ? (e) Repeat part (d) but use the semi-empirical binding energy formula rather than the known atomic masses.
9. (a) Calculate the binding energies per nucleon of the isobars (defined in Exercise 8) boron-12, carbon-12, and nitrogen-12. (b) In which of the terms of the semi-empirical binding energy formula do these binding en-

ergies differ, and how should these differences affect the binding energies per nucleon? (c) Determine the binding energies per nucleon using the semi-empirical binding energy formula and discuss the results.
10. Obtain a semi-empirical binding energy *per nucleon* formula. Using this as a guide, explain why the Coulomb force, only about $\frac{1}{100}$ as strong as the internucleon attraction for two protons "in contact" (cf. Table 10.2), would eventually *have to become* a dominant factor in large nuclei. Assume that $Z$, $N$, and $A$ increase in rough proportion to one another.

### Section 10.5

11. What is the daughter nucleus and how much kinetic energy is released in the $\alpha$ decay of polonium-210?
12. What is the daughter nucleus and how much kinetic energy is released in the $\beta^+$ decay of nitrogen-13?
13. Oxygen-19 $\beta^-$-decays. What is the daughter nucleus, and what may be said of the kinetic energy of the emitted $\beta^-$-particle?
14. Polonium-207 may undergo three kinds of radioactive decay: $\beta^+$, $\alpha$, and electron capture. But it does not $\beta^-$-decay. Given its position relative to the curve of stability, explain why this is not unexpected.
15. Potassium-40 ($Z = 19$, $N = 21$) is a radioactive isotope rare but not unknown in nature. It is particularly interesting in that it lies along the curve of stability yet decays by *both* $\beta^+$ and $\beta^-$—that is, in both directions *away* from the curve of stability. (a) Identify the daughter nuclei for both decays.

    Many factors governing nuclear stability are discussed in the chapter (e.g., those in the semi-empirical binding energy formula, magic numbers, and even numbers); identify those that would argue (b) *only* for $\beta^-$ decay, (c) *only* for $\beta^+$ decay, and (d) for either $\beta^-$ or $\beta^+$ decay.
16. Find $Q$ for the decay of beryllium-10.
17. Calcium-41 decays by electron capture. (a) Find the $Q$-value for the decay. (b) Show that calcium-41 cannot decay by $\beta^+$ emission.
18. Using the semi-empirical binding energy formula to determine whether energy would likely be released, would

you expect the hypothetical nucleus $^{288}_{119}X$ to (a) $\alpha$-decay, (b) $\beta^+$-decay, or (c) $\beta^-$-decay?

19. What is the recoil speed of the daughter nucleus when $^{152}_{67}Ho$ $\alpha$-decays? (Treat all motion as nonrelativistic.)

**Section 10.6**

20. Given an initial amount of 100 g of plutonium-239, how much time must pass for the amount to drop to 1 g?

21. The initial decay rate of a sample of a certain radioactive isotope is $2.00 \times 10^{11}$ s$^{-1}$. After half an hour, the decay rate is $6.42 \times 10^{10}$ s$^{-1}$. Determine the half-life of the isotope.

22. The half-life $T_{1/2}$ is not the average lifetime $\tau$ of a radioactive nucleus. The average lifetime is the time multiplied by the probability per unit time $P(t)$ that the nucleus will "live" that long, then summed (integrated) over all time. (a) Show that $P(t)$ should be given by $\lambda e^{-\lambda t}$. (*Hint:* What must be the *total* probability?) (b) Show that $\tau = T_{1/2}/\ln 2$.

23. Eighty centuries after its death, what will be the decay rate of 1 g of carbon from the thigh bone of an animal?

24. A fossil specimen has a carbon-14 decay rate of 3.0 s$^{-1}$. (a) How many carbon-14 nuclei are present? (b) If this number is one-tenth the number that must have been present when the animal died, how old is the fossil?

25. A fossil specimen has a $^{14}C$ decay rate of 5.0 s$^{-1}$. (a) How many carbon-14 nuclei are present? (b) If the specimen is 20,000 years old, how many carbon-14 nuclei were present when the animal died? (c) How much kinetic energy (in MeV) is released in each $\beta$ decay, and what is the total amount released in all $\beta$ decays since the animal died?

26. A bone of an animal contains $\frac{1}{10}$ mole of carbon when it dies. (a) How many carbon-14 atoms would be left after 200,000 yr? (b) Is carbon dating useful to predict the age of such an old bone? Explain.

27. Potassium-40 has a half-life of $1.26 \times 10^9$ yr, decaying to calcium-40 and argon-40 in ratio 8.54 to 1. A rock sample, which contained no argon when it formed a solid, now contains one argon-40 atom for every potassium-40 atom. How old is the rock?

28. (a) Determine the total amount of energy released in the complete decay of 1 mg of tritium. (b) According to the law of radioactive decay, how much time would this release of energy span? (c) In a practical sense, how much time will it span?

29. Given initially 40 mg of radium-226 (one of the decay products of uranium-238), determine (a) the amount that will be left after 500 yr, (b) the number of $\alpha$-particles the radium will have emitted during this time, and

(c) the amount of kinetic energy that will have been released. (d) Find the decay rate of the radium at the end of the 500 yr.

30. Ten milligrams of pure polonium-210 is placed in 500 g of water. If no heat is allowed to escape to the surroundings, how much will the temperature rise in 1 hr?

**Section 10.7**

31. Determine the value of $Q$ for the reaction:

$$^7_3Li + ^1_1H \rightarrow ^7_4Be + ^1_0n$$

32. Calculate the net amount of energy released in the deuterium–tritium reaction:

$$^2_1D + ^3_1T \rightarrow ^4_2He + ^1_0n$$

33. In an assembly of fissionable material, the larger the surface area per fissioning nucleus (i.e., per unit volume), the more likely is the escape of valuable neutrons. (a) What is the surface-to-volume ratio of a sphere of radius $r_0$? (b) What is the surface-to-volume ratio of a cube of the same volume? (c) What is the surface-to-volume ratio of a sphere of twice the volume?

34. If all the nuclei in a pure sample of uranium-235 were to fission, yielding about 200 MeV each, what is the kinetic energy yield in joules per kilogram of fuel?

35. (a) To release 100 MW of power, approximately how many uranium fissions must occur every second? (b) How many kilograms of U-235 would have to fission in 1 yr?

36. In both of the D–D fusion reactions noted in the discussion of the fusion reactor, two deuterons fuse to produce *two* particles, a nucleus of $A = 3$ and a free nucleon. Mass decreases because the binding energy of the $A = 3$ nucleus is greater than the combined binding energies of the two deuterons. The binding energy of helium-4 is even greater still. Why can't the deuterons simply fuse into a helium-4 nucleus and nothing else; why must *multiple* particles be produced?

37. Assume that the fusion of D–D fuel liberates approximately 2 MeV of kinetic energy per deuteron. (a) What is the yield, in joules of kinetic energy liberated per kilogram of fuel? (b) How does this compare to a typical yield of $10^6$ J/kg for chemical fuels?

38. (a) How much energy can be extracted from a cup (240 mL) of sea water? Assume that the hydrogen in sea water is 0.015% deuterium and that an average D–D fusion yield is about 2 MeV per atom. (The density of sea water is $\sim 10^3$ kg/m$^3$. Ignore impurities.) (b) A modern supertanker can hold $9 \times 10^7$ gal. How many "water tankers" would be needed to supply the energy needs of greater Los Angeles, consuming electricity at a rate of about 20 GW, for 1 yr? Assume that only 20% of the available energy actually becomes electrical energy.

## General Exercises

39. As noted in Section 10.5, carbon-11 decays to boron-11. What factor in nuclear stability argues that such a decay is favorable?

40. The binding energy per nucleon in helium-3 is 2.57 MeV/nucleon. Assuming a nucleon separation of 2.5 fm, determine (a) the gravitational potential energy *per nucleon,* and (b) the electrostatic energy *per proton* between the protons. (c) What is the approximate value of the internucleon potential energy per nucleon? (d) Do these results agree qualitatively with Table 10.2?

41. For very light nuclei, stability cannot be understood solely on the basis of binding energy per nucleon; there is no nucleon binding at all for a single proton or neutron, yet one is stable and the other is not. (a) Helium-3 and hydrogen-3, tritium, differ only in the switch of a nucleon—between neutron and proton. Which has the higher binding energy per nucleon? (b) Helium-3 is stable, while tritium is not (in fact decaying into He-3). Why?

42. You occupy a one-dimensional world in which beads, of mass $m_0$ when isolated, attract each other when, but only when, in contact. Were the beads to interact solely by this attraction, it would take energy $H$ to break the contact. Consequently, we could extract this much energy by sticking two together. However, they also share a repulsive force no matter what their separation, for which the potential energy is $U(r) = 0.85Ha/r$, where $a$ is a bead's radius and $r$ is center-to-center separation. The closer the beads, the higher is this energy. (a) For one stationary bead, by how much does the energy differ from $m_0c^2$? (b) For two stationary beads in contact, by how much does the energy differ from $2m_0c^2$? (c) For three beads in contact (in a line, of course, since this world is one-dimensional), by how much does the energy differ from $3m_0c^2$? (d) For four beads in contact, by how much does the energy differ from $4m_0c^2$? (e) If you had 12 isolated beads and wished to extract the most energy by sticking them together (in linear groupings), into sets of what number would you group them? (f) Sets of what number would be suitable fuel for the release of fusion energy? Of fission energy?

# Fundamental Particles
# and Interactions

I t is fitting that our study culminate in a discussion of fundamental particles. In physics, we seek to understand why our seemingly complex universe behaves as it does. Invariably, we must concentrate on some small part of it, and in so doing we are apt to discover what appear to be building blocks of nature. But as we look closer, what might have seemed to be fundamental particles turn out to be collections of even more basic building blocks. Ancient philosophers postulated that the world comprised simply air, water, earth, and fire. Scientific study later revealed a more complex structure; it came to be accepted that the world was composed of atoms, of which there seemed to be dozens of distinct types—that is, the "elements." A brief return to a simpler view occurred when it was discovered that all the elements appeared to be built of the same three "fundamental" particles: protons, neutrons, and electrons. However, as ever greater energy became available to scientists, a vast host of "new" subatomic particles appeared. Even more recently, research has again reduced the number of particles suspected of being fundamental; protons, neutrons, and many of the "new" particles have internal structure and appear to be merely different combinations of a limited number of more-basic constituents. In this chapter, we discuss the prevailing theory of the building blocks of nature—**fundamental particles.** Inextricably related to this, however, is the question of how fundamental particles interact with each other—**fundamental interactions**—and it is here that we begin.

# 11.1 How Forces Act

In the most basic view, the physical universe consists of elementary particles that interact in only a few distinct ways. But this simple view is the product of an evolution of understanding much like the path taken by each student of physics.

The beginning physics student is apt to view forces as requiring physical contact (e.g., tension, normal forces, friction), with the curious exception of gravity, which appears to act through a vacuum. Later, in a deeper study of gravitation and electromagnetism, the student learns that force without physical contact is more the rule than the exception. Forces of contact, for instance, are really coarse manifestations of microscopic electromagnetic forces. However, rather than particles simply exerting forces on one another at a distance, the student is taught that a force is really conveyed by a pervasive field, gravitational or electromagnetic as the case may be. In elementary particle physics, we take the view of how a force is conveyed one step further.

In the modern view, a fundamental force[1] between particles is conveyed by the exchange of a **mediating particle** whose role is special to that force. The mediating particle is also known as a **field quantum,** meaning a particle of field, which gives the theory its name, **quantum field theory.** There are two classes of particles, then—those that experience the force and those that are exchanged to convey it.[2] An analogy: Two physics students slide along a frictionless frozen pond, each helpless to change his velocity without engaging in a mutual force with some other object. As shown in Figure 11.1, one of the students has a snowball, which he throws to the other. The exchange of the snowball has the effect of a mutual repulsive force between the students. The first student's momentum is changed by $-\mathbf{p}_{snowball}$ as the snowball is thrown and the second student's is changed by $+\mathbf{p}_{snowball}$ as the snowball is caught. We might term this the "snowball-mediated force": a force between physics students (the particles experiencing the force), mediated by a snowball (the particle conveying it). The net effect is represented in the lower diagram in Figure 11.1. The exchange of the mediating particle is the cause of the momentum changes experienced by the particles engaging in the force. We shall have more to say about diagrams of this sort in Section 11.5.

While clarifying how the exchange of a particle might convey a force, as an analogy for the workings of a fundamental force, the snowball example has several shortcomings. One is that it would seem to be able to convey only a *repulsive* force. Both thrower and catcher experience momentum changes *away* from the mediating particle. To convey an attraction, the throwing of the snowball would have to force the thrower *toward* the snowball, and the catching of the snowball would have to cause the catcher to move in the direction from which the snowball came! The total momentum after the catch might well equal that before the throw, but momentum conservation would surely be violated *during* the mediating particle's exchange. Another deficiency is that for a real fundamental force *the mediating particle exists only during its exchange.* The "snowball" is created solely for the occasion, disappearing just as mysteriously afterward. This raises concerns about energy conservation; although energy might be conserved overall,

[1]Following common habit, we use the terms "force" and "interaction" interchangeably, although the latter is a better choice.

[2]As we shall see, this is too restrictive. Often, the particle exchanged is also capable of experiencing the force it conveys.

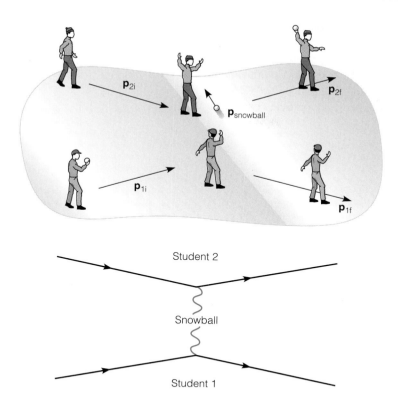

Figure 11.1 A force between students, conveyed by exchange of a snowball.

what is the source of this temporary fluctuation? In one of the great wonders of physics, the answers to these questions are found in the quantum-mechanical limitations of the conservation laws.

## Range and the Mass of the Mediating Particle

Quantum-mechanically, it is impossible to know that energy is conserved during an arbitrarily small interval of time; the uncertainty principle forbids such knowledge. Considered for an interval of time $\Delta t$, there is an inherent uncertainty $\Delta E$ in the energy of a system, governed by

$$\Delta t \; \Delta E \; \approx \; \hbar$$

Therefore, in an interval $\Delta t$, energy conservation might be violated by as much as $\Delta E \approx \hbar/\Delta t$ and it would be theoretically impossible to prove otherwise. This very fact makes it possible for a force to be conveyed by a particle of energy $\Delta E$ appearing out of nowhere—so long as it exists for no more than a time interval $\Delta t \approx \hbar/\Delta E$.[3] Existing only within this constraint, the particle exchanged is often referred to as a **virtual** particle. (The term evokes an image of the particle's lot in life: fated to skulking along with its unlawful energy cache in the nether world allowed by the uncertainty principle.)

[3]The possible violation of *momentum* conservation is essentially the same problem. We might say that the creation and exchange of the mediating particle implies a fluctuation in momentum-energy, but one that is unverifiable if occurring over a short enough space-time interval.

By these arguments, a force whose mediating particle has mass must be a "short-range" force. If a massive particle must be created, there is an absolute lower limit on the amount by which the total energy must fluctuate: the mass/internal energy of that particle. This in turn implies that the maximum time it could survive without verifiably upsetting energy conservation is $\Delta t \approx \hbar/mc^2$. Since the speed of light is an absolute maximum speed, the particle could then travel at most

$$\Delta x \approx c \; \Delta t \approx \frac{\hbar}{mc}$$

or

$$\text{range} \approx \frac{\hbar}{c} \frac{1}{m} \qquad\qquad (11\text{-}1)$$

If the mediating particle has mass, the range of the force is limited!

Conversely, we see that an infinite-range force, such as the electromagnetic force, would have to be conveyed by massless particles. With no mass, there would be no lower limit on the energy of the mediating particle; its energy would be all kinetic, which can be arbitrarily small. The time and range could then be arbitrarily large. Indeed, it is no coincidence that electromagnetic forces are conveyed by the exchange of massless photons. An electrostatic repulsion between electrons, for instance, is conveyed when one electron emits a virtual photon that is later absorbed by the second electron.

We shall return to equation (11-1) in Section 11.3 when we discuss specific fundamental forces. Here, we merely note that it has been used in several instances to predict fairly accurately the mediating particle's mass from knowledge of the force's range *before* the mediating particle was first detected! This is one of the most convincing arguments for treating forces as the exchange of particles.

## 11.2 Antiparticles

Before delving further into fundamental interactions between particles, we must understand antiparticles. For each kind of particle, there is an antiparticle, which shares essentially all the properties of the particle, except being of opposite charge. Section 2.4 introduced us to the positron—the antiparticle of the electron. It has the same mass and spin as an electron, but is of positive charge. Many other antiparticles have been found: antiprotons (negatively charged), antineutrons (uncharged), and so on.[4] A particle and its antiparticle can be created in the process known as pair production, and when they meet, pair annihilation may follow, in which they disappear and their mass energy is converted to photon energy. The symbol for an antiparticle is the same as the particle's but with an overbar. The antiproton is thus $\bar{p}$, the antineutron $\bar{n}$. The positron, however, is usually represented $e^+$ rather than $\overline{e^-}$. (There are many exceptions to the convention, e.g. the $\pi^+$ and $\pi^-$.)

The existence of antiparticles is an experimental fact, but there is a theoretical basis: relativistic quantum mechanics. Let us pursue just enough of this advanced

[4]It might seem that an uncharged particle such as a neutron has no property to distinguish it from its antiparticle. However, the distinct identity of the antineutron is confirmed by the evidence that it does annihilate with the neutron, whereas two neutrons do not annihilate. Furthermore, as we soon see, the neutron has internal structure that distinguishes particle from antiparticle. Uncharged particles for which this is not the case are indeed their own antiparticles (e.g., the photon, the $\pi^0$).

theory to gain some idea of how combining relativity with quantum mechanics might suggest the existence of antiparticles.

For a free particle in one dimension, the Schrödinger equation is

$$-\frac{\hbar^2}{2m}\frac{\partial^2}{\partial x^2}\Psi(x,\ t) = i\hbar\frac{\partial}{\partial t}\Psi(x,\ t) \tag{11-2}$$

or, expressed in terms of operators $[\hat{p} = -i\hbar(\partial/\partial x),$ and $\hat{E} = i\hbar(\partial/\partial t)]$,

$$\frac{1}{2m}\hat{p}^2\Psi(x,\ t) = \hat{E}\Psi(x,\ t)$$

As we know, this equation is based on energy; in the absence of external potential energies, the kinetic energy of a particle, $p^2/2m$, equals its total energy $E$. However, since $p^2/2m$ is not the relativistically correct expression for kinetic energy, $p^2/2m = E$ cannot serve as the basis of a relativistic replacement for the Schrödinger equation. A logical basis is equation (1-26), the relativistically correct relationship between energy, momentum, and mass:

$$p^2c^2 + m^2c^4 = E^2 \tag{11-3}$$

To obtain a relativistic matter-wave equation, we might try inserting the appropriate operators and then have each term operate on a wave function:

$$c^2\hat{p}^2\Psi(x,\ t) + m^2c^4\Psi(x,\ t) = \hat{E}^2\Psi(x,\ t)$$

or

$$-c^2\hbar^2\frac{\partial^2}{\partial x^2}\Psi(x,\ t) + m^2c^4\Psi(x,\ t) = -\hbar^2\frac{\partial^2}{\partial t^2}\Psi(x,\ t) \tag{11-4}$$

Klein–Gordon equation

This relativistically correct analog to the Schrödinger equation (11-2) is known as the **Klein–Gordon equation.** It has been shown to yield correct predictions about the behavior of spinless massive particles at all speeds.[5]

The main difficulty with basing a relativistic matter-wave equation on (11-3) is that it involves $E$ *squared*. If $E = p^2/2m$, then $E$ is positive. Solutions to the free-particle Schrödinger equation are thus always of positive energy. However, equation (11-3) is equally well satisfied whether $E$ is positive or negative. There are perfectly valid mathematical solutions of (11-4) for which the total energy is negative (see Exercise 5). The question is how to interpret these mathematical solutions *physically*.

There is another difficulty with (11-4). It can be shown that the integral $\int \Psi^* \Psi\, dV$ does not change with time if $\Psi$ is a solution of the Schrödinger equation, but it may change with time if $\Psi$ is instead a solution of the Klein–Gordon equation (see Exercises 6 and 8). Were we to interpret the integral in the usual way—as the total probability of finding the particle—the particle might either appear or disappear as time passes! There *is* a density that doesn't change in time, formed from solutions of the Klein–Gordon equation,[6] but it cannot be interpreted as a probability density, for it can be either positive or negative.

A conspicuous link between the problems of negative energy and negative density is that the sign of the density is correlated with the sign of the energy. Positive-energy solutions correspond to a density of one sign and negative-energy solutions

[5]An example of a zero-spin massive particle is the pion, discussed in Section 11.3. Particles with nonzero spin, such as electrons, obey the related but more complicated **Dirac equation.** A discussion of this equation would not be appropriate at this level. However, it is in the Dirac equation that spin is included naturally, not in the ad hoc fashion required when using the Schrödinger equation. So far as antiparticles are concerned, the most important point is that the Dirac equation is also based upon $E^2 = p^2c^2 + m^2c^4$, and thus shares the "problem" of negative-energy solutions.

[6]The density is

$$i\Psi^*(\partial\Psi/\partial t) - i\Psi(\partial\Psi^*/\partial t)$$

and the integral

$$\int [i\Psi^*(\partial\Psi/\partial t) - i\Psi(\partial\Psi^*/\partial t)\, dV$$

is real and constant in time.

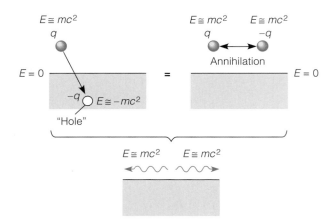

**Figure 11.2** Whether a positive-energy particle drops into an empty negative-energy state, or two positive-energy particles annihilate, the result is the same.

to a density of the opposite sign. This clue serves as the basis for the following interpretation: The new density we identify as *charge* density, whose integral over space must indeed be a constant in time, but may be of either positive or negative sign. A negative-energy solution describes a particle identical to that described by the positive-energy solution, except that its charge is of the opposite sign—the antiparticle.

The idea of negative energy is still troubling. Without going too deeply into relativistic quantum mechanics, it might be said that, as depicted in Figure 11.2, a negative-energy *antiparticle* state behaves as an oppositely-charged "hole" in the allowed states of the *particle*. A particle may fall into the hole (thus disappearing), but only if the hole already exists—when there exists an antiparticle. In filling the hole, the particle surrenders an energy equal to the difference between the positive initial and negative final energies—twice the energy of the particle—usually resulting in the production of two photons. Because equal energy would be released in either view, we see that it is equivalent to regard the filling of the hole as the annihilation of two particles, each of (positive) energy $mc^2$.

## 11.3  Forces and Particles: How Many?

The process of "unifying forces," that is, of showing the essential equivalence of seemingly diverse forces, has been going on for a long time. For instance, electric and magnetic phenomena were once thought to be completely distinct, but by the early twentieth century were conclusively shown to be simply different faces of the single "electromagnetic" interaction.[7] Our present knowledge indicates that all forces are manifestations of no more than three distinct fundamental interactions: the **gravitational,** the **electroweak,** and the **strong.** Until the 1970s, four were accepted. But in the 1979 Nobel prize–winning theoretical work of Sheldon Glashow, Abdus Salam, and Steven Weinberg (all working separately) it was shown that the weak and electromagnetic interactions, formerly considered distinct, are in fact merely two aspects of the same "electroweak" interaction. Nevertheless, the two are still often referred to separately, since the "weak part" is

[7] An important piece of evidence was the finding that a static electric field in one frame of reference (perhaps due to a stationary point charge) is, in a frame moving relative to the first, a combination of electric and magnetic fields. The two "mix together" in a well-defined way.

manifest only in certain reactions among subatomic particles. It is not revealed in ordinary electromagnetic phenomena.

Much research continues to be devoted to showing the equivalence of the three remaining interactions. A theory in which all interactions appear as merely different aspects of a single fundamental interaction would be the ultimate "unified theory." Alas, such a theory has eluded many of the brightest minds in physics to this day. Unified theories are discussed in Section 11.7.

Certain characteristics are common to all fundamental forces—surely an enticing bit of evidence that they are somehow related. First, fundamental fermions engage in a force by exchanging mediating particles that are bosons. Secondly, for a fermion to engage in a force, it must possess the property associated with that force. These properties are mass/energy, for gravitation; charge (or weak charge), for the electroweak; and color-charge, for the strong interaction. As a crude example (involving particles *not* truly fundamental), two neutrons interact via the gravitational force, since they possess the required property of mass, but not via the electrostatic force, because they don't possess the property of charge. Table 11.1 summarizes the **standard model**—the state of the art in particle physics. Although a full appreciation will unfold only gradually, we present it here so

**Table 11.1   The Standard Model**

| Force | Gravitation | Electroweak | | | Strong | Residual |
|---|---|---|---|---|---|---|
| Property | Mass/energy | Charge/weak-charge | | | Color-charge | |
| Strength | $\sim 10^{-39}$ | $\sim 10^{-2}$ | | $\sim 10^{-6}$ | 1 | |
| Range | $1/r^2$ | $1/r^2$ | | $10^{-3}$fm | complex | 1fm |
| **Mediating Bosons** | graviton? | photon $\gamma$ | $W^+, W^-$ | $Z^0$ | gluon | $\pi^\pm, \pi^0$ |
| spin | 2? | 1 | 1 | 1 | 1 | 0 |
| mass | 0? | $<6 \times 10^{-22}$ | $80.3 \times 10^3$ | $91.2 \times 10^3$ | $<10$ | 140, 135 |
| charge | — | 0 | $+1, -1$ | 0 | 0 | $\pm 1, 0$ |
| color-charge | — | — | — | — | r, g, or b + $\bar{\text{r}}$, $\bar{\text{g}}$, or $\bar{\text{b}}$ | neutral |

| Leptons | | | | Quarks | | | |
|---|---|---|---|---|---|---|---|
| Participants in Gravitation and Electroweak | | | | Participants in Gravitation, Electroweak, and Strong | | | |
| | Spin | Mass | Charge | | Spin | Mass | Charge | Color-charge |
| electron, e | $\frac{1}{2}$ | 0.511 | $-1$ | up, u | $\frac{1}{2}$ | $\sim 5$ | $+\frac{2}{3}$ | r,g,b |
| e-neutrino, $\nu_e$ | $\frac{1}{2}$ | $<10^{-6}$ | 0 | down, d | $\frac{1}{2}$ | $\sim 10$ | $-\frac{1}{3}$ | r,g,b |
| muon, $\mu$ | $\frac{1}{2}$ | 106 | $-1$ | strange, s | $\frac{1}{2}$ | $\sim 200$ | $-\frac{1}{3}$ | r,g,b |
| $\mu$-neutrino, $\nu_\mu$ | $\frac{1}{2}$ | $<0.17$ | 0 | charm, c | $\frac{1}{2}$ | $\sim 1.5 \times 10^3$ | $+\frac{2}{3}$ | r,g,b |
| tauon, $\tau$ | $\frac{1}{2}$ | $1.78 \times 10^3$ | $-1$ | bottom, b | $\frac{1}{2}$ | $\sim 4.5 \times 10^3$ | $-\frac{1}{3}$ | r,g,b |
| $\tau$-neutrino, $\nu_\tau$ | $\frac{1}{2}$ | $<24$ | 0 | top, t | $\frac{1}{2}$ | $\sim 180 \times 10^3$ | $+\frac{2}{3}$ | r,g,b |

Spins given in units of $\hbar$, masses in units of MeV/$c^2$, and charges in units of $e$. All particles have antiparticles, of opposite charge and color-charge.

that the reader may see how the pieces fall into place. Let us now study qualitatively the presently accepted fundamental interactions, in order of decreasing strength, and introduce the particles that engage in them.

## The Strong Force

Neither the proton nor the neutron is fundamental. We now know that each is a combination of more-basic particles—**quarks**—bound together by their mutual strong-force attraction (Figure 11.3). At present, it is believed that quarks are the fundamental fermions that engage in the strong force. Six different types are known, and they have been given the names up, down, strange, charm, top, and bottom. (No meaning should be attached to these names. In the early days of elementary particle physics, frivolity reigned, and particles and properties were often given fanciful names.) Their properties are summarized in Table 11.1. The proton and the neutron comprise three quarks each, with **quark content** uud and udd, respectively. From the tabulated charges of these quarks, $+\frac{2}{3}$ for the up and $-\frac{1}{3}$ for the down, we obtain the proper charges of the proton and neutron: $+\frac{2}{3} + \frac{2}{3} - \frac{1}{3} = +1$ and $-\frac{1}{3} - \frac{1}{3} + \frac{2}{3} = 0$. Particles composed of quarks are known by the generic name **hadrons.** We shall discuss hadrons other than the proton and neutron later.

Perhaps the most remarkable property of quarks is that they have nonintegral charge. This would appear to dash our time-honored belief that any amount of charge is always an integral multiple of the "fundamental charge" $e = 1.6 \times 10^{-19}$ C. The belief is safe for the time being, however, for no means has yet been found of isolating a quark! Moreover, there is reason to believe that it will never be done. Because the electrostatic force falls off as $1/r^2$, the energy required to pull opposite charges apart, even infinitely far, is finite. The strong force, on the other hand, does *not* decrease with separation.[8] Consequently, infinite energy would be required to separate quarks. **Quark confinement** is the term used to describe the refusal of quarks to be separated. Certainly the attempt can be made, but what we invariably find is that the energy expended is sufficient, even at very small separations, to create quark–antiquark pairs. We end up not with *separated* quarks, but simply a greater *number* of quarks, which form multiquark hadrons of integral charge. Even so, this explanation would seem to raise more questions than it answers: If quarks cannot be isolated, how do we know they are of nonintegral charge? Indeed, how do we know they even exist?

It is helpful to note that the atom was shown to possess a concentrated positive nucleus, without actually tearing it apart, by probing with high-energy alpha-particles. In a similar way, much information about the internal structure of hadrons is gleaned from experiments in which they are struck by high-energy electrons.[9] Fundamental point electrons are sent in to probe for other fundamental

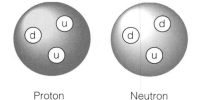

Proton            Neutron

**Figure 11.3** The proton and neutron are not fundamental; they contain quarks.

Hadron: A bound state of multiple quarks

---

[8]It is probably worthwhile to reiterate a point made in Chapter 10: The force responsible for holding the *nucleons* together in the nucleus—as opposed to holding quarks together in a nucleon—is an *aspect* of the strong force that does indeed diminish rapidly with $r$. This "residual strong force" is discussed later in the section.

[9]It is because essentially all information about fundamental particles and interactions is gained in this way that the terms "elementary particle physics" and "high-energy physics" are used interchangeably.

point particles, that is, quarks. In **deep inelastic scattering experiments,** electrons of energy greater than 20 GeV and momentum 20 GeV/c have been smashed into protons.[10] The wavelength corresponding to such a momentum is on the order of $10^{-16}$ m. Since this is no more than about a tenth the nucleon radius, the electron should indeed be able to resolve structural details within a nucleon. Analysis of the scattering in these experiments has provided clear evidence: The nucleon is indeed composed of point particles of fractional charge.

The mediating boson of the strong force is the **gluon.** Thus, hadrons may be thought of as quarks bound together by continuous exchange of gluons. Gluons appear to have zero intrinsic mass. This fits with the observation that the strong force does not decrease with quark separation: According to equation (11-1), an infinite-range force would imply a massless mediating particle. However, as noted previously, attempting to pull quarks apart even small distances simply generates more hadrons. This makes defining a range for the strong force rather difficult.

The property a particle must possess to engage in the strong force (the property to which the gluon "couples") is known as **color-charge.** While similar in many respects, it is distinct from electric charge. Whereas electric charge can be of two kinds—plus and minus—color-charge can be of three kinds, styled red, green, and blue. (They are not actual colors of light.) Quarks may be any of the three "colors" of color-charge, while antiquarks may be antired, antigreen, or antiblue. Gluons also carry color. (For reasons discussed in Section 11.5, gluons carry a color-anticolor pair, e.g., blue-antigreen.) Being the only particles possessing color, quarks and gluons are the only ones able to engage in the strong interaction.[11] A rationale for the term "color" is that red, green, and blue light add to neutral light, and all evidence indicates that hadrons must be "color neutral." In electrostatics, we know that two electrons, with a nonzero net *electric* charge, repel one another, while there is a net attraction in a charge-neutral system such as a hydrogen atom. Similarly, zero net color leads to attraction. To be color neutral, hadrons can be bound states of three quarks, one of each color (red + green + blue = neutral), or bound states of two quarks, a quark of one color and an antiquark of its anticolor (red + antired = neutral).[12]

Just as for quarks, it has not been possible to isolate gluons or net color-charge, but there is ample evidence of their presence. High-energy collisions often produce **jets** (isolated sectors) of "extra" hadrons, perfectly explained as decay products of gluons emitted immediately after the collision. With regard to color, the $\Delta^{++}$ and $\Delta^{-}$ hadrons comprise three identical quarks, uuu and ddd, respectively, and they are spin-$\frac{3}{2}$, meaning that the quark spins are aligned; all three quarks are

---

[10]The collisions are inelastic because some of the available kinetic energy is "lost," producing new massive particles. Also, with the electron's total energy so much larger than its mass energy (511 keV), the relationship $E^2 = p^2c^2 + m^2c^4$ becomes essentially $E = pc$, just as for a photon. Thus, the simple conversion from energy to momentum.

[11]This is one of the important differences between the strong force and the more familiar electromagnetic. Photons, the mediating particles of the electromagnetic interaction, do not themselves possess electric charge. Thus, while photon emission and absorption are the means by which charged fundamental fermions interact, photons cannot interact with each other. But gluons do possess color-charge. Not only are they emitted and absorbed by colored fundamental fermions (i.e., quarks), they also interact with each other. We see then that it is not quite general to say that fundamental forces are conveyed by the exchange of mediating bosons *between fermions.*

[12]Antiparticles of three-quark hadrons are composed of antiquarks (e.g., the $\overline{uud}$ antiproton) and are color-neutral by virtue of comprising antired, antigreen, and antiblue antiquarks.

in the same spin state. Without another property to distinguish their quarks, these hadrons would violate the exclusion principle. Different colors make the quarks distinguishable. Indicating the importance of color, the theory of the strong interaction is known as **quantum chromodynamics** ("chroma" is Greek for color), or **QCD.**

A quick glance at Table 11.1 shows that the masses of the up and down quarks are far too small to account for the mass of a three-quark nucleon ($\sim$940 MeV/$c^2$). The masses given are often called "bare masses" and are at best only rough values. Scattering experiments show that a nucleon is not simply three spheres stuck together, but rather three small quarks surrounded by a sizable cloud of energetic gluons. This cloud constitutes the majority of the nucleon mass and makes assigning definite values to quark masses problematic.

## Categories of Hadrons

With six quarks and six antiquarks, a multitude of different hadrons is possible. Many of the possible combinations have been produced (discovered) in high-energy inelastic collisions in which kinetic energy is converted to mass. Table 11.2 lists some commonly produced hadrons, with their quark content, masses, and other properties. (All have antiparticles in which each quark is replaced by its

### Table 11.2　Commonly Produced Hadrons

| Baryons | Mass (MeV/$c^2$) | Spin$^P$ | Strange-ness | $I, I_3$ | Lifetime, $\tau$ (or width $\hbar/\tau$) | Mesons | Mass (MeV/$c^2$) | Spin$^P$ | Strange-ness | $I, I_3$ | Lifetime, $\tau$ (or width $\hbar/\tau$) |
|---|---|---|---|---|---|---|---|---|---|---|---|
| p (uud) | 938 | $\frac{1}{2}^+$ | 0 | $\frac{1}{2}, +\frac{1}{2}$ | $> 10^{32}$ yr | $\pi^+$ (u$\bar{\text{d}}$) | 140 | $0^-$ | 0 | $1, +1$ | $2.6 \times 10^{-8}$ s |
| n (udd) | 940 | $\frac{1}{2}^+$ | 0 | $\frac{1}{2}, -\frac{1}{2}$ | 889 s | $\pi^0$ (u$\bar{\text{u}}$ + d$\bar{\text{d}}$) | 135 | $0^-$ | 0 | $1, 0$ | $8.4 \times 10^{-17}$ s |
| $\Sigma^+$ (uus) | 1189 | $\frac{1}{2}^+$ | $-1$ | $1, +1$ | $8.0 \times 10^{-11}$ s | $\pi^-$ (d$\bar{\text{u}}$) | 140 | $0^-$ | 0 | $1, -1$ | $2.6 \times 10^{-8}$ s |
| $\Sigma^0$ (uds) | 1193 | $\frac{1}{2}^+$ | $-1$ | $1, 0$ | $7.4 \times 10^{-20}$ s | $K^+$ (u$\bar{\text{s}}$) | 494 | $0^-$ | $+1$ | $\frac{1}{2}, +\frac{1}{2}$ | $1.2 \times 10^{-8}$ s |
| $\Lambda^0$ (uds) | 1116 | $\frac{1}{2}^+$ | $-1$ | $0, 0$ | $2.6 \times 10^{-10}$ s | $K_S^0$ (d$\bar{\text{s}}$, s$\bar{\text{d}}$) | 498 | $0^-$ | mix | $\frac{1}{2}$, mix | $8.9 \times 10^{-11}$ s |
| $\Sigma^-$ (dds) | 1197 | $\frac{1}{2}^+$ | $-1$ | $1, -1$ | $1.5 \times 10^{-10}$ s | $K_L^0$ (d$\bar{\text{s}}$, s$\bar{\text{d}}$) | 498 | $0^-$ | mix | $\frac{1}{2}$, mix | $5.2 \times 10^{-8}$ s |
| $\Xi^0$ (uss) | 1315 | $\frac{1}{2}^+$ | $-2$ | $\frac{1}{2}, +\frac{1}{2}$ | $2.9 \times 10^{-10}$ s | $K^-$ (s$\bar{\text{u}}$) | 494 | $0^-$ | $-1$ | $\frac{1}{2}, -\frac{1}{2}$ | $1.2 \times 10^{-8}$ s |
| $\Xi^-$ (dss) | 1321 | $\frac{1}{2}^+$ | $-2$ | $\frac{1}{2}, -\frac{1}{2}$ | $1.6 \times 10^{-10}$ s | $\rho^+$ (u$\bar{\text{d}}$) | 769 | $1^-$ | 0 | $1, +1$ | 151 MeV |
| $\Delta^{++}$ (uuu) | 1232 | $\frac{3}{2}^+$ | 0 | $\frac{3}{2}, +\frac{3}{2}$ | 120 MeV | $\rho^0$ (u$\bar{\text{u}}$ + d$\bar{\text{d}}$) | 769 | $1^-$ | 0 | $1, 0$ | 151 MeV |
| $\Delta^+$ (uud) | 1232 | $\frac{3}{2}^+$ | 0 | $\frac{3}{2}, +\frac{1}{2}$ | 120 MeV | $\rho^-$ (d$\bar{\text{u}}$) | 769 | $1^-$ | 0 | $1, -1$ | 151 MeV |
| $\Delta^0$ (udd) | 1232 | $\frac{3}{2}^+$ | 0 | $\frac{3}{2}, -\frac{1}{2}$ | 120 MeV | $K^{*+}$ (u$\bar{\text{s}}$) | 892 | $1^-$ | $+1$ | $\frac{1}{2}, +\frac{1}{2}$ | 50 MeV |
| $\Delta^-$ (ddd) | 1232 | $\frac{3}{2}^+$ | 0 | $\frac{3}{2}, -\frac{3}{2}$ | 120 MeV | $K^{*0}$ (d$\bar{\text{s}}$) | 896 | $1^-$ | $+1$ | $\frac{1}{2}, -\frac{1}{2}$ | 51 MeV |
| $\Sigma^{*+}$ (uus) | 1383 | $\frac{3}{2}^+$ | $-1$ | $1, +1$ | $\sim$40 MeV | $\overline{K^{*0}}$ (s$\bar{\text{d}}$) | 896 | $1^-$ | $-1$ | $\frac{1}{2}, +\frac{1}{2}$ | 51 MeV |
| $\Sigma^{*0}$ (uds) | 1384 | $\frac{3}{2}^+$ | $-1$ | $1, 0$ | $\sim$40 MeV | $K^{*-}$ (s$\bar{\text{u}}$) | 892 | $1^-$ | $-1$ | $\frac{1}{2}, -\frac{1}{2}$ | 50 MeV |
| $\Sigma^{*-}$ (dds) | 1387 | $\frac{3}{2}^+$ | $-1$ | $1, -1$ | $\sim$40 MeV | | | | | | |
| $\Xi^{*0}$ (uss) | 1532 | $\frac{3}{2}^+$ | $-2$ | $\frac{1}{2}, +\frac{1}{2}$ | $\sim$10 MeV | Heavy mesons—containing quarks beyond the strange | | | | | |
| $\Xi^{*-}$ (dss) | 1535 | $\frac{3}{2}^+$ | $-2$ | $\frac{1}{2}, -\frac{1}{2}$ | $\sim$10 MeV | $J/\psi$ (c$\bar{\text{c}}$) | 3097 | $1^-$ | 0 | $0, 0$ | 87 keV |
| $\Omega^-$ (sss) | 1672 | $\frac{3}{2}^+$ | $-3$ | $0, 0$ | $8.2 \times 10^{-11}$ s | $\Upsilon$ (b$\bar{\text{b}}$) | 9460 | $1^-$ | 0 | $0, 0$ | $\sim$50 keV |

antiquark.) Three-quark hadrons are known as **baryons** (from the Greek for heavy) and two-quark hadrons as **mesons** (from the Greek for intermediate). The neutron (udd), for instance, is a baryon. Although it is impossible to know which quark is which of the three colors, the requirement of color neutrality demands that there be one quark of each. An example of a meson is the **pion**, which comes in three varieties. The $\pi^+$ is of quark content $u\bar{d}$. Its $+1$ charge is accounted for by the $+\frac{2}{3}$ charge of the up quark and the $+\frac{1}{3}$ of the *anti*down (the antiparticle of the down quark). The up and antidown quarks might be red and antired, green and antigreen, or blue and antiblue. The antiparticle of the $\pi^+$ is the $\pi^-$, of quark content $\bar{u}d$ and therefore of charge $-1$. The $\pi^0$ is of zero electric charge; it is a mixture of the charge- and color-neutral quark combinations $\bar{u}u$ and $\bar{d}d$.

From the table we see that many hadrons, particularly baryons, are much heavier than nucleons. Greater mass output requires greater kinetic energy input, so heavier ones tend to be more recent discoveries, made via higher-energy accelerators. We also see, from the average lifetime data, that all nonnucleon hadrons are fairly short-lived. In cases of very short-lived particles, it is common to specify an energy **width** rather than a lifetime. According to the uncertainty principle, a particle's energy can be known to no greater precision than $\Delta E \cong \hbar/\tau$, where $\tau$ is the time interval during which the particle exists, that is, its lifetime. The lifetime corresponding to a width of 50 Mev is about $10^{-23}$ s.

### Intrinsic Properties

Table 11.2 gives the spins of the hadrons, which follow from the spins of the quarks. All quarks are spin-$\frac{1}{2}$, so by the rules of angular momentum addition, baryons are either spin-$\frac{3}{2}$ or spin-$\frac{1}{2}$ (i.e., all three spins aligned, $\frac{1}{2} + \frac{1}{2} + \frac{1}{2}$, or two aligned and one antialigned, $\frac{1}{2} + \frac{1}{2} - \frac{1}{2}$). Baryons are fermions. Similarly, mesons are bosons of spin-0 or spin-1. (The term "spin" here really means *total* angular momentum, which includes any *orbital* angular momentum the quarks might have within a hadron. The ground state hadrons of Table 11.2 have no orbital angular momentum.)

The column giving spins in Table 11.2 also gives another property—**intrinsic parity** (symbol: $P$). Quarks are endowed with a host of "new" intrinsic properties. With mass, charge, and spin as examples, we know that a particle's intrinsic properties are those that are immutable, not subject to change. They are independent of any spatial extent and so are often called "point properties." We may understand intrinsic parity via an analogy: The spin of an orbiting electron is the same physical property as its spread-out orbital angular momentum (with which it may therefore add), but of an intrinsic, point nature. Similarly, intrinsic parity is a point property essentially the same as the parity of a spread-out wave function. A wave function that changes sign if $(x, y, z)$ is replaced by $(-x, -y, -z)$ has negative parity; if it does not change sign, it is of positive parity. In one dimension, such wave functions would be called "odd" and "even" functions of $x$, respectively. Thus, intrinsic parity is the point analog of odd and even functions in one dimension. The intrinsic parity is positive for quarks and negative for antiquarks. Total parity is the product of individual parities, so the three-quark baryons of Table 11.2 are of positive parity $(+1) \times (+1) \times (+1)$ and the quark–antiquark mesons are of negative parity $(+1) \times (-1)$. (As in the case of spin, *total* parity must also include the parities of the orbiting quarks' wave functions. The hadrons of Table 11.2 have their quarks in lowest-energy, positive-parity spatial states.)

Another property of quarks underlies a curious feature in Table 11.2: Hadrons may have the same quark content and spin, yet still be different. For instance, the $\Lambda^0$ and $\Sigma^0$ are both uds and spin-$\frac{1}{2}$. The distinction lies in the intrinsic property called **isospin,** given the symbol $I$. It is not an angular momentum, as is spin, but does obey the same addition rules—hence the name. The up and down quark each have isospin $I = \frac{1}{2}$, while the strange quark has isospin $I = 0$. In the $\Sigma^0$, the isospins of the up and down quark are aligned, giving a total isospin $I = 1$, while in the $\Lambda^0$ they are antialigned, yielding $I = 0$. The distinction between the $K_S^0$ and $K_L^0$ (the different subscripts referring to their lifetimes—"short" and "long") is not based on isospin, but on an unusual mixing of the $K^0$ particle and its antiparticle $\overline{K^0}$. A discussion of this mixing may be found in more-advanced texts on particle physics. Both isospin and parity are discussed further in Section 11.6.

Yet another intrinsic property given in Table 11.2 is **strangeness.** Sensibly, it is the strange quark that endows a particle with strangeness. By arbitrary sign choice, possession of one gives a strangeness of $-1$, of two a strangeness $-2$, of an antistrange a strangeness $+1$, and so forth. We discuss strangeness further in Section 11.5. Here we note simply that the hadrons containing only u, d, and s quarks can be grouped into several "multiplets" according to spin, and an individual member of a multiplet is distinguished by its strangeness and isospin. (Four such groups are set apart in the table.)

### The Residual Strong Force

Neutral atoms may attract each other electrostatically: Each may have, or acquire, an electric dipole moment, and if these are properly oriented, a net attraction results. Limited though it may be, this attraction is how some substances form solids, notably water (cf. Section 9.4). In a similar effect, color-neutral particles may attract each other via the strong interaction. It is this attraction that holds the nucleons together in the nucleus. Called the **residual strong force,** it is not a distinct force, but rather a coarse manifestation of the true strong interaction. Still, it too can be viewed as being conveyed by exchange of a particle: Color-neutral uud protons and udd neutrons may exchange a two-quark, color-neutral combination of u, d, $\overline{u}$, and $\overline{d}$ quarks—a pion. Although the lightest of hadrons, the pion does have mass, and it is because of this that the residual strong force is short-range. In fact, rough knowledge of the force's range led, through equation (11-1), to a reasonably accurate prediction of the pion's mass before it was ever detected (see Exercise 1).

## The Electroweak Force

The electroweak force is engaged in by particles that possess the property of charge and/or **weak-charge.** Besides quarks, this includes the class of particles known as **leptons** (from the Greek for light), which lack color-charge and so are "blind" to the strong force. As shown in Table 11.1, there are six known leptons (plus their antiparticles), in three pairs: the electron and electron-neutrino, the muon and mu-neutrino, and the tauon and tau-neutrino. Although of increasing mass, the electron, muon, and tauon are otherwise very much alike, and many physicists wonder why there "need" to be three different particles.

The electroweak force is conveyed by the exchange of any of four mediating bosons: the photon and the so-called **weak bosons** $W^+$, $W^-$, and $Z^0$. A cursory examination of their properties quickly reveals the complexity of the electroweak interaction: The photon, exchanged in the "electromagnetic part" of the interaction, is massless, while the $W^\pm$ and $Z^0$ particles, exchanged in the "weak part," are among the heaviest fundamental particles yet detected! If there is indeed only one underlying interaction, these four are all field quanta *of the same field.* Yet how?

The schism is known as a **spontaneously broken symmetry.** The asymmetry—the different behaviors of the electromagnetic and weak forces—occurs at low particle energies. As a child's toy top, oriented symmetrically relative to the ground at high rotational energy, at lower energy spontaneously trades its top-heavy state for an asymmetric slumber on its side, so does the asymmetry in the electroweak force arise at low energies. The true underlying symmetry of the electroweak interaction becomes apparent only at particle energies much higher than those commonly accessible to observation. At very high energies, the intrinsic mass of a particle becomes irrelevant. The massive-particle momentum-energy relation, $E^2 = p^2c^2 + m^2c^4$, becomes that for a massless particle, $E = pc,$ and the behaviors of the mediating particles converge. The precise mechanism by which the symmetry is broken at low energies is still not fully understood, but current theory suggests that the picture would be completed by the existence of one more field particle—the **Higgs boson.** Its production is expected to require particle accelerators at least at the $\sim$1-TeV limit of existing machines. Discovery of the Higgs boson would be a great step in solving the mysteries of the electroweak force.

All leptons are spin-$\frac{1}{2}$ fermions. Unlike quarks, they possess neither intrinsic parity nor isospin. Neutrinos are uncharged and therefore do not engage in *strictly electromagnetic* interactions. But all leptons engage in weak interactions; all possess weak-charge. We won't study the relationship between charge and weak-charge, except to note that the greater strength of the electromagnetic relative to the weak force is not because weak-*charge* is weaker than electric charge; it is because of the yet mysterious *mass* difference of the mediating bosons. Nonzero mass is an impediment, responsible for both the weakness and the short range of the weak force relative to the electromagnetic.

## The Gravitational Force

General relativity tells us that all particles with mass *or any other form of energy* attract one another via the gravitational force. Obviously, this broad categorization includes all the fundamental particles, for even massless particles move at $c$ and so have kinetic energy. However, to date, no quantum aspects of the gravitational force have been observed! Primarily this is due to its relative weakness. By analogy with the two other interactions, we assume that things attract one another gravitationally via exchange of a field quantum, which, since the force is of infinite range, should be massless. Moreover, given gravity's attractive-only nature, the theory of gravitation suggests that this mediating boson should be of spin-2. It has even been given a name—**graviton.** Yet it has never been detected. We shall return to this quantum-mechanically intransigent force when we discuss unified theories.

# 11.4 Particle Production and Detection

As noted earlier, our view of which particles are truly fundamental has changed much through the years. We now briefly survey the progress and experimental means by which our knowledge has been gained.

## A Brief Chronology

Perhaps not surprisingly, the first subatomic particle to be discovered was the electron, in 1897. Proof of a compact positively charged nucleus came in 1911, followed in 1919 by the discovery of the proton. Despite mounting interest and rapidly advancing technology, the neutron was not discovered until 1932. (This gives some idea of the difficulty of detecting uncharged particles. Even today's particle detectors detect uncharged particles only indirectly, by analyzing the behavior of the charged particles with which they interact.)

The first source of high energy for particle creation was natural: cosmic rays reaching Earth from space. Cosmic-ray photons often have energies greater than 1 GeV, so they are energetic enough to produce particles whose mass energies are comparable to the nucleon's ~1 GeV. Besides the positron (Section 2.4), cosmic rays served as the source for the discovery in the 1930s and 1940s of the muon ($m_\mu = 0.11$ GeV), the pion ($m_\pi = 0.14$ GeV), and the kaon ($m_K = 0.49$ GeV). After the K-meson, a great multitude of other hadrons were identified. But cosmic rays were soon no longer the best energy source—the particle accelerator had arrived.

Although not realized at the time, the discovery of the kaon had brought to three the number of quarks subject to study. "Early" hadrons were merely different combinations of the same three quarks: the up, the down, and the strange. Indeed, after a decade or so of finding recurring patterns, strong suspicions arose that the known hadrons were composed of a smaller number of more fundamental particles. The three-quark theory that came to be accepted is due to Murry Gell-Mann, who was awarded the 1969 Nobel prize for his work (and by whom the name "quark" was chosen). The theory was widely embraced, but it was not until the deep inelastic scattering experiments (see Section 11.3) of Friedman, Kendall, and Taylor in 1967 that the first direct experimental evidence of quarks was obtained. Still, other more massive quarks lay undiscovered.

As the energy available to us has increased, so has our ability to produce massive particles. The appearance of much more powerful accelerators since the 1970s has led to regular successes in the production of many theoretically predicted particles. Consider a few examples of the progression. In 1974, the 3.1-GeV J/$\psi$ particle, a $\bar{c}c$ meson, was discovered, adding a fourth quark. Discovery of the third massive lepton (joining the electron and muon), the 1.8-GeV tauon, came in 1975. The bottom quark, the fifth, was identified in 1977 via production of the 9.5-GeV $\Upsilon$ meson, a bound $\bar{b}b$. The weak bosons, $W^{\pm}$ (80 GeV) and $Z^0$ (91 GeV), were found in 1983. And in 1994, the sixth quark, the 180-GeV top, was positively identified.

## Accelerators and Detectors

Today's accelerators may be categorized in several ways: (1) They may have two colliding beams, or one beam and a stationary target. A collider is inherently more difficult to align, but technology has surmounted this disadvantage, making its advantages decisive.[13] Accordingly, most accelerators nowadays are colliders, and several older ones are being converted from stationary targets to colliders. (2) The accelerated particles are usually either protons or electrons, including $\bar{p}$ and $e^+$. (In some experiments, heavy ions are accelerated.) (3) The particles may be linearly accelerated, or may follow a circular path. In all cases, electric fields speed up the particles, while turning is done by magnetic fields, in some cases produced by superconducting magnets. Data on a few of the many colliding-beam facilities are given in Table 11.3. In a **synchrotron,** the magnetic field is constantly adjusted so that while being "kicked" by electric fields the particles still move in a circle of constant radius; they circle in a narrow tube. In older "cyclotrons," they spiral outward from the center. Figures 11.4 through 11.7 show features of several of the larger accelerators.

It is useless to produce particles if their presence can't be verified; reliable detectors are crucial. Of the many types, a few of the more important are the

[13]Recall from Section 1.9: In a fixed-target collision, there is a large initial momentum, so there must be a large final momentum and thus kinetic energy. Not all the kinetic energy is available to become mass. In a colliding beam set-up, the initial total momentum is zero. Were a single particle the result (unrealistic though it might be), the particle would be stationary and all of the initial kinetic energy would have been converted to mass. A collider "wastes" no energy.

**Table 11.3**

| Facility | Location | Type | Energy |
|----------|----------|------|--------|
| SLC | Stanford, California USA | Linear $e^+e^-$ | 50 GeV ($\times 2$) |
| LEP—CERN | Geneva, Switzerland | Synchrotron $e^+e^-$ | 85+ GeV ($\times 2$) |
| TEVATRON—Fermilab | Batavia, Illinois USA | Synchrotron $\bar{p}p$ | 1 TeV ($\times 2$) |

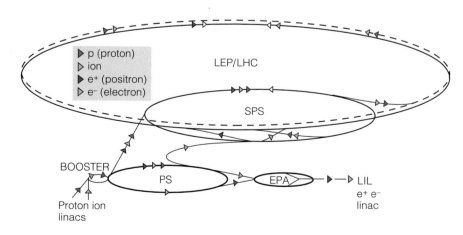

**Figure 11.4** The accelerator complex at CERN, covering about 70 km$^2$ in Switzerland and France. The large dashed circle is the planned Large Hadron Collider (LHC). Linear accelerators (linacs) are the sources for various particles.

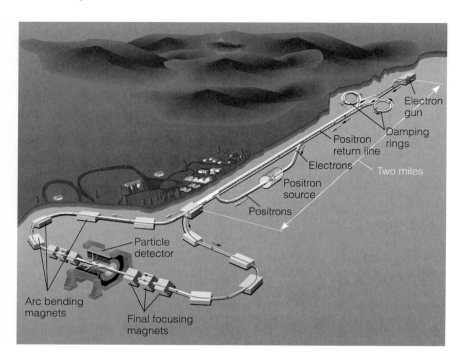

**Figure 11.5** The Stanford Linear Accelerator Center (SLAC), surrounding the Stanford Linear Collider (SLC).

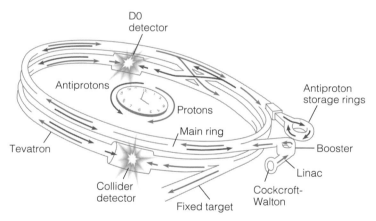

**Figure 11.6** Fermilab's $1\frac{1}{4}$-mile-diameter Tevatron.

scintillation counter, the multiwire proportional counter, the Cerenkov counter, the bubble chamber, and the calorimeter. The **scintillation counter** houses one of many substances (e.g., thallium-doped NaI) known to produce a flash of light when struck by a charged particle. This light is easily observed with electronic sensors and response time is short.

A **multiwire proportional counter** contains a grid of conductors at alternating electric potential within a gas-filled region. When a charged particle enters the region, it ionizes the gas around it, which then causes a brief current between nearby conductors. Determining which conductors are conducting at a given time establishes the particle's location, and computer analysis of the signals provides

**Figure 11.7** The Tevatron's antiproton storage rings.

an excellent picture of its motion. The "proportional" refers to the proportion between the energy deposited by the particle and the electric response of the detector. (Simpler gas discharge detectors, like the Geiger counter, lack these desirable properties; they give only a "yes" or "no" response and do not track a moving particle.)

In a **Cerenkov detector,** a charged particle enters a medium at a speed greater than light moves in the medium.[14] It rapidly slows down while emitting electromagnetic radiation called Cerenkov radiation, particularly identifiable by its direction and coherence. From this, the particle's speed can be determined.

A **bubble chamber** contains a liquid at a pressure and temperature such that it is very near the transition point to a gas. In fact, it is beyond that point, in an unstable superheated state in which virtually any disturbance would cause a rapid change to the gas phase. The passage of a charged particle is such a disturbance, and its path leaves a clear trail of bubbles through the chamber. Photographs are taken of the bubble trail from various aspects, producing an image of the motion in three dimensions. In most applications, bubble chambers are immersed in an external magnetic field, so that the curvature of path indicates the speed of the particle. Afterward, an increase in pressure erases the trail (recondenses the gas), and the chamber is ready for another "shot."

When a particle enters a **calorimeter,** it adds energy, which is then shed in numerous ways—creating showers of other particles, for instance. Analysis of these secondary occurrences determines the incoming particle's energy.

An important result of our discoveries to date is that we have identified six quarks and six leptons, occurring in three increasingly massive tiers whose charges are $\left(+\frac{2}{3}e, -\frac{1}{3}e\right)$ and $(-e, 0)$: (u, d) and $(e^-, \nu_e)$; (c, s) and $(\mu^-, \nu_\mu)$; and

[14]The particle, of course, never moves faster than the speed of light *in a vacuum.*

(t, b) and $(\tau^-, \nu_\tau)$. This is a pleasing symmetry, and though their interactions are still not fully understood—unified theories are still being developed—many physicists speculate that these 12 particles may complete the list of fundamental fermions.

# 11.5 Interactions: Decay Modes and Conservation Rules

Besides simple elastic interactions, in which particles interact with each other but are not fundamentally changed, there are many transmutations possible among subatomic particles. Transmutations may be deliberately induced in a collision or may be the result of the natural decay of an unstable particle. Collisions allow us to produce heavier particles, but even simple decays provide valuable insight into all particle interactions.

It might well be argued that there are few stable particles. The $\mu$ and $\tau$ leptons are unstable, and all hadrons but the proton are known to decay.[15] Decays are classified as strong, electromagnetic, or weak depending on whether the interaction involves the exchange of gluons, photons, or weak bosons, respectively. (We ignore the very weak force of gravity, whose field quantum, the graviton, has not even been found.) In all cases, though, conservation rules apply. Perhaps the most obvious is energy conservation: A spontaneous decay, in which kinetic energy increases, must be accompanied by a decrease in mass. This alone suggests that *all* massive particles should be liable to decay to the lightest stable particle. But other rules restrict the possibilities.

Historically, it has often come to the attention of scientists that one quantity or another seems to be conserved. After standing the test of laborious scrutiny, some of these are officially proclaimed "conserved quantities" and their conservation accepted as a law of physics. Mistakes have been made. Mass was once thought to be conserved, but we now know that it is merely one kind of energy, and may change while total energy is still conserved. As we discuss various decays, we will refer to the rules summarized in Table 11.4 pertaining to quantities that must be conserved. The reader should bear in mind that, just as for "mass conservation," these rules are often based upon not much more than a pattern of which no viola-

[15] We refer to *free* particles. Nucleons behave completely differently when subject to the complex forces within the nucleus. There, neutrons need not decay (helium, two neutrons and two protons, is stable) and protons may decay (e.g., in carbon-11 decay, where a proton becomes a neutron plus a positron).

**Table 11.4   Conservation Rules**

| Interaction | Momentum, Energy, Angular Momentum, Charge | Lepton Numbers $(L_e, L_\mu, L_\tau)$ | Baryon Number $(B)$ | Strangeness Charm, Bottomness, Topness | Isospin $(I)$ | Parity $(P)$ | Charge Conjugation $(C)$ | Time Reversal $(T)$ |
|---|---|---|---|---|---|---|---|---|
| Strong | Yes | Yes | Yes | Yes | Yes | Yes | Yes | Yes |
| Electromagnetic | Yes | Yes | Yes | Yes | No | Yes | Yes | Yes |
| Weak | Yes | Yes | Yes | No | No | No | No | No |

tion has *yet* been observed. They undergo continual reassessment. In many cases, a conservation rule has been found to be obeyed only for a particular type of decay. For instance, strangeness and the analogously defined (see Section 11.3) **charm, bottomness,** and **topness** are found to be conserved in strong and electromagnetic interactions but not in weak ones (another example of the asymmetry in the electroweak interaction).

However, certain quantities appear to be conserved in all interactions. In addition to the usual momentum, energy, angular momentum, and charge, **lepton numbers** and **baryon number** are always conserved. Lepton number is simply the number of leptons. But there are three such numbers, one for each family ($e$-$\nu_e$, $\mu$-$\nu_\mu$, and $\tau$-$\nu_\tau$), and they are conserved independently. (See Progress and Applications, page 542, for a newly found exception.) The *electron* lepton number $L_e$ is $+1$ for the electron and electron neutrino, $-1$ for their antiparticles, and zero for all other particles. The muon lepton number $L_\mu$ and tau lepton number $L_\tau$ are defined analogously. Similarly, the baryon number $B$ is $+1$ for baryons, $-1$ for antibaryons, and zero for nonbaryons.

Let us now study particle decays and conservation rules[16] by considering interactions on the simplest possible level—the interaction of a mediating boson and a fundamental fermion. This we do via **Feynman diagrams.** These are very convenient devices, for they show what is occurring without going through a complex calculation.

## Feynman Diagrams

The basic element of a Feynman diagram is a **vertex** representing the interaction of a fundamental fermion with a mediating boson. The axes are space and time, and the paths of fermions and mediating bosons are represented, respectively, by straight and wavy lines. Figure 11.8 shows the vertices for emission of a boson by a fundamental fermion. A few points of note are as follows:

1.  In the quark–gluon vertex, the quark may change color. To conserve color in the interaction, then, the gluon carries a color-anticolor charge. In the case shown, the initial quark is red, the final green, and the gluon therefore red-antigreen.

2.  Among the quarks and leptons, only neutrinos are uncharged. Thus, all fundamental particles except neutrinos may interact electromagnetically. The strictly weak interaction of neutrinos makes them very difficult to detect.

3.  All fundamental fermions may interact via the $Z^0$ and $W^\pm$ bosons, but in the $W^\pm$ case the charge of the fermion is changed and a transmutation of particles takes place. Thus, $W^+$ emission changes a $+\frac{2}{3}$ charge quark (u, c, or t) to a $-\frac{1}{3}$ charge quark (d, s, or b), or a neutrino to the other member of its lepton pair. Emission of a $W^-$ has the opposite effect.

To obtain the vertices for absorption, we would simply draw the boson as incoming—coming into rather than leaving the vertex—and switch its charge or color-charge (i.e., if emission of a positive conserves charge, so would the absorption of a negative). Replacing all particles with their antiparticles would give the antiparticle vertices.

[16]The rules pertaining to isospin, parity, charge conjugation, and time reversal are deferred to Section 11.6.

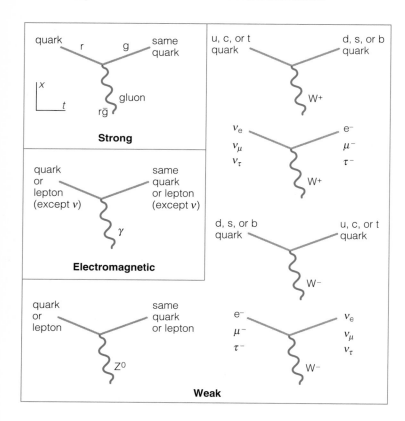

Figure 11.8  Emission vertices.

As shown in Figure 11.9, if the final fermion of an emission vertex is replaced with its antiparticle and the diagram rotated counterclockwise, a diagram for the destruction of a particle and an antiparticle results. Reversal along the time axis then represents the creation of such a pair. That it is always a particle and an antiparticle explains why $\beta^-$ decay creates an electron (particle) and *anti*neutrino (antiparticle) while $\beta^+$ decay creates a positron (antiparticle) and neutrino (particle); both maintain $L_e$ at zero.

Figure 11.10 is a simple Feynman diagram representing not a decay but the elastic interaction of two electrons. One electron emits a photon, both propagate

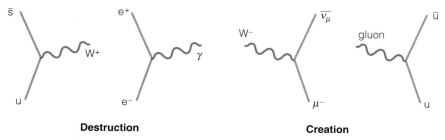

Figure 11.9  Representative destruction and creation vertices.

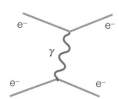

Figure 11.10  Electrons repel by exchanging a photon.

forward in time, then the photon is absorbed by the other electron. Note the similarity to the "snowball-mediated" repulsion of Figure 11.1.

Diagrams representing the various types of spontaneous decay are shown in Figure 11.11. Diagram (a) shows the $\beta$ decay of a free neutron, a weak decay because it involves the exchange of a weak boson. Diagram (b) shows the weak $\pi^-$ decay. Its lifetime of $2.6 \times 10^{-8}$ s is typical for a weak decay. (The neutron's 15-min lifetime is unusually long.) Diagrams (c) and (d) show weak $\Sigma^-$ and $\mu^+$ decays. An example of an electromagnetic decay is the decay of the $\Sigma^0$, shown in diagram (e). Involving the stronger part of the electroweak force, electromagnetic decays occur in a much shorter time than weak ones, typically $10^{-17}$ s. A gluon-mediated strong decay is the $K^{*0}$ decay shown in diagram (f). Involving a yet stronger force, strong decay lifetimes are typically only $10^{-23}$ s. Once created, a particle able to decay strongly does not survive long. Indeed, such a short-lived entity is often referred to as a **resonance** rather than a particle; its constituents "stay in tune" for scarcely longer than it takes them to pass one another at high speed. (A radius of $10^{-15}$ m divided by a speed of $v \cong 10^8$ m/s gives a time of approximately $10^{-23}$ s.) Lasting so short a time, proof of the occurrence of these

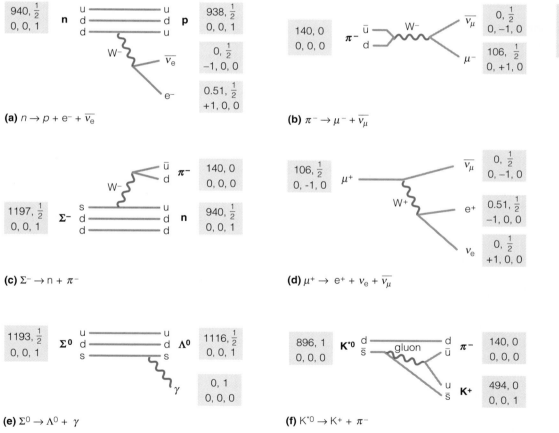

**(a)** $n \rightarrow p + e^- + \overline{v}_e$

**(b)** $\pi^- \rightarrow \mu^- + \overline{v}_\mu$

**(c)** $\Sigma^- \rightarrow n + \pi^-$

**(d)** $\mu^+ \rightarrow e^+ + v_e + \overline{v}_\mu$

**(e)** $\Sigma^0 \rightarrow \Lambda^0 + \gamma$

**(f)** $K^{*0} \rightarrow K^+ + \pi^-$

**Figure 11.11** Feyman diagrams for various spontaneous decays.

resonances must rest on indirect evidence, such as the abrupt appearance of their longer-lived decay products.

Momentum, energy, angular momentum, charge, and lepton and baryon numbers are conserved in all interactions. The values given in the shaded rectangles in Figure 11.11 illustrate how some of these properties are conserved: The final mass is always less than the initial (kinetic energy increases in a spontaneous explosion); lepton and baryon numbers are conserved; and by the usual angular momentum addition rules, the final angular momenta are able to add to the initial. Note that the *spins* don't add up in the $K^{*0}$ decay of diagram (f), in which a spin-1 particle becomes two spinless particles. But the decay is allowed because the kaon and pion depart in a state of unit *orbital* angular momentum; *total* angular momentum is still conserved.

The conservation rules for strangeness are borne out in diagrams (c), (e), and (f). While conserved in electromagnetic and strong decays, (e) and (f), strangeness is not conserved in weak decay (c). The W emission changes the strange quark to one of another kind.

---

### Example II.I

By considering the rules for quantities that are always conserved—charge, energy (i.e., mass decrease), angular momentum, and lepton and baryon numbers—indicate whether each of the proposed decays is possible. If it is not possible, indicate which rules are violated. If it is possible, indicate whether it is a strong, electromagnetic or weak decay, and sketch a Feynman diagram. (Use data from Tables 11.1 and 11.2.)

(a) $\pi^0 \rightarrow \mu^+ + e^-$

(b) $\tau^- \rightarrow \mu^- + \nu_\tau + \overline{\nu}_\mu$

(c) $\Xi^0 \rightarrow K^+ + K^-$

(d) $\Delta^+ \rightarrow n + K^+$

(e) $\overline{\Lambda^0} \rightarrow \overline{p} + \pi^+$

### Solution

Charge is conserved in all the decays.

(a) The final mass (106 MeV + 0.511 MeV) is less than the initial (135 MeV), and no baryons are involved. But the $\mu^+$ has *muon* lepton number $-1$ and the electron *electron* lepton number $+1$. These do *not* cancel, so the decay is forbidden. A spinless particle may decay into two spin-$\frac{1}{2}$ particles; spin is not a problem.

(b) The initial mass is $1.78 \times 10^3$ MeV, the final is 106 MeV, and no baryons are involved. The initial lepton numbers are $(L_e, L_\mu, L_\tau) = (0, 0, +1)$, and those of the product particles are $(0, +1, 0)$, $(0, 0, +1)$, and $(0, -1, 0)$. Lepton numbers are conserved. Three spin-$\frac{1}{2}$ particles may add up to a spin of $\frac{1}{2}$, so angular momentum may be conserved. The decay is allowed. Since neutrinos interact only via the weak force this must be a weak decay. The

Feynman diagram is shown in the margin. The boson must be a W; a Z could not "erase" the $\tau^-$.

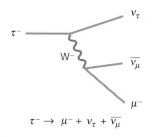

$\tau^- \rightarrow \mu^- + \nu_\tau + \overline{\nu}_\mu$

(c) This is forbidden on two accounts. First, although the final mass of 2 × 494 MeV is less than the initial 1315 MeV, and no leptons are involved, it begins with a baryon number of +1 and ends with no baryons at all. Secondly, angular momentum cannot be conserved. The initial spin is $\frac{1}{2}$, while the product particles are spinless. Orbital angular momentum, which comes in integral steps, cannot make up the difference.

(d) Since the $\Delta$ and n are both baryons, and the kaon a meson, this decay would conserve baryon number. No leptons are involved. A final orbital angular momentum might allow the spin-$\frac{3}{2}$ particle to decay to a spin-$\frac{1}{2}$ and a spinless pion. However, since the final mass of 940 MeV + 494 MeV is greater than the initial 1232 MeV, the $\Delta^+$ cannot decay this way.

(e) The initial and final masses are 1116 MeV and 938 MeV + 140 MeV. The $\overline{\Lambda}^0$ and antiproton are both of baryon number −1 and the pion has baryon number zero, so baryon number is conserved. The $\overline{\Lambda}^0$ and antiproton are spin-$\frac{1}{2}$, while the pion is spinless. This decay is allowed, and it does occur. (There is a guiding principle in particle physics that says that if a decay *can* occur—if it violates no conservation rules—then it *must* occur.) Since the $\overline{\Lambda}^0$ contains a strange quark, which is absent in the products, strangeness is not conserved. This must be a weak decay. The Feynman diagram is shown in the margin. A W boson changes the strange quark to a $\overline{u}$.

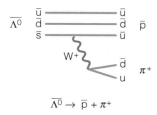

$\overline{\Lambda}^0 \rightarrow \overline{p} + \pi^+$

It should be noted that, just as some nuclei are capable of decaying in multiple ways (e.g., via $\alpha$ or $\beta^+$), many hadrons and leptons have multiple decay modes. The $\tau^-$, for example, may decay via $\tau^- \rightarrow e^- + \overline{\nu}_e + \nu_\tau$, and the $\overline{\Lambda}^0$ via $\overline{\Lambda}^0 \rightarrow \overline{p} + e^+ + \nu_e$.

## ♦ 11.6  Isospin and P, C, and T

Let us briefly study the remaining conservation rules of Table 11.4. They touch upon fundamental ideas of space and time, and raise difficult questions.

As noted in Section 11.3, isospin obeys the same addition rules as spin. Interestingly, though, hadrons have not only intrinsic isospin $I$, but an intrinsic "third component" $I_3$. The proton, for instance, with $(I, I_3) = (\frac{1}{2}, +\frac{1}{2})$, is isospin-up while the neutron, $(\frac{1}{2}, -\frac{1}{2})$, is isospin-down. Strong interactions conserve isospin, but the others do not. In the strong decay of Figure 11.11(f), $(I, I_3)$ goes from $(\frac{1}{2}, -\frac{1}{2})$ for the K$^{*0}$ to $(\frac{1}{2}, +\frac{1}{2})$ and $(1, -1)$ for the K$^+$ and $\pi^-$. In the weak decay of diagram (a), the neutron is $(\frac{1}{2}, -\frac{1}{2})$ and the proton is $(\frac{1}{2}, +\frac{1}{2})$, but electrons and neutrinos (and all other leptons) do not possess isospin, so isospin changes in the process. In the weak pion decay of diagram (b), a particle *with* isospin becomes particles without. The weak decay of the $\Sigma^-$ in diagram (c) begins with isospin $(1, -1)$ and ends with $(\frac{1}{2}, -\frac{1}{2})$ and $(1, -1)$. These do not add up, and isospin thus changes. Photons do not possess isospin, so we see that isospin changes from

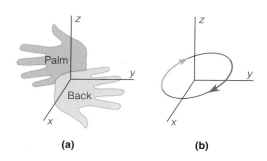

**Figure 11.12** Parity inversion changes a right hand into a left, but does not change the direction of angular momentum.

**(a)**          **(b)**

(1, 0) to (0, 0) in the electromagnetic decay of the $\Sigma^0$. (It happens, however, that the third component of isospin is conserved in electromagnetic interactions.)

**Parity inversion** changes the sign of all spatial coordinates: $(x, y, z) \rightarrow (-x, -y, -z)$. Among other things, it changes a right hand into a left hand, as shown in Figure 11.12(a). Strong and electromagnetic interactions appear identical from a parity-inverted view, but not the weak interaction. The simplest example is $\beta$ decay, whose parity violation we see by combining two observations. The first is that if its coordinates at all points in time changed according to $(x, y, z) \rightarrow (-x, -y, -z)$, a point particle orbiting in a clockwise sense about the $z$-axis would still orbit clockwise, Figure 11.12(b). By the same token, the direction of spin (intrinsic angular momentum) is unaffected by a parity inversion. The second observation has to do with a special property of neutrinos.

Massless particles may possess definite **helicity.** Helicity refers to the orientation of a particle's spin relative to its momentum. Electrons, for instance, do not have a definite helicity. Suppose an electron moves in the positive $z$-direction at 500 m/s with spin up, that is, spin and momentum aligned. In a frame moving at 1000 m/s in the positive $z$-direction relative to the first, the electron is moving at 500 m/s in the *negative* direction and is spin-up, that is, spin and momentum antialigned. The spin and momentum of the same electron can be either aligned or antialigned depending on the frame of reference. A massless particle, on the other hand, moves at $c$ in any frame. Since it cannot be overtaken, if its spin and momentum are aligned in one frame, they must be aligned in all frames. Although recent evidence suggests that neutrinos have a small mass (see Progress and Applications), it appears that they always have their spin and momentum antialigned (Figure 11.13). They are said to be left-handed, meaning that they "rotate" counterclockwise when viewed along their direction of motion, as does a left-hand screw. Antineutrinos are always right-handed, rotating clockwise when viewed along their direction of motion, as does a right-hand screw—spin and momentum are aligned.

Let us now combine our two observations. Parity inversion, changing the direction of motion but not the spin, would change a right-handed antineutrino into a left-handed one—which does not exist! Similarly, a left-handed neutrino would become a nonexistent right-handed neutrino. Since it is via the weak interaction alone that neutrinos are created, we say that *the weak interaction is not invariant under a parity inversion.*

As parity inversion replaces all spatial coordinates with their opposite, **charge conjugation** replaces all particles with their antiparticles. We begin to see its role by "applying" charge conjugation in a special case: a $\beta^-$ decay in which the

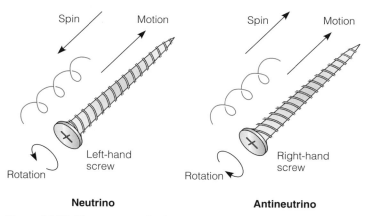

**Figure 11.13** Momentum and spin are opposite for a neutrino and parallel for an antineutrino.

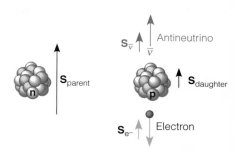

**Figure 11.14** A nucleus losing one unit of spin in $\beta$ decay.

antineutrino and electron move in opposite directions and carry one unit of angular momentum ($\hbar$) away from the nucleus, shown in Figure 11.14. To conserve angular momentum, the antineutrino and electron spins, each $\frac{1}{2}$, must both be in the direction of the parent's spin. However, since antineutrinos are always right-handed, this requires that the antineutrino be the particle that actually moves in that direction.[17] We already know what parity inversion would do. Shown in Figure 11.15, it would yield a nonexistent left-handed antineutrino. Charge conjugation (including exchanging protons for antiprotons and neutrons for antineutrons in the nucleus) would change the right-handed antineutrino into a right-handed neutrino—something else that doesn't exist. Thus, it may also be said that *the weak interaction is not invariant under charge conjugation.* Again it differs from the strong and electromagnetic interactions, for which replacing particles with antiparticles yields processes that can and do occur.

Figure 11.15 also shows the *combined* operation of parity inversion and charge conjugation, changing a right-handed antineutrino into a left-handed neutrino. Interestingly, the result is allowed. Thus, the weak interaction seems to be invariant under the product of the two. This is *almost* always the case. But there is a famous counterexample: In 0.3% of instances, the $K_L^0$ meson decays weakly in a way that is not invariant under the product of charge conjugation and parity inversion. Although the effect was discovered in 1964 (Nobel prize 1980; James Cronin and Val Fitch), its true nature is one of the persistent mysteries of physics.

Finally, **time reversal** replaces *t* with $-t$. In effect, to view the time-reversed result of a given process would be to view a motion picture of the process run backward. *Almost* without exception, microscopic processes are time-reversal invariant. They obey the same laws of physics whether time moves forward or backward. A general result of quantum field theory is the **CPT theorem,** which states that *under the combined operations of charge conjugation (C), parity inversion (P), and time reversal (T), all interactions are invariant.* Thus, if processes are invariant under the combined operation *CP,* they *must* be invariant under *T.* On the other hand, if a process is not invariant under *CP,* it cannot be under *T,* and the $K_L^0$ decay is therefore an exception to microscopic time-reversal invariance. The

[17]This existence of a preferred direction for beta emission—that is, opposite the nuclear spin—was the most striking early evidence of the violation of invariance under a parity inversion. It was first demonstrated in 1957 by C. S. Wu et al., studying the $\beta^-$ decay of Co-60 to Ni-60.

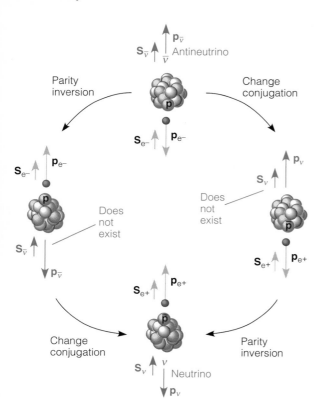

**Figure 11.15** Applied to $\beta$ decay, neither parity inversion nor charge conjugation alone gives a process that occurs—but the combination does.

belief that microscopic laws should be obeyed equally well backward as forward in time is a strong one in physicists, so this enigmatic violation is particularly troubling.

## 11.7  Unified Theories

The low-energy symmetry-breaking responsible for the schism in the electroweak force may shed light on the pattern for the unification of *all* forces. It is believed by many physicists that the strong and electroweak interactions are really asymmetric manifestations of the same interaction, but that their equivalence becomes apparent only at particle energies even higher than the level at which the electromagnetic and weak converge. Similarly, this unified interaction may converge with the gravitational interaction at yet higher energies.

A theory in which the strong and electroweak forces appear as one is known as a **grand unified theory,** or **GUT.** At present, various competing theories are being pursued. To establish its validity, a new theory must, of course, agree with existing evidence, but it must also either explain the previously unexplainable or make new predictions that can be experimentally verified. Present GUTs have not been extremely successful in either regard, so they remain "mere" theories.

One of the most provocative predictions of GUTs is proton decay. As we know, conservation of baryon number is presently accepted as applying to all decays. If true, the proton must be stable, for there is no lighter baryon into which it might decay. But in GUTs, the barrier between quark and lepton is broken down, and the proton is liable to decay into electrons plus pions, photons, or other light particles. However, the predicted lifetime is invariably greater than $10^{30}$ s, or $10^{22}$ years. To put this in perspective, many cosmologists speculate that the age of the universe is $10^{10}$ years. Thus, to have any hope of succeeding, experiments dedicated to finding proton decay have employed gigantic quantities of matter, are maintained for many years, and are buried deep below the ground, to reduce cosmic ray "noise." Several throughout the world have been in continuous operation since the early 1980s. Despite the hopes of theorists and vigilance of experimentalists, no proton decay has yet been substantiated. Rather, the value for the minimum possible lifetime of the proton has progressively increased—now known to be greater than $10^{32}$ s. Another prediction of GUTs is the existence of magnetic monopoles, which should be very massive. Efforts to find these have also failed. Nevertheless, buoyed by similarities between them, the quest to unify the strong and electroweak forces continues. As the yet mysterious asymmetry that breaks the electroweak into electromagnetic and weak parts is expected to disappear at particle energies above about 1 TeV, it is believed that the strong and electroweak forces converge at around $10^{13}$ TeV (see Figure 11.16). Note that this is a fantastically high energy—a megajoule (a car at 100 mph) *per elementary particle!*

Recently, the prospect of unifying gravity with the other forces, long gloomy, has brightened somewhat. Still, it is a particularly tough nut to crack, because it is weak yet it pulls on *everything*—all forms of energy, including its own potential energy. To give some measure of the difficulty, Einstein himself, driven by tantalizing similarities, spent much of his later years attempting to unify gravitation with electromagnetism—without success. While quantum gravity still lacks experimental evidence such as the graviton, recent theoretical advances offer a reasonable hope of future success (see Progress and Applications). A theory that goes beyond a GUT, by also incorporating gravity, is often referred to as a "theory of everything." At what particle energy might gravitation unite with the strong-electroweak force? There is much speculation. Perhaps $10^{16}$ TeV.

## Back to Cosmology

We now come full circle, and return to a topic introduced in Chapter 1. Cosmology—the study of the behavior and evolution of the universe—is not only related to general relativity, but also to unified theories. There is good reason: The equivalence of the fundamental interactions is expected to be manifest under conditions of extremely high particle energies, and cosmological evidence suggests that our universe is expanding from an ancient cataclysmic explosion in which it was so small and the energy density so high that this very condition existed.

Because gravity is the only fundamental interaction that is both long range and of one sign (attractive only), it is by far the most important force in the motion of the things we see in the heavens today—planets, stars, galaxies, and so on. But things would have been quite different in the early moments following a Big Bang. At such an unimaginably high energy-density and temperature, one superunified

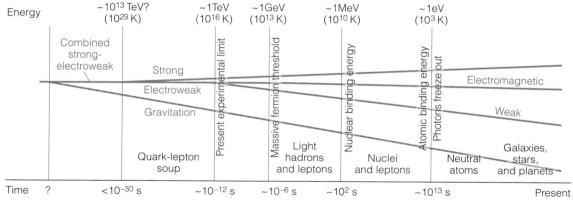

**Figure 11.16** The genealogy of our forces?

force would have governed all. Only later, with decreasing density and lower average particle energy, could asymmetries creep into the picture. Symmetries might be broken, and the one force manifest itself as many.

Although cosmologists still debate about many of the steps, particularly at times earlier than the first microsecond, there is some consensus on a rough chronology of events beginning with the Big Bang. Immediately after $t = 0$, the universe was likely just an angry firestorm of high-energy fundamental particles interacting at a feverish pace via their lone unified interaction. It would have been unimaginably small—orders of magnitude smaller than a nucleon. Predictions about the universe in such a state are necessarily the most speculative of all. In a very short time, average particle energies would have so diminished that gravitation would sever its relationship with the strong-electroweak force, striking out on its own at an early age. Later, when typical energies drop to approximately $10^{13}$ TeV, the strong and electroweak forces would go their separate ways. Even so, energies would still be high enough to ensure an abundance of all three tiers of fundamental fermions—that is, the least massive tier (u, d, e, $\nu_e$), the middle (s, c, $\mu$, $\nu_\mu$), and the heaviest tier (b, t, $\tau$, $\nu_\tau$). This compact hot mixture is often called the **quark-lepton soup.**

By perhaps $10^{-12}$ s, average energies would have dropped to approximately 1 TeV, whereupon the electromagnetic and weak forces would part company, thus beginning the four-force, asymmetric behavior we observe today. From this point onward, roughly at the limit of experimentally attainable energies, the evidence is more clear. At around 1 $\mu$s and $\sim$1 GeV, the heavier fermions would have become increasingly scarce, for they are liable to decay to lighter particles, and available energies would be insufficient to renew the supply. Below this limit, the universe would consist mostly of nucleons and light leptons. After several minutes, average energy would have dropped to $\sim$1 MeV. Since this is roughly the binding energy per nucleon, nuclei would form.

Atoms, however, could not have formed until the average energy dropped to the limit of *atomic* binding, a matter of electronvolts, at around $10^{13}$ s, or 500,000 years. This point, when the temperature was approximately 3000 K, marks a significant event in the history of the universe, for it can be readily verified by experiment. Photons, which had been participating in energy exchange with the cooling plasma of separated *charged* particles, would no longer interact vigorously with *neutral* atoms. It is said that the photons were "frozen out." The weaker, infrequent remaining interactions would no longer have been able to keep them in equilibrium with the neutral matter. Thus, the photons embarked upon their random, independent journeys, contributing to the universe a blackbody spectrum characteristic of the 3000 K temperature at which they ceased to interact. This is known as the **cosmic background radiation.**

Such radiation seen today was produced in equilibrium with the plasma very far away, very long ago. But the universe was expanding even then, and this distant plasma would have been receding from Earth at a significant fraction of *c*. Accordingly, the radiation received now is Doppler red-shifted; it has the equilibrium blackbody spectrum of an object at only 2.7 K. The agreement between experimental observations and the theory of an expanding universe is excellent, and the discovery of the cosmic background (1978 Nobel prize; A. A. Penzias and R. W. Wilson) was thus a great step forward in understanding the history of the universe.

## Conclusion

At present, we have a fairly pleasing view of fundamental particles and interactions: six quarks and six leptons, and three forces seemingly destined to reveal themselves as one. However, given the lessons of physics past—with blunders like the predictions of virtual stagnation made near the end of the nineteenth century—it would be folly to discount the possibility of a watershed discovery that would turn this pleasant view upside-down. Perhaps there are particles more fundamental than quarks and leptons. Perhaps interactions yet unimagined will be found. These possibilities might well be tied to enduring problems in our present view: Just how does the asymmetry in the electroweak interaction come about? Will a theory encompassing all forces be established, and what will it tell us about the universe? Why do the fundamental fermions have the charges and masses that they do? Indeed, why do we have all these particles?

## Progress and Applications

- While we continue to investigate the unification of the strong and electroweak forces, work proceeds toward fully characterizing them separately. The Higgs boson is the next expected "big discovery" in particle physics and anticipation builds as the effort accelerates, for with it would come much clarification of the electroweak force. The strong force is also still a mystery of sorts. Among the difficulties is that, while studies intended to produce a clear, self-consistent picture at the highest energies, where quark-gluon interactions predominate, meet with considerable success, clarifying the transition to lower energies of hadron binding has yet proved elusive.

- A persistent puzzle in high-energy physics is the so-called **spin crisis.** Although its three spin-$\frac{1}{2}$ quarks contribute to the overall spin of the proton, experimental evidence indicates they account for less than half. Plausible candidates for making up the difference include orbital motion, angular momenta of the spin-1 gluons, and hidden quark–antiquark pairs appearing and disappearing within the stew of quarks and gluons, but the question is still open.

- The search for neutrino mass, important to predicting the future of our expanding universe, centers now on attempts to observe **neutrino oscillations.** If an electron neutrino may change back and forth (oscillate) into a mu- or tau-neutrino, and vice versa, then neutrinos must have mass. Specially designed experiments are scouring neutrino "events" for evidence of oscillations. The statistics show a considerable likelihood, but the evidence is not yet conclusive.

■ The search for neutrino mass has in recent years centered on finding **neutrino oscillations**. If an electron-neutrino can change back and forth (oscillate) into a mu-neutrino or tau-neutrino, and vice versa, then neutrinos must have mass. Specially designed experiments have been scouring neutrino "events" for oscillations, and success finally seems to have been achieved. In June 1998, scientists working at the Super Kamiokande neutrino detector (Figure 11.17), located a kilometer underground in Japan and operated jointly by Japan and the United States, reported the first convincing evidence for neutrino oscillations. Unfortunately, while the data gives a rough value for the *difference* between the electron- and mu-neutrino masses (implying that at least one kind of neutrino has mass), it does not yield actual masses. Still, the discovery is of great importance to particle physics. One implication is that the three lepton numbers are not strictly conserved independently: another is that neutrinos don't have definite helicity (though right- versus left-handedness still plays a role, via the related property of chirality). Moreover, the discovery may solve the **solar neutrino problem**. Existing theories predict one value for the electron-neutrino flux reaching Earth from the sun, but the actual value is much smaller. The deficiency may be due to electron-neutrinos changing along the way into mu-neutrinos, which escape detection. Neutrino mass may also answer the cosmological question of whether our universe will or will not continue to expand forever.

■ The $K_L^0$ meson's violation of time-reversal symmetry has been known for decades. It is now widely believed that oscillations between particle and antiparticle are at the root of the phenomenon. Yet, given the profoundness of the very idea, progress in its understanding has been painfully slow. Recently, though,

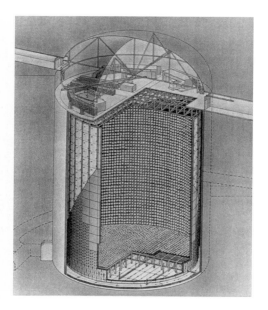

**Figure 11.17** The 1-km-deep Super Kamiokande neutrino detector in Japan, 40 m in diameter and over 50 m tall.

time-reversal asymmetry has been studied from new directions. One involves searching for electric dipole moments in certain particles. For instance, if the electron were found to have an intrinsic electric dipole moment, both parity and time reversal would be violated. Another route looks for CP-violation in the decays of $B^0$ ($b\bar{d}$) and $\overline{B}^0$ mesons, which should behave much the same as neutral K mesons, but with fewer unrelated decays to obscure the results. Until these studies progress further, however, we are still left with only the one conclusive example of CP-violation.

- In the 1980s, several seemingly different **string theories** gained much popularity in attempts to combine gravity with the other forces, with modest success. In early string theories, the fundamental particles of the universe were extremely short one-dimensional "strings." More recently it has been found that all string theories could be understood as parts of a larger, all-encompassing **M-theory,** in which higher-dimensional "membranes" are the fundamental stuff of which all things are made. Still they would be extraordinarily small, on the order of $10^{-33}$ m, the so-called **Planck length.** Among the predictions of M-theory is **supersymmetry,** a theory that had already been put forth and was garnering much attention in the grand unification scheme. The symmetry in supersymmetry is that for each fundamental fermion there would be a bosonic supersymmetric partner, and vice versa. For the electron there would be the (boson) selectron, for each quark a (boson) squark, for the photon the (fermion) photino, for the graviton the (fermion) gravitino, and so forth. In this scheme, there would also be multiple Higgs particles. The great appeal of supersymmetry is that grand unification would be a built-in consequence. Alas, no supersymmetric partners have yet been found.

## Summary

The study of fundamental particles and their interactions, known as "particle" or "high-energy" physics, is the quest to understand the physical universe on the most basic level possible. At this level, interactions/forces between particles occur in discrete events: Fundamental fermions share a force when one emits a mediating boson, or field quantum, which is then absorbed by another.

For each fundamental particle there is an antiparticle, which has the same mass and spin as the particle but is of opposite charge. Antiparticles are an experimental fact, and a theoretical prediction of relativistic quantum mechanics.

At present, the standard model of particle physics recognizes three fundamental forces. Each involves a property required to participate in that force, and each has its own mediating boson(s). In order of decreasing strength, they are the strong force, the electroweak force, and the gravitational force.

To engage in the strong force requires the property of color-charge, which comes in three kinds: red, green, and blue (plus anticolors). The fundamental fermions with this property are called quarks, and come in three pairs: the u and d, the c and s, and the t and b. The mediating boson is the gluon. Separated individual quarks have never been found. The smallest gluon-bound units are quarks–antiquark mesons and three-quark baryons, known collectively as hadrons. Hadrons are color-neutral; mesons are a color-anticolor pair, while baryons comprise one quark of each color. The proton and neutron are baryons, of quark content uud and udd, respectively. In a nucleus, all nucleons attract each other via a coarse manifestation of the strong force known as the residual strong force.

The property of charge/weak-charge must be possessed for a particle to engage in the electroweak force. Besides quarks, leptons possess this property. There are six known leptons, in three pairs: $(e^-, \nu_e)$, $(\mu^-, \nu_\mu)$, and $(\tau^-, \nu_\tau)$. There are four mediating bosons: the massless photon, associated with the electromagnetic part, and three very massive weak bosons $W^\pm$ and $Z^0$, associated with the weak part. The asymmetry between the two parts of the interaction is not yet fully understood, but it is expected to disappear at particle energies somewhat above approximately 1 TeV.

Gravitation is an attractive force that pulls on all energy and so is felt by all particles. The mediating boson, the graviton, has not yet been found. Although its long-range attraction is of primary importance in the cosmological realm, gravity's extreme weakness among the fundamental forces makes detection of quantum gravitational effects very difficult.

Many physicists believe that today's three forces will one day be revealed as merely different aspects of one fundamental interaction. The unification is expected to become apparent at very high particle energies. Much work has yet to be done in developing a theory of unified forces, and data gained from experiments at higher energies than presently attainable will be crucial. Help may come from the area of cosmology—the study of the heavens—for it is believed by many that the universe began with a "bang" so energetic as to realize the unification of the fundamental interactions.

## EXERCISES

### Section 11.1

1. From the experimental evidence that the force between nucleons has a range of about 1 fm, obtain a rough value (in MeV/$c^2$) for the mass of the particle exchanged to convey this force, the pion.

2. From the masses of the weak bosons given in Table 11.1, show that the range of the weak part of the electroweak force should be about $10^{-3}$ fm.

3. The electrostatic force between two charges of $+q$ a distance $L$ apart is equal to that between two charges of $+2q$ a distance $2L$ apart. What would be a typical energy of a virtual photon in each case? If these energies are not equal, how can the forces be equal?

4. Charges of $+q$ and $-q$ are separated by a distance $L$.
(a) What would be a typical energy of a virtual photon they exchange? To what wavelength does this correspond? (b) Evaluate these for a separation of 1 m.

### Section 11.2

5. To show that the Klein–Gordon equation has valid solutions for negative values of $E$, verify that equation (11-4) is satisfied by a wave function of the form

$$\Psi(x,\ t) = A \exp\left[i\left(\frac{\pm|p|}{\hbar}x - \frac{\pm|E|}{\hbar}t\right)\right]$$

6. (a) Show that $\Psi_1(x, t) = Ae^{ikx - i\omega t}$ is a solution of both the Klein–Gordon and the Schrödinger equations.
(b) Show that $\Psi_2(x, t) = Ae^{ikx}\cos(\omega t)$ is a solution of the Klein–Gordon but not of the Schrödinger equation.
(c) Show that $\Psi_2$ is a combination of positive and negative energy solutions of the Klein–Gordon equation (see Exercise 5). (d) Compare the time dependence of $|\Psi|^2$ for $\Psi_1$ and $\Psi_2$.

7. For solutions of the Klein–Gordon equation, the quantity

$$i\Psi^* \frac{\partial}{\partial t}\Psi - i\Psi\frac{\partial}{\partial t}\Psi^*$$

is interpreted as charge-density. Show that for a positive-energy plane-wave solution (see Exercise 5) it is a real constant and for a negative-energy plane-wave solution it is the negative of that constant.

8. In nonrelativistic quantum mechanics, governed by the Schrödinger equation, the total probability of finding a particle does not change with time.

(a) Prove it! Begin with the time derivative of the total probability:

$$\frac{d}{dt} \int \Psi^*(x,\ t)\Psi(x,\ t)\ dx$$

$$= \int \left( \Psi(x,\ t)\frac{\partial}{\partial t}\Psi^*(x,\ t)\ +\ \Psi^*(x,\ t)\frac{\partial}{\partial t}\Psi(x,\ t) \right) dx$$

Then use the Schrödinger equation to eliminate the partial time derivatives in favor of partial spatial derivatives, integrate by parts, and show that the result is zero. (*Note:* It is assumed that the particle described is reasonably well localized, so that $\Psi$ and $\partial\Psi/\partial x$ are zero when evaluated at $\pm\infty$.)

(b) May this procedure be used to draw the same conclusion if the wave function must obey the Klein–Gordon rather than the Schrödinger equation? Why or why not?

## Section 11.3

9. For which particles does Table 11.2 show both particle and antiparticle? Which particles are their own antiparticle?

10. We have noted that photons cannot interact with photons but that gluons can interact with gluons. (a) Can the $W^\pm$ weak bosons self-interact? Explain. (b) Should gravitons self-interact? Explain.

## Section 11.4

11. The classical magnetic force formula $F = qvB$ is correct relativistically. But if a magnetic field is to keep a high-energy charged particle moving in a circle, it must satisfy the relativistically correct relationship between force and centripetal acceleration, $F = \gamma_v m(v^2/r)$.
(a) To keep a 1-TeV proton in a 1-km-radius circle, as is done at the Tevatron, how strong must be the magnetic field? (b) How large would the radius have to be for magnets of the same strength to keep a 20-TeV proton in a circle?

## Section 11.5

12. (a) What is the quark content of the antineutron? (b) Sketch the Feyman diagram for its $\beta$ decay.

**In the following exercises, indicate whether the proposed decay is possible. If it is not possible, indicate which rules are violated. Consider only those obeyed in all interactions: charge, energy, angular momentum, and lepton and baryon numbers. If it is possible, indicate whether it is a strong, electromag-netic, or weak decay, and sketch a Feynman diagram.**

13. $\tau^+ \rightarrow e^+ + \nu_e + \overline{\nu}_\tau$
14. $\mu^+ \rightarrow e^+ + \nu_e$
15. $\overline{n} \rightarrow p + e^- + \overline{\nu}_e$
16. $\Xi^- \rightarrow \Lambda^0 + \pi^-$
17. $\Xi^{*-} \rightarrow \Sigma^0 + e^- + \nu_e$
18. $\Lambda^0 \rightarrow p + \pi^-$
19. $K^+ \rightarrow \mu^+ + \overline{\nu}_\mu$
20. $\Omega^- \rightarrow \Xi^0 + \pi^-$
21. $\Delta^+ \rightarrow \Sigma^+ + \pi^0 + \gamma$
22. $\pi^0 \rightarrow \mu^- + \overline{\nu}_\mu$

**In the following exercises, two protons are smashed together in an attempt to convert kinetic energy into mass—and new particles. Indicate whether the proposed reaction is possible. If it is not possible, indicate which rules are violated. (Consider only those for charge, angular momentum, and baryon number.) If it is possible, calculate the minimum kinetic energy required of the colliding protons.**

23. $p + p \rightarrow p + p + p + p$
24. $p + p \rightarrow p + p + p + \overline{p}$
25. $p + p \rightarrow p + p + n + \overline{n}$
26. $p + p \rightarrow p + \overline{p} + n + \overline{n}$
27. $p + p \rightarrow p + p + \pi^+ + \pi^-$
28. $p + p \rightarrow p + K^+$
29. $p + p \rightarrow p + K^+ + \Lambda^0$
30. $p + p \rightarrow p + \Sigma^+ + K^0$

## Section 11.6

31. You are a promising young theoretical physicist who does not believe that gravity is a distinct fundamental force separate unto itself, but is instead related to the other forces by an all-encompassing relativistic, quantum-mechanical theory. In particular, you do not believe that the universal gravitational constant $G$ is really one of nature's elite set of fundamental constants. Rather, you believe that $G$ can be *derived* from more-basic constants: the fundamental constant of quantum mechanics, Planck's constant $h$; the fundamental speed limiting the propagation of any force, the speed of light $c$; and one other—a fundamental length $\ell$, important to the one unified force. Using simply dimensional analysis, find a formula for $G$, then an order-of-magnitude value for the fundamental length. (*Note:* This is known as the Planck length, and is indeed an element in M-theory, intended to unite gravity with the other forces.)

# The Michelson–Morley Experiment

The Michelson–Morley experiment was intended to determine the speed at which Earth in its orbital motion moves through the aether, the medium of light propagation *presumably* permeating space. It was based on the fact that if light were constrained to move through an aether at a fixed speed, while the aether moves relative to Earth, round-trips parallel and perpendicular to the aether would take different amounts of time. Measuring the time difference would then give the aether's speed relative to Earth. The analysis is identical to that in a situation often considered in classical physics.

Consider a swimmer who always swims relative to water at speed $v_s$, swimming in a river (the medium) flowing at speed $v_r$ relative to the ground. An observer on a bridge overhead watches the swimmer perform two different tasks: (1) a round-trip from just below the observer, upstream a distance of $L$, then back downstream to the point under the observer, and (2) a round-trip from just below the observer, a distance $L$ perpendicular to the bank, then in the opposite direction back to the point under the observer. For both legs of both round-trips, the velocity of the swimmer relative to the ground is the vector sum of the velocities of the swimmer relative to the river and of the river relative to the ground. From Figure A.1 we see that the results of these four vector additions are

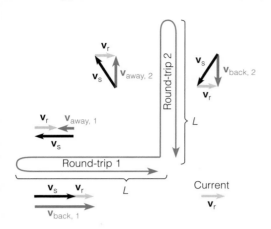

**Figure A.1** Perpendicular round-trips through a moving medium.

**Round-trip 1:**   $v_{away,1} = v_s - v_r$       $v_{back,1} = v_s + v_r$

**Round-trip 2:**   $v_{away,2} = \sqrt{v_s^2 - v_r^2}$       $v_{back,2} = \sqrt{v_s^2 - v_r^2}$

In both cases, the time required for a leg of the round-trip is the distance divided by the speed. The total times are thus

**Round-trip 1:**   $t_{\parallel} = \dfrac{L}{v_s - v_r} + \dfrac{L}{v_s + v_r} = \dfrac{2Lv_s}{v_s^2 - v_r^2}$     **(A-1)**

**Round-trip 2:**   $t_{\perp} = \dfrac{L}{\sqrt{v_s^2 - v_r^2}} + \dfrac{L}{\sqrt{v_s^2 - v_r^2}} = \dfrac{2L\sqrt{v_s^2 - v_r^2}}{v_s^2 - v_r^2}$     **(A-2)**

Since $v_s > \sqrt{v_s^2 - v_r^2}$, we see that round-trip 1 takes more time than round-trip 2.

Michelson and Morley applied the same analysis to light. If Earth in its orbital motion about the Sun passes through the aether that permeates space, then the aether moves relative to Earth. Thus, Earth is analogous to the bridge and the aether to the river. If there really is a medium for light, an aether, a light beam would always have to move through it at $c$, so the light beam is analogous to the swimmer.

Speed of light beam (swimmer) relative to aether (river):   $v_s \rightarrow c$

Speed of aether (river) relative to Earth (ground):   $v_r \rightarrow v_{aether}$

With these replacements, equations (A-1) and (A-2) give the time difference between round-trips of light beams on Earth:

$$\Delta t = \frac{2Lc}{c^2 - v_{aether}^2} - \frac{2L\sqrt{c^2 - v_{aether}^2}}{c^2 - v_{aether}^2} \qquad \textbf{(A-3)}$$

At this point, the reader may wonder why we bother with *two* round-trips; if $L$ and $v_s$ ($c$) are known and a round-trip time could be measured, either (A-1) or (A-2) alone would suffice to determine $v_r$ ($v_{aether}$). A far-fetched but useful example illustrates the difficulty: If the speed $v_r$ of the river current were 1 m/s, but the swimmer swims through water at $v_s = 1000$ m/s, the round-trip time would differ very little from what it would be if the river's speed were zero; considerable precision would be needed to distinguish between a current of 1 m/s and of 0 m/s (about 1 part in $10^6$). In the case of light, the swimmer is a light beam and the river current is the aether moving past Earth at earth's orbital speed. Because the "swimmer swims" roughly 10,000 times faster than the current, very great precision would be required. A round-trip of a light beam on Earth would be very short in any case, so a slight increase caused by Earth's motion would have been beyond the capabilities of early instruments to measure. *Timing* two different round-trips would not solve the problem, since both times would still be minuscule. However, having light beams travel two slightly different round trips is the perfect application of the Michelson interferometer. By observing *interference* between the beams, exceptionally small differences may be detected.

In an interferometer, depicted in Figure A.2, a light beam is split in two by a half-silvered mirror, the beams take different but equal-length paths, are reunited at the half-silvered mirror, and the interference between them is then observed. If the beams travel at the same speed, constructive interference is observed. But if the beams move at different speeds, they will take different times to traverse the distance $2L$ through the apparatus, and will have passed through a different number of periods; a different interference condition will generally result. Were there an aether moving relative to Earth, an interferometer would have a "slow arm" and a "fast arm" for round-trips of light. Equation (A-3) would give the difference between the two times, and from this could be calculated the number of fringes (cycles of light and dark) the pattern should be shifted from an equal-times condition. However, we can't observe the interference condition and then ask that the world be slowly brought to a halt, so as to count the number of fringes until the equal-times condition is reached.

The trick is to begin with the interferometer oriented, as shown in Figure A.2, with one slow arm and one fast arm, observe the initial interference condition, then rotate the interferometer by 90°, meanwhile counting fringes. The interference condition would start at some arbitrary state, pass through a symmetric, equal-times condition, in which the aether "flows" at 45° to both arms, then continue to a final interference condition in which the former slow arm becomes the fast arm, and vice versa. The number of fringes counted should then correspond to *twice* the difference in times between the two round-trips. The number of fringes (cycles) that the asymmetry would shift the interference condition from

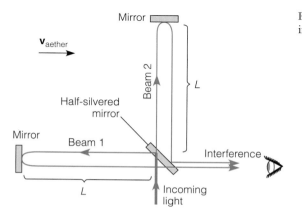

**Figure A.2**  The Michelson interferometer.

the *equal* times constructive condition is found simply by dividing the time difference (A-3) by the period $T$ of the light wave.

$$\text{Number of fringes}_{\text{from equal times}} = \frac{2L(c - \sqrt{c^2 - v_{\text{aether}}^2})}{T(c^2 - v_{\text{aether}}^2)}$$

Thus, when the apparatus is rotated 90°, the number of fringes seen should be

$$\text{Number of fringes}_{\text{through 90°}} = \frac{4L(c - \sqrt{c^2 - v_{\text{aether}}^2})}{T(c^2 - v_{\text{aether}}^2)} = \frac{4L\left[1 - \sqrt{1 - \dfrac{v_{\text{aether}}^2}{c^2}}\right]}{Tc\left[1 - \dfrac{v_{\text{aether}}^2}{c^2}\right]}$$

In the original Michelson–Morley experiment, $L$ was approximately 11 m and the wavelength 550 nm, corresponding to a period of $1.83 \times 10^{-15}$ s. The orbital speed of Earth (the presumed speed of the aether relative to Earth) is approximately $3.0 \times 10^4$ m/s. Thus,[1]

$$\text{Number of fringes}_{\text{through 90°}} =$$

$$\frac{4(11 \text{ m})\left[1 - \sqrt{1 - \dfrac{(3 \times 10^4)^2}{(3 \times 10^8)^2}}\right]}{(1.83 \times 10^{-15} \text{ s})(3 \times 10^8 \text{ m/s})\left[1 - \dfrac{(3 \times 10^4)^2}{(3 \times 10^8)^2}\right]} = 0.4$$

Perhaps 40% of a fringe does not seem like much, but it should definitely have been detectable! The finding of Michelson and Morley as the apparatus was rotated 90° was *no fringe shift whatsoever*. This is what would be observed if $v_{\text{aether}}$ were zero, if the "river" were not in motion! Hence the mystery: If aether exists and permeates space, why is it not in motion relative to the orbiting Earth?

---

[1] Because $3 \times 10^4$, Earth's orbital speed, is so much smaller than $3 \times 10^8$, the speed of light, the calculation requires either a calculator of considerable precision or the approximation of equation (1–14), discussed in Example 1.3.

# Lorentz Transformation: Plotting Events

To gain more insight into the mathematical nature of the Lorentz transformation, let us consider a related topic: rotation of axes. Suppose we have two reference frames, $F$ and $F'$. Neither frame is moving relative to the other; they differ only in that frame $F'$ is rotated by an angle $\theta$ about the $z$-axis relative to frame $F$. Figure B.1 shows the two frames. While perhaps not seeming terribly significant at this point, note that lines of constant $y$ ($y'$) are parallel to the $x$ ($x'$) axis and lines of constant $x$ ($x'$) are parallel to the $y$ ($y'$) axis. A little thought shows that this must always be the case: If a line of constant $y$, where $y \neq 0$, were *not* parallel to the $x$-axis, it would at some point cross that axis, *which defines $y = 0$.* Nonsense!

Now consider an arbitrary two-dimensional position vector **r.** It has different components in the two frames, either $x$ and $y$ or $x'$ and $y'$, but there are two points of particular interest: (1) $x$ and $y$ are combinations of $x'$ and $y'$, and vice versa, and (2) the length of the vector is the same in both frames. Figure B.2 illustrates the first point. From the diagram,

$$x = x' \cos \theta - y' \sin \theta \qquad y = y' \cos \theta + x' \sin \theta \qquad \text{(B-1a)}$$

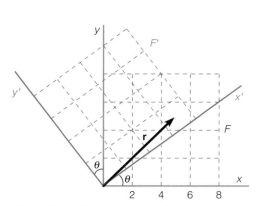

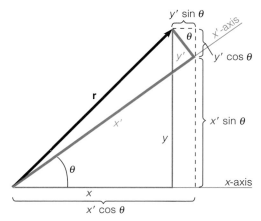

**Figure B.1** A frame rotation and a vector **r**, whose length is invariant.

**Figure B.2** The components of a vector in two frames.

or, written another way,

$$\begin{bmatrix} x \\ y \end{bmatrix} = \begin{bmatrix} \cos\theta & -\sin\theta \\ \sin\theta & \cos\theta \end{bmatrix} \begin{bmatrix} x' \\ y' \end{bmatrix}$$    (B-1b)

The second point becomes clear by squaring and adding both sides of equations (B-1a), yielding $x^2 + y^2 = x'^2 + y'^2$; the length of **r** is invariant. In particular, the vector shown in Figure B.1 has components $(x, y) = (5, 5)$ and $(x', y') = (7, 1)$, for a length of $\sqrt{50}$ in both frames.

Let us now see what a plot of the $x'$-axis would be in terms of $x$ and $y$. It would be all points for which $y' = 0$. Inserting $y' = 0$ in equation (B-1a),

$$x = x' \cos\theta \quad \text{and} \quad y = x' \sin\theta \quad \Rightarrow \quad \frac{y}{x} = \frac{x' \sin\theta}{x' \cos\theta} \quad \Rightarrow \quad y = \tan\theta \, x$$

This is a straight line with positive slope. Similarly, a plot of the $y'$-axis would be all points for which $x' = 0$. Again using (B-1a),

$$x = -y' \sin\theta \quad \text{and} \quad y = y' \cos\theta \quad \Rightarrow \quad \frac{y}{x} = \frac{y' \cos\theta}{-y' \sin\theta} \quad \Rightarrow \quad y = -\cot\theta \, x$$

This is a straight line with negative slope. These results are borne out in Figure B.1; the $x'$-axis indeed has positive slope, and the $y'$-axis negative slope.

Now consider the Lorentz transformation equations (1-9).

$$x = \gamma_v x' + \left(\gamma_v \frac{v}{c}\right)(ct') \qquad ct = \left(\gamma_v \frac{v}{c}\right)x' + \gamma_v(ct')$$    (B-2a)

or

$$\begin{bmatrix} x \\ ct \end{bmatrix} = \begin{bmatrix} \gamma_v & \gamma_v \dfrac{v}{c} \\ \gamma_v \dfrac{v}{c} & \gamma_v \end{bmatrix} \begin{bmatrix} x' \\ ct' \end{bmatrix}$$    (B-2b)

(*Note:* We have multiplied and divided by $c$ at various places, so that the equations are expressed not in terms of $t$, but in terms of $ct$, which has the same units as $x$.) As discussed in Section 1.11, these equations may be used to show that $(ct)^2 - x^2 = (ct')^2 - x'^2$, which implies that the proper time for a series of events is invariant, the same when determined from any frame of reference.

   Transformations (B-1b) and (B-2b) are in two important ways conspicuously similar: (1) The top-left and bottom-right elements of both $2 \times 2$ matrices are equal, while the top-right and bottom-left elements are of equal *magnitude*. (2) There is some quantity that is invariant—the length of the vector in the case of rotation, and the proper time in the case of the Lorentz transformation.[1] There is one important difference: a different sign in one element of the transformation matrix. This accounts for the different sign in the expression for the invariant and also (as we shall see) for a difference in the appearance of $S$ and $S'$ plotted on the same graph compared to $F$ and $F'$.[2] Mathematically expressed,

> Both the rotation transformation and the Lorentz transformation are linear transformations; that is, $x'$ and $y'$ depend on $x$ and $y$ to the first power, and $t'$ and $x'$ depend on $t$ and $x$ to the first power. However, while the rotation leaves $x^2 + y^2$ invariant, the Lorentz transformation leaves $(ct)^2 - x^2$, rather than $(ct)^2 + x^2$, invariant.

   Suppose we plot the $t'$-axis, all points for which $x' = 0$, in terms of $t$ and $x$. From equations (B-2a),

$$x = \left(\gamma_v \frac{v}{c}\right)(ct') \quad \text{and} \quad ct = \gamma_v(ct') \;\Rightarrow\; \frac{x}{ct} = \frac{\left(\gamma_v \frac{v}{c}\right)(ct')}{\gamma_v(ct')} \;\Rightarrow\; x = \frac{v}{c}(ct)$$

This is a straight line with positive slope ($v/c$). A plot of the $x'$-axis, all points for which $t' = 0$, follows in a similar way.

$$x = \gamma_v x' \quad \text{and} \quad (ct) = \left(\gamma_v \frac{v}{c}\right)x' \;\Rightarrow\; \frac{x}{ct} = \frac{\gamma_v x'}{\left(\gamma_v \frac{v}{c}\right)x'} \;\Rightarrow\; x = \frac{c}{v}(ct)$$

This is also a straight line with positive slope ($c/v$). *Both axes have positive slope.* Obviously, a plot of $S$ and $S'$ together should look distinctly different from a plot of $F$ and $F'$ together.

   Figure B.3 shows $x'$- and $t'$-axes plotted on the $x$-$t$ axes of frame $S$ for $v = \frac{\sqrt{3}}{2}c$. The $t'$- and $x'$-axes have slopes of $\frac{\sqrt{3}}{2}$ and $\frac{2}{\sqrt{3}}$, respectively. Lines of constant position are parallel to the time axes, and those of constant time are parallel to the position axes. Since the transformation from $x$ and $t$ to $x'$ and $t'$ differs from the reverse only in the sign of $v$, if $x$- and $t$-axes are plotted on the $x'$-$t'$ axes of frame $S'$, both would be of negative slope, as shown in Figure B.4.

---

   [1] Both matrices also have a determinant of unity.
   [2] Had we instead used the Lorentz transformation giving $x'$ and $t'$ in terms of $x$ and $t$, rather than $x$ and $t$ in terms of $x'$ and $t'$, the situation would not have been different; whether there are zero or two negative elements depends only on the direction chosen positive, but having only one, as in the rotation matrix, is qualitatively different.

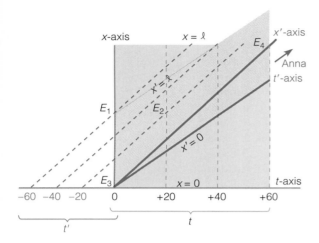

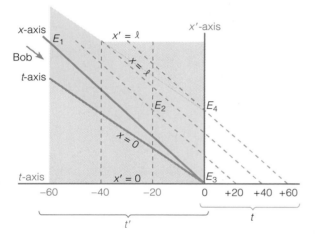

In particular, Figures B.3 and B.4 show the events of Example 1.4. The shaded regions represent the space covered by Anna's (blue) and Bob's (gray) ships over the 60-ns interval in which events $E_1$ through $E_4$ occur. To see how these **space-time plots** describe the events (which are also shown in Figure B.5), consider a few points.

In Figure B.3, Anna's ship has a positive velocity relative to Bob, while in B.4, Bob's has a negative velocity relative to Anna.

While the sequence of events according to Bob is $E_1$ and $E_3$ simultaneously, then $E_2$, then $E_4$, according to Anna it is $E_1$, then $E_2$, then $E_3$ and $E_4$ simultaneously, and this may be seen from *either* plot.

In either plot, we see that $E_1$ and $E_3$ occur on the same $t = 0$ line, but on different $t'$ lines—relative simultaneity. Similarly, $E_3$ and $E_4$ occur on the same $t' = 0$ line, but different $t$ lines.

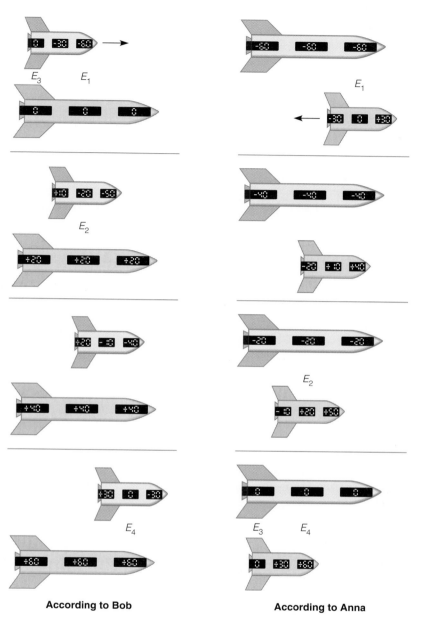

**Figure B.5** The passing of two ships, seen from both frames.

**According to Bob**

**According to Anna**

In both plots, at Bob's $t = 0$, Anna's ship, bounded by events $E_1$ and $E_3$, fits between one end and the middle of Bob's ($x = 0$ and $x = \frac{1}{2}\ell$); according to Bob, Anna's ship is contracted. Similarly, at Anna's $t' = 0$, Bob's ship, bounded by events $E_3$ and $E_4$, is contracted.

# C

# Planck's Blackbody Radiation Law

A  ll materials emit electromagnetic radiation, because all materials contain charged particles that move about randomly. Whenever a charged particle accelerates, it radiates electromagnetic energy. The amount of energy radiated is related to the average energy of the random motion, which is in turn related to the temperature. For example, both human beings and red-hot coals radiate electromagnetic energy in the invisible infrared portion of the spectrum; but the higher-temperature coals emit more radiation per unit of surface area, much of which is in the red end of the visible spectrum. Red-hot coals glow in the dark; human beings do not. Most materials also *reflect* electromagnetic energy. Mirrors reflect electromagnetic radiation in the visible range quite well. Coal reflects poorly. A **blackbody** is defined to be any object from which electromagnetic radiation emanates solely due to the thermal motion of the charges within it. It must reflect no light, hence the name. It is well established that the electromagnetic radiation emitted by a blackbody is independent of specific properties of the material. That is, so far as spectral emissions go, all blackbodies of temperature $T$ are identical.

Coal is a reasonable approximation of a blackbody, emitting much radiation when heated, while reflecting little at any temperature.[1] Still, fabricating a true blackbody—absolutely no reflection of any wavelengths—would seem to be

[1]The term *black*body must not be taken too literally; the surface of the Sun is a blackbody! The electromagnetic energy it *reflects* is a truly insignificant fraction of the energy it emits due to the thermal motion of its charges.

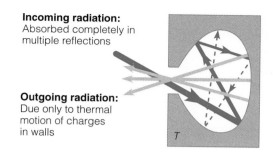

**Incoming radiation:**
Absorbed completely in
multiple reflections

**Outgoing radiation:**
Due only to thermal
motion of charges
in walls

**Figure C.1** Electromagnetic radiation incident on and emanating from
a cavity.

problematic. Imagine, however, an object with an interior cavity and a small hole
connecting the cavity to the exterior, as depicted in Figure C.1. Any radiation
entering the hole would reflect from the cavity's inner surface many times, losing
energy to the object at each reflection. Very little would reflect back through the
hole. On the other hand, all areas of the inner surface consist of charges in thermal
motion, and so constantly emit radiation in all directions, some of which exits the
hole. Therefore, the hole behaves as a blackbody of temperature $T$. (Accordingly,
the terms "blackbody radiation" and "cavity radiation" are synonymous.)

Early experiments showed that, although a blackbody emits some radiation at
all frequencies, the energy emitted is maximum at a certain frequency, which increases proportionally with the object's temperature. Correspondingly, the wavelength of maximum emission is inversely proportional to $T$. For example, as a
piece of iron is heated, not only does it emit more energy overall, but there is a
shift in its predominant wavelengths from the far infrared (longer than visible)
toward visible red.[2]

Classical theory, however, based upon the assumption that electromagnetic radiation is strictly a wave, failed miserably to predict what experiment had established as fact. It indicated that the energy should be a monatonically increasing
function of frequency, diverging as frequency increases without bound. All materials should radiate infinite energy per unit time! The theory did agree with experiment at low frequencies, but it failed at high/ultraviolet frequencies. The
discrepancy was termed the "ultraviolet catastrophe." Of course, there was no
*real* catastrophe; it was simply a case where the classical theory was inadequate.

Max Planck solved the problem. As we outline Planck's work, it will be necessary to invoke some principles of statistical mechanics (the topic of Chapter 8)
that may be unfamiliar. Still the reader will be able to discern one of the most
important features of Planck's work—its departure from classical theory.

Our goal is to calculate the **spectral energy density:** the electromagnetic energy inside a cavity in a narrow frequency range between $f$ and $f + df$.

$dE_f$ = energy in frequency range $f$ to $f + df$

Two factors go into this: First, assuming that the energy may be treated as standing
waves of certain wavelengths, there will be a certain *number* of standing waves
whose frequencies are in the range $f$ to $f + df$.

$dn_f \equiv$ number of waves in frequency range $f$ to $f + df$

[2]Iron vaporizes before the temperature is reached at which the maximum
would actually occur in the visible range.

Secondly, although the energy in a wave of frequency $f$ might fluctuate as electromagnetic energy is constantly emitted and absorbed, it should have a predictable average, which we designate as $\overline{E(f)}$.

$\overline{E(f)}$ ≡ average energy expected for a wave of frequency $f$

Putting these together, and assuming that $\overline{E(f)}$ does not vary too rapidly with $f$, the total energy in the range of frequencies around $f$ should be

$$dE_f = \overline{E(f)}\, dn_f \qquad\qquad (C\text{-}1)$$

Let us determine $\overline{E(f)}$ and $dn_f$ separately, combining them afterward to obtain $dE_f$.

## $\overline{E(f)}$: Average Energy of Wave of Frequency $f$

In a system as complicated as a cavity whose walls constantly emit and absorb electromagnetic energy, the energy of a wave of frequency $f$ is a matter of probabilities. Its average we obtain by multiplying a *possible* energy $E$ by the *probability, $P(E)$,* that the wave would have that energy, then summing over all possible energies.

$$\overline{E(f)} = \Sigma_E\, E\, P(E) \qquad\qquad (C\text{-}2)$$

Of course, a central question is, What is the probability? We assume that the electromagnetic radiation in the cavity is in equilibrium with the randomly moving charges in the object's walls, that there is no *net* heat flow between walls and radiation. (Radiation "leaks" out the hole so slowly as to have little effect on this energy balance.) Thus, the *radiation itself* has the same temperature as the walls, $T$. Statistical mechanics then gives us the probability, known as the Boltzmann probability. In a system of distinct elements—in this case, standing waves—at temperature $T$, the probability that a given element/wave would have energy $E$ is

$$P(E) = Ae^{-E/k_B T}$$

where $k_B$ is the Boltzmann constant, $1.38 \times 10^{-23}$ J/K. We see that the probability decreases exponentially as the energy increases. The value of the multiplicative constant $A$ must be such that the total probability is unity.

$$\Sigma_E\, Ae^{-E/k_B T} = 1 \quad\Rightarrow\quad A = \frac{1}{\Sigma_E\, e^{-E/k_B T}}$$

Thus, we may write the probability as

$$P(E) = \frac{e^{-E/k_B T}}{\Sigma_E\, e^{-E/k_B T}}$$

and equation (C-2) then becomes

$$\overline{E(f)} = \Sigma_E\, E\frac{e^{-E/k_B T}}{\Sigma_E\, e^{-E/k_B T}} = \frac{\Sigma_E\, Ee^{-E/k_B T}}{\Sigma_E\, e^{-E/k_B T}}$$

At this point, the classical theory and Planck's diverge. To clarify the differences, physical and mathematical, we follow them side by side.

Classically, a wave can have any energy in the continuum zero to infinity. The sum becomes an integral.

$$\overline{E(f)} = \frac{\displaystyle\int_0^\infty E e^{-E/k_B T}\, dE}{\displaystyle\int_0^\infty e^{-E/k_B T}\, dE}$$

Carrying out the integration,

$$\overline{E(f)} = k_B T$$

Planck assumed[3] that electromagnetic radiation of frequency $f$ is allowed only those energies given by $E = nhf$, where $n$ is an integer and $h$ a constant.

$$\overline{E(f)} = \frac{\displaystyle\sum_{n=0}^\infty nhf e^{-nhf/k_B T}}{\displaystyle\sum_{n=0}^\infty e^{-nhf/k_B T}}$$

Carrying out the summation,[4]

$$\overline{E(f)} = \frac{hf}{e^{hf/k_B T} - 1}$$

Although the results appear quite different, they are equivalent at low frequency. To see this, consider a power series expansion of the exponential.

$$e^{hf/k_B T} = 1 + \frac{hf}{k_B T} + \frac{1}{2!}\left(\frac{hf}{k_B T}\right)^2 + \frac{1}{3!}\left(\frac{hf}{k_B T}\right)^3 + \cdots$$

The expansion is valid so long as $hf \ll k_B T$. In this low-frequency limit, the first term in the denominator of Planck's result ($e^{hf/k_B T} - 1$) would be $hf/k_B T$, and $\overline{E(f)}$ would then be $k_B T$, the same as the classical. As frequency increases, however, the classical result remains $k_B T$ while Planck's approaches zero. The two are compared in Figure C.2.

From our understanding of light's particle nature, we interpret Planck's result as follows: The total energy at frequency $f$, $nhf$, comprises $n$ photons each of energy $hf$. The average energy is $k_B T$ only if the thermal energy of the charges in the cavity's walls is sufficient to create *many* photons. At high frequencies, where the energy per photon $hf$ is much greater than $k_B T$, it is unlikely that random thermal energy would create even a single photon. Thus, we expect to find no photon energy at arbitrarily high frequencies. Classically, a wave of frequency $f$ has an energy related to its amplitude, which may take on any value, from very small to very large, but Planck's theory allows only the energies 0, $hf$, $2hf$, and so on.

---

[3]In the interest of historical accuracy, it should be noted that Planck originally proposed that the condition $E = nhf$ be applied not to the radiation in the cavity, but only to the *charges oscillating in the walls*. These "resonators," as he called them, could thus change energy only in discrete jumps. Given the assumption of thermal equilibrium between walls and radiation, the two views are equivalent.

[4]With the definition $x \equiv e^{-\beta hf}$ where $\beta \equiv 1/k_B T$, the denominator becomes $\sum_{n=0}^\infty x^n$, which equals $1/(1 - x)$ or $1/(1 - e^{-hf/k_B T})$. The numerator is $-(\partial/\partial\beta) \sum_{n=0}^\infty e^{-n\beta hf} = -(\partial/\partial\beta) \sum_{n=0}^\infty x^n = -(\partial/\partial\beta)(1 - x)^{-1} = -(1 - x)^{-2}\, \partial x/\partial\beta = (1 - x)^{-2} hf e^{-\beta hf} = hf e^{-\beta hf}/(1 - e^{-hf/k_B T})^2$. Dividing numerator by denominator, and then multiplying top and bottom by $e^{hf/k_B T}$, yields Planck's result.

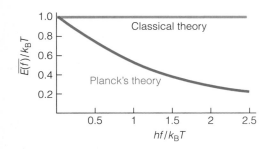

**Figure C.2** Average energy at frequency $f$, classically and according to Planck.

# $dn_f$: Number of Waves in Range $f$ to $f + df$

The quantity $dn_f$ is the same quantum-mechanically as classically. As such, it is less important if our goal is merely to investigate the distinction between the classical and quantum-mechanical predictions. Nevertheless, Planck's triumphant theory would be incomplete without it.

With no loss of generality we assume that the cavity is a cube and, again, that the energy inside is in the form of electromagnetic standing waves. Because there are a few twists in three dimensions, let us introduce some of the ideas in the simpler context of one-dimensional waves on a string.

## One Dimension

If a string stretching from $x = 0$ to $x = L$ is deformed in some way then released, a wave will propagate. A possible standing wave would be described by

$$W(x, t) = A \sin \frac{m\pi x}{L} \cos 2\pi ft \tag{C-3}$$

where $W(x, t)$ is the string's displacement from the flat equilibrium state. This function is zero—has nodes—at $x = 0$ and $x = L$, has $m$ antinodes, and oscillates with frequency $f$. But what really governs the behavior of disturbances on the string?

All waves on the string must obey the **wave equation** for the string, which is of the form

$$\frac{\partial^2}{\partial t^2} W(x, t) = v^2 \frac{\partial^2}{\partial x^2} W(x, t) \tag{C-4}$$

The reason this must be obeyed is that it is simply the second law of motion rearranged in a form most readily applicable to disturbances on a stretched string. (The proof is given in most introductory calculus-based physics texts.) We verify that (C-3) does indeed obey the second law by showing that it is a solution of the wave equation. Inserting (C-3) in (C-4) and carrying out the partial derivatives gives

$$-(2\pi f)^2 A \sin \frac{m\pi x}{L} \cos 2\pi ft = -v^2 \left(\frac{m\pi}{L}\right)^2 A \sin \frac{m\pi x}{L} \cos 2\pi ft$$

The function $W(x, t)$ itself cancels, leaving $(2\pi f)^2 = v^2(m\pi/L)^2$. We conclude, then, that the second law of motion is obeyed for all values of $x$ and $t$ so long as

$$f = v\frac{m}{2L}$$

Since $f = v/\lambda$, this merely states that the allowed wavelengths are given by $\lambda_m = 2L/m$, where $m$ is the number of antinodes, as expected. More important for our purpose, though, is that allowed *frequencies* are separated by $v/2L$. Figure C.3 represents the allowed standing waves by points on a one-dimensional axis. There is one wave per change in frequency of $v/2L$, so the number of waves $dn_f$ per frequency is

$$\frac{dn_f}{df} = \frac{1}{v/2L} = \frac{2L}{v} \qquad \textbf{(One dimension)}$$

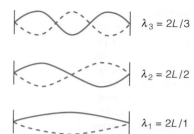

$\lambda_3 = 2L/3$

$\lambda_2 = 2L/2$

$\lambda_1 = 2L/1$

$dm = 1$
$df = v/2L$

**Figure C.3** Number of antinodes for different standing waves, represented as points on a line.

## Three Dimensions

Whereas mechanical waves obey a wave equation derived from Newton's second law of motion, light waves obey one derived from the fundamental laws of electromagnetism—Maxwell's equations. Despite the *waves'* different natures, the *wave equations* are very similar. Light waves obey:[5]

$$\frac{\partial^2}{\partial t^2} W(x, y, z, t) = v^2\left(\frac{\partial^2}{\partial x^2} + \frac{\partial^2}{\partial y^2} + \frac{\partial^2}{\partial z^2}\right) W(x, y, z, t) \qquad \textbf{(C-5)}$$

The three-dimensional standing wave is

$$W(x, y, z, t) = A \sin\frac{m_x\pi x}{L} \sin\frac{m_y\pi y}{L} \sin\frac{m_z\pi z}{L} \cos 2\pi ft \qquad \textbf{(C-6)}$$

where $m_x$, $m_y$, and $m_z$ are the number of antinodes along the three axes. Again, we find the relationship between frequency and the allowed shapes of the waves, that is, the numbers of antinodes, by inserting $W(x, y, z, t)$ into the wave equation, (C-5):

$$-(2\pi f)^2 A \sin\frac{m_x\pi x}{L} \sin\frac{m_y\pi y}{L} \sin\frac{m_z\pi z}{L} \cos 2\pi ft$$

$$= v^2\left[\left(\frac{m_x\pi}{L}\right)^2 + \left(\frac{m_y\pi}{L}\right)^2 + \left(\frac{m_z\pi}{L}\right)^2\right] A \sin\frac{m_x\pi x}{L} \sin\frac{m_y\pi y}{L} \sin\frac{m_z\pi z}{L} \cos 2\pi ft$$

Once again $W(x, y, z, t)$ cancels and we obtain

$$f = \frac{v}{2L} m \quad \text{where} \quad m \equiv \sqrt{m_x^2 + m_y^2 + m_z^2}$$

We see now that *different* standing waves may have the *same* frequency. For instance, the frequency $f = (c/2L)\sqrt{27}$ is shared by four different waves: $(m_x, m_y, m_z) = (5, 1, 1)$, $(1, 5, 1)$, $(1, 1, 5)$, and $(3, 3, 3)$. How then do we determine the number of waves per unit frequency?

[5] The equation actually applies to any given *component* of the electric or magnetic field, for example, $W = E_x$.

As does Figure C.3, Figure C.4 represents possible standing waves as separate points. Here, though, they are points in a three-dimensional "antinode-number space." The point labeled (3, 1, 2), for instance, is the standing wave with three antinodes along the x-axis, one along the y, and two along the z. All points with the same "radius," $m = \sqrt{m_x^2 + m_y^2 + m_z^2}$, have the same frequency, and the number of points/waves in a region is merely the volume of that region. Thus, for a change $dm$ in "radius," $df$ would be $(v/2L)\,dm$ and the corresponding number of waves $dn_f$ would be the volume of a one-eighth section of a spherical shell: $\frac{1}{8}4\pi m^2 \times dm$. The number of waves per frequency is thus

$$\frac{dn_f}{df} = \frac{\frac{1}{2}\pi m^2\,dm}{(v/2L)\,dm} = \frac{\pi m^2 L}{v} \quad \textbf{(Three dimensions)} \tag{C-7}$$

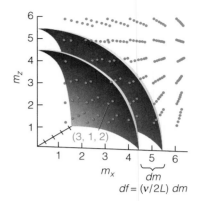

**Figure C.4** Three-dimensional "antinode-number space."

In three dimensions, the number of waves per frequency is an increasing function of $m$ and thus $f$, because the number of waves *at a given frequency* is an increasing function of $f$. Clearly, the geometrical calculation used here is not absolutely precise; a *number* of waves is not a continuous function. But, except at very low energies, the "radius" $m$ is so large that the change in $m$ is only a small fraction of $m$, so that $m$ is fairly smooth. Now, using $f = (v/2L)m$ to put (C-7) in terms of $f$ rather than $m$, we have

$$dn_f = \frac{4\pi L^3}{v^3}f^2\,df$$

Finally, we note that electromagnetic radiation moves at speed $c$ and, moreover, that it is transverse. Consequently, there are two independent possible orientations (polarizations) of the electric field vector for any given wave. The number of waves at any frequency is thus actually twice what is indicated above.

$$dn_f = \frac{8\pi L^3}{c^3}f^2\,df \tag{C-8}$$

It is worth reiterating that while looking at the frequencies allowed, we have not in any way addressed the question of how much *energy* would be expected at a given frequency. That was the point of the previous calculation and is where the classical and Planck's theory differ.

## $dE_f$: Spectral Energy Density

We now find the spectral energy density $dE_f$ via equation (C-1). Since they differ in $\overline{E(f)}$, we again consider the classical theory and Planck's side by side.

| **Classical Wave Theory** | **Planck's Theory** |
|---|---|
| $dE_f = k_\text{B}T\dfrac{8\pi L^3}{c^3}f^2\,df$ | $dE_f = \dfrac{hf}{e^{hf/k_\text{B}T}-1}\dfrac{8\pi L^3}{c^3}f^2\,df$ |

The two are plotted in Figure C.5. Both initially increase because the number of states $dn_f$ increases as $f^2$. But Planck's ultimately falls to zero because at very high frequencies there isn't enough thermal energy to produce even one photon—*and*

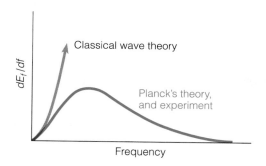

**Figure C.5** Energy per unit frequency, classical versus Planck.

*it is not possible to produce less than one.* Most importantly, it is Planck's theory that agrees with the experimental evidence.

$$dE_f = \frac{hf}{e^{hf/k_BT} - 1}\frac{8\pi L^3}{c^3}f^2\,df$$

Spectral energy density

It should be borne in mind that although Planck derived the spectral energy density from his quantization assumption, *he did not derive the value of h.* His value was simply the one that gave the best fit with the experimental data. Still, though varying *h* would change the energy density's *magnitude* at all frequencies, it would not affect its basic functional form, its "shape." Had there been no merit in Planck's assumption, it would have been the most bizarre coincidence for the shape of his energy density to match the experiment data. In fact, we now realize that *h cannot* be theoretically predicted, for it is one of nature's fundamental constants (e.g., the universal gravitational constant *G*), all of which must be determined experimentally.

Often, we are more interested in the amount of energy in a given range of wavelengths, rather than frequencies. The distinction rests upon the relationship

$$df = -\frac{c}{\lambda^2}\,d\lambda$$

which follows by differentiating $f = c/\lambda$. A fixed range of frequencies *df* is not directly proportional to a fixed range of wavelengths *dλ*. (We ignore the negative sign; we seek merely the energy in a range of wavelengths of a certain magnitude.) Substituting into the preceding spectral energy densities gives

| **Classical Wave Theory** | **Planck's Theory** |
|---|---|
| $k_BT\dfrac{8\pi L^3}{c^3}\left(\dfrac{c}{\lambda}\right)^2\dfrac{c}{\lambda^2}\,d\lambda$ | $\dfrac{h(c/\lambda)}{e^{hc/\lambda k_BT} - 1}\dfrac{8\pi L^3}{c^3}\left(\dfrac{c}{\lambda}\right)^2\dfrac{c}{\lambda^2}\,d\lambda$ |
| $dE_\lambda = 8\pi L^3 k_BT\dfrac{1}{\lambda^4}\,d\lambda$ | $dE_\lambda = \dfrac{8\pi L^3 hc}{e^{hc/\lambda k_BT} - 1}\dfrac{1}{\lambda^5}\,d\lambda$ |

Figure C.6 shows energy density versus wavelength in the two theories. While they agree at long wavelengths—low frequencies, low energy per photon—only Planck's agrees with experiment over the entire spectrum. The diagram also illustrates two important predictions of Planck's theory of blackbody radiation: As temperature increases, (1) the total energy radiated by a blackbody increases much

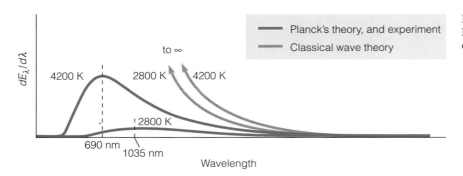

**Figure C.6** Energy per unit wavelength at two different temperatures, classical versus Planck.

faster, (the Stefan-Boltzmann law), and (2) the wavelength of maximum radiated energy decreases inversely (Wien's law). As to the first, the area under the curve, $\int (dE_\lambda/d\lambda)\, d\lambda$, is the total energy and is clearly many times larger for the 4200 K than for the 2800 K curve, though the temperature ratio is only $\frac{3}{2}$. As to the second, the maximum of the 4200 K curve does occur at a wavelength $\frac{2}{3}$ that for the 2800 K curve. These phenomena were accepted *experimental* truths—"laws" of thermal radiation—before Planck's work; they simply lacked a theoretical basis. According to classical theory, the spectral energy density has no maximum, and diverges as $\lambda \to 0$, implying infinite total energy!

In conclusion, not only did Planck originate the notion of energy quantization and obtain the first numerical value for the fundamental constant that now bears his name, his theory explains aspects of thermal radiation that could not be explained previously.

## EXERCISES

1. Derive the Stefan–Boltzmann law: The total electromagnetic intensity of all wavelengths thermally radiated by a body of temperature $T$ obeys $I = \sigma T^4$ where $\sigma = 5.67 \times 10^{-8}$ W/m$^2$·K$^4$. First show that the total energy in a volume of electromagnetic radiation at temperature $T$ is $E = \frac{8}{15}(\pi^5 k_B^4/h^3 c^3)VT^4$. Note that

$$\int_0^\infty \frac{dx}{x^5(e^{a/x} - 1)} = \frac{1}{15}\left(\frac{\pi}{a}\right)^4$$

Intensity, power per unit area, is then the product of the energy per unit volume and distance per unit time. However, since intensity is a flow in a given direction away from the blackbody, the correct speed is not $c$; for radiation moving uniformly in all directions, the average *component* of velocity in a given direction is $\frac{1}{4}c$.

2. Derive Wien's law: $\lambda_{max}T = 2.898 \times 10^{-3}$ m·K, where $\lambda_{max}$ is the wavelength at which $dE_\lambda$ is maximum. To do this, obtain from Planck's $dE_\lambda$ an equation that when solved would yield the wavelength at which $dE_\lambda$ is maximum. This transcendental equation cannot be solved exactly, so it is enough to show that it is solved to a reasonable degree of precision by $\lambda = 2.898 \times 10^{-3}$ m·K/$T$.

3. At what wavelength does the human body emit the maximum electromagnetic radiation? Use Wien's law from the previous exercise and assume a skin temperature of $70°$F.

# Solving for Spectral Content

O ur goal here is find the spectral content $\tilde{f}(k)$ of function $f(x)$ from the expression in which $f(x)$ is represented as a sum of complex exponential plane waves:

$$f(x) = \int_{-\infty}^{+\infty} \tilde{f}(k')e^{ik'x}\,dk'$$

Strictly speaking, this cannot be "solved" for $\tilde{f}(k')$, since $k'$ is a dummy variable of integration. (We have chosen to use $k'$ rather than $k$ only for convenience.) But we may solve for the *functional form* of $\tilde{f}$. We begin by multiplying both sides of the equation by $e^{-ikx}$, then integrating over $x$ from $-b$ to $+b$, eventually intending to let $b$ approach infinity.

$$\int_{-b}^{+b} f(x)e^{-ikx}\,dx = \int_{-b}^{+b}\int_{-\infty}^{+\infty} \tilde{f}(k')e^{ik'x}\,dk'\,e^{-ikx}\,dx$$

Rearranging a bit, we have

$$\int_{-b}^{+b} f(x)e^{-ikx}\,dx = \int_{-\infty}^{+\infty} \tilde{f}(k')\left[\int_{-b}^{+b} e^{i(k'-k)x}\,dx\right]dk' \qquad \text{(D-I)}$$

The integral in brackets is

$$\int_{-b}^{+b} e^{i(k'-k)x}\, dx = \frac{e^{+i(k'-k)b} - e^{-i(k'-k)b}}{i(k'-k)}$$

$$= \frac{2}{k'-k} \frac{e^{+i(k'-k)b} - e^{-i(k'-k)b}}{2i}$$

$$= \frac{2\,\sin[(k'-k)b]}{k'-k}$$

so that (D-1) becomes

$$\int_{-b}^{+b} f(x) e^{-ikx}\, dx = \int_{-\infty}^{+\infty} \tilde{f}(k') \frac{2\,\sin[(k'-k)b]}{k'-k}\, dk' \qquad \textbf{(D-2)}$$

Our goal is near, but to reach it we must take the limit as $b$ approaches infinity. The integral on the right-hand side of the equation, with this limit taken, comes up often in physics. Let us consider it separately.

## Integrals Involving the Dirac Delta Function

Here we concentrate on integrals of the form

$$\lim_{b\to\infty} \int_{-\infty}^{+\infty} g(z') \frac{2\,\sin[(z'-z)b]}{z'-z}\, dz' \qquad \textbf{(D-3)}$$

where $g(z')$ is an arbitrary function. Figure D.1 plots an arbitrary $g(z')$ and the function $2\,\sin[(z'-z)b]/(z'-z)$. The latter function, the focus of our attention, is shown in "close-up" in Figure D.2. It is $2/(z'-z)$ modulated by $\sin[(z'-z)b]$. The $\sin[(z'-z)b]$ oscillates at a frequency proportional to $b$, so that as $b$ approaches infinity, it would oscillate between $+1$ and $-1$ infinitely rapidly. Multiplying any other smooth finite functions in an integral, it would give an area under

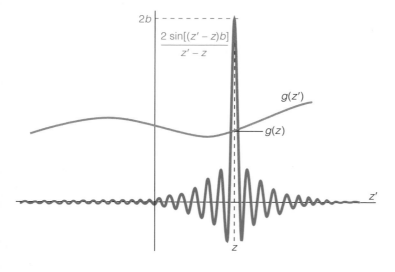

Figure D.1  An arbitrary smooth function $g(z')$, and one sharply peaked at $z' = z$.

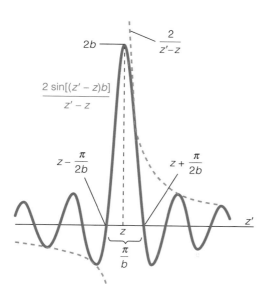

**Figure D.2** As $b$ becomes large, $2 \sin[(z' - z)b]/(z' - z)$ becomes a narrow but tall peak, with rapidly oscillating "wings."

the graph of zero in any interval along the $z$-axis. However, the $2/(z' - z)$ is *not* finite everywhere; it diverges at $z' = z$. In the region $z - (\pi/2b) < z' < z + (\pi/2b)$ the product $\sin[(z' - z)b] \cdot 2/(z' - z)$ is not only strictly positive; at $z' = z$ it is $2b$,[1] and so becomes infinitely tall as $b \to \infty$. On the other hand, this region is of width $\pi/b$, becoming infinitesimally narrow as $b \to \infty$. All things considered, then, we see that, whatever its value might be, the integral in (D-3) is obtained fully in an infinitesimal region surrounding $z' = z$. This allows a great if surprising simplification: Since $2 \sin[(z' - z)b]/(z' - z)$ is *effectively* zero everywhere but at $z' = z$, the values of the integrand's other function, $g(z')$, at $z'$-values *not* equal to $z$ are irrelevant. Thus, $g(z')$ might as well be $g(z)$ everywhere; we may call it a constant $g(z)$ and take it out of the integral!

$$\lim_{b \to \infty} \int_{-\infty}^{+\infty} g(z') \frac{2 \, \sin[(z' - z)b]}{z' - z} \, dz' = g(z) \lim_{b \to \infty} \int_{-\infty}^{+\infty} \frac{2 \, \sin[(z' - z)b]}{z' - z} \, dz'$$

The remaining integral is a standard one, found in any table of integrals. Its value is $2\pi$, *independent of $z$ and $b$*. [Changing variables, with $\theta \equiv (z' - z)b$, "erases" $z$ and $b$.] Finally, then

$$\lim_{b \to \infty} \int_{-\infty}^{+\infty} g(z') \frac{2 \, \sin[(z' - z)b]}{z' - z} \, dz' = g(z)2\pi \qquad \text{(D-4)}$$

The function responsible for this surprising result is called the **Dirac delta function**. In our derivation, it is the $2 \sin[(z' - z)b]/(z' - z)$ in the limit $b \to \infty$. However, there are numerous seemingly different functions that, in a certain limit, have exactly the same characteristics: (1) effectively zero everywhere except at $z' = z$, where it is infinite, and, consequently, (2) when multiplying a smooth function $g(z')$ in an integral over all $z'$, it leaves merely $2\pi$ times the function $g$ evaluated at $z' = z$. We leave further discussion of this peculiar function to a higher-level course.

[1] Which may be shown via L'Hopital's rule.

At last, having dealt with the limit $b \to \infty$ on the right side of equation (D-2), we may now let $b \to \infty$ on the left side, so that

$$\int_{-\infty}^{+\infty} f(x)e^{-ikx}\,dx = \tilde{f}(k)2\pi$$

or

$$\tilde{f}(k) = \frac{1}{2\pi}\int_{-\infty}^{+\infty} f(x)e^{-ikx}\,dx$$

# The Momentum Operator

The derivation of the momentum operator rests upon the assertion that, as the probability amplitude for position $x$ is $\psi(x)$, the probability amplitude for wave number $k$ is the Fourier transform $\tilde{\psi}(k)$. The considerable symmetry in the way $x$ and $k$ appear in the Fourier transform and its inverse is certainly suggestive:

$$\psi(x) = \int_{-\infty}^{+\infty} \tilde{\psi}(k)e^{ikx}\,dk \qquad\qquad \tilde{\psi}(k) = \frac{1}{2\pi}\int_{-\infty}^{+\infty} \psi(x)e^{-ikx}\,dx$$

**Inverse Fourier transform** $(k \rightarrow x)$       **Fourier transform** $(x \rightarrow k)$

(3-11)             (3-12)

What $\psi(x)$ is to $x$, $\tilde{\psi}(k)$ should be to $k$. Furthermore, much of what Section 3.6 tells us about the Fourier transform also agrees: Only a $\psi(x)$ that is a spike at $x_0$ has a well-defined position of $x_0$. Only an infinite plane wave can have a well-defined wave number $k_0$, and the Fourier transform of such a plane wave is a spike at $k_0$. Just as $|\psi(x)|^2$ is the probability per distance of finding the particle at $x$, $|\tilde{\psi}(k)|^2$ is the probability per wave number of finding the particle to have wave number $k$. It follows that if we write

$$\bar{x} = \int_{-\infty}^{+\infty} x|\psi(x)|^2\,dx$$

then we must write

$$\bar{k} = \frac{1}{2\pi} \int_{-\infty}^{+\infty} k |\tilde{\psi}(k)|^2 \, dk \tag{E-1}$$

(Later we will see why the multiplicative $1/2\pi$ is needed.) Since $p = \hbar k$, to know $\bar{k}$ is to know $\bar{p}$.

However, we wish to be able to calculate $\bar{k}$ using $\psi(x)$, not its Fourier transform. Thus, using the definition of the Fourier transform,

$$\bar{k} = \frac{1}{2\pi} \int_{-\infty}^{+\infty} k \left( \int_{-\infty}^{+\infty} \psi(x')e^{-ikx'} \, dx' \right)^* \left( \int_{-\infty}^{+\infty} \psi(x)e^{-ikx} \, dx \right) dk$$

or, rearranging,

$$\bar{k} = \frac{1}{2\pi} \int\int\int_{-\infty}^{+\infty} k \, \psi^*(x')e^{+ikx'}(\psi(x)e^{-ikx}) \, dx' \, dx \, dk \tag{E-2}$$

For reasons to become clear, let us integrate by parts the term in parentheses involving $x$.

$$\int_{-\infty}^{+\infty} \psi(x)e^{-ikx} \, dx = \int_{-\infty}^{+\infty} \psi(x) \, d\left( \frac{i}{k} e^{-ikx} \right)$$

$$= \left( \psi(x) \frac{i}{k} e^{-ikx} \right) \Bigg|_{-\infty}^{+\infty} - \int_{-\infty}^{+\infty} \frac{d\psi(x)}{dx} \frac{i}{k} e^{-ikx} \, dx$$

Assuming that $\psi(x)$ is well behaved, falling to zero at $x = \pm\infty$, the first term on the right is zero. The benefit of the integration by parts is that when the second term is reinserted in (E-2), $k$ cancels. (We wish to eliminate $k$ from the integral.) Thus,

$$\bar{k} = \frac{1}{2\pi} \int\int\int_{-\infty}^{+\infty} \psi^*(x')e^{+ikx'}(-i)\frac{\partial\psi(x)}{\partial x} e^{-ikx} \, dx' \, dx \, dk$$

or

$$\bar{k} = \frac{1}{2\pi} \int\int_{-\infty}^{+\infty} \psi^*(x') \left[ \int_{-\infty}^{+\infty} e^{+ik(x'-x)} \, dk \right] (-i)\frac{\partial\psi(x)}{\partial x} \, dx' \, dx \tag{E-3}$$

Now the integration over $k$ may be carried out, but here we must be careful. Writing it explicitly as a limit, we have

$$\lim_{b\to\infty} \int_{-b}^{+b} e^{+ik(x'-x)} \, dk = \lim_{b\to\infty} \frac{e^{+ib(x'-x)} - e^{-ib(x'-x)}}{i(x'-x)} = \lim_{b\to\infty} \frac{2 \, \sin[(x' - x)b]}{x' - x}$$

Thus, (E-3) becomes

$$\bar{k} = \frac{1}{2\pi} \int\int_{-\infty}^{+\infty} \psi^*(x') \left\{ \lim_{b\to\infty} \frac{2 \, \sin[(x' - x)b]}{x' - x} \right\} (-i)\frac{\partial\psi(x)}{\partial x} \, dx' \, dx$$

and grouping together the functions of $x'$ then gives

$$\overline{k} = \frac{1}{2\pi} \int_{-\infty}^{+\infty} \left( \int_{-\infty}^{+\infty} \psi^*(x') \left\{ \lim_{b \to \infty} \frac{2 \, \sin[(x' - x)b]}{x' - x} \right\} dx' \right) (-i) \frac{\partial \psi(x)}{\partial x} \, dx$$

(E-4)

Integrals having the form of the $x'$ integration in parentheses are discussed in Appendix D. Using equation (D-4), we arrive at

$$\overline{k} = \int_{-\infty}^{+\infty} \psi^*(x)(-i) \frac{\partial \psi(x)}{\partial x} \, dx$$

(E-5)

We now address the multiplicative $1/2\pi$ in (E-1). For this average to make sense, the average of unity should be unity. That is, we must have

$$\frac{1}{2\pi} \int_{-\infty}^{+\infty} 1 \, |\tilde{\psi}(k)|^2 \, dk = 1$$

To carry out this integral, we would follow somewhat the same course as before: We would use the Fourier transform definition to trade $\tilde{\psi}(k)$ for $\psi(x)$, but we would skip the integration by parts over $x$; no $-(i/k)(\partial/\partial x)$ would be needed to cancel a factor of $k$. The integration over $x'$ would again replace $\psi^*(x')$ by $2\pi\psi^*(x)$, canceling the multiplicative $1/2\pi$. Thus,

$$\frac{1}{2\pi} \int_{-\infty}^{+\infty} 1 \, |\tilde{\psi}(k)|^2 \, dk = \int_{-\infty}^{+\infty} \psi^*(x)\psi(x) \, dx$$

Since $\psi(x)$ is assumed normalized, the result is indeed unity. In short, $\int |\psi(x)|^2 \, dx = 1$ implies that $\int |\tilde{\psi}(k)|^2 \, dk = 2\pi$, and it is this different normalization of the Fourier transform that requires the multiplicative factor.

Finally, multiplying both sides of (E-5) by $\hbar$ and using $p = \hbar k$, we have

$$\overline{p} = \int_{-\infty}^{+\infty} \psi^*(x)(-i\hbar) \frac{\partial \psi(x)}{\partial x} \, dx$$

We conclude that to calculate the expectation value of the momentum, we must apply the operator $-i\hbar(\partial/\partial x)$ to the wave function $\psi(x)$ in the probability integral.

# Finite-Well Wave Functions

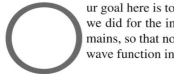

ur goal here is to take the finite-well wave function to the point, as we did for the infinite well, where only one arbitrary constant remains, so that normalization can be carried out. We begin with the wave function in the form given in (4-13).

$$\psi(x) = \begin{cases} Ce^{+\alpha x} & x < 0 \\ A\sin kx + B\cos kx & 0 < x < L \\ Ge^{-\alpha x} & x > L \end{cases} \qquad \text{(F-I)}$$

We may use three of the smoothness conditions to express $C$, $A$, and $G$ in terms of $B$:

$$C = B$$

$$\alpha C = kA \qquad \text{or} \qquad A = \frac{\alpha}{k}B$$

$$A\sin kL + B\cos kL = Ge^{-\alpha L} \qquad \text{or} \qquad G = \left(\frac{\alpha}{k}\sin kL + \cos kL\right)e^{\alpha L}B$$

Inserting these in (F-1) gives

$$\psi(x) = B \times \begin{cases} e^{+\alpha x} & x < 0 \\ \dfrac{\alpha}{k}\sin kx + \cos kx & 0 < x < L \\ \left(\dfrac{\alpha}{k}\sin kL + \cos kL\right)e^{\alpha L}e^{-\alpha x} & x > L \end{cases} \qquad \text{(F-2)}$$

Now, rather than use the fourth condition, it will be easier to use the quantization conditions (4-23b) we obtained by combining all four: $\alpha = k\tan(\frac{1}{2}kL)$ and $\alpha = -k\cot(\frac{1}{2}kL)$. However, we cannot handle both at once. Inserting the first gives a function that, for reasons soon to be clear, we refer to as the even function $\psi_e(x)$.

$$\psi_e(x) = B \times \begin{cases} e^{+\alpha x} & x < 0 \\ \tan(\frac{1}{2}kL)\sin kx + \cos kx & 0 < x < L \\ [\tan(\frac{1}{2}kL)\sin kL + \cos kL]e^{\alpha L}e^{-\alpha x} & x > L \end{cases} \qquad \text{(F-3)}$$

Writing $\tan(\frac{1}{2}kL)$ as $\sin(\frac{1}{2}kL)/\cos(\frac{1}{2}kL)$, unity as $\cos(\frac{1}{2}kL)/\cos(\frac{1}{2}kL)$, and rearranging a bit gives

$$\psi_e(x) = B \times \begin{cases} e^{+\alpha x} & x < 0 \\ \dfrac{\sin(\frac{1}{2}kL)\sin kx}{\cos(\frac{1}{2}kL)} + \dfrac{\cos(\frac{1}{2}kL)\cos kx}{\cos(\frac{1}{2}kL)} & 0 < x < L \\ \left[\dfrac{\sin(\frac{1}{2}kL)\sin kL}{\cos(\frac{1}{2}kL)} + \dfrac{\cos(\frac{1}{2}kL)\cos kL}{\cos(\frac{1}{2}kL)}\right]e^{\alpha L}e^{-\alpha x} & x > L \end{cases}$$

The identity $\cos(\alpha - \beta) = \sin\beta\sin\alpha + \cos\beta\cos\alpha$ may be used in two places. Thus,

$$\psi_e(x) = B \times \begin{cases} e^{+\alpha x} & x < 0 \\ \dfrac{\cos(kx - \frac{1}{2}L)}{\cos(\frac{1}{2}kL)} & 0 < x < L \\ \dfrac{\cos(kL - \frac{1}{2}kL)}{\cos(\frac{1}{2}kL)}e^{\alpha L}e^{-\alpha x} & x > L \end{cases}$$

$$= B \times \begin{cases} e^{+\alpha x} & x < 0 \\ \dfrac{\cos(kx - \frac{1}{2}kL)}{\cos(\frac{1}{2}kL)} & 0 < x < L \\ e^{-\alpha(x - L)} & x > L \end{cases}$$

If we now change variables, with $z \equiv x - \frac{1}{2}L$, the result is

$$\psi_e(z) = B \times \begin{cases} e^{+\alpha(z+\frac{1}{2}L)} & z < -\frac{1}{2}L \\[2mm] \dfrac{\cos(kz)}{\cos(\frac{1}{2}kL)} & -\frac{1}{2}L < z < +\frac{1}{2}L \\[3mm] e^{-\alpha(z-\frac{1}{2}L)} & z > +\frac{1}{2}L \end{cases}$$

This function is symmetric about the center of the well, $z = 0$ or $x = \frac{1}{2}L$, and is in fact an even function of $z$. If the other quantization condition, $\alpha = -k \cot(\frac{1}{2}kL)$, is inserted into (F-2) and similar steps taken, the result is

$$\psi_o(z) = B \times \begin{cases} e^{+\alpha(z+\frac{1}{2}L)} & z < -\frac{1}{2}L \\[2mm] -\dfrac{\sin(kz)}{\sin(\frac{1}{2}kL)} & -\frac{1}{2}L < z < +\frac{1}{2}L \\[3mm] -e^{-\alpha(z-\frac{1}{2}L)} & z > +\frac{1}{2}L \end{cases}$$

This is an odd function about the center of the well. In either case, only one arbitrary constant remains. To find it, we would have to solve (4-23) for an allowed value of $k$ (which would also determine $E$ and $\alpha$), then apply normalization. Since this would require numerical means, we go no further. Still, in the preceding forms it is easy to see that the functions are continuous at the walls, $z = \pm\frac{1}{2}L$. (The derivative is, of course, also still continuous.) The even wave functions include the ground-state ($n = 1$) as well as the third, the fifth, . . . , while the second, fourth, sixth, . . . are described by the odd functions.

# The Kinetic Energy Operator in Spherical Polar Coordinates

In this appendix, we have two related goals: (1) to find $\hat{L}^2$, the operator for the square of the angular momentum, and (2) to identify the part of the kinetic energy operator in spherical polar coordinates that corresponds to radial kinetic energy.

Classically, angular momentum is $\mathbf{L} = \mathbf{r} \times \mathbf{p}$ and its components are

$$L_x = yp_z - zp_y \qquad L_y = zp_x - xp_z \qquad L_z = xp_y - yp_x$$

We determine the quantum-mechanical operators for the components by inserting the basic operators for momentum and position. This is done for $L_z$ in Section 6.4, and the result expressed in spherical polar coordinates is

$$\hat{L}_z = -i\hbar \frac{\partial}{\partial \phi}$$

Let us do the same for $L_x$ and $L_y$.

$$\hat{L}_x = y\left(-i\hbar \frac{\partial}{\partial z}\right) - z\left(-i\hbar \frac{\partial}{\partial y}\right) \qquad \hat{L}_y = z\left(-i\hbar \frac{\partial}{\partial x}\right) - x\left(-i\hbar \frac{\partial}{\partial z}\right)$$

To express these in spherical polar coordinates, we transform the partial derivatives using standard rules of partial differentiation.

$$\hat{L}_x = -i\hbar \left\{ y\frac{\partial}{\partial z} - z\frac{\partial}{\partial y} \right\}$$

$$= -i\hbar \left\{ y\left( \frac{\partial r}{\partial z}\frac{\partial}{\partial r} + \frac{\partial \theta}{\partial z}\frac{\partial}{\partial \theta} + \frac{\partial \phi}{\partial z}\frac{\partial}{\partial \phi} \right) - z\left( \frac{\partial r}{\partial y}\frac{\partial}{\partial r} + \frac{\partial \theta}{\partial y}\frac{\partial}{\partial \theta} + \frac{\partial \phi}{\partial y}\frac{\partial}{\partial \phi} \right) \right\}$$

$$\hat{L}_y = -i\hbar \left\{ z\frac{\partial}{\partial x} - x\frac{\partial}{\partial z} \right\}$$

$$= -i\hbar \left\{ z\left( \frac{\partial r}{\partial x}\frac{\partial}{\partial r} + \frac{\partial \theta}{\partial x}\frac{\partial}{\partial \theta} + \frac{\partial \phi}{\partial x}\frac{\partial}{\partial \phi} \right) - x\left( \frac{\partial r}{\partial z}\frac{\partial}{\partial r} + \frac{\partial \theta}{\partial z}\frac{\partial}{\partial \theta} + \frac{\partial \phi}{\partial z}\frac{\partial}{\partial \phi} \right) \right\}$$

Partial derivatives such as $\partial r/\partial z$, $\partial \theta/\partial z$, and so on, are found in terms of $r$, $\theta$, and $\phi$ by expressing $dr$, $d\theta$, and $d\phi$ in terms of $dx$, $dy$, and $dz$. This is done by implicit differentiation of the formulas for $r$, $\theta$, and $\phi$ in terms of $x$, $y$, and $z$ (given in Table G.1), followed by reexpression of all coefficients in terms of $r$, $\theta$, and $\phi$. The mathematical details are left to the concerned student. The results are as follows:

$$dr = \frac{\partial r}{\partial x}dx + \frac{\partial r}{\partial y}dy + \frac{\partial r}{\partial z}dz = \sin\theta\cos\phi\,dx + \sin\theta\sin\phi\,dy + \cos\theta\,dz$$

$$d\theta = \frac{\partial \theta}{\partial x}dx + \frac{\partial \theta}{\partial y}dy + \frac{\partial \theta}{\partial z}dz = \frac{\cos\theta\cos\phi}{r}dx + \frac{\cos\theta\sin\phi}{r}dy - \frac{\sin\theta}{r}dz$$

$$d\phi = \frac{\partial \phi}{\partial x}dx + \frac{\partial \phi}{\partial y}dy + \frac{\partial \phi}{\partial z}dz = -\frac{\sin\phi}{r\sin\theta}dx + \frac{\cos\phi}{r\sin\theta}dy + 0\,dz$$

Inserting these in the previous expressions gives

$$\hat{L}_x = -i\hbar \left\{ r\sin\theta\sin\phi\left( \cos\theta\frac{\partial}{\partial r} - \frac{\sin\theta}{r}\frac{\partial}{\partial \theta} + 0\frac{\partial}{\partial \phi} \right) \right.$$

$$\left. - r\cos\theta\left( \sin\theta\sin\phi\frac{\partial}{\partial r} + \frac{\cos\theta\sin\phi}{r}\frac{\partial}{\partial \theta} + \frac{\cos\phi}{r\sin\theta}\frac{\partial}{\partial \phi} \right) \right\}$$

$$= -i\hbar \left\{ -\sin\phi\frac{\partial}{\partial \theta} - \cot\theta\cos\phi\frac{\partial}{\partial \phi} \right\}$$

**Table G.1**

| Rectangular–Polar Transformation | |
| --- | --- |
| $x = r\sin\theta\cos\phi$ | $r = \sqrt{x^2 + y^2 + z^2}$ |
| $y = r\sin\theta\sin\phi$ | $\theta = \cos^{-1}\dfrac{z}{\sqrt{x^2 + y^2 + z^2}}$ |
| $z = r\cos\theta$ | $\phi = \tan^{-1}\dfrac{y}{x}$ |

$$\hat{L}_y = -i\hbar \left\{ r \cos\theta \left( \sin\theta \cos\phi \frac{\partial}{\partial r} + \frac{\cos\theta \cos\phi}{r} \frac{\partial}{\partial\theta} - \frac{\sin\phi}{r \sin\theta} \frac{\partial}{\partial\phi} \right) \right.$$

$$\left. - r \sin\theta \cos\phi \left( \cos\theta \frac{\partial}{\partial r} - \frac{\sin\theta}{r} \frac{\partial}{\partial\theta} + 0 \frac{\partial}{\partial\phi} \right) \right\}$$

$$= -i\hbar \left\{ \cos\phi \frac{\partial}{\partial\theta} - \cot\theta \sin\phi \frac{\partial}{\partial\phi} \right\}$$

Now squaring these, we obtain

$$\hat{L}_x^2 = -\hbar^2 \left( -\sin\phi \frac{\partial}{\partial\theta} - \cot\theta \cos\phi \frac{\partial}{\partial\phi} \right) \left( -\sin\phi \frac{\partial}{\partial\theta} - \cot\theta \cos\phi \frac{\partial}{\partial\phi} \right)$$

$$= -\hbar^2 \left[ \sin^2\phi \frac{\partial^2}{\partial\theta^2} + \cot\theta \cos\phi \frac{\partial}{\partial\phi} \left( \sin\phi \frac{\partial}{\partial\theta} \right) \right.$$

$$\left. + \sin\phi \cos\phi \frac{\partial}{\partial\theta} \left( \cot\theta \frac{\partial}{\partial\phi} \right) + \cot^2\theta \cos\phi \frac{\partial}{\partial\phi} \left( \cos\phi \frac{\partial}{\partial\phi} \right) \right]$$

$$\hat{L}_y^2 = -\hbar^2 \left( \cos\phi \frac{\partial}{\partial\theta} - \cot\theta \sin\phi \frac{\partial}{\partial\phi} \right) \left( \cos\phi \frac{\partial}{\partial\theta} - \cot\theta \sin\phi \frac{\partial}{\partial\phi} \right)$$

$$= -\hbar^2 \left[ \cos^2\phi \frac{\partial^2}{\partial\theta^2} - \cot\theta \sin\phi \frac{\partial}{\partial\phi} \left( \cos\phi \frac{\partial}{\partial\theta} \right) \right.$$

$$\left. - \cos\phi \sin\phi \frac{\partial}{\partial\theta} \left( \cot\theta \frac{\partial}{\partial\phi} \right) + \cot^2\theta \sin\phi \frac{\partial}{\partial\phi} \left( \sin\phi \frac{\partial}{\partial\phi} \right) \right]$$

When these are added, the first terms add to $-\hbar^2(\partial^2/\partial\theta^2)$ and the third terms add to zero. Using the product rule on the quantities in parentheses shows that the second terms add to $-\hbar^2 \cot\theta(\partial/\partial\theta)$ and the fourth to $-\hbar^2 \cot^2\theta(\partial^2/\partial\phi^2)$. Thus,

$$\hat{L}_x^2 + \hat{L}_y^2 = -\hbar^2 \left[ \frac{\partial^2}{\partial\theta^2} + \cot\theta \frac{\partial}{\partial\theta} + \cot^2\theta \frac{\partial^2}{\partial\phi^2} \right]$$

Together with $\hat{L}_z^2 = -\hbar^2(\partial^2/\partial\phi^2)$, we then have

$$\hat{L}^2 = \hat{L}_x^2 + \hat{L}_y^2 + \hat{L}_z^2 = -\hbar^2 \left[ \frac{\partial^2}{\partial\theta^2} + \cot\theta \frac{\partial}{\partial\theta} + \cot^2\theta \frac{\partial^2}{\partial\phi^2} \right] - \hbar^2 \frac{\partial^2}{\partial\phi^2}$$

$$= -\hbar^2 \left[ \frac{\partial^2}{\partial\theta^2} + \cot\theta \frac{\partial}{\partial\theta} + (\cot^2\theta + 1) \frac{\partial^2}{\partial\phi^2} \right] = -\hbar^2 \left[ \frac{\partial^2}{\partial\theta^2} + \cot\theta \frac{\partial}{\partial\theta} + \csc^2\theta \frac{\partial^2}{\partial\phi^2} \right]$$

or

$$\hat{L}^2 = -\hbar^2 \left[ \csc\theta \frac{\partial}{\partial\theta} \left( \sin\theta \frac{\partial}{\partial\theta} \right) + \csc^2\theta \frac{\partial^2}{\partial\phi^2} \right]$$

Our first goal has been accomplished. Now let us use this to identify radial kinetic energy. In spherical polar coordinates, the Schrödinger equation is

$$-\frac{\hbar^2}{2m} \frac{1}{r^2} \left[ \frac{\partial}{\partial r} \left( r^2 \frac{\partial}{\partial r} \right) + \csc\theta \frac{\partial}{\partial\theta} \left( \sin\theta \frac{\partial}{\partial\theta} \right) + \csc^2\theta \frac{\partial^2}{\partial\phi^2} \right] \psi(r, \theta, \phi) + U(r, \theta, \phi)\psi(r, \theta, \phi) = E\psi(r, \theta, \phi)$$

$$\underbrace{\phantom{-\frac{\hbar^2}{2m}}}_{\text{Kinetic}} \qquad\qquad \underbrace{\phantom{U(r,\theta,\phi)}}_{\text{Potential}} \qquad \underbrace{\phantom{E\psi}}_{\text{Total}}$$

The first term on the left involves the kinetic energy operator.

$$\hat{KE}\psi(r, \theta, \phi) = -\frac{\hbar^2}{2mr^2}\frac{\partial}{\partial r}\left(r^2\frac{\partial}{\partial r}\right)\psi(r, \theta, \phi) - \frac{\hbar^2}{2mr^2}\left[\csc\theta\frac{\partial}{\partial\theta}\left(\sin\theta\frac{\partial}{\partial\theta}\right) + \csc^2\theta\frac{\partial^2}{\partial\phi^2}\right]\psi(r, \theta, \phi)$$

With our result for $\hat{L}^2$, we see that

$$\hat{KE}\psi(r, \theta, \phi) = -\frac{\hbar^2}{2mr^2}\frac{\partial}{\partial r}\left(r^2\frac{\partial}{\partial r}\right)\psi(r, \theta, \phi) + \frac{1}{2mr^2}\hat{L}^2\psi(r, \theta, \phi)$$

Since $KE_{rot} = L^2/2I$ (just as $KE_{trans} = p^2/2m$) and the moment of inertia $I$ for a point particle is $mr^2$, we see that the second term involves the rotational kinetic energy operator. The first term is then the kinetic energy associated strictly with radial motion.

$$\hat{KE}_{radial} = -\frac{\hbar^2}{2mr^2}\frac{\partial}{\partial r}\left(r^2\frac{\partial}{\partial r}\right) \quad \text{(G-1)} \qquad \hat{KE}_{rotational} = \frac{1}{2mr^2}\hat{L}^2 \quad \text{(G-2)}$$

We have deduced that (G-1) gives radial kinetic, but let us show it directly. First, we need to prove the following theorem:

$$\frac{\partial}{\partial r} = \frac{x}{r}\frac{\partial}{\partial x} + \frac{y}{r}\frac{\partial}{\partial y} + \frac{z}{r}\frac{\partial}{\partial z}$$

Using standard properties of partial derivatives, the partial derivative with respect to $r$ of a function of $(x, y, z)$ is

$$\frac{\partial}{\partial r}f = \left\{\frac{\partial x}{\partial r}\frac{\partial}{\partial x} + \frac{\partial y}{\partial r}\frac{\partial}{\partial y} + \frac{\partial z}{\partial r}\frac{\partial}{\partial z}\right\}f(x, y, z)$$

Using the $(x, y, z) \leftrightarrow (r, \theta, \phi)$ transformations of Table G.1, it may easily be shown that

$$\frac{\partial x}{\partial r} = \frac{x}{r} \qquad \frac{\partial y}{\partial r} = \frac{y}{r} \qquad \frac{\partial z}{\partial r} = \frac{z}{r}$$

Thus,

$$\frac{\partial}{\partial r}f = \left\{\frac{x}{r}\frac{\partial}{\partial x} + \frac{y}{r}\frac{\partial}{\partial y} + \frac{z}{r}\frac{\partial}{\partial z}\right\}f(x, y, z)$$

Now, using the result of the theorem,

$$\frac{-\hbar^2}{2mr^2}\frac{\partial}{\partial r}\left(r^2\frac{\partial}{\partial r}\right) = \frac{-\hbar^2}{2mr^2}\left\{\frac{x}{r}\frac{\partial}{\partial x} + \frac{y}{r}\frac{\partial}{\partial y} + \frac{z}{r}\frac{\partial}{\partial z}\right\}\left(r^2\left\{\frac{x}{r}\frac{\partial}{\partial x} + \frac{y}{r}\frac{\partial}{\partial y} + \frac{z}{r}\frac{\partial}{\partial z}\right\}\right)$$

To find a quantum-mechanical operator from a classical property, we replace $x$, $y$, $z$, $p_x$, $p_y$, and $p_z$ with their operators. Here we wish to do the opposite: to identify the classical property to which a quantum-mechanical operator corresponds. Thus, we make the following "backward" associations:

$$\frac{\partial}{\partial x} \rightarrow \frac{p_x}{-i\hbar} \qquad \frac{\partial}{\partial y} \rightarrow \frac{p_y}{-i\hbar} \qquad \frac{\partial}{\partial z} \rightarrow \frac{p_z}{-i\hbar}$$

whereupon

$$\frac{-\hbar^2}{2mr^2}\frac{\partial}{\partial r}\left(r^2\frac{\partial}{\partial r}\right) \quad\rightarrow\quad \frac{-\hbar^2}{2mr^2}\frac{1}{(-i\hbar)^2}\left\{\frac{x}{r}p_x + \frac{y}{r}p_y + \frac{z}{r}p_z\right\}\left(r^2\left\{\frac{x}{r}p_x + \frac{y}{r}p_y + \frac{z}{r}p_z\right\}\right)$$

We now have classical properties rather than differential operators, so we may cancel powers of r, leaving

$$\frac{-\hbar^2}{2mr^2}\frac{\partial}{\partial r}\left(r^2\frac{\partial}{\partial r}\right) \quad\rightarrow\quad \frac{1}{2mr^2}(xp_x + yp_y + zp_z)^2$$

With a few self-explanatory steps, we now show what we set out to show:

$$\frac{-\hbar^2}{2mr^2}\frac{\partial}{\partial r}\left(r^2\frac{\partial}{\partial r}\right) \quad\rightarrow\quad \frac{1}{2mr^2}(xp_x + yp_y + zp_z)^2 = \frac{1}{2mr^2}\left(x\,m\,\frac{dx}{dt} + y\,m\,\frac{dy}{dt} + z\,m\,\frac{dz}{dt}\right)^2$$

$$= \frac{1}{2mr^2}\left(m\frac{1}{2}\frac{dx^2}{dt} + m\frac{1}{2}\frac{dy^2}{dt} + m\frac{1}{2}\frac{dz^2}{dt}\right)^2$$

$$= \frac{1}{2mr^2}\left(\frac{1}{2}m\,\frac{d(x^2 + y^2 + z^2)}{dt}\right)^2$$

$$= \frac{1}{2mr^2}\left(\frac{1}{2}m\,\frac{dr^2}{dt}\right)^2 = \frac{1}{2mr^2}\left(\frac{1}{2}m2r\,\frac{dr}{dt}\right)^2 = \frac{1}{2}m\left(\frac{dr}{dt}\right)^2$$

Using a dot to indicate a total derivative with respect to time,

$$\hat{\mathrm{KE}}_{\mathrm{radial}} = -\frac{\hbar^2}{2mr^2}\frac{\partial}{\partial r}\left(r^2\frac{\partial}{\partial r}\right) \quad\rightarrow\quad \frac{1}{2}m\,\dot{r}^2$$

The operator in (G-1) is indeed the kinetic energy for solely radial motion.

By a similar procedure it may be shown that the operator in (G-2) corresponds to classical quantities as follows:

$$\hat{\mathrm{KE}}_{\mathrm{rotational}} = \frac{1}{2mr^2}\hat{L}^2 \quad\rightarrow\quad \frac{1}{2}m\,r^2\,\dot{\theta}^2 + \frac{1}{2}mr^2\,\sin^2\theta\,\dot{\phi}^2$$

This fits perfectly, for the classical kinetic energy may be written as

$$\mathrm{KE} = \frac{1}{2}m(v_x^2 + v_y^2 + v_z^2) = \frac{1}{2}m(\dot{x}^2 + \dot{y}^2 + \dot{z}^2)$$

$$= \frac{1}{2}m\,\dot{r}^2 + \left(\frac{1}{2}mr^2\,\dot{\theta}^2 + \frac{1}{2}mr^2\,\sin^2\theta\,\dot{\phi}^2\right)$$

The $\mathrm{KE}_{\mathrm{radial}}$ term is nonzero if and only if r varies with time, while the $\mathrm{KE}_{\mathrm{rotational}}$ terms are nonzero if and only if the angles vary with time.

# Quantum
# Distributions

ur goal here is to determine for a general quantum thermodynamic system the average number of particles $\overline{N}(E_n)$ that will occupy individual-particle state $n$ whose energy is $E_n$.[1] Written most simply, this average should be

$$\overline{N}(E_n) = \frac{\Sigma_{N_n} N_n W_n}{W} = \frac{\Sigma_{N_n} N_n W_n}{\Sigma_{N_n} W_n}$$

where $N_n$ is a *possible* number of particles in that state; $W$ is the total number of ways of distributing the system's energy $E$ among its $N$ particles; and $W_n$ (i.e., with $N_n$ fixed) is the restricted number of ways of distributing among the $N - N_n$ particles *not* in state $n$ the energy $E - N_n E_n$ remaining for those particles. While $W$ is a function of $N$ and $E$, $W_n$ is a function of $N - N_n$ and $E - N_n E_n$. Thus, a more complete way of writing the average is

$$\overline{N}(E_n) = \frac{\Sigma_{N_n} N_n W_n(N - N_n, \ E - E_n N_n)}{W(N, \ E)} = \frac{\Sigma_{N_n} N_n W_n(N - N_n, \ E - E_n N_n)}{\Sigma_{N_n} W_n(N - N_n, \ E - E_n N_n)}$$

(H-1)

For the time being, we allow for an unspecified number of particles in state $n$ by leaving off the limits on the sums.

[1] The bar over the $N$ is used only in this appendix. An "occupation number," such as we derive here, is indeed an average number of particles in state $n$ (although it is a very reliable one if the system is truly thermodynamic/large), and the bar is here to distinguish this *average* value from the various *possible* values, $N_n$.

It would certainly seem that $W_n$ cannot be known unless the specific quantum system is known. However, with just one assumption and one observation, equation (H-1) assumes a form that does not depend on the system in any detailed way, but only upon two of its overall properties, one of which is temperature. The assumption is that $N_n$ is a small fraction of $N$, so that the $N - N_n$ other particles constitute the vastly larger part of the system, called the **reservoir.**[2] Thus, a minimal (unit) change in $N_n$ would have only small (differential) effects on $N - N_n$ and $E - N_n E_n$. The observation is that numbers of ways always involve "combinatorial factors" such as $a^N$ and $N!$, which, though not smooth functions, become progressively smooth in the limit of large $N$ (cf. Figure 8.1). Moreover, when the arguments of such functions are $X - x$, where $X$ is a large quantity and $x$ a small quantity, they decrease *exponentially* with $x$.[3] Here, the small quantity is $N_n$. Putting this all together, we expect $W_n$ to be a smooth exponentially decreasing function of $N_n$. Therefore, we may expand the *logarithm* of $W_n$ to the first power in $N_n$, which will give us the coefficient of $N_n$ in the exponential's argument.

$$\ln(W_n(N - N_n, E - N_n E_n)) \cong$$

$$\ln(W_n(N - N_n, E - N_n E_n))\bigg|_{N_n=0} + N_n \left[ \frac{d}{dN_n} \ln(W_n(N - N_n, E - N_n E_n)) \right]\bigg|_{N_n=0} \tag{H-2}$$

Consider the derivative in the second term. By the chain rule,

$$\frac{d}{dN_n} \ln(W_n(N - N_n, E - N_n E_n))$$

$$= \frac{\partial(N - N_n)}{\partial N_n} \frac{\partial}{\partial(N - N_n)} \ln(W_n(N - N_n, E - N_n E_n))$$

$$+ \frac{\partial(E - N_n E_n)}{\partial N_n} \frac{\partial}{\partial(E - N_n E_n)} \ln(W_n(N - N_n, E - N_n E_n))$$

$$= -\frac{\partial}{\partial(N - N_n)} \ln(W_n(N - N_n, E - N_n E_n))$$

$$- E_n \frac{\partial}{\partial(E - N_n E_n)} \ln(W_n(N - N_n, E - N_n E_n))$$

Evaluating at $N_n = 0$,

$$\left[ \frac{d}{dN_n} \ln(W_n(N - N_n, E - N_n E_n)) \right]\bigg|_{N_n=0} = -\frac{\partial}{\partial N} \ln(W_n(N, E)) - E_n \frac{\partial}{\partial E} \ln(W_n(N, E))$$

---

[2]If the distributions we derive are to be reliable, there must be a "vastly larger part" of the overall system to serve as reservoir. In many cases, the system of interest is indeed a small part of a larger system; it exchanges energy with its environment, which is then the reservoir. In cases where the system of interest encompasses everything, either we assume that $N_n \ll N - N_n$, with the $N - N_n$ serving as reservoir, or we consider $n$ to be a state within only a subvolume of the larger system, whose remainder is the reservoir, and assume that the distribution in the subvolume is representative of the whole system.

[3]The reader may verify that $(X - x)! \cong X! e^{-x \ln X}$ by plotting the two functions versus $x$, using a large value, perhaps 100, for $X$. Note also that $a^{X-x} = a^X a^{-x} = a^X e^{-x \ln a}$.

Now, reinserting in (H-2) then exponentiating, we have our exponential dependence on $N_n$:

$$\ln(W_n(N - N_n, E - N_nE_n)) \cong \ln(W_n(N, E)) + N_n\left[-\frac{\partial}{\partial N}\ln(W_n(N, E)) - E_n\frac{\partial}{\partial E}\ln(W_n(N, E))\right]$$

$$W_n(N - N_n, E - N_nE_n) \cong W_n(N, E)\exp\left\{-N_n\left[\frac{\partial}{\partial N}\ln(W_n(N, E)) + E_n\frac{\partial}{\partial E}\ln(W_n(N, E))\right]\right\} \quad \text{(H-3)}$$

From definition (8-2), we know that $\ln W = S/k_B$, where $S$ is the entropy. In equation (H-3), however, we have $\ln W_n(N, E)$, where $N_n$ has been evaluated at zero. The distinction is not significant. So long as $N_n$ is only a small fraction of $N$, the entropy with $N_n$ fixed at zero differs negligibly from what it would be if $N_n$ were unrestricted. Thus,

$$W_n(N - N_n, E - N_nE_n) \cong W_n(N, E)\exp\left\{-\frac{N_n}{k_B}\left[\frac{\partial S}{\partial N} + E_n\frac{\partial S}{\partial E}\right]\right\}$$

Let us take stock of how our calculation has been simplified. The quantities $\partial S/\partial N$ and $\partial S/\partial E$ depend only on the reservoir; they are properties of the overall system. We will not attempt to characterize $\partial S/\partial N$, but merely state that it is proportional to the "chemical potential," used much in chemistry. We shall see that it is simply related to the constant $B$ in equations (8-30) and (8-31). It best serves our purposes merely to redefine it: $\partial S/\partial N \equiv k_B\alpha$. The other property, though, is quite familiar; from Section 8.2, $\partial S/\partial E \equiv 1/T$. Do not overlook this important step: Energy distributions depend on temperature, and it is here that it enters the picture. Substituting,

$$W_n(N - N_n, E - N_nE_n) \cong W_n(N, E)\exp\left(-\alpha N_n - \frac{N_nE_n}{k_BT}\right)$$

Now reinserting in (H-1),

$$\overline{N}(E_n) = \frac{\sum_{N_n} N_nW_n(N, E)\exp\left(-\alpha N_n - \frac{N_nE_n}{k_BT}\right)}{\sum_{N_n} W_n(N, E)\exp\left(-\alpha N_n - \frac{N_nE_n}{k_BT}\right)} = \frac{\sum_{N_n} N_n\exp\left(-\alpha N_n - \frac{N_nE_n}{k_BT}\right)}{\sum_{N_n} \exp\left(-\alpha N_n - \frac{N_nE_n}{k_BT}\right)} \quad \text{(H-4)}$$

Because it was already evaluated at $N_n = 0$, $W_n(N, E)$ cancels. This step too is easily overlooked but of great importance, for $W_n(N, E)$ contains all the information about distributing the remaining energy among the states other than $n$. That it need not even be known means that *the Bose–Einstein and Fermi–Dirac distributions are applicable regardless of the specific system involved.*

To simplify the calculation, we resort to a "standard mathematical trick." That (H-4) leads to the following is left as a mental exercise.

$$\overline{N}(E_n) = -\frac{\partial}{\partial\alpha}\ln[\sum_{N_n}(e^{-\alpha - E_n/k_BT})^{N_n}] \quad \text{(H-5)}$$

Only now must we distinguish between fermions and bosons. For fermions, there cannot be multiple particles in the same state; $N_n$ is restricted to the values 0 and 1. For bosons, the number of particles in any given state is arbitrary; we allow

$N_n$ to range from zero to infinity. Of course, we have assumed that $N_n$ is much less than $N$, yet it may still be a large number. The important point is that the summand $\exp[-N_n(\alpha + E_n/k_BT)]$ falls to practically zero long before the assumption is violated. Extending the sum to infinity merely makes it easier to see its value: It is of the form $\sum_{m=0}^{\infty} x^m = 1/(1-x)$.

For bosons,

$$\overline{N}(E_n)_{\text{B-E}} = -\frac{\partial}{\partial\alpha}\ln\left[\sum_{N_n=0}^{N_n=\infty}(e^{-\alpha-E_n/k_BT})^{N_n}\right] = -\frac{\partial}{\partial\alpha}\ln\left(\frac{1}{1-e^{-\alpha-E_n/k_BT}}\right) = \frac{\partial}{\partial\alpha}\ln(1-e^{-\alpha-E_n/k_BT})$$

$$\overline{N}(E_n)_{\text{B-E}} = \frac{1}{e^{\alpha+E_n/k_BT}-1}$$

For fermions,

$$\overline{N}(E_n)_{\text{F-D}} = -\frac{\partial}{\partial\alpha}\ln\left[\sum_{N_n=0}^{N_n=1}(e^{-\alpha-E_n/k_BT})^{N_n}\right] = -\frac{\partial}{\partial\alpha}\ln(1+e^{-\alpha-E_n/k_BT})$$

$$\overline{N}(E_n)_{\text{F-D}} = \frac{1}{e^{\alpha+E_n/k_BT}+1}$$

With $e^{\alpha}$ defined to be $B$, these are expressions (8-30) and (8-31).

## The Boltzmann Distribution

From the preceding derivation it might appear that only the Bose–Einstein and Fermi–Dirac distributions result naturally from our careful statistical accounting. This would be troubling, because we know that there are distinguishable quantum systems for which the Boltzmann distribution is correct. The resolution of this seeming difficulty is that early in the derivation we did assume the particles to be indistinguishable. We assumed that $W_n$ is the number of ways of distributing energy with $N_n$ fixed, irrespective of *which* particles were in state $n$. However, there are $N!/N_n!$ ways of switching specific particles/labels between state $n$ and the rest of the states without actually changing $N_n$, and *these are indeed different ways if (but only if) the particles are distinguishable.* We may account for the increased permutations of particle labels by including $N!/N_n!$ within all sums over $N_n$. It carries through to (H-4), whereupon $N!$ cancels top and bottom, so that (H-5) becomes

$$\overline{N}(E_n)_{\text{Boltz}} = -\frac{\partial}{\partial\alpha}\ln\left[\sum_{N_n}\frac{(e^{-\alpha-E_n/k_BT})^{N_n}}{N_n!}\right]$$

Allowing $N_n$ to range over all possible values, the sum becomes $\sum_{m=0}^{\infty}(x^m/m!) = e^x$. Thus,

$$\overline{N}(E_n)_{\text{Boltz}} = -\frac{\partial}{\partial\alpha}\ln[\exp(e^{-\alpha-E_n/k_BT})] = -\frac{\partial}{\partial\alpha}e^{-\alpha-E_n/k_BT}$$

$$\overline{N}(E_n)_{\text{Boltz}} = \frac{1}{e^{\alpha+E_n/k_BT}}$$

Again identifying $e^{\alpha}$ with $B$, this is expression (8-29), the Boltzmann distribution.

# I

# The Diatomic-Molecule Schrödinger Equation

Classically, the nontranslational energy of the two-particle system depicted in Figure I.1 is given by

$$E = E_{\text{rot}} + E_{\text{vib}} = \frac{1}{2}(I_1 + I_2)\omega^2 + \left(\frac{1}{2}m_2v_{1x}^2 + \frac{1}{2}m_2v_{2x}^2 + \frac{1}{2}\kappa x_r^2\right)$$

$$\underbrace{\qquad\qquad}_{\textbf{Rotational}} \qquad \underbrace{\qquad\qquad\qquad\qquad\qquad}_{\textbf{Vibrational}}$$

This accounts for (1) the particles' rotation at common angular frequency $\omega$ about the system's center of mass, with moments of inertia $I_1$ and $I_2$, and (2) their vibrational motion along the molecular axis, each having kinetic energy and exchanging energy with the spring. However, rotational and vibrational energies are not independent of each other. As the atoms move apart in their vibrational motion, the system's moment of inertia increases and its rotational kinetic energy decreases. (We may write $E_{\text{rot}}$ as $L^2/2I$, and if $I$ increases, $E_{\text{rot}}$ decreases, because $L$ cannot change.) To conserve energy, the vibrational energy would have to increase correspondingly—due to the action, one might argue, of "centrifugal forces." Similarly, when the atoms move toward each other, rotational energy increases and vibrational decreases.

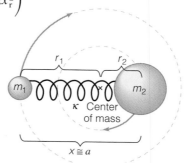

**Figure I.1** In a diatomic molecule, atoms rotate about their common center of mass.

This interdependence would make analysis of the problem extremely difficult, but we will assume that the two energies *are independent,* that although the separation between the atoms does vary periodically, it is not so severe as to significantly alter the system's moment of inertia. In other words, we treat the separation as constant for rotational motion, but allow it to vary so far as vibration is concerned. This may strike the reader a very shaky premise, but it is often a good approximation. (See Exercise 11 in Chapter 9.)

We seek an equivalent one-particle system. Let us consider rotation and vibration separately. The rotational energy is

$$E_{\text{rot}} = \frac{1}{2}(I_1 + I_2)\omega^2 = \frac{1}{2}(m_1 r_1^2 + m_1 r_2^2)\omega^2$$

where the atoms are treated as point masses (i.e., $I_{\text{point mass}} = mr^2$). Since the rotation is about an axis at the center of mass, we must have

$$m_1 r_1 = m_2 r_2$$

But under the assumption (for rotation only) of fixed separation, we also have

$$r_1 + r_2 = a$$

Solving for $r_1$ and $r_2$ between these equations, we find that

$$r_1 = \frac{m_2}{m_1 + m_2}a \quad \text{and} \quad r_2 = \frac{m_1}{m_1 + m_2}a$$

Inserting these in the expression for $E_{\text{rot}}$ and rearranging a bit gives

$$E_{\text{rot}} = \frac{1}{2}(\mu a^2)\omega^2 \quad \text{where} \quad \mu \equiv \frac{m_1 m_2}{m_1 + m_2}$$

This is an expression for the rotational energy of a single particle of mass $\mu$ rotating at a distance $a$ about a fixed point.

The vibrational energy is

$$E_{\text{vib}} = \frac{1}{2}m_1 v_{1x}^2 + \frac{1}{2}m_2 v_{2x}^2 + \frac{1}{2}\kappa x_r^2$$

In the reference frame where the center of mass is at rest, the masses must have opposite momenta.

$$m_1 v_{1x} = -m_2 v_{2x}$$

If we define the relative velocity by

$$v_r \equiv v_{2x} - v_{1x}$$

the individual velocities may then be expressed as

$$v_{1x} = -\frac{m_2}{m_1 + m_2}v_r \quad \text{and} \quad v_{2x} = \frac{m_1}{m_1 + m_2}v_r$$

Inserting these in $E_{\text{vib}}$ then rearranging a bit, we obtain

$$E_{\text{vib}} = \frac{1}{2}\mu v_r^2 + \frac{1}{2}\kappa x_r^2$$

Once again, we have an expression for the energy of a single particle of mass $\mu$. In this case, it is connected to a spring of force constant $\kappa$.

Putting rotation and vibration together, we conclude that analysis of the two-particle system is equivalent to analysis of the one-particle system shown in Figure I.2.

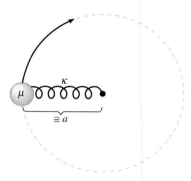

$$E = \frac{1}{2}(\mu a^2)\omega^2 + \left(\frac{1}{2}\mu v_r^2 + \frac{1}{2}\kappa x_r^2\right)$$

       **Rotational**       **Vibrational**

**Figure I.2**  One-particle equivalent of a diatomic molecule.

The quantity $\mu = m_1 m_2/(m_1 + m_2)$ appearing in both terms is the so-called reduced mass.

Solution of the one-particle Schrödinger equation corresponding to our classical model is straightforward: It is a central force problem, identical to the hydrogen atom except that a spring, rather than electrostatic attraction, now provides the force. This affects only the radial part. In spherical polar coordinates, the Schrödinger equation is

$$-\frac{\hbar^2}{2m}\frac{1}{r^2}\left[\frac{\partial}{\partial r}\left(r^2\frac{\partial}{\partial r}\right) + \csc\theta\frac{\partial}{\partial\theta}\left(\sin\theta\frac{\partial}{\partial\theta}\right) + \csc^2\theta\frac{\partial^2}{\partial\phi^2}\right]\psi(r, \theta, \phi) + U(r, \theta, \phi)\psi(r, \theta, \phi) = E\psi(r, \theta, \phi)$$

Inserting the potential energy $\frac{1}{2}\kappa x_r^2$, substituting $\mu$ for $m$, then rearranging, gives

$$\left\{-\frac{\hbar^2}{2\mu}\frac{1}{r^2}\left[\csc\theta\frac{\partial}{\partial\theta}\left(\sin\theta\frac{\partial}{\partial\theta}\right) + \csc^2\theta\frac{\partial^2}{\partial\phi^2}\right]\right\}\psi + \left\{-\frac{\hbar^2}{2\mu}\frac{1}{r^2}\left[\frac{\partial}{\partial r}\left(r^2\frac{\partial}{\partial r}\right)\right] + \frac{1}{2}\kappa x_r^2\right\}\psi = E\psi$$

                **Rotational kinetic**                       **Vibrational: radial kinetic and potential**

As usual, we separate variables by substituting $R\Theta\Phi$ for $\psi$, then dividing by $R\Theta\Phi$.

$$\frac{1}{\Theta\Phi}\left\{-\frac{\hbar^2}{2\mu}\frac{1}{r^2}\left[\csc\theta\frac{\partial}{\partial\theta}\left(\sin\theta\frac{\partial}{\partial\theta}\right) + \csc^2\theta\frac{\partial^2}{\partial\phi^2}\right]\right\}\Theta\Phi + \frac{1}{R}\left\{-\frac{\hbar^2}{2\mu}\frac{1}{r^2}\left[\frac{\partial}{\partial r}\left(r^2\frac{\partial}{\partial r}\right)\right] + \frac{1}{2}\kappa x_r^2\right\}R = E$$

                **Rotational kinetic**                       **Vibrational: radial kinetic and potential**     (I-1)

As noted earlier, we assume that in the rotational term, $r$ is a constant—the equilibrium atomic separation $a$. Solution of this purely angular part is then the same as for all central forces. According to equation (6-20),

$$\frac{1}{2\mu a^2}\frac{1}{\Theta\Phi}\left\{-\hbar^2\left[\csc\theta\frac{\partial}{\partial\theta}\left(\sin\theta\frac{\partial}{\partial\theta}\right) + \csc^2\theta\frac{\partial^2}{\partial\phi^2}\right]\right\}\Theta\Phi = \frac{1}{2\mu a^2}\frac{1}{\Theta\Phi}\hat{L}^2\Theta\Phi$$

Equation (6-21) then gives us the same sort of angular momentum quantization as in the hydrogen atom:

$$\frac{1}{2\mu a^2}\frac{1}{\Theta\Phi}\left\{-\hbar^2\left[\csc\theta\frac{\partial}{\partial\theta}\left(\sin\theta\frac{\partial}{\partial\theta}\right) + \csc^2\theta\frac{\partial^2}{\partial\phi^2}\right]\right\}\Theta\Phi = \frac{1}{2\mu a^2}\hbar^2\ell(\ell + 1)$$

Thus, we write

$$E_{\text{rot}} = \frac{\hbar^2\ell(\ell + 1)}{2\mu a^2} \quad \text{where} \quad \ell = 0, 1, 2, \ldots$$

A molecule may not possess any arbitrary amount of rotational kinetic energy, but only certain discrete values.

Equation (I-1) now reduces to

$$E_{\text{rot}} + \frac{1}{R}\left\{ -\frac{\hbar^2}{2\mu}\frac{1}{r^2}\left[\frac{d}{dr}\left(r^2\frac{d}{dr}\right)\right] + \frac{1}{2}\kappa x_r^2 \right\} R = E$$

or, identifying $E - E_{\text{rot}}$ as $E_{\text{vib}}$,

$$\frac{1}{R}\left\{ -\frac{\hbar^2}{2\mu}\frac{1}{r^2}\left[\frac{d}{dr}\left(r^2\frac{d}{dr}\right)\right] + \frac{1}{2}\kappa x_r^2 \right\} R = E_{\text{vib}}$$

It is left to the concerned student to show that this may be rewritten as

$$-\frac{\hbar^2}{2\mu}\frac{d^2}{dr^2}(rR) + \frac{1}{2}\kappa x_r^2(rR) = E_{\text{vib}}(rR)$$

Finally, with the definition

$$f(x_r) \equiv rR$$

we obtain

$$-\frac{\hbar^2}{2\mu}\frac{d^2}{dx_r^2}f(x_r) + \frac{1}{2}\kappa x_r^2 f(x_r) = E_{\text{vib}}f(x_r)$$

(*Note:* The transformation of the derivative is valid since $x_r$ and $r$ differ only by a constant; i.e., $x_r \equiv r - a$.) This is simply the one-dimensional harmonic oscillator Schrödinger equation, with energy levels given by

$$E_{\text{vib}} = \left(n + \frac{1}{2}\right)\hbar\sqrt{\frac{\kappa}{\mu}} \quad \text{where} \quad n = 0, 1, 2, \ldots$$

Combining vibration and rotation, we arrive at a total energy dependent on two quantum numbers, $n$ and $\ell$:[1]

$$E_{n,\ell} = E_{\text{vib}} + E_{\text{rot}} = \left(n + \frac{1}{2}\right)\hbar\sqrt{\frac{\kappa}{\mu}} + \frac{\hbar^2\ell(\ell+1)}{2\mu a^2} \quad \begin{array}{l} n = 0, 1, 2, \ldots \\ \ell = 0, 1, 2, \ldots \end{array}$$

---

[1] In the hydrogen atom, $n$ is a "principal" quantum number, related to the *total* energy, both rotational and radial forms. The condition restricting $\ell$ to values no greater than $n - 1$ arises because rotational energy cannot exceed the total energy. Here, however, $n$ is not a principal quantum number; it determines only vibrational energy. Consequently, there is no condition linking $\ell$ to $n$.

# Properties of Isotopes

| Z | Element | Symbol | A | Atomic Mass | % Natural Abundance* | Half-life (Decay Modes)† |
|---|---------|--------|---|-------------|----------------------|--------------------------|
| 1 | Hydrogen | H | 1 | 1.007825 | 99.985 | |
|   |          |   | 2 | 2.014102 | 0.015 | |
|   |          |   | 3 | 3.016049 |  | 12.32 yr ($\beta^-$) |
| 2 | Helium | He | 3 | 3.016029 | $1.37 \times 10^{-6}$ | |
|   |        |    | 4 | 4.002603 | $\approx$100 | |
| 3 | Lithium | Li | 6 | 6.015121 | 7.5 | |
|   |         |    | 7 | 7.016003 | 92.5 | |
| 4 | Beryllium | Be | 7 | 7.016928 |  | 53.28 d (E.C.) |
|   |           |    | 8 | 8.005305 |  | $\approx$0.07 fs ($2\alpha$) |
|   |           |    | 9 | 9.012182 | 100 | |
|   |           |    | 10 | 10.013534 |  | 1.52 Myr ($\beta^-$) |
| 5 | Boron | B | 10 | 10.012937 | 19.9 | |
|   |       |   | 11 | 11.009305 | 80.1 | |
|   |       |   | 12 | 12.014352 |  | 20.2 ms ($\beta^-$) |
| 6 | Carbon | C | 11 | 11.01143 |  | 20.3 min (E.C., $\beta^+$) |
|   |        |   | 12 | 12 | 98.9 | |
|   |        |   | 13 | 13.003355 | 1.1 | |
|   |        |   | 14 | 14.003241 |  | 5730 yr ($\beta^-$) |
| 7 | Nitrogen | N | 12 | 12.018613 |  | 11.00 ms ($\beta^+$) |
|   |          |   | 13 | 13.005738 |  | 9.97 min ($\beta^+$) |
|   |          |   | 14 | 14.003074 | 99.63 | |
|   |          |   | 15 | 15.000108 | 0.37 | |
| 8 | Oxygen | O | 15 | 15.003065 |  | 122.2 s ($\beta^+$) |
|   |        |   | 16 | 15.994915 | 99.76 | |
|   |        |   | 17 | 16.999131 | 0.04 | |
|   |        |   | 18 | 17.99916 | 0.30 | |
|   |        |   | 19 | 19.003577 |  | 26.9 s ($\beta^-$) |
| 9 | Fluorine | F | 19 | 18.998403 | 100 | |
| 10 | Neon | Ne | 20 | 19.992435 | 90.48 | |
|    |      |    | 21 | 20.993843 | 0.27 | |
|    |      |    | 22 | 21.991383 | 0.25 | |
| 11 | Sodium | Na | 23 | 22.989767 | 100 | |
| 12 | Magnesium | Mg | 24 | 23.985042 | 78.99 | |
|    |           |    | 25 | 24.98537 | 10.00 | |
|    |           |    | 26 | 25.982593 | 11.01 | |
| 13 | Aluminum | Al | 27 | 26.981539 | 100 | |
| 14 | Silicon | Si | 28 | 27.976927 | 92.23 | |
|    |         |    | 29 | 28.976495 | 4.67 | |
|    |         |    | 30 | 29.97377 | 3.10 | |
| 15 | Phosphorus | P | 31 | 30.973762 | 100 | |
|    |            |   | 32 | 31.973907 |  | 14.28 d ($\beta^-$) |
| 16 | Sulfur | S | 32 | 31.97207 | 95.02 | |
|    |        |   | 33 | 32.97146 | 0.75 | |
|    |        |   | 34 | 33.967866 | 4.21 | |
|    |        |   | 35 | 34.969031 |  | 87.2 d ($\beta^-$) |
|    |        |   | 36 | 35.96708 | 0.02 | |

*Note: Isotopes with a percent natural abundance normally do not have a half-life, and vice versa. Potassium-40, thorium-232 and uranium-234, -235, and -238 are exceptions.
†E.C. stands for electron capture.

| Z | Element | Symbol | A | Atomic Mass | % Natural Abundance* | Half-life (Decay Modes)† |
|---|---------|--------|---|-------------|----------------------|--------------------------|
| 17 | Chlorine | Cl | 35 | 34.968852 | 75.77 | |
| | | | 37 | 36.965903 | 24.23 | |
| 18 | Argon | Ar | 36 | 35.967545 | 0.337 | |
| | | | 38 | 37.962732 | 0.063 | |
| | | | 40 | 39.962384 | 99.600 | |
| 19 | Potassium | K | 39 | 38.963707 | 93.2581 | |
| | | | 40 | 39.963999 | 0.0117 | 1.26 Gyr ($\beta^-, \beta^+$, E.C.) |
| | | | 41 | 40.961825 | 6.7302 | |
| 20 | Calcium | Ca | 35 | 34.99523 | | 50 ms ($\beta^+$) |
| | | | 40 | 39.962591 | 96.941 | |
| | | | 41 | 40.962278 | | 0.103 Myr (E.C.) |
| | | | 42 | 41.958618 | 0.647 | |
| | | | 43 | 42.958766 | 0.135 | |
| | | | 44 | 43.95548 | 2.086 | |
| | | | 46 | 45.953689 | 0.004 | |
| | | | 48 | 47.952533 | 0.187 | |
| 21 | Scandium | Sc | 45 | 44.95591 | 100 | |
| 22 | Titanium | Ti | 46 | 45.952629 | 8.0 | |
| | | | 47 | 46.951764 | 7.3 | |
| | | | 48 | 47.947947 | 73.8 | |
| | | | 49 | 48.947871 | 5.5 | |
| | | | 50 | 49.944792 | 5.4 | |
| 23 | Vanadium | V | 50 | 49.947161 | 0.250 | |
| | | | 51 | 50.943962 | 99.750 | |
| 24 | Chromium | Cr | 50 | 49.946046 | 4.345 | |
| | | | 52 | 51.940509 | 83.79 | |
| | | | 53 | 52.940651 | 9.50 | |
| | | | 54 | 53.938882 | 2.365 | |
| 25 | Manganese | Mn | 55 | 54.938047 | 100 | |
| 26 | Iron | Fe | 54 | 53.939612 | 5.9 | |
| | | | 56 | 55.934939 | 91.72 | |
| | | | 57 | 56.935396 | 2.1 | |
| | | | 58 | 57.933277 | 0.28 | |
| 27 | Cobalt | Co | 59 | 58.933198 | 100 | |
| | | | 60 | 59.933819 | | 5.271 yr ($\beta^-$) |
| 28 | Nickel | Ni | 58 | 57.935346 | 68.077 | |
| | | | 60 | 59.930788 | 26.223 | |
| | | | 61 | 60.931058 | 1.140 | |
| | | | 62 | 61.928346 | 3.634 | |
| | | | 64 | 63.927968 | 0.926 | |
| 29 | Copper | Cu | 63 | 62.939598 | 69.17 | |
| | | | 65 | 64.927793 | 30.83 | |
| 30 | Zinc | Zn | 64 | 63.929145 | 48.6 | |
| | | | 66 | 65.926304 | 27.9 | |
| | | | 67 | 66.927129 | 4.1 | |
| | | | 68 | 67.924846 | 18.8 | |
| | | | 70 | 69.925325 | 0.6 | |
| 31 | Gallium | Ga | 69 | 68.92558 | 60.108 | |
| | | | 71 | 70.9247 | 39.892 | |
| 32 | Germanium | Ge | 70 | 69.92425 | 21.24 | |
| | | | 72 | 71.922079 | 27.66 | |
| | | | 73 | 72.923463 | 7.72 | |
| | | | 74 | 73.921177 | 35.94 | |
| | | | 76 | 75.921401 | 7.44 | |
| 33 | Arsenic | As | 75 | 74.921594 | 100 | |
| 34 | Selenium | Se | 74 | 73.922475 | 0.89 | |

| Z | Element | Symbol | A | Atomic Mass | % Natural Abundance* | Half-life (Decay Modes)† |
|---|---------|--------|---|-------------|----------------------|--------------------------|
| | | | 76 | 75.919212 | 9.36 | |
| | | | 77 | 76.919912 | 7.63 | |
| | | | 78 | 77.917308 | 23.77 | |
| | | | 80 | 79.91625 | 49.61 | |
| | | | 82 | 81.916698 | 8.74 | |
| 35 | Bromine | Br | 79 | 78.918336 | 50.69 | |
| | | | 81 | 80.916289 | 49.61 | |
| 36 | Krypton | Kr | 78 | 77.9204 | 0.35 | |
| | | | 80 | 79.91638 | 2.25 | |
| | | | 82 | 81.913482 | 11.6 | |
| | | | 83 | 82.914135 | 11.5 | |
| | | | 84 | 83.911507 | 57.0 | |
| | | | 86 | 85.910616 | 17.3 | |
| | | | 92 | 91.926270 | | 1.84 s ($\beta^-$) |
| 37 | Rubidium | Rb | 85 | 84.911794 | 72.17 | |
| | | | 87 | 86.909187 | 27.83 | |
| 38 | Strontium | Sr | 84 | 83.91343 | 0.56 | |
| | | | 86 | 85.909267 | 9.86 | |
| | | | 87 | 86.908884 | 7.00 | |
| | | | 88 | 87.905619 | 82.58 | |
| | | | 94 | 93.915367 | | 1.27 min ($\beta^-$) |
| 39 | Yttrium | Y | 89 | 88.905849 | 100 | |
| 40 | Zirconium | Zr | 90 | 89.904703 | 51.45 | |
| | | | 91 | 90.905644 | 11.22 | |
| | | | 92 | 91.905039 | 17.15 | |
| | | | 94 | 93.906314 | 17.38 | |
| | | | 96 | 95.908275 | 2.80 | |
| 41 | Niobium | Nb | 93 | 92.906377 | 100 | |
| 42 | Molybdenum | Mo | 92 | 91.906808 | 14.84 | |
| | | | 94 | 93.905085 | 9.25 | |
| | | | 95 | 94.90584 | 15.92 | |
| | | | 96 | 95.904678 | 16.68 | |
| | | | 97 | 96.90602 | 9.55 | |
| | | | 98 | 97.905406 | 24.13 | |
| | | | 100 | 99.907477 | 9.63 | |
| | | | 101 | 100.910345 | | 14.6 min ($\beta^-$) |
| 43 | Technetium | Tc | 98 | 97.907215 | | 4.2 Myr ($\beta^-$) |
| 44 | Ruthenium | Ru | 96 | 95.907599 | 5.54 | |
| | | | 98 | 97.905287 | 1.86 | |
| | | | 99 | 98.905939 | 12.7 | |
| | | | 100 | 99.904219 | 12.6 | |
| | | | 101 | 100.905582 | 17.1 | |
| | | | 102 | 101.904348 | 31.6 | |
| | | | 104 | 103.905424 | 18.6 | |
| 45 | Rhodium | Rh | 103 | 102.9055 | 100 | |
| 46 | Palladium | Pd | 102 | 101.905634 | 1.02 | |
| | | | 104 | 103.904029 | 11.14 | |
| | | | 105 | 104.905079 | 22.33 | |
| | | | 106 | 105.903478 | 27.33 | |
| | | | 108 | 107.903895 | 26.46 | |
| | | | 110 | 109.905167 | 11.72 | |
| 47 | Silver | Ag | 107 | 106.905092 | 51.839 | |
| | | | 109 | 108.904757 | 48.161 | |
| 48 | Cadmium | Cd | 106 | 105.906461 | 1.25 | |
| | | | 108 | 107.90418 | 0.89 | |
| | | | 110 | 109.903005 | 12.49 | |
| | | | 111 | 110.904182 | 12.8 | |
| | | | 112 | 111.902758 | 24.13 | |
| | | | 113 | 112.9044 | 12.22 | |

| Z | Element | Symbol | A | Atomic Mass | % Natural Abundance* | Half-life (Decay Modes)[†] | Z | Element | Symbol | A | Atomic Mass | % Natural Abundance* | Half-life (Decay Modes)[†] |
|---|---|---|---|---|---|---|---|---|---|---|---|---|---|
| | | | 114 | 113.903357 | 28.73 | | 61 | Promethium | Pm | 145 | 144.912743 | | 5.98 h ($\beta^-$) |
| | | | 116 | 115.904754 | 7.49 | | 62 | Samarium | Sm | 144 | 143.911998 | 3.1 | |
| 49 | Indium | In | 113 | 112.904061 | 4.3 | | | | | 147 | 146.914895 | 15.0 | |
| | | | 115 | 114.90388 | 95.7 | | | | | 148 | 147.91482 | 11.3 | |
| 50 | Tin | Sn | 112 | 111.904826 | 0.97 | | | | | 149 | 148.917181 | 13.8 | |
| | | | 114 | 113.902784 | 0.65 | | | | | 150 | 149.917273 | 7.4 | |
| | | | 115 | 114.903348 | 0.36 | | | | | 152 | 151.919729 | 26.7 | |
| | | | 116 | 115.901747 | 14.53 | | | | | 154 | 153.922206 | 22.7 | |
| | | | 117 | 116.902956 | 7.68 | | 63 | Europium | Eu | 151 | 150.919847 | 47.8 | |
| | | | 118 | 117.901609 | 24.22 | | | | | 153 | 152.921225 | 52.2 | |
| | | | 119 | 118.90331 | 8.58 | | 64 | Gadolinium | Gd | 152 | 151.919786 | 0.20 | |
| | | | 120 | 119.9022 | 32.59 | | | | | 154 | 153.920861 | 2.18 | |
| | | | 122 | 121.90344 | 4.63 | | | | | 155 | 154.922618 | 14.80 | |
| | | | 124 | 123.905274 | 5.79 | | | | | 156 | 155.922118 | 20.47 | |
| | | | 132 | 131.917760 | | 40 s ($\beta^-$) | | | | 157 | 156.923956 | 15.65 | |
| 51 | Antimony | Sb | 121 | 120.903821 | 57.36 | | | | | 158 | 157.924099 | 24.84 | |
| | | | 123 | 122.904216 | 42.64 | | | | | 160 | 159.927049 | 21.86 | |
| 52 | Tellurium | Te | 120 | 119.904048 | 0.095 | | 65 | Terbium | Tb | 148 | 147.924140 | | 1.0 h ($\beta^+$, E.C.) |
| | | | 122 | 121.903504 | 2.59 | | | | | 159 | 158.925342 | 100 | |
| | | | 123 | 122.904271 | 0.905 | | 66 | Dysprosium | Dy | 152 | 151.924716 | | 2.37 h (E.C., $\alpha$) |
| | | | 124 | 123.902823 | 4.79 | | | | | 156 | 155.925277 | 0.06 | |
| | | | 125 | 124.904433 | 7.12 | | | | | 158 | 157.924403 | 0.10 | |
| | | | 126 | 125.903314 | 18.93 | | | | | 160 | 159.925193 | 2.34 | |
| | | | 128 | 127.904463 | 31.70 | | | | | 161 | 160.92693 | 18.9 | |
| | | | 130 | 129.906229 | 33.87 | | | | | 162 | 161.926795 | 25.5 | |
| 53 | Iodine | I | 125 | 124.90462 | | 59.4 d (E.C.) | | | | 163 | 162.928728 | 24.9 | |
| | | | 127 | 126.904473 | 100 | | | | | 164 | 163.929171 | 28.2 | |
| 54 | Xenon | Xe | 124 | 123.905894 | 0.10 | | 67 | Holmium | Ho | 152 | 151.931580 | | 2.4 min ($\beta^+$, $\alpha$) |
| | | | 126 | 125.904281 | 0.09 | | | | | 165 | 164.930319 | 100 | |
| | | | 128 | 127.903531 | 1.91 | | 68 | Erbium | Er | 162 | 161.928775 | 0.14 | |
| | | | 129 | 128.90478 | 26.4 | | | | | 164 | 163.929198 | 1.61 | |
| | | | 130 | 129.903509 | 4.1 | | | | | 166 | 165.93029 | 33.6 | |
| | | | 131 | 130.905072 | 21.2 | | | | | 167 | 166.932046 | 22.95 | |
| | | | 132 | 131.904144 | 26.9 | | | | | 168 | 167.932368 | 26.8 | |
| | | | 134 | 133.905395 | 10.4 | | | | | 170 | 169.935461 | 14.9 | |
| | | | 136 | 135.907214 | 8.9 | | 69 | Thulium | Tm | 169 | 168.934212 | 100 | |
| | | | 140 | 139.921620 | | 13.6 s ($\beta^-$) | 70 | Ytterbium | Yb | 168 | 167.933894 | 0.13 | |
| 55 | Cesium | Cs | 133 | 132.905429 | 100 | | | | | 170 | 169.934759 | 3.05 | |
| | | | 135 | 134.905885 | | 2.3 Myr ($\beta^-$) | | | | 171 | 170.936323 | 14.3 | |
| 56 | Barium | Ba | 130 | 129.906282 | 0.106 | | | | | 172 | 171.936378 | 21.9 | |
| | | | 132 | 131.905042 | 0.101 | | | | | 173 | 172.938208 | 16.12 | |
| | | | 134 | 133.904486 | 2.42 | | | | | 174 | 173.938859 | 31.8 | |
| | | | 135 | 134.905665 | 6.593 | | | | | 176 | 175.942564 | 12.7 | |
| | | | 136 | 135.904553 | 7.85 | | 71 | Lutetium | Lu | 175 | 174.94077 | 97.41 | |
| | | | 137 | 136.905812 | 11.23 | | | | | 176 | 175.942679 | 2.59 | |
| | | | 138 | 137.905232 | 71.70 | | 72 | Hafnium | Hf | 174 | 173.940044 | 0.162 | |
| | | | 141 | 140.914363 | | 18.3 min ($\beta^-$) | | | | 176 | 175.941406 | 5.206 | |
| 57 | Lanthanum | La | 138 | 137.90711 | 0.0902 | | | | | 177 | 176.943217 | 18.606 | |
| | | | 139 | 138.906347 | 99.9098 | | | | | 178 | 177.943696 | 27.297 | |
| 58 | Cerium | Ce | 136 | 135.90714 | 0.19 | | | | | 179 | 178.945812 | 13.629 | |
| | | | 138 | 137.905985 | 0.25 | | | | | 180 | 179.946545 | 35.100 | |
| | | | 140 | 139.905433 | 88.43 | | 73 | Tantalum | Ta | 180 | 179.947462 | 0.012 | |
| | | | 142 | 141.909241 | 11.13 | | | | | 181 | 180.947992 | 99.988 | |
| 59 | Praseodymium | Pr | 141 | 140.907647 | 100 | | 74 | Tungsten | W | 180 | 179.946701 | 0.12 | |
| 60 | Neodymium | Nd | 142 | 141.907719 | 27.13 | | | | | 182 | 181.948202 | 26.3 | |
| | | | 143 | 142.90981 | 12.18 | | | | | 183 | 182.95022 | 14.28 | |
| | | | 144 | 143.910083 | 23.80 | | | | | 184 | 183.950928 | 30.7 | |
| | | | 145 | 144.91257 | 8.30 | | | | | 186 | 185.954357 | 28.6 | |
| | | | 146 | 145.913113 | 17.19 | | 75 | Rhenium | Re | 185 | 184.952951 | 37.40 | |
| | | | 148 | 147.916889 | 5.76 | | | | | | | | |
| | | | 150 | 149.920887 | 5.64 | | | | | | | | |

| Z | Element | Sym-bol | A | Atomic Mass | % Natural Abundance* | Half-life (Decay Modes)† |
|---|---------|---------|---|-------------|----------------------|--------------------------|
| | | | 187 | 186.955744 | 62.60 | |
| 76 | Osmium | Os | 184 | 183.952488 | 0.02 | |
| | | | 186 | 185.95383 | 1.58 | |
| | | | 187 | 186.955741 | 1.6 | |
| | | | 188 | 187.95586 | 13.3 | |
| | | | 189 | 188.958137 | 16.1 | |
| | | | 190 | 189.958436 | 26.4 | |
| | | | 192 | 191.961467 | 41.0 | |
| 77 | Iridium | Ir | 191 | 190.960584 | 37.3 | |
| | | | 193 | 192.962917 | 62.7 | |
| 78 | Platinum | Pt | 190 | 189.959917 | 0.01 | |
| | | | 192 | 191.961019 | 0.79 | |
| | | | 194 | 193.962655 | 32.9 | |
| | | | 195 | 194.964766 | 33.8 | |
| | | | 196 | 195.964926 | 25.3 | |
| | | | 198 | 197.967869 | 7.2 | |
| 79 | Gold | Au | 197 | 196.966543 | 100 | |
| 80 | Mercury | Hg | 196 | 195.965807 | 0.15 | |
| | | | 198 | 197.966743 | 9.97 | |
| | | | 199 | 198.968254 | 16.87 | |
| | | | 200 | 199.9683 | 23.10 | |
| | | | 201 | 200.970277 | 13.18 | |
| | | | 202 | 201.970617 | 29.86 | |
| | | | 204 | 203.973467 | 6.87 | |
| 81 | Thallium | Tl | 203 | 202.97232 | 29.524 | |
| | | | 205 | 204.974401 | 70.476 | |
| | | | 208 | 207.981988 | | 3.053 min $(\beta^-)$ |
| 82 | Lead | Pb | 204 | 203.97302 | 1.4 | |
| | | | 206 | 205.97444 | 24.1 | |
| | | | 207 | 206.975872 | 22.1 | |
| | | | 208 | 207.976627 | 52.4 | |
| | | | 210 | 209.984163 | | 22.6 yr $(\beta^-)$ |
| | | | 212 | 211.991871 | | 10.64 h $(\beta^-)$ |
| | | | 214 | 213.999798 | | 27 min $(\beta^-)$ |
| 83 | Bismuth | Bi | 209 | 208.980374 | 100 | |
| | | | 210 | 209.984095 | | 5.01 d $(\beta^-)$ |
| | | | 212 | 211.991255 | | 1.009 h $(\beta^-)$ |
| | | | 214 | 213.998691 | | 19.9 min $(\beta^-)$ |
| 84 | Polonium | Po | 209 | 208.982404 | | 102 yr $(\alpha)$ |
| | | | 210 | 209.982848 | | 138.38 d $(\alpha)$ |
| | | | 212 | 211.988842 | | 298 ns $(\alpha)$ |
| | | | 214 | 213.995176 | | 163.7 $\mu$s $(\alpha)$ |
| | | | 216 | 216.001889 | | 145 ms $(\alpha)$ |
| 85 | Astatine | At | 210 | 209.987126 | | 8.1 h (E.C., $\alpha$) |
| 86 | Radon | Rn | 220 | 220.017570 | | 55.6 s $(\alpha)$ |

| Z | Element | Sym-bol | A | Atomic Mass | % Natural Abundance* | Half-life (Decay Modes)† |
|---|---------|---------|---|-------------|----------------------|--------------------------|
| | | | 222 | 222.01757 | | 3.8235 d $(\alpha, \beta^-)$ |
| 87 | Francium | Fr | 223 | 223.019733 | | 21.8 min $(\beta^-)$ |
| 88 | Radiun | Ra | 224 | 224.020186 | | 3.66 d $(\alpha)$ |
| | | | 226 | 226.025402 | | 1599 yr $(\alpha)$ |
| | | | 228 | 228.031064 | | 5.76 yr $(\beta^-)$ |
| 89 | Actinium | Ac | 227 | 227.02775 | | 21.77 yr $(\beta^-, \alpha)$ |
| | | | 228 | 228.031015 | | 6.15 h $(\beta^-)$ |
| | | | 229 | 229.032980 | | 1.04 h $(\beta^-)$ |
| | | | 230 | 230.038550 | | 7.5 min $(\beta^-)$ |
| 90 | Thorium | Th | 232 | 232.038054 | 100 | 14 Gyr $(\alpha)$ |
| | | | 234 | 234.043593 | | 24.1 d $(\beta^-)$ |
| 91 | Protactinium | Pa | 231 | 231.03588 | | 32.5 kyr $(\alpha)$ |
| 92 | Uranium | U | 234 | 234.040946 | 0.0055 | 245 kyr $(\alpha)$ |
| | | | 235 | 235.043924 | 0.720 | 704 Myr $(\alpha)$ |
| | | | 238 | 238.050784 | 99.2745 | 4.46 Gyr $(\alpha)$ |
| 93 | Neptunium | Np | 234 | 234.042888 | | 4.4 d $(\beta^+, E.C.)$ |
| | | | 237 | 237.048167 | | 214 Myr $(\alpha)$ |
| | | | 238 | 238.050941 | | 2.117 d $(\beta^-)$ |
| 94 | Plutonium | Pu | 239 | 239.052157 | | 24.11 kyr $(\alpha)$ |
| | | | 244 | 244.064199 | | 82 Myr $(\alpha, S.F.)$ |
| 95 | Americium | Am | 243 | 243.061375 | | 7.37 kyr $(\alpha)$ |
| 96 | Curium | Cm | 247 | 247.070347 | | 15.6 Myr $(\alpha)$ |
| 97 | Berkelium | Bk | 247 | 247.070300 | | 1.4 kyr $(\alpha)$ |
| 98 | Californium | Cf | 251 | 251.079580 | | 0.90 kyr $(\alpha)$ |
| 99 | Einsteinium | Es | 252 | 252.082944 | | 1.29 yr $(\alpha, E.C.)$ |
| 100 | Fermium | Fm | 257 | 257.075099 | | 100.5 d $(\alpha, S.F.)$ |
| 101 | Mendelevium | Md | 258 | 258.098570 | | 51.5 d $(\alpha)$ |
| 102 | Nobelium | No | 259 | 259.100931 | | 58 min $(\alpha, E.C.)$ |
| 103 | Lawrencium | Lr | 260 | 260.105320 | | 3 min $(\alpha)$ |
| 104 | Rutherfordium | Rf | 261 | 261.108690 | | 65 s $(\alpha)$ |
| 105 | Dubnium | Db | 262 | 262.113760 | | 34 s (S.F., $\alpha$) |
| 106 | Seaborgium | Sg | 263 | 263.1182 | | 0.8 s (S.F., $\alpha$) |
| 107 | Bohrium | Bh | 262 | 262.1231 | | 0.10 s $(\alpha)$ |
| 108 | Hassium | Hs | 265 | 265.1300 | | 2 ms $(\alpha)$ |
| 109 | Meitnerium | Mt | 266 | 266.1378 | | $\approx$3.4 ms $(\alpha)$ |

# K

# Related Math

## Complex Numbers

A complex number has two parts: a real number, and another real number multiplied by $i$, where $i$ is defined by $i^2 = -1$. For example, in $X = 3 + i4$ the **real part** is 3 and the **imaginary part** (which is a real number) is 4. Complex numbers include real numbers as a subset—that is, when the imaginary part is zero—and may be thought of as a compact way of conveying twice the information of a real number. However, they also obey special rules of arithmetic, based on keeping $i$ (to the first power) separate, while $i^2$ becomes $-1$:

Given: $\quad U = u_1 + iu_2 \quad$ and $\quad V = v_1 + iv_2$

$$U + V = (u_1 + iu_2) + (v_1 + iv_2)$$

$$= (u_1 + v_1) + i(u_2 + v_2)$$

$$UV = (u_1 + iu_2)(v_1 + iv_2)$$

$$= u_1v_1 + iu_1v_2 + iu_2v_1 + i^2u_2v_2$$

$$= (u_1v_1 - u_2v_2) + i(u_1v_2 + u_2v_1)$$

The two parts of a complex number may each be functions, for example, $U(x) = u_1(x) + iu_2(x)$.

A function whose *argument* is complex may often be (and usually is) broken into its real and imaginary parts. A common function is the exponential. Consider first an exponential whose argument has no real part: $e^{ix}$, where $x$ is real. With the power series, $e^b = \sum_0^\infty (b^n/n!)$, we may write

$$e^{ix} = \sum_0^\infty \frac{(ix)^n}{n!} = 1 + ix + i^2\frac{1}{2!}x^2 + i^3\frac{1}{3!}x^3 + i^4\frac{1}{4!}x^4 + i^5\frac{1}{5!}x^5 + i^6\frac{1}{6!}x^6 + i^7\frac{1}{7!}x^7 + i^8\frac{1}{8!}x^8 + \cdots$$

Using $i^3 = i^2 i = -i$, $i^4 = i^3 i = 1$, $i^5 = i^4 i = i$, . . . , this naturally breaks into two series.

$$e^{ix} = \left(1 - \frac{1}{2!}x^2 + \frac{1}{4!}x^4 - \frac{1}{6!}x^6 + \frac{1}{8!}x^8 - \cdots\right) + i\left(x - \frac{1}{3!}x^3 + \frac{1}{5!}x^5 - \frac{1}{7!}x^7 + \cdots\right)$$

The sums in parentheses are the power series for cosine and sine, and the result is an identity used very often in physics, known as the **Euler formula:**

$$e^{ix} = \cos x + i \sin x$$

If the argument also has a real part, then $e^X = e^{(x_1 + ix_2)} = e^{x_1}e^{ix_2} = (e^{x_1}\cos x_2) + i(e^{x_1}\sin x_2)$.

Useful formulas following directly from the Euler formula are

$$\cos x = \frac{e^{+ix} + e^{-ix}}{2} \qquad \sin x = \frac{e^{+ix} - e^{-ix}}{2i}$$

In the same way as a two-dimensional vector (by which it can be represented and which shares the same addition rule), a complex number has a real, positive magnitude defined as the square root of the sum of the squares of its two parts.

$$|U| = \sqrt{u_1^2 + u_2^2}$$

Given a complex number or function $X$, its **complex conjugate,** denoted $X^*$, is obtained by replacing $i$ by $-i$ wherever it occurs. Examples:

If $U = u_1 + iu_2$       then $U^* = u_1 - iu_2$
If $f(x) = \cos x - i \sin x$   then $f^*(x) = \cos x + i \sin x$
If $f(x) = e^{-x^2}e^{+ix}$   then $f^*(x) = e^{-x^2}e^{-ix}$
If $f(x) = \tanh^{-1}(3 - i4)$   then $f^*(x) = \tanh^{-1}(3 + i4)$

The product of a quantity and its complex conjugate gives the square of the quantity's magnitude.

$$U^*U = (u_1 - iu_2)(u_1 + iu_2) = u_1^2 + u_2^2 = |U|^2$$

An important special function is the **complex exponential** $e^{ikx} = \cos(kx) + i\sin(kx)$, whose behavior is represented in Figure K.1. It is oscillatory because both its real and imaginary parts are oscillatory, of angular frequency $k$. (With $x$ measured in meters, $k$ would have to be in rad/m. An equivalent function is $e^{i\omega t}$, where $\omega$ would be measured in rad/s.) Moreover, its magnitude is unity:

$$f(x) = e^{ikx} \quad\Rightarrow\quad f^*(x)f(x) = e^{-ikx}e^{+ikx} = e^{-ikx+ikx} = e^0 = 1$$

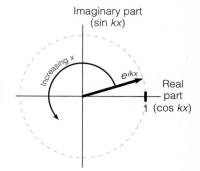

**Figure K.1** The complex exponential $e^{ikx}$ has unit magnitude and "rotates" in the real-complex plane at frequency $k$.

or

$$f^*(x)f(x) = [\cos(kx) - i\,\sin(kx)][\cos(kx) + i\,\sin(kx)] = \cos^2(kx) + \sin^2(kx) = 1$$

Thus, it rotates in the real-imaginary plane at frequency $k$ while maintaining unit length.

## Differential Equations

A differential equation is one involving derivatives of a function. To be **$m$th-order** is to involve derivatives as high as the $m$th. A **partial,** as opposed to **ordinary,** differential equation is one in which the function has multiple independent variables and the derivatives are partial derivatives with respect to those variables. A **linear,** as opposed to **nonlinear,** differential equation is one in which each term involves either the function or one of its derivatives to the *first* power only. Examples:

$$\frac{df(x)}{dx} = bf(x) \qquad\qquad \text{First-order, ordinary, linear}$$

$$\frac{d^2f(t)}{dt^2} = bf(t) \qquad\qquad \text{Second-order, ordinary, linear}$$

$$\frac{\partial^2 f(x,\,y,\,z)}{\partial x^2} + \frac{\partial^2 f(x,\,y,\,z)}{\partial y^2} + \frac{\partial^2 f(x,\,y,\,z)}{\partial z^2} = bf(x,\,y,\,z) \qquad \text{Second-order, partial, linear}$$

$$\frac{\partial^2 f(x,\,t)}{\partial x^2} = b\frac{\partial f(x,\,t)}{\partial t} \qquad\qquad \text{Second-order, partial, linear}$$

$$\frac{df(t)}{dt} = bf^2(t) \qquad\qquad \text{First-order, ordinary, nonlinear}$$

Different kinds of differential equations demand different techniques to solve for $f$. The easiest to solve are the ordinary, linear ones. Partial differential equations are often converted to separate ordinary differential equations, one for each independent variable, by the technique of **separation of variables:** For a function such as $f(x,\,y,\,z)$ the product $f_1(x)f_2(y)f_3(z)$ is substituted. The differential equation is then rearranged so that dependence on each variable is isolated, resulting in three separate ordinary differential equations, which when solved give $f_1(x), f_2(y)$, and $f_3(z)$. For many differential equations, particularly nonlinear ones, no technique exists to find a closed-form solution. Numerical approximation techniques are required.

In this text, the reader need be familiar with only two differential equations, summarized below; partial differential equations will be converted to one of these forms. Readers unfamiliar with the techniques by which the function $f$ is actually solved for in these cases are strongly encouraged to consult a math text. Our interest here is in the results. The reader should at the very least *verify* by substitution that the given functions $f$ do indeed solve the differential equations.

First-order, ordinary, linear:

$$\frac{df(x)}{dx} = bf(x) \quad \Rightarrow \quad f(x) = Ae^{bx}$$

Solution of a first-order equation yields one arbitrary constant (essentially a constant of integration), here $A$, which can take on any value while still solving the equation. *Note:* The technique to solve this basic differential equation is the simplest — merely rearrange $(df/f = b\,dx)$ and integrate. Others are more complicated.

Second-order, ordinary, linear:

$$\frac{d^2 f(x)}{dx^2} = bf(x)$$

$$\Rightarrow \quad f(x) = \begin{cases} A\,\sin(\sqrt{|b|}\,x) + B\,\cos(\sqrt{|b|}\,x) \quad \text{or} \quad Ae^{+i\sqrt{|b|}x} + Be^{-i\sqrt{|b|}x} & b < 0 \\ Ae^{+\sqrt{b}x} + Be^{-\sqrt{b}x} \quad \text{or} \quad A\,\sinh(\sqrt{b}x) + B\,\cosh(\sqrt{b}x) & b > 0 \\ Ax + B & b = 0 \end{cases}$$

A second-order equation yields two arbitrary constants (of integration). *Note:* The sign of the constant $b$, which is dependent on the physical application, is *crucial;* if negative, the function $f$ is oscillatory; if positive, exponentially growing or decaying (or a combination); if zero, a straight line.

## Useful Integrals

$$\int \sin^2\left(\frac{n\pi}{L}x\right) dx = \frac{x}{2} - \frac{L}{4n\pi}\sin\left(\frac{2n\pi}{L}x\right)$$

$$\int x\,\sin^2\left(\frac{n\pi}{L}x\right) dx = \frac{x^2}{4} - \frac{Lx}{4n\pi}\sin\left(\frac{2n\pi}{L}x\right) - \frac{L^2}{8n^2\pi^2}\cos\left(\frac{2n\pi}{L}x\right)$$

$$\int x^2\,\sin^2\left(\frac{n\pi}{L}x\right) dx = \frac{x^3}{6} - \frac{Lx^2}{4n\pi}\sin\left(\frac{2n\pi}{L}x\right) - \frac{L^2 x}{4n^2\pi^2}\cos\left(\frac{2n\pi}{L}x\right) + \frac{L^3}{8n^3\pi^3}\sin\left(\frac{2n\pi}{L}x\right)$$

$$\int_0^\infty x^m e^{-bx}\,dx = \frac{m!}{b^{m+1}}$$

## Gaussian Integrals

$$\int_{-\infty}^{+\infty} e^{-a(z-b)^2}\,dz = \sqrt{\frac{\pi}{a}} \qquad \int_{-\infty}^{+\infty} z\,e^{-a(z-b)^2}\,dz = b\sqrt{\frac{\pi}{a}}$$

$$\int_{-\infty}^{+\infty} e^{-az^2 + bz}\,dz = e^{b^2/4a}\sqrt{\frac{\pi}{a}} \qquad \int_{-\infty}^{+\infty} z^2\,e^{-az^2}\,dz = \frac{1}{2}\sqrt{\frac{\pi}{a^3}}$$

$$\int_{-\infty}^{+\infty} z^{n+2}\,e^{-az^2}\,dz = -\frac{d}{da}\int_{-\infty}^{+\infty} z^n\,e^{-az^2}\,dz$$

# Answers to Selected Exercises

## Chapter I

**1.** $0.14c$ **3.** $43.75$ m **5.** later, $0.8$ m **7.** $60$ m, $-2.67 \times 10^{-7}$ s **9.** $v/c = 0.781$, $1.04 \times 10^{-7}$ s, $24.375$ m **11.** $24$ m,

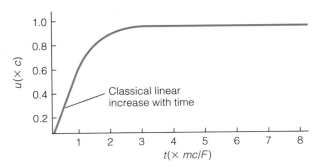

$v/c = 0.6$ **13.** $0.0067$ s behind **15.** $9.8$ ps earlier
**17.** $c/\sqrt{2}$ **19.** it must; top passes through $(v/c^2)L_0 \cos \theta_0$
earlier **21.** $1549$ m **23.** (a) $-100$ ns; (b) $141$ ns; (c) $100$ ns

zero **25.** Bob is $60$ yr, Al is $52$ yr **27.** (a) jumps ahead
$128$ days; (b) $250$ ns; (c) behind by same amounts **31.** (a) toward, $v/c = 0.25$; (b) $687$ nm; (c) $549$ nm **33.** yes
**35.** $\dfrac{\sqrt{3}\,k_\mathrm{B}T/m}{c}\,\lambda$, $0.11$ nm **37.** $0.385c$, $c$ **41.** $\gamma_u$
**43.** $9 \times 10^{16}$ J, $6 \times 10^{16}$ J, $1.5 \times 10^{17}$ J **45.** $2.5 \times 10^{-17}$ kg
**47.** (a) $2.19 \times 10^{-26}$ kg·m/s; (b) $3.64 \times 10^{-22}$ kg·m/s;
(c) $3 \times 10^{-7}\%$ low, $40\%$ low **49.** $c/\sqrt{2}$ **51.** $183$ kg/day
**53.** $4.71 \times 10^6$ kg/s **55.** $\frac{\sqrt{3}}{2}c$ **57.** $25.1$ MV
**59.** (a) $0.759c$; (b) $1.52c$; (c) $0.948c$ **61.** $u_\mathrm{e} = 0.99971c$,
$u_\mathrm{C} = -1.62 \times 10^{-3}\,c$, $20.6$ MeV, $17.1$ keV **63.** $m_2 = 6.43$ kg, $m_1 = 1.43$ kg, $1.93 \times 10^{17}$ J **65.** (a) $1.29m_0c^2$;
(b) $0.795c$; (c) $2.57m_0$, $3.29m_0$, exp. B
**69.** $\begin{bmatrix} 1 & 0 & 0 & -v/c \\ 0 & 1 & 0 & 0 \\ 0 & 0 & 1 & 0 \\ -v/c & 0 & 0 & 1 \end{bmatrix}$

**81.**

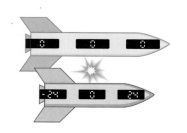
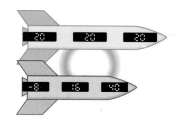
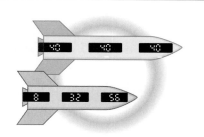
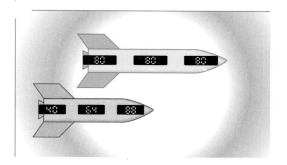
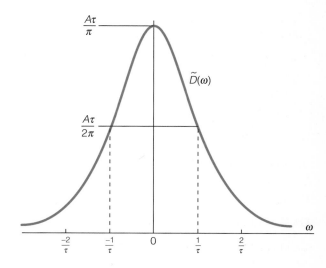

## Chapter 2

**1.** 3.12 eV, 399 nm    **3.** 82.4 nm    **5.** 6.42 × 10³¹ photons per sec    **7.** 1.22 × 10⁶ m/s    **11.** 8.15 × 10⁷ m/s    **17.** 130.5°, 23.9°    **21.** 0.00659 nm    **25.** (b) opposite, 1.18 × 10⁻¹⁴ m    **27.** 60°, 5.73 × 10⁻³ degrees, visible light    **29.** 7.38 × 10⁻¹⁰ m, 600    **31.** 5 × 10⁻⁹ Pa, 6.37 × 10⁵ N    **33.** $\frac{1}{2}\epsilon_0 E^2 \lambda/h$    **35.** 2.91 × 10⁻¹² m    **37.** $\frac{3}{4}m_0 c^2$, $\frac{7}{4}m_0$

**37.** 95 nm    **39.** $\dfrac{L}{2\pi}\dfrac{\sin[(k_0 - k)\frac{1}{2}L]}{(k_0 - k)\frac{1}{2}L}$    **41.** $2n\pi\hbar/w$

**43.** $\dfrac{A\tau}{2\pi}\dfrac{2}{1 + (\omega\tau)^2}$

## Chapter 3

**1.** 40    **3.** 2.43 × 10⁻¹² m    **5.** 728 m/s    **7.** (a) 6.29 nm; (b) 0.147 nm    **9.** (a) 1.46 × 10⁻¹⁰ m to 1.32 × 10⁻¹³ m; (b) 6.23 × 10⁻⁹ m to 2.43 × 10⁻¹⁰ m    **11.** 2.2 × 10⁻¹⁰ m    **13.** cannot be treated classically    **15.** (a) 0.21 mm; (b) 1600; (c) 400    **21.** 1.56 × 10⁻²⁵ kg    **23.** 3.6 × 10⁻²⁸ m/s    **25.** 6.3 × 10⁶ m/s    **27.** $\Delta x \sim 10^{-26} \times 25$ cm    **29.** (a) 0.0167°; (b) 1.05 × 10⁻³¹ degrees = 1.83 × 10⁻³³ rad    **31.** $r < 10^{-5}$ m    **33.** (b) $\frac{3}{2}\sqrt[3]{b/2m^2\hbar}$    **35.** 0.3 fs

**45.** $\dfrac{1}{2\pi} \dfrac{(A + B)\alpha + (A - B)ik}{\alpha^2 + k^2}$, $A = B$, it falls off more rapidly

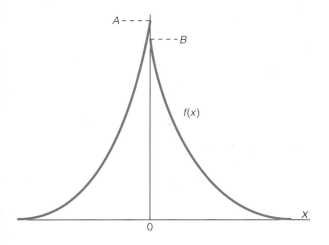

**47.** $e^2/(4\pi\epsilon_0)\hbar n$, $2.2 \times 10^6$ m/s   **49.** $7.64 \times 10^{-16}$ m, particle

## Chapter 4

**1.** $2 \pm \sqrt{2}$, yes, $\sqrt{2} - 1$, no

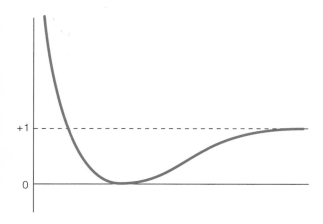

**3.** $1.41 \times 10^4$ m$^{-1/2}$ sin$(9.42 \times 10^8$ m$^{-1}$ $x)e^{-i(5.14 \times 10^{13}\text{ s}^{-1})t}$

**5.** 0.64 nm   **15.** 0.038 eV   **25.** zero   **29.** $\sqrt{\dfrac{3}{2}\left(\dfrac{\hbar^2}{m\kappa}\right)^{1/4}}$,

$\sqrt{\dfrac{3\hbar}{2}}(m\kappa)^{1/4}, \frac{3}{2}\hbar$   **31.** $\pm\left(\dfrac{\hbar^2}{m\kappa}\right)^{1/4}$   **33.** $C + D = B$,

$\alpha(C - D) = kA$, $A \sin kL + B \cos kL = Fe^{\alpha L} + Ge^{-\alpha L}$, $k(A \cos kL - B \sin kL) = \alpha(Fe^{\alpha L} - Ge^{-\alpha L})$   **39.** (a) and (b) closer together; (c) Yes; (d) energy cannot be arbitrarily high, infinite number of states still possible   **41.** (b) no

**43.**

**45.** 0.323   **47.** 0.866/$a$   **49.** $a\hbar$   **53.** (b) $\frac{1}{2}$; (c) 0.798   **55.** $a$

## Chapter 5

**1.** 0.00704   **3.** $1.1 \times 10^{-54}$   **5.** No   **7.** (a) $2t/m$; (b) $2L/n$

**13.** $\dfrac{i - \frac{1}{2}}{i + \frac{1}{2}} e^{-ikx}, \dfrac{2i}{i + \frac{1}{2}} e^{-\alpha x}$   **15.** (b) 0, 0°, 2$A$, 90°

**17.** zero, $e^{-4}$, yes

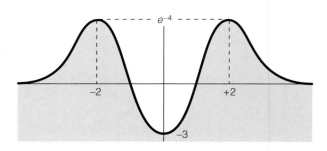

**19.** 1.8%   **23.** $v_{\text{phase}} = 0.30$ m/s, $v_{\text{group}} = 0.45$ m/s
**25.** $v_{\text{phase}} = 2.79$ m/s, $v_{\text{group}} = 1.40$ m/s, yes   **27.** 0.043 s

## Chapter 6

**3.** 0.64 nm   **5.** 0.609, 0.196, 0.609   **7.** 107 $\pi^2\hbar^2/2mL^2$, at center   **9.** yes, 45°, 135°   **13.** 150°, 125.3°, 106.8°, 90°, 73.2°, 54.7°, 30°   **15.** $-0.85$ eV, magnitude of angular momentum: $0\sqrt{2}\hbar, \sqrt{6}\hbar, \sqrt{12}\hbar$, $z$-component of angular momentum: $-3\hbar, -2\hbar, -\hbar, 0, +\hbar, +2\hbar, +3\hbar$   **21.** (a) yes; (b) no   **23.** $(\pi a^3)^{-1/2}e^{-r/a_0}$   **25.** (a) $\frac{1}{2}$; (b) $\frac{11}{16}$; (c) $\frac{203}{256}$   **27.** 0.212

**29.** 0.238

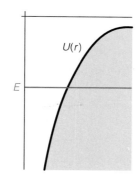

$U(r)$

$E$

**31.** ellipse   **33.** no   **37.** $n_i = 6$, $n_f = 3$, $1.1 \times 10^{-6}$ m
**39.** $n = 6$, 13.2 eV, $10^5$ K   **41.** $5.5 \times 10^{-8}$ eV, $5 \times 10^{-9}$
**43.** 13.5 nm   **47.** $2 \times 10^{-7}$ s   **49.** 0.0012 s, $1.9 \times 10^{-6}$ m
**51.** (a) $re^{-r/2a_0} \sin\theta\, 2\cos\phi$, $4r^2 e^{-r/a_0} \sin^2\theta \cos^2\phi$; (b) angular probabilities; (d) $\psi_{2,1,+1} - \psi_{2,1,-1}$   **53.** $E_n = n^2 \pi^2 \hbar^2 / 2ma^2$, $L = 0$

## Chapter 7

**1.** $9.11 \times 10^{-63}$ kg·m², $10^{28}$ rad/s, $3000c$   **5.** spin-down, $1.2 \times 10^{10}$ Hz   **7.** sym. 0.43, antisym. 0.07   **9.** $19\,\pi^2\hbar^2/2mL^2$, $5\,\pi^2\hbar^2/2mL^2$, $8\,\pi^2\hbar^2/2mL^2$   **13.** (b) symmetric triplet spin state   **15.** phosphorus: $1s^2 2s^2 2p^6 3s^2 3p^3$; germanium: $1s^2 2s^2 2p^6 3s^2 3p^6 3d^{10} 4s^2 4p^2$; cesium: $1s^2 2s^2 2p^6 3s^2 3p^6 3d^{10} 4s^2 4p^6 4d^{10} 5s^2 5p^6 6s^1$
**17.** $1s^2 2s^2 2p^6 3s^2 3p^6 3d^{10} 4s^2 4p^6 4d^{10} 4f^{14}$ $5s^2 5p^6 5d^{10} 5f^{14} 6s^2 6p^6 6d^{10} 7s^2 7p^6 8s^1$
**19.** (a) $4s \to 3d$ is smaller; (b) yes; (c) yes   **23.** (a) 66°; (b) 145°   **27.** (a) $\frac{\sqrt{63}}{2}\hbar$, $\frac{\sqrt{35}}{2}\hbar$, $\frac{\sqrt{15}}{2}\hbar$, $\frac{\sqrt{3}}{2}\hbar$; (b) 20; (c) $(\frac{7}{2}, +\frac{7}{2})$, $(\frac{7}{2}, +\frac{5}{2})$, $(\frac{7}{2}, +\frac{3}{2})$, $(\frac{7}{2}, +\frac{1}{2})$, $(\frac{7}{2}, -\frac{1}{2})$, $(\frac{7}{2}, -\frac{3}{2})$, $(\frac{7}{2}, -\frac{5}{2})$, $(\frac{7}{2}, -\frac{7}{2})$, $(\frac{5}{2}, +\frac{5}{2})$, $(\frac{5}{2}, +\frac{3}{2})$, $(\frac{5}{2}, +\frac{1}{2})$, $(\frac{5}{2}, -\frac{1}{2})$, $(\frac{5}{2}, -\frac{3}{2})$, $(\frac{5}{2}, -\frac{5}{2})$, $(\frac{3}{2}, +\frac{3}{2})$, $(\frac{3}{2}, +\frac{1}{2})$, $(\frac{3}{2}, -\frac{1}{2})$, $(\frac{3}{2}, -\frac{3}{2})$, $(\frac{1}{2}, +\frac{1}{2})$, $(\frac{1}{2}, -\frac{1}{2})$
**29.** 167°   **35.** 3.4 eV/$hc$ $(3Z^2 + 5Z - 19.25)$

## Chapter 8

**1.** (a) 60; (b) 20
**3.**

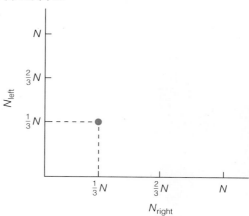

**5.** $\Delta S_1$: 0.20 J/K, $\Delta S_2$: $-0.15$ J/K, $+0.05$ J/K, $e^{3.6 \times 10^{21}}$
**7.** (a) (3, 0, 0, 0), (0, 3, 0, 0), (0, 0, 3, 0), (0, 0, 0, 3), (2, 1, 0, 0), (2, 0, 1, 0), (2, 0, 0, 1), (1, 2, 0, 0), (1, 0, 2, 0), (1, 0, 0, 2), (0, 2, 1, 0), (0, 2, 0, 1), (0, 1, 2, 0), (0, 1, 0, 2), (0, 0, 2, 1), (0, 0, 1, 2), (1, 1, 1, 0), (1, 1, 0, 1), (1, 0, 1, 1), (0, 1, 1, 1); (b) 0.5

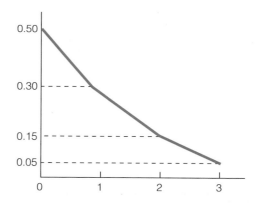

**9.** $N\hbar\omega/k_B$   **19.** $\sqrt{2k_B T/m}$, smaller
**21.** (a) $(mg/k_B T)\, e^{-mgy/k_B T}$; (b) 8.4% less; (c) 9.6% less
**27.** 7.0 eV   **29.** 1 K   **31.** (a) $3.71 \times 10^{-9}$ m; (b) $2.76 \times 10^{-11}$ m; (c) classically   **35.** $10^{-11}$ C, $10^{-16}$   **39.** $9.85 \times 10^{-6}$ m   **41.** $(32\pi^5 k_B^4/15h^3 c^3)T^3$, $8.12 \times 10^{-8}$ J/K·m³ **45.** $\frac{3}{2}k_B T$   **51.** (a) 21.8 MeV; (b) 16.5 MeV   **53.** $7.17 \times 10^6$ m

## Chapter 9

**3.** 1.53 eV must be put in, 0.94 nm, 4.46 eV lower
**7.** $\frac{1}{2} - \frac{3}{8}\sin(2\tau)$, $\frac{3\pi}{4}$, $-1$, $\frac{7}{8}$

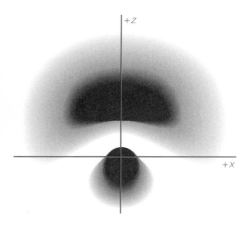

**9.** 2.94, $1.2 \times 10^{-5}$   **11.** (a) 0.179 eV, 0.537 eV;
(b) 0.011 nm, 0.019 nm; (c) 8.5%, 14.6%; (d) $5.4 \times 10^{14}$ s$^{-1}$, $5.4 \times 10^{12}$ s$^{-1}$   **13.** 33   **15.** vibrational levels
would be spaced more closely, rotational levels more closely
spaced   **19.** (a) $2.6 \times 10^{-10}$ m; (b) $5.3 \times 10^{-8}$ m;
(c) no   **25.** $\sim\frac{1}{10}$ eV   **27.** $10^{-6}$ m   **31.** positive bias,
bipolar: low-input impedance, MOSFET: high-input
impedance

**33.**

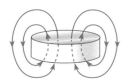

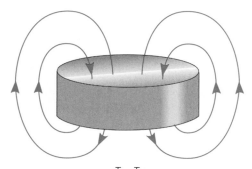

$T < T_c$

**35.** 100 A

## Chapter 10

**1.** $3 \times 10^{-14}$   **3.** 10.811 u   **5.** 7.68 MeV/nuc
**7.** 8.75 MeV/nuc   **9.** (a) 6.63 MeV/nuc, 7.68 MeV/nuc,
6.17 MeV/nuc; (c) 6.85 MeV/nuc, 7.25 MeV/nuc,
6.28 MeV/nuc   **11.** lead-206, 5.41 MeV   **13.** fluorine-19,
4.82 MeV, a maximum KE, since shared with antineutrino
**15.** (a) Ar-40, Ca-40; (b) magic numbers, asymmetry;
(c) Coulomb repulsion; (d) even numbers   **17.** 0.713 MeV
**19.** $3.94 \times 10^5$ m/s   **21.** 18.3 min   **23.** 5.7 per min
**25.** $1.30 \times 10^{12}$, $1.47 \times 10^{13}$, 0.156 MeV, 0.332 J
**27.** $4.28 \times 10^9$ yr   **29.** 32.2 mg, $2.1 \times 10^{19}$, $1.6 \times 10^7$ J,
$1.2 \times 10^9$ s$^{-1}$   **31.** $-1.64$ MeV   **33.** (a) $3/r_0$; (b) $3.72/r_0$;
(c) $2.38/r_0$   **35.** $3.13 \times 10^{18}$ fissions/s, 38.5 kg/yr
**37.** $9.57 \times 10^{13}$ J/kg, about eight orders of magnitude larger
**39.** reduced Coulomb repulsion   **41.** hydrogen-3

## Chapter 11

**1.** 197 MeV/$c^2$   **3.** $\hbar c/L$, $\hbar c/2L$   **9.** mesons, $\pi^0$ and $\rho^0$
**11.** 3.4 T, 20 km   **13.** weak

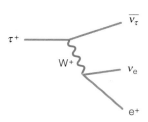

**15.** baryon number not conserved   **17.** lepton number not
conserved   **19.** lepton number not conserved   **21.** energy
conservation would be violated   **23.** baryon number and
charge not conserved   **25.** 1880 MeV   **27.** 280 MeV
**29.** 672 MeV   **31.** $c^3\ell^2/h$, $4 \times 10^{-35}$ m

# Credits

# Index

# Useful Equations (from Appendix $K$)

## Complex Numbers

$$e^{+ix} = \cos x + i \sin x$$

$$\cos x = \frac{e^{+ix} + e^{-ix}}{2} \qquad \sin x = \frac{e^{+ix} - e^{-ix}}{2i}$$

$$U^*U = (u_1 - iu_2)(u_1 + iu_2) = u_1^2 + u_2^2 = |U|^2$$

$$f(x) = e^{ikx} \quad \Rightarrow \quad f^*(x)f(x) = e^{-ikx}e^{+ikx} = e^{-ikx+ikx} = e^0 = 1$$

$$|U| = \sqrt{u_1^2 + u_2^2}$$

## Differential Equations

$$\frac{df(x)}{dx} = bf(x) \quad \Rightarrow \quad f(x) = Ae^{bx}$$

$$\frac{d^2f(x)}{dx^2} = bf(x)$$

$$\Rightarrow \quad f(x) = \begin{cases} A \sin(\sqrt{|b|}x) + B \cos(\sqrt{|b|}x) \quad \text{or} \quad Ae^{+i\sqrt{|b|}x} + Be^{-i\sqrt{|b|}x} & b < 0 \\ Ae^{+\sqrt{b}x} + Be^{-\sqrt{b}x} \quad \text{or} \quad A \sinh(\sqrt{b}x) + B \cosh(\sqrt{b}x) & b > 0 \\ Ax + B & b = 0 \end{cases}$$